DISCRETE-TIME
SIGNAL
PROCESSING

SECOND EDITION

DISCRETE-TIME
SIGNAL
PROCESSING

ALAN V. OPPENHEIM
MASSACHUSETTS INSTITUTE OF TECHNOLOGY

RONALD W. SCHAFER
GEORGIA INSTITUTE OF TECHNOLOGY

WITH

JOHN R. BUCK
UNIVERSITY OF MASSACHUSETTS DARTMOUTH

Prentice Hall International, Inc.

Acquisitions Editor: **Tom Robbins**
Production service: **Interactive Composition Corporation**
Editorial/production supervision: **Sharyn Vitrano**
Copy editor: **Brian Baker**
Art Director: **Amy Rosen**
Managing editor: **Eileen Clark**
Editor-in-Chief: **Marcia Horton**
Director of production and manufacturing: **David W. Riccardi**
Manufacturing buyer: **Pat Brown**
Editorial assistant: **Dan De Pasquale**

Printed in the United States of America

10 9 8 7 6 5 4

ISBN 0-13-083443-2

Prentice-Hall International (UK) Limited, London
Prentice-Hall of Australia Pty. Limited, Sydney
Prentice-Hall Canada Inc., Toronto
Prentice-Hall Hispanoamericana, S. A., Mexico
Prentice-Hall of India Private Limited, New Delhi
Prentice-Hall of Japan, Inc., Tokyo
Simon & Schuster Asia Pte. Ltd., Singapore
Editora Prentice-Hall do Brasil, Ltda., Rio de Janeiro
Prentice-Hall, Inc., Upper Saddle River, New Jersey

To Phyllis, Justine and Jason

To Dorothy, Bill, Tricia, Ken and Kate
and in memory of John

To Susan

CONTENTS

LIST OF EXAMPLES   XV

PREFACE   XIX

ACKNOWLEDGMENTS   XXV

1   INTRODUCTION   1

2   DISCRETE-TIME SIGNALS AND SYSTEMS   8

2.0   **Introduction   8**
2.1   **Discrete-Time Signals: Sequences   9**
   2.1.1   Basic Sequences and Sequence Operations   11
2.2   **Discrete-Time Systems   16**
   2.2.1   Memoryless Systems   18
   2.2.2   Linear Systems   18
   2.2.3   Time-Invariant Systems   20
   2.2.4   Causality   21
   2.2.5   Stability   21
2.3   **Linear Time-Invariant Systems   22**
2.4   **Properties of Linear Time-Invariant Systems   28**
2.5   **Linear Constant-Coefficient Difference Equations   34**
2.6   **Frequency-Domain Representation of Discrete-Time Signals and Systems   40**
   2.6.1   Eigenfunctions for Linear Time-Invariant Systems   40
   2.6.2   Suddenly Applied Complex Exponential Inputs   46
2.7   **Representation of Sequences by Fourier Transforms   48**
2.8   **Symmetry Properties of the Fourier Transform   55**
2.9   **Fourier Transform Theorems   58**
   2.9.1   Linearity of the Fourier Transform   59
   2.9.2   Time Shifting and Frequency Shifting   59
   2.9.3   Time Reversal   60
   2.9.4   Differentiation in Frequency   60
   2.9.5   Parseval's Theorem   60
   2.9.6   The Convolution Theorem   60
   2.9.7   The Modulation or Windowing Theorem   61
2.10   **Discrete-Time Random Signals   65**
2.11   **Summary   70**
   Problems   70

3   THE z-TRANSFORM   94

3.0   **Introduction   94**
3.1   **z-Transform   94**

3.2   Properties of the Region of Convergence for the z-Transform   105
3.3   The Inverse z-Transform   111
    3.3.1   Inspection Method   111
    3.3.2   Partial Fraction Expansion   112
    3.3.3   Power Series Expansion   116
3.4   z-Transform Properties   119
    3.4.1   Linearity   119
    3.4.2   Time Shifting   120
    3.4.3   Multiplication by an Exponential Sequence   121
    3.4.4   Differentiation of $X(z)$ 122
    3.4.5   Conjugation of a Complex Sequence   123
    3.4.6   Time Reversal   123
    3.4.7   Convolution of Sequences   124
    3.4.8   Initial-Value Theorem   126
    3.4.9   Summary of Some z-Transform Properties   126
3.5   Summary   126
    Problems   127

4   SAMPLING OF CONTINUOUS-TIME SIGNALS   140

4.0   Introduction   140
4.1   Periodic Sampling   140
4.2   Frequency-Domain Representation of Sampling   142
4.3   Reconstruction of a Bandlimited Signal from Its Samples   150
4.4   Discrete-Time Processing of Continuous-Time Signals   153
    4.4.1   Linear Time-Invariant Discrete-Time Systems   154
    4.4.2   Impulse Invariance   160
4.5   Continuous-Time Processing of Discrete-Time Signals   163
4.6   Changing the Sampling Rate Using Discrete-Time Processing   167
    4.6.1   Sampling Rate Reduction by an Integer Factor   167
    4.6.2   Increasing the Sampling Rate by an Integer Factor   172
    4.6.3   Changing the Sampling Rate by a Noninteger Factor   176
4.7   Multirate Signal Processing   179
    4.7.1   Interchange of Filtering and Downsampling/Upsampling   179
    4.7.2   Polyphase Decompositions   180
    4.7.3   Polyphase Implementation of Decimation Filters   182
    4.7.4   Polyphase Implementation of Interpolation Filters   183
4.8   Digital Processing of Analog Signals   185
    4.8.1   Prefiltering to Avoid Aliasing   185
    4.8.2   Analog-to-Digital (A/D) Conversion   187
    4.8.3   Analysis of Quantization Errors   193
    4.8.4   D/A Conversion   197
4.9   Oversampling and Noise Shaping in A/D and D/A Conversion   201
    4.9.1   Oversampled A/D Conversion with Direct Quantization   201
    4.9.2   Oversampled A/D Conversion with Noise Shaping   206
    4.9.3   Oversampling and Noise Shaping in D/A Conversion   210

4.10 Summary   213
       Problems   214

5   Transform Analysis of Linear Time-Invariant Systems   240

5.0 Introduction   240
5.1 The Frequency Response of LTI Systems   241
       5.1.1 Ideal Frequency-Selective Filters   241
       5.1.2 Phase Distortion and Delay   242
5.2 System Functions for Systems Characterized by Linear Constant-Coefficient Difference Equations   245
       5.2.1 Stability and Causality   247
       5.2.2 Inverse Systems   248
       5.2.3 Impulse Response for Rational System Functions   250
5.3 Frequency Response for Rational System Functions   253
       5.3.1 Frequency Response of a Single Zero or Pole   258
       5.3.2 Examples with Multiple Poles and Zeros   265
5.4 Relationship between Magnitude and Phase   270
5.5 All-Pass Systems   274
5.6 Minimum-Phase Systems   280
       5.6.1 Minimum-Phase and All-Pass Decomposition   280
       5.6.2 Frequency-Response Compensation   282
       5.6.3 Properties of Minimum-Phase Systems   287
5.7 Linear Systems with Generalized Linear Phase   291
       5.7.1 Systems with Linear Phase   292
       5.7.2 Generalized Linear Phase   295
       5.7.3 Causal Generalized Linear-Phase Systems   297
       5.7.4 Relation of FIR Linear-Phase Systems to Minimum-Phase Systems   308
5.8 Summary   311
       Problems   312

6   Structures for Discrete-Time Systems   340

6.0 Introduction   340
6.1 Block Diagram Representation of Linear Constant-Coefficient Difference Equations   341
6.2 Signal Flow Graph Representation of Linear Constant-Coefficient Difference Equations   348
6.3 Basic Structures for IIR Systems   354
       6.3.1 Direct Forms   354
       6.3.2 Cascade Form   356
       6.3.3 Parallel Form   359
       6.3.4 Feedback in IIR Systems   361
6.4 Transposed Forms   363
6.5 Basic Network Structures for FIR Systems   366

6.5.1   Direct Form   367
6.5.2   Cascade Form   367
6.5.3   Structures for Linear-Phase FIR Systems   368
6.6   Overview of Finite-Precision Numerical Effects   370
6.6.1   Number Representations   371
6.6.2   Quantization in Implementing Systems   374
6.7   The Effects of Coefficient Quantization   377
6.7.1   Effects of Coefficient Quantization in IIR Systems   377
6.7.2   Example of Coefficient Quantization in an Elliptic Filter   379
6.7.3   Poles of Quantized Second-Order Sections   382
6.7.4   Effects of Coefficient Quantization in FIR Systems   384
6.7.5   Example of Quantization of an Optimum FIR Filter   386
6.7.6   Maintaining Linear Phase   390
6.8   Effects of Round-off Noise in Digital Filters   391
6.8.1   Analysis of the Direct-Form IIR Structures   391
6.8.2   Scaling in Fixed-Point Implementations of IIR Systems   399
6.8.3   Example of Analysis of a Cascade IIR Structure   403
6.8.4   Analysis of Direct-Form FIR Systems   410
6.8.5   Floating-Point Realizations of Discrete-Time Systems   412
6.9   Zero-Input Limit Cycles in Fixed-Point Realizations of IIR Digital Filters   413
6.9.1   Limit Cycles due to Round-off and Truncation   414
6.9.2   Limit Cycles Due to Overflow   416
6.9.3   Avoiding Limit Cycles   417
6.10   Summary   418
Problems   419

7   FILTER DESIGN TECHNIQUES   439
7.0   Introduction   439
7.1   Design of Discrete-Time IIR Filters from Continuous-Time Filters   442
7.1.1   Filter Design by Impulse Invariance   443
7.1.2   Bilinear Transformation   450
7.1.3   Examples of Bilinear Transformation Design   454
7.2   Design of FIR Filters by Windowing   465
7.2.1   Properties of Commonly Used Windows   467
7.2.2   Incorporation of Generalized Linear Phase   469
7.2.3   The Kaiser Window Filter Design Method   474
7.2.4   Relationship of the Kaiser Window to Other Windows   478
7.3   Examples of FIR Filter Design by the Kaiser Window Method   478
7.3.1   Highpass Filter   479
7.3.2   Discrete-Time Differentiators   482
7.4   Optimum Approximations of FIR Filters   486
7.4.1   Optimal Type I Lowpass Filters   491
7.4.2   Optimal Type II Lowpass Filters   497
7.4.3   The Parks–McClellan Algorithm   498

7.4.4   Characteristics of Optimum FIR Filters   501
7.5   Examples of FIR Equiripple Approximation   503
7.5.1   Lowpass Filter   503
7.5.2   Compensation for Zero-Order Hold   506
7.5.3   Bandpass Filter   507
7.6   Comments on IIR and FIR Discrete-Time Filters   510
7.7   Summary   511
Problems   511

8   THE DISCRETE FOURIER TRANSFORM   541

8.0   Introduction   541
8.1   Representation of Periodic Sequences: The Discrete Fourier Series   542
8.2   Properties of the Discrete Fourier Series   546
8.2.1   Linearity   546
8.2.2   Shift of a Sequence   546
8.2.3   Duality   547
8.2.4   Symmetry Properties   547
8.2.5   Periodic Convolution   548
8.2.6   Summary of Properties of the DFS Representation of Periodic Sequences   550
8.3   The Fourier Transform of Periodic Signals   551
8.4   Sampling the Fourier Transform   555
8.5   Fourier Representation of Finite-Duration Sequences: The Discrete Fourier Transform   559
8.6   Properties of the Discrete Fourier Transform   564
8.6.1   Linearity   564
8.6.2   Circular Shift of a Sequence   564
8.6.3   Duality   567
8.6.4   Symmetry Properties   568
8.6.5   Circular Convolution   571
8.6.6   Summary of Properties of the Discrete Fourier Transform   575
8.7   Linear Convolution Using the Discrete Fourier Transform   576
8.7.1   Linear Convolution of Two Finite-Length Sequences   577
8.7.2   Circular Convolution as Linear Convolution with Aliasing   577
8.7.3   Implementing Linear Time-Invariant Systems Using the DFT   582
8.8   The Discrete Cosine Transform (DCT)   589
8.8.1   Definitions of the DCT   589
8.8.2   Definition of the DCT-1 and DCT-2   590
8.8.3   Relationship between the DFT and the DCT-1   593
8.8.4   Relationship between the DFT and the DCT-2   594
8.8.5   Energy Compaction Property of the DCT-2   595
8.8.6   Applications of the DCT   598
8.9   Summary   599
Problems   600

9 Computation of the Discrete Fourier Transform   629

9.0   **Introduction**   629
9.1   **Efficient Computation of the Discrete Fourier Transform**   630
9.2   **The Goertzel Algorithm**   633
9.3   **Decimation-in-Time FFT Algorithms**   635
   9.3.1   In-Place Computations   640
   9.3.2   Alternative Forms   643
9.4   **Decimation-in-Frequency FFT Algorithms**   646
   9.4.1   In-Place Computation   650
   9.4.2   Alternative Forms   650
9.5   **Practical Considerations**   652
   9.5.1   Indexing   652
   9.5.2   Coefficients   654
   9.5.3   Algorithms for More General Values of N 655
9.6   **Implementation of the DFT Using Convolution**   655
   9.6.1   Overview of the Winograd Fourier Transform Algorithm   655
   9.6.2   The Chirp Transform Algorithm   656
9.7   **Effects of Finite Register Length**   661
9.8   **Summary**   669
   Problems   669

10 Fourier Analysis of Signals Using the Discrete Fourier Transform   693

10.0   **Introduction**   693
10.1   **Fourier Analysis of Signals Using the DFT**   694
10.2   **DFT Analysis of Sinusoidal Signals**   697
   10.2.1   The Effect of Windowing   698
   10.2.2   The Effect of Spectral Sampling   703
10.3   **The Time-Dependent Fourier Transform**   714
   10.3.1   The Effect of the Window   717
   10.3.2   Sampling in Time and Frequency   718
10.4   **Block Convolution Using the Time-Dependent Fourier Transform**   722
10.5   **Fourier Analysis of Nonstationary Signals**   723
   10.5.1   Time-Dependent Fourier Analysis of Speech Signals   724
   10.5.2   Time-Dependent Fourier Analysis of Radar Signals   728
10.6   **Fourier Analysis of Stationary Random Signals: The Periodogram**   730
   10.6.1   The Periodogram   731
   10.6.2   Properties of the Periodogram   733
   10.6.3   Periodogram Averaging   737
   10.6.4   Computation of Average Periodograms Using the DFT   739
   10.6.5   An Example of Periodogram Analysis   739

10.7 **Spectrum Analysis of Random Signals Using Estimates of the Autocorrelation Sequence** **743**

    10.7.1 Computing Correlation and Power Spectrum Estimates Using the DFT 746

    10.7.2 An Example of Power Spectrum Estimation Based on Estimation of the Autocorrelation Sequence 748

10.8 **Summary** **754**

    Problems **755**

11 **DISCRETE HILBERT TRANSFORMS** **775**

11.0 **Introduction** **775**

11.1 **Real- and Imaginary-Part Sufficiency of the Fourier Transform for Causal Sequences** **777**

11.2 **Sufficiency Theorems for Finite-Length Sequences** **782**

11.3 **Relationships Between Magnitude and Phase** **788**

11.4 **Hilbert Transform Relations for Complex Sequences** **789**

    11.4.1 Design of Hilbert Transformers 792

    11.4.2 Representation of Bandpass Signals 796

    11.4.3 Bandpass Sampling 799

11.5 **Summary** **801**

    Problems **802**

APPENDIX A RANDOM SIGNALS **811**

A.1 **Discrete-Time Random Processes** **811**

A.2 **Averages** **813**

    A.2.1 Definitions 813

    A.2.2 Time Averages 815

A.3 **Properties of Correlation and Covariance Sequences** **817**

A.4 **Fourier Transform Representation of Random Signals** **818**

A.5 **Use of the *z*-Transform in Average Power Computations** **820**

APPENDIX B CONTINUOUS-TIME FILTERS **824**

B.1 **Butterworth Lowpass Filters** **824**

B.2 **Chebyshev Filters** **826**

B.3 **Elliptic Filters** **828**

APPENDIX C ANSWERS TO SELECTED BASIC PROBLEMS **830**

BIBLIOGRAPHY **851**

INDEX **859**

LIST OF EXAMPLES

Example 2.1	Combining Basic Sequences	13
Example 2.2	Periodic and Aperiodic Discrete-Time Sinusoids	15
Example 2.3	The Ideal Delay System	17
Example 2.4	Moving Average	17
Example 2.5	A Memoryless System	18
Example 2.6	The Accumulator System	19
Example 2.7	A Nonlinear System	19
Example 2.8	The Accumulator as a Time-Invariant System	20
Example 2.9	The Compressor System	20
Example 2.10	The Forward and Backward Difference Systems	21
Example 2.11	Testing for Stability or Instability	22
Example 2.12	Computation of the Convolution Sum	25
Example 2.13	Analytical Evaluation of the Convolution Sum	26
Example 2.14	Difference Equation Representation of the Accumulator	34
Example 2.15	Difference Equation Representation of the Moving-Average System	35
Example 2.16	Recursive Computation of Difference Equations	37
Example 2.17	Frequency Response of the Ideal Delay System	41
Example 2.18	Sinusoidal Response of LTI Systems	42
Example 2.19	Ideal Frequency-Selective Filters	43
Example 2.20	Frequency Response of the Moving-Average System	44
Example 2.21	Absolute Summability for a Suddenly-Applied Exponential	51
Example 2.22	Square-Summability for the Ideal Lowpass Filter	52
Example 2.23	Fourier Transform of a Constant	53
Example 2.24	Fourier Transform of Complex Exponential Sequences	54
Example 2.25	Illustration of Symmetry Properties	57
Example 2.26	Determining a Fourier Transform Using Tables 2.2 and 2.3	63
Example 2.27	Determining an Inverse Fourier Transform Using Tables 2.2 and 2.3	63
Example 2.28	Determining the Impulse Response from the Frequency Response	64
Example 2.29	Determining the Impulse Response for a Difference Equation	64
Example 2.30	White Noise	69
Example 3.1	Right-Sided Exponential Sequence	98
Example 3.2	Left-Sided Exponential Sequence	99
Example 3.3	Sum of Two Exponential Sequences	100
Example 3.4	Sum of Two Exponentials (Again)	101
Example 3.5	Two-Sided Exponential Sequence	102
Example 3.6	Finite-Length Sequence	103
Example 3.7	Stability, Causality, and the ROC	110

...ters ....on with Aliasing 572
...onship Using the DFT 573
...onship Between DFT Values 579
.... 596
.... 661
.... 697
.... 697

Example 3.8
Example 3.9   Second-Order z-Transf
Example 3.10   Inverse by Par
Example 3.11   Finite L

Example 10.3   Effect of Windowing on Fourier Analysis of Sinusoidal
                 Signals ....

Example 10.4   Illustration of the Effect of Spectral Sampling ....

Example 10.5   Spectral Sampling with Frequencies Matching DFT
                 Frequencies ....

Example 10.6   DFT Analysis of Sinusoidal Signals Using a Kaiser Window .....

Example 10.7   DFT Analysis with 32-point Kaiser Window and
                 Zero-Padding ....

Example 10.8   Oversampling and Linear Interpolation for Frequency
                 Estimation ....

Example 10.9   Time-Dependent Fourier Transform of a Linear Chirp Signal ...

Example 10.10   Spectrogram Display of the Time-Dependent Fourier
                 Transform of Speech ....

Example 11.1   Finite-Length Sequence ....

Example 11.2   Exponential Sequence ....

Example 11.3   Periodic Sequence ....

Example 11.4   Kaiser Window Design of Hilbert Transformers ....

Example A.1   Noise Power Output of Ideal Lowpass Filter......................

Example A.2   Noise Power Output of a Second-Order IIR Filter ....

E
E
Ex
Exa
Exan
Exam

LIST OF EXAMPLES

Example 2.1	Combining Basic Sequences	13
Example 2.2	Periodic and Aperiodic Discrete-Time Sinusoids	15
Example 2.3	The Ideal Delay System	17
Example 2.4	Moving Average	17
Example 2.5	A Memoryless System	18
Example 2.6	The Accumulator System	19
Example 2.7	A Nonlinear System	19
Example 2.8	The Accumulator as a Time-Invariant System	20
Example 2.9	The Compressor System	20
Example 2.10	The Forward and Backward Difference Systems	21
Example 2.11	Testing for Stability or Instability	22
Example 2.12	Computation of the Convolution Sum	25
Example 2.13	Analytical Evaluation of the Convolution Sum	26
Example 2.14	Difference Equation Representation of the Accumulator	34
Example 2.15	Difference Equation Representation of the Moving-Average System	35
Example 2.16	Recursive Computation of Difference Equations	37
Example 2.17	Frequency Response of the Ideal Delay System	41
Example 2.18	Sinusoidal Response of LTI Systems	42
Example 2.19	Ideal Frequency-Selective Filters	43
Example 2.20	Frequency Response of the Moving-Average System	44
Example 2.21	Absolute Summability for a Suddenly-Applied Exponential	51
Example 2.22	Square-Summability for the Ideal Lowpass Filter	52
Example 2.23	Fourier Transform of a Constant	53
Example 2.24	Fourier Transform of Complex Exponential Sequences	54
Example 2.25	Illustration of Symmetry Properties	57
Example 2.26	Determining a Fourier Transform Using Tables 2.2 and 2.3	63
Example 2.27	Determining an Inverse Fourier Transform Using Tables 2.2 and 2.3	63
Example 2.28	Determining the Impulse Response from the Frequency Response	64
Example 2.29	Determining the Impulse Response for a Difference Equation	64
Example 2.30	White Noise	69
Example 3.1	Right-Sided Exponential Sequence	98
Example 3.2	Left-Sided Exponential Sequence	99
Example 3.3	Sum of Two Exponential Sequences	100
Example 3.4	Sum of Two Exponentials (Again)	101
Example 3.5	Two-Sided Exponential Sequence	102
Example 3.6	Finite-Length Sequence	103
Example 3.7	Stability, Causality, and the ROC	110

Example 3.8   Second-Order z-Transform 113
Example 3.9   Inverse by Partial Fractions.....................................115
Example 3.10   Finite-Length Sequence 117
Example 3.11   Inverse Transform by Power Series Expansion 117
Example 3.12   Power Series Expansion by Long Division 118
Example 3.13   Power Series Expansion for a Left-Sided Sequence 118
Example 3.14   Shifted Exponential Sequence....................................120
Example 3.15   Exponential Multiplication 121
Example 3.16   Inverse of Non-Rational z-Transform 122
Example 3.17   Second-Order Pole..123
Example 3.18   Time-Reversed Exponential Sequence............................124
Example 3.19   Evaluating a Convolution Using the z-Transform 125
Example 4.1   Sampling and Reconstruction of a Sinusoidal Signal 147
Example 4.2   Aliasing in the Reconstruction of an Undersampled
                Sinusoidal Signal..148
Example 4.3   A Second Example of Aliasing 149
Example 4.4   Ideal Continuous-Time Lowpass Filtering Using a
                Discrete-Time Lowpass Filter 155
Example 4.5   Discrete-Time Implementation of an Ideal
                Continuous-Time Bandlimited Differentiator 158
Example 4.6   Illustration of Example 4.5 with a Sinusoidal Input..............159
Example 4.7   A Discrete-Time Lowpass Filter Obtained By Impulse
                Invariance 162
Example 4.8   Impulse Invariance Applied to Continuous-Time Systems
                with Rational System Functions 162
Example 4.9   Noninteger Delay...164
Example 4.10   Moving-Average System with Noninteger Delay 165
Example 4.11   Sampling Rate Conversion by a Noninteger Rational Factor.....177
Example 4.12   Quantization Error For a Sinusoidal Signal 194
Example 5.1   Effects of Attenuation and Group Delay 243
Example 5.2   Second-Order System...246
Example 5.3   Determining the ROC 247
Example 5.4   Inverse System for First-Order System...........................249
Example 5.5   Inverse for System with a Zero in the ROC.......................250
Example 5.6   A First-Order IIR System 251
Example 5.7   A Simple FIR System...252
Example 5.8   Second-Order IIR System 265
Example 5.9   Second-Order FIR System..268
Example 5.10   Third-Order IIR System 268
Example 5.11   Systems with the Same $C(z)$......................................271
Example 5.13   First- and Second-Order All-Pass Systems.......................275
Example 5.14   Minimum-Phase/All-Pass Decomposition 281
Example 5.15   Compensation of an FIR System 283
Example 5.16   Ideal Lowpass with Linear Phase.................................293
Example 5.17   Type I Linear-Phase System 300
Example 5.18   Type II Linear-Phase System 302
Example 5.19   Type III Linear-Phase System 302

Example 5.20   Type IV Linear-Phase System 302
Example 5.21   Decomposition of a Linear-Phase System 308
Example 6.1   Block Diagram Representation of a Difference Equation 342
Example 6.2   Direct Form I and Direct Form II Implementation of an LTI
                System 347
Example 6.3   Determination of the System Function from a Flow Graph 352
Example 6.4   Illustration of Direct Form I and Direct Form II Structures 355
Example 6.5   Illustration of Cascade Structures 358
Example 6.6   Illustration of Parallel-Form Structures 360
Example 6.7   Transposed Form for a First-Order System with No Zeroes 363
Example 6.8   Transposed Form for a Basic Second-Order Section 364
Example 6.9   Round-off Noise in a First-Order System 396
Example 6.10   Round-off Noise in a Second-Order System 397
Example 6.11   Interaction Between Scaling and Round-off Noise 402
Example 6.12   Scaling Considerations for the FIR System in Section 6.7.5 411
Example 6.13   Limit Cycle Behavior in a First-Order System 414
Example 6.14   Overflow Oscillations in a Second-Order System 416
Example 7.1   Determining Specifications for a Discrete-Time Filter 440
Example 7.2   Impulse Invariance with a Butterworth Filter 446
Example 7.3   Bilinear Transformation of a Butterworth Filter 454
Example 7.4   Butterworth Approximation 458
Example 7.5   Chebyshev Approximation 460
Example 7.6   Elliptic Approximation 463
Example 7.7   Linear-Phase Lowpass Filter 472
Example 7.8   Kaiser Window Design of a Lowpass Filter 476
Example 7.9   Kaiser Window Design of a Highpass Filter 479
Example 7.10   Kaiser Window Design of a Differentiator 483
Example 7.11   Alternation Theorem and Polynomials 490
Example 8.1   Discrete Fourier Series of a Periodic Impulse Train 544
Example 8.2   Duality in the Discrete Fourier Series 544
Example 8.3   The Discrete Fourier Series of a Periodic Rectangular Pulse
                Train 545
Example 8.4   Periodic Convolution 549
Example 8.5   The Fourier Transform of a Periodic Impulse Train 552
Example 8.6   Relationship Between the Fourier Series Coefficients and
                the Fourier Transform of One Period 554
Example 8.7   The DFT of a Rectangular Pulse 561
Example 8.8   Circular Shift of a Sequence 566
Example 8.9   The Duality Relationship for the DFT 568
Example 8.10   Circular Convolution with a Delayed Impulse Sequence 572
Example 8.11   Circular Convolution of Two Rectangular Pulses 573
Example 8.12   Circular Convolution as Linear Convolution with Aliasing 579
Example 8.13   Energy Compaction in the DCT-2 596
Example 9.1   Chirp Transform Parameters 661
Example 10.1   Fourier Analysis Using the DFT 697
Example 10.2   Relationship Between DFT Values 697

Example 10.3   Effect of Windowing on Fourier Analysis of Sinusoidal
Signals 698

Example 10.4   Illustration of the Effect of Spectral Sampling 703

Example 10.5   Spectral Sampling with Frequencies Matching DFT
Frequencies 706

Example 10.6   DFT Analysis of Sinusoidal Signals Using a Kaiser Window 708

Example 10.7   DFT Analysis with 32-point Kaiser Window and
Zero-Padding 711

Example 10.8   Oversampling and Linear Interpolation for Frequency
Estimation 713

Example 10.9   Time-Dependent Fourier Transform of a Linear Chirp Signal ... 715

Example 10.10   Spectrogram Display of the Time-Dependent Fourier
Transform of Speech 725

Example 11.1   Finite-Length Sequence 779

Example 11.2   Exponential Sequence 779

Example 11.3   Periodic Sequence 787

Example 11.4   Kaiser Window Design of Hilbert Transformers 793

Example A.1   Noise Power Output of Ideal Lowpass Filter..................... 820

Example A.2   Noise Power Output of a Second-Order IIR Filter 823

PREFACE

This text is a second generation descendent of our text, *Digital Signal Processing*, which was published in 1975. At that time, the technical field of digital signal processing was in its infancy, but certain basic principles had emerged and could be organized into a coherent presentation. Although courses existed at a few schools, they were almost exclusively at the graduate level. The original text was designed for such courses.

By 1985, the pace of research and integrated circuit technology made it clear that digital signal processing would realize the potential that had been evident in the 1970s. The burgeoning importance of DSP clearly justified a revision and updating of the original text. However, in organizing that revision, it was clear that so many changes had occurred that it was most appropriate to develop a new textbook, strongly based on our original text, while keeping the original text in print. We titled the new book *Discrete-Time Signal Processing* to emphasize that most of the theory and design techniques discussed in the text apply to discrete-time systems in general.

By the time *Discrete-Time Signal Processing* was published in 1989, the basic principles of DSP were commonly taught at the undergraduate level, sometimes even as part of a first course on linear systems, or at a somewhat more advanced level in third-year, fourth-year, or beginning graduate subjects. Therefore, it was appropriate to expand considerably the treatment of such topics as linear systems, sampling, multirate signal processing, applications, and spectral analysis. In addition, more examples were included to emphasize and illustrate important concepts. We also removed and condensed some topics that time had shown were not fundamental to the understanding of discrete-time signal processing. Consistent with the importance that we placed on well constructed examples and homework problems, the new text contained more than 400 problems.

In the decade or so since *Discrete-Time Signal Processing* was published, some important new concepts have been developed, the capability of digital integrated circuits has grown exponentially, and an increasing number of applications have emerged. However, the underlying basics and fundamentals remain largely the same albeit with a refinement of emphasis, understanding and pedagogy. Consequently when we looked at what was needed to keep *Discrete-Time Signal Processing* up-to-date as a textbook emphasizing the fundamentals of DSP, we found that the changes needed were far less drastic than before. In planning this current revision we were guided by the principle that the main objective of a fundamental textbook is to uncover a subject rather than to cover it. Consequently, our goal in this current revision is to make the subject of discrete-time signal processing even more accessible to students and practicing engineers, without compromising on coverage of what we consider to be the essential concepts that define the field. Toward this end we have considerably expanded our coverage of multi-rate signal processing due to its importance in oversampled A-to-D and D-to-A conversion and digital filter implementation. We have added a discussion of the cosine transform, which plays a central role in data compression standards. We have also removed some material that we judged to be of lesser importance in the present context, or more appropriate for advanced textbooks and upper level graduate courses. Many of the concepts that were removed from the text (such as basic results on the cepstrum) have reappeared in some of the new homework problems.

A major part of our emphasis in this revision has been directed toward the homework problems and examples. We have significantly increased the number of examples which are important in illustrating and understanding the basic concepts, and we have increased the number of homework problems. Furthermore, the homework problems have been reorganized according to their level of difficulty and sophistication, and answers are provided to a selected set of problems. The instructor's manual available from the publisher contains updated solutions for all of the problems in the book. These solutions were prepared by Li Lee and Maya Said of MIT and Jordan Rosenthal and Greg Slabaugh of Georgia Tech. This manual also contains some suggested exam problems based on our courses at MIT, Georgia Tech and the University of Massachusetts Dartmouth.

As in the earlier texts, it is assumed that the reader has a background of advanced calculus, along with a good understanding of the elements of complex numbers and variables. In this edition, we have refrained from the use of complex contour integration in order to make the discussion accessible to a wider audience. An exposure to linear system theory for continuous-time signals, including Laplace and Fourier transforms, as taught in most undergraduate electrical and mechanical engineering curricula is still a basic prerequisite. With this background, the book is self-contained. In particular, no prior experience with discrete-time signals, z-transforms, discrete Fourier transforms, and the like is assumed. In later sections of some chapters, some topics such as quantization noise are included that assume a basic background in stochastic signals. A brief review of the background for these sections is included in Chapter 2 and in Appendix A.

It has become common in many signal processing courses to include exercises to be done on a computer, and many of the homework problems in this book are easily turned into problems to be solved with the aid of a computer. As in the first edition, we have purposely avoided providing special software to implement algorithms described in this book, for a variety of reasons. Foremost among them is that there are a variety of inexpensive signal processing software packages readily available for demonstrating and implementing signal processing on any of the popular personal computers and workstations. These packages are well documented and have excellent technical support, and many of them have excellent user interfaces that make them easily accessible to students. Furthermore, they are in a constant state of evolution, which strongly suggests that available software for classroom use should be constantly reviewed and updated. We share the enthusiasm of many for MATLAB, which an increasing number of students are learning at early stages of their education. However, we continue to prefer a presentation that utilizes the power of computational tools such as MATLAB to create examples and illustrations of the theory and fundamentals for use in the text, but does not let issues of programming syntax and functionality of the software environment detract from the emphasis on the concepts and the way that they are used. We firmly believe that there is enormous value in hands-on experience. Indeed, software tools such as MATLAB allow students to implement sophisticated signal processing systems on their own personal computers, and we feel that there is great benefit to this once the student is confident of the fundamentals and is capable of sorting out programming mistakes from conceptual errors. For this reason, the instructor's manual contains a section of suggestions for assignments in the inexpensive texts *Computer-Based Exercises for Signal Processing Using Matlab 5* by McClellan, et al., and *Computer Explorations in Signals and Systems Using Matlab* by Buck, Daniel and Singer, both of which are also available from Prentice-Hall, Inc. These suggestions link projects in these computer exercise books to specific sections, examples and problems in this textbook. This will allow instructors to design computer assignments which are related to the material and examples they have covered in class, and to link these computer assignments to traditional analytic homework problems to reinforce the concepts demonstrated there.

The material in this book is organized in a way that provides considerable flexibility in its use at both the undergraduate and graduate level. A typical one-semester undergraduate elective might cover in depth Chapter 2, Sections 2.0–2.9; Chapter 3; Chapter 4, Sections 4.0–4.6; Chapter 5, Sections 5.0–5.3; Chapter 6, Sections 6.0–6.5; Chapter 7, Sections 7.0–7.3 and a brief overview of Sections 7.4–7.5. If students have studied discrete-time signals and systems in a general signals and systems course, it would be possible to move more quickly through the material of Chapters 2, 3, and 4, thus freeing time for covering Chapter 8. A first-year graduate course could augment the above topics with the remaining topics in Chapter 5, a discussion of multirate signal processing (Section 4.7) an exposure to some of the quantization issues introduced in Section 4.8 and perhaps an introduction to noise shaping in A/D and D/A converters as discussed in Section 4.9. A first-year graduate course should also include exposure to some of the quantization issues addressed in Sections 6.6–6.9, to a discussion of optimal FIR filters as incorporated in Sections 7.4 and 7.5, and a thorough treatment of the discrete Fourier transform (Chapter 8) and its computation using the FFT (Chapter 9). The discussion of the DFT can be effectively augmented with many of the examples in Chapter 10. In a two-semester graduate course, the entire text together with a number of additional advanced topics can be covered.

In Chapter 2, we introduce the basic class of discrete-time signals and systems and define basic system properties such as linearity, time invariance, stability, and causality. The primary focus of the book is on linear time-invariant systems because of the rich set of tools available for designing and analyzing this class of systems. In particular, in Chapter 2 we develop the time-domain representation of linear time-invariant systems through the convolution sum and introduce the class of linear time-invariant systems represented by linear constant-coefficient difference equations. In Chapter 6, we develop this class of systems in considerably more detail. Also in Chapter 2 we introduce the frequency-domain representation of signals and systems through the Fourier transform. The primary focus in Chapter 2 is on the representation of sequences in terms of the Fourier transform, i.e., as a linear combination of complex exponentials, and the development of the basic properties of the Fourier transform.

In Chapter 3, we develop the z-transform as a generalization of the Fourier transform. This chapter focuses on developing the basic theorems and properties of the z-transform and the development of the partial fraction expansion method for the inverse transform operation. In Chapter 5, the results developed in Chapters 3 and 4 are used extensively in a detailed discussion of the representation and analysis of linear time-invariant systems.

In Chapter 4, we carry out a detailed discussion of the relationship between continuous-time and discrete-time signals when the discrete-time signals are obtained through periodic sampling of continuous-time signals. This includes a development of the Nyquist sampling theorem. In addition, we discuss upsampling and downsampling of discrete-time signals, as used, for example, in multirate signal processing systems and for sampling rate conversion. The chapter concludes with a discussion of some of the practical issues encountered in conversion from continuous time to discrete time including prefiltering to avoid aliasing, modeling the effects of amplitude quantization when the discrete-time signals are represented digitally, and the use of oversampling in simplifying the A-to-D and D-to-A conversion processes.

In Chapter 5 we apply the concepts developed in the previous chapters to a detailed study of the properties of linear time-invariant systems. We define the class of ideal, frequency-selective filters and develop the system function and pole-zero representation for systems described by linear constant-coefficient difference equations, a class of systems whose implementation is considered in detail in Chapter 6. Also in Chapter 5, we define and discuss group delay, phase response and phase distortion, and the relationships between the magnitude response and the phase response of systems, including a discussion of minimum-phase, allpass, and generalized linear phase systems.

In Chapter 6, we focus specifically on systems described by linear constant-coefficient difference equations and develop their representation in terms of block diagrams and linear signal flow graphs. Much of this chapter is concerned with developing a variety of the important system structures and comparing some of their properties. The importance of this discussion and the variety of filter structures relate to the fact that in a practical implementation of a discrete-time system, the effects of coefficient inaccuracies and arithmetic error can be very dependent on the specific structure used. While these basic issues are similar whether the technology used for implementation is digital or discrete-time analog, we illustrate them in this chapter in the context of a digital implementation through a discussion of the effects of coefficient quantization and arithmetic roundoff noise for digital filters.

While Chapter 6 is concerned with the representation and implementation of linear constant-coefficient difference equations, Chapter 7 is a discussion of the procedures for obtaining the coefficients of this class of difference equations to approximate a desired system response. The design techniques separate into those used for infinite impulse response (IIR) filters and those used for finite impulse response (FIR) filters.

In continuous-time linear system theory, the Fourier transform is primarily an analytical tool for representing signals and systems. In contrast, in the discrete-time case, many signal processing systems and algorithms involve the explicit computation of the Fourier transform. While the Fourier transform itself cannot be computed, a sampled version of it, the discrete Fourier transform (DFT), can be computed, and for finite-length signals the DFT is a complete Fourier representation of the signal. In Chapter 8, the discrete Fourier transform is introduced and its properties and relationship to the discrete-time Fourier transform are developed in detail. In this chapter we also provide an introduction to the discrete cosine transform which is playing an increasingly important role in many applications including audio and video compression. In Chapter 9, the rich and important variety of algorithms for computing or generating the discrete Fourier transform is introduced and discussed, including the Goertzel algorithm, the fast Fourier transform (FFT) algorithms, and the chirp transform.

With the background developed in the earlier chapters and particularly Chapters 2, 3, 5, and 8, we focus in Chapter 10 on Fourier analysis of signals using the discrete Fourier transform. Without a careful understanding of the issues involved and the relationship between the DFT and the Fourier transform, using the DFT for practical signal analysis can often lead to confusions and misinterpretations. We address a number of these issues in Chapter 10. We also consider in some detail the Fourier analysis of signals with time-varying characteristics by means of the time-dependent Fourier transform.

In Chapter 11, we introduce the discrete Hilbert transform. This transform arises in a variety of practical applications, including inverse filtering, complex representations for real bandpass signals, single-sideband modulation techniques, and many others.

With this edition we thank and welcome Professor John Buck. John has been a long time contributor to this book through his teaching of the subject while a student at MIT and more recently as a member of the faculty at the University of Massachusetts Dartmouth. In this edition he has taken the major responsibility for a total reworking and reorganization of the homework problems and many of the examples in the book. His insight and dedication to the task are obvious in the final result.

Alan V. Oppenheim
Ronald W. Schafer

ACKNOWLEDGMENTS

In preparing the two editions of this book, we have been fortunate to receive valuable assistance, suggestions, and support from numerous colleagues, students, and friends. Over the years a number of our colleagues have taught the material with us at MIT and Georgia Tech, and we have benefited greatly from their perspectives and input. These colleagues include Professors Arthur Baggeroer, Sidney Burrus, Meir Feder, Jae Lim, Bruce Musicus, Hamid Nawab, Gregory Wornell and Victor Zue at MIT and Professors Tom Barnwell, Mark Clements, Monty Hayes, Jim McClellan, Russ Mersereau, David Schwartz, Mark Smith, Vijay Madisetti, Doug Williams, and Tong Zhou at Georgia Tech.

MIT and Georgia Tech have provided us with a stimulating environment for research and teaching throughout a major part of our technical careers and have provided significant encouragement and support for this project. In addition RWS particularly thanks W. Kelley Mosley for his friendship and support, and the John and Mary Franklin Foundation for many years of support through the John and Marilu McCarty Chair. AVO expresses deep appreciation to Mr. Ray Stata and Analog Devices, Inc. and to the Ford Foundation for their generous and continued support of the signal processing activities at MIT including the funding of the Distinguished Professor Chair in Electrical Engineering and the Ford Chair in Engineering.

We feel extremely fortunate to have worked with Prentice Hall. Our relationship with Prentice Hall spans many years and many writing projects. The encouragement and support provided for this current edition by Eileen Clark, Marcia Horton, Tom Robbins, Amy Rosen, and Sharyn Vitrano at Prentice Hall enhance the enjoyment of writing and completing this one.

In producing this second edition, we were fortunate to receive the help of many colleagues, students, and friends. We greatly appreciate their generosity in devoting their time to help us with this project. Specifically, we express our thanks to:

Li Lee and Maya Said of MIT and Jordan Rosenthal and Greg Slabaugh of Georgia Tech for preparing the solution manual for the homework problems, and Hu Dou of the University of Massachusetts Dartmouth for his work on the answers to basic problems.

Wade Torres, Akmal Butt and Faramarz Fekri for their assistance in updating the bibliography.

Vivian Berman for her help in designing the new cover.

Darla Chupp, Stacy Schultz and Kay Gilstrap for their assistance with preparation of this revision and continued support of our teaching activities.

Matthew Secor and Giovanni Aliberti for their help with the many computer issues related to preparation of the manuscript.

And to all who helped in careful reviewing of the manuscript and page proofs:

Susan Alderman, Jon Arrowood, Joe Arrowood, Chalee Asavathiratham, Halük Aydınoğlu, Ali Behboodian, Albert Chan, Matthew Cobb, Yonina Eldar,

Christoforos Hadjicostis, Chris Lanciani, Nicholas Laneman, Li Lee, Michael Lopez, Fernando Mujica, Burhan Necioğlu, Ara Nefian, Eric Reed, Andrew Russell, Maya Said, and Trevor Trinkaus.

1
INTRODUCTION

Signal processing has a long and rich history. It is a technology that spans an immense set of disciplines including entertainment, communications, space exploration, medicine, and archaeology, just to name a few. Sophisticated signal processing algorithms and hardware are prevalent in a wide range of systems, from highly specialized military systems through industrial applications to low-cost, high-volume consumer electronics. Although we routinely take for granted the performance of home entertainment systems such as television and high-fidelity audio, these systems have always relied heavily on state-of-the-art signal processing. This is even more true today with the emergence of advanced television and multimedia entertainment and information systems. Furthermore, as communication systems become increasingly wireless, mobile, and multifunctional, the importance of sophisticated signal processing in these systems continues to grow. Overall, as we look to the future, it is clear that the role of signal processing in our society is accelerating, driven in part by the convergence of communications, computers and signal processing in both the consumer arena and in advanced industrial and government applications.

The field of signal processing has always benefited from a close coupling between theory, applications, and technologies for implementing signal processing systems. The growing number of applications and demand for increasingly sophisticated algorithms goes hand-in-hand with the rapid pace of device technology for implementing signal processing systems. By some estimates the processing capability of signal processing microprocessors is likely to increase by a factor of 200 or more over the next ten years. It seems clear that in many ways the importance and role of signal processing is accelerating and expanding.

Signal processing is concerned with the representation, transformation, and manipulation of signals and the information they contain. For example, we may wish to

1

separate two or more signals that have somehow been combined, or we may want to enhance some signal component or some parameter of a signal model. In communications systems, it is generally necessary to do pre-processing such as modulation, signal conditioning, and compression prior to transmission over a channel and then to post process at the receiver. Prior to the 1960s, the technology for signal processing was almost exclusively continuous-time analog technology.[1] The rapid evolution of digital computers and microprocessors together with some important theoretical developments such as the fast Fourier transform algorithm (FFT) caused a major shift to digital technologies, giving rise to the field of digital signal processing. A fundamental aspect of digital signal processing is that it is based on processing sequences of samples. This discrete-time nature of digital signal processing technology is also characteristic of other signal processing technologies such as surface acoustic wave (SAW) devices, charge-coupled devices (CCDs), charge transport devices (CTDs), and switched-capacitor technologies. In digital signal processing, signals are represented by sequences of finite-precision numbers, and processing is implemented using digital computation. The more general term *discrete-time signal processing* includes digital signal processing as a special case, but also includes the possibility that sequences of samples (sampled data) are processed with other discrete-time technologies. Often the distinction between the terms discrete-time signal processing and digital signal processing is of minor importance, since both are concerned with discrete-time signals. While there are many examples in which signals to be processed are inherently sequences, most applications involve the use of discrete-time technology for processing continuous-time signals. In this case, a continuous-time signal is converted into a sequence of samples, i.e., a discrete-time signal. After discrete-time processing, the output sequence is converted back to a continuous-time signal. Real-time operation is often desirable for such systems, meaning that the discrete-time system is implemented so that samples of the output are computed at the same rate at which the continuous-time signal is sampled. Discrete-time processing of continuous-time signals in real time is commonplace in communication systems, radar and sonar, speech and video coding and enhancement, and biomedical engineering to name just a few. The compact disc player is a somewhat different example in which a processed form of the input is stored (on the compact disc) and final processing is carried out in real time when the output is desired. The compact disc recording and playback system relies on many of the signal processing concepts which we discuss in this book.

Much of traditional signal processing involves processing one signal to obtain another signal. Another important class of signal processing problems is *signal interpretation*. In such problems the objective of the processing is not to obtain an output signal but to obtain a characterization of the input signal. For example, in a speech recognition or understanding system, the objective is to interpret the input signal or extract information from it. Typically, such a system will apply digital preprocessing (filtering, parameter estimation, etc.) followed by a pattern recognition system to produce a symbolic representation such as a phonemic transcription of the speech. This symbolic output can in turn be the input to a symbolic processing system, such as a rule-based expert system, to

---

[1] In a general context, we typically refer to the independent variable as "time" even though in specific contexts the independent variable may take on any of a broad range of possible dimensions. Consequently, continuous time and discrete time should be thought of as generic terms referring to a continuous independent variable and a discrete independent variable, respectively.

provide the final signal interpretation. Still another and relatively new category of signal processing involves the symbolic manipulation of signal processing expressions. This type of processing is particularly useful in signal processing workstations and for the computer-aided design of signal processing systems. In this class of processing, signals and systems are represented and manipulated as abstract data objects. Object-oriented programming languages provide a convenient environment for manipulating signals, systems, and signal processing expressions without explicitly evaluating the data sequences and provide the basis for this class of processing. The sophistication of systems designed to do signal expression processing is directly influenced by the incorporation of fundamental signal processing concepts, theorems, and properties such as those that form the basis for this book. For example, a signal processing environment that incorporates the property that convolution in the time domain corresponds to multiplication in the frequency domain can explore a variety of rearrangements of filtering structures, including those involving the direct use of the discrete Fourier transform and the fast Fourier transform algorithm. Similarly, environments that incorporate the relationship between sampling rate and aliasing can make effective use of decimation and interpolation strategies for filter implementation. Similar ideas are currently being explored for implementing signal processing in network environments. In this type of environment, data can potentially be tagged with a high-level description of the processing to be done and the details of the implementation can be based dynamically on the resources available on the network.

The development of object-oriented environments for computer-aided system design and for signal processing on dynamically changing networks is still in its very early stages and any detailed discussion of it is beyond the scope of this text. However, it is important to recognize that the basic concepts that are the subject of this book should not be viewed as just theoretical in nature; they are likely to become an explicit integral part of computer-aided signal processing environments, workstations, and networks.

Many of the concepts and design techniques discussed in this text are now incorporated into the structure of sophisticated software systems such as Matlab. In many cases where discrete-time signals are acquired and stored in computers, these tools allow extremely sophisticated signal processing operations to be formed from basic functions. In such cases, it is not generally necessary to know the details of the underlying algorithm that implements the computation of an operation like the FFT, but it is essential to understand what is computed and how it should be interpreted. In other words, a good understanding of the concepts considered in this text is essential for intelligent use of the signal processing software tools that are now widely available.

Signal processing problems are not confined, of course, to one-dimensional signals. Although there are some fundamental differences in the theories for one-dimensional and multidimensional signal processing, much of the material that we discuss in this text has a direct counterpart in multidimensional systems. The theory of multidimensional digital signal processing is presented in detail in Dudgeon and Mersereau (1984), Lim (1989), and Bracewell (1986).[2] Many image processing applications require the use of two-dimensional signal processing techniques. This is the case in such areas as video coding, medical imaging, enhancement and analysis of aerial photographs, analysis

---

[2]Authors names and dates are used throughout the text to refer to books and papers listed in the Bibliography at the end of the book.

of satellite weather photos, and enhancement of video transmissions from lunar and deep-space probes. Applications of multidimensional digital signal processing to image processing are discussed in Andrews and Hunt (1977), Macovski (1983), Pratt (1991), Castleman (1996), Jain (1989), and Chellappa et al. (1998). Seismic data analysis as required in oil exploration, earthquake measurement, and nuclear test monitoring also utilizes multidimensional signal processing techniques. Seismic applications are discussed in Robinson and Treitel (1980) and Robinson and Durrani (1985).

Multidimensional signal processing is only one of many advanced and specialized topics that build on the fundamentals covered in this text. Spectral analysis based on the use of the discrete Fourier transform and the use of signal modeling is another particularly rich and important aspect of signal processing. We introduce many facets of this topic, focusing on the basic concepts and techniques relating to the use of the discrete Fourier transform. In addition to these techniques, a variety of spectral analysis methods rely in one way or another on specific signal models. For example, a class of high-resolution spectral analysis methods referred to as maximum entropy methods (MEM spectral analysis) is based on representing the signal to be analyzed as the response of a discrete-time linear time-invariant filter to either an impulse or to white noise. Spectral analysis is achieved by estimating the parameters (e.g., the difference equation coefficients) of the system and then evaluating the magnitude squared of the frequency response of the model filter. A thorough and detailed treatment of the issues and techniques of this approach to signal modeling and spectral analysis builds directly from the fundamentals in this text. Detailed discussions can be found in the texts by Kay (1988), Marple (1987), and Hayes (1996). Signal modeling also plays an important role in data compression and coding, and again the fundamentals of difference equations provide the basis for understanding many of these techniques. For example, one class of signal coding techniques, referred to as linear predictive coding (LPC), exploits the notion that if a signal is the response of a certain class of discrete-time filters, the signal value at any time index is a linear function of (and thus linearly predictable from) previous values. Consequently, efficient signal representations can be obtained by estimating these prediction parameters and using them along with the prediction error to represent the signal. The signal can then be regenerated when needed from the model parameters. This class of signal coding techniques has been particularly effective in speech coding and is described in considerable detail in Jayant and Noll (1984), Markel and Gray (1976), Rabiner and Schafer (1978), and Deller et al. (1993).

Another advanced topic of considerable importance is adaptive signal processing. Adaptive systems represent a particular class of time-varying and, in some sense, non-linear systems with broad application and with established and effective techniques for their design and analysis. Again, many of these techniques build from the fundamentals of discrete-time signal processing covered in this text. Details of adaptive signal processing are given by Haykin (1996), and Widrow and Stearns (1985).

These represent only a few of the many advanced topics that extend from the topics covered in this text. Others include advanced and specialized filter design procedures, a variety of specialized algorithms for evaluation of the Fourier transform, specialized filter structures, and various advanced multirate signal processing techniques including filter banks and wavelet transforms.

It is often said that the purpose of a fundamental textbook should be to uncover rather than cover a subject, and in choosing the topics and depth of coverage in this book we have been guided by this philosophy. The preceding brief discussion of advanced topics and the Bibliography at the end of the book should be strongly suggestive of the rich variety of directions that these fundamentals begin to uncover.

HISTORICAL PERSPECTIVE

Discrete-time signal processing has advanced in uneven steps over a long period of time. Looking back at the development of the field provides a valuable perspective on fundamentals that will remain central to the field long into the future. Since the invention of calculus in the 17th century, scientists and engineers have developed models to represent physical phenomena in terms of functions of continuous variables and differential equations. Numerical techniques have been used to solve these equations when analytical solutions are not possible. Indeed, Newton used finite-difference methods that are special cases of some of the discrete-time systems that we present in this text. Mathematicians of the 18th century, such as Euler, Bernoulli, and Lagrange, developed methods for numerical integration and interpolation of functions of a continuous variable. Interesting historical research by Heideman, Johnson, and Burrus (1984) showed that Gauss discovered the fundamental principle of the fast Fourier transform (discussed in Chapter 9) as early as 1805–even before the publication of Fourier's treatise on harmonic series representation of functions.

Until the early 1950s, signal processing as we have defined it was typically done with analog systems that were implemented with electronic circuits or even with mechanical devices. Even though digital computers were becoming available in business environments and in scientific laboratories, they were expensive and had relatively limited capabilities. About that time, the need for more sophisticated signal processing in some application areas created considerable interest in discrete-time signal processing. One of the first uses of digital computers in digital signal processing was in oil prospecting, where seismic data could be recorded on magnetic tape for later processing. This type of signal processing could not generally be done in real time; minutes or even hours of computer time were often required to process only seconds of data. Even so, the flexibility of the digital computer and the potential payoffs made this alternative extremely inviting.

Also in the 1950s, the use of digital computers in signal processing arose in a different way. Because of the flexibility of digital computers, it was often useful to simulate a signal processing system on a digital computer before implementing it in analog hardware. In this way, a new signal processing algorithm or system could be studied in a flexible experimental environment before committing economic and engineering resources to constructing it. Typical examples of such simulations were the vocoder simulations carried out at Lincoln Laboratory and Bell Laboratories. In the implementation of an analog channel vocoder, for example, the filter characteristics affected the perceived quality of the coded speech signal in ways that were difficult to quantify objectively. Through computer simulations, these filter characteristics could be adjusted and the perceived quality of a speech coding system evaluated prior to construction of the analog equipment.

In all of these examples of signal processing using digital computers, the com offered tremendous advantages in flexibility. However, the processing could not b in real time. Consequently, a prevalent attitude was that the digital computer was being used to *approximate,* or *simulate,* an analog signal processing system. In keeping with that style, early work on digital filtering was very much concerned with ways in which a filter could be programmed on a digital computer so that with analog-to-digital conversion of the signal, followed by digital filtering, followed by digital-to-analog conversion, the overall system approximated a good analog filter. The notion that digital systems might, in fact, be practical for the actual real-time implementation of signal processing in speech communication, radar processing, or any of a variety of other applications seemed at the most optimistic times to be highly speculative. Speed, cost, and size were, of course, three of the important factors in favor of the use of analog components.

As signals were being processed on digital computers, researchers had a natural tendency to experiment with increasingly sophisticated signal processing algorithms. Some of these algorithms grew out of the flexibility of the digital computer and had no apparent practical implementation in analog equipment. Thus, many of these algorithms were treated as interesting, but somewhat impractical, ideas. The development of such signal processing algorithms made the notion of all-digital implementation of signal processing systems even more tempting. Active work began on the investigation of digital vocoders, digital spectrum analyzers, and other all-digital systems, with the hope that eventually such systems would become practical.

The evolution of a new point of view toward discrete-time signal processing was further accelerated by the disclosure by Cooley and Tukey (1965) of an efficient algorithm for computation of Fourier transforms. This class of algorithms has come to be known as the fast Fourier transform, or FFT. The FFT was significant for several reasons. Many signal processing algorithms that had been developed on digital computers required processing times several orders of magnitude greater than real time. Often this was because spectrum analysis was an important component of the signal processing and no efficient means were available for implementing it. The fast Fourier transform algorithm reduced the computation time of the Fourier transform by orders of magnitude, permitting the implementation of increasingly sophisticated signal processing algorithms with processing times that allowed interactive experimentation with the system. Furthermore, with the realization that the fast Fourier transform algorithms might, in fact, be implementable in special-purpose digital hardware, many signal processing algorithms that previously had appeared to be impractical began to appear to have practical implementations.

Another important implication of the fast Fourier transform algorithm was that it was an inherently discrete-time concept. It was directed toward the computation of the Fourier transform of a discrete-time signal or sequence and involved a set of properties and mathematics that was exact in the discrete-time domain–it was not simply an approximation to a continuous-time Fourier transform. This had the effect of stimulating a reformulation of many signal processing concepts and algorithms in terms of discrete-time mathematics, and these techniques then formed an exact set of relationships in the discrete-time domain. Following this shift away from the notion that signal processing on a digital computer was merely an approximation to analog signal processing techniques, there emerged a strong interest in discrete-time signal processing as an important field of investigation in its own right.

Another major development in the history of discrete-time signal processing occurred in the field of microelectronics. The invention and subsequent proliferation of the microprocessor paved the way for low-cost implementations of discrete-time signal processing systems. Although the first microprocessors were too slow to implement most discrete-time systems in real time, by the mid-1980s integrated circuit technology had advanced to a level that permitted the implementation of very fast fixed-point and floating-point microcomputers with architectures specially designed for implementing discrete-time signal processing algorithms. With this technology came, for the first time, the possibility of widespread application of discrete-time signal processing techniques.

FUTURE PROMISE

Microelectronics engineers continue to strive for increased circuit densities and production yields, and as a result, the complexity and sophistication of microelectronic systems are continually increasing. Indeed, complexity and capability of DSP chips have grown exponentially since the early 1980s and show no sign of slowing down. As wafer-scale integration techniques become highly developed, very complex discrete-time signal processing systems will be implemented with low cost, miniature size, and low power consumption. Consequently, the importance of discrete-time signal processing will almost certainly continue to increase and the future development of the field is likely to be even more dramatic than the course of development that we have just described. Discrete-time signal processing techniques are already promoting revolutionary advances in some fields of application. A notable example is in the area of telecommunications, where discrete-time signal processing techniques, microelectronic technology, and fiber optic transmission combine to change the nature of communication systems in truly revolutionary ways. A similar impact can be expected in many other areas of technology.

While discrete-time signal processing is a dynamic, rapidly growing field, its fundamentals are well formulated. Our goal in this book is to provide a coherent treatment of the theory of discrete-time linear systems, filtering, sampling, and discrete-time Fourier analysis. The topics presented should provide the reader with the knowledge necessary for an appreciation of the wide scope of applications for discrete-time signal processing and a foundation for contributing to future developments in this exciting field of technology.

2

DISCRETE-TIME SIGNALS AND SYSTEMS

2.0 INTRODUCTION

The term *signal* is generally applied to something that conveys information. Signals generally convey information about the state or behavior of a physical system, and often, signals are synthesized for the purpose of communicating information between humans or between humans and machines. Although signals can be represented in many ways, in all cases the information is contained in some pattern of variations. Signals are represented mathematically as functions of one or more independent variables. For example, a speech signal is represented mathematically as a function of time, and a photographic image is represented as a brightness function of two spatial variables. A common convention—and one that usually will be followed in this book—is to refer to the independent variable of the mathematical representation of a signal as time, although in specific examples the independent variable may in fact not represent time.

The independent variable in the mathematical representation of a signal may be either continuous or discrete. *Continuous-time signals* are defined along a continuum of times and thus are represented by a continuous independent variable. Continuous-time signals are often referred to as *analog signals*. *Discrete-time signals* are defined at discrete times, and thus, the independent variable has discrete values; i.e., discrete-time signals are represented as sequences of numbers. Signals such as speech or images may have either a continuous- or a discrete-variable representation, and if certain conditions hold, these representations are entirely equivalent. Besides the independent variables being either continuous or discrete, the signal amplitude may be either continuous or discrete. *Digital signals* are those for which both time and amplitude are discrete.

Signal-processing systems may be classified along the same lines as signals. That is, continuous-time systems are systems for which both the input and the output are continuous-time signals, and discrete-time systems are those for which both the input and the output are discrete-time signals. Similarly, a digital system is a system for which both the input and the output are digital signals. Digital signal processing, then, deals with the transformation of signals that are discrete in both amplitude and time. The principal focus in this book is on discrete-time (rather than digital) signals and systems. However, the theory of discrete-time signals and systems is also exceedingly useful for digital signals and systems, particularly if the signal amplitudes are finely quantized. The effects of signal amplitude quantization are considered in Sections 4.8, 6.7–6.9, and 9.7.

Discrete-time signals may arise by sampling a continuous-time signal, or they may be generated directly by some discrete-time process. Whatever the origin of the discrete-time signals, discrete-time signal-processing systems have many attractive features. They can be realized with great flexibility with a variety of technologies, such as charge transport devices, surface acoustic wave devices, general-purpose digital computers, or high-speed microprocessors. Complete signal-processing systems can be implemented using VLSI techniques. Discrete-time systems can be used to simulate analog systems or, more importantly, to realize signal transformations that cannot be implemented with continuous-time hardware. Thus, discrete-time representations of signals are often desirable when sophisticated and flexible signal processing is required.

In this chapter, we consider the fundamental concepts of discrete-time signals and signal-processing systems for one-dimensional signals. We emphasize the class of linear time-invariant discrete-time systems. Many of the properties and results that we derive in this and subsequent chapters will be similar to properties and results for linear time-invariant continuous-time systems, as presented in a variety of texts. (See, for example, Oppenheim and Willsky, 1997.) In fact, it is possible to approach the discussion of discrete-time systems by treating sequences as analog signals that are impulse trains. This approach, if implemented carefully, can lead to correct results and has formed the basis for much of the classical discussion of sampled data systems. (See, for example, Phillips and Nagle, 1995.) However, not all sequences arise from sampling a continuous-time signal, and many discrete-time systems are not simply approximations to corresponding analog systems. Furthermore, there are important and fundamental differences between discrete- and continuous-time systems. Therefore, rather than attempt to force results from continuous-time system theory into a discrete-time framework, we will derive parallel results starting within a framework and with notation that is suitable to discrete-time systems. Discrete-time signals will be related to continuous-time signals only when it is necessary and useful to do so.

2.1 DISCRETE-TIME SIGNALS: SEQUENCES

Discrete-time signals are represented mathematically as sequences of numbers. A sequence of numbers x, in which the nth number in the sequence is denoted $x[n]$,[1] is formally written as

$$x = \{x[n]\}, \qquad -\infty < n < \infty, \tag{2.1}$$

where n is an integer. In a practical setting, such sequences can often arise from periodic

---

[1]A sequence is simply a function whose domain is the set of integers. Note that we use [ ] to enclose the independent variable of such functions, and we use ( ) to enclose the independent variable of continuous-variable functions.

sampling of an analog signal. In this case, the numeric value of the nth number in the
sequence is equal to the value of the analog signal, $x_a(t)$, at time nT; i.e.,

$$x[n] = x_a(nT), \qquad -\infty < n < \infty. \tag{2.2}$$

The quantity T is called the *sampling period,* and its reciprocal is the *sampling frequency.* Although sequences do not always arise from sampling analog waveforms, it is
convenient to refer to $x[n]$ as the "nth sample" of the sequence. Also, although, strictly
speaking, $x[n]$ denotes the nth number in the sequence, the notation of Eq. (2.1) is often
unnecessarily cumbersome, and it is convenient and unambiguous to refer to "the sequence $x[n]$" when we mean the entire sequence, just as we referred to the "analog signal
$x_a(t)$." Discrete-time signals (i.e., sequences) are often depicted graphically as shown in
Figure 2.1. Although the abscissa is drawn as a continuous line, it is important to recognize that $x[n]$ is defined only for integer values of n. It is not correct to think of $x[n]$ as
being zero for n is not an integer; $x[n]$ is simply undefined for noninteger values of n.

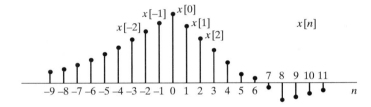

Figure 2.1 Graphical representation of a discrete-time signal.

As an example, Figure 2.2(a) shows a segment of a speech signal corresponding to
acoustic pressure variation as a function of time, and Figure 2.2(b) presents a sequence

Figure 2.2 (a) Segment of a continuous-time speech signal. (b) Sequence of samples
obtained from part (a) with $T = 125\ \mu s$.

of samples of the speech signal. Although the original speech signal is defined at all values of time t, the sequence contains information about the signal only at discrete instants. From the sampling theorem, discussed in Chapter 4, the original signal can be reconstructed as accurately as desired from a corresponding sequence of samples if the samples are taken frequently enough.

2.1.1 Basic Sequences and Sequence Operations

In the analysis of discrete-time signal-processing systems, sequences are manipulated in several basic ways. The product and sum of two sequences $x[n]$ and $y[n]$ are defined as the sample-by-sample product and sum, respectively. Multiplication of a sequence $x[n]$ by a number α is defined as multiplication of each sample value by α. A sequence $y[n]$ is said to be a delayed or shifted version of a sequence $x[n]$ if

$$y[n] = x[n - n_0], \tag{2.3}$$

where n_0 is an integer.

In discussing the theory of discrete-time signals and systems, several basic sequences are of particular importance. These sequences are shown in Figure 2.3 and are discussed next.

The *unit sample sequence* (Figure 2.3a) is defined as the sequence

$$\delta[n] = \begin{cases} 0, & n \neq 0, \\ 1, & n = 0. \end{cases} \tag{2.4}$$

As we will see, the unit sample sequence plays the same role for discrete-time signals and systems that the unit impulse function (Dirac delta function) does for continuous-time signals and systems. For convenience, the unit sample sequence is often referred to as a *discrete-time impulse* or simply as an *impulse*. It is important to note that a discrete-time impulse does not suffer from the mathematical complications of the continuous-time impulse; its definition is simple and precise.

As we will see in the discussion of linear systems, one of the important aspects of the impulse sequence is that an arbitrary sequence can be represented as a sum of scaled, delayed impulses. For example, the sequence $p[n]$ in Figure 2.4 can be expressed as

$$p[n] = a_{-3}\delta[n + 3] + a_1\delta[n - 1] + a_2\delta[n - 2] + a_7\delta[n - 7]. \tag{2.5}$$

More generally, any sequence can be expressed as

$$x[n] = \sum_{k=-\infty}^{\infty} x[k]\delta[n - k]. \tag{2.6}$$

We will make specific use of Eq. (2.6) in discussing the representation of discrete-time linear systems.

The *unit step sequence* (Figure 2.3b) is given by

$$u[n] = \begin{cases} 1, & n \geq 0, \\ 0, & n < 0. \end{cases} \tag{2.7}$$

(a)

(b)

(c)

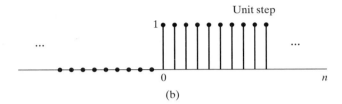

(d)

Figure 2.3   Some basic sequences. The sequences shown play important roles in the analysis and representation of discrete-time signals and systems.

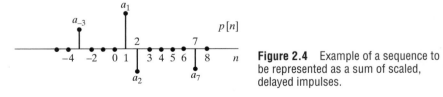

Figure 2.4   Example of a sequence to be represented as a sum of scaled, delayed impulses.

The unit step is related to the impulse by

$$u[n] = \sum_{k=-\infty}^{n} \delta[k]; \qquad (2.8)$$

that is, the value of the unit step sequence at (time) index n is equal to the accumulated sum of the value at index n and all previous values of the impulse sequence. An alternative representation of the unit step in terms of the impulse is obtained by interpreting the unit step in Figure 2.3(b) in terms of a sum of delayed impulses as in Eq. (2.6). In this case, the nonzero values are all unity, so

$$u[n] = \delta[n] + \delta[n-1] + \delta[n-2] + \cdots \qquad (2.9a)$$

or

$$u[n] = \sum_{k=0}^{\infty} \delta[n-k]. \qquad (2.9b)$$

Conversely, the impulse sequence can be expressed as the first backward difference of the unit step sequence, i.e.,

$$\delta[n] = u[n] - u[n-1]. \qquad (2.10)$$

Exponential sequences are extremely important in representing and analyzing linear time-invariant discrete-time systems. The general form of an exponential sequence is

$$x[n] = A\alpha^n. \qquad (2.11)$$

If A and α are real numbers, then the sequence is real. If $0 < \alpha < 1$ and A is positive, then the sequence values are positive and decrease with increasing n, as in Figure 2.3(c). For $-1 < \alpha < 0$, the sequence values alternate in sign, but again decrease in magnitude with increasing n. If $|\alpha| > 1$, then the sequence grows in magnitude as n increases.

Example 2.1   Combining Basic Sequences

We often combine basic sequences to form simple representations of other sequences. If we want an exponential sequence that is zero for $n < 0$, we can write this as the somewhat cumbersome expression

$$x[n] = \begin{cases} A\alpha^n, & n \geq 0, \\ 0, & n < 0. \end{cases} \qquad (2.12)$$

A much simpler expression is $x[n] = A\alpha^n u[n]$.

Sinusoidal sequences are also very important. A sinusoidal sequence has the general form

$$x[n] = A\cos(\omega_0 n + \phi), \qquad \text{for all } n, \qquad (2.13)$$

with A and ϕ real constants, and is illustrated in Figure 2.3(d).

The exponential sequence $A\alpha^n$ with complex α has real and imaginary parts that are exponentially weighted sinusoids. Specifically, if $\alpha = |\alpha|e^{j\omega_0}$ and $A = |A|e^{j\phi}$, the sequence $A\alpha^n$ can be expressed in any of the following ways:

$$x[n] = A\alpha^n = |A|e^{j\phi}|\alpha|^n e^{j\omega_0 n}$$

$$= |A|\,|\alpha|^n e^{j(\omega_0 n + \phi)} \qquad (2.14)$$

$$= |A|\,|\alpha|^n \cos(\omega_0 n + \phi) + j|A|\,|\alpha|^n \sin(\omega_0 n + \phi).$$

The sequence oscillates with an exponentially growing envelope if $|\alpha| > 1$ or with an exponentially decaying envelope if $|\alpha| < 1$. (As a simple example, consider the case $\omega_0 = \pi$.)

When $|\alpha| = 1$, the sequence is referred to as a *complex exponential sequence* and has the form

$$x[n] = |A|e^{j(\omega_0 n + \phi)} = |A|\cos(\omega_0 n + \phi) + j|A|\sin(\omega_0 n + \phi); \qquad (2.15)$$

that is, the real and imaginary parts of $e^{j\omega_0 n}$ vary sinusoidally with n. By analogy with the continuous-time case, the quantity ω_0 is called the *frequency* of the complex sinusoid or complex exponential, and ϕ is called the *phase*. However, note that n is a dimensionless integer. Thus, the dimension of ω_0 must be radians. If we wish to maintain a closer analogy with the continuous-time case, we can specify the units of ω_0 to be radians per sample and the units of n to be samples.

The fact that n is always an integer in Eq. (2.15) leads to some important differences between the properties of discrete-time and continuous-time complex exponential sequences and sinusoidal sequences. An important difference between continuous-time and discrete-time complex sinusoids is seen when we consider a frequency $(\omega_0 + 2\pi)$. In this case,

$$\begin{aligned} x[n] &= Ae^{j(\omega_0 + 2\pi)n} \\ &= Ae^{j\omega_0 n}e^{j2\pi n} = Ae^{j\omega_0 n}. \end{aligned} \qquad (2.16)$$

More generally, we can easily see that complex exponential sequences with frequencies $(\omega_0 + 2\pi r)$, where r is an integer, are indistinguishable from one another. An identical statement holds for sinusoidal sequences. Specifically, it is easily verified that

$$\begin{aligned} x[n] &= A\cos[(\omega_0 + 2\pi r)n + \phi] \\ &= A\cos(\omega_0 n + \phi). \end{aligned} \qquad (2.17)$$

The implications of this property for sequences obtained by sampling sinusoids and other signals will be discussed in Chapter 4. For now, we simply conclude that, when discussing complex exponential signals of the form $x[n] = Ae^{j\omega_0 n}$ or real sinusoidal signals of the form $x[n] = A\cos(\omega_0 n + \phi)$, we need only consider frequencies in an interval of length 2π, such as $-\pi < \omega_0 \leq \pi$ or $0 \leq \omega_0 < 2\pi$.

Another important difference between continuous-time and discrete-time complex exponentials and sinusoids concerns their periodicity. In the continuous-time case, a sinusoidal signal and a complex exponential signal are both periodic, with the period equal to 2π divided by the frequency. In the discrete-time case, a periodic sequence is a sequence for which

$$x[n] = x[n + N], \qquad \text{for all } n, \qquad (2.18)$$

where the period N is necessarily an integer. If this condition for periodicity is tested for the discrete-time sinusoid, then

$$A\cos(\omega_0 n + \phi) = A\cos(\omega_0 n + \omega_0 N + \phi), \qquad (2.19)$$

which requires that

$$\omega_0 N = 2\pi k, \qquad (2.20)$$

where k is an integer. A similar statement holds for the complex exponential sequence

$Ce^{j\omega_0 n}$; that is, periodicity with period N requires that

$$e^{j\omega_0(n+N)} = e^{j\omega_0 n}, \tag{2.21}$$

which is true only for $\omega_0 N = 2\pi k$, as in Eq. (2.20). Consequently, complex exponential and sinusoidal sequences are not necessarily periodic in n with period $(2\pi/\omega_0)$ and, depending on the value of ω_0, may not be periodic at all.

Example 2.2   Periodic and Aperiodic Discrete-Time Sinusoids

Consider the signal $x_1[n] = \cos(\pi n/4)$. This signal has a period of $N = 8$. To show this, note that $x[n + 8] = \cos(\pi(n + 8)/4) = \cos(\pi n/4 + 2\pi) = \cos(\pi n/4) = x[n]$, satisfying the definition of a discrete-time periodic signal. Contrary to our intuition from continuous-time sinusoids, increasing the frequency of a discrete-time sinusoid does not necessarily decrease the period of the signal. Consider the discrete-time sinusoid $x_2[n] = \cos(3\pi n/8)$, which has a higher frequency than $x_1[n]$. However, $x_2[n]$ is not periodic with period 8, since $x_2[n+8] = \cos(3\pi(n+8)/8) = \cos(3\pi n/8 + 3\pi) = -x_2[n]$. Using an argument analogous to the one for $x_1[n]$, we can show that $x_2[n]$ has a period of $N = 16$. Thus, increasing the frequency from $\omega_0 = 2\pi/8$ to $\omega_0 = 3\pi/8$ also increases the period of the signal. This occurs because discrete-time signals are defined only for integer indices n.

The integer restriction on n causes some sinusoidal signals not to be periodic at all. For example, there is no integer N such that the signal $x_3[n] = \cos(n)$ satisfies the condition $x_3[n + N] = x_3[n]$ for all n. These and other properties of discrete-time sinusoids that run counter to their continuous-time counterparts are caused by the limitation of the time index n to integers for discrete-time signals and systems.

When we combine the condition of Eq. (2.20) with our previous observation that ω_0 and $(\omega_0 + 2\pi r)$ are indistinguishable frequencies, it becomes clear that there are N distinguishable frequencies for which the corresponding sequences are periodic with period N. One set of frequencies is $\omega_k = 2\pi k/N$, $k = 0, 1, \ldots, N - 1$. These properties of complex exponential and sinusoidal sequences are basic to both the theory and the design of computational algorithms for discrete-time Fourier analysis, and they will be discussed in more detail in Chapters 8 and 9.

Related to the preceding discussion is the fact that the interpretation of high and low frequencies is somewhat different for continuous-time and discrete-time sinusoidal and complex exponential signals. For a continuous-time sinusoidal signal $x(t) = A\cos(\Omega_0 t + \phi)$, as Ω_0 increases, $x(t)$ oscillates more and more rapidly. For the discrete-time sinusoidal signal $x[n] = A\cos(\omega_0 n + \phi)$, as ω_0 increases from $\omega_0 = 0$ toward $\omega_0 = \pi$, $x[n]$ oscillates more and more rapidly. However, as ω_0 increases from $\omega_0 = \pi$ to $\omega_0 = 2\pi$, the oscillations become slower. This is illustrated in Figure 2.5. In fact, because of the periodicity in ω_0 of sinusoidal and complex exponential sequences, $\omega_0 = 2\pi$ is indistinguishable from $\omega_0 = 0$, and, more generally, frequencies around $\omega_0 = 2\pi$ are indistinguishable from frequencies around $\omega_0 = 0$. As a consequence, for sinusoidal and complex exponential signals, values of ω_0 in the vicinity of $\omega_0 = 2\pi k$ for any integer value of k are typically referred to as low frequencies (relatively slow oscillations), while values of ω_0 in the vicinity of $\omega_0 = (\pi + 2\pi k)$ for any integer value of k are typically referred to as high frequencies (relatively rapid oscillations).

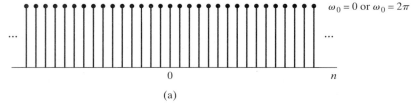

$\omega_0 = 0$ or $\omega_0 = 2\pi$

(a)

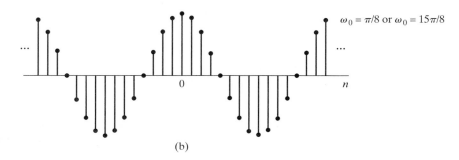

$\omega_0 = \pi/8$ or $\omega_0 = 15\pi/8$

(b)

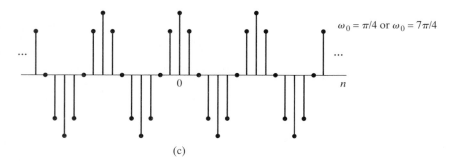

$\omega_0 = \pi/4$ or $\omega_0 = 7\pi/4$

(c)

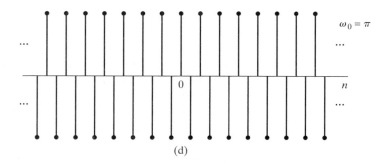

$\omega_0 = \pi$

(d)

Figure 2.5 $\cos\omega_0 n$ for several different values of ω_0. As ω_0 increases from zero toward π (parts a–d), the sequence oscillates more rapidly. As ω_0 increases from π to 2π (parts d–a), the oscillations become slower.

2.2 DISCRETE-TIME SYSTEMS

A discrete-time system is defined mathematically as a transformation or operator that maps an input sequence with values $x[n]$ into an output sequence with values $y[n]$. This can be denoted as

$$y[n] = T\{x[n]\} \tag{2.22}$$

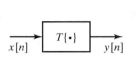

and is indicated pictorially in Figure 2.6. Equation (2.22) represents a rule or formula for computing the output sequence values from the input sequence values. It should be emphasized that the value of the output sequence at each value of the index n may depend on $x[n]$ for all values of n. The following examples illustrate some simple and useful systems.

Example 2.3   The Ideal Delay System

The ideal delay system is defined by the equation

$$y[n] = x[n - n_d], \qquad -\infty < n < \infty, \tag{2.23}$$

where n_d is a fixed positive integer called the delay of the system. In words, the ideal delay system simply shifts the input sequence to the right by n_d samples to form the output. If, in Eq. (2.23), n_d is a fixed negative integer, then the system would shift the input to the left by $|n_d|$ samples, corresponding to a time advance.

In Example 2.3, only one sample of the input sequence is involved in determining a certain output sample. In the following example, this is not the case.

Example 2.4   Moving Average

The general moving-average system is defined by the equation

$$y[n] = \frac{1}{M_1 + M_2 + 1} \sum_{k=-M_1}^{M_2} x[n - k]$$

$$= \frac{1}{M_1 + M_2 + 1} \{x[n + M_1] + x[n + M_1 - 1] + \cdots + x[n] \tag{2.24}$$

$$+ x[n - 1] + \cdots + x[n - M_2]\}.$$

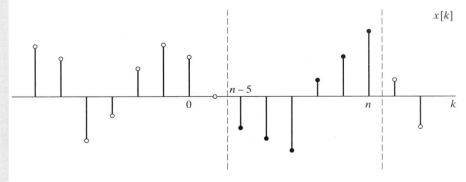

Figure 2.7   Sequence values involved in computing a causal moving average.

This system computes the nth sample of the output sequence as the average of ($M_1 + M_2 + 1$) samples of the input sequence around the nth sample. Figure 2.7 shows an input sequence plotted as a function of a dummy index k and the samples involved in the computation of the output sample $y[n]$ for $n = 7$, $M_1 = 0$, and $M_2 = 5$. The output sample $y[7]$ is equal to one-sixth of the sum of all the samples between the vertical dotted lines. To compute $y[8]$, both dotted lines would move one sample to the right.

Classes of systems are defined by placing constraints on the properties of the transformation $T\{\cdot\}$. Doing so often leads to very general mathematical representations, as we will see. Of particular importance are the system constraints and properties, discussed in Sections 2.2.1–2.2.5.

2.2.1 Memoryless Systems

A system is referred to as memoryless if the output $y[n]$ at every value of n depends only on the input $x[n]$ at the same value of n.

Example 2.5 A Memoryless System

An example of a memoryless system is a system for which $x[n]$ and $y[n]$ are related by

$$y[n] = (x[n])^2, \qquad \text{for each value of } n. \tag{2.25}$$

The system in Example 2.3 is not memoryless unless $n_d = 0$; in particular, this system is referred to as having "memory" whether n_d is positive (a time delay) or negative (a time advance). The system in Example 2.4 is not memoryless unless $M_1 = M_2 = 0$.

2.2.2 Linear Systems

The class of *linear systems* is defined by the principle of superposition. If $y_1[n]$ and $y_2[n]$ are the responses of a system when $x_1[n]$ and $x_2[n]$ are the respective inputs, then the system is linear if and only if

$$T\{x_1[n] + x_2[n]\} = T\{x_2[n]\} + T\{x_2[n]\} = y_1[n] + y_2[n] \tag{2.26a}$$

and

$$T\{ax[n]\} = a T\{x[n]\} = ay[n], \tag{2.26b}$$

where a is an arbitrary constant. The first property is called the *additivity property,* and the second is called the *homogeneity* or *scaling property*. These two properties can be combined into the principle of superposition, stated as

$$T\{ax_1[n] + bx_2[n]\} = a T\{x_1[n]\} + b T\{x_2[n]\} \tag{2.27}$$

for arbitrary constants a and b. This equation can be generalized to the superposition of many inputs. Specifically, if

$$x[n] = \sum_k a_k x_k[n], \tag{2.28a}$$

then the output of a linear system will be

$$y[n] = \sum_k a_k y_k[n], \tag{2.28b}$$

where $y_k[n]$ is the system response to the input $x_k[n]$.

By using the definition of the principle of superposition, we can easily show that the systems of Examples 2.3 and 2.4 are linear systems. (See Problem 2.23.) An example of a nonlinear system is the system in Example 2.5.

Example 2.6   The Accumulator System

The system defined by the input–output equation

$$y[n] = \sum_{k=-\infty}^{n} x[k] \tag{2.29}$$

is called the *accumulator* system, since the output at time n is just the sum of the present and all previous input samples. The accumulator system is a linear system. In order to prove this, we must show that it satisfies the superposition principle for all inputs, not just any specific set of inputs. We begin by defining two arbitrary inputs $x_1[n]$ and $x_2[n]$ and their corresponding outputs

$$y_1[n] = \sum_{k=-\infty}^{n} x_1[k], \tag{2.30}$$

$$y_2[n] = \sum_{k=-\infty}^{n} x_2[k]. \tag{2.31}$$

When the input is $x_3[n] = ax_1[n] + bx_2[n]$, the superposition principle requires the output $y_3[n] = ay_1[n] + by_2[n]$ for all possible choices of a and b. We can show this by starting from Eq. (2.29):

$$y_3[n] = \sum_{k=-\infty}^{n} x_3[k], \tag{2.32}$$

$$= \sum_{k=-\infty}^{n} (ax_1[k] + bx_2[k]), \tag{2.33}$$

$$= a \sum_{k=-\infty}^{n} x_1[k] + b \sum_{k=-\infty}^{n} x_2[k], \tag{2.34}$$

$$= ay_1[n] + by_2[n]. \tag{2.35}$$

Thus, the accumulator system of Eq. (2.29) satisfies the superposition principle for all inputs and is therefore linear.

In general, it may be simpler to prove that a system is not linear (if it is not) than to prove that it is linear (if it is). We simply must find an input or set of inputs for which the system does not satisfy the conditions of linearity.

Example 2.7   A Nonlinear System

Consider the system defined by

$$w[n] = \log_{10}(|x[n]|). \tag{2.36}$$

This system is not linear. In order to prove this, we only need to find one counterexample—that is, one set of inputs and outputs which demonstrates that the system violates the superposition principle, Eq. (2.27). The inputs $x_1[n] = 1$ and $x_2[n] = 10$ are a counterexample. The output for the first signal is $w_1[n] = 0$, while for the second, $w_2[n] = 1$. The scaling property of linear systems requires that, since $x_2[n] = 10x_1[n]$, if the system is linear, it must be true that $w_2[n] = 10w_1[n]$. Since this is not so for Eq. (2.36) for this set of inputs and outputs, the system is *not* linear.

2.2.3 Time-Invariant Systems

A time-invariant system (often referred to equivalently as a shift-invariant system) is a system for which a time shift or delay of the input sequence causes a corresponding shift in the output sequence. Specifically, suppose that a system transforms the input sequence with values $x[n]$ into the output sequence with values $y[n]$. Then the system is said to be time invariant if, for all n_0, the input sequence with values $x_1[n] = x[n - n_0]$ produces the output sequence with values $y_1[n] = y[n - n_0]$.

   As in the case of linearity, proving that a system is time invariant requires a general proof making no specific assumptions about the input signals. All of the systems in Examples 2.3–2.7 are time invariant. The style of proof for time invariance is illustrated in Examples 2.8 and 2.9.

Example 2.8   The Accumulator as a Time-Invariant System

Consider the accumulator from Example 2.6. We define $x_1[n] = x[n - n_0]$. To show time invariance, we solve for both $y[n - n_0]$ and $y_1[n]$ and compare them to see whether they are equal. First,

$$y[n - n_0] = \sum_{k=-\infty}^{n-n_0} x[k]. \tag{2.37}$$

Next, we find

$$y_1[n] = \sum_{k=-\infty}^{n} x_1[k] \tag{2.38}$$

$$= \sum_{k=-\infty}^{n} x[k - n_0]. \tag{2.39}$$

Substituting the change of variables $k_1 = k - n_0$ into the summation gives

$$y_1[n] = \sum_{k_1=-\infty}^{n-n_0} x[k_1] = y[n - n_0]. \tag{2.40}$$

Thus, the accumulator is a time-invariant system.

The following example illustrates a system that is not time invariant.

Example 2.9   The Compressor System

The system defined by the relation

$$y[n] = x[Mn], \qquad -\infty < n < \infty, \tag{2.41}$$

with M a positive integer, is called a *compressor*. Specifically, it discards $(M - 1)$ samples out of M; i.e., it creates the output sequence by selecting every Mth sample. This system is not time invariant. We can show that it is not by considering the response $y_1[n]$ to the input $x_1[n] = x[n - n_0]$. In order for the system to be time invariant, the output of the system when the input is $x_1[n]$ must be equal to $y[n - n_0]$. The output $y_1[n]$ that results from the input $x_1[n]$ can be directly computed from Eq. (2.41) to be

$$y_1[n] = x_1[Mn] = x[Mn - n_0]. \tag{2.42}$$

Delaying the output $y[n]$ by n_0 samples yields

$$y[n - n_0] = x[M(n - n_0)]. \qquad (2.43)$$

Comparing these two outputs, we see that $y[n - n_0]$ is not equal to $y_1[n]$ for all M and n_0, and therefore, the system is not time invariant.

It is also possible to prove that a system is not time invariant by finding a single counterexample that violates the time-invariance property. For instance, a counterexample for the compressor is the case when $M = 2$, $x[n] = \delta[n]$, and $x_1[n] = \delta[n - 1]$. For this choice of inputs and M, $y[n] = \delta[n]$, but $y_1[n] = 0$; thus, it is clear that $y_1[n] \neq y[n - 1]$ for this system.

2.2.4 Causality

A system is causal if, for every choice of n_0, the output sequence value at the index $n = n_0$ depends only on the input sequence values for $n \leq n_0$. This implies that if $x_1[n] = x_2[n]$ for $n \leq n_0$, then $y_1[n] = y_2[n]$ for $n \leq n_0$. That is, the system is *nonanticipative*. The system of Example 2.3 is causal for $n_d \geq 0$ and is noncausal for $n_d < 0$. The system of Example 2.4 is causal if $-M_1 \geq 0$ and $M_2 \geq 0$; otherwise it is noncausal. The system of Example 2.5 is causal, as is the accumulator of Example 2.6 and the nonlinear system in Example 2.7. However, the system of Example 2.9 is noncausal if $M > 1$, since $y[1] = x[M]$. Another noncausal system is given in the following example.

Example 2.10   The Forward and Backward Difference Systems

Consider the *forward difference system* defined by the relationship

$$y[n] = x[n + 1] - x[n]. \qquad (2.44)$$

This system is not causal, since the current value of the output depends on a future value of the input. The violation of causality can be demonstrated by considering the two inputs $x_1[n] = \delta[n - 1]$ and $x_2[n] = 0$ and their corresponding outputs $y_1[n] = \delta[n] - \delta[n - 1]$ and $y_2[n] = 0$. Note that $x_1[n] = x_2[n]$ for $n \leq 0$, so the definition of causality requires that $y_1[n] = y_2[n]$ for $n \leq 0$, which is clearly not the case for $n = 0$. Thus, by this counterexample, we have shown that the system is not causal.

The *backward difference system*, defined as

$$y[n] = x[n] - x[n - 1], \qquad (2.45)$$

has an output that depends only on the present and past values of the input. Because there is no way for the output at a specific time $y[n_0]$ to incorporate values of the input for $n > n_0$, the system is causal.

2.2.5 Stability

A system is stable in the bounded-input, bounded-output (BIBO) sense if and only if every bounded input sequence produces a bounded output sequence. The input $x[n]$ is bounded if there exists a fixed positive finite value B_x such that

$$|x[n]| \leq B_x < \infty, \qquad \text{for all } n. \qquad (2.46)$$

Stability requires that, for every bounded input, there exist a fixed positive finite value B_y such that

$$|y[n]| \le B_y < \infty, \qquad \text{for all } n. \tag{2.47}$$

It is important to emphasize that the properties we have defined in this section are properties of *systems,* not of the inputs to a system. That is, we may be able to find inputs for which the properties hold, but the existence of the property for some inputs does not mean that the system has the property. For the system to have the property, it must hold for *all* inputs. For example, an unstable system may have some bounded inputs for which the output is bounded, but for the system to have the property of stability, it must be true that for *all* bounded inputs, the output is bounded. If we can find just one input for which the system property does not hold, then we have shown that the system does *not* have that property. The following example illustrates the testing of stability for several of the systems that we have defined.

Example 2.11   Testing for Stability or Instability

The system of Example 2.5 is stable. To see this, assume that the input $x[n]$ is bounded such that $|x[n]| \le B_x$ for all n. Then $|y[n]| = |x[n]|^2 \le B_x^2$. Thus, we can choose $B_y = B_x^2$ and prove that $y[n]$ is bounded.

Likewise, we can see that the system defined in Example 2.7 is unstable, since $y[n] = \log_{10}(|x[n]|) = -\infty$ for any values of the time index n at which $x[n] = 0$, even though the output will be bounded for any input samples that are not equal to zero.

The accumulator, as defined in Example 2.6 by Eq. (2.29), is also not stable. For example, consider the case when $x[n] = u[n]$, which is clearly bounded by $B_x = 1$. For this input, the output of the accumulator is

$$y[n] = \sum_{k=-\infty}^{n} u[k] \tag{2.48}$$

$$= \begin{cases} 0, & n < 0, \\ (n+1), & n \ge 0. \end{cases} \tag{2.49}$$

There is no finite choice for B_y such that $(n+1) \le B_y < \infty$ for all n; thus, the system is unstable.

Using similar arguments, it can be shown that the systems in Examples 2.3, 2.4, 2.9 and 2.10 are all stable.

2.3 LINEAR TIME-INVARIANT SYSTEMS

A particularly important class of systems consists of those that are linear and time invariant. These two properties in combination lead to especially convenient representations for such systems. Most important, this class of systems has significant signal-processing applications. The class of linear systems is defined by the principle of superposition in Eq. (2.27). If the linearity property is combined with the representation of a general sequence as a linear combination of delayed impulses as in Eq. (2.6), it follows that a linear system can be completely characterized by its impulse response. Specifically, let $h_k[n]$ be the response of the system to $\delta[n - k]$, an impulse occurring at $n = k$. Then, from Eq. (2.6),

$$y[n] = T\left\{\sum_{k=-\infty}^{\infty} x[k]\delta[n-k]\right\}. \tag{2.50}$$

From the principle of superposition in Eq. (2.27), we can write

$$y[n] = \sum_{k=-\infty}^{\infty} x[k]T\{\delta[n-k]\} = \sum_{k=-\infty}^{\infty} x[k]h_k[n]. \tag{2.51}$$

According to Eq. (2.51), the system response to any input can be expressed in terms of the responses of the system to the sequences $\delta[n-k]$. If only linearity is imposed, $h_k[n]$ will depend on both n and k, in which case the computational usefulness of Eq. (2.51) is limited. We obtain a more useful result if we impose the additional constraint of time invariance.

The property of time invariance implies that if $h[n]$ is the response to $\delta[n]$, then the response to $\delta[n-k]$ is $h[n-k]$. With this additional constraint, Eq. (2.51) becomes

$$y[n] = \sum_{k=-\infty}^{\infty} x[k]h[n-k]. \tag{2.52}$$

As a consequence of Eq. (2.52), a linear time-invariant system (which we will sometimes abbreviate as LTI) is completely characterized by its impulse response $h[n]$ in the sense that, given $h[n]$, it is possible to use Eq. (2.52) to compute the output $y[n]$ due to *any* input $x[n]$.

Equation (2.52) is commonly called the *convolution sum*. If $y[n]$ is a sequence whose values are related to the values of two sequences $h[n]$ and $x[n]$ as in Eq. (2.52), we say that $y[n]$ is the convolution of $x[n]$ with $h[n]$ and represent this by the notation

$$y[n] = x[n] * h[n]. \tag{2.53}$$

The operation of discrete-time convolution takes two sequences $x[n]$ and $h[n]$ and produces a third sequence $y[n]$. Equation (2.52) expresses each sample of the output sequence in terms all of the samples of the input and impulse response sequences.

The derivation of Eq. (2.52) suggests the interpretation that the input sample at $n = k$, represented as $x[k]\delta[n-k]$, is transformed by the system into an output sequence $x[k]h[n-k]$, for $-\infty < n < \infty$, and that, for each k, these sequences are superimposed to form the overall output sequence. This interpretation is illustrated in Figure 2.8, which shows an impulse response, a simple input sequence having three nonzero samples, the individual outputs due to each sample, and the composite output due to all the samples in the input sequence. Specifically, $x[n]$ can be decomposed as the sum of the three sequences $x[-2]\delta[n+2]$, $x[0]\delta[n]$, and $x[3]\delta[n-3]$ representing the three nonzero values in the sequence $x[n]$. The sequences $x[-2]h[n+2]$, $x[0]h[n]$, and $x[3]h[n-3]$ are the system responses to $x[-2]\delta[n+2]$, $x[0]\delta[n]$, and $x[3]\delta[n-3]$, respectively. The response to $x[n]$ is then the sum of these three individual responses.

Although the convolution-sum expression is analogous to the convolution integral of continuous-time linear system theory, the convolution sum should not be thought of as an approximation to the convolution integral. The convolution integral plays mainly a theoretical role in continuous-time linear system theory; we will see that the convolution sum, in addition to its theoretical importance, often serves as an explicit realization of a discrete-time linear system. Thus, it is important to gain some insight into the properties of the convolution sum in actual calculations.

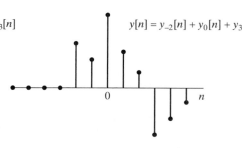

Figure 2.8   Representation of the output of a linear time-invariant system as the superposition of responses to individual samples of the input.

The preceding interpretation of Eq. (2.52) emphasizes that the convolution sum is a direct result of linearity and time invariance. However, a slightly different way of looking at Eq. (2.52) leads to a particularly useful computational interpretation. When viewed as a formula for computing a single value of the output sequence, Eq. (2.52) dictates that $y[n]$ (i.e., the nth value of the output) is obtained by multiplying the input sequence (expressed as a function of k) by the sequence whose values are $h[n - k]$, $-\infty < k < \infty$, and then, for any fixed value of n, summing all the values of the products $x[k]h[n-k]$, with k a counting index in the summation process. Therefore, the operation of convolving two sequences involves doing the computation for all values of n, thus generating the complete output sequence $y[n]$, $-\infty < n < \infty$. The key to carrying out the computations of Eq. (2.52) to obtain $y[n]$ is understanding how to form the sequence $h[n - k]$, $-\infty < k < \infty$, for all values of n that are of interest. To this end, it is useful to note that

$$h[n - k] = h[-(k - n)]. \tag{2.54}$$

The interpretation of Eq. (2.54) is best done with an example.

Example 2.12   Computation of the Convolution Sum

Suppose $h[k]$ is the sequence shown in Figure 2.9(a) and we wish to find $h[n - k] = h[-(k - n)]$. Define $h_1[k]$ to be $h[-k]$, which is shown in Figure 2.9(b). Next, define

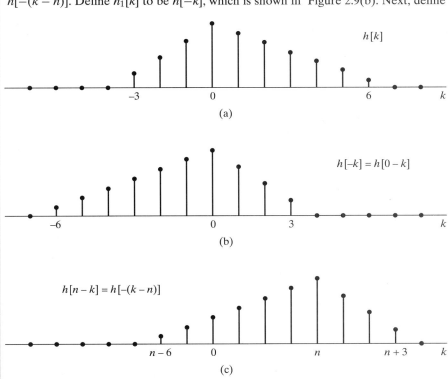

Figure 2.9   Forming the sequence $h[n - k]$. (a) The sequence $h[k]$ as a function of k. (b) The sequence $h[-k]$ as a function of k. (c) The sequence $h[n - k] = h[-(k - n)]$ as a function of k for $n = 4$.

$h_2[k]$ to be $h_1[k]$, delayed, by n samples on the k axis, i.e., $h_2[k] = h_1[k-n]$. Figure 2.9(c) shows the sequence that results from delaying the sequence in Figure 2.9(b) by n samples. Using the relationship between $h_1[k]$ and $h[k]$, we can show that $h_2[k] = h_1[k-n] = h[-(k-n)] = h[n-k]$, and thus, the bottom figure is the desired signal. To summarize, to compute $h[n-k]$ from $h[k]$, we first reverse $h[k]$ in time about $k=0$ and then delay the time-reversed signal by n samples.

From Example 2.3, it should be clear that, in general, the sequence $h[n-k]$, $-\infty < k < \infty$, is obtained by

1. reflecting $h[k]$ about the origin to obtain $h[-k]$;

2. shifting the origin of the reflected sequence to $k = n$.

To implement discrete-time convolution, the two sequences $x[k]$ and $h[n-k]$ are multiplied together for $-\infty < k < \infty$, and the products are summed to compute the output sample $y[n]$. To obtain another output sample, the origin of the sequence $h[-k]$ is shifted to the new sample position, and the process is repeated. This computational procedure applies whether the computations are carried out numerically on sampled data or analytically with sequences for which the sample values have simple formulas. The following example illustrates discrete-time convolution for the latter case.

Example 2.13   Analytical Evaluation of the Convolution Sum

Consider a system with impulse response

$$h[n] = u[n] - u[n - N]$$

$$= \begin{cases} 1, & 0 \le n \le N - 1, \\ 0, & \text{otherwise.} \end{cases}$$

The input is

$$x[n] = a^n u[n].$$

To find the output at a particular index n, we must form the sums over all k of the product $x[k]h[n-k]$. In this case, we can find formulas for $y[n]$ for different sets of values of n. For example, Figure 2.10(a) shows the sequences $x[k]$ and $h[n-k]$, plotted for n a negative integer. Clearly, all negative values of n give a similar picture; i.e., the nonzero portions of the sequences $x[k]$ and $h[n-k]$ do not overlap, so

$$y[n] = 0, \qquad n < 0.$$

Figure 2.10(b) illustrates the two sequences when $0 \le n$ and $n - N + 1 \le 0$. These two conditions can be combined into the single condition $0 \le n \le N - 1$. By considering Figure 2.10(b), we see that, since

$$x[k]h[n-k] = a^k,$$

it follows that

$$y[n] = \sum_{k=0}^{n} a^k, \qquad \text{for } 0 \le n \le N - 1. \tag{2.55}$$

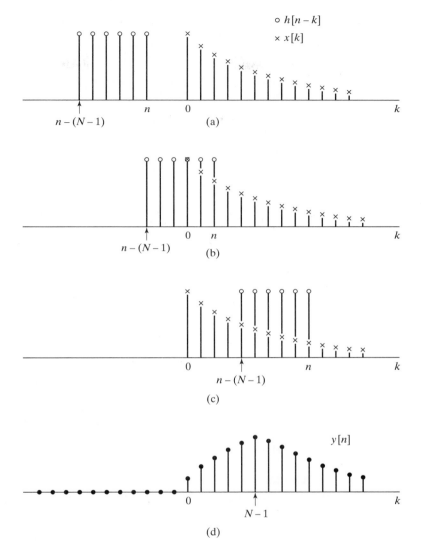

Figure 2.10   Sequence involved in computing a discrete convolution. (a)–(c) The sequences $x[k]$ and $h[n - k]$ as a function of k for different values of n. (Only nonzero samples are shown.) (d) Corresponding output sequence as a function of n.

The limits on the sum are determined directly from Figure 2.10(b). Equation (2.55) shows that $y[n]$ is the sum of $n + 1$ terms of a geometric series in which the ratio of terms is a. This sum can be expressed in closed form using the general formula

$$\sum_{k=N_1}^{N_2} \alpha^k = \frac{\alpha^{N_1} - \alpha^{N_2+1}}{1 - \alpha}, \qquad N_2 \geq N_1. \tag{2.56}$$

Applying this formula to Eq. (2.55), we obtain

$$y[n] = \frac{1 - a^{n+1}}{1 - a}, \qquad 0 \le n \le N - 1. \tag{2.57}$$

Finally, Figure 2.10(c) shows the two sequences when $0 < n - N + 1$ or $N - 1 < n$. As before,

$$x[k]h[n - k] = a^k, \qquad n - N + 1 < k \le n,$$

but now the lower limit on the sum is $n - N + 1$, as seen in Figure 2.10(c). Thus,

$$y[n] = \sum_{k=n-N+1}^{n} a^k, \qquad \text{for } N - 1 < n. \tag{2.58}$$

Using Eq. (2.56), we obtain

$$y[n] = \frac{a^{n-N+1} - a^{n+1}}{1 - a},$$

or

$$y[n] = a^{n-N+1} \left(\frac{1 - a^N}{1 - a} \right). \tag{2.59}$$

Thus, because of the piecewise-exponential nature of both the input and the unit sample response, we have been able to obtain the following closed-form expression for $y[n]$ as a function of the index n:

$$y[n] = \begin{cases} 0, & n < 0, \\ \dfrac{1 - a^{n+1}}{1 - a}, & 0 \le n \le N - 1, \\ a^{n-N+1} \left(\dfrac{1 - a^N}{1 - a} \right), & N - 1 < n. \end{cases} \tag{2.60}$$

This sequence is shown in Figure 2.10(d).

Example 2.13 illustrates how the convolution sum can be computed analytically when the input and the impulse response are given by simple formulas. In such cases, the sums may have a compact form that may be derived using the formula for the sum of a geometric series or other "closed-form" formulas.[2] When no simple form is available, the convolution sum can still be evaluated numerically using the technique illustrated in Example 2.13 whenever the sums are finite, which will be the case if either the input sequence or the impulse response is of finite length, i.e., has a finite number of nonzero samples.

2.4 PROPERTIES OF LINEAR TIME-INVARIANT SYSTEMS

Since all linear time-invariant systems are described by the convolution sum of Eq. (2.52), the properties of this class of systems are defined by the properties of discrete-time convolution. Therefore, the impulse response is a complete characterization of the properties of a specific linear time-invariant system.

[2]Such results are discussed, for example, in Grossman (1992).

Some general properties of the class of linear time-invariant systems can be found by considering properties of the convolution operation. For example, the convolution operation is commutative:

$$x[n] * h[n] = h[n] * x[n]. \tag{2.61}$$

This can be shown by applying a substitution of variables to Eq. (2.52). Specifically, with $m = n - k$,

$$y[n] = \sum_{m=\infty}^{-\infty} x[n-m]h[m] = \sum_{m=-\infty}^{\infty} h[m]x[n-m] = h[n] * x[n], \tag{2.62}$$

so the roles of $x[n]$ and $h[n]$ in the summation are interchanged. That is, the order of the sequences in a convolution is unimportant, and hence, the system output is the same if the roles of the input and impulse response are reversed. Accordingly, a linear time-invariant system with input $x[n]$ and impulse response $h[n]$ will have the same output as a linear time-invariant system with input $h[n]$ and impulse response $x[n]$. The convolution operation also distributes over addition; i.e.,

$$x[n] * (h_1[n] + h_2[n]) = x[n] * h_1[n] + x[n] * h_2[n].$$

This follows in a straightforward way from Eq. (2.52) and is a direct result of the linearity and commutativity of convolution.

In a *cascade connection* of systems, the output of the first system is the input to the second, the output of the second is the input to the third, etc. The output of the last system is the overall output. Two linear time-invariant systems in cascade correspond to a linear time-invariant system with an impulse response that is the convolution of the impulse responses of the two systems. This is illustrated in Figure 2.11. In the upper block diagram, the output of the first system will be $h_1[n]$ if $x[n] = \delta[n]$. Thus, the output of the second system (and, by definition, the impulse response of the overall system) will be

$$h[n] = h_1[n] * h_2[n]. \tag{2.63}$$

As a consequence of the commutative property of convolution, the impulse response of a cascade combination of linear time-invariant systems is independent of the order in which they are cascaded. This result is summarized in Figure 2.11, where the three systems all have the same impulse response.

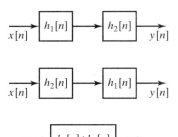

Figure 2.11   Three linear time-invariant systems with identical impulse responses.

(a)

(b)

Figure 2.12 (a) Parallel combination of linear time-invariant systems. (b) An equivalent system.

In a *parallel connection,* the systems have the same input, and their outputs are summed to produce an overall output. It follows from the distributive property of convolution that the connection of two linear time-invariant systems in parallel is equivalent to a single system whose impulse response is the sum of the individual impulse responses; i.e.,

$$h[n] = h_1[n] + h_2[n]. \tag{2.64}$$

This is depicted in Figure 2.12.

The constraints of linearity and time invariance define a class of systems with very special properties. Stability and causality represent additional properties, and it is often important to know whether a linear time-invariant system is stable and whether it is causal. Recall from Section 2.2.5 that a stable system is a system for which every bounded input produces a bounded output. Linear time-invariant systems are stable if and only if the impulse response is absolutely summable, i.e., if

$$S = \sum_{k=-\infty}^{\infty} |h[k]| < \infty. \tag{2.65}$$

This can be shown as follows. From Eq. (2.62),

$$|y[n]| = \left| \sum_{k=-\infty}^{\infty} h[k]x[n-k] \right| \le \sum_{k=-\infty}^{\infty} |h[k]| \, |x[n-k]|. \tag{2.66}$$

If $x[n]$ is bounded, so that

$$|x[n]| \le B_x,$$

then substituting B_x for $|x[n-k]|$ can only strengthen the inequality. Hence,

$$|y[n]| \le B_x \sum_{k=-\infty}^{\infty} |h[k]|. \tag{2.67}$$

Thus, $y[n]$ is bounded if Eq. (2.65) holds; in other words, Eq. (2.65) is a sufficient condition for stability. To show that it is also a necessary condition, we must show that if $S = \infty$, then a bounded input can be found that will cause an unbounded output. Such an input is the sequence with values

$$x[n] = \begin{cases} \dfrac{h^*[-n]}{|h[-n]|}, & h[n] \ne 0, \\ 0, & h[n] = 0, \end{cases} \tag{2.68}$$

where $h^*[n]$ is the complex conjugate of $h[n]$. The sequence $x[n]$ is clearly bounded by unity. However, the value of the output at $n = 0$ is

$$y[0] = \sum_{k=-\infty}^{\infty} x[-k]h[k] = \sum_{k=-\infty}^{\infty} \frac{|h[k]|^2}{|h[k]|} = S. \tag{2.69}$$

Therefore, if $S = \infty$, it is possible for a bounded input sequence to produce an unbounded output sequence.

The class of causal systems was defined in Section 2.2.4 as those systems for which the output $y[n_0]$ depends only on the input samples $x[n]$, for $n \leq n_0$. It follows from Eq. (2.52) or Eq. (2.62) that this definition implies the condition

$$h[n] = 0, \qquad n < 0, \tag{2.70}$$

for causality of linear time-invariant systems. (See Problem 2.62.) For this reason, it is sometimes convenient to refer to a sequence that is zero for $n < 0$ as a *causal sequence*, meaning that it could be the impulse response of a causal system.

To illustrate how the properties of linear time-invariant systems are reflected in the impulse response, let us consider again some of the systems defined in Examples 2.3–2.10. First note that only the systems of Examples 2.3, 2.4, 2.6, and 2.10 are linear and time invariant. Although the impulse response of nonlinear or time-varying systems can be found, it is generally of limited interest, since the convolution-sum formula and Eqs. (2.65) and (2.70), expressing stability and causality, do not apply to such systems.

First, let us find the impulse responses of the systems in Examples 2.3, 2.4, 2.6, and 2.10. We can do this by simply computing the response of each system to $\delta[n]$, using the defining relationship for the system. The resulting impulse responses are as follows:

Ideal Delay (Example 2.3)

$$h[n] = \delta[n - n_d], \qquad n_d \text{ a positive fixed integer.} \tag{2.71}$$

Moving Average (Example 2.4)

$$h[n] = \frac{1}{M_1 + M_2 + 1} \sum_{k=-M_1}^{M_2} \delta[n - k]$$

$$= \begin{cases} \dfrac{1}{M_1 + M_2 + 1}, & -M_1 \leq n \leq M_2, \\ 0, & \text{otherwise.} \end{cases} \tag{2.72}$$

Accumulator (Example 2.6)

$$h[n] = \sum_{k=-\infty}^{n} \delta[k]$$

$$= \begin{cases} 1, & n \geq 0, \\ 0, & n < 0, \end{cases} \tag{2.73}$$

$$= u[n].$$

Forward Difference (Example 2.10)

$$h[n] = \delta[n + 1] - \delta[n]. \tag{2.74}$$

Backward Difference (Example 2.10)

$$h[n] = \delta[n] - \delta[n-1]. \tag{2.75}$$

Given the impulse responses of these basic systems [Eqs. (2.71)–(2.75)], we can test the stability of each one by computing the sum

$$S = \sum_{n=-\infty}^{\infty} |h[n]|.$$

For the ideal delay, moving-average, forward difference, and backward difference examples, it is clear that $S < \infty$, since the impulse response has only a finite number of nonzero samples. Such systems are called *finite-duration impulse response* (FIR) systems. Clearly, FIR systems will always be stable, as long as each of the impulse response values is finite in magnitude. The accumulator, however, is unstable because

$$S = \sum_{n=0}^{\infty} u[n] = \infty.$$

In Section 2.2.5, we also demonstrated the instability of the accumulator by giving an example of a bounded input (the unit step) for which the output is unbounded.

The impulse response of the accumulator is infinite in duration. This is an example of the class of systems referred to as *infinite-duration impulse response* (IIR) systems. An example of an IIR system that is stable is a system whose impulse response is $h[n] = a^n u[n]$ with $|a| < 1$. In this case,

$$S = \sum_{n=0}^{\infty} |a|^n. \tag{2.76}$$

If $|a| < 1$, the formula for the sum of the terms of an infinite geometric series gives

$$S = \frac{1}{1-|a|} < \infty. \tag{2.77}$$

If, on the other hand, $|a| \geq 1$, the sum is infinite and the system is unstable.

To test causality of the linear time-invariant systems in Examples 2.3, 2.4, 2.6, and 2.10, we can check to see whether $h[n] = 0$ for $n < 0$. As discussed in Section 2.2.4, the ideal delay [$n_d \geq 0$ in Eq. (2.23)] is causal. If $n_d < 0$, the system is noncausal. For the moving average, causality requires that $-M_1 \geq 0$ and $M_2 \geq 0$. The accumulator and backward difference systems are causal, and the forward difference system is noncausal.

The concept of convolution as an operation between two sequences leads to the simplification of many problems involving systems. A particularly useful result can be stated for the ideal delay system. Since the output of the delay system is $y[n] = x[n-n_d]$, and since the delay system has impulse response $h[n] = \delta[n - n_d]$, it follows that

$$x[n] * \delta[n - n_d] = \delta[n - n_d] * x[n] = x[n - n_d]. \tag{2.78}$$

That is, the convolution of a shifted impulse sequence with any signal $x[n]$ is easily evaluated by simply shifting $x[n]$ by the displacement of the impulse.

Since delay is a fundamental operation in the implementation of linear systems, the preceding result is often useful in the analysis and simplification of interconnections of linear time-invariant systems. As an example, consider the system of Figure 2.13(a),

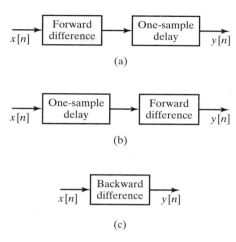

Figure 2.13   Equivalent systems found by using the commutative property of convolution.

which consists of a forward difference system cascaded with an ideal delay of one sample. According to the commutative property of convolution, the order in which systems are cascaded does not matter, as long as they are linear and time invariant. Therefore, we obtain the same result when we compute the forward difference of a sequence and delay the result (Figure 2.13a) as when we delay the sequence first and then compute the forward difference (Figure 2.13b). Also, it follows from Eq. (2.63) that the overall impulse response of each cascade system is the convolution of the individual impulse responses. Consequently,

$$h[n] = (\delta[n+1] - \delta[n]) * \delta[n-1]$$
$$= \delta[n-1] * (\delta[n+1] - \delta[n]) \qquad (2.79)$$
$$= \delta[n] - \delta[n-1].$$

Thus, $h[n]$ is identical to the impulse response of the backward difference system; that is, the cascaded systems of Figures 2.13(a) and 2.13(b) can be replaced by a backward difference system, as shown in Figure 2.13(c).

Note that the noncausal forward difference systems in Figures 2.13(a) and (b) have been converted to causal systems by cascading them with a delay. In general, any noncausal FIR system can be made causal by cascading it with a sufficiently long delay.

Another example of cascaded systems introduces the concept of an *inverse system*. Consider the cascade of systems in Figure 2.14. The impulse response of the cascade system is

$$h[n] = u[n] * (\delta[n] - \delta[n-1])$$
$$= u[n] - u[n-1] \qquad (2.80)$$
$$= \delta[n].$$

That is, the cascade combination of an accumulator followed by a backward difference (or vice versa) yields a system whose overall impulse response is the impulse. Thus, the output of the cascade combination will always be equal to the input, since $x[n] * \delta[n] = x[n]$. In this case, the backward difference system compensates exactly for (or inverts) the effect of the accumulator; that is, the backward difference system is the *inverse*

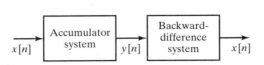

Figure 2.14 An accumulator in cascade with a backward difference. Since the backward difference is the inverse system for the accumulator, the cascade combination is equivalent to the identity system.

system for the accumulator. From the commutative property of convolution, the accumulator is likewise the inverse system for the backward difference system. Note that this example provides a system interpretation of Eqs. (2.8) and (2.10). In general, if a linear time-invariant system has impulse response $h[n]$, then its inverse system, if it exists, has impulse response $h_i[n]$ defined by the relation

$$h[n] * h_i[n] = h_i[n] * h[n] = \delta[n]. \tag{2.81}$$

Inverse systems are useful in many situations in which it is necessary to compensate for the effects of a linear system. In general, it is difficult to solve Eq. (2.81) directly for $h_i[n]$, given $h[n]$. However, in Chapter 3 we will see that the z-transform provides a straightforward method of finding an inverse system.

2.5 LINEAR CONSTANT-COEFFICIENT DIFFERENCE EQUATIONS

An important subclass of linear time-invariant systems consists of those systems for which the input $x[n]$ and the output $y[n]$ satisfy an Nth-order linear constant-coefficient difference equation of the form

$$\sum_{k=0}^{N} a_k y[n-k] = \sum_{m=0}^{M} b_m x[n-m]. \tag{2.82}$$

The properties discussed in Section 2.4 and some of the analysis techniques introduced there can be used to find difference equation representations for some of the linear time-invariant systems that we have defined.

Example 2.14   Difference Equation Representation of the Accumulator

An example of the class of linear constant-coefficient difference equations is the accumulator system defined by

$$y[n] = \sum_{k=-\infty}^{n} x[k]. \tag{2.83}$$

To show that the input and output satisfy a difference equation of the form of Eq. (2.82), note that we can write the output for $n-1$ as

$$y[n-1] = \sum_{k=-\infty}^{n-1} x[k]. \tag{2.84}$$

By separating the term $x[n]$ from the sum, we can rewrite Eq. (2.83) as

$$y[n] = x[n] + \sum_{k=-\infty}^{n-1} x[k]. \tag{2.85}$$

Substituting Eq. (2.84) into Eq. (2.85) yields

$$y[n] = x[n] + y[n-1], \tag{2.86}$$

from which the desired form of the difference equation can be obtained by grouping all the input and output terms on separate sides of the equation:

$$y[n] - y[n-1] = x[n]. \tag{2.87}$$

Thus, we have shown that, in addition to satisfying the defining relationship of Eq. (2.83), the input and output satisfy a linear constant-coefficient difference equation of the form Eq. (2.82), with $N = 1$, $a_0 = 1$, $a_1 = -1$, $M = 0$, and $b_0 = 1$.

The difference equation in the form of Eq. (2.86) gives us a better understanding of how we could implement the accumulator system. According to Eq. (2.86), for each value of n, we add the current input value $x[n]$ to the previously accumulated sum $y[n-1]$. This interpretation of the accumulator is represented in block diagram form in Figure 2.15.

Figure 2.15   Block diagram of a recursive difference equation representing an accumulator.

Equation (2.86) and the block diagram in Figure 2.15 are referred to as a *recursive representation* of the system, since each value is computed using previously computed values. This general notion will be explored in more detail later in the section.

Example 2.15   Difference Equation Representation of the Moving-Average System

Consider the moving-average system of Example 2.4, with $M_1 = 0$ so that the system is causal. In this case, from Eq. (2.72), the impulse response is

$$h[n] = \frac{1}{(M_2 + 1)} (u[n] - u[n - M_2 - 1]), \tag{2.88}$$

from which it follows that

$$y[n] = \frac{1}{(M_2 + 1)} \sum_{k=0}^{M_2} x[n - k], \tag{2.89}$$

which is a special case of Eq. (2.82), with $N = 0$, $a_0 = 1$, $M = M_2$, and $b_k = 1/(M_2 + 1)$ for $0 \leq k \leq M_2$.

Also, the impulse response can be expressed as

$$h[n] = \frac{1}{(M_2 + 1)} (\delta[n] - \delta[n - M_2 - 1]) * u[n], \tag{2.90}$$

which suggests that the causal moving-average system can be represented as the cascade system of Figure 2.16. We can obtain a difference equation for this block diagram

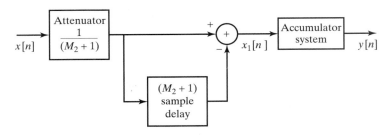

Figure 2.16 Block diagram of the recursive form of a moving-average system.

by noting first that

$$x_1[n] = \frac{1}{(M_2 + 1)}(x[n] - x[n - M_2 - 1]). \tag{2.91}$$

From Eq. (2.87) of Example 2.14, the output of the accumulator satisfies the difference equation

$$y[n] - y[n - 1] = x_1[n],$$

so that

$$y[n] - y[n - 1] = \frac{1}{(M_2 + 1)}(x[n] - x[n - M_2 - 1]). \tag{2.92}$$

Again, we have a difference equation in the form of Eq. (2.82), but this time $N = 1$, $a_0 = 1$, $a_1 = -1$, $M = M_2$ and $b_0 = -b_{M_2+1} = 1/(M_2 + 1)$, and $b_k = 0$ otherwise.

In Example 2.15, we showed two different difference equation representations of the moving-average system. In Chapter 6 we will see that an unlimited number of distinct difference equations can be used to represent a given linear time-invariant input–output relation.

Just as in the case of linear constant-coefficient differential equations for continuous-time systems, without additional constraints or information a linear constant-coefficient difference equation for discrete-time systems does not provide a unique specification of the output for a given input. Specifically, suppose that, for a given input $x_p[n]$, we have determined by some means one output sequence $y_p[n]$, so that an equation of the form of Eq. (2.82) is satisfied. Then the same equation with the same input is satisfied by any output of the form

$$y[n] = y_p[n] + y_h[n], \tag{2.93}$$

where $y_h[n]$ is any solution to Eq. (2.82) with $x[n] = 0$, i.e., to the equation

$$\sum_{k=0}^{N} a_k y_h[n - k] = 0. \tag{2.94}$$

Equation (2.94) is referred to as the *homogeneous equation* and $y_h[n]$ the *homogeneous solution*. The sequence $y_h[n]$ is in fact a member of a family of solutions of the form

$$y_h[n] = \sum_{m=1}^{N} A_m z_m^n. \tag{2.95}$$

Substituting Eq. (2.95) into Eq. (2.94) shows that the complex numbers z_m must be roots of the polynomial

$$\sum_{k=0}^{N} a_k z^{-k} = 0. \tag{2.96}$$

Equation (2.95) assumes that all N roots of the polynomial in Eq. (2.96) are distinct. The form of terms associated with multiple roots is slightly different, but there are always N undetermined coefficients. An example of the homogeneous solution with multiple roots is considered in Problem 2.38.

Since $y_h[n]$ has N undetermined coefficients, a set of N auxiliary conditions is required for the unique specification of $y[n]$ for a given $x[n]$. These auxiliary conditions might consist of specifying fixed values of $y[n]$ at specific values of n, such as $y[-1]$, $y[-2], \ldots, y[-N]$, and then solving a set of N linear equations for the N undetermined coefficients.

Alternatively, if the auxiliary conditions are a set of auxiliary values of $y[n]$, the other values of $y[n]$ can be generated by rewriting Eq. (2.82) as a recurrence formula, i.e., in the form

$$y[n] = -\sum_{k=1}^{N} \frac{a_k}{a_0} y[n-k] + \sum_{k=0}^{M} \frac{b_k}{a_0} x[n-k]. \tag{2.97}$$

If the input $x[n]$, together with a set of auxiliary values, say, $y[-1], y[-2], \ldots, y[-N]$, is specified, then $y[0]$ can be determined from Eq. (2.97). With $y[0], y[-1], \ldots, y[-N+1]$ available, $y[1]$ can then be calculated, and so on. When this procedure is used, $y[n]$ is said to be computed *recursively;* i.e., the output computation involves not only the input sequence, but also previous values of the output sequence.

To generate values of $y[n]$ for $n < -N$ (again assuming that the values $y[-1]$, $y[-2], \ldots, y[-N]$ are given as auxiliary conditions), we can rearrange Eq. (2.82) in the form

$$y[n-N] = -\sum_{k=0}^{N-1} \frac{a_k}{a_N} y[n-k] + \sum_{k=0}^{M} \frac{b_k}{a_N} x[n-k], \tag{2.98}$$

from which $y[-N-1], y[-N-2], \ldots$ can be computed recursively. The following example illustrates this recursive procedure.

Example 2.16   Recursive Computation of Difference Equations

The difference equation satisfied by the input and output of a system is

$$y[n] = ay[n-1] + x[n]. \tag{2.99}$$

Consider the input $x[n] = K\delta[n]$, where K is an arbitrary number, and the auxiliary condition $y[-1] = c$. Beginning with this value, the output for $n > -1$ can be computed recursively as follows:

$$y[0] = ac + K,$$
$$y[1] = ay[0] + 0 = a(ac + K) = a^2 c + aK,$$

$$y[2] = ay[1] + 0 = a(a^2c + aK) = a^3c + a^2K,$$

$$y[3] = ay[2] + 0 = a(a^3c + a^2K) = a^4c + a^3K,$$

$$\vdots \qquad\qquad \vdots$$

For this simple case, we can see that for $n \geq 0$,

$$y[n] = a^{n+1}c + a^nK, \qquad \text{for } n \geq 0. \tag{2.100}$$

To determine the output for $n < 0$, we express the difference equation in the form

$$y[n-1] = a^{-1}(y[n] - x[n]), \tag{2.101a}$$

or

$$y[n] = a^{-1}(y[n+1] - x[n+1]). \tag{2.101b}$$

Using the auxiliary condition $y[-1] = c$, we can compute $y[n]$ for $n < -1$ as follows:

$$y[-2] = a^{-1}(y[-1] - x[-1]) = a^{-1}c,$$

$$y[-3] = a^{-1}(y[-2] - x[-2]) = a^{-1}a^{-1}c = a^{-2}c,$$

$$y[-4] = a^{-1}(y[-3] - x[-3]) = a^{-1}a^{-2}c = a^{-3}c,$$

It then follows that

$$y[n] = a^{n+1}c \qquad \text{for } n \leq -1. \tag{2.102}$$

In sum, combining Eqs. (2.100) and (2.102), we obtain, as the result of the recursive computation,

$$y[n] = a^{n+1}c + Ka^nu[n], \qquad \text{for all } n. \tag{2.103}$$

Several important points are illustrated by the solution of Example 2.16. First, note that we implemented the system by recursively computing the output in both the positive and the negative direction, beginning with $n = -1$. Clearly, this procedure is noncausal. Also, note that when $K = 0$, the input is zero, but $y[n] = a^{n+1}c$. A linear system requires that the output be zero for all time when the input is zero for all time. (See Problem 2.21.) Consequently, this system is not linear. Furthermore, if the input were shifted by n_0 samples, i.e., $x_1[n] = K\delta[n - n_0]$, the output would be

$$y_1[n] = a^{n+1}c + Ka^{n-n_0}u[n - n_0], \tag{2.104}$$

and the system is therefore not time invariant.

Our principal interest in this text is in systems that are linear and time invariant, in which case the auxiliary conditions must be consistent with these additional requirements. In Chapter 3, when we discuss the solution of difference equations using the z-transform, we implicitly incorporate conditions of linearity and time invariance. As we will see in that discussion, even with the additional constraints of linearity and time invariance, the solution to the difference equation, and therefore the system, is not uniquely specified. In particular, there are, in general, both causal and noncausal linear time-invariant systems consistent with a given difference equation.

If a system is characterized by a linear constant-coefficient difference equation and is further specified to be linear, time invariant, and causal, the solution is unique. In this case, the auxiliary conditions are often stated as *initial-rest conditions*. In other words, the auxiliary information is that if the input $x[n]$ is zero for n less than some time

n_0, then the output $y[n]$ is constrained to be zero for n less than n_0. This then provides sufficient initial conditions to obtain $y[n]$ for $n \geq n_0$ recursively using Eq. (2.97).

To summarize, for a system for which the input and output satisfy a linear constant-coefficient difference equation:

- The output for a given input is not uniquely specified. Auxiliary information or conditions are required.

- If the auxiliary information is in the form of N sequential values of the output, later values can be obtained by rearranging the difference equation as a recursive relation running forward in n, and prior values can be obtained by rearranging the difference equation as a recursive relation running backward in n.

- Linearity, time invariance, and causality of the system will depend on the auxiliary conditions. If an additional condition is that the system is initially at rest, then the system will be linear, time invariant, and causal.

With the preceding discussion in mind, let us now consider again Example 2.16, but with initial-rest conditions. With $x[n] = K\delta[n]$, $y[-1] = 0$, since $x[n] = 0, n < 0$. Consequently, from Eq. (2.103),

$$y[n] = Ka^n u[n]. \tag{2.105}$$

If the input is instead $K\delta[n - n_0]$, again with initial-rest conditions, then the recursive solution is carried out using the initial condition $y[n] = 0, n < n_0$. Note that for $n_0 < 0$, initial rest implies that $y[-1] \neq 0$. That is, initial rest does not always mean $y[-1] = \cdots = y[-N] = 0$. It does mean that $y[n_0 - 1] = \cdots = y[n_0 - N] = 0$ if $x[n] = 0$ for $n < n_0$. Note also that the impulse response for the example is $h[n] = a^n u[n]$; i.e., $h[n]$ is zero for $n < 0$, consistent with the causality imposed by the assumption of initial rest.

The preceding discussion assumed that $N \geq 1$ in Eq. (2.82). If, instead, $N = 0$, no recursion is required to use the difference equation to compute the output, and therefore, no auxiliary conditions are required. That is,

$$y[n] = \sum_{k=0}^{M} \left(\frac{b_k}{a_0} \right) x[n - k]. \tag{2.106}$$

Equation (2.106) is in the form of a convolution, and by setting $x[n] = \delta[n]$, we see that the impulse response is

$$h[n] = \sum_{k=0}^{M} \left(\frac{b_k}{a_0} \right) \delta[n - k],$$

or

$$h[n] = \begin{cases} \left(\dfrac{b_n}{a_0} \right), & 0 \leq n \leq M, \\ 0, & \text{otherwise.} \end{cases} \tag{2.107}$$

The impulse response is obviously finite in duration. Indeed, the output of any FIR system can be computed nonrecursively using the difference equation of Eq. (2.106), where the coefficients are the values of the impulse response sequence. The moving-average system of Example 2.15 with $M_1 = 0$ is an example of a causal FIR system. An interesting feature of that system was that we also found a recursive equation for the output. In

Chapter 6 we will show that there are many possible ways of implementing a desired signal transformation using difference equations. Advantages of one method over another depend on practical considerations such as numerical accuracy, data storage, and the number of multiplications and additions required to compute each sample of the output.

2.6 FREQUENCY-DOMAIN REPRESENTATION OF DISCRETE-TIME SIGNALS AND SYSTEMS

In the previous sections, we have introduced some of the fundamental concepts of the theory of discrete-time signals and systems. For linear time-invariant systems, we saw that a representation of the input sequence as a weighted sum of delayed impulses leads to a representation of the output as a weighted sum of delayed impulse responses. As with continuous-time signals, discrete-time signals may be represented in a number of different ways. For example, sinusoidal and complex exponential sequences play a particularly important role in representing discrete-time signals. This is because complex exponential sequences are eigenfunctions of linear time-invariant systems and the response to a sinusoidal input is sinusoidal with the same frequency as the input and with amplitude and phase determined by the system. This fundamental property of linear time-invariant systems makes representations of signals in terms of sinusoids or complex exponentials (i.e., Fourier representations) very useful in linear system theory.

2.6.1 Eigenfunctions for Linear Time-Invariant Systems

To demonstrate the eigenfunction property of complex exponentials for discrete-time systems, consider an input sequence $x[n] = e^{j\omega n}$ for $-\infty < n < \infty$, i.e., a complex exponential of radian frequency ω. From Eq. (2.62), the corresponding output of a linear time-invariant system with impulse response $h[n]$ is

$$y[n] = \sum_{k=-\infty}^{\infty} h[k]e^{j\omega(n-k)}$$

$$= e^{j\omega n} \left(\sum_{k=-\infty}^{\infty} h[k]e^{-j\omega k} \right).$$

(2.108)

If we define

$$H(e^{j\omega}) = \sum_{k=-\infty}^{\infty} h[k]e^{-j\omega k},$$

(2.109)

Eq. (2.108) becomes

$$y[n] = H(e^{j\omega})e^{j\omega n}.$$

(2.110)

Consequently, $e^{j\omega n}$ is an eigenfunction of the system, and the associated eigenvalue is $H(e^{j\omega})$. From Eq. (2.110), we see that $H(e^{j\omega})$ describes the change in complex amplitude of a complex exponential input signal as a function of the frequency ω. The eigenvalue $H(e^{j\omega})$ is called the *frequency response* of the system. In general, $H(e^{j\omega})$ is complex and can be expressed in terms of its real and imaginary parts as

$$H(e^{j\omega}) = H_R(e^{j\omega}) + jH_I(e^{j\omega})$$

(2.111)

or in terms of magnitude and phase as

$$H(e^{j\omega}) = |H(e^{j\omega})|e^{j \sphericalangle H(e^{j\omega})}. \tag{2.112}$$

Example 2.17   Frequency Response of the Ideal Delay System

As a simple example of how we can find the frequency response of a linear time-invariant system, consider the ideal delay system defined by

$$y[n] = x[n - n_d], \tag{2.113}$$

where n_d is a fixed integer. If we consider $x[n] = e^{j\omega n}$ as input to this system, then, from Eq. (2.113), we have

$$y[n] = e^{j\omega(n-n_d)} = e^{-j\omega n_d}e^{j\omega n}.$$

Thus, for any given value of ω, we obtain an output that is the input multiplied by a complex constant, the value of which depends on the frequency ω and the delay n_d. The frequency response of the ideal delay is therefore

$$H(e^{j\omega}) = e^{-j\omega n_d}. \tag{2.114}$$

As an alternative method of obtaining the frequency response, recall that $h[n] = \delta[n - n_d]$ for the ideal delay system. Using Eq. (2.109), we obtain

$$H(e^{j\omega}) = \sum_{n=-\infty}^{\infty} \delta[n - n_d]e^{-j\omega n} = e^{-j\omega n_d}.$$

From the Euler relation, the real and imaginary parts of the frequency response are

$$H_R(e^{j\omega}) = \cos(\omega n_d), \tag{2.115a}$$

$$H_I(e^{j\omega}) = -\sin(\omega n_d). \tag{2.115b}$$

The magnitude and phase are

$$|H(e^{j\omega})| = 1, \tag{2.116a}$$

$$\sphericalangle H(e^{j\omega}) = -\omega n_d. \tag{2.116b}$$

In Section 2.7 we will show that a broad class of signals can be represented as a linear combination of complex exponentials in the form

$$x[n] = \sum_k \alpha_k e^{j\omega_k n}. \tag{2.117}$$

From the principle of superposition, the corresponding output of a linear time-invariant system is

$$y[n] = \sum_k \alpha_k H(e^{j\omega_k})e^{j\omega_k n}. \tag{2.118}$$

Thus, if we can find a representation of $x[n]$ as a superposition of complex exponential sequences, as in Eq. (2.117), then we can find the output using Eq. (2.118) if we know the frequency response of the system. The following simple example illustrates this fundamental property of linear time-invariant systems.

Example 2.18   Sinusoidal Response of LTI Systems

Since it is simple to express a sinusoid as a linear combination of complex exponentials, let us consider a sinusoidal input

$$x[n] = A\cos(\omega_0 n + \phi) = \frac{A}{2}e^{j\phi}e^{j\omega_0 n} + \frac{A}{2}e^{-j\phi}e^{-j\omega_0 n}. \tag{2.119}$$

From Eq. (2.110), the response to $x_1[n] = (A/2)e^{j\phi}e^{j\omega_0 n}$ is

$$y_1[n] = H(e^{j\omega_0})\frac{A}{2}e^{j\phi}e^{j\omega_0 n}. \tag{2.120a}$$

The response to $x_2[n] = (A/2)e^{-j\phi}e^{-j\omega_0 n}$ is

$$y_2[n] = H(e^{-j\omega_0})\frac{A}{2}e^{-j\phi}e^{-j\omega_0 n}. \tag{2.120b}$$

Thus, the total response is

$$y[n] = \frac{A}{2}[H(e^{j\omega_0})e^{j\phi}e^{j\omega_0 n} + H(e^{-j\omega_0})e^{-j\phi}e^{-j\omega_0 n}]. \tag{2.121}$$

If $h[n]$ is real, it can be shown (see Problem 2.71) that $H(e^{-j\omega_0}) = H^*(e^{j\omega_0})$. Consequently,

$$y[n] = A|H(e^{j\omega_0})|\cos(\omega_0 n + \phi + \theta), \tag{2.122}$$

where $\theta = \sphericalangle H(e^{j\omega_0})$ is the phase of the system function at frequency ω_0.

For the simple example of the ideal delay, $|H(e^{j\omega_0})| = 1$ and $\theta = -\omega_0 n_d$, as we determined in Example 2.17. Therefore,

$$\begin{aligned} y[n] &= A\cos(\omega_0 n + \phi - \omega_0 n_d) \\ &= A\cos[\omega_0(n - n_d) + \phi], \end{aligned} \tag{2.123}$$

which is consistent with what we would obtain directly using the definition of the ideal delay system.

The concept of the frequency response of linear time-invariant systems is essentially the same for continuous-time and discrete-time systems. However, an important distinction arises because the frequency response of discrete-time linear time-invariant systems is *always* a periodic function of the frequency variable ω with period 2π. To show this, we substitute $\omega + 2\pi$ into Eq. (2.109) to obtain

$$H(e^{j(\omega+2\pi)}) = \sum_{n=-\infty}^{\infty} h[n]e^{-j(\omega+2\pi)n}. \tag{2.124}$$

Using the fact that $e^{\pm j2\pi n} = 1$ for n an integer, we have

$$e^{-j(\omega+2\pi)n} = e^{-j\omega n}e^{-j2\pi n} = e^{-j\omega n}.$$

Therefore,

$$H(e^{j(\omega+2\pi)}) = H(e^{j\omega}), \tag{2.125}$$

and, more generally,

$$H(e^{j(\omega+2\pi r)}) = H(e^{j\omega}), \qquad \text{for } r \text{ an integer.} \tag{2.126}$$

That is, $H(e^{j\omega})$ is periodic with period 2π. Note that this is obviously true for the ideal delay system, since $e^{-j(\omega+2\pi)n_d} = e^{-j\omega n_d}$ when n_d is an integer.

The reason for this periodicity is related directly to our earlier observation that the sequence

$$\{e^{j\omega n}\}, \qquad -\infty < n < \infty,$$

is indistinguishable from the sequence

$$\{e^{j(\omega+2\pi)n}\}, \qquad -\infty < n < \infty.$$

Because these two sequences have identical values for all n, the system must respond identically to both input sequences. This condition requires that Eq. (2.125) hold.

Since $H(e^{j\omega})$ is periodic with period 2π, and since the frequencies ω and $\omega + 2\pi$ are indistinguishable, it follows that we need only specify $H(e^{j\omega})$ over an interval of length 2π, e.g., $0 \leq \omega \leq 2\pi$ or $-\pi < \omega \leq \pi$. The inherent periodicity defines the frequency response everywhere outside the chosen interval. For simplicity and for consistency with the continuous-time case, it is generally convenient to specify $H(e^{j\omega})$ over the interval $-\pi < \omega \leq \pi$. With respect to this interval, the "low frequencies" are frequencies close to zero, while the "high frequencies" are frequencies close to $\pm\pi$. Recalling that frequencies differing by an integer multiple of 2π are indistinguishable, we might generalize the preceding statement as follows: The "low frequencies" are those that are close to an even multiple of π, while the "high frequencies" are those that are close to an odd multiple of π, consistent with our earlier discussion in Section 2.1.

Example 2.19   Ideal Frequency-Selective Filters

An important class of linear time-invariant systems includes those systems for which the frequency response is unity over a certain range of frequencies and is zero at the remaining frequencies. These correspond to *ideal frequency-selective filters*. The frequency response of an ideal lowpass filter is shown in Figure 2.17(a). Because of the inherent periodicity of the discrete-time frequency response, it has the appearance of

(a)

(b)

Figure 2.17   Ideal lowpass filter showing (a) periodicity of the frequency response and (b) one period of the periodic frequency response.

(a)

(b)

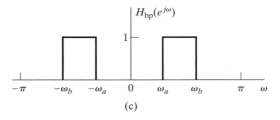

(c)

Figure 2.18   Ideal frequency-selective filters. (a) Highpass filter. (b) Bandstop filter. (c) Bandpass filter. In each case, the frequency response is periodic with period 2π. Only one period is shown.

a multiband filter, since frequencies around $\omega = 2\pi$ are indistinguishable from frequencies around $\omega = 0$. In effect, however, the frequency response passes only low frequencies and rejects high frequencies. Since the frequency response is completely specified by its behavior over the interval $-\pi < \omega \le \pi$, the ideal lowpass filter frequency response is more typically shown only in the interval $-\pi < \omega \le \pi$, as in Figure 2.17(b). It is understood that the frequency response repeats periodically with period 2π outside the plotted interval. The frequency responses for ideal highpass, bandstop, and bandpass filters are shown in Figures 2.18(a), (b), and (c), respectively.

Example 2.20   Frequency Response of the Moving-Average System

The impulse response of the moving-average system of Example 2.4 is

$$
h[n] = \begin{cases} \dfrac{1}{M_1 + M_2 + 1}, & -M_1 \le n \le M_2, \\ \\ 0, & \text{otherwise.} \end{cases}
$$

Therefore, the frequency response is

$$H(e^{j\omega}) = \frac{1}{M_1 + M_2 + 1} \sum_{n=-M_1}^{M_2} e^{-j\omega n}. \tag{2.127}$$

Equation (2.127) can be expressed in closed form by using Eq. (2.56), so that

$$H(e^{j\omega}) = \frac{1}{M_1 + M_2 + 1} \frac{e^{j\omega M_1} - e^{-j\omega(M_2+1)}}{1 - e^{-j\omega}}$$

$$= \frac{1}{M_1 + M_2 + 1} \frac{e^{j\omega(M_1+M_2+1)/2} - e^{-j\omega(M_1+M_2+1)/2}}{1 - e^{-j\omega}} e^{-j\omega(M_2-M_1+1)/2}$$

$$\tag{2.128}$$

$$= \frac{1}{M_1 + M_2 + 1} \frac{e^{j\omega(M_1+M_2+1)/2} - e^{-j\omega(M_1+M_2+1)/2}}{e^{j\omega/2} - e^{-j\omega/2}} e^{-j\omega(M_2-M_1)/2}$$

$$= \frac{1}{M_1 + M_2 + 1} \frac{\sin[\omega(M_1 + M_2 + 1)/2]}{\sin(\omega/2)} e^{-j\omega(M_2-M_1)/2}.$$

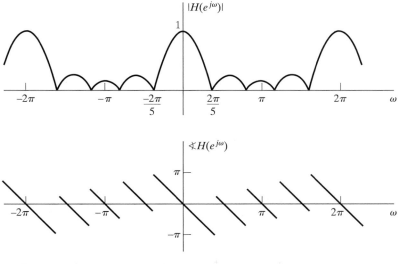

Figure 2.19 (a) Magnitude and (b) phase of the frequency response of the moving-average system for the case $M_1 = 0$ and $M_2 = 4$.

The magnitude and phase of $H(e^{j\omega})$ are plotted in Figure 2.19 for $M_1 = 0$ and $M_2 = 4$. Note that $H(e^{j\omega})$ is periodic, as is required of the frequency response of a discrete-time system. Note also that $|H(e^{j\omega})|$ falls off at "high frequencies" and $\sphericalangle H(e^{j\omega})$, i.e., the phase of $H(e^{j\omega})$, varies linearly with ω. This attenuation of the high frequencies suggests that the system will smooth out rapid variations in the input sequence; in other words, the system is a rough approximation to a lowpass filter. This is consistent with what we would intuitively expect about the behavior of the moving-average system.

2.6.2 Suddenly Applied Complex Exponential Inputs

We have seen that complex exponential inputs of the form $e^{j\omega n}$ for $-\infty < n < \infty$ produce outputs of the form $H(e^{j\omega})e^{j\omega n}$ for linear time-invariant systems. Such inputs, nonzero over a doubly infinite domain, may seem to be impractical models of signals; however, as we will see in the next section, models of this kind are crucial to the mathematical representation of a wide range of signals, even those that exist only over a finite domain. Even so, we can gain additional insight into linear time-invariant systems by considering more practical-appearing inputs of the form

$$x[n] = e^{j\omega n}u[n],$$

i.e., complex exponentials that are suddenly applied at an arbitrary time, which for convenience here we choose as $n = 0$. Using the convolution sum in Eq. (2.62), the corresponding output of a causal linear time-invariant system with impulse response $h[n]$ is

$$y[n] = \begin{cases} 0, & n < 0, \\ \left(\sum_{k=0}^{n} h[k]e^{-j\omega k} \right) e^{j\omega n}, & n \geq 0. \end{cases}$$

If we consider the output for $n \geq 0$, we can write

$$y[n] = \left(\sum_{k=0}^{\infty} h[k]e^{-j\omega k} \right) e^{j\omega n} - \left(\sum_{k=n+1}^{\infty} h[k]e^{-j\omega k} \right) e^{j\omega n} \tag{2.129}$$

$$= H(e^{j\omega})e^{j\omega n} - \left(\sum_{k=n+1}^{\infty} h[k]e^{-j\omega k} \right) e^{j\omega n}. \tag{2.130}$$

From Eq. (2.130), we see that the output consists of the sum of two terms, i.e., $y[n] = y_{ss}[n] + y_t[n]$. The first term,

$$y_{ss}[n] = H(e^{j\omega})e^{j\omega n},$$

is called the *steady-state response*. It is identical to the response of the system when the input is $e^{j\omega n}$ for all n. In a sense, the second term,

$$y_t[n] = - \sum_{k=n+1}^{\infty} h[k]e^{-j\omega k}e^{j\omega n},$$

is the amount by which the output differs from the eigenfunction result. This part is called the *transient response*, because it is clear that in some cases it may approach zero. To see the conditions for which this is true, let us consider the size of the second term. Its magnitude is bounded as follows:

$$|y_t[n]| = \left| \sum_{k=n+1}^{\infty} h[k]e^{-j\omega k}e^{j\omega n} \right| \leq \sum_{k=n+1}^{\infty} |h[k]|. \tag{2.131}$$

From Eq. (2.131), it should be clear that if the impulse response has finite length, so that $h[n] = 0$ except for $0 \leq n \leq M$, then the term $y_t[n] = 0$ for $n + 1 > M$, or $n > M - 1$. In this case,

$$y[n] = y_{ss}[n] = H(e^{j\omega})e^{j\omega n}, \qquad \text{for } n > M - 1.$$

When the impulse response has infinite duration, the transient response does not disappear abruptly, but if the samples of the impulse response approach zero with increasing n, then $y_t[n]$ will approach zero. Note that Eq. (2.131) can be written

$$|y_t[n]| = \left| \sum_{k=n+1}^{\infty} h[k]e^{-j\omega k}e^{j\omega n} \right| \leq \sum_{k=n+1}^{\infty} |h[k]| \leq \sum_{k=0}^{\infty} |h[k]|. \qquad (2.132)$$

That is, the transient response is bounded by the sum of the absolute values of *all* of the impulse response samples. If the right-hand side of Eq. (2.132) is bounded, so that

$$\sum_{k=0}^{\infty} |h[k]| < \infty,$$

then the system is stable. From Eq. (2.132), it follows that, for stable systems, the transient response must become increasingly smaller as $n \to \infty$. Thus, a sufficient condition for the transient response to die out is that the system be stable.

Figure 2.20 shows the real part of a complex exponential signal with frequency $\omega = 2\pi/10$. The solid dots indicate the samples $x[k]$ of the suddenly applied complex exponential, while the open circles indicate the samples of the complex exponential that are "missing." The shaded dots indicate the samples of the impulse response $h[n - k]$ as a function of k for $n = 8$. In the finite-length case shown in Figure 2.20(a), it is clear that the output would consist only of the steady-state component for $n \geq 8$, while in the infinite-length case, it is clear that the missing samples have less and less effect as n increases, due to the decaying nature of the impulse response.

(a)

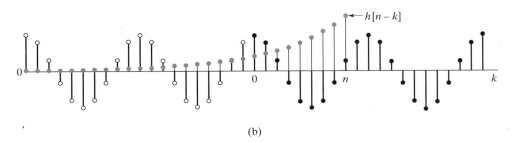

(b)

Figure 2.20 Illustration of real part of suddenly applied complex exponential input with (a) finite-length impulse response and (b) infinite-length impulse response.

The condition for stability is also a sufficient condition for the existence of the frequency response function. To see this, note that, in general,

$$|H(e^{j\omega})| = \left| \sum_{k=-\infty}^{\infty} h[k]e^{-j\omega k} \right| \le \sum_{k=-\infty}^{\infty} |h[k]e^{-j\omega k}| \le \sum_{k=-\infty}^{\infty} |h[k]|,$$

so the general condition

$$\sum_{k=-\infty}^{\infty} |h[k]| < \infty$$

ensures that $H(e^{j\omega})$ exists. It is no surprise that the condition for existence of the frequency response is the same as the condition for dominance of the steady-state solution. Indeed, a complex exponential that exists for all n can be thought of as one that is applied at $n = -\infty$. The eigenfunction property of complex exponentials depends on stability of the system, since at finite n, the transient response must have become zero, so that we only see the steady-state response $H(e^{j\omega})e^{j\omega n}$ for all finite n.

2.7 REPRESENTATION OF SEQUENCES BY FOURIER TRANSFORMS

One of the advantages of the frequency-response representation of a linear time-invariant system is that interpretations of system behavior such as the one we made in Example 2.20 often follow easily. We will elaborate on this point in considerably more detail in Chapter 5. At this point, however, let us return to the question of how we may find representations of the form of Eq. (2.117) for an arbitrary input sequence.

Many sequences can be represented by a Fourier integral of the form

$$x[n] = \frac{1}{2\pi} \int_{-\pi}^{\pi} X(e^{j\omega})e^{j\omega n} d\omega, \tag{2.133}$$

where

$$X(e^{j\omega}) = \sum_{n=-\infty}^{\infty} x[n]e^{-j\omega n}. \tag{2.134}$$

Equations (2.133) and (2.134) together form a *Fourier representation* for the sequence. Equation (2.133), the *inverse Fourier transform,* is a *synthesis* formula. That is, it represents $x[n]$ as a superposition of infinitesimally small complex sinusoids of the form

$$\frac{1}{2\pi} X(e^{j\omega})e^{j\omega n} d\omega,$$

with ω ranging over an interval of length 2π and with $X(e^{j\omega})$ determining the relative amount of each complex sinusoidal component. Although, in writing Eq. (2.133), we have chosen the range of values for ω between $-\pi$ and $+\pi$, any interval of length 2π can be used. Equation (2.134), the *Fourier transform,*[3] is an expression for computing $X(e^{j\omega})$ from the sequence $x[n]$, i.e., for *analyzing* the sequence $x[n]$ to determine how much of each frequency component is required to synthesize $x[n]$ using Eq. (2.133).

[3]Sometimes we will refer to Eq. (2.134) more explicitly as the discrete-time Fourier transform, or DTFT, particularly when it is important to distinguish it from the continuous-time Fourier transform.

In general, the Fourier transform is a complex-valued function of ω. As with the frequency response, we may either express $X(e^{j\omega})$ in rectangular form as

$$X(e^{j\omega}) = X_R(e^{j\omega}) + j X_I(e^{j\omega}) \tag{2.135a}$$

or in polar form as

$$X(e^{j\omega}) = |X(e^{j\omega})|e^{j \sphericalangle X(e^{j\omega})}. \tag{2.135b}$$

The quantities $|X(e^{j\omega})|$ and $\sphericalangle X(e^{j\omega})$ are the *magnitude* and *phase,* respectively, of the Fourier transform. The Fourier transform is sometimes referred to as the *Fourier spectrum* or, simply, the *spectrum*. Also, the terminology *magnitude spectrum* or *amplitude spectrum* is sometimes used to refer to $|X(e^{j\omega})|$, and the angle or phase $\sphericalangle X(e^{j\omega})$ is sometimes called the *phase spectrum.*

The phase $\sphericalangle X(e^{j\omega})$ is not uniquely specified by Eq. (2.135b), since any integer multiple of 2π may be added to $\sphericalangle X(e^{j\omega})$ at any value of ω without affecting the result of the complex exponentiation. When we specifically want to refer to the principal value, i.e., $\sphericalangle X(e^{j\omega})$ restricted to the range of values between $-\pi$ and $+\pi$, we will denote this as $\text{ARG}[X(e^{j\omega})]$. If we want to refer to a phase function that is a continuous function of ω for $0 < \omega < \pi$, we will use the notation $\arg[X(e^{j\omega})]$.

By comparing Eqs. (2.109) and (2.134), we can see that the frequency response of a linear time-invariant system is simply the Fourier transform of the impulse response and that, therefore, the impulse response can be obtained from the frequency response by applying the inverse Fourier transform integral; i.e.,

$$h[n] = \frac{1}{2\pi} \int_{-\pi}^{\pi} H(e^{j\omega})e^{j\omega n} d\omega. \tag{2.136}$$

As discussed previously, the frequency response is a periodic function. Likewise, the Fourier transform is periodic with period 2π. Indeed, Eq. (2.134) is of the form of a Fourier series for the continuous-variable periodic function $X(e^{j\omega})$, and Eq. (2.133), which expresses the sequence values $x[n]$ in terms of the periodic function $X(e^{j\omega})$, is of the form of the integral that would be used to obtain the coefficients in the Fourier series. Our use of Eqs. (2.133) and (2.134) focuses on the representation of the sequence $x[n]$. Nevertheless, it is useful to be aware of the equivalence between the Fourier series representation of continuous-variable periodic functions and the Fourier transform representation of discrete-time signals, since all the familiar properties of Fourier series can be applied, with appropriate interpretation of variables, to the Fourier transform representation of a sequence.

We have not yet shown explicitly that Eqs. (2.133) and (2.134) are inverses of each other, nor have we considered the question of how broad a class of signals can be represented in the form of Eq. (2.133). To demonstrate that Eq. (2.133) is the inverse of Eq. (2.134), we can find $X(e^{j\omega})$ using Eq. (2.134) and then substitute the result into

Eq. (2.133). Specifically, consider

$$\frac{1}{2\pi} \int_{-\pi}^{\pi} \left(\sum_{m=-\infty}^{\infty} x[m]e^{-j\omega m} \right) e^{j\omega n} d\omega = \hat{x}[n], \tag{2.137}$$

where we have tentatively used $\hat{x}[n]$ to denote the result of the Fourier synthesis. We wish to show that $\hat{x}[n] = x[n]$ if $X(e^{j\omega})$ can be found using Eq. (2.134). Note that the "dummy index" of summation has been changed to m to distinguish it from n, the variable index in Eq. (2.133). If the infinite sum converges uniformly for all ω, then we can interchange the order of integration and summation to obtain

$$\hat{x}[n] = \sum_{m=-\infty}^{\infty} x[m] \left(\frac{1}{2\pi} \int_{-\pi}^{\pi} e^{j\omega(n-m)} d\omega \right). \tag{2.138}$$

Evaluating the integral within the parentheses gives

$$\frac{1}{2\pi} \int_{-\pi}^{\pi} e^{j\omega(n-m)} d\omega = \frac{\sin \pi(n-m)}{\pi(n-m)}$$

$$= \begin{cases} 1, & m = n, \\ 0, & m \neq n, \end{cases}$$

$$= \delta[n-m].$$

Thus,

$$\hat{x}[n] = \sum_{m=-\infty}^{\infty} x[m]\delta[n-m] = x[n],$$

which is what we set out to show.

Determining the class of signals that can be represented by Eq. (2.133) is equivalent to considering the convergence of the infinite sum in Eq. (2.134). That is, we are concerned with the conditions that must be satisfied by the terms in the sum in Eq. (2.134) such that

$$|X(e^{j\omega})| < \infty \qquad \text{for all } \omega,$$

where $X(e^{j\omega})$ is the limit as $M \to \infty$ of the finite sum

$$X_M(e^{j\omega}) = \sum_{n=-M}^{M} x[n]e^{-j\omega n}. \tag{2.139}$$

A sufficient condition for convergence can be found as follows:

$$|X(e^{j\omega})| = \left| \sum_{n=-\infty}^{\infty} x[n]e^{-j\omega n} \right|$$

$$\leq \sum_{n=-\infty}^{\infty} |x[n]| \, |e^{-j\omega n}|$$

$$\leq \sum_{n=-\infty}^{\infty} |x[n]| < \infty.$$

Thus, if $x[n]$ is *absolutely summable,* then $X(e^{j\omega})$ exists. Furthermore, in this case, the series can be shown to converge uniformly to a continuous function of ω.

Since a stable sequence is, by definition, absolutely summable, all stable sequences have Fourier transforms. It also follows, then, that any stable *system* will have a finite and continuous frequency response.

Absolute summability is a sufficient condition for the existence of a Fourier transform representation. In Examples 2.17 and 2.20, we computed the Fourier transforms of the sequences $\delta[n - n_d]$ and $[1/(M_1 + M_2 + 1)](u[n + M_1] - u[n - M_2 - 1])$. These sequences are absolutely summable, since they are finite in length. Clearly, any finite-length sequence is absolutely summable and thus will have a Fourier transform representation. In the context of linear time-invariant systems, any FIR system will be stable and therefore will have a finite, continuous frequency response. When a sequence has infinite length, we must be concerned about convergence of the infinite sum. The following example illustrates this case.

Example 2.21   Absolute Summability for a Suddenly-Applied Exponential

Let $x[n] = a^n u[n]$. The Fourier transform of this sequence is

$$X(e^{j\omega}) = \sum_{n=0}^{\infty} a^n e^{-j\omega n} = \sum_{n=0}^{\infty} (ae^{-j\omega})^n$$

$$= \frac{1}{1 - ae^{-j\omega}} \qquad \text{if } |ae^{-j\omega}| < 1 \quad \text{or} \quad |a| < 1.$$

Clearly, the condition $|a| < 1$ is the condition for the absolute summability of $x[n]$; i.e.,

$$\sum_{n=0}^{\infty} |a|^n = \frac{1}{1 - |a|} < \infty \qquad \text{if } |a| < 1. \tag{2.140}$$

Absolute summability is a *sufficient* condition for the existence of a Fourier transform representation, and it also guarantees uniform convergence. Some sequences are not absolutely summable, but are square summable, i.e.,

$$\sum_{n=-\infty}^{\infty} |x[n]|^2 < \infty. \tag{2.141}$$

Such sequences can be represented by a Fourier transform if we are willing to relax the condition of uniform convergence of the infinite sum defining $X(e^{j\omega})$. Specifically, in this case we have mean-square convergence; that is, with

$$X(e^{j\omega}) = \sum_{n=-\infty}^{\infty} x[n]e^{-j\omega n} \tag{2.142a}$$

and

$$X_M(e^{j\omega}) = \sum_{n=-M}^{M} x[n]e^{-j\omega n}, \tag{2.142b}$$

it follows that

$$\lim_{M \to \infty} \int_{-\pi}^{\pi} |X(e^{j\omega}) - X_M(e^{j\omega})|^2 d\omega = 0. \tag{2.143}$$

In other words, the error $|X(e^{j\omega}) - X_M(e^{j\omega})|$ may not approach zero at each value of ω as $M \to \infty$, but the total "energy" in the error does. Example 2.22 illustrates this case.

Example 2.22   Square-Summability for the Ideal Lowpass Filter

Let us determine the impulse response of the ideal lowpass filter discussed in Example 2.19. The frequency response is

$$H_{\mathrm{lp}}(e^{j\omega}) = \begin{cases} 1, & |\omega| < \omega_c, \\ 0, & \omega_c < |\omega| \le \pi, \end{cases} \tag{2.144}$$

with periodicity 2π also understood. The impulse response $h_{\mathrm{lp}}[n]$ can be found using the Fourier transform synthesis equation (2.133):

$$\begin{aligned} h_{\mathrm{lp}}[n] &= \frac{1}{2\pi} \int_{-\omega_c}^{\omega_c} e^{j\omega n} d\omega \\ &= \frac{1}{2\pi jn} \left[e^{j\omega n} \right]_{-\omega_c}^{\omega_c} = \frac{1}{2\pi jn} (e^{j\omega_c n} - e^{-j\omega_c n}) \\ &= \frac{\sin \omega_c n}{\pi n}, \qquad -\infty < n < \infty. \end{aligned} \tag{2.145}$$

We note that, since $h_{\mathrm{lp}}[n]$ is nonzero for $n < 0$, the ideal lowpass filter is noncausal. Also, $h_{\mathrm{lp}}[n]$ is *not* absolutely summable. The sequence values approach zero as $n \to \infty$, but only as $1/n$. This is because $H_{\mathrm{lp}}(e^{j\omega})$ is discontinuous at $\omega = \omega_c$. Since $h_{\mathrm{lp}}[n]$ is not absolutely summable, the infinite sum

$$\sum_{n=-\infty}^{\infty} \frac{\sin \omega_c n}{\pi n} e^{-j\omega n}$$

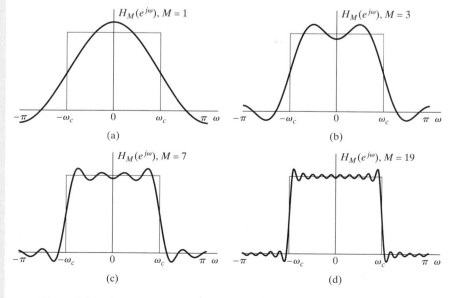

Figure 2.21   Convergence of the Fourier transform. The oscillatory behavior at $\omega = \omega_c$ is often called the Gibbs phenomenon.

does not converge uniformly for all values of ω. To obtain an intuitive feeling for this, let us consider $H_M(e^{j\omega})$ as the sum of a finite number of terms:

$$H_M(e^{j\omega}) = \sum_{n=-M}^{M} \frac{\sin \omega_c n}{\pi n} e^{-j\omega n}. \tag{2.146}$$

We can show that $H_M(e^{j\omega})$ can be expressed as

$$H_M(e^{j\omega}) = \frac{1}{2\pi} \int_{-\omega_c}^{\omega_c} \frac{\sin[(2M+1)(\omega-\theta)/2]}{\sin[(\omega-\theta)]/2} d\theta.$$

The function $H_M(e^{j\omega})$ is evaluated in Figure 2.21 for several values of M. Note that as M increases, the oscillatory behavior at $\omega = \omega_c$ (often referred to as the Gibbs phenomenon) is more rapid, but the size of the ripples does not decrease. In fact, it can be shown that as $M \to \infty$, the maximum amplitude of the oscillations does not approach zero, but the oscillations converge in location toward the point $\omega = \omega_c$. Thus, the infinite sum does not converge uniformly to the discontinuous function $H_{lp}(e^{j\omega})$ of Eq. (2.144). However, $h_{lp}[n]$, as given in Eq. (2.145), is square summable, and correspondingly, $H_M(e^{j\omega})$ converges in the mean-square sense to $H_{lp}(e^{j\omega})$; i.e.,

$$\lim_{M\to\infty} \int_{-\pi}^{\pi} |H_{lp}(e^{j\omega}) - H_M(e^{j\omega})|^2 d\omega = 0.$$

Although the error between $\lim_{M\to\infty} H_M(e^{j\omega})$ and $H_{lp}(e^{j\omega})$ might seem unimportant because the two functions differ only at $\omega = \omega_c$, we will see in Chapter 7 that the behavior of finite sums has important implications in the design of discrete-time systems for filtering.

It is sometimes useful to have a Fourier transform representation for certain sequences that are neither absolutely summable nor square summable. We illustrate several of these in the following examples.

Example 2.23   Fourier Transform of a Constant

Consider the sequence $x[n] = 1$ for all n. This sequence is neither absolutely summable nor square summable, and Eq. (2.134) does not converge in either the uniform or mean-square sense for this case. However, it is possible and useful to define the Fourier transform of the sequence $x[n]$ to be the periodic impulse train[4]

$$X(e^{j\omega}) = \sum_{r=-\infty}^{\infty} 2\pi \delta(\omega + 2\pi r). \tag{2.147}$$

The impulses in this case are functions of a continuous variable and therefore are of "infinite height, zero width, and unit area," consistent with the fact that Eq. (2.134) does not converge. The use of Eq. (2.147) as a Fourier representation of the sequence $x[n] = 1$ is justified principally because formal substitution of Eq. (2.147) into Eq. (2.133) leads to the correct result. Example 2.24 represents a generalization of this example.

---

[4]The impulse function is defined as that "function" that has the following properties: $\delta(\omega) = 0$ for $\omega \neq 0$; $X(e^{j\omega})\delta(\omega) = X(e^{j0})\delta(\omega)$; $\int_{-\infty}^{\infty} \delta(\omega)d\omega = 1$; and $\delta(\omega) * X(e^{j\omega}) = X(e^{j\omega})$, where $*$ denotes continuous-variable convolution. See Oppenheim and Willsky (1997) for a discussion of the impulse function.

Example 2.24   Fourier Transform of Complex Exponential Sequences

Consider a sequence $x[n]$ whose Fourier transform is the periodic impulse train

$$X(e^{j\omega}) = \sum_{r=-\infty}^{\infty} 2\pi\delta(\omega - \omega_0 + 2\pi r). \qquad (2.148)$$

We show in this example that $x[n]$ is the complex exponential sequence $e^{j\omega_0 n}$.

We can safely assume that $-\pi < \omega_0 \leq \pi$ in this problem. If the chosen value of ω_0 does not satisfy this requirement, there is a choice of ω_0 in the interval which produces the same $X(e^{j\omega})$, since the impulses repeat periodically every 2π. Thus, we can redefine ω_0 to be the frequency of the impulse in the summation of Eq. (2.148), which falls in the interval between $-\pi$ and π without any change in the spectrum $X(e^{j\omega})$.

We can determine $x[n]$ by substituting $X(e^{j\omega})$ into the inverse Fourier transform integral of Eq. (2.133). Because the integration of $X(e^{j\omega})$ extends only over one period, from $-\pi < \omega < \pi$, we need include only the $r = 0$ term from Eq. (2.148). Consequently, we can write

$$x[n] = \frac{1}{2\pi} \int_{-\pi}^{\pi} 2\pi\delta(\omega - \omega_0)e^{j\omega n}d\omega. \qquad (2.149)$$

From the definition of the impulse function, it follows that

$$x[n] = e^{j\omega_0 n} \qquad \text{for any } n.$$

For $\omega_0 = 0$, this reduces to the sequence considered in Example 2.23.

Clearly, $x[n]$ in Example 2.24 is not absolutely summable, nor is it square summable, and $|X(e^{j\omega})|$ is not finite for all ω. Thus, the mathematical statement

$$\sum_{n=-\infty}^{\infty} e^{j\omega_0 n}e^{-j\omega n} = \sum_{r=-\infty}^{\infty} 2\pi\delta(\omega - \omega_0 + 2\pi r) \qquad (2.150)$$

must be interpreted in a special way. Such an interpretation is provided by the theory of generalized functions (Lighthill, 1958). Using that theory, we can rigorously extend the concept of a Fourier transform representation to the class of sequences that can be expressed as a sum of discrete frequency components, such as

$$x[n] = \sum_{k} a_k e^{j\omega_k n}, \qquad -\infty < n < \infty. \qquad (2.151)$$

From the result of Example 2.24, it follows that

$$X(e^{j\omega}) = \sum_{r=-\infty}^{\infty} \sum_{k} 2\pi a_k \delta(\omega - \omega_k + 2\pi r) \qquad (2.152)$$

is a consistent Fourier transform representation of $x[n]$ in Eq. (2.151).

Another sequence that is neither absolutely summable nor square summable is the unit step sequence $u[n]$. Although it is not completely straightforward to show, this sequence can be represented by the following Fourier transform:

$$U(e^{j\omega}) = \frac{1}{1 - e^{-j\omega}} + \sum_{r=-\infty}^{\infty} \pi\delta(\omega + 2\pi r). \qquad (2.153)$$

2.8 SYMMETRY PROPERTIES OF THE FOURIER TRANSFORM

In using Fourier transforms, it is useful to have a detailed knowledge of the way that properties of the sequence manifest themselves in the Fourier transform and vice versa. In this section and Section 1.9, we discuss and summarize a number of such properties.

Symmetry properties of the Fourier transform are often very useful for simplifying the solution of problems. The following discussion presents these properties, and the proofs are considered in Problems 1.72 and 1.73. Before presenting the properties, however, we begin with some definitions.

A *conjugate-symmetric sequence* $x_e[n]$ is defined as a sequence for which $x_e[n] = x_e^*[-n]$, and a *conjugate-antisymmetric sequence* $x_o[n]$ is defined as a sequence for which $x_o[n] = -x_o^*[-n]$, where * denotes complex conjugation. Any sequence $x[n]$ can be expressed as a sum of a conjugate-symmetric and conjugate-antisymmetric sequence. Specifically,

$$x[n] = x_e[n] + x_o[n], \qquad (2.154a)$$

where

$$x_e[n] = \tfrac{1}{2}(x[n] + x^*[-n]) = x_e^*[-n] \qquad (2.154b)$$

and

$$x_o[n] = \tfrac{1}{2}(x[n] - x^*[-n]) = -x_o^*[-n]. \qquad (2.154c)$$

A real sequence that is conjugate symmetric such that $x_e[n] = x_e[-n]$ is called an *even sequence*, and a real sequence that is conjugate antisymmetric such that $x_o[n] = -x_o[-n]$ is called an *odd sequence*.

A Fourier transform $X(e^{j\omega})$ can be decomposed into a sum of conjugate-symmetric and conjugate-antisymmetric functions as

$$X(e^{j\omega}) = X_e(e^{j\omega}) + X_o(e^{j\omega}), \qquad (2.155a)$$

where

$$X_e(e^{j\omega}) = \tfrac{1}{2}[X(e^{j\omega}) + X^*(e^{-j\omega})] \qquad (2.155b)$$

and

$$X_o(e^{j\omega}) = \tfrac{1}{2}[X(e^{j\omega}) - X^*(e^{-j\omega})]. \qquad (2.155c)$$

By substituting $-\omega$ for ω in Eqs. (2.155b) and (2.155c), it follows that $X_e(e^{j\omega})$ is conjugate symmetric and $X_o(e^{j\omega})$ is conjugate antisymmetric; i.e.,

$$X_e(e^{j\omega}) = X_e^*(e^{-j\omega}) \qquad (2.156a)$$

and

$$X_o(e^{j\omega}) = -X_o^*(e^{-j\omega}). \qquad (2.156b)$$

If a real function of a continuous variable is conjugate symmetric, it is referred to as an *even function,* and a real conjugate-antisymmetric function of a continuous variable is referred to as an *odd function.*

The symmetry properties of the Fourier transform are summarized in Table 1.1. The first six properties apply for a general complex sequence $x[n]$ with Fourier transform $X(e^{j\omega})$. Properties 1 and 2 are considered in Problem 1.72. Property 3 follows from

TABLE 2.1   SYMMETRY PROPERTIES OF THE FOURIER TRANSFORM

Sequence $x[n]$	Fourier Transform $X(e^{j\omega})$
1. $x^*[n]$	$X^*(e^{-j\omega})$
2. $x^*[-n]$	$X^*(e^{j\omega})$
3. $\mathcal{R}e\{x[n]\}$	$X_e(e^{j\omega})$ (conjugate-symmetric part of $X(e^{j\omega})$)
4. $j\mathcal{I}m\{x[n]\}$	$X_o(e^{j\omega})$ (conjugate-antisymmetric part of $X(e^{j\omega})$)
5. $x_e[n]$ (conjugate-symmetric part of $x[n]$)	$X_R(e^{j\omega}) = \mathcal{R}e\{X(e^{j\omega})\}$
6. $x_o[n]$ (conjugate-antisymmetric part of $x[n]$)	$j X_I(e^{j\omega}) = j\mathcal{I}m\{X(e^{j\omega})\}$
The following properties apply only when $x[n]$ is real:	
7. Any real $x[n]$	$X(e^{j\omega}) = X^*(e^{-j\omega})$ (Fourier transform is conjugate symmetric)
8. Any real $x[n]$	$X_R(e^{j\omega}) = X_R(e^{-j\omega})$ (real part is even)
9. Any real $x[n]$	$X_I(e^{j\omega}) = -X_I(e^{-j\omega})$ (imaginary part is odd)
10. Any real $x[n]$	$|X(e^{j\omega})| = |X(e^{-j\omega})|$ (magnitude is even)
11. Any real $x[n]$	$\sphericalangle X(e^{j\omega}) = -\sphericalangle X(e^{-j\omega})$ (phase is odd)
12. $x_e[n]$ (even part of $x[n]$)	$X_R(e^{j\omega})$
13. $x_o[n]$ (odd part of $x[n]$)	$j X_I(e^{j\omega})$

properties 1 and 2, together with the fact that the Fourier transform of the sum of two sequences is the sum of their Fourier transforms. Specifically, the Fourier transform of $\mathcal{R}e\{x[n]\} = \frac{1}{2}(x[n] + x^*[n])$ is the conjugate-symmetric part of $X(e^{j\omega})$, or $X_e(e^{j\omega})$. Similarly, $j\mathcal{I}m\{x[n]\} = \frac{1}{2}(x[n] - x^*[n])$, or equivalently, $j\mathcal{I}m\{x[n]\}$ has a Fourier transform that is the conjugate-antisymmetric component $X_o(e^{j\omega})$ corresponding to property 4. By considering the Fourier transform of $x_e[n]$ and $x_o[n]$, the conjugate-symmetric and conjugate-antisymmetric components, respectively, of $x[n]$, it can be shown that properties 5 and 6 follow.

If $x[n]$ is a real sequence, these symmetry properties become particularly straightforward and useful. Specifically, for a real sequence, the Fourier transform is conjugate symmetric; i.e., $X(e^{j\omega}) = X^*(e^{-j\omega})$ (property 7). Expressing $X(e^{j\omega})$ in terms of its real and imaginary parts as

$$X(e^{j\omega}) = X_R(e^{j\omega}) + j X_I(e^{j\omega}), \tag{2.157}$$

we can derive properties 8 and 9—specifically,

$$X_R(e^{j\omega}) = X_R(e^{-j\omega}) \tag{2.158a}$$

and

$$X_I(e^{j\omega}) = -X_I(e^{-j\omega}). \tag{2.158b}$$

In other words, the real part of the Fourier transform is an even function, and the imaginary part is an odd function, if the sequence is real. In a similar manner, by expressing $X(e^{j\omega})$ in polar form as

$$X(e^{j\omega}) = |X(e^{j\omega})|e^{j\sphericalangle X(e^{j\omega})}, \tag{2.159}$$

we can show that, for a real sequence $x[n]$, the magnitude of the Fourier transform, $|X(e^{j\omega})|$, is an even function of ω and the phase, $\sphericalangle X(e^{j\omega})$, can be chosen to be an odd function of ω (properties 10 and 11). Also, for a real sequence, the even part of $x[n]$ transforms to $X_R(e^{j\omega})$, and the odd part of $x[n]$ transforms to $j X_I(e^{j\omega})$ (properties 12 and 13).

Example 2.25   Illustration of Symmetry Properties

Let us return to the sequence of Example 2.21, where we showed that the Fourier transform of the real sequence $x[n] = a^n u[n]$ is

$$X(e^{j\omega}) = \frac{1}{1 - ae^{-j\omega}} \qquad \text{if } |a| < 1. \tag{2.160}$$

Then, from the properties of complex numbers, it follows that

$$X(e^{j\omega}) = \frac{1}{1 - ae^{-j\omega}} = X^*(e^{-j\omega}) \qquad \text{(property 7)},$$

$$X_R(e^{j\omega}) = \frac{1 - a\cos\omega}{1 + a^2 - 2a\cos\omega} = X_R(e^{-j\omega}) \qquad \text{(property 8)},$$

$$X_I(e^{j\omega}) = \frac{-a\sin\omega}{1 + a^2 - 2a\cos\omega} = -X_I(e^{-j\omega}) \qquad \text{(property 9)},$$

$$|X(e^{j\omega})| = \frac{1}{(1 + a^2 - 2a\cos\omega)^{1/2}} = |X(e^{-j\omega})| \qquad \text{(property 10)},$$

$$\sphericalangle X(e^{j\omega}) = \tan^{-1}\left(\frac{-a\sin\omega}{1 - a\cos\omega}\right) = -\sphericalangle X(e^{-j\omega}) \qquad \text{(property 11)}.$$

Figure 2.22   Frequency response for a system with impulse response $h[n] = a^n u[n]$. (a) Real part. $a > 0$; $a = 0.9$ (solid curve) and $a = 0.5$ (dashed curve). (b) Imaginary part.

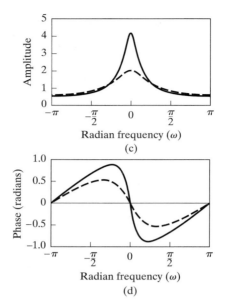

Figure 2.22 (*Continued*) (c) Magnitude. $a > 0$; $a = 0.9$ (solid curve) and $a = 0.5$ (dashed curve). (d) Phase.

These functions are plotted in Figure 2.22 for $a > 0$, specifically, $a = 0.9$ (solid curve) and $a = 0.5$ (dashed curve). In Problem 2.43, we consider the corresponding plots for $a < 0$.

2.9 FOURIER TRANSFORM THEOREMS

In addition to the symmetry properties, a variety of theorems (presented in Sections 2.9.1–2.9.7) relate operations on the sequence to operations on the Fourier transform. We will see that these theorems are quite similar in most cases to corresponding theorems for continuous-time signals and their Fourier transforms. To facilitate the statement of the theorems, we introduce the following operator notation:

$$X(e^{j\omega}) = \mathcal{F}\{x[n]\},$$
$$x[n] = \mathcal{F}^{-1}\{X(e^{j\omega})\},$$
$$x[n] \overset{\mathcal{F}}{\longleftrightarrow} X(e^{j\omega}).$$

That is, $\mathcal{F}$ denotes the operation of "taking the Fourier transform of $x[n]$," and $\mathcal{F}^{-1}$ is the inverse of that operation. Most of the theorems will be stated without proof. The proofs, which are left as exercises (Problem 2.74), generally involve only simple manipulations of variables of summation or integration. The theorems in this section are summarized in Table 2.2.

TABLE 2.2   FOURIER TRANSFORM THEOREMS

Sequence $x[n]$ $y[n]$	Fourier Transform $X(e^{j\omega})$ $Y(e^{j\omega})$
1. $ax[n] + by[n]$	$aX(e^{j\omega}) + bY(e^{j\omega})$
2. $x[n - n_d]$ (n_d an integer)	$e^{-j\omega n_d} X(e^{j\omega})$
3. $e^{j\omega_0 n}x[n]$	$X(e^{j(\omega-\omega_0)})$
4. $x[-n]$	$X(e^{-j\omega})$ $X^*(e^{j\omega})$ if $x[n]$ real.
5. $nx[n]$	$j\dfrac{dX(e^{j\omega})}{d\omega}$
6. $x[n] * y[n]$	$X(e^{j\omega})Y(e^{j\omega})$
7. $x[n]y[n]$	$\dfrac{1}{2\pi}\displaystyle\int_{-\pi}^{\pi} X(e^{j\theta})Y(e^{j(\omega-\theta)})d\theta$

Parseval's theorem:

8. $\displaystyle\sum_{n=-\infty}^{\infty} |x[n]|^2 = \dfrac{1}{2\pi}\int_{-\pi}^{\pi} |X(e^{j\omega})|^2 d\omega$

9. $\displaystyle\sum_{n=-\infty}^{\infty} x[n]y^*[n] = \dfrac{1}{2\pi}\int_{-\pi}^{\pi} X(e^{j\omega})Y^*(e^{j\omega})d\omega$

2.9.1 Linearity of the Fourier Transform

If

$$x_1[n] \xleftrightarrow{\mathcal{F}} X_1(e^{j\omega})$$

and

$$x_2[n] \xleftrightarrow{\mathcal{F}} X_2(e^{j\omega}),$$

then it follows by substitution into the definition of the discrete-time Fourier transform that

$$ax_1[n] + bx_2[n] \xleftrightarrow{\mathcal{F}} aX_1(e^{j\omega}) + bX_2(e^{j\omega}). \tag{2.161}$$

2.9.2 Time Shifting and Frequency Shifting

If

$$x[n] \xleftrightarrow{\mathcal{F}} X(e^{j\omega}),$$

then, for the time-shifted sequence, a simple transformation of the index of summation in the discrete-time Fourier transform yields

$$x[n - n_d] \xleftrightarrow{\mathcal{F}} e^{-j\omega n_d} X(e^{j\omega}). \tag{2.162}$$

Direct substitution proves the following result for the frequency-shifted Fourier transform:

$$e^{j\omega_0 n}x[n] \xleftrightarrow{\mathcal{F}} X(e^{j(\omega-\omega_0)}). \tag{2.163}$$

2.9.3 Time Reversal

If

$$x[n] \xleftrightarrow{\mathcal{F}} X(e^{j\omega}),$$

then if the sequence is time reversed,

$$x[-n] \xleftrightarrow{\mathcal{F}} X(e^{-j\omega}). \tag{2.164}$$

If $x[n]$ is real, this theorem becomes

$$x[-n] \xleftrightarrow{\mathcal{F}} X^*(e^{j\omega}). \tag{2.165}$$

2.9.4 Differentiation in Frequency

If

$$x[n] \xleftrightarrow{\mathcal{F}} X(e^{j\omega}),$$

then, by differentiating the discrete-time Fourier transform, it is seen that

$$nx[n] \xleftrightarrow{\mathcal{F}} j\frac{dX(e^{j\omega})}{d\omega}. \tag{2.166}$$

2.9.5 Parseval's Theorem

If

$$x[n] \xleftrightarrow{\mathcal{F}} X(e^{j\omega}),$$

then

$$E = \sum_{n=-\infty}^{\infty} |x[n]|^2 = \frac{1}{2\pi} \int_{-\pi}^{\pi} |X(e^{j\omega})|^2 d\omega. \tag{2.167}$$

The function $|X(e^{j\omega})|^2$ is called the *energy density spectrum,* since it determines how the energy is distributed in the frequency domain. Necessarily, the energy density spectrum is defined only for finite-energy signals. A more general form of Parseval's theorem is shown in Problem 2.77.

2.9.6 The Convolution Theorem

If

$$x[n] \xleftrightarrow{\mathcal{F}} X(e^{j\omega})$$

and

$$h[n] \xleftrightarrow{\mathcal{F}} H(e^{j\omega}),$$

and if

$$y[n] = \sum_{k=-\infty}^{\infty} x[k]h[n-k] = x[n] * h[n], \tag{2.168}$$

then

$$Y(e^{j\omega}) = X(e^{j\omega})H(e^{j\omega}). \tag{2.169}$$

Thus, convolution of sequences implies multiplication of the corresponding Fourier transforms. Note that the time-shifting property is a special case of the convolution property, since

$$\delta[n - n_d] \xleftrightarrow{\mathcal{F}} e^{-j\omega n_d} \tag{2.170}$$

and if $h[n] = \delta[n - n_d]$, then $y[n] = x[n] * \delta[n - n_d] = x[n - n_d]$. Therefore,

$$H(e^{j\omega}) = e^{-j\omega n_d} \quad \text{and} \quad Y(e^{j\omega}) = e^{-j\omega n_d} X(e^{j\omega}).$$

A formal derivation of the convolution theorem is easily achieved by applying the definition of the Fourier transform to $y[n]$ as expressed in Eq. (2.168). This theorem can also be interpreted as a direct consequence of the eigenfunction property of complex exponentials for linear time-invariant systems. Recall that $H(e^{j\omega})$ is the frequency response of the linear time-invariant system whose impulse response is $h[n]$. Recall also that if

$$x[n] = e^{j\omega n},$$

then

$$y[n] = H(e^{j\omega})e^{j\omega n}.$$

That is, complex exponentials are *eigenfunctions* of linear time-invariant systems, where $H(e^{j\omega})$, the Fourier transform of $h[n]$, is the eigenvalue. From the definition of integration, the Fourier transform synthesis equation corresponds to the representation of a sequence $x[n]$ as a superposition of complex exponentials of infinitesimal size; that is,

$$x[n] = \frac{1}{2\pi} \int_{-\pi}^{\pi} X(e^{j\omega})e^{j\omega n} d\omega = \lim_{\Delta\omega \to 0} \frac{1}{2\pi} \sum_{k} X(e^{jk\Delta\omega})e^{jk\Delta\omega n} \Delta\omega.$$

By the eigenfunction property of linear systems and by the principle of superposition, the corresponding output will be

$$y[n] = \lim_{\Delta\omega \to 0} \frac{1}{2\pi} \sum_{k} H(e^{jk\Delta\omega})X(e^{jk\Delta\omega})e^{jk\Delta\omega n} \Delta\omega = \frac{1}{2\pi} \int_{-\pi}^{\pi} H(e^{j\omega})X(e^{j\omega})e^{j\omega n} d\omega.$$

Thus, we conclude that

$$Y(e^{j\omega}) = H(e^{j\omega})X(e^{j\omega}),$$

as in Eq. (2.169).

2.9.7 The Modulation or Windowing Theorem

If

$$x[n] \xleftrightarrow{\mathcal{F}} X(e^{j\omega})$$

and

$$w[n] \xleftrightarrow{\mathcal{F}} W(e^{j\omega}),$$

and if

$$y[n] = x[n]w[n], \tag{2.171}$$

then

$$Y(e^{j\omega}) = \frac{1}{2\pi} \int_{-\pi}^{\pi} X(e^{j\theta})W(e^{j(\omega-\theta)})d\theta. \tag{2.172}$$

Equation (2.172) is a periodic convolution, i.e., a convolution of two periodic functions with the limits of integration extending over only one period. The duality inherent in most Fourier transform theorems is evident when we compare the convolution and modulation theorems. However, in contrast to the continuous-time case, where this duality is complete, in the discrete-time case fundamental differences arise because the Fourier transform is a sum while the inverse transform is an integral with a periodic integrand. Although for continuous time we can state that convolution in the time domain is represented by multiplication in the frequency domain and vice versa, in discrete time this statement must be modified somewhat. Specifically, discrete-time convolution of sequences (the convolution sum) is equivalent to multiplication of corresponding periodic Fourier transforms, and multiplication of sequences is equivalent to *periodic* convolution of corresponding Fourier transforms.

The theorems of this section and a number of fundamental Fourier transform pairs are summarized in Tables 2.2 and 2.3, respectively. One of the ways that knowledge of

TABLE 2.3   FOURIER TRANSFORM PAIRS

Sequence	Fourier Transform
1. $\delta[n]$	1
2. $\delta[n - n_0]$	$e^{-j\omega n_0}$
3. 1 $(-\infty < n < \infty)$	$\displaystyle\sum_{k=-\infty}^{\infty} 2\pi\delta(\omega + 2\pi k)$
4. $a^n u[n]$ $(|a| < 1)$	$\dfrac{1}{1 - ae^{-j\omega}}$
5. $u[n]$	$\dfrac{1}{1 - e^{-j\omega}} + \displaystyle\sum_{k=-\infty}^{\infty} \pi\delta(\omega + 2\pi k)$
6. $(n + 1)a^n u[n]$ $(|a| < 1)$	$\dfrac{1}{(1 - ae^{-j\omega})^2}$
7. $\dfrac{r^n \sin\omega_p(n + 1)}{\sin\omega_p}u[n]$ $(|r| < 1)$	$\dfrac{1}{1 - 2r\cos\omega_p e^{-j\omega} + r^2 e^{-j2\omega}}$
8. $\dfrac{\sin\omega_c n}{\pi n}$	$X(e^{j\omega}) = \begin{cases} 1, & |\omega| < \omega_c, \\ 0, & \omega_c < |\omega| \le \pi \end{cases}$
9. $x[n] = \begin{cases} 1, & 0 \le n \le M \\ 0, & \text{otherwise} \end{cases}$	$\dfrac{\sin[\omega(M + 1)/2]}{\sin(\omega/2)}e^{-j\omega M/2}$
10. $e^{j\omega_0 n}$	$\displaystyle\sum_{k=-\infty}^{\infty} 2\pi\delta(\omega - \omega_0 + 2\pi k)$
11. $\cos(\omega_0 n + \phi)$	$\displaystyle\sum_{k=-\infty}^{\infty} [\pi e^{j\phi}\delta(\omega - \omega_0 + 2\pi k) + \pi e^{-j\phi}\delta(\omega + \omega_0 + 2\pi k)]$

Fourier transform theorems and properties is useful is in determining Fourier transforms or inverse transforms. Often, by using the theorems and known transform pairs, it is possible to represent a sequence in terms of operations on other sequences for which the transform is known, thereby simplifying an otherwise difficult or tedious problem. Examples 2.26–2.30 illustrate this approach.

Example 2.26   Determining a Fourier Transform using Tables 2.2 and 2.3

Suppose we wish to find the Fourier transform of the sequence $x[n] = a^n u[n - 5]$. This transform can be computed by exploiting Theorems 1 and 2 of Table 2.2 and transform pair 4 of Table 2.3. Let $x_1[n] = a^n u[n]$. We start with this signal because it is the most similar signal to $x[n]$ in Table 2.3. The table states that

$$X_1(e^{j\omega}) = \frac{1}{1 - ae^{-j\omega}}. \tag{2.173}$$

To obtain $x[n]$ from $x_1[n]$, we first delay $x_1[n]$ by 5 samples, i.e., $x_2[n] = x_1[n-5]$. Theorem 2 of Table 2.2 gives the corresponding frequency-domain relationship, $X_2(e^{j\omega}) = e^{-j5\omega} X_1(e^{j\omega})$, so

$$X_2(e^{j\omega}) = \frac{e^{-j5\omega}}{1 - ae^{-j\omega}}. \tag{2.174}$$

In order to get from $x_2[n]$ to the desired $x[n]$, we need only multiply by the constant a^5, i.e., $x[n] = a^5 x_2[n]$. The linearity property of the Fourier transform, Theorem 1 of Table 2.2, then yields the desired Fourier transform,

$$X(e^{j\omega}) = \frac{a^5 e^{-j5\omega}}{1 - ae^{-j\omega}}. \tag{2.175}$$

Example 2.27   Determining an Inverse Fourier Transform Using Tables 2.2 and 2.3

Suppose that

$$X(e^{j\omega}) = \frac{1}{(1 - ae^{-j\omega})(1 - be^{-j\omega})}. \tag{2.176}$$

Direct substitution of $X(e^{j\omega})$ into Eq. (2.133) leads to an integral that is difficult to evaluate by ordinary real integration techniques. However, using the technique of partial fraction expansion, which we discuss in detail in Chapter 3, we can expand $X(e^{j\omega})$ into the form

$$X(e^{j\omega}) = \frac{a/(a - b)}{1 - ae^{-j\omega}} - \frac{b/(a - b)}{1 - be^{-j\omega}}. \tag{2.177}$$

From Theorem 1 of Table 2.2 and transform pair 4 of Table 2.3, it follows that

$$x[n] = \left(\frac{a}{a - b}\right) a^n u[n] - \left(\frac{b}{a - b}\right) b^n u[n]. \tag{2.178}$$

Example 2.28   Determining the Impulse Response from the Frequency Response

The frequency response of a highpass filter with delay is

$$H(e^{j\omega}) = \begin{cases} e^{-j\omega n_d}, & \omega_c < |\omega| < \pi, \\ 0, & |\omega| < \omega_c, \end{cases} \tag{2.179}$$

where a period of 2π is understood. This frequency response can be expressed as

$$H(e^{j\omega}) = e^{-j\omega n_d}(1 - H_{lp}(e^{j\omega})) = e^{-j\omega n_d} - e^{-j\omega n_d} H_{lp}(e^{j\omega}),$$

where $H_{lp}(e^{j\omega})$ is periodic with period 2π and

$$H_{lp}(e^{j\omega}) = \begin{cases} 1, & |\omega| < \omega_c, \\ 0, & \omega_c < |\omega| < \pi. \end{cases}$$

Using the result of Example 2.22 to obtain the inverse transform of $H_{lp}(e^{j\omega})$, together with properties 1 and 2 of Table 2.2, we have

$$h[n] = \delta[n - n_d] - r[n - n_d]$$

$$= \delta[n - n_d] - \frac{\sin \omega_c(n - n_d)}{\pi(n - n_d)}.$$

Example 2.29   Determining the Impulse Response for a Difference Equation

In this example we determine the impulse response for a stable linear time-invariant system for which the input $x[n]$ and output $y[n]$ satisfy the linear constant-coefficient difference equation

$$y[n] - \tfrac{1}{2}y[n-1] = x[n] - \tfrac{1}{4}x[n-1]. \tag{2.180}$$

In Chapter 3 we will see that the z-transform is more useful than the Fourier transform for dealing with difference equations. However, this example offers a hint of the utility of transform methods in the analysis of linear systems. To find the impulse response, we set $x[n] = \delta[n]$; with $h[n]$ denoting the impulse response, Eq. (2.180) becomes

$$h[n] - \tfrac{1}{2}h[n-1] = \delta[n] - \tfrac{1}{4}\delta[n-1]. \tag{2.181}$$

Applying the Fourier transform to both sides of Eq. (2.181) and using properties 1 and 2 of Table 2.2, we obtain

$$H(e^{j\omega}) - \tfrac{1}{2}e^{-j\omega}H(e^{j\omega}) = 1 - \tfrac{1}{4}e^{-j\omega}, \tag{2.182}$$

or

$$H(e^{j\omega}) = \frac{1 - \tfrac{1}{4}e^{-j\omega}}{1 - \tfrac{1}{2}e^{-j\omega}}. \tag{2.183}$$

To obtain $h[n]$, we want to determine the inverse Fourier transform of $H(e^{j\omega})$. Toward this end, we rewrite Eq. (2.183) as

$$H(e^{j\omega}) = \frac{1}{1 - \tfrac{1}{2}e^{-j\omega}} - \frac{\tfrac{1}{4}e^{-j\omega}}{1 - \tfrac{1}{2}e^{-j\omega}}. \tag{2.184}$$

From transform 4 of Table 2.3,

$$\left(\tfrac{1}{2}\right)^n u[n] \xleftrightarrow{\mathcal{F}} \frac{1}{1 - \tfrac{1}{2}e^{-j\omega}}.$$

Combining this transform with property 3 of Table 2.2, we obtain

$$-\left(\tfrac{1}{4}\right)\left(\tfrac{1}{2}\right)^{n-1} u[n-1] \xleftrightarrow{\mathcal{F}} -\frac{\tfrac{1}{4}e^{-j\omega}}{1 - \tfrac{1}{2}e^{-j\omega}}. \tag{2.185}$$

Based on property 1 of Table 2.2, then,

$$h[n] = \left(\tfrac{1}{2}\right)^{n} u[n] - \left(\tfrac{1}{4}\right)\left(\tfrac{1}{2}\right)^{n-1} u[n-1]. \tag{2.186}$$

2.10 DISCRETE-TIME RANDOM SIGNALS

The preceding sections have focused on mathematical representations of discrete-time signals and systems and the insights that derive from such mathematical representations. We have seen that discrete-time signals and systems have both a time-domain and a frequency-domain representation, each with an important place in the theory and design of discrete-time signal-processing systems. Until now, we have assumed that the signals are deterministic, i.e., that each value of a sequence is uniquely determined by a mathematical expression, a table of data, or a rule of some type.

In many situations, the processes that generate signals are so complex as to make precise description of a signal extremely difficult or undesirable, if not impossible. In such cases, modeling the signal as a stochastic process is analytically useful. As an example, we will see in Chapter 6 that many of the effects encountered in implementing digital signal-processing algorithms with finite register length can be represented by additive noise, i.e., a stochastic sequence. Many mechanical systems generate acoustic or vibratory signals that can be processed to diagnose potential failure; again, signals of this type are often best modeled in terms of stochastic signals. Speech signals to be processed for automatic recognition or bandwidth compression and music to be processed for quality enhancement are two more of many examples.

A stochastic signal is considered to be a member of an ensemble of discrete-time signals that is characterized by a set of probability density functions. More specifically, for a particular signal at a particular time, the amplitude of the signal sample at that time is assumed to have been determined by an underlying scheme of probabilities. That is, each individual sample $x[n]$ of a particular signal is assumed to be an outcome of some underlying random variable $\mathbf{x}_n$. The entire signal is represented by a collection of such random variables, one for each sample time, $-\infty < n < \infty$. This collection of random variables is called a *random process,* and we assume that a particular sequence of samples $x[n]$ for $-\infty < n < \infty$ has been generated by the random process that underlies the signal. To completely describe the random process, we need to specify the individual and joint probability distributions of all the random variables.

The key to obtaining useful results from such models of signals lies in their description in terms of averages that can be computed from assumed probability laws or estimated from specific signals. While stochastic signals are not absolutely summable or square summable and, consequently, do not directly have Fourier transforms, many (but not all) of the properties of such signals can be summarized in terms of averages such as the *autocorrelation* or *autocovariance sequence,* for which the Fourier transform often exists. As we will discuss in this section, the Fourier transform of the autocovariance sequence has a useful interpretation in terms of the frequency distribution of the power in the signal. The use of the autocovariance sequence and its transform has another important advantage: The effect of processing stochastic signals with a discrete-time linear system can be conveniently described in terms of the effect of the system on the autocovariance sequence.

In the following discussion, we assume that the reader is familiar with the basic concepts of stochastic processes, such as averages, correlation and covariance functions, and the power spectrum. A brief review and summary of notation and concepts is provided in Appendix A. A more detailed presentation of the theory of random signals can be found in a variety of excellent texts, such as Davenport (1970) and Papoulis (1984).

Our primary objective in this section is to present a specific set of results that will be useful in subsequent chapters. Therefore, we focus on wide-sense stationary random signals and their representation in the context of processing with linear time-invariant systems. Although, for simplicity, we assume that $x[n]$ and $h[n]$ are real valued, the results can be generalized to the complex case.

Consider a stable linear time-invariant system with real impulse response $h[n]$. Let $x[n]$ be a real-valued sequence that is a sample sequence of a wide-sense stationary discrete-time random process. Then the output of the linear system is also a sample function of a random process related to the input process by the linear transformation

$$y[n] = \sum_{k=-\infty}^{\infty} h[n-k]x[k] = \sum_{k=-\infty}^{\infty} h[k]x[n-k].$$

As we have shown, since the system is stable, $y[n]$ will be bounded if $x[n]$ is bounded. We will see shortly that if the input is stationary,[5] then so is the output. The input signal may be characterized by its mean m_x and its autocorrelation function $\phi_{xx}[m]$, or we may also have additional information about first- or even second-order probability distributions. In characterizing the output random process $y[n]$ we desire similar information. For many applications, it is sufficient to characterize both the input and output in terms of simple averages, such as the mean, variance, and autocorrelation. Therefore, we will derive input–output relationships for these quantities.

The means of the input and output processes are, respectively,

$$m_{x_n} = \mathcal{E}\{\mathbf{x}_n\}, \qquad m_{y_n} = \mathcal{E}\{\mathbf{y}_n\}, \tag{2.187}$$

where $\mathcal{E}\{\cdot\}$ denotes the expected value. In most of our discussion, it will not be necessary to carefully distinguish between the random variables $\mathbf{x}_n$ and $\mathbf{y}_n$ and their specific values $x[n]$ and $y[n]$. This will simplify the mathematical notation significantly. For example, Eqs. (2.187) will alternatively be written

$$m_x[n] = \mathcal{E}\{x[n]\}, \qquad m_y[n] = \mathcal{E}\{y[n]\}. \tag{2.188}$$

If $x[n]$ is stationary, then $m_x[n]$ is independent of n and will be written as m_x, with similar notation for $m_y[n]$ if $y[n]$ is stationary.

---

[5]In the remainder of the text, we will use the term *stationary* to mean "wide-sense stationary."

The mean of the output process is

$$m_y[n] = \mathcal{E}\{y[n]\} = \sum_{k=-\infty}^{\infty} h[k]\mathcal{E}\{x[n-k]\},$$

where we have used the fact that the expected value of a sum is the sum of the expected values. Since the input is stationary, $m_x[n-k] = m_x$, and consequently,

$$m_y[n] = m_x \sum_{k=-\infty}^{\infty} h[k]. \tag{2.189}$$

From Eq. (2.189), we see that the mean of the output is also constant. An equivalent expression to Eq. (2.189) in terms of the frequency response is

$$m_y = H(e^{j0})m_x. \tag{2.190}$$

Assuming temporarily that the output is nonstationary, the autocorrelation function of the output process for a real input is

$$\phi_{yy}[n, n+m] = \mathcal{E}\{y[n]y[n+m]\}$$

$$= \mathcal{E}\left\{\sum_{k=-\infty}^{\infty}\sum_{r=-\infty}^{\infty} h[k]h[r]x[n-k]x[n+m-r]\right\}$$

$$= \sum_{k=-\infty}^{\infty} h[k]\sum_{r=-\infty}^{\infty} h[r]\mathcal{E}\{x[n-k]x[n+m-r]\}.$$

Since $x[n]$ is assumed to be stationary, $\mathcal{E}\{x[n-k]x[n+m-r]\}$ depends only on the time difference $m+k-r$. Therefore,

$$\phi_{yy}[n, n+m] = \sum_{k=-\infty}^{\infty} h[r]\sum_{r=-\infty}^{\infty} h[r]\phi_{xx}[m+k-r] = \phi_{yy}[m]. \tag{2.191}$$

That is, the output autocorrelation sequence also depends only on the time difference m. Thus, for a linear time-invariant system having a wide-sense stationary input, the output is also wide-sense stationary.

By making the substitution $\ell = r - k$, we can express Eq. (2.191) as

$$\phi_{yy}[m] = \sum_{\ell=-\infty}^{\infty} \phi_{xx}[m-\ell]\sum_{k=-\infty}^{\infty} h[k]h[\ell+k]$$

$$= \sum_{\ell=-\infty}^{\infty} \phi_{xx}[m-\ell]c_{hh}[\ell], \tag{2.192}$$

where we have defined

$$c_{hh}[\ell] = \sum_{k=-\infty}^{\infty} h[k]h[\ell+k]. \tag{2.193}$$

A sequence of the form of $c_{hh}[\ell]$ is called a *deterministic autocorrelation sequence* or, simply, the *autocorrelation sequence of $h[n]$*. It should be emphasized that $c_{hh}[\ell]$ is the autocorrelation of an aperiodic—i.e., finite-energy—sequence and should not be confused with the autocorrelation of an infinite-energy random sequence. Indeed, it can be seen that $c_{hh}[\ell]$ is simply the discrete convolution of $h[n]$ with $h[-n]$. Equation (2.192), then, can be interpreted to mean that the autocorrelation of the output of a linear system is the convolution of the autocorrelation of the input with the aperiodic autocorrelation of the system impulse response.

Equation (2.192) suggests that Fourier transforms may be useful in characterizing the response of a linear time-invariant system to a stochastic input. Assume, for convenience, that $m_x = 0$; i.e., the autocorrelation and autocovariance sequences are identical. Then, with $\Phi_{xx}(e^{j\omega})$, $\Phi_{yy}(e^{j\omega})$, and $C_{hh}(e^{j\omega})$ denoting the Fourier transforms of $\phi_{xx}[m]$, $\phi_{yy}[m]$, and $c_{hh}[\ell]$, respectively, from Eq. (2.192),

$$\Phi_{yy}(e^{j\omega}) = C_{hh}(e^{j\omega})\Phi_{xx}(e^{j\omega}). \tag{2.194}$$

Also, from Eq. (2.193),

$$C_{hh}(e^{j\omega}) = H(e^{j\omega})H^*(e^{j\omega})$$
$$= |H(e^{j\omega})|^2,$$

so

$$\Phi_{yy}(e^{j\omega}) = |H(e^{j\omega})|^2 \Phi_{xx}(e^{j\omega}). \tag{2.195}$$

Equation (2.195) provides the motivation for the term *power density spectrum*. Specifically,

$$\mathcal{E}\{y^2[n]\} = \phi_{yy}[0] = \frac{1}{2\pi} \int_{-\pi}^{\pi} \Phi_{yy}(e^{j\omega})\, d\omega \tag{2.196}$$
$$= \text{total average power in output.}$$

Substituting Eq. (2.195) into Eq. (2.196), we have

$$\mathcal{E}\{y^2[n]\} = \phi_{yy}[0] = \frac{1}{2\pi} \int_{-\pi}^{\pi} |H(e^{j\omega})|^2 \Phi_{xx}(e^{j\omega})\, d\omega. \tag{2.197}$$

Suppose that $H(e^{j\omega})$ is an ideal bandpass filter, as shown in Figure 2.18(c). We recall that $\phi_{xx}[m]$ is an even sequence, so

$$\Phi_{xx}(e^{j\omega}) = \Phi_{xx}(e^{-j\omega}).$$

Likewise, $|H(e^{j\omega})|^2$ is an even function of ω. Therefore, we can write

$$\phi_{yy}[0] = \text{average power in output}$$

$$= \frac{1}{2\pi} \int_{\omega_a}^{\omega_b} \Phi_{xx}(e^{j\omega})\, d\omega + \frac{1}{2\pi} \int_{-\omega_b}^{-\omega_a} \Phi_{xx}(e^{j\omega})\, d\omega. \tag{2.198}$$

Thus, the area under $\Phi_{xx}(e^{j\omega})$ for $\omega_a \leq |\omega| \leq \omega_b$ can be taken to represent the mean-square value of the input in that frequency band. We observe that the output power must remain nonnegative, so

$$\lim_{(\omega_b - \omega_a) \to 0} \phi_{yy}[0] \geq 0.$$

This result, together with Eq. (2.198) and the fact that the band $\omega_a \leq \omega \leq \omega_b$ can be arbitrarily small, implies that

$$\Phi_{xx}(e^{j\omega}) \geq 0 \qquad \text{for all } \omega. \tag{2.199}$$

Hence, we note that the power density function of a real signal is real, even, and nonnegative.

Example 2.30   White Noise

The concept of white noise is exceedingly useful in quantization error analysis. A white-noise signal is a signal for which $\phi_{xx}[m] = \sigma_x^2 \delta[m]$. We assume in this example that the signal has zero mean. The power spectrum of a white noise signal is a constant, i.e.,

$$\Phi_{xx}(e^{j\omega}) = \sigma_x^2 \qquad \text{for all } \omega.$$

The average power of a white-noise signal is therefore

$$\phi_{xx}[0] = \frac{1}{2\pi} \int_{-\pi}^{\pi} \Phi_{xx}(e^{j\omega})\, d\omega = \frac{1}{2\pi} \int_{-\pi}^{\pi} \sigma_x^2\, d\omega = \sigma_x^2.$$

The concept of white noise is also useful in the representation of random signals whose power spectra are not constant with frequency. For example, a random signal $y[n]$ with power spectrum $\Phi_{yy}(e^{j\omega})$ can be assumed to be the output of a linear time-invariant system with a white-noise input. That is, we use Eq. (2.195) to define a system with frequency response $H(e^{j\omega})$ to satisfy the equation

$$\Phi_{yy}(e^{j\omega}) = |H(e^{j\omega})|^2 \sigma_x^2,$$

where σ_x^2 is the average power of the assumed white-noise input signal. We adjust the average power of this input signal to give the correct average power for $y[n]$. For example, suppose that $h[n] = a^n u[n]$. Then

$$H(e^{j\omega}) = \frac{1}{1 - ae^{-j\omega}},$$

and we can represent all random signals whose power spectra are of the form

$$\Phi_{yy}(e^{j\omega}) = \left| \frac{1}{1 - ae^{-j\omega}} \right|^2 \sigma_x^2 = \frac{\sigma_x^2}{1 + a^2 - 2a \cos \omega}.$$

Another important result concerns the cross-correlation between the input and output of a linear time-invariant system:

$$\phi_{xy}[m] = \mathcal{E}\{x[n]y[n+m]\}$$

$$= \mathcal{E}\left\{ x[n] \sum_{k=-\infty}^{\infty} h[k]x[n+m-k] \right\} \qquad (2.200)$$

$$= \sum_{k=-\infty}^{\infty} h[k]\phi_{xx}[m-k].$$

In this case, we note that the cross-correlation between input and output is the convolution of the impulse response with the input autocorrelation sequence.

The Fourier transform of Eq. (2.200) is

$$\Phi_{xy}(e^{j\omega}) = H(e^{j\omega})\Phi_{xx}(e^{j\omega}). \qquad (2.201)$$

This result has a useful application when the input is white noise, i.e., when $\phi_{xx}[m] = \sigma_x^2 \delta[m]$. Substituting into Eq. (2.198), we note that

$$\phi_{xy}[m] = \sigma_x^2 h[m]. \qquad (2.202)$$

That is, for a zero-mean white-noise input, the cross-correlation between input and output of a linear system is proportional to the impulse response of the system. Similarly, the power spectrum of a white-noise input is

$$\Phi_{xx}(e^{j\omega}) = \sigma_x^2, \qquad -\pi \leq \omega \leq \pi. \tag{2.203}$$

Thus, from Eq. (2.201),

$$\Phi_{xy}(e^{j\omega}) = \sigma_x^2 H(e^{j\omega}). \tag{2.204}$$

In other words, the cross power spectrum is in this case proportional to the frequency response of the system. Equations (2.202) and (2.204) may serve as the basis for estimating the impulse response or frequency response of a linear time-invariant system if it is possible to observe the output of the system in response to a white-noise input.

2.11 SUMMARY

In this chapter, we have considered a number of basic definitions relating to discrete-time signals and systems. We considered the definition of a set of basic sequences, the definition and representation of linear time-invariant systems in terms of the convolution sum, and some implications of stability and causality. The class of systems for which the input and output satisfy a linear constant-coefficient difference equation with initial rest conditions was shown to be an important subclass of linear time-invariant systems. The recursive solution of such difference equations was discussed and the classes of FIR and IIR systems defined.

An important means for the analysis and representation of linear time-invariant systems lies in their frequency-domain representation. The response of a system to a complex exponential input was considered, leading to the definition of the frequency response. The relation between impulse response and frequency response was then interpreted as a Fourier transform pair.

We called attention to many properties of Fourier transform representations and discussed a variety of useful Fourier transform pairs. Tables 2.1 and 2.2 summarize the properties and theorems, and Table 2.3 contains some useful Fourier transform pairs.

The chapter concludes with an introduction to discrete-time random signals. These basic ideas and results will be developed further and used in later chapters.

Although the material in this chapter was presented without direct reference to continuous-time signals, an important class of discrete-time signal-processing problems arises from sampling such signals. In Chapter 4 we consider the relationship between continuous-time signals and sequences obtained by periodic sampling.

PROBLEMS

Basic Problems with Answers

2.1. For each of the following systems, determine whether the system is (1) stable, (2) causal, (3) linear, (4) time invariant, and (5) memoryless:

(a) $T(x[n]) = g[n]x[n]$ with $g[n]$ given

(b) $T(x[n]) = \sum_{k=n_0}^{n} x[k]$

(c) $T(x[n]) = \sum_{k=n-n_0}^{n+n_0} x[k]$

(d) $T(x[n]) = x[n - n_0]$

(e) $T(x[n]) = e^{x[n]}$

(f) $T(x[n]) = ax[n] + b$

(g) $T(x[n]) = x[-n]$

(h) $T(x[n]) = x[n] + 3u[n + 1]$

2.2. (a) The impulse response $h[n]$ of a linear time-invariant system is known to be zero, except in the interval $N_0 \le n \le N_1$. The input $x[n]$ is known to be zero, except in the interval $N_2 \le n \le N_3$. As a result, the output is constrained to be zero, except in some interval $N_4 \le n \le N_5$. Determine N_4 and N_5 in terms of N_0, N_1, N_2, and N_3.

(b) If $x[n]$ is zero, except for N consecutive points, and $h[n]$ is zero, except for M consecutive points, what is the maximum number of consecutive points for which $y[n]$ can be nonzero?

2.3. By direct evaluation of the convolution sum, determine the step response of a linear time-invariant system whose impulse response is

$$h[n] = a^{-n}u[-n], \qquad 0 < a < 1.$$

2.4. Consider the linear constant-coefficient difference equation

$$y[n] - \tfrac{3}{4}y[n - 1] + \tfrac{1}{8}y[n - 2] = 2x[n - 1].$$

Determine $y[n]$ for $n \ge 0$ when $x[n] = \delta[n]$ and $y[n] = 0, n < 0$.

2.5. A causal linear time-invariant system is described by the difference equation

$$y[n] - 5y[n - 1] + 6y[n - 2] = 2x[n - 1].$$

(a) Determine the homogeneous response of the system, i.e., the possible outputs if $x[n] = 0$ for all n.

(b) Determine the impulse response of the system.

(c) Determine the step response of the system.

2.6. (a) Find the frequency response $H(e^{j\omega})$ of the linear time-invariant system whose input and output satisfy the difference equation

$$y[n] - \tfrac{1}{2}y[n - 1] = x[n] + 2x[n - 1] + x[n - 2].$$

(b) Write a difference equation that characterizes a system whose frequency response is

$$H(e^{j\omega}) = \frac{1 - \tfrac{1}{2}e^{-j\omega} + e^{-j3\omega}}{1 + \tfrac{1}{2}e^{-j\omega} + \tfrac{3}{4}e^{-j2\omega}}.$$

2.7. Determine whether each of the following signals is periodic. If the signal is periodic, state its period.

(a) $x[n] = e^{j(\pi n/6)}$

(b) $x[n] = e^{j(3\pi n/4)}$

(c) $x[n] = [\sin(\pi n/5)]/(\pi n)$

(d) $x[n] = e^{j\pi n/\sqrt{2}}$

2.8. An LTI system has impulse response $h[n] = 5(-1/2)^n u[n]$. Use the Fourier transform to find the output of this system when the input is $x[n] = (1/3)^n u[n]$.

2.9. Consider the difference equation

$$y[n] - \frac{5}{6}y[n - 1] + \frac{1}{6}y[n - 2] = \frac{1}{3}x[n - 1].$$

(a) What are the impulse response, frequency response, and step response for the causal LTI system satisfying this difference equation.

(b) What is the general form of the homogeneous solution of the difference equation?

(c) Consider a different system satisfying the difference equation that is neither causal nor LTI, but that has $y[0] = y[1] = 1$. Find the response of this system to $x[n] = \delta[n]$.

2.10. Determine the output of a linear time-invariant system if the impulse response $h[n]$ and the input $x[n]$ are as follows:

(a) $x[n] = u[n]$ and $h[n] = a^n u[-n-1]$, with $a > 1$.

(b) $x[n] = u[n-4]$ and $h[n] = 2^n u[-n-1]$.

(c) $x[n] = u[n]$ and $h[n] = (0.5)2^n u[-n]$.

(d) $h[n] = 2^n u[-n-1]$ and $x[n] = u[n] - u[n-10]$

Use your knowledge of linearity and time invariance to minimize the work in Parts (b)–(d).

2.11. Consider an LTI system with frequency response

$$H(e^{j\omega}) = \frac{1 - e^{-j2\omega}}{1 + \frac{1}{2}e^{-j4\omega}}, \qquad -\pi < \omega \leq \pi.$$

Determine the output $y[n]$ for all n if the input $x[n]$ for all n is

$$x[n] = \sin\left(\frac{\pi n}{4}\right).$$

2.12. Consider a system with input $x[n]$ and output $y[n]$ that satisfy the difference equation

$$y[n] = ny[n-1] + x[n].$$

The system is causal and satisfies initial-rest conditions; i.e., if $x[n] = 0$ for $n < n_0$, then $y[n] = 0$ for $n < n_0$.

(a) If $x[n] = \delta[n]$, determine $y[n]$ for all n.

(b) Is the system linear? Justify your answer.

(c) Is the system time invariant? Justify your answer.

2.13. Indicate which of the following discrete-time signals are eigenfunctions of stable, linear time-invariant discrete-time systems:

(a) $e^{j2\pi n/3}$

(b) 3^n

(c) $2^n u[-n-1]$

(d) $\cos(\omega_0 n)$

(e) $(1/4)^n$

(f) $(1/4)^n u[n] + 4^n u[-n-1]$

2.14. A single input–output relationship is given for each of the following three systems:

(a) System A: $x[n] = (1/3)^n$, $\quad y[n] = 2(1/3)^n$.

(b) System B: $x[n] = (1/2)^n$, $\quad y[n] = (1/4)^n$.

(c) System C: $x[n] = (2/3)^n u[n]$, $\quad y[n] = 4(2/3)^n u[n] - 3(1/2)^n u[n]$.

Based on this information, pick the strongest possible conclusion that you can make about each system from the following list of statements:

(i) The system cannot possibly be LTI.

(ii) The system must be LTI.

(iii) The system can be LTI, and there is only one LTI system that satisfies this input–output constraint.

(iv) The system can be LTI, but cannot be uniquely determined from the information in this input–output constraint.

If you chose option (iii) from this list, specify either the impulse response $h[n]$ or the frequency response $H(e^{j\omega})$ for the LTI system.

2.15. Consider the system illustrated in Figure P2.15-1. The ouput of an LTI system with an impulse response $h[n] = \left(\frac{1}{4}\right)^n u[n + 10]$ is multiplied by a unit step function $u[n]$ to yield the output of the overall system. Answer each of the following questions, and briefly justify your answers:

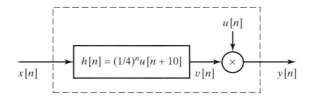

Figure P2.15-1

(a) Is the overall system LTI?
(b) Is the overall system causal?
(c) Is the overall system stable in the BIBO sense?

2.16. Consider the following difference equation:

$$y[n] - \frac{1}{4}y[n - 1] - \frac{1}{8}y[n - 2] = 3x[n].$$

(a) Determine the general form of the homogeneous solution to this difference equation.
(b) Both a causal and an anticausal LTI system are characterized by this difference equation. Find the impulse responses of the two systems.
(c) Show that the causal LTI system is stable and the anticausal LTI system is unstable.
(d) Find a particular solution to the difference equation when $x[n] = (1/2)^n u[n]$.

2.17. (a) Determine the Fourier transform of the sequence

$$r[n] = \begin{cases} 1, & 0 \le n \le M, \\ 0, & \text{otherwise.} \end{cases}$$

(b) Consider the sequence

$$w[n] = \begin{cases} \frac{1}{2}\left[1 - \cos\left(\frac{2\pi n}{M}\right)\right], & 0 \le n \le M, \\ 0, & \text{otherwise.} \end{cases}$$

Sketch $w[n]$ and express $W(e^{j\omega})$, the Fourier transform of $w[n]$, in terms of $R(e^{j\omega})$, the Fourier transform of $r[n]$. (*Hint:* First express $w[n]$ in terms of $r[n]$ and the complex exponentials $e^{j(2\pi n/M)}$ and $e^{-j(2\pi n/M)}$.)
(c) Sketch the magnitude of $R(e^{j\omega})$ and $W(e^{j\omega})$ for the case when $M = 4$.

2.18. For each of the following impulse responses of LTI systems, indicate whether or not the system is causal:
(a) $h[n] = (1/2)^n u[n]$
(b) $h[n] = (1/2)^n u[n - 1]$
(c) $h[n] = (1/2)^{|n|}$
(d) $h[n] = u[n + 2] - u[n - 2]$
(e) $h[n] = (1/3)^n u[n] + 3^n u[-n - 1]$

2.19. For each of the following impulse responses of LTI systems, indicate whether or not the system is stable:
(a) $h[n] = 4^n u[n]$
(b) $h[n] = u[n] - u[n - 10]$
(c) $h[n] = 3^n u[-n - 1]$

  (d) $h[n] = \sin(\pi n/3)u[n]$

  (e) $h[n] = (3/4)^{|n|}\cos(\pi n/4 + \pi/4)$

  (f) $h[n] = 2u[n+5] - u[n] - u[n-5]$

2.20. Consider the difference equation representing a causal LTI system

$$y[n] + (1/a)y[n-1] = x[n-1].$$

  (a) Find the impulse response of the system, $h[n]$, as a function of the constant a.

  (b) For what range of values of a will the system be stable?

Basic Problems

2.21. Consider an arbitrary linear system with input $x[n]$ and output $y[n]$. Show that if $x[n] = 0$ for all n, then $y[n]$ must also be zero for all n.

2.22. For each of the pairs of sequences in Figure P2.22-1, use discrete convolution to find the response to the input $x[n]$ of the linear time-invariant system with impulse response $h[n]$.

(a)

(b)

(c)

(d)

Figure P2.22-1

2.23. Using the definition of linearity (Eqs. (2.26a)–(2.26b)), show that the ideal delay system (Example 2.3) and the moving-average system (Example 2.4) are both linear systems.

2.24. The impulse response of a linear time-invariant system is shown in Figure P2.24-1. Determine and carefully sketch the response of this system to the input $x[n] = u[n-4]$.

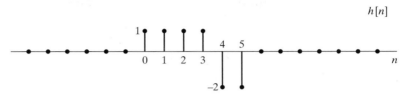

$h[n]$

Figure P2.24-1

2.25. A linear time-invariant system has impulse response $h[n] = u[n]$. Determine the response of this system to the input $x[n]$ shown in Figure P2.25-1 and described as

$$x[n] = \begin{cases} 0, & n < 0, \\ a^n, & 0 \le n \le N_1, \\ 0, & N_1 < n < N_2, \\ a^{n-N_2}, & N_2 \le n \le N_2 + N_1, \\ 0, & N_2 + N_1 < n, \end{cases}$$

where $0 < a < 1$.

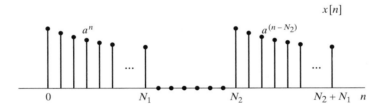

$x[n]$

Figure P2.25-1

2.26. Which of the following discrete-time signals could be eigenfunctions of any stable LTI system?
(a) $5^n u[n]$
(b) $e^{j2\omega n}$
(c) $e^{j\omega n} + e^{j2\omega n}$
(d) 5^n
(e) $5^n \cdot e^{j2\omega n}$

2.27. Three systems A, B, and C have the inputs and outputs indicated in Figure P2.27-1. Determine whether each system could be LTI. If your answer is yes, specify whether there could be more than one LTI system with the given input–output pair. Explain your answer.

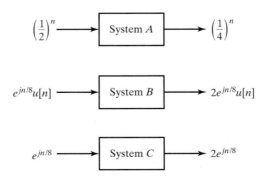

Figure P2.27-1

2.28. Determine which of the following signals is periodic. If a signal is periodic, determine its period.
  (a) $x[n] = e^{j(2\pi n/5)}$
  (b) $x[n] = \sin(\pi n/19)$
  (c) $x[n] = ne^{j\pi n}$
  (d) $x[n] = e^{jn}$

2.29. A discrete-time signal $x[n]$ is shown in Figure P2.29-1.

Figure P2.29-1

Sketch and label carefully each of the following signals:
  (a) $x[n - 2]$
  (b) $x[4 - n]$
  (c) $x[2n]$
  (d) $x[n]u[2 - n]$
  (e) $x[n - 1]\delta[n - 3]$

2.30. For each of the following systems, determine whether the system is (1) stable, (2) causal, (3) linear, and (4) time invariant.
  (a) $T(x[n]) = (\cos \pi n)x[n]$
  (b) $T(x[n]) = x[n^2]$
  (c) $T(x[n]) = x[n] \sum_{k=0}^{\infty} \delta[n - k]$
  (d) $T(x[n]) = \sum_{k=n-1}^{\infty} x[k]$

2.31. Consider the difference equation

$$y[n] + \frac{1}{15} y[n - 1] - \frac{2}{5} y[n - 2] = x[n].$$

  (a) Determine the general form of the homogeneous solution to this equation.
  (b) Both a causal and an anticausal LTI system are characterized by the given difference equation. Find the impulse responses of the two systems.
  (c) Show that the causal LTI system is stable and the anticausal LTI system is unstable.
  (d) Find a particular solution to the difference equation when $x[n] = (3/5)^n u[n]$.

2.32. Consider an LTI system with frequency response

$$H(e^{j\omega}) = e^{-j(\omega - \frac{\pi}{4})} \left(\frac{1 + e^{-j2\omega} + 4e^{-j4\omega}}{1 + \frac{1}{2}e^{-j2\omega}} \right), \qquad -\pi < \omega \le \pi.$$

Determine the output $y[n]$ for all n if the input for all n is

$$x[n] = \cos \left(\frac{\pi n}{2} \right).$$

2.33. Consider an LTI system with $|H(e^{j\omega})| = 1$, and let $\arg[H(e^{j\omega})]$ be as shown in Figure P2.33-1. If the input is

$$x[n] = \cos \left(\frac{3\pi}{2} n + \frac{\pi}{4} \right),$$

determine the output $y[n]$.

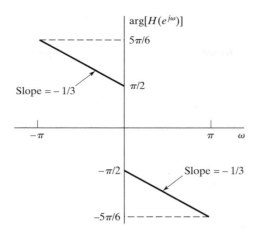

Figure P2.33-1

2.34. The input–output pair shown in Figure P2.34-1 is given for a stable LTI system.

Figure P2.34-1

(a) Determine the response to the input $x_1[n]$ in Figure P2.34-2.

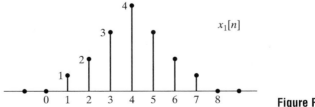

Figure P2.34-2

(b) Determine the impulse response of the system.

Advanced Problems

2.35. The system T in Figure P2.35-1 is known to be *time invariant*. When the inputs to the system are $x_1[n]$, $x_2[n]$, and $x_3[n]$, the responses of the system are $y_1[n]$, $y_2[n]$, and $y_3[n]$, as shown.

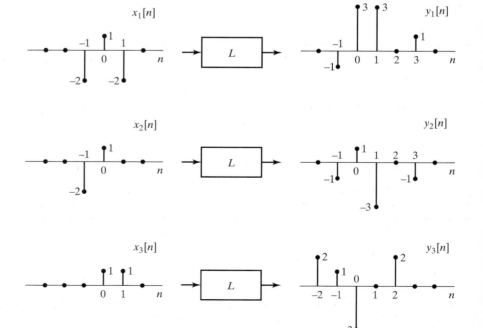

Figure P2.35-1

(a) Determine whether the system T could be linear.

(b) If the input $x[n]$ to the system T is $\delta[n]$, what is the system response $y[n]$?

(c) What are all possible inputs $x[n]$ for which the response of the system T can be determined from the given information alone?

2.36. The system L in Figure P2.36-1 is known to be *linear*. Shown are three output signals $y_1[n]$, $y_2[n]$, and $y_3[n]$ in response to the input signals $x_1[n]$, $x_2[n]$, and $x_3[n]$, respectively.

Figure P2.36-1

(a) Determine whether the system L could be time invariant.

(b) If the input $x[n]$ to the system L is $\delta[n]$, what is the system response $y[n]$?

2.37. Consider a discrete-time linear time-invariant system with impulse response $h[n]$. If the input $x[n]$ is a periodic sequence with period N (i.e., if $x[n] = x[n + N]$), show that the output $y[n]$ is also a periodic sequence with period N.

2.38. In Section 2.5, we stated that the solution to the homogeneous difference equation

$$\sum_{k=0}^{N} a_k y_h[n - k] = 0 \qquad \text{(P2.38-1)}$$

is of the form

$$y_h[n] = \sum_{m=1}^{N} A_m z_m^n, \qquad \text{(P2.38-2)}$$

with the A_m's arbitrary and the z_m's the N roots of the polynomial

$$\sum_{k=0}^{N} a_k z^{-k} = 0; \qquad \text{(P2.38-3)}$$

i.e.,

$$\sum_{k=0}^{N} a_k z^{-k} = \prod_{m=1}^{N} (1 - z_m z^{-1}). \qquad \text{(P2.38-4)}$$

(a) Determine the general form of the homogeneous solution to the difference equation

$$y[n] - \tfrac{3}{4} y[n - 1] + \tfrac{1}{8} y[n - 2] = 2x[n - 1]. \qquad \text{(P2.38-5)}$$

(b) Determine the coefficients A_m in the homogeneous solution if $y[-1] = 1$ and $y[0] = 0$.

(c) Now consider the difference equation

$$y[n] - y[n - 1] + \tfrac{1}{4} y[n - 2] = 2y[n - 1]. \qquad \text{(P2.38-6)}$$

If the homogeneous solution contains only terms of the form of Eq. (P2.38-2), show that the initial conditions $y[-1] = 1$ and $y[0] = 0$ cannot be satisfied.

(d) If Eq. (P2.38-3) has two roots that are identical, then, in place of Eq. (P2.38-2), $y_h[n]$ will take the form

$$y_h[n] = \sum_{m=1}^{N-1} A_m z_m^n + n B_1 z_1^n, \qquad \text{(P2.38-7)}$$

where we have assumed that the double root is z_1. Using Eq. (P2.38-7), determine the general form of $y_h[n]$ for Eq. (P2.38-6). Verify explicitly that your answer satisfies Eq. (P2.38-6) with $x[n] = 0$.

(e) Determine the coefficients A_1 and B_1 in the homogeneous solution obtained in Part (d) if $y[-1] = 1$ and $y[0] = 0$.

2.39. Consider a system with input $x[n]$ and output $y[n]$. The input–output relation for the system is defined by the following two properties:

1. $y[n] - ay[n - 1] = x[n]$,

2. $y[0] = 1$.

(a) Determine whether the system is time invariant.

(b) Determine whether the system is linear.

(c) Assume that the difference equation (property 1) remains the same, but the value $y[0]$ is specified to be zero. Does this change your answer to either Part (a) or Part (b)?

2.40. Consider the linear time-invariant system with impulse response

$$h[n] = \left(\frac{j}{2}\right)^n u[n], \qquad \text{where } j = \sqrt{-1}.$$

Determine the steady-state response, i.e., the response for large n, to the excitation

$$x[n] = \cos(\pi n)u[n].$$

2.41. A linear time-invariant system has frequency response

$$H(e^{j\omega}) = \begin{cases} e^{-j\omega 3}, & |\omega| < \dfrac{2\pi}{16}\left(\dfrac{3}{2}\right), \\[2mm] 0, & \dfrac{2\pi}{16}\left(\dfrac{3}{2}\right) \le |\omega| \le \pi. \end{cases}$$

The input to the system is a periodic unit-impulse train with period $N = 16$; i.e.,

$$x[n] = \sum_{k=-\infty}^{\infty} \delta[n + 16k].$$

Find the output of the system.

2.42. Consider the system in Figure P2.42-1.

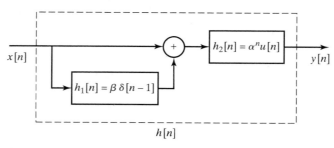

$$h[n]$$

Figure P2.42-1

(a) Find the impulse response $h[n]$ of the overall system.
(b) Find the frequency response of the overall system.
(c) Specify a difference equation that relates the output $y[n]$ to the input $x[n]$.
(d) Is this system causal? Under what condition would the system be stable?

2.43. For $X(e^{j\omega}) = 1/(1 - ae^{-j\omega})$, with $-1 < a < 0$, determine and sketch the following as a function of ω:
(a) $\mathcal{R}e\{X(e^{j\omega})\}$
(b) $\mathcal{I}m\{X(e^{j\omega})\}$
(c) $|X(e^{j\omega})|$
(d) $\sphericalangle X(e^{j\omega})$

2.44. Let $X(e^{j\omega})$ denote the Fourier transform of the signal $x[n]$ shown in Figure P2.44-1. Perform the following calculations without explicitly evaluating $X(e^{j\omega})$:

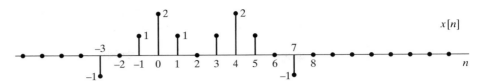

Figure P2.44-1

(a) Evaluate $X(e^{j\omega})|_{\omega=0}$.
(b) Evaluate $X(e^{j\omega})|_{\omega=\pi}$.
(c) Find $\sphericalangle X(e^{j\omega})$.
(d) Evaluate $\int_{-\pi}^{\pi} X(e^{j\omega})d\omega$.
(e) Determine and sketch the signal whose Fourier transform is $X(e^{-j\omega})$.
(f) Determine and sketch the signal whose Fourier transform is $\mathcal{R}e\{X(e^{j\omega})\}$.

2.45. For the system in Figure P2.45-1, determine the output $y[n]$ when the input $x[n]$ is $\delta[n]$ and $H(e^{j\omega})$ is an ideal lowpass filter as indicated, i.e.,

$$H(e^{j\omega}) = \begin{cases} 1, & |\omega| < \pi/2, \\ 0, & \pi/2 < |\omega| \le \pi. \end{cases}$$

Figure P2.45-1

2.46. A sequence has the discrete-time Fourier transform

$$X(e^{j\omega}) = \frac{1-a^2}{(1-ae^{-j\omega})(1-ae^{j\omega})}, \qquad |a| < 1.$$

(a) Find the sequence $x[n]$.
(b) Calculate $\int_{-\pi}^{\pi} X(e^{j\omega})\cos(\omega)d\omega/2\pi$.

2.47. A linear time-invariant system is described by the input–output relation

$$y[n] = x[n] + 2x[n-1] + x[n-2].$$

(a) Determine $h[n]$, the impulse response of the system.
(b) Is this a stable system?
(c) Determine $H(e^{j\omega})$, the frequency response of the system. Use trigonometric identities to obtain a simple expression for $H(e^{j\omega})$.

(d) Plot the magnitude and phase of the frequency response.

(e) Now consider a new system whose frequency response is $H_1(e^{j\omega}) = H(e^{j(\omega+\pi)})$. Determine $h_1[n]$, the impulse response of the new system.

2.48. Let the real discrete-time signal $x[n]$ with Fourier transform $X(e^{j\omega})$ be the input to a system with the output defined by

$$y[n] = \begin{cases} x[n], & \text{if } n \text{ is even,} \\ 0, & \text{otherwise.} \end{cases}$$

(a) Sketch the discrete-time signal $s[n] = 1 + \cos(\pi n)$ and its (generalized) Fourier transform $S(e^{j\omega})$.

(b) Express $Y(e^{j\omega})$, the Fourier transform of the output, as a function of $X(e^{j\omega})$ and $S(e^{j\omega})$.

(c) You would like to approximate $x[n]$ by the interpolated signal $w[n] = y[n] + (1/2)(y[n+1] + y[n-1])$. Determine the Fourier transform $W(e^{j\omega})$ as a function of $Y(e^{j\omega})$.

(d) Sketch $X(e^{j\omega})$, $Y(e^{j\omega})$, and $W(e^{j\omega})$ for the case when $x[n] = \sin(\pi n/a)/(\pi n/a)$ and $a > 1$. Under what conditions is the proposed interpolated signal $w[n]$ a good approximation for the original $x[n]$.

2.49. Consider a discrete-time LTI system with frequency response $H(e^{j\omega})$ and corresponding impulse response $h[n]$.

(a) We are first given the following three clues about the system:

    (i) The system is causal.

    (ii) $H(e^{j\omega}) = H^*(e^{-j\omega})$.

    (iii) The DTFT of the sequence $h[n+1]$ is real.

    Using these three clues, show that the system has an impulse response of finite duration.

(b) In addition to the preceding three clues, we are now given two more clues:

    (iv) $\dfrac{1}{2\pi}\displaystyle\int_{-\pi}^{\pi} H(e^{j\omega})d\omega = 2.$

    (v) $H(e^{j\pi}) = 0.$

    Is there enough information to identify the system uniquely? If so, determine the impulse response $h[n]$. If not, specify as much as you can about the sequence $h[n]$.

2.50. Consider the three sequences

$$v[n] = u[n] - u[n-6],$$
$$w[n] = \delta[n] + 2\delta[n-2] + \delta[n-4],$$
$$q[n] = v[n] * w[n].$$

(a) Find and sketch the sequence $q[n]$.

(b) Find and sketch the sequence $r[n]$ such that $r[n] * v[n] = \sum_{k=-\infty}^{n-1} q[k]$.

(c) Is $q[-n] = v[-n] * w[-n]$? Justify your answer.

2.51. A linear time-invariant system has impulse response $h[n] = a^n u[n]$.

(a) Determine $y_1[n]$, the response of the system to the input $x_1[n] = e^{j(\pi/2)n}$.

(b) Use the result of Part (a) to help to determine $y_2[n]$, the response of the system to the input $x_2[n] = \cos(\pi n/2)$.

(c) Determine $y_3[n]$, the response of the system to the input $x_3[n] = e^{j(\pi/2)n}u[n]$.

(d) Compare $y_3[n]$ with $y_1[n]$ for large n.

2.52. The frequency response of an LTI system is

$$H(e^{j\omega}) = e^{-j\omega/4}, \qquad -\pi < \omega \leq \pi.$$

Determine the output of the system, $y[n]$, when the input is $x[n] = \cos(5\pi n/2)$. Express your answer in as simple a form as you can.

2.53. Consider the cascade of LTI discrete-time systems shown in Figure P2.53-1.

Figure P2.53-1

The first system is described by the equation

$$H_1(e^{j\omega}) = \begin{cases} 1, & |\omega| < 0.5\pi, \\ 0, & 0.5\pi \leq |\omega| < \pi, \end{cases}$$

and the second system is described by the equation

$$y[n] = w[n] - w[n-1].$$

The input to this system is

$$x[n] = \cos(0.6\pi n) + 3\delta[n-5] + 2.$$

Determine the output $y[n]$. With careful thought, you will be able to use the properties of LTI systems to write down the answer by inspection.

2.54. Consider an LTI system with frequency response

$$H(e^{j\omega}) = e^{-j[(\omega/2) + (\pi/4)]}, \qquad -\pi < \omega \leq \pi.$$

Determine $y[n]$, the output of this system, if the input is

$$x[n] = \cos\left(\frac{15\pi n}{4} - \frac{\pi}{3}\right)$$

for all n.

2.55. For the system shown in Figure P2.55-1, System 1 is a memoryless nonlinear system. System 2 determines the value of A according to the relation

$$A = \sum_{n=0}^{100} y[n].$$

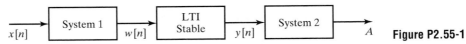

Figure P2.55-1

Specifically, consider the class of inputs of the form $x[n] = \cos(\omega n)$, with ω a real finite number. Varying the value of ω at the input will change A; i.e., A will be a function of ω. In general, will A be periodic in ω? Justify your answer.

2.56. Consider a system S with input $x[n]$ and output $y[n]$ related according to the block diagram in Figure P2.56-1.

Figure P2.56-1

The input $x[n]$ is multiplied by $e^{-j\omega_0 n}$, and the product is passed through a stable LTI system with impulse response $h[n]$.

(a) Is the system S linear? Justify your answer.

(b) Is the system S time invariant? Justify your answer.

(c) Is the system S stable? Justify your answer.

(d) Specify a system C such that the block diagram in Figure P2.56-2 represents an alternative way of expressing the input–output relationship of the system S. (*Note:* The system C does not have to be an LTI system.)

Figure P2.56-2

2.57. An ideal lowpass filter with zero delay has impulse response $h_{lp}[n]$ and frequency response

$$H_{lp}(e^{j\omega}) = \begin{cases} 1, & |\omega| < 0.2\pi, \\ 0, & 0.2\pi \leq |\omega| \leq \pi. \end{cases}$$

(a) A new filter is defined by the equation $h_1[n] = (-1)^n h_{lp}[n] = e^{j\pi n} h_{lp}[n]$. Determine an equation for the frequency response of $H_1(e^{j\omega})$, and plot the equation for $|\omega| < \pi$. What kind of filter is this?

(b) A second filter is defined by the equation $h_2[n] = 2h_{lp}[n]\cos(0.5\pi n)$. Determine the equation for the frequency response $H_2(e^{j\omega})$, and plot the equation for $|\omega| < \pi$. What kind of filter is this?

(c) A third filter is defined by the equation

$$h_3[n] = \frac{\sin(0.1\pi n)}{\pi n} h_{lp}[n].$$

Determine the equation for the frequency response $H_3(e^{j\omega})$, and plot the equation for $|\omega| < \pi$. What kind of filter is this?

2.58. The LTI system

$$H(e^{j\omega}) = \begin{cases} -j, & 0 < \omega < \pi, \\ j, & -\pi < \omega < 0, \end{cases}$$

is referred to as a $90°$ phase shifter and is used to generate what is referred to as an analytic signal $w[n]$ as shown in Figure P2.58-1. Specifically, the analytic signal $w[n]$ is a complex-valued signal for which

$$\mathcal{Re}\{w[n]\} = x[n],$$

$$\mathcal{Im}\{w[n]\} = y[n].$$

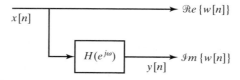

Figure P2.58-1

If $X(e^{j\omega})$ is as shown in Figure P2.58-2, determine and sketch $W(e^{j\omega})$, the Fourier transform of the analytic signal $w[n] = x[n] + jy[n]$.

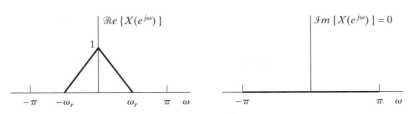

Figure P2.58-2

2.59. The autocorrelation sequence of a signal $x[n]$ is defined as

$$R_x[n] = \sum_{k=-\infty}^{\infty} x^*[k]x[n+k].$$

(a) Show that for an appropriate choice of the signal $g[n]$, $R_x[n] = x[n] * g[n]$, and identify the proper choice for $g[n]$.

(b) Show that the Fourier transform of $R_x[n]$ is equal to $|X(e^{j\omega})|^2$.

2.60. The signals $x[n]$ and $y[n]$ shown in Figure P2.60-1 are the input and corresponding output for an LTI system.

Figure P2.60-1

(a) Find the response of the system to the sequence $x_2[n]$ in Figure P2.60-2.

Figure P2.60-2

(b) Find the impulse response $h[n]$ for this LTI system.

2.61. Consider a system for which the input $x[n]$ and output $y[n]$ satisfy the difference equation

$$y[n] - \tfrac{1}{2}y[n-1] = x[n]$$

and for which $y[-1]$ is constrained to be zero for every input. Determine whether or not the system is stable. If you conclude that the system is stable, show your reasoning. If you conclude that the system is **not** stable, give an example of a bounded input that results in an unbounded output.

Extension Problems

2.62. The causality of a system was defined in Section 2.2.4. From this definition, show that, for a linear time-invariant system, causality implies that the impulse response $h[n]$ is zero for $n < 0$. One approach is to show that if $h[n]$ is *not* zero for $n < 0$, then the system *cannot* be causal. Show also that if the impulse response is zero for $n < 0$, then the system will necessarily be causal.

2.63. Consider a discrete-time system with input $x[n]$ and output $y[n]$. When the input is

$$x[n] = \left(\frac{1}{4}\right)^n u[n],$$

the output is

$$y[n] = \left(\frac{1}{2}\right)^n \qquad \text{for all } n.$$

Determine which of the following statements is correct:

- The system must be LTI.
- The system could be LTI.
- The system cannot be LTI.

If your answer is that the system must or could be LTI, give a possible impulse response. If your answer is that the system could not be LTI, explain clearly why not.

2.64. Consider an LTI system whose frequency response is

$$H(e^{j\omega}) = e^{-j\omega/2}, \qquad |\omega| < \pi.$$

Determine whether or not the system is causal. Show your reasoning.

2.65. In Figure P2.65-1, two sequences $x_1[n]$ and $x_2[n]$ are shown. Both sequences are zero for all n outside the regions shown. The Fourier transforms of these sequences are $X_1(e^{j\omega})$ and $X_2(e^{j\omega})$, which, in general, can be expected to be complex and can be written in the form

$$X_1(e^{j\omega}) = A_1(\omega)e^{j\theta_1(\omega)},$$
$$X_2(e^{j\omega}) = A_2(\omega)e^{j\theta_2(\omega)},$$

where $A_1(\omega)$, $\theta_1(\omega)$, $A_2(\omega)$, and $\theta_2(\omega)$ are all real functions chosen so that both $A_1(\omega)$ and $A_2(\omega)$ are nonnegative at $\omega = 0$, but otherwise can take on both positive and negative values. Determine appropriate choices for $\theta_1(\omega)$ and $\theta_2(\omega)$, and sketch these two phase functions in the range $0 < \omega < 2\pi$.

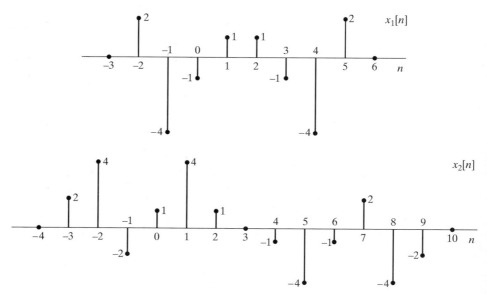

Figure P2.65-1

2.66. Consider the cascade of discrete-time systems in Figure P1.66-1. The time-reversal systems are defined by the equations $f[n] = e[-n]$ and $y[n] = g[-n]$. Assume throughout the problem that $x[n]$ and $h_1[n]$ are real sequences.

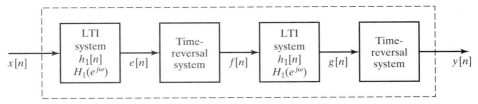

Figure P2.66-1

(a) Express $E(e^{j\omega})$, $F(e^{j\omega})$, $G(e^{j\omega})$, and $Y(e^{j\omega})$ in terms of $X(e^{j\omega})$ and $H_1(e^{j\omega})$.

(b) The result from Part (a) should convince you that the overall system is LTI. Find the frequency response $H(e^{j\omega})$ of the overall system.

(c) Determine an expression for the impulse response $h[n]$ of the overall system in terms of $h_1[n]$.

2.67. The overall system in the dotted box in Figure P1.67-1 can be shown to be linear and time invariant.

(a) Determine an expression for $H(e^{j\omega})$, the frequency response of the overall system from the input $x[n]$ to the output $y[n]$, in terms of $H_1(e^{j\omega})$, the frequency response of the internal LTI system. Remember that $(-1)^n = e^{j\pi n}$.

(b) Plot $H(e^{j\omega})$ for the case when the frequency response of the internal LTI system is

$$H_1(e^{j\omega}) = \begin{cases} 1, & |\omega| < \omega_c, \\ 0, & \omega_c < |\omega| \le \pi. \end{cases}$$

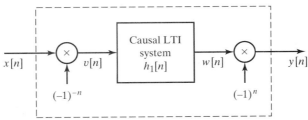

Figure P2.67-1

2.68. Figure P1.68-1 shows the input–output relationships of Systems A and B, while Figure P1.68-2 contains two possible cascade combinations of these systems.

Figure P2.68-1

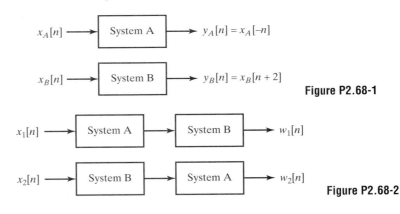

Figure P2.68-2

If $x_1[n] = x_2[n]$, will $w_1[n]$ and $w_2[n]$ necessarily be equal? If your answer is *yes*, clearly and concisely explain why and demonstrate with an example. If your answer is *not necessarily*, demonstrate with a counterexample.

2.69. Consider the system in Figure P2.69-1, where the subsystems S_1 and S_2 are LTI.

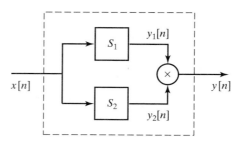

Figure P2.69-1

(a) Is the overall system enclosed by the dashed box, with input $x[n]$ and output $y[n]$ equal to the product of $y_1[n]$ and $y_2[n]$, guaranteed to be an LTI system? If so, explain your reasoning. If not, provide a counterexample.

(b) Suppose S_1 and S_2 have frequency responses $H_1(e^{j\omega})$ and $H_2(e^{j\omega})$ that are known to be zero over certain regions. Let

$$H_1(e^{j\omega}) = \begin{cases} 0, & |\omega| \le 0.2\pi, \\ \text{unspecified}, & 0.2\pi < |\omega| \le \pi, \end{cases}$$

$$H_2(e^{j\omega}) = \begin{cases} \text{unspecified}, & |\omega| \le 0.4\pi, \\ 0, & 0.4\pi < |\omega| \le \pi. \end{cases}$$

Suppose also that the input $x[n]$ is known to be bandlimited to 0.3π, i.e.,

$$X(e^{j\omega}) = \begin{cases} \text{unspecified}, & |\omega| < 0.3\pi, \\ 0, & 0.3\pi \le |\omega| \le \pi. \end{cases}$$

Over what region of $-\pi \le \omega < \pi$ is $Y(e^{j\omega})$, the DTFT of $y[n]$, guaranteed to be zero?

2.70. A commonly used numerical operation called the *first backward difference* is defined as

$$y[n] = \nabla(x[n]) = x[n] - x[n-1],$$

where $x[n]$ is the input and $y[n]$ is the output of the first-backward-difference system.

(a) Show that this system is linear and time invariant.

(b) Find the impulse response of the system.

(c) Find and sketch the frequency response (magnitude and phase).

(d) Show that if

$$x[n] = f[n] * g[n],$$

then

$$\nabla(x[n]) = \nabla(f[n]) * g[n] = f[n] * \nabla(g[n]),$$

where $*$ denotes discrete convolution.

(e) Find the impulse response of a system that could be cascaded with the first-difference system to recover the input; i.e., find $h_i[n]$, where

$$h_i[n] * \nabla(x[n]) = x[n].$$

2.71. Let $H(e^{j\omega})$ denote the frequency response of an LTI system with impulse response $h[n]$, where $h[n]$ is, in general, complex.

(a) Using Eq. (2.109), show that $H^*(e^{-j\omega})$ is the frequency response of a system with impulse response $h^*[n]$, where $*$ denotes complex conjugation.

(b) Show that if $h[n]$ is real, the frequency response is conjugate symmetric, i.e., $H(e^{-j\omega}) = H^*(e^{j\omega})$.

2.72. Let $X(e^{j\omega})$ denote the Fourier transform of $x[n]$. Using the Fourier transform synthesis or analysis equations (Eqs. (2.133) and (2.134)), show that

(a) the Fourier transform of $x^*[n]$ is $X^*(e^{-j\omega})$,

(b) the Fourier transform of $x^*[-n]$ is $X^*(e^{j\omega})$.

2.73. Show that for $x[n]$ real, property 7 in Table 2.1 follows from property 1 and that properties 8–11 follow from property 7.

2.74. In Section 2.9, we stated a number of Fourier transform theorems without proof. Using the Fourier synthesis or analysis equations (Eqs. (2.133) and (2.134)), demonstrate the validity of Theorems 1–5 in Table 2.2.

2.75. In Section 2.9.6, it was argued intuitively that

$$Y(e^{j\omega}) = H(e^{j\omega})X(e^{j\omega}), \tag{P2.75-1}$$

when $Y(e^{j\omega})$, $H(e^{j\omega})$, and $X(e^{j\omega})$ are, respectively, the Fourier transforms of the output $y[n]$, impulse response $h[n]$, and input $x[n]$ of a linear time-invariant system; i.e.,

$$y[n] = \sum_{k=-\infty}^{\infty} x[k]h[n-k]. \tag{P2.75-2}$$

Verify Eq. (P2.75-1) by applying the Fourier transform to the convolution sum given in Eq. (P2.75-2).

2.76. By applying the Fourier synthesis equation (Eq. (2.133)) to Eq. (2.172) and using Theorem 3 in Table 2.2, demonstrate the validity of the modulation theorem (Theorem 7, Table 2.2).

2.77. Let $x[n]$ and $y[n]$ denote complex sequences and $X(e^{j\omega})$ and $Y(e^{j\omega})$ their respective Fourier transforms.

(a) By using the convolution theorem (Theorem 6 in Table 2.2) and appropriate properties from Table 2.2, determine, in terms of $x[n]$ and $y[n]$, the sequence whose Fourier transform is $X(e^{j\omega})Y^*(e^{j\omega})$.

(b) Using the result in Part (a), show that

$$\sum_{n=-\infty}^{\infty} x[n]y^*[n] = \frac{1}{2\pi}\int_{-\pi}^{\pi} X(e^{j\omega})Y^*(e^{j\omega})d\omega. \tag{P2.77-1}$$

Equation (P2.77-1) is a more general form of Parseval's theorem, as given in Section 2.9.5.

(c) Using Eq. (P2.77-1), determine the numerical value of the sum

$$\sum_{n=-\infty}^{\infty} \frac{\sin(\pi n/4)}{2\pi n}\frac{\sin(\pi n/6)}{5\pi n}.$$

2.78. Let $x[n]$ and $X(e^{j\omega})$ represent a sequence and its Fourier transform, respectively. Determine, in terms of $X(e^{j\omega})$, the transforms of $y_s[n]$, $y_d[n]$, and $y_e[n]$. In each case, sketch $Y(e^{j\omega})$ for $X(e^{j\omega})$ as shown in Figure P2.78-1.

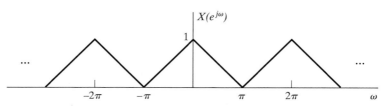

Figure P2.78-1

(a) Sampler:

$$y_s[n] = \begin{cases} x[n], & n \text{ even}, \\ 0, & n \text{ odd}. \end{cases}$$

Note that $y_s[n] = \frac{1}{2}\{x[n] + (-1)^n x[n]\}$ and $-1 = e^{j\pi}$.

(b) Compressor:

$$y_d[n] = x[2n].$$

(c) Expander:

$$y_e[n] = \begin{cases} x[n/2], & n \text{ even}, \\ 0, & n \text{ odd}. \end{cases}$$

2.79. The two-frequency correlation function $\Phi_x(N, \omega)$ is often used in radar and sonar to evaluate the frequency and travel-time resolution of a signal. For discrete-time signals, we define

$$\Phi_x(N, \omega) = \sum_{n=-\infty}^{\infty} x[n + N]x^*[n - N]e^{-j\omega n}.$$

(a) Show that

$$\Phi_x(-N, -\omega) = \Phi_x^*(N, \omega).$$

(b) If

$$x[n] = Aa^n u[n], \qquad 0 < a < 1,$$

find $\Phi_x(N, \omega)$. (Assume that $N \geq 0$.)

(c) The function $\Phi_x(N, \omega)$ has a frequency domain dual. Show that

$$\Phi_x(N, \omega) = \frac{1}{2\pi} \int_{-\pi}^{\pi} X\left(e^{j[v+(\omega/2)]}\right) X^*\left(e^{j[v-(\omega/2)]}\right) e^{j2vN} dv.$$

2.80. Let $x[n]$ and $y[n]$ be stationary, uncorrelated random signals. Show that if

$$w[n] = x[n] + y[n],$$

then

$$m_w = m_x + m_y \quad \text{and} \quad \sigma_w^2 = \sigma_x^2 + \sigma_y^2.$$

2.81. Let $e[n]$ denote a white-noise sequence, and let $s[n]$ denote a sequence that is uncorrelated with $e[n]$. Show that the sequence

$$y[n] = s[n]e[n]$$

is white, i.e., that

$$E\{y[n]y[n+m]\} = A\delta[m],$$

where A is a constant.

2.82. Consider a random signal $x[n] = s[n] + e[n]$, where both $s[n]$ and $e[n]$ are independent zero-mean stationary random signals with autocorrelation functions $\phi_{ss}[m]$ and $\phi_{ee}[m]$
respectively.
  (a) Determine expressions for $\phi_{xx}[m]$ and $\Phi_{xx}(e^{j\omega})$.
  (b) Determine expressions for $\phi_{xe}[m]$ and $\Phi_{xe}(e^{j\omega})$.
  (c) Determine expressions for $\phi_{xs}[m]$ and $\Phi_{xs}(e^{j\omega})$.

2.83. Consider an LTI system with impulse response $h[n] = a^n u[n]$ with $|a| < 1$.
  (a) Compute the deterministic autocorrelation function $\phi_{hh}[m]$ for this impulse response.
  (b) Determine the energy density function $|H(e^{j\omega})|^2$ for the system.
  (c) Use Parseval's theorem to evaluate the integral

$$\frac{1}{2\pi} \int_{-\pi}^{\pi} |H(e^{j\omega})|^2 \, d\omega$$

  for the system.

2.84. The input to the first-backward-difference system (Example 2.10) is a zero-mean white-
noise signal whose autocorrelation function is $\phi_{xx}[m] = \sigma_x^2 \delta[m]$.
  (a) Determine and plot the autocorrelation function and the power spectrum of the corresponding output of the system.
  (b) What is the average power of the output of the system?
  (c) What does this problem tell you about the first backward difference of a noisy signal?

2.85. Let $x[n]$ be a real, stationary, white-noise process, with zero mean and variance σ_x^2. Let $y[n]$
be the corresponding output when $x[n]$ is the input to a linear time-invariant system with
impulse response $h[n]$. Show that
  (a) $E\{x[n]y[n]\} = h[0]\sigma_x^2$,
  (b) $\sigma_y^2 = \sigma_x^2 \sum_{n=-\infty}^{\infty} h^2[n]$.

2.86. Let $x[n]$ be a real stationary white-noise sequence, with zero mean and variance σ_x^2. Let
$x[n]$ be the input to the cascade of two causal linear time-invariant discrete-time systems,
as shown in Figure P1.86-1.

Figure P2.86-1

  (a) Is $\sigma_y^2 = \sigma_x^2 \sum_{k=0}^{\infty} h_1^2[k]$?
  (b) Is $\sigma_w^2 = \sigma_y^2 \sum_{k=0}^{\infty} h_2^2[k]$?
  (c) Let $h_1[n] = a^n u[n]$ and $h_2[n] = b^n u[n]$. Determine the impulse response of the overall
system in Figure P1.86-1, and, from this, determine σ_w^2. Are your answers to parts (b)
and (c) consistent?

2.87. Sometimes we are interested in the statistical behavior of a linear time-invariant system
when the input is a suddenly applied random signal. Such a situation is depicted in Figure P1.87-1.

(switch closed at $n = 0$)   **Figure P2.87–1**

Let $x[n]$ be a stationary white-noise process. The input to the system, $w[n]$, given by

$$w[n] = \begin{cases} x[n], & n \geq 0, \\ 0, & n < 0, \end{cases}$$

is a nonstationary process, as is the output $y[n]$.

(a) Derive an expression for the mean of the output in terms of the mean of the input.

(b) Derive an expression for the autocorrelation sequence $\phi_{yy}[n_1, n_2]$ of the output.

(c) Show that, for large n, the formulas derived in parts (a) and (b) approach the results for stationary inputs.

(d) Assume that $h[n] = a^n u[n]$. Find the mean and mean-square values of the output in terms of the mean and mean-square values of the input. Sketch these parameters as a function of n.

2.88. Let $x[n]$ and $y[n]$ respectively denote the input and output of a system. The input–output relation of a system sometimes used for the purpose of noise reduction in images is given by

$$y[n] = \frac{\sigma_s^2[n]}{\sigma_x^2[n]}(x[n] - m_x[n]) + m_x[n],$$

where

$$\sigma_x^2[n] = \frac{1}{3}\sum_{k=n-1}^{n+1}(x[k] - m_x[n])^2,$$

$$m_x[n] = \frac{1}{3}\sum_{k=n-1}^{n+1}x[k],$$

$$\sigma_s^2[n] = \begin{cases} \sigma_x^2[n] - \sigma_w^2, & \sigma_x^2[n] \geq \sigma_w^2, \\ 0, & \text{otherwise,} \end{cases}$$

and σ_w^2 is a known constant proportional to the noise power.

(a) Is the system linear?

(b) Is the system shift invariant?

(c) Is the system stable?

(d) Is the system causal?

(e) For a fixed $x[n]$, determine $y[n]$ when σ_w^2 is very large (large noise power) and when σ_w^2 is very small (small noise power). Does $y[n]$ make sense for these extreme cases?

2.89. Consider a random process $x[n]$ that is the response of the linear time-invariant system shown in Figure P2.89-1. In the figure, $w[n]$ represents a real zero-mean stationary white-noise process with $E\{w^2[n]\} = \sigma_w^2$.

$$w[n] \longrightarrow \boxed{H(e^{j\omega}) = \frac{1}{1 - 0.5\,e^{-j\omega}}} \longrightarrow x[n] \qquad \textbf{Figure P2.89-1}$$

(a) Express $\mathcal{E}\{x^2[n]\}$ in terms of $\phi_{xx}[n]$ or $\Phi_{xx}(e^{j\omega})$.

(b) Determine $\Phi_{xx}(e^{j\omega})$, the power density spectrum of $x[n]$.

(c) Determine $\phi_{xx}[n]$, the correlation function of $x[n]$.

2.90. Consider a linear time-invariant system whose impulse response is real and is given by $h[n]$. Suppose the responses of the system to the two inputs $x[n]$ and $v[n]$ are, respectively, $y[n]$ and $z[n]$, as shown in Figure P2.90-1.

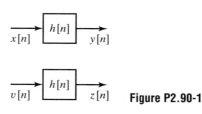

Figure P2.90-1

The inputs $x[n]$ and $v[n]$ in the figure represent real zero-mean stationary random processes with autocorrelation functions $\phi_{xx}[n]$ and $\phi_{vv}[n]$, cross-correlation function $\phi_{xv}[n]$, power spectra $\Phi_{xx}(e^{j\omega})$ and $\Phi_{vv}(e^{j\omega})$, and cross power spectrum $\Phi_{xv}(e^{j\omega})$.

(a) Given $\phi_{xx}[n]$, $\phi_{vv}[n]$, $\phi_{xv}[n]$, $\Phi_{xx}(e^{j\omega})$, $\Phi_{vv}(e^{j\omega})$, and $\Phi_{xv}(e^{j\omega})$, determine $\Phi_{yz}(e^{j\omega})$, the cross power spectrum of $y[n]$ and $z[n]$, where $\Phi_{yz}(e^{j\omega})$ is defined by

$$\phi_{yz}[n] \xleftrightarrow{\;\mathcal{F}\;} \Phi_{yz}(e^{j\omega}),$$

with $\phi_{yz}[n] = E\{y[k]z[k-n]\}$.

(b) Is the cross power spectrum $\Phi_{xv}(e^{j\omega})$ always nonnegative; i.e., is $\Phi_{xv}(e^{j\omega}) \geq 0$ for all ω? Justify your answer.

2.91. Consider the LTI system shown in Figure P2.91-1. The input to this system, $e[n]$, is a stationary zero-mean white-noise signal with average power σ_e^2. The first system is a backward-difference system as defined in Eq. 2.45 with $f[n] = e[n] - e[n-1]$. The second system is an ideal lowpass filter with frequency response

$$H_2(e^{j\omega}) = \begin{cases} 1, & |\omega| < \omega_c, \\ 0, & \omega_c < |\omega| \leq \pi. \end{cases}$$

$e[n]$ → LTI system #1 → $f[n]$ → LTI system #2 → $g[n]$ **Figure P2.91-1**

(a) Determine an expression for $\Phi_{ff}(e^{j\omega})$, the power spectrum of $f[n]$, and plot this expression for $-2\pi < \omega < 2\pi$.

(b) Determine an expression for $\phi_{ff}[m]$, the autocorrelation function of $f[n]$.

(c) Determine an expression for $\Phi_{gg}(e^{j\omega})$, the power spectrum of $g[n]$, and plot this expression for $-2\pi < \omega < 2\pi$.

(d) Determine an expression for σ_g^2, the average power of the output.

3

THE Z-TRANSFORM

3.0 INTRODUCTION

We have seen that the Fourier transform plays a key role in representing and analyzing discrete-time signals and systems. In this chapter, we develop the z-transform representation of a sequence and study how the properties of a sequence are related to the properties of its z-transform. The z-transform for discrete-time signals is the counterpart of the Laplace transform for continuous-time signals, and they each have a similar relationship to the corresponding Fourier transform. One motivation for introducing this generalization is that the Fourier transform does not converge for all sequences and it is useful to have a generalization of the Fourier transform that encompasses a broader class of signals. A second advantage is that in analytical problems the z-transform notation is often more convenient than the Fourier transform notation.

3.1 z-TRANSFORM

The Fourier transform of a sequence $x[n]$ was defined in Chapter 2 as

$$X(e^{j\omega}) = \sum_{n=-\infty}^{\infty} x[n]e^{-j\omega n}. \tag{3.1}$$

The z-transform of a sequence $x[n]$ is defined as

$$X(z) = \sum_{n=-\infty}^{\infty} x[n]z^{-n}. \tag{3.2}$$

This equation is, in general, an infinite sum or infinite power series, with z being a complex variable. Sometimes it is useful to consider Eq. (3.2) as an operator that transforms a sequence into a function, and we will refer to the *z-transform operator* $\mathcal{Z}\{\cdot\}$, defined as

$$\mathcal{Z}\{x[n]\} = \sum_{n=-\infty}^{\infty} x[n]z^{-n} = X(z). \tag{3.3}$$

With this interpretation, the z-transform operator is seen to transform the sequence $x[n]$ into the function $X(z)$, where z is a continuous complex variable. The correspondence between a sequence and its z-transform is indicated by the notation

$$x[n] \overset{\mathcal{Z}}{\longleftrightarrow} X(z). \tag{3.4}$$

The z-transform, as we have defined it in Eq. (3.2), is often referred to as the *two-sided* or *bilateral z-transform,* in contrast to the *one-sided* or *unilateral z-transform,* which is defined as

$$\mathcal{X}(z) = \sum_{n=0}^{\infty} x[n]z^{-n}. \tag{3.5}$$

Clearly, the bilateral and unilateral transforms are equivalent only if $x[n] = 0$ for $n < 0$. In this book, we focus on the bilateral transform exclusively.

It is evident from a comparison of Eqs. (3.1) and (3.2) that there is a close relationship between the Fourier transform and the z-transform. In particular, if we replace the complex variable z in Eq. (3.2) with the complex variable $e^{j\omega}$, then the z-transform reduces to the Fourier transform. This is one motivation for the notation $X(e^{j\omega})$ for the Fourier transform; when it exists, the Fourier transform is simply $X(z)$ with $z = e^{j\omega}$. This corresponds to restricting z to have unity magnitude; i.e., for $|z| = 1$, the z-transform corresponds to the Fourier transform. More generally, we can express the complex variable z in polar form as

$$z = re^{j\omega}.$$

With z expressed in this form, Eq. (3.2) becomes

$$X(re^{j\omega}) = \sum_{n=-\infty}^{\infty} x[n](re^{j\omega})^{-n},$$

or

$$X(re^{j\omega}) = \sum_{n=-\infty}^{\infty} (x[n]r^{-n})e^{-j\omega n}. \tag{3.6}$$

Equation (3.6) can be interpreted as the Fourier transform of the product of the original sequence $x[n]$ and the exponential sequence r^{-n}. Obviously, for $r = 1$, Eq. (3.6) reduces to the Fourier transform of $x[n]$.

Since the z-transform is a function of a complex variable, it is convenient to describe and interpret it using the complex z-plane. In the z-plane, the contour corresponding to $|z| = 1$ is a circle of unit radius, as illustrated in Figure 3.1. This contour is referred to as the *unit circle*. The z-transform evaluated on the unit circle corresponds to the Fourier transform. Note that ω is the angle between the vector to a point z on the unit circle and the real axis of the complex z-plane. If we evaluate $X(z)$ at points on the

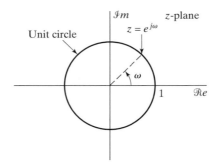

Figure 3.1 The unit circle in the complex z-plane.

unit circle in the z-plane beginning at $z = 1$ (i.e., $\omega = 0$) through $z = j$ (i.e., $\omega = \pi/2$) to $z = -1$ (i.e., $\omega = \pi$), we obtain the Fourier transform for $0 \leq \omega \leq \pi$. Continuing around the unit circle would correspond to examining the Fourier transform from $\omega = \pi$ to $\omega = 2\pi$ or, equivalently, from $\omega = -\pi$ to $\omega = 0$. In Chapter 2, the Fourier transform was displayed on a linear frequency axis. Interpreting the Fourier transform as the z-transform on the unit circle in the z-plane corresponds conceptually to wrapping the linear frequency axis around the unit circle with $\omega = 0$ at $z = 1$ and $\omega = \pi$ at $z = -1$. With this interpretation, the inherent periodicity in frequency of the Fourier transform is captured naturally, since a change of angle of 2π radians in the z-plane corresponds to traversing the unit circle once and returning to exactly the same point.

As we discussed in Chapter 2, the power series representing the Fourier transform does not converge for all sequences; i.e., the infinite sum may not always be finite. Similarly, the z-transform does not converge for all sequences or for all values of z. For any given sequence, the set of values of z for which the z-transform converges is called the *region of convergence,* which we abbreviate ROC. As we stated in Sec. 2.7, if the sequence is absolutely summable, the Fourier transform converges to a continuous function of ω. Applying this criterion to Eq. (3.6) leads to the condition

$$\sum_{n=-\infty}^{\infty} |x[n]r^{-n}| < \infty \tag{3.7}$$

for convergence of the z-transform. It should be clear from Eq. (3.7) that, because of the multiplication of the sequence by the real exponential r^{-n}, it is possible for the z-transform to converge even if the Fourier transform does not. For example, the sequence $x[n] = u[n]$ is not absolutely summable, and therefore, the Fourier transform does not converge absolutely. However, $r^{-n}u[n]$ is absolutely summable if $r > 1$. This means that the z-transform for the unit step exists with a region of convergence $|z| > 1$.

Convergence of the power series of Eq. (3.2) depends only on $|z|$, since $|X(z)| < \infty$ if

$$\sum_{n=-\infty}^{\infty} |x[n]|\,|z|^{-n} < \infty, \tag{3.8}$$

i. e., the region of convergence of the power series in Eq. (3.2) consists of all values of z such that the inequality in Eq. (3.8) holds. Thus, if some value of z, say, $z = z_1$, is in the ROC, then all values of z on the circle defined by $|z| = |z_1|$ will also be in the ROC. As one consequence of this, the region of convergence will consist of a ring in the z-plane

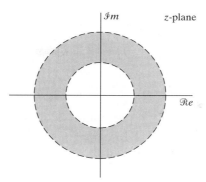

Figure 3.2   The region of convergence (ROC) as a ring in the z-plane. For specific cases, the inner boundary can extend inward to the origin, and the ROC becomes a disc. For other cases, the outer boundary can extend outward to infinity.

centered about the origin. Its outer boundary will be a circle (or the ROC may extend outward to infinity), and its inner boundary will be a circle (or it may extend inward to include the origin). This is illustrated in Figure 3.2. If the ROC includes the unit circle, this of course implies convergence of the z-transform for $|z| = 1$, or equivalently, the Fourier transform of the sequence converges. Conversely, if the ROC does not include the unit circle, the Fourier transform does not converge absolutely.

A power series of the form of Eq. (3.2) is a Laurent series. Therefore, a number of elegant and powerful theorems from the theory of functions of a complex variable can be employed in the study of the z-transform. (See, for example, Churchill and Brown, 1990.) A Laurent series, and therefore the z-transform, represents an analytic function at every point inside the region of convergence; hence, the z-transform and all its derivatives must be continuous functions of z within the region of convergence. This implies that if the region of convergence includes the unit circle, then the Fourier transform and all its derivatives with respect to ω must be continuous functions of ω. Also, from the discussion in Section 2.7, the sequence must be absolutely summable, i.e., a stable sequence.

Uniform convergence of the z-transform requires absolute summability of the exponentially weighted sequence, as stated in Eq. (3.7). Neither of the sequences

$$x_1[n] = \frac{\sin \omega_c n}{\pi n}, \qquad -\infty < n < \infty,$$

and

$$x_2[n] = \cos \omega_0 n, \qquad -\infty < n < \infty,$$

is absolutely summable. Furthermore, neither of these sequences multiplied by r^{-n} would be absolutely summable for any value of r. Thus, these sequences do not have a z-transform that converges absolutely for any z. However, we showed in Section 2.7 that even though sequences such as $x_1[n]$ are not absolutely summable, they do have finite energy, and the Fourier transform converges in the mean-square sense to a discontinuous periodic function. Similarly, the sequence $x_2[n]$ is neither absolutely nor square summable, but a useful Fourier transform for $x_2[n]$ can be defined using impulses. In both cases the Fourier transforms are not continuous, infinitely differentiable functions, so they cannot result from evaluating a z-transform on the unit circle. Thus, in such cases it is not strictly correct to think of the Fourier transform as being the z-transform evaluated on the unit circle, although we nevertheless use the notation $X(e^{j\omega})$ that implies this.

The z-transform is most useful when the infinite sum can be expressed in closed form, i.e., when it can be "summed" and expressed as a simple mathematical formula.

Among the most important and useful z-transforms are those for which $X(z)$ is a rational function inside the region of convergence, i.e.,

$$X(z) = \frac{P(z)}{Q(z)}, \tag{3.9}$$

where $P(z)$ and $Q(z)$ are polynomials in z. The values of z for which $X(z) = 0$ are called the *zeros* of $X(z)$, and the values of z for which $X(z)$ is infinite are referred to as the *poles* of $X(z)$. The poles of $X(z)$ for finite values of z are the roots of the denominator polynomial. In addition, poles may occur at $z = 0$ or $z = \infty$. For rational z-transforms, a number of important relationships exist between the locations of poles of $X(z)$ and the region of convergence of the z-transform. We discuss these more specifically in Section 3.2. First, however, we illustrate the z-transform with several examples.

Example 3.1   Right-Sided Exponential Sequence

Consider the signal $x[n] = a^n u[n]$. Because it is nonzero only for $n \geq 0$, this is an example of a *right-sided* sequence. From Eq. (3.2),

$$X(z) = \sum_{n=-\infty}^{\infty} = a^n u[n] z^{-n} = \sum_{n=0}^{\infty} (az^{-1})^n.$$

For convergence of $X(z)$, we require that

$$\sum_{n=0}^{\infty} |az^{-1}|^n < \infty.$$

Thus, the region of convergence is the range of values of z for which $|az^{-1}| < 1$ or, equivalently, $|z| > |a|$. Inside the region of convergence, the infinite series converges to

$$X(z) = \sum_{n=0}^{\infty} (az^{-1})^n = \frac{1}{1 - az^{-1}} = \frac{z}{z - a}, \qquad |z| > |a|. \tag{3.10}$$

Here we have used the familiar formula for the sum of terms of a geometric series. The z-transform has a region of convergence for any finite value of $|a|$. The Fourier transform of $x[n]$, on the other hand, converges only if $|a| < 1$. For $a = 1$, $x[n]$ is the unit step sequence with z-transform

$$X(z) = \frac{1}{1 - z^{-1}}, \qquad |z| > 1. \tag{3.11}$$

In Example 3.1, the infinite sum is equal to a rational function of z inside the region of convergence; for most purposes, this rational function is a much more convenient representation than the infinite sum. We will see that any sequence that can be represented as a sum of exponentials can equivalently be represented by a rational z-transform. Such a z-transform is determined to within a constant multiplier by its zeros and its poles. For this example, there is one zero, at $z = 0$, and one pole, at $z = a$. The pole–zero plot and the region of convergence for Example 3.1 are shown in Figure 3.3 where a "∘" denotes the zero and an "×" the pole. For $|a| > 1$, the ROC does not include the unit circle, consistent with the fact that, for these values of a, the Fourier transform of the exponentially growing sequence $a^n u[n]$ does not converge.

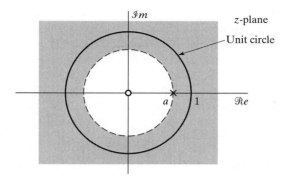

Figure 3.3 Pole–zero plot and region of convergence for Example 3.1.

Example 3.2   Left-Sided Exponential Sequence

Now let $x[n] = -a^n u[-n - 1]$. Since the sequence is nonzero only for $n \leq -1$, this is a *left-sided* sequence. Then

$$X(z) = -\sum_{n=-\infty}^{\infty} a^n u[-n-1] z^{-n} = -\sum_{n=-\infty}^{-1} a^n z^{-n}$$

$$= -\sum_{n=1}^{\infty} a^{-n} z^n = 1 - \sum_{n=0}^{\infty} (a^{-1} z)^n. \tag{3.12}$$

If $|a^{-1}z| < 1$ or, equivalently, $|z| < |a|$, the sum in Eq. (3.12) converges, and

$$X(z) = 1 - \frac{1}{1 - a^{-1}z} = \frac{1}{1 - az^{-1}} = \frac{z}{z - a}, \qquad |z| < |a|. \tag{3.13}$$

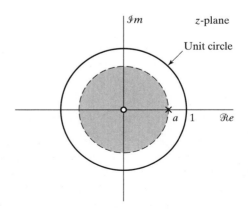

Figure 3.4 Pole–zero plot and region of convergence for Example 3.2.

The pole–zero plot and region of convergence for this example are shown in Figure 3.4. Note that for $|a| < 1$, the sequence $-a^n u[-n - 1]$ grows exponentially as $n \to -\infty$, and thus, the Fourier transform does not exist.

Comparing Eqs. (3.10) and (3.13) and Figures 3.3 and 3.4, we see that the sequences and, therefore, the infinite sums are different; however, the algebraic expressions for $X(z)$ and the corresponding pole–zero plots are identical in Examples 3.1 and 3.2. The z-transforms differ only in the region of convergence. This emphasizes the need for specifying both the algebraic expression and the region of convergence for the z-transform of a given sequence. Also, in both examples, the sequences were exponentials and the resulting z-transforms were rational. In fact, as is further suggested by the next example, $X(z)$ will be rational whenever $x[n]$ is a linear combination of real or complex exponentials.

Example 3.3   Sum of Two Exponential Sequences

Consider a signal that is the sum of two real exponentials:

$$x[n] = \left(\tfrac{1}{2}\right)^n u[n] + \left(-\tfrac{1}{3}\right)^n u[n]. \tag{3.14}$$

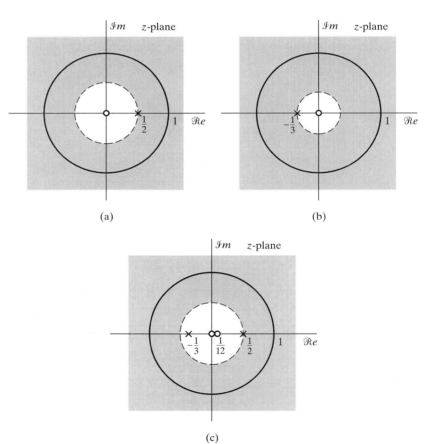

(a)   (b)

(c)

Figure 3.5 Pole–zero plot and region of convergence for the individual terms and the sum of terms in Examples 3.3 and 3.4. (a) $1/(1 - \tfrac{1}{2}z^{-1})$, $|z| > \tfrac{1}{2}$. (b) $1/(1 + \tfrac{1}{3}z^{-1})$, $|z| > \tfrac{1}{3}$. (c) $1/(1 - \tfrac{1}{2}z^{-1}) + 1/(1 + \tfrac{1}{3}z^{-1})$, $|z| > \tfrac{1}{2}$.

The z-transform is then

$$X(z) = \sum_{n=-\infty}^{\infty} \left\{ \left(\tfrac{1}{2}\right)^n u[n] + \left(-\tfrac{1}{3}\right)^n u[n] \right\} z^{-n}$$

$$= \sum_{n=-\infty}^{\infty} \left(\tfrac{1}{2}\right)^n u[n] z^{-n} + \sum_{n=-\infty}^{\infty} \left(-\tfrac{1}{3}\right)^n u[n] z^{-n} \qquad (3.15)$$

$$= \sum_{n=0}^{\infty} \left(\tfrac{1}{2} z^{-1}\right)^n + \sum_{n=0}^{\infty} \left(-\tfrac{1}{3} z^{-1}\right)^n$$

$$= \frac{1}{1 - \tfrac{1}{2} z^{-1}} + \frac{1}{1 + \tfrac{1}{3} z^{-1}} = \frac{2\left(1 - \tfrac{1}{12} z^{-1}\right)}{\left(1 - \tfrac{1}{2} z^{-1}\right)\left(1 + \tfrac{1}{3} z^{-1}\right)}$$

$$= \frac{2z\left(z - \tfrac{1}{12}\right)}{\left(z - \tfrac{1}{2}\right)\left(z + \tfrac{1}{3}\right)}. \qquad (3.16)$$

For convergence of $X(z)$, both sums in Eq. (3.15) must converge, which requires that both $\left|\tfrac{1}{2} z^{-1}\right| < 1$ and $\left|\left(-\tfrac{1}{3}\right) z^{-1}\right| < 1$ or, equivalently, $|z| > \tfrac{1}{2}$ and $|z| > \tfrac{1}{3}$. Thus, the region of convergence is the region of overlap, $|z| > \tfrac{1}{2}$. The pole–zero plot and ROC for the z-transform of each of the individual terms and for the combined signal are shown in Figure 3.5.

In each of the preceding examples, we started with the definition of the sequence and manipulated each of the infinite sums into a form whose sum could be recognized. When the sequence is recognized as a sum of exponential sequences of the form of Examples 3.1 and 3.2, the z-transform can be obtained much more simply using the fact that the z-transform operator is linear. Specifically, from the definition of the z-transform, Eq. (3.2), if $x[n]$ is the sum of two terms, then $X(z)$ will be the sum of the corresponding z-transforms of the individual terms. The ROC will be the intersection of the individual regions of convergence, i.e., the values of z for which both individual sums converge. We have already demonstrated this fact in obtaining Eq. (3.15) in Example 3.3. Example 3.4 shows how the z-transform in Example 3.3 can be obtained in a much more straightforward manner.

Example 3.4   Sum of Two Exponentials (Again)

Again, let $x[n]$ be given by Eq. (3.14). Then using the general result of Example 3.1 with $a = \tfrac{1}{2}$ and $a = -\tfrac{1}{3}$, the z-transforms of the two individual terms are easily seen to be

$$\left(\tfrac{1}{2}\right)^n u[n] \overset{z}{\longleftrightarrow} \frac{1}{1 - \tfrac{1}{2} z^{-1}}, \qquad |z| > \tfrac{1}{2}, \qquad (3.17)$$

$$\left(-\tfrac{1}{3}\right)^n u[n] \overset{z}{\longleftrightarrow} \frac{1}{1 + \tfrac{1}{3} z^{-1}}, \qquad |z| > \tfrac{1}{3}, \qquad (3.18)$$

and, consequently,

$$\left(\tfrac{1}{2}\right)^n u[n] + \left(-\tfrac{1}{3}\right)^n u[n] \overset{z}{\longleftrightarrow} \frac{1}{1 - \tfrac{1}{2} z^{-1}} + \frac{1}{1 + \tfrac{1}{3} z^{-1}}, \qquad |z| > \tfrac{1}{2}, \qquad (3.19)$$

as we had determined in Example 3.3. The pole–zero plot and ROC for the z-transform of each of the individual terms and for the combined signal are shown in Figure 3.5.

All the major points of Examples 3.1–3.4 are summarized in Example 3.5.

Example 3.5   Two-Sided Exponential Sequence

Consider the sequence

$$x[n] = \left(-\tfrac{1}{3}\right)^n u[n] - \left(\tfrac{1}{2}\right)^n u[-n-1]. \qquad (3.20)$$

Note that this sequence grows exponentially as $n \to -\infty$. Using the general result of Example 3.1 with $a = -\tfrac{1}{3}$, we obtain

$$\left(-\tfrac{1}{3}\right)^n u[n] \overset{\mathcal{Z}}{\longleftrightarrow} \frac{1}{1 + \tfrac{1}{3}z^{-1}}, \qquad |z| > \tfrac{1}{3},$$

and using the result of Example 3.2 with $a = \tfrac{1}{2}$ yields

$$-\left(\tfrac{1}{2}\right)^n u[-n-1] \overset{\mathcal{Z}}{\longleftrightarrow} \frac{1}{1 - \tfrac{1}{2}z^{-1}}, \qquad |z| < \tfrac{1}{2}.$$

Thus, by the linearity of the z-transform,

$$X(z) = \frac{1}{1 + \tfrac{1}{3}z^{-1}} + \frac{1}{1 - \tfrac{1}{2}z^{-1}}, \qquad \tfrac{1}{3} < |z|, \quad |z| < \tfrac{1}{2},$$

$$= \frac{2\left(1 - \tfrac{1}{12}z^{-1}\right)}{\left(1 + \tfrac{1}{3}z^{-1}\right)\left(1 - \tfrac{1}{2}z^{-1}\right)} = \frac{2z\left(z - \tfrac{1}{12}\right)}{\left(z + \tfrac{1}{3}\right)\left(z - \tfrac{1}{2}\right)}. \qquad (3.21)$$

In this case, the ROC is the annular region $\tfrac{1}{3} < |z| < \tfrac{1}{2}$. Note that the rational function in this example is identical to the rational function in Examples 3.3 and 3.4, but the ROC is different in the three cases. The pole–zero plot and the ROC for this example are shown in Figure 3.6.

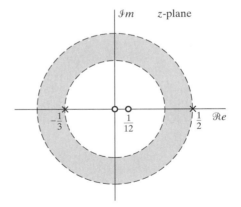

Figure 3.6   Pole–zero plot and region of convergence for Example 3.5.

Note that the ROC does not contain the unit circle, so the sequence in Eq. (3.20) does not have a Fourier transform.

In each of the preceding examples, we expressed the z-transform both as a ratio of polynomials in z and as a ratio of polynomials in z^{-1}. From the form of the definition of the z-transform as given in Eq. (3.2), we see that, for sequences that are zero for

$n < 0$, $X(z)$ involves only negative powers of z. Thus, for this class of signals, it is particularly convenient for $X(z)$ to be expressed in terms of polynomials in z^{-1} rather than z; however, even when $x[n]$ is nonzero for $n < 0$, $X(z)$ can still be expressed in terms of factors of the form $(1 - az^{-1})$. It should be remembered that such a factor introduces both a pole and a zero, as illustrated by the algebraic expressions in the preceding examples.

From these examples, it is easily seen that infinitely long exponential sequences have z-transforms that can be expressed as rational functions of either z or z^{-1}. The case where the sequence has finite length also has a rather simple form. If the sequence is nonzero only in the interval $N_1 \le n \le N_2$, the z-transform

$$X(z) = \sum_{n=N_1}^{N_2} x[n]z^{-n} \tag{3.22}$$

has no problems of convergence, as long as each of the terms $|x[n]z^{-n}|$ is finite. In general, it may not be possible to express the sum of a finite set of terms in a closed form, but in such cases it may be unnecessary. For example, it is easily seen that if $x[n] = \delta[n] + \delta[n-5]$, then $X(z) = 1 + z^{-5}$, which is finite for $|z| > 0$. An example of a case where a finite number of terms can be summed to produce a more compact representation of the z-transform is given in Example 3.6.

Example 3.6   Finite-Length Sequence

Consider the signal

$$x[n] = \begin{cases} a^n, & 0 \le n \le N-1, \\ 0, & \text{otherwise.} \end{cases}$$

Then

$$X(z) = \sum_{n=0}^{N-1} a^n z^{-n} = \sum_{n=0}^{N-1} (az^{-1})^n$$

$$= \frac{1 - (az^{-1})^N}{1 - az^{-1}} = \frac{1}{z^{N-1}} \frac{z^N - a^N}{z - a}, \tag{3.23}$$

where we have used the general formula in Eq. (2.56) to sum the finite series. The ROC is determined by the set of values of z for which

$$\sum_{n=0}^{N-1} |az^{-1}|^n < \infty.$$

Since there are only a finite number of nonzero terms, the sum will be finite as long as az^{-1} is finite, which in turn requires only that $|a| < \infty$ and $z \ne 0$. Thus, assuming that $|a|$ is finite, the ROC includes the entire z-plane, with the exception of the origin ($z = 0$). The pole–zero plot for this example, with $N = 16$ and a real and between zero and unity, is shown in Figure 3.7. Specifically, the N roots of the numerator polynomial are at

$$z_k = ae^{j(2\pi k/N)}, \qquad k = 0, 1, \ldots, N-1. \tag{3.24}$$

(Note that these values satisfy the equation $z^N = a^N$, and when $a = 1$, these complex values are the Nth roots of unity.) The zero at $k = 0$ cancels the pole at $z = a$.

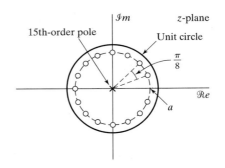

Figure 3.7 Pole–zero plot for Example 3.6 with $N = 16$ and a real such that $0 < a < 1$. The region of convergence in this example consists of all values of z except $z = 0$.

Consequently, there are no poles other than at the origin. The remaining zeros are at

$$z_k = ae^{j(2\pi k/N)}, \qquad k = 1, \ldots, N - 1. \tag{3.25}$$

TABLE 3.1   SOME COMMON z-TRANSFORM PAIRS

Sequence	Transform	ROC
1. $\delta[n]$	1	All z
2. $u[n]$	$\dfrac{1}{1 - z^{-1}}$	$|z| > 1$
3. $-u[-n - 1]$	$\dfrac{1}{1 - z^{-1}}$	$|z| < 1$
4. $\delta[n - m]$	z^{-m}	All z except 0 (if $m > 0$) or ∞ (if $m < 0$)
5. $a^n u[n]$	$\dfrac{1}{1 - az^{-1}}$	$|z| > |a|$
6. $-a^n u[-n - 1]$	$\dfrac{1}{1 - az^{-1}}$	$|z| < |a|$
7. $na^n u[n]$	$\dfrac{az^{-1}}{(1 - az^{-1})^2}$	$|z| > |a|$
8. $-na^n u[-n - 1]$	$\dfrac{az^{-1}}{(1 - az^{-1})^2}$	$|z| < |a|$
9. $[\cos \omega_0 n]u[n]$	$\dfrac{1 - [\cos \omega_0]z^{-1}}{1 - [2\cos \omega_0]z^{-1} + z^{-2}}$	$|z| > 1$
10. $[\sin \omega_0 n]u[n]$	$\dfrac{[\sin \omega_0]z^{-1}}{1 - [2\cos \omega_0]z^{-1} + z^{-2}}$	$|z| > 1$
11. $[r^n \cos \omega_0 n]u[n]$	$\dfrac{1 - [r\cos \omega_0]z^{-1}}{1 - [2r\cos \omega_0]z^{-1} + r^2 z^{-2}}$	$|z| > r$
12. $[r^n \sin \omega_0 n]u[n]$	$\dfrac{[r\sin \omega_0]z^{-1}}{1 - [2r\cos \omega_0]z^{-1} + r^2 z^{-2}}$	$|z| > r$
13. $\begin{cases} a^n, & 0 \le n \le N - 1, \\ 0, & \text{otherwise} \end{cases}$	$\dfrac{1 - a^N z^{-N}}{1 - az^{-1}}$	$|z| > 0$

The transform pairs corresponding to some of the preceding examples, as well as a number of other commonly encountered z-transform pairs, are summarized in Table 3.1. We will see that these basic transform pairs are very useful in finding z-transforms given a sequence or, conversely, in finding the sequence corresponding to a given z-transform.

3.2 PROPERTIES OF THE REGION OF CONVERGENCE FOR THE z-TRANSFORM

The examples of the previous section suggest that the properties of the region of convergence depend on the nature of the signal. These properties are summarized next, followed by some discussion and intuitive justification. We assume specifically that the algebraic expression for the z-transform is a rational function and that $x[n]$ has finite amplitude, except possibly at $n = \infty$ or $n = -\infty$.

PROPERTY 1:   The ROC is a ring or disk in the z-plane centered at the origin; i.e., $0 \leq r_R < |z| < r_L \leq \infty$.

PROPERTY 2:   The Fourier transform of $x[n]$ converges absolutely if and only if the ROC of the z-transform of $x[n]$ includes the unit circle.

PROPERTY 3:   The ROC cannot contain any poles.

PROPERTY 4:   If $x[n]$ is a *finite-duration sequence,* i.e., a sequence that is zero except in a finite interval $-\infty < N_1 \leq n \leq N_2 < \infty$, then the ROC is the entire z-plane, except possibly $z = 0$ or $z = \infty$.

PROPERTY 5:   If $x[n]$ is a *right-sided sequence,* i.e., a sequence that is zero for $n < N_1 < \infty$, the ROC extends outward from the *outermost* (i.e., largest magnitude) finite pole in $X(z)$ to (and possibly including) $z = \infty$.

PROPERTY 6:   If $x[n]$ is a *left-sided sequence,* i.e., a sequence that is zero for $n > N_2 > -\infty$, the ROC extends inward from the *innermost* (smallest magnitude) nonzero pole in $X(z)$ to (and possibly including) $z = 0$.

PROPERTY 7:   A *two-sided sequence* is an infinite-duration sequence that is neither right sided nor left sided. If $x[n]$ is a two-sided sequence, the ROC will consist of a ring in the z-plane, bounded on the interior and exterior by a pole and, consistent with property 3, not containing any poles.

PROPERTY 8:   The ROC must be a connected region.

As discussed in Section 3.1, property 1 results from the fact that convergence of Eq. (3.2) for a given $x[n]$ is dependent only on $|z|$, and property 2 is a consequence of the fact that Eq. (3.2) reduces to the Fourier transform when $|z| = 1$. Property 3 follows from the recognition that $X(z)$ is infinite at a pole and therefore, by definition, does not converge.

Properties 4 through 7 can all be developed more or less directly from the interpretation of the z-transform as the Fourier transform of the original sequence, modified by an exponential weighting. Let us first consider property 4. Figure 3.8 shows a finite-duration sequence and the exponential sequence r^{-n} for $1 < r$ (a decaying exponential) and for $0 < r < 1$ (a growing exponential). Convergence of the z-transform is implied by absolute summability of the sequence $x[n]|z|^{-n}$ or, equivalently, $x[n]r^{-n}$. It should

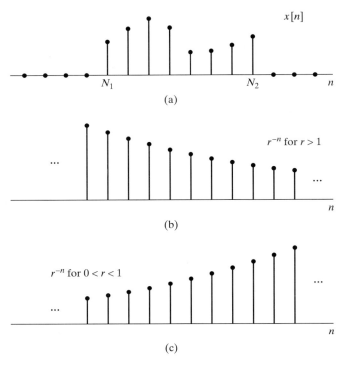

$x[n]$

(a)

r^{-n} for $r > 1$

(b)

r^{-n} for $0 < r < 1$

(c)

Figure 3.8   Finite-length sequence and weighting sequences implicit in convergence of the z-transform. (a) The finite-length sequence $x[n]$. (b) Weighting sequence r^{-n} for $1 < r$. (c) Weighting sequence r^{-n} for $0 < r < 1$.

be evident from Figure 3.8 that, since $x[n]$ has only a finite number of nonzero values, as long as each of these values is finite, $x[n]$ will be absolutely summable. Furthermore, this will not be affected by the exponential weighting if the weighting sequence has finite amplitude in the interval where $x[n]$ is nonzero, i.e., $N_1 \leq n \leq N_2$. Therefore, for a finite-duration sequence, $x[n]r^{-n}$ will be absolutely summable for $0 < r < \infty$. The only possible complication arises for $r = 0$ or for $r = \infty$. If $x[n]$ is nonzero for any positive values of n (i.e., if $N_2 > 0$), and if r, or, equivalently, $|z|$, is zero, then $x[n]r^{-n}$ will be infinite for $0 < n \leq N_2$. Correspondingly, if $x[n]$ is nonzero for any negative values of n (i.e., if $N_1 < 0$), then $x[n]r^{-n}$ will be infinite for $N_1 \leq n < 0$ if r, or, equivalently, $|z|$, is infinite.

Property 5 can be interpreted in a somewhat similar manner. Figure 3.9 illustrates a right-sided sequence and the exponential sequence r^{-n} for two different values of r. A right-sided sequence is zero prior to some value of n, say, N_1. If the circle $|z| = r_0$ is in the ROC, then $x[n]r_0^{-n}$ is absolutely summable, or equivalently, the Fourier transform of $x[n]r_0^{-n}$ converges. Since $x[n]$ is right sided, the sequence $x[n]r_1^{-n}$ will also be absolutely summable if r_1^{-n} decays faster than r_0^{-n}. Specifically, as illustrated in Figure 3.9, this more rapid exponential decay will further attenuate sequence values for positive values of n and cannot cause sequence values for negative values of n to become unbounded, since $x[n]z^{-n} = 0$ for $n < N_1$. Based on this property, we can conclude that, for a right-sided sequence, the ROC extends outward from some circle in the z-plane, concentric with the origin. This circle, in fact, is at the outermost pole in $X(z)$. To see this, assume that the poles occur at $z = d_1, \ldots, d_N$, with d_1 having the smallest magnitude, i.e., corresponding

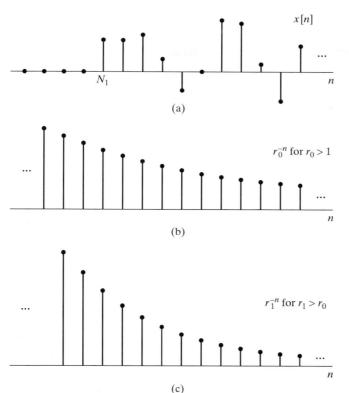

Figure 3.9   Right-sided sequence and weighting sequences implicit in convergence of the z-transform. (a) The right-sided sequence $x[n]$.
(b) Weighting sequence r_0^{-n} for $1 < r_0$.
(c) Weighting sequence r_1^{-n} for $r_1 > r_0$.

to the innermost pole, and d_N having the largest magnitude, i.e., corresponding to the outermost pole. To simplify the argument, we will assume that all the poles are distinct, although the argument can be easily generalized for multiple-order poles. As we will see in Section 3.3, for $N_1 \leq n$, $x[n]$ will consist of a sum of exponentials of the form

$$x[n] = \sum_{k=1}^{N} A_k(d_k)^n, \qquad n \geq N_1. \tag{3.26}$$

The least rapidly increasing of these exponentials, as n increases, is the one corresponding to the innermost pole, i.e., d_1, and the most slowly decaying (or most rapidly growing) is the one corresponding to the outermost pole, i.e., d_N. Now let us consider $x[n]$ with the exponential weighting r^{-n} applied, i.e.,

$$x[n]r^{-n} = r^{-n} \sum_{k=1}^{N} A_k(d_k)^n, \qquad n \geq N_1, \tag{3.27}$$

$$= \sum_{k=1}^{N} A_k(d_k r^{-1})^n, \qquad n \geq N_1. \tag{3.28}$$

Absolute summability of $x[n]r^{-n}$ requires that each exponential in Eq. (3.28) be absolutely summable; i.e.,

$$\sum_{n=N_1}^{\infty} |d_k r^{-1}|^n < \infty, \qquad k = 1, \ldots, N, \tag{3.29}$$

or equivalently,

$$|r| > |d_k|, \qquad k = 1, \ldots, N. \tag{3.30}$$

Since the outermost pole, d_N, is the one with the largest absolute value,

$$|r| > |d_N|; \tag{3.31}$$

i.e., the ROC is outside the outermost pole, extending to infinity. If $N_1 < 0$, the ROC will not include $|z| = \infty$, since r^{-n} is infinite for r infinite and n negative.

As suggested by the preceding discussion, it is possible to be very precise about property 5 (as well as the associated properties 6 and 7). The essence of the argument, however, is that for a sum of right-sided exponential sequences with an exponential weighting applied, the exponential weighting must be restricted so that all of the exponentially weighted terms decay with increasing n.

For property 6, which is concerned with left-sided sequences, an exactly parallel argument can be carried out. Here, however, $x[n]$ will consist of a sum of exponentials of the same form as Eq. (3.28), but for $n \leq N_2$; i.e.,

$$x[n] = \sum_{k=1}^{N} A_k (d_k)^n, \qquad n \leq N_2, \tag{3.32}$$

or, with exponential weighting,

$$x[n]r^{-n} = \sum_{k=1}^{N} A_k (d_k r^{-1})^n, \qquad n \leq N_2. \tag{3.33}$$

Since $x[n]$ now extends to $-\infty$ along the negative n-axis, r must be restricted so that for each d_k, the exponential sequence $(d_k r^{-1})^n$ decays to zero as n *decreases* toward $-\infty$. Equivalently,

$$|r| < |d_k|, \qquad k = 1, \ldots, N,$$

or, since d_1 has the smallest magnitude,

$$|r| < |d_1|; \tag{3.34}$$

i.e., the ROC is inside the innermost pole. If the left-sided sequence has nonzero values for positive values of n, then the ROC will not include the origin, $z = 0$.

For right-sided sequences, the ROC is dictated by the exponential weighting required to have all exponential terms decay to zero for increasing n; for left-sided sequences, the exponential weighting must be such that all exponential terms decay to zero for decreasing n. For two-sided sequences, the exponential weighting needs to be balanced, since if it decays too fast for increasing n, it may grow too quickly for decreasing n and vice versa. More specifically, for two-sided sequences, some of the poles contribute only for $n > 0$ and the rest only for $n < 0$. The region of convergence is

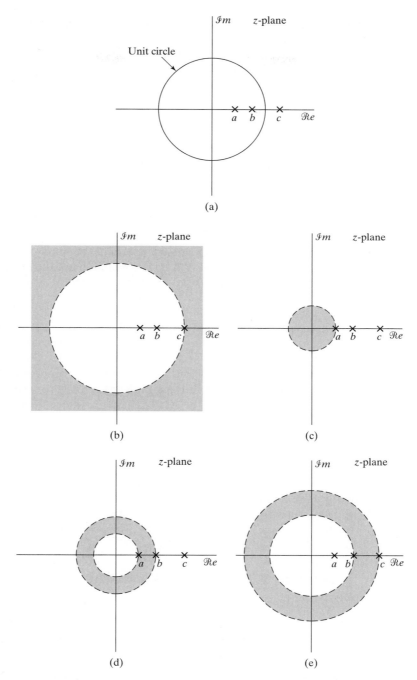

Figure 3.10   Examples of four z- transforms with the same pole–zero locations, illustrating the different possibilities for the region of convergence. Each ROC corresponds to a different sequence: (b) to a right-sided sequence, (c) to a left-sided sequence, (d) to a two-sided sequence, and (e) to a two-sided sequence.

bounded on the inside by the pole with the largest magnitude that contributes for $n > 0$ and on the outside by the pole with the smallest magnitude that contributes for $n < 0$.

Property 8 is somewhat more difficult to develop formally, but, at least intuitively, it is strongly suggested by our discussion of properties 4 through 7. Any infinite two-sided sequence can be represented as a sum of a right-sided part (say, for $n \geq 0$) and a left-sided part that includes everything not included in the right-sided part. The right-sided part will have an ROC given by Eq. (3.31), while the ROC of the left-sided part will be given by Eq. (3.34). The ROC of the entire two-sided sequence must be the intersection of these two regions. Thus, if such an intersection exists, it will always be a simply connected annular region of the form

$$r_R < |z| < r_L.$$

There is a possibility of no overlap between the regions of convergence of the right- and left-sided parts; i.e., $r_L < r_R$. An example is the sequence $x[n] = \left(\frac{1}{2}\right)^n u[n] - \left(-\frac{1}{3}\right)^n u[-n-1]$. In this case, the z-transform of the sequence simply does not exist. If such cases arise, however, it may still be possible to use the z-transform by considering the sequence to be the sum of two sequences, each of which has a z-transform, but the two transforms cannot be combined in algebraic expressions, since they have no common ROC.

As we indicated in comparing Examples 3.1 and 3.2, the algebraic expression or pole–zero pattern does not completely specify the z-transform of a sequence; i.e., the ROC must also be specified. The properties considered in this section limit the possible ROC's that can be associated with a given pole–zero pattern. To illustrate, consider the pole–zero pattern shown in Figure 3.10(a). From properties 1, 3, and 8, there are only four possible choices for the ROC. These are indicated in Figures 3.10(b), (c), (d), and (e), each being associated with a different sequence. Specifically, Figure 3.10(b) corresponds to a right-sided sequence, Figure 3.10(c) to a left-sided sequence, and Figures 3.10(d) and 3.10(e) to two different two-sided sequences. If we assume, as indicated in Figure 3.10(a), that the unit circle falls between the pole at $z = b$ and the pole at $z = c$, then the only one of the four cases for which the Fourier transform would converge is that in Figure 3.10(e).

In representing a sequence through its z-transform, it is sometimes convenient to specify the ROC implicitly through an appropriate time-domain property of the sequence. This is illustrated in Example 3.7.

Example 3.7   Stability, Causality, and the ROC

Consider a system with impulse response $h[n]$ for which the z-transform $H(z)$ has the pole–zero plot shown in Figure 3.11. There are three possible ROC's consistent with properties 1–8 that can be associated with this pole-zero plot. However, if we state in addition that the system is stable (or equivalently, that $h[n]$ is absolutely summable and therefore has a Fourier transform), then the ROC must include the unit circle. Thus, stability of the system and properties 1–8 imply that the ROC is the region $\frac{1}{2} < |z| < 2$. Note that as a consequence, $h[n]$ is two sided, and therefore, the system is not causal.

If we state instead that the system is causal, and therefore that $h[n]$ is right sided, then property 5 would require that the ROC be the region $|z| > 2$. Under this

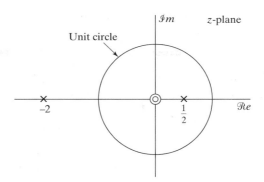

Figure 3.11   Pole–zero plot for the system function in Example 3.7.

condition, the system would not be stable; i.e., for this specific pole–zero plot, there is no ROC that would imply that the system is both stable and causal.

3.3 THE INVERSE z-TRANSFORM

One of the important roles of the z-transform is in the analysis of discrete-time linear systems. Often, this analysis involves finding the z-transform of sequences and, after some manipulation of the algebraic expressions, finding the inverse z-transform. There are a number of formal and informal ways of determining the inverse z-transform from a given algebraic expression and associated region of convergence. There is a formal inverse z-transform expression that is based on the Cauchy integral theorem (Churchill and Brown, 1990). However, for the typical kinds of sequences and z-transforms that we will encounter in the analysis of discrete linear time-invariant systems, less formal procedures are sufficient and preferable. In Sections 3.3.1–3.3.3 we consider some of these procedures, specifically the inspection method, partial fraction expansion, and power series expansion.

3.3.1 Inspection Method

The inspection method consists simply of becoming familiar with, or recognizing "by inspection," certain transform pairs. For example, in Section 3.1, we evaluated the z-transform for sequences of the form $x[n] = a^n u[n]$, where a can be either real or complex. Sequences of this form arise quite frequently, and consequently, it is particularly useful to make direct use of the transform pair

$$a^n u[n] \overset{z}{\longleftrightarrow} \frac{1}{1 - az^{-1}}, \qquad |z| > |a|. \tag{3.35}$$

If we need to find the inverse z-transform of

$$X(z) = \left(\frac{1}{1 - \frac{1}{2}z^{-1}} \right), \qquad |z| > \tfrac{1}{2}, \tag{3.36}$$

and we recall the z-transform pair of Eq. (3.35), we would recognize "by inspection" the associated sequence as $x[n] = \left(\frac{1}{2}\right)^n u[n]$. If the ROC associated with $X(z)$ in Eq. (3.36) had been $|z| < \frac{1}{2}$, we can recall transform pair 6 in Table 3.1 to find by inspection that $x[n] = -\left(\frac{1}{2}\right)^n u[-n-1]$.

Tables of z-transforms, such as Table 3.1, are invaluable in applying the inspection method. If the table is extensive, it may be possible to express a given z-transform as a sum of terms, each of whose inverse is given in the table. If so, the inverse transform (i.e., the corresponding sequence) can be written from the table.

3.3.2 Partial Fraction Expansion

As already described, inverse z-transforms can be found by inspection if the z-transform expression is recognized or tabulated. Sometimes $X(z)$ may not be given explicitly in an available table, but it may be possible to obtain an alternative expression for $X(z)$ as a sum of simpler terms, each of which is tabulated. This is the case for any rational function, since we can obtain a partial fraction expansion and easily identify the sequences corresponding to the individual terms.

To see how to obtain a partial fraction expansion, let us assume that $X(z)$ is expressed as a ratio of polynomials in z^{-1}; i.e.,

$$X(z) = \frac{\displaystyle\sum_{k=0}^{M} b_k z^{-k}}{\displaystyle\sum_{k=0}^{N} a_k z^{-k}}. \tag{3.37}$$

Such z-transforms arise frequently in the study of linear time-invariant systems. An equivalent expression is

$$X(z) = \frac{z^N \displaystyle\sum_{k=0}^{M} b_k z^{M-k}}{z^M \displaystyle\sum_{k=0}^{N} a_k z^{N-k}}. \tag{3.38}$$

Equation (3.38) explicitly shows that for such functions, there will be M zeros and N poles at nonzero locations in the z-plane. In addition, there will be either $M - N$ poles at $z = 0$ if $M > N$ or $N - M$ zeros at $z = 0$ if $N > M$. In other words, z-transforms of the form of Eq. (3.37) always have the same number of poles and zeros in the finite z-plane, and there are no poles or zeros at $z = \infty$. To obtain the partial fraction expansion of $X(z)$ in Eq. (3.37), it is most convenient to note that $X(z)$ could be expressed in the form

$$X(z) = \frac{b_0}{a_0} \frac{\displaystyle\prod_{k=1}^{M}(1 - c_k z^{-1})}{\displaystyle\prod_{k=1}^{N}(1 - d_k z^{-1})}, \tag{3.39}$$

where the c_k's are the nonzero zeros of $X(z)$ and the d_k's are the nonzero poles of $X(z)$. If $M < N$ and the poles are all first order, then $X(z)$ can be expressed as

$$X(z) = \sum_{k=1}^{N} \frac{A_k}{1 - d_k z^{-1}}. \tag{3.40}$$

Obviously, the common denominator of the fractions in Eq. (3.40) is the same as the denominator in Eq. (3.39). Multiplying both sides of Eq. (3.40) by $(1 - d_k z^{-1})$ and evaluating for $z = d_k$ shows that the coefficients, A_k, can be found from

$$A_k = (1 - d_k z^{-1}) X(z) \big|_{z=d_k}. \tag{3.41}$$

Example 3.8   Second-Order z-Transform

Consider a sequence $x[n]$ with z-transform

$$X(z) = \frac{1}{\left(1 - \frac{1}{4} z^{-1}\right) \left(1 - \frac{1}{2} z^{-1}\right)}, \qquad |z| > \frac{1}{2}. \tag{3.42}$$

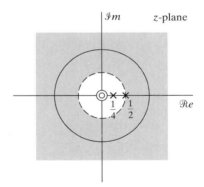

Figure 3.12   Pole-zero plot and ROC for Example 3.8.

The pole–zero plot for $X(z)$ is shown in Figure 3.12. From the region of convergence and property 5, Section 3.2, we see that $x[n]$ is a right-sided sequence. Since the poles are both first order, $X(z)$ can be expressed in the form of Eq. (3.40); i.e.,

$$X(z) = \frac{A_1}{\left(1 - \frac{1}{4} z^{-1}\right)} + \frac{A_2}{\left(1 - \frac{1}{2} z^{-1}\right)}.$$

From Eq. (3.41),

$$A_1 = \left(1 - \frac{1}{4} z^{-1}\right) X(z) \big|_{z=1/4} = -1,$$

$$A_2 = \left(1 - \frac{1}{2} z^{-1}\right) X(z) \big|_{z=1/2} = 2.$$

Therefore,

$$X(z) = \frac{-1}{\left(1 - \frac{1}{4} z^{-1}\right)} + \frac{2}{\left(1 - \frac{1}{2} z^{-1}\right)}.$$

Since $x[n]$ is right sided, the ROC for each term extends outward from the outermost pole. From Table 3.1 and the linearity of the z-transform, it then follows that

$$x[n] = 2\left(\frac{1}{2}\right)^n u[n] - \left(\frac{1}{4}\right)^n u[n].$$

Clearly, the numerator that would result from adding the terms in Eq. (3.40) would be at most of degree $(N-1)$ in the variable z^{-1}. If $M \geq N$, then a polynomial must be added to the right-hand side of Eq. (3.40), the order of which is $(M-N)$. Thus, for $M \geq N$, the complete partial fraction expansion would have the form

$$X(z) = \sum_{r=0}^{M-N} B_r z^{-r} + \sum_{k=1}^{N} \frac{A_k}{1 - d_k z^{-1}}. \tag{3.43}$$

If we are given a rational function of the form of Eq. (3.37), with $M \geq N$, the B_r's can be obtained by long division of the numerator by the denominator, with the division process terminating when the remainder is of lower degree than the denominator. The A_k's can still be obtained with Eq. (3.41).

If $X(z)$ has multiple-order poles and $M \geq N$, Eq. (3.43) must be further modified. In particular, if $X(z)$ has a pole of order s at $z = d_i$ and all the other poles are first-order, then Eq. (3.43) becomes

$$X(z) = \sum_{r=0}^{M-N} B_r z^{-r} + \sum_{k=1, k \neq i}^{N} \frac{A_k}{1 - d_k z^{-1}} + \sum_{m=1}^{s} \frac{C_m}{(1 - d_i z^{-1})^m}. \tag{3.44}$$

The coefficients A_k and B_r are obtained as before. The coefficients C_m are obtained from the equation

$$C_m = \frac{1}{(s-m)!(-d_i)^{s-m}} \left\{ \frac{d^{s-m}}{dw^{s-m}} [(1 - d_i w)^s X(w^{-1})] \right\}_{w=d_i^{-1}}. \tag{3.45}$$

Equation (3.44) gives the most general form for the partial fraction expansion of a rational z-transform expressed as a function of z^{-1} for the case $M \geq N$ and for d_i a pole of order s. If there are several multiple-order poles, there will be a term like the third sum in Eq. (3.44) for each multiple-order pole. If there are no multiple-order poles, Eq. (3.44) reduces to Eq. (3.43). If the order of the numerator is less than the order of the denominator ($M < N$), then the polynomial term disappears from Eqs. (3.43) and (3.44).

It should be emphasized that we could have achieved the same results by assuming that the rational z-transform was expressed as a function of z instead of z^{-1}. That is, instead of factors of the form $(1 - az^{-1})$, we could have considered factors of the form $(z - a)$. This would lead to a set of equations similar in form to Eqs. (3.39)–(3.45) that would be convenient for use with a table of z-transforms expressed in terms of z. Since Table 3.1 is expressed in terms of z^{-1}, the development we pursued is more useful.

To see how to find the sequence corresponding to a given rational z-transform, let us suppose that $X(z)$ has only first-order poles, so that Eq. (3.43) is the most general form of the partial fraction expansion. To find $x[n]$, we first note that the z-transform operation is linear, so that the inverse transform of individual terms can be found and then added together to form $x[n]$.

The terms $B_r z^{-r}$ correspond to shifted and scaled impulse sequences, i.e., terms of the form $B_r \delta[n-r]$. The fractional terms correspond to exponential sequences. To decide whether a term

$$\frac{A_k}{1 - d_k z^{-1}}$$

corresponds to $(d_k)^n u[n]$ or $-(d_k)^n u[-n-1]$, we must use the properties of the region of convergence that were discussed in Section 3.2. From that discussion, it follows that if $X(z)$ has only simple poles and the ROC is of the form $r_R < |z| < r_L$, then a given pole d_k will correspond to a right-sided exponential $(d_k)^n u[n]$ if $|d_k| < r_R$, and it will correspond to a left-sided exponential if $|d_k| > r_L$. Thus, the region of convergence can be used to sort the poles. Multiple-order poles also are divided into left-sided and right-sided contributions in the same way. The use of the region of convergence in finding inverse z-transforms from the partial fraction expansion is illustrated by the following examples.

Example 3.9   Inverse by Partial Fractions

To illustrate the case in which the partial fraction expansion has the form of Eq. (3.43), consider a sequence $x[n]$ with z-transform

$$X(z) = \frac{1 + 2z^{-1} + z^{-2}}{1 - \frac{3}{2}z^{-1} + \frac{1}{2}z^{-2}} = \frac{(1 + z^{-1})^2}{\left(1 - \frac{1}{2}z^{-1}\right)\left(1 - z^{-1}\right)}, \qquad |z| > 1. \qquad (3.46)$$

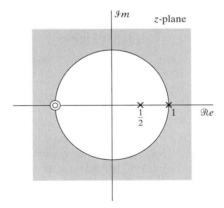

Figure 3.13   Pole–zero plot for the z-transform in Example 3.9.

The pole–zero plot for $X(z)$ is shown in Figure 3.13. From the region of convergence and property 5, Section 3.2, it is clear that $x[n]$ is a right-sided sequence. Since $M = N = 2$ and the poles are all first order, $X(z)$ can be represented as

$$X(z) = B_0 + \frac{A_1}{1 - \frac{1}{2}z^{-1}} + \frac{A_2}{1 - z^{-1}}.$$

The constant B_0 can be found by long division:

$$\frac{1}{2}z^{-2} - \frac{3}{2}z^{-1} + 1 \,\overline{\big)\, z^{-2} + 2z^{-1} + 1} \quad \overset{\displaystyle 2}{}$$

$$\frac{z^{-2} - 3z^{-1} + 2}{5z^{-1} - 1}$$

Since the remainder after one step of long division is of degree 1 in the variable z^{-1}, it is not necessary to continue to divide. Thus, $X(z)$ can be expressed as

$$X(z) = 2 + \frac{-1 + 5z^{-1}}{\left(1 - \frac{1}{2}z^{-1}\right)\left(1 - z^{-1}\right)}. \tag{3.47}$$

Now the coefficients A_1 and A_2 can be found by applying Eq. (3.41) to Eq. (3.46) or, equivalently, Eq. (3.47). Using Eq. (3.47), we obtain

$$A_1 = \left[\left(2 + \frac{-1 + 5z^{-1}}{\left(1 - \frac{1}{2}z^{-1}\right)\left(1 - z^{-1}\right)}\right)\left(1 - \frac{1}{2}z^{-1}\right)\right]_{z=1/2} = -9,$$

$$A_2 = \left[\left(2 + \frac{-1 + 5z^{-1}}{\left(1 - \frac{1}{2}z^{-1}\right)\left(1 - z^{-1}\right)}\right)\left(1 - z^{-1}\right)\right]_{z=1} = 8.$$

Therefore,

$$X(z) = 2 - \frac{9}{1 - \frac{1}{2}z^{-1}} + \frac{8}{1 - z^{-1}}. \tag{3.48}$$

From Table 3.1, we see that since the ROC is $|z| > 1$,

$$2 \overset{z}{\longleftrightarrow} 2\delta[n],$$

$$\frac{1}{1 - \frac{1}{2}z^{-1}} \overset{z}{\longleftrightarrow} \left(\tfrac{1}{2}\right)^n u[n],$$

$$\frac{1}{1 - z^{-1}} \overset{z}{\longleftrightarrow} u[n].$$

Thus, from the linearity of the z-transform,

$$x[n] = 2\delta[n] - 9\left(\tfrac{1}{2}\right)^n u[n] + 8u[n].$$

In Section 3.4 we will discuss and illustrate a number of properties of the z-transform that, in combination with the partial fraction expansion, provide a means for determining the inverse z-transform from a given rational algebraic expression and associated ROC, even when $X(z)$ is not exactly in the form of Eq. (3.39).

3.3.3 Power Series Expansion

The defining expression for the z-transform is a Laurent series where the sequence values $x[n]$ are the coefficients of z^{-n}. Thus, if the z-transform is given as a power series in the form

$$X(z) = \sum_{n=-\infty}^{\infty} x[n]z^{-n} \tag{3.49}$$

$$= \cdots + x[-2]z^2 + x[-1]z + x[0] + x[1]z^{-1} + x[2]z^{-2} + \cdots,$$

we can determine any particular value of the sequence by finding the coefficient of the appropriate power of z^{-1}. We have already used this approach in finding the inverse transform of the polynomial part of the partial fraction expansion when $M \geq N$. This approach is also very useful for finite-length sequences where $X(z)$ may have no simpler form than a polynomial in z^{-1}.

Example 3.10   Finite-Length Sequence

Suppose $X(z)$ is given in the form

$$X(z) = z^2 \left(1 - \tfrac{1}{2}z^{-1}\right)(1 + z^{-1})(1 - z^{-1}). \tag{3.50}$$

Although $X(z)$ is obviously a rational function, its only poles are at $z = 0$, so a partial fraction expansion according to the technique of Section 3.3.2 is not appropriate. However, by multiplying the factors of Eq. (3.50), we can express $X(z)$ as

$$X(z) = z^2 - \tfrac{1}{2}z - 1 + \tfrac{1}{2}z^{-1}.$$

Therefore, by inspection, $x[n]$ is seen to be

$$x[n] = \begin{cases} 1, & n = -2, \\ -\tfrac{1}{2}, & n = -1, \\ -1, & n = 0, \\ \tfrac{1}{2}, & n = 1, \\ 0, & \text{otherwise.} \end{cases}$$

Equivalently,

$$x[n] = \delta[n+2] - \tfrac{1}{2}\delta[n+1] - \delta[n] + \tfrac{1}{2}\delta[n-1].$$

In finding z-transforms of a sequence, we generally seek to sum the power series of Eq. (3.49) to obtain a simpler mathematical expression, e.g., a rational function. If we wish to use the power series to find the sequence corresponding to a given $X(z)$ expressed in closed form, we must expand $X(z)$ back into a power series. Many power series have been tabulated for transcendental functions such as log, sin, sinh, etc. In some cases such power series can have a useful interpretation as z-transforms, as we illustrate in Example 3.11. For rational z-transforms, a power series expansion can be obtained by long division, as illustrated in Examples 3.12 and 3.13.

Example 3.11   Inverse Transform by Power Series Expansion

Consider the z-transform

$$X(z) = \log(1 + az^{-1}), \qquad |z| > |a|. \tag{3.51}$$

Using the power series expansion for $\log(1 + x)$, with $|x| < 1$, we obtain

$$X(z) = \sum_{n=1}^{\infty} \frac{(-1)^{n+1} a^n z^{-n}}{n}.$$

Therefore,

$$x[n] = \begin{cases} (-1)^{n+1}\dfrac{a^n}{n}, & n \geq 1, \\ 0, & n \leq 0. \end{cases} \qquad (3.52)$$

Example 3.12   Power Series Expansion by Long Division

Consider the z-transform

$$X(z) = \frac{1}{1 - az^{-1}}, \qquad |z| > |a|. \qquad (3.53)$$

Since the region of convergence is the exterior of a circle, the sequence is a right-sided one. Furthermore, since $X(z)$ approaches a finite constant as z approaches infinity, the sequence is causal. Thus, we divide, so as to obtain a series in powers of z^{-1}. Carrying out the long division, we obtain

$$
\begin{array}{r}
1 + az^{-1} + a^2z^{-2} + \cdots \\
1 - az^{-1} \,\overline{\big)\, 1 } \\
\underline{1 - az^{-1}} \\
az^{-1} \\
\underline{az^{-1} - a^2z^{-2}} \\
a^2z^{-2} \quad \cdots
\end{array}
$$

or

$$\frac{1}{1 - az^{-1}} = 1 + az^{-1} + a^2z^{-2} + \cdots.$$

Hence, $x[n] = a^n u[n]$.

Example 3.13   Power Series Expansion for a Left-Sided Sequence

As another example, we can consider the same ratio of polynomials as in Eq. (3.53), but with a different region of convergence:

$$X(z) = \frac{1}{1 - az^{-1}}, \qquad |z| < |a|. \qquad (3.54)$$

Because of the region of convergence, the sequence is a left-sided one, and since $X(z)$ at $z = 0$ is finite, the sequence is zero for $n > 0$. Thus, we divide, so as to obtain a series in powers of z as follows:

$$
\begin{array}{r}
-a^{-1}z - a^{-2}z^{-2} - \cdots \\
-a + z \,\overline{\big)\, z } \\
\underline{z - a^{-1}z^2} \\
a^{-1}z^2
\end{array}
$$

Therefore, $x[n] = -a^n u[-n - 1]$.

3.4 z-TRANSFORM PROPERTIES

Many of the properties of the z-transform are particularly useful in studying discrete-time signals and systems. For example, these properties are often used in conjunction with the inverse z-transform techniques discussed in Sec. 3.3 to obtain the inverse z-transform of more complicated expressions. In Chapter 5 we will see that the properties also form the basis for transforming linear constant-coefficient difference equations to algebraic equations in terms of the transform variable z, the solution to which can then be obtained using the inverse z-transform. In this section, we consider some of the most frequently used properties. In the following discussion, $X(z)$ denotes the z-transform of $x[n]$, and the ROC of $X(z)$ is indicated by R_x; i.e.,

$$x[n] \overset{z}{\longleftrightarrow} X(z), \qquad \text{ROC} = R_x.$$

As we have seen, R_x represents a set of values of z such that $r_R < |z| < r_L$. For properties that involve two sequences and associated z-transforms, the transform pairs will be denoted as

$$x_1[n] \overset{z}{\longleftrightarrow} X_1(z), \qquad \text{ROC} = R_{x_1},$$
$$x_2[n] \overset{z}{\longleftrightarrow} X_2(z), \qquad \text{ROC} = R_{x_2}.$$

3.4.1 Linearity

The linearity property states that

$$ax_1[n] + bx_2[n] \overset{z}{\longleftrightarrow} aX_1(z) + bX_2(z), \qquad \text{ROC contains } R_{x_1} \cap R_{x_2},$$

and follows directly from the z-transform definition, Eq. (3.2). As indicated, the region of convergence is at least the intersection of the individual regions of convergence. For sequences with rational z-transforms, if the poles of $aX_1(z) + bX_2(z)$ consist of all the poles of $X_1(z)$ and $X_2(z)$ (i.e., if there is no pole–zero cancellation), then the region of convergence will be exactly equal to the overlap of the individual regions of convergence. If the linear combination is such that some zeros are introduced that cancel poles, then the region of convergence may be larger. A simple example of this occurs when $x_1[n]$ and $x_2[n]$ are of infinite duration, but the linear combination is of finite duration. In this case the region of convergence of the linear combination is the entire z-plane, with the possible exception of $z = 0$ or $z = \infty$. An example was given in Example 3.6, where $x[n]$ can be expressed as

$$x[n] = a^n u[n] - a^n u[n - N].$$

Both $a^n u[n]$ and $a^n u[n - N]$ are infinite-extent right-sided sequences, and their z-transforms have a pole at $z = a$. Therefore, their individual regions of convergence would both be $|z| > |a|$. However, as shown in Example 3.6, the pole at $z = a$ is canceled by a zero at $z = a$, and therefore, the ROC extends to the entire z-plane, with the exception of $z = 0$.

We have already exploited the linearity property in our previous discussion of the use of the partial fraction expansion for evaluating the inverse z-transform. With that procedure, $X(z)$ is expanded into a sum of simpler terms, and through linearity, the inverse z-transform is the sum of the inverse transforms of each of these terms.

3.4.2 Time Shifting

According to the time-shifting property,

$$x[n - n_0] \overset{z}{\longleftrightarrow} z^{-n_0} X(z), \qquad \text{ROC} = R_x (\text{except for the possible addition or deletion of } z = 0 \text{ or } z = \infty).$$

The quantity n_0 is an integer. If n_0 is positive, the original sequence $x[n]$ is shifted right, and if n_0 is negative, $x[n]$ is shifted left. As in the case of linearity, the ROC can be changed, since the factor z^{-n_0} can alter the number of poles at $z = 0$ or $z = \infty$.

The derivation of this property follows directly from the z-transform expression in Eq. (3.2). Specifically, if $y[n] = x[n - n_0]$, the corresponding z-transform is

$$Y(z) = \sum_{n=-\infty}^{\infty} x[n - n_0] z^{-n}.$$

With the substitution of variables $m = n - n_0$,

$$Y(z) = \sum_{m=-\infty}^{\infty} x[m] z^{-(m+n_0)}$$

$$= z^{-n_0} \sum_{m=-\infty}^{\infty} x[m] z^{-m},$$

or

$$Y(z) = z^{-n_0} X(z).$$

The time-shifting property is often useful, in conjunction with other properties and procedures, for obtaining the inverse z-transform. We illustrate with an example.

Example 3.14   Shifted Exponential Sequence

Consider the z-transform

$$X(z) = \frac{1}{z - \frac{1}{4}}, \qquad |z| > \tfrac{1}{4}.$$

From the ROC, we identify this as corresponding to a right-sided sequence. We can first rewrite $X(z)$ in the form

$$X(z) = \frac{z^{-1}}{1 - \frac{1}{4} z^{-1}}, \qquad |z| > \tfrac{1}{4}. \tag{3.55}$$

This z-transform is of the form of Eq. (3.39) with $M = N = 1$, and its expansion in the form of Eq. (3.43) is

$$X(z) = -4 + \frac{4}{1 - \frac{1}{4} z^{-1}}. \tag{3.56}$$

From Eq. (3.56), it follows that $x[n]$ can be expressed as

$$x[n] = -4\delta[n] + 4 \left(\tfrac{1}{4}\right)^n u[n]. \tag{3.57}$$

An expression for $x[n]$ can be obtained more directly by applying the time-shifting property. First, $X(z)$ can be written as

$$X(z) = z^{-1}\left(\frac{1}{1 - \frac{1}{4}z^{-1}}\right), \qquad |z| > \tfrac{1}{4}. \tag{3.58}$$

From the time-shifting property, we recognize the factor z^{-1} in Eq. (3.58) as being associated with a time shift of one sample to the right of the sequence $\left(\frac{1}{4}\right)^n u[n]$; i.e.,

$$x[n] = \left(\tfrac{1}{4}\right)^{n-1} u[n - 1]. \tag{3.59}$$

It is easily verified that Eqs. (3.57) and (3.59) are the same for all values of n; i.e., they represent the same sequence.

3.4.3 Multiplication by an Exponential Sequence

The exponential multiplication property is expressed mathematically as

$$z_0^n x[n] \xleftrightarrow{\;\mathcal{Z}\;} X(z/z_0), \qquad \text{ROC} = |z_0| R_x.$$

The notation $\text{ROC} = |z_0| R_x$ denotes that the ROC is R_x scaled by $|z_0|$; i.e., if R_x is the set of values of z such that $r_R < |z| < r_L$, then $|z_0| R_x$ is the set of values of z such that $|z_0| r_R < |z| < |z_0| r_L$.

This property is easily shown simply by substituting $z_0^n x[n]$ into Eq. (3.2). As a consequence of the exponential multiplication property, all the pole–zero locations are scaled by a factor z_0, since, if $X(z)$ has a pole at $z = z_1$, then $X(z_0^{-1}z)$ will have a pole at $z = z_0 z_1$. If z_0 is a positive real number, the scaling can be interpreted as a shrinking or expanding of the z-plane; i.e., the pole and zero locations change along radial lines in the z-plane. If z_0 is complex with unity magnitude, so that $z_0 = e^{j\omega_0}$, the scaling corresponds to a rotation in the z-plane by an angle of ω_0; i.e., the pole and zero locations change in position along circles centered at the origin. This in turn can be interpreted as a frequency shift or translation, associated with the modulation in the time domain by the complex exponential sequence $e^{j\omega_0 n}$. That is, if the Fourier transform exists, this property has the form

$$e^{j\omega_0 n} x[n] \xleftrightarrow{\;\mathcal{F}\;} X(e^{j(\omega - \omega_0)}).$$

Example 3.15   Exponential Multiplication

Starting with the transform pair

$$u[n] \xleftrightarrow{\;\mathcal{Z}\;} \frac{1}{1 - z^{-1}}, \qquad |z| > 1, \tag{3.60}$$

we can use the exponential multiplication property to determine the z-transform of

$$x[n] = r^n \cos(\omega_0 n) u[n]. \tag{3.61}$$

First, $x[n]$ is expressed as

$$x[n] = \tfrac{1}{2}(re^{j\omega_0})^n u[n] + \tfrac{1}{2}(re^{-j\omega_0})^n u[n].$$

Then, using Eq. (3.60) and the exponential multiplication property, we see that

$$\tfrac{1}{2}(re^{j\omega_0})^n u[n] \xleftrightarrow{\mathcal{Z}} \frac{\tfrac{1}{2}}{1 - re^{j\omega_0}z^{-1}}, \qquad |z| > r,$$

$$\tfrac{1}{2}(re^{-j\omega_0})^n u[n] \xleftrightarrow{\mathcal{Z}} \frac{\tfrac{1}{2}}{1 - re^{-j\omega_0}z^{-1}}, \qquad |z| > r.$$

From the linearity property, it follows that

$$X(z) = \frac{\tfrac{1}{2}}{1 - re^{j\omega_0}z^{-1}} + \frac{\tfrac{1}{2}}{1 - re^{-j\omega_0}z^{-1}}, \qquad |z| > r$$

$$= \frac{(1 - r\cos\omega_0 z^{-1})}{1 - 2r\cos\omega_0 z^{-1} + r^2 z^{-2}}, \qquad |z| > r. \tag{3.62}$$

3.4.4 Differentiation of $X(z)$

The differentiation property states that

$$nx[n] \xleftrightarrow{\mathcal{Z}} -z\frac{dX(z)}{dz}, \qquad \text{ROC} = R_x.$$

This property is verified by differentiating the z-transform expression of Eq. (3.2); i.e.,

$$X(z) = \sum_{n=-\infty}^{\infty} x[n]z^{-n},$$

$$-z\frac{dX(z)}{dz} = -z \sum_{n=-\infty}^{\infty} (-n)x[n]z^{-n-1}$$

$$= \sum_{n=-\infty}^{\infty} nx[n]z^{-n} = \mathcal{Z}\{nx[n]\}.$$

We illustrate the use of the differentiation property with two examples.

Example 3.16   Inverse of Non-Rational z-Transform

In this example, we use the differentiation property together with the time-shifting property to determine the inverse z-transform considered in Example 3.11. With

$$X(z) = \log(1 + az^{-1}), \qquad |z| > |a|,$$

we first differentiate to obtain a rational expression:

$$\frac{dX(z)}{dz} = \frac{-az^{-2}}{1 + az^{-1}}.$$

From the differentiation property,

$$nx[n] \xleftrightarrow{\mathcal{Z}} -z\frac{dX(z)}{dz} = \frac{az^{-1}}{1 + az^{-1}}, \qquad |z| > |a|. \tag{3.63}$$

The inverse transform of Eq. (3.63) can be obtained by the combined use of the z-transform pair of Example 3.1, the linearity property, and the time-shifting property. Specifically, we can express $nx[n]$ as

$$nx[n] = a(-a)^{n-1}u[n-1].$$

Therefore,

$$x[n] = (-1)^{n+1}\frac{a^n}{n}u[n-1] \overset{z}{\longleftrightarrow} \log(1 + az^{-1}), \qquad |z| > |a|.$$

Example 3.17   Second-Order Pole

As another example of the use of the differentiation property, let us determine the z-transform of the sequence

$$x[n] = na^nu[n] = n(a^nu[n]).$$

From the z-transform pair of Example 3.1 and the differentiation property, it follows that

$$X(z) = -z\frac{d}{dz}\left(\frac{1}{1 - az^{-1}}\right), \qquad |z| > |a|$$

$$= \frac{az^{-1}}{(1 - az^{-1})^2}, \qquad |z| > |a|.$$

Therefore,

$$na^nu[n] \overset{z}{\longleftrightarrow} \frac{az^{-1}}{(1 - az^{-1})^2}, \qquad |z| > |a|.$$

3.4.5 Conjugation of a Complex Sequence

The conjugation property is expressed as

$$x^*[n] \overset{z}{\longleftrightarrow} X^*(z^*), \qquad \text{ROC} = R_x.$$

This property follows in a straightforward manner from the definition of the z-transform, the details of which are left as an exercise (Problem 3.51).

3.4.6 Time Reversal

By the time-reversal property,

$$x^*[-n] \overset{z}{\longleftrightarrow} X^*(1/z^*), \qquad \text{ROC} = \frac{1}{R_x}.$$

The notation $\text{ROC} = 1/R_x$ implies that R_x is inverted; i.e., if R_x is the set of values of z such that $r_R < |z| < r_L$, then the ROC is the set of values of z such that $1/r_L < |z| < 1/r_R$. Thus, if z_0 is in the ROC for $x[n]$, then $1/z_0^*$ is in the ROC for the z-transform of $x^*[-n]$. If the sequence $x[n]$ is real or we do not conjugate a complex sequence, the result becomes

$$x[-n] \overset{z}{\longleftrightarrow} X(1/z), \qquad \text{ROC} = \frac{1}{R_x}.$$

As with the conjugation property, the time-reversal property follows easily from the definition of the z-transform, and the details are left as an exercise (Problem 3.51).

Example 3.18   Time-Reversed Exponential Sequence

As an example of the use of the property of time reversal, consider the sequence

$$x[n] = a^{-n}u[-n],$$

which is a time-reversed version of $a^n u[n]$. From the time-reversal property, it follows that

$$X(z) = \frac{1}{1-az} = \frac{-a^{-1}z^{-1}}{1-a^{-1}z^{-1}}, \qquad |z| < |a^{-1}|.$$

3.4.7 Convolution of Sequences

According to the convolution property,

$$x_1[n] * x_2[n] \overset{z}{\longleftrightarrow} X_1(z)X_2(z), \qquad \text{ROC contains } R_{x_1} \cap R_{x_2}.$$

To derive this property formally, we consider

$$y[n] = \sum_{k=-\infty}^{\infty} x_1[k]x_2[n-k],$$

so that

$$Y(z) = \sum_{n=-\infty}^{\infty} y[n]z^{-n}$$

$$= \sum_{n=-\infty}^{\infty} \left\{ \sum_{k=-\infty}^{\infty} x_1[k]x_2[n-k] \right\} z^{-n}.$$

If we interchange the order of summation,

$$Y(z) = \sum_{k=-\infty}^{\infty} x_1[k] \sum_{n=-\infty}^{\infty} x_2[n-k]z^{-n}.$$

Changing the index of summation in the second sum from n to $m = n - k$, we obtain

$$Y(z) = \sum_{k=-\infty}^{\infty} x_1[k] \left\{ \sum_{m=-\infty}^{\infty} x_2[m]z^{-m} \right\} z^{-k}.$$

Thus, for values of z inside the regions of convergence of both $X_1(z)$ and $X_2(z)$, we can write

$$Y(z) = X_1(z)X_2(z),$$

where the region of convergence includes the intersection of the regions of convergence of $X_1(z)$ and $X_2(z)$. If a pole that borders on the region of convergence of one of the z-transforms is canceled by a zero of the other, then the region of convergence of $Y(z)$ may be larger. As we develop and exploit it in Chapter 5, the convolution property plays a particularly important role in the analysis of LTI systems. Specifically, as a consequence of this property, the z-transform of the output of an LTI system is the product of the z-transform of the input and the z-transform of the system impulse response. The z-transform of the impulse response of an LTI system is typically referred to as the system function.

Example 3.19   Evaluating a Convolution Using the z-Transform

Let $x_1[n] = a^n u[n]$ and $x_2[n] = u[n]$. The corresponding z-transforms are

$$X_1(z) = \sum_{n=0}^{\infty} a^n z^{-n} = \frac{1}{1 - az^{-1}}, \qquad |z| > |a|,$$

and

$$X_2(z) = \sum_{n=0}^{\infty} z^{-n} = \frac{1}{1 - z^{-1}}, \qquad |z| > 1.$$

If $|a| < 1$, the z-transform of the convolution of $x_1[n]$ with $x_2[n]$ is then

$$Y(z) = \frac{1}{(1 - az^{-1})(1 - z^{-1})} = \frac{z^2}{(z - a)(z - 1)}, \qquad |z| > 1. \qquad (3.64)$$

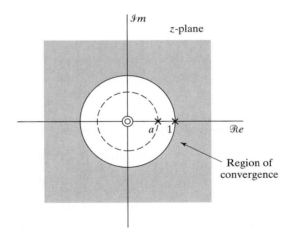

Figure 3.14 Pole–zero plot for the z-transform of the convolution of the sequences $u[n]$ and $a^n u[n]$.

The poles and zeros of $Y(z)$ are plotted in Figure 3.14, and the region of convergence is seen to be the overlap region. The sequence $y[n]$ can be obtained by determining the inverse z-transform. Expanding $Y(z)$ in Eq. (3.64) in a partial fraction expansion, we get

$$Y(z) = \frac{1}{1 - a}\left(\frac{1}{1 - z^{-1}} - \frac{a}{1 - az^{-1}}\right), \qquad |z| > 1.$$

Therefore,

$$y[n] = \frac{1}{1 - a}(u[n] - a^{n+1}u[n]).$$

TABLE 3.2   SOME z-TRANSFORM PROPERTIES

Property Number	Section Reference	Sequence	Transform	ROC
		$x[n]$	$X(z)$	R_x
		$x_1[n]$	$X_1(z)$	R_{x_1}
		$x_2[n]$	$X_2(z)$	R_{x_2}
1	3.4.1	$ax_1[n] + bx_2[n]$	$aX_1(z) + bX_2(z)$	Contains $R_{x_1} \cap R_{x_2}$
2	3.4.2	$x[n - n_0]$	$z^{-n_0} X(z)$	R_x, except for the possible addition or deletion of the origin or ∞
3	3.4.3	$z_0^n x[n]$	$X(z/z_0)$	$|z_0| R_x$
4	3.4.4	$nx[n]$	$-z\dfrac{dX(z)}{dz}$	R_x, except for the possible addition or deletion of the origin or ∞
5	3.4.5	$x^*[n]$	$X^*(z^*)$	R_x
6		$\mathcal{R}e\{x[n]\}$	$\dfrac{1}{2}[X(z) + X^*(z^*)]$	Contains R_x
7		$\mathcal{I}m\{x[n]\}$	$\dfrac{1}{2j}[X(z) - X^*(z^*)]$	Contains R_x
8	3.4.6	$x^*[-n]$	$X^*(1/z^*)$	$1/R_x$
9	3.4.7	$x_1[n] * x_2[n]$	$X_1(z)X_2(z)$	Contains $R_{x_1} \cap R_{x_2}$
10	3.4.8	Initial-value theorem:		
		$x[n] = 0, \quad n < 0 \qquad \lim\limits_{z\to\infty} X(z) = x[0]$		

3.4.8 Initial-Value Theorem

If $x[n]$ is zero for $n < 0$ (i.e., if $x[n]$ is causal), then

$$x[0] = \lim_{z\to\infty} X(z).$$

This theorem is shown by considering the limit of each term in the series of Eq. (3.2). (See Problem 3.56.)

3.4.9 Summary of Some z-Transform Properties

We have presented and discussed a number of the theorems and properties of z-transforms, many of which are useful in manipulating z-transforms. These properties and a number of others are summarized for convenient reference in Table 3.2.

3.5 SUMMARY

In this chapter, we have defined the z-transform of a sequence and shown that it is a generalization of the Fourier transform. The discussion focused on the properties of the z-transform and techniques for obtaining the z-transform of a sequence and vice versa. Specifically, we showed that the defining power series of the z-transform may converge when the Fourier transform does not. We explored in detail the dependence of the shape of the region of convergence on the properties of the sequence. A full understanding of the properties of the region of convergence is essential for successful use of the z-transform. This is particularly true in developing techniques for finding the sequence that corresponds to a given z-transform, i.e., finding inverse z-transforms. Much of the discussion focused on z-transforms that are rational functions in their region of convergence. For such functions, we described a technique of inverse transformation based on the partial fraction expansion of $X(z)$. We also discussed other techniques for inverse transformation, such as the use of tabulated power series expansions and long division.

An important part of the chapter was a discussion of some of the many properties of the z-transform that make it useful in analyzing discrete-time signals and systems. A variety of examples demonstrated how these properties can be used to find direct and inverse z-transforms.

PROBLEMS

Basic Problems with Answers

3.1. Determine the z-transform, including the region of convergence, for each of the following sequences:

(a) $\left(\frac{1}{2}\right)^n u[n]$

(b) $-\left(\frac{1}{2}\right)^n u[-n-1]$

(c) $\left(\frac{1}{2}\right)^n u[-n]$

(d) $\delta[n]$

(e) $\delta[n-1]$

(f) $\delta[n+1]$

(g) $\left(\frac{1}{2}\right)^n (u[n] - u[n-10])$

3.2. Determine the z-transform of the sequence

$$x[n] = \begin{cases} n, & 0 \le n \le N-1, \\ N, & N \le n. \end{cases}$$

3.3. Determine the z-transform of each of the following sequences. Include with your answer the region of convergence in the z-plane and a sketch of the pole–zero plot. Express all sums in closed form; α can be complex.

(a) $x_a[n] = \alpha^{|n|}, \qquad 0 < |\alpha| < 1.$

(b) $x_b[n] = \begin{cases} 1, & 0 \le n \le N-1, \\ 0, & \text{otherwise.} \end{cases}$

(c) $x_c[n] = \begin{cases} n, & 0 \le n \le N, \\ 2N - n, & N+1 \le n \le 2N, \\ 0, & \text{otherwise.} \end{cases}$

Hint: Note that $x_b[n]$ is a rectangular sequence and $x_c[n]$ is a triangular sequence. First express $x_c[n]$ in terms of $x_b[n]$.

3.4. Consider the z-transform $X(z)$ whose pole–zero plot is as shown in Figure P3.4-1.

(a) Determine the region of convergence of $X(z)$ if it is known that the Fourier transform exists. For this case, determine whether the corresponding sequence $x[n]$ is right sided, left sided, or two sided.

(b) How many possible two-sided sequences have the pole–zero plot shown in Figure P3.4-1?

(c) Is it possible for the pole–zero plot in Figure P3.4-1 to be associated with a sequence that is both stable and causal? If so, give the appropriate region of convergence.

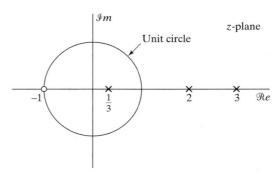

Figure P3.4-1

3.5. Determine the sequence $x[n]$ with z-transform

$$X(z) = (1 + 2z)(1 + 3z^{-1})(1 - z^{-1}).$$

3.6. Following are several z-transforms. For each, determine the inverse z-transform using both methods—partial fraction expansion and power series expansion—discussed in Section 3.3. In addition, indicate in each case whether the Fourier transform exists.

(a) $X(z) = \dfrac{1}{1 + \frac{1}{2}z^{-1}}, \qquad |z| > \frac{1}{2}$

(b) $X(z) = \dfrac{1}{1 + \frac{1}{2}z^{-1}}, \qquad |z| < \frac{1}{2}$

(c) $X(z) = \dfrac{1 - \frac{1}{2}z^{-1}}{1 + \frac{3}{4}z^{-1} + \frac{1}{8}z^{-2}}, \qquad |z| > \frac{1}{2}$

(d) $X(z) = \dfrac{1 - \frac{1}{2}z^{-1}}{1 - \frac{1}{4}z^{-2}}, \qquad |z| > \frac{1}{2}$

(e) $X(z) = \dfrac{1 - az^{-1}}{z^{-1} - a}, \qquad |z| > |1/a|$

3.7. The input to a causal linear time-invariant system is

$$x[n] = u[-n - 1] + \left(\tfrac{1}{2}\right)^{n} u[n].$$

The z-transform of the output of this system is

$$Y(z) = \frac{-\frac{1}{2}z^{-1}}{\left(1 - \frac{1}{2}z^{-1}\right)\left(1 + z^{-1}\right)}.$$

(a) Determine $H(z)$, the z-transform of the system impulse response. Be sure to specify the region of convergence.

(b) What is the region of convergence for $Y(z)$?

(c) Determine $y[n]$.

3.8. The system function of a causal linear time-invariant system is

$$H(z) = \frac{1 - z^{-1}}{1 + \frac{3}{4}z^{-1}}.$$

The input to this system is

$$x[n] = \left(\tfrac{1}{3}\right)^n u[n] + u[-n-1].$$

(a) Find the impulse response of the system, $h[n]$.
(b) Find the output $y[n]$.
(c) Is the system stable? That is, is $h[n]$ absolutely summable?

3.9. A causal LTI system has impulse response $h[n]$, for which the z-transform is

$$H(z) = \frac{1 + z^{-1}}{\left(1 - \frac{1}{2}z^{-1}\right)\left(1 + \frac{1}{4}z^{-1}\right)}.$$

(a) What is the region of convergence of $H(z)$?
(b) Is the system stable? Explain.
(c) Find the z-transform $X(z)$ of an input $x[n]$ that will produce the output

$$y[n] = -\tfrac{1}{3}\left(-\tfrac{1}{4}\right)^n u[n] - \tfrac{4}{3}(2)^n u[-n-1].$$

(d) Find the impulse response $h[n]$ of the system.

3.10. Without explicitly solving for $X(z)$, find the region of convergence of the z-transform of each of the following sequences, and determine whether the Fourier transform converges:

(a) $x[n] = \left[\left(\tfrac{1}{2}\right)^n + \left(\tfrac{3}{4}\right)^n\right] u[n - 10]$

(b) $x[n] = \begin{cases} 1, & -10 \leq n \leq 10, \\ 0, & \text{otherwise,} \end{cases}$

(c) $x[n] = 2^n u[-n]$

(d) $x[n] = \left[\left(\tfrac{1}{4}\right)^{n+4} - (e^{j\pi/3})^n\right] u[n - 1]$

(e) $x[n] = u[n + 10] - u[n + 5]$

(f) $x[n] = \left(\tfrac{1}{2}\right)^{n-1} u[n] + (2 + 3j)^{n-2} u[-n - 1]$

3.11. Following are four z-transforms. Determine which ones *could* be the z-transform of a *causal* sequence. Do not evaluate the inverse transform. You should be able to give the answer by inspection. Clearly state your reasons in each case.

(a) $\dfrac{(1 - z^{-1})^2}{\left(1 - \frac{1}{2}z^{-1}\right)}$

(b) $\dfrac{(z - 1)^2}{\left(z - \frac{1}{2}\right)}$

(c) $\dfrac{\left(z - \frac{1}{4}\right)^5}{\left(z - \frac{1}{2}\right)^6}$

(d) $\dfrac{\left(z - \frac{1}{4}\right)^6}{\left(z - \frac{1}{2}\right)^5}$

3.12. Sketch the pole–zero plot for each of the following z-transforms and shade the region of convergence:

(a) $X_1(z) = \dfrac{1 - \frac{1}{2}z^{-1}}{1 + 2z^{-1}}$,   ROC: $|z| < 2$

(b) $X_2(z) = \dfrac{1 - \frac{1}{3}z^{-1}}{\left(1 + \frac{1}{2}z^{-1}\right)\left(1 - \frac{2}{3}z^{-1}\right)}$, $x_2[n]$ causal

(c) $X_3(z) = \dfrac{1 + z^{-1} - 2z^{-2}}{1 - \frac{13}{6}z^{-1} + z^{-2}}$, $x_3[n]$ absolutely summable.

3.13. A causal sequence $g[n]$ has the z-transform

$$G(z) = \sin(z^{-1})(1 + 3z^{-2} + 2z^{-4}).$$

Find $g[11]$.

3.14. If $H(z) = \dfrac{1}{1 - \frac{1}{4}z^{-2}}$ and $h[n] = A_1\alpha_1^n u[n] + A_2\alpha_2^n u[n]$, determine the values of A_1, A_2, α_1, and α_2.

3.15. If $H(z) = \dfrac{1 - \frac{1}{1024}z^{-10}}{1 - \frac{1}{2}z^{-1}}$ for $|z| > 0$, is the corresponding LTI system causal? Justify your answer.

3.16. When the input to an LTI system is

$$x[n] = \left(\frac{1}{3}\right)^n u[n] + (2)^n u[-n-1],$$

the corresponding output is

$$y[n] = 5\left(\frac{1}{3}\right)^n u[n] - 5\left(\frac{2}{3}\right)^n u[n].$$

(a) Find the system function $H(z)$ of the system. Plot the pole(s) and zero(s) of $H(z)$ and indicate the region of convergence.
(b) Find the impulse response $h[n]$ of the system.
(c) Write a difference equation that is satisfied by the given input and output.
(d) Is the system stable? Is it causal?

3.17. Consider an LTI system with input $x[n]$ and output $y[n]$ that satisfies the difference equation

$$y[n] - \frac{5}{2}y[n-1] + y[n-2] = x[n] - x[n-1].$$

Determine all possible values for the system's impulse response $h[n]$ at $n = 0$.

3.18. A causal LTI system has the system function

$$H(z) = \frac{1 + 2z^{-1} + z^{-2}}{\left(1 + \frac{1}{2}z^{-1}\right)(1 - z^{-1})}.$$

(a) Find the impulse response of the system, $h[n]$.
(b) Find the output of this system, $y[n]$, for the input

$$x[n] = e^{j(\pi/2)n}.$$

3.19. For each of the following pairs of input z-transform $X(z)$ and system function $H(z)$, determine the region of convergence for the output z-transform $Y(z)$:

(a)

$$X(z) = \frac{1}{1 + \frac{1}{2}z^{-1}}, \qquad |z| > \frac{1}{2}$$

$$H(z) = \frac{1}{1 - \frac{1}{4}z^{-1}}, \qquad |z| > \frac{1}{4}$$

(b)

$$X(z) = \frac{1}{1 - 2z^{-1}}, \qquad |z| < 2$$

$$H(z) = \frac{1}{1 - \frac{1}{3}z^{-1}}, \qquad |z| > \frac{1}{3}$$

(c)

$$X(z) = \frac{1}{\left(1 - \frac{1}{5}z^{-1}\right)\left(1 + 3z^{-1}\right)}, \qquad \frac{1}{5} < |z| < 3$$

$$H(z) = \frac{1 + 3z^{-1}}{1 + \frac{1}{3}z^{-1}}, \qquad |z| > \frac{1}{3}$$

3.20. For each of the following pairs of input and output z-transforms $X(z)$ and $Y(z)$, determine the region of convergence for the system function $H(z)$:

(a)

$$X(z) = \frac{1}{1 - \frac{3}{4}z^{-1}}, \qquad |z| > \frac{3}{4}$$

$$Y(z) = \frac{1}{1 + \frac{2}{3}z^{-1}}, \qquad |z| > \frac{2}{3}$$

(b)

$$X(z) = \frac{1}{1 + \frac{1}{3}z^{-1}}, \qquad |z| < \frac{1}{3}$$

$$Y(z) = \frac{1}{\left(1 - \frac{1}{6}z^{-1}\right)\left(1 + \frac{1}{3}z^{-1}\right)}, \qquad \frac{1}{6} < |z| < \frac{1}{3}$$

Basic Problems

3.21. Consider a linear time-invariant system with impulse response

$$h[n] = \begin{cases} a^n, & n \geq 0, \\ 0, & n < 0, \end{cases}$$

and input

$$x[n] = \begin{cases} 1, & 0 \leq n \leq (N - 1), \\ 0, & \text{otherwise.} \end{cases}$$

(a) Determine the output $y[n]$ by explicitly evaluating the discrete convolution of $x[n]$ and $h[n]$.

(b) Determine the output $y[n]$ by computing the inverse z-transform of the product of the z-transforms of $x[n]$ and $h[n]$.

3.22. Consider an LTI system that is stable and for which $H(z)$, the z-transform of the impulse response, is given by

$$H(z) = \frac{3}{1 + \frac{1}{3}z^{-1}}.$$

Suppose $x[n]$, the input to the system, is a unit step sequence.

(a) Find the output $y[n]$ by evaluating the discrete convolution of $x[n]$ and $h[n]$.

(b) Find the output $y[n]$ by computing the inverse z-transform of $Y(z)$.

3.23. An LTI system is characterized by the system function

$$H(z) = \frac{\left(1 - \frac{1}{2}z^{-2}\right)}{\left(1 - \frac{1}{2}z^{-1}\right)\left(1 - \frac{1}{4}z^{-1}\right)}, \qquad |z| > \frac{1}{2}.$$

(a) Determine the impulse response of the system.

(b) Determine the difference equation relating the system input $x[n]$ and the system output $y[n]$.

3.24. Sketch each of the following sequences and determine their z-transforms, including the region of convergence:

(a) $\displaystyle\sum_{k=-\infty}^{\infty} \delta[n - 4k]$

(b) $\dfrac{1}{2}\left[e^{j\pi n} + \cos\left(\dfrac{\pi}{2}n\right) + \sin\left(\dfrac{\pi}{2} + 2\pi n\right) \right] u[n]$

3.25. Consider a right-sided sequence $x[n]$ with z-transform

$$X(z) = \frac{1}{(1 - az^{-1})(1 - bz^{-1})} = \frac{z^2}{(z - a)(z - b)}.$$

In Section 3.3 we considered the determination of $x[n]$ by carrying out a partial fraction expansion, with $X(z)$ considered as a ratio of polynomials in z^{-1}. Carry out a partial fraction expansion of $X(z)$, considered as a ratio of polynomials in z, and determine $x[n]$ from this expansion.

Advanced Problems

3.26. Determine the inverse z-transform of each of the following. In Parts (a)–(c), use the methods specified. In Part (d), use any method you prefer.

(a) Long division:

$$X(z) = \frac{1 - \frac{1}{3}z^{-1}}{1 + \frac{1}{3}z^{-1}}, \qquad x[n] \text{ a right-sided sequence}$$

(b) Partial fraction:

$$X(z) = \frac{3}{z - \frac{1}{4} - \frac{1}{8}z^{-1}}, \qquad x[n] \text{ stable}$$

(c) Power series:

$$X(z) = \ln(1 - 4z), \qquad |z| < \tfrac{1}{4}$$

(d) $X(z) = \dfrac{1}{1 - \frac{1}{3}z^{-3}}, \qquad |z| > (3)^{-1/3}$

3.27. Using any method, determine the inverse z-transform for each of the following:

(a) $X(z) = \dfrac{1}{\left(1 + \frac{1}{2}z^{-1}\right)^2 (1 - 2z^{-1})(1 - 3z^{-1})}$,   stable sequence

(b) $X(z) = e^{z^{-1}}$

(c) $X(z) = \dfrac{z^3 - 2z}{z - 2}$,   left-sided sequence

3.28. Determine the inverse z-transform of each of the following. You should find the z-transform properties in Section 3.4 helpful.

(a) $X(z) = \dfrac{3z^{-3}}{\left(1 - \frac{1}{4}z^{-1}\right)^2}$, $x[n]$ left sided

(b) $X(z) = \sin(z)$,   ROC includes $|z| = 1$

(c) $X(z) = \dfrac{z^7 - 2}{1 - z^{-7}}$, $|z| > 1$

3.29. Determine a sequence $x[n]$ whose z-transform is $X(z) = e^z + e^{1/z}$, $z \neq 0$.

3.30. Determine the inverse z-transform of

$$X(z) = \log 2\left(\tfrac{1}{2} - z\right), \qquad |z| < \tfrac{1}{2},$$

by

(a) using the power series

$$\log(1 - x) = -\sum_{i=1}^{\infty} \frac{x^i}{i}, \qquad |x| < 1;$$

(b) first differentiating $X(z)$ and then using the derivative to recover $x[n]$.

3.31. For each of the following sequences, determine the z-transform and region of convergence, and sketch the pole–zero diagram:

(a) $x[n] = a^n u[n] + b^n u[n] + c^n u[-n - 1]$, $|a| < |b| < |c|$

(b) $x[n] = n^2 a^n u[n]$

(c) $x[n] = e^{n^4}\left[\cos\left(\frac{\pi}{12}n\right)\right] u[n] - e^{n^4}\left[\cos\left(\frac{\pi}{12}n\right)\right] u[n - 1]$

3.32. The pole–zero diagram in Figure P3.32-1 corresponds to the z-transform $X(z)$ of a causal sequence $x[n]$. Sketch the pole–zero diagram of $Y(z)$, where $y[n] = x[-n+3]$. Also, specify the region of convergence for $Y(z)$.

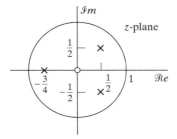

Figure P3.32-1

3.33. Let $x[n]$ be the sequence with the pole–zero plot shown in Figure P3.33-1. Sketch the pole–zero plot for:

(a) $y[n] = \left(\frac{1}{2}\right)^n x[n]$

(b) $w[n] = \cos\left(\dfrac{\pi n}{2}\right) x[n]$

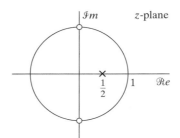

Figure P3.33-1

3.34. Consider an LTI system that is stable and for which $H(z)$, the z-transform of the impulse response, is given by

$$H(z) = \frac{3 - 7z^{-1} + 5z^{-2}}{1 - \frac{5}{2}z^{-1} + z^{-2}}.$$

Suppose $x[n]$, the input to the system, is a unit step sequence.
(a) Find the output $y[n]$ by evaluating the discrete convolution of $x[n]$ and $h[n]$.
(b) Find the output $y[n]$ by computing the inverse z-transform of $Y(z)$.

3.35. Determine the unit step response of the causal system for which the z-transform of the impulse response is

$$H(z) = \frac{1 - z^3}{1 - z^4}.$$

3.36. If the input $x[n]$ to an LTI system is $x[n] = u[n]$, the output is

$$y[n] = \left(\tfrac{1}{2}\right)^{n-1} u[n+1].$$

(a) Find $H(z)$, the z-transform of the system impulse response, and plot its pole–zero diagram.
(b) Find the impulse response $h[n]$.
(c) Is the system stable?
(d) Is the system causal?

3.37. Consider a sequence $x[n]$ for which the z-transform is

$$X(z) = \frac{\frac{1}{3}}{1 - \frac{1}{2}z^{-1}} + \frac{\frac{1}{4}}{1 - 2z^{-1}}$$

and for which the region of convergence includes the unit circle. Determine $x[0]$ using the initial-value theorem.

3.38. Consider a stable linear time-invariant system. The z-transform of the impulse response is

$$H(z) = \frac{z^{-1} + z^{-2}}{\left(1 - \frac{1}{2}z^{-1}\right)\left(1 + \frac{1}{3}z^{-1}\right)}.$$

Suppose $x[n]$, the input to the system, is $2u[n]$. Determine $y[n]$ at $n = 1$.

3.39. Suppose the z-transform of $x[n]$ is

$$X(z) = \frac{z^{10}}{\left(z - \frac{1}{2}\right)\left(z - \frac{3}{2}\right)^{10}\left(z + \frac{3}{2}\right)^2\left(z + \frac{5}{2}\right)\left(z + \frac{7}{2}\right)}.$$

It is also known that $x[n]$ is a stable sequence.
(a) Determine the region of convergence of $X(z)$.
(b) Determine $x[n]$ at $n = -8$.

3.40. In Figure P3.40-1, $H(z)$ is the system function of a causal LTI system.
   (a) Using z-transforms of the signals shown in the figure, obtain an expression for $W(z)$ in the form

$$W(z) = H_1(z)X(z) + H_2(z)E(z),$$

   where both $H_1(z)$ and $H_2(z)$ are expressed in terms of $H(z)$.
   (b) For the special case $H(z) = z^{-1}/(1 - z^{-1})$, determine $H_1(z)$ and $H_2(z)$.
   (c) Is the system $H(z)$ stable? Are the systems $H_1(z)$ and $H_2(z)$ stable?

Figure P3.40-1

3.41. In Figure P3.41-1, $h[n]$ is the impulse response of the LTI system within the inner box. The input to system $h[n]$ is $v[n]$, and the output is $w[n]$. The z-transform of $h[n]$, $H(z)$, exists in the following region of convergence:

$$0 < r_{min} < |z| < r_{max} < \infty.$$

   (a) Can the LTI system with impulse response $h[n]$ be BIBO stable? If so, determine inequality constraints on r_{min} and r_{max} such that it is stable. If not, briefly explain why.
   (b) Is the overall system (in the large box, with input $x[n]$ and output $y[n]$) LTI? If so, find its impulse response $g[n]$. If not, briefly explain why.
   (c) Can the overall system be BIBO stable? If so, determine inequality constraints relating α, r_{min}, and r_{max} such that it is stable. If not, briefly explain why.

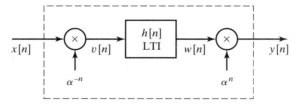

Figure P3.41-1

3.42. A causal and stable LTI system S has its input $x[n]$ and output $y[n]$ related by the linear constant-coefficient difference equation

$$y[n] + \sum_{k=1}^{10} \alpha_k y[n - k] = x[n] + \beta x[n - 1].$$

Let the impulse response of S be the sequence $h[n]$.
   (a) Show that $h[0]$ must be nonzero.
   (b) Show that α_1 can be determined from the knowledge of $h[0]$ and $h[1]$.
   (c) If $h[n] = (0.9)^n \cos(\pi n/4)$ for $0 \le n \le 10$, sketch the pole–zero plot for the system function of S, and indicate the region of convergence.

3.43. When the input to an LTI system is

$$x[n] = \left(\frac{1}{2}\right)^n u[n] + 2^n u[-n-1],$$

the output is

$$y[n] = 6\left(\frac{1}{2}\right)^n u[n] - 6\left(\frac{3}{4}\right)^n u[n].$$

(a) Find the system function $H(z)$ of the system. Plot the poles and zeros of $H(z)$, and indicate the region of convergence.
(b) Find the impulse response $h[n]$ of the system.
(c) Write the difference equation that characterizes the system.
(d) Is the system stable? Is it causal?

3.44. When the input to a causal LTI system is

$$x[n] = -\frac{1}{3}\left(\frac{1}{2}\right)^n u[n] - \frac{4}{3}2^n u[-n-1],$$

the z-transform of the output is

$$Y(z) = \frac{1 + z^{-1}}{\left(1 - z^{-1}\right)\left(1 + \frac{1}{2}z^{-1}\right)\left(1 - 2z^{-1}\right)}.$$

(a) Find the z-transform of $x[n]$.
(b) What is the region of convergence of $Y(z)$?
(c) Find the impulse response of the system.
(d) Is the system stable?

3.45. Let $x[n]$ be a discrete-time signal with $x[n] = 0$ for $n \le 0$ and z-transform $X(z)$. Furthermore, given $x[n]$, let the discrete-time signal $y[n]$ be defined by

$$y[n] = \begin{cases} \frac{1}{n}x[n], & n > 0, \\ 0, & \text{otherwise.} \end{cases}$$

(a) Compute $Y(z)$ in terms of $X(z)$.
(b) Using the result of Part (a), find the z-transform of

$$w[n] = \frac{1}{n + \delta[n]}u[n-1].$$

3.46. The signal $y[n]$ is the output of an LTI system with impulse response $h[n]$ for a given input $x[n]$. Throughout the problem, assume that $y[n]$ is stable and has a z-transform $Y(z)$ with the pole–zero diagram shown in Figure P3.46-1. The signal $x[n]$ is stable and has the pole–zero diagram shown in Figure P3.46-2.

Figure P3.46-1

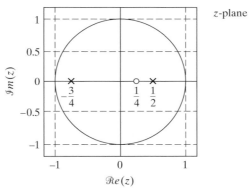

Figure P3.46-2

(a) What is the region of convergence, $Y(z)$?
(b) Is $y[n]$ left sided, right sided, or two sided?
(c) What is the ROC of $X(z)$?
(d) Is $x[n]$ a causal sequence? That is, does $x[n] = 0$ for $n < 0$?
(e) What is $x[0]$?
(f) Draw the pole–zero plot of $H(z)$, and specify its ROC.
(g) Is $h[n]$ anticausal? That is, does $h[n] = 0$ for $n > 0$?

Extension Problems

3.47. Let $x[n]$ denote a causal sequence; i.e., $x[n] = 0, n < 0$. Furthermore, assume that $x[0] \neq 0$.
  (a) Show that there are no poles or zeros of $X(z)$ at $z = \infty$, i.e., that $\lim_{z \to \infty} X(z)$ is nonzero and finite.
  (b) Show that the number of poles in the finite z-plane equals the number of zeros in the finite z-plane. (The finite z-plane excludes $z = \infty$.)

3.48. Consider a sequence with z-transform $X(z) = P(z)/Q(z)$, where $P(z)$ and $Q(z)$ are polynomials in z. If the sequence is absolutely summable and if all the roots of $Q(z)$ are inside the unit circle, is the sequence necessarily causal? If your answer is yes, clearly explain. If your answer is no, give a counterexample.

3.49. Let $x[n]$ be a causal stable sequence with z-transform $X(z)$. The *complex cepstrum* $\hat{x}[n]$ is defined as the inverse transform of the logarithm of $X(z)$; i.e.,

$$\hat{X}(z) = \log X(z) \overset{\mathcal{Z}}{\longleftrightarrow} \hat{x}[n],$$

where the ROC of $\hat{X}(z)$ includes the unit circle. (Strictly speaking, taking the logarithm of a complex number requires some careful considerations. Furthermore, the logarithm of a valid z-transform may not be a valid z-transform. For now, we assume that this operation is valid.)

Determine the complex cepstrum for the sequence

$$x[n] = \delta[n] + a\delta[n - N], \qquad \text{where } |a| < 1.$$

3.50. Assume that $x[n]$ is real and even; i.e., $x[n] = x[-n]$. Further, assume that z_0 is a zero of $X(z)$; i.e., $X(z_0) = 0$.
(a) Show that $1/z_0$ is also a zero of $X(z)$.
(b) Are there other zeros of $X(z)$ implied by the information given?

3.51. Using the definition of the z-transform in Eq. (3.2), show that if $X(z)$ is the z-transform of $x[n] = x_R[n] + jx_I[n]$, then

(a) $x^*[n] \overset{\mathcal{Z}}{\longleftrightarrow} X^*(z^*)$

(b) $x[-n] \overset{\mathcal{Z}}{\longleftrightarrow} X(1/z)$

(c) $x_R[n] \overset{\mathcal{Z}}{\longleftrightarrow} \frac{1}{2}[X(z) + X^*(z^*)]$

(d) $x_I[n] \overset{\mathcal{Z}}{\longleftrightarrow} \frac{1}{2j}[X(z) - X^*(z^*)]$

3.52. Consider a *real* sequence $x[n]$ that has all the poles and zeros of its z-transform inside the unit circle. Determine, in terms of $x[n]$, a *real* sequence $x_1[n]$ not equal to $x[n]$, but for which $x_1[0] = x[0]$, $|x_1[n]| = |x[n]|$, and the z-transform of $x_1[n]$ has all its poles and zeros inside the unit circle.

3.53. A real finite-duration sequence whose z-transform has no zeros at conjugate reciprocal pair locations and no zeros on the unit circle is uniquely specified to within a positive scale factor by its Fourier transform phase (Hayes et al., 1980).

An example of zeros at conjugate reciprocal pair locations is $z = a$ and $(a^*)^{-1}$. Even though we can generate sequences that do not satisfy the preceding set of conditions, almost any sequence of practical interest satisfies the conditions and therefore is uniquely specified to within a positive scale factor by the phase of its Fourier transform.

Consider a sequence $x[n]$ that is real, that is zero outside $0 \leq n \leq N - 1$, and whose z-transform has no zeros at conjugate reciprocal pair locations and no zeros on the unit circle. We wish to develop an algorithm that reconstructs $cx[n]$ from $\sphericalangle X(e^{j\omega})$, the Fourier transform phase of $x[n]$, where c is a positive scale factor.
(a) Specify a set of $(N-1)$ linear equations, the solution to which will provide the recovery of $x[n]$ to within a positive or negative scale factor from $\tan\{\sphericalangle X(e^{j\omega})\}$. You do not have to prove that the set of $(N-1)$ linear equations has a unique solution. Further, show that if we know $\sphericalangle X(e^{j\omega})$ rather than just $\tan\{\sphericalangle X(e^{j\omega})\}$, the sign of the scale factor can also be determined.
(b) Suppose

$$x[n] = \begin{cases} 0, & n < 0, \\ 1, & n = 0, \\ 2, & n = 1, \\ 3, & n = 2, \\ 0, & n \geq 3. \end{cases}$$

Using the approach developed in Part (a), demonstrate that $cx[n]$ can be determined from $\sphericalangle X(e^{j\omega})$, where c is a positive scale factor.

3.54. For a sequence $x[n]$ that is zero for $n < 0$, use Eq. (3.2) to show that

$$\lim_{z \to \infty} X(z) = x[0].$$

What is the corresponding theorem if the sequence is zero for $n > 0$?

3.55. The aperiodic autocorrelation function for a real-valued stable sequence $x[n]$ is defined as

$$c_{xx}[n] = \sum_{k=-\infty}^{\infty} x[k]x[n+k].$$

(a) Show that the z-transform of $c_{xx}[n]$ is

$$C_{xx}(z) = X(z)X(z^{-1}).$$

Determine the region of convergence for $C_{xx}(z)$.

(b) Suppose that $x[n] = a^n u[n]$. Sketch the pole–zero plot for $C_{xx}(z)$, including the region of convergence. Also, find $c_{xx}[n]$ by evaluating the inverse z-transform of $C_{xx}(z)$.

(c) Specify another sequence, $x_1[n]$, that is not equal to $x[n]$ in Part (b), but that has the same autocorrelation function, $c_{xx}[n]$, as $x[n]$ in Part (b).

(d) Specify a third sequence, $x_2[n]$, that is not equal to $x[n]$ or $x_1[n]$, but that has the same autocorrelation function as $x[n]$ in Part (b).

3.56. Determine whether or not the function $X(z) = z^*$ can correspond to the z-transform of a sequence. Clearly explain your reasoning.

3.57. Let $X(z)$ denote a ratio of polynomials in z; i.e.,

$$X(z) = \frac{B(z)}{A(z)}.$$

Show that if $X(z)$ has a first-order pole at $z = z_0$, then the residue of $X(z)$ at $z = z_0$ is equal to

$$\frac{B(z_0)}{A'(z_0)},$$

where $A'(z_0)$ denotes the derivative of $A(z)$ evaluated at $z = z_0$.

4

SAMPLING OF CONTINUOUS-TIME SIGNALS

4.0 INTRODUCTION

Discrete-time signals can arise in many ways, but they most commonly occur as representations of sampled continuous-time signals. It is remarkable that under reasonable constraints, a continuous-time signal can be quite accurately represented by samples taken at discrete points in time. In this chapter we discuss the process of periodic sampling in some detail, including the phenomenon of aliasing, which occurs when the signal is not bandlimited or when the sampling rate is too low. Of particular importance is the fact that continuous-time signal processing can be implemented through a process of sampling, discrete-time processing, and the subsequent reconstruction of a continuous-time signal.

4.1 PERIODIC SAMPLING

Although other possibilities exist (see Steiglitz, 1965; Oppenheim and Johnson, 1972), the typical method of obtaining a discrete-time representation of a continuous-time signal is through periodic sampling, wherein a sequence of samples, $x[n]$, is obtained from a continuous-time signal $x_c(t)$ according to the relation

$$x[n] = x_c(nT), \qquad -\infty < n < \infty. \tag{4.1}$$

In Eq. (4.1), T is the *sampling period*, and its reciprocal, $f_s = 1/T$, is the *sampling*

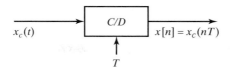

Figure 4.1 Block diagram representation of an ideal continuous-to-discrete-time (C/D) converter.

frequency, in samples per second. We also express the sampling frequency as $\Omega_s = 2\pi / T$ when we want to use frequencies in radians per second.

We refer to a system that implements the operation of Eq. (4.1) as an *ideal continuous-to-discrete-time (C/D) converter,* and we depict it in block diagram form as indicated in Figure 4.1. As an example of the relationship between $x_c(t)$ and $x[n]$, in Figure 2.2 we illustrated a continuous-time speech waveform and the corresponding sequence of samples.

In a practical setting, the operation of sampling is implemented by an analog-to-digital (A/D) converter. Such systems can be viewed as approximations to the ideal C/D converter. Important considerations in the implementation or choice of an A/D converter include quantization of the output samples, linearity of quantization steps, the need for sample-and-hold circuits, and limitations on the sampling rate. The effects of quantization are discussed in Sections 4.8.2 and 4.8.3. Other practical issues of A/D conversion are electronic circuit concerns that are outside the scope of this text.

The sampling operation is generally not invertible; i.e., given the output $x[n]$, it is not possible in general to reconstruct $x_c(t)$, the input to the sampler, since many continuous-time signals can produce the same output sequence of samples. The inherent ambiguity in sampling is a fundamental issue in signal processing. Fortunately, it is possible to remove the ambiguity by restricting the input signals that go into the sampler.

It is convenient to represent the sampling process mathematically in the two stages depicted in Figure 4.2(a). The stages consist of an impulse train modulator followed by conversion of the impulse train to a sequence. Figure 4.2(b) shows a continuous-time signal $x_c(t)$ and the results of impulse train sampling for two different sampling rates. Figure 4.2(c) depicts the corresponding output sequences. The essential difference between $x_s(t)$ and $x[n]$ is that $x_s(t)$ is, in a sense, a continuous-time signal (specifically, an impulse train) that is zero except at integer multiples of T. The sequence $x[n]$, on the other hand, is indexed on the integer variable n, which in effect introduces a time normalization; i.e., the sequence of numbers $x[n]$ contains no explicit information about the sampling rate. Furthermore, the samples of $x_c(t)$ are represented by finite numbers in $x[n]$ rather than as the areas of impulses, as with $x_s(t)$.

It is important to emphasize that Figure 4.2(a) is strictly a mathematical representation that is convenient for gaining insight into sampling in both the time domain and frequency domain. It is not a close representation of any physical circuits or systems designed to implement the sampling operation. Whether a piece of hardware can be construed to be an approximation to the block diagram of Figure 4.2(a) is a secondary issue at this point. We have introduced this representation of the sampling operation because it leads to a simple derivation of a key result and because the approach leads to a number of important insights that are difficult to obtain from a more formal derivation based on manipulation of Fourier transform formulas.

(a)

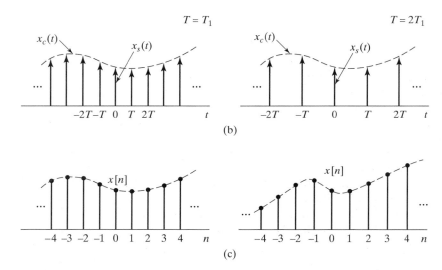

(b)

(c)

Figure 4.2 Sampling with a periodic impulse train followed by conversion to a discrete-time sequence. (a) Overall system. (b) $x_s(t)$ for two sampling rates. (c) The output sequence for the two different sampling rates.

4.2 FREQUENCY-DOMAIN REPRESENTATION OF SAMPLING

To derive the frequency-domain relation between the input and output of an ideal C/D converter, let us first consider the conversion of $x_c(t)$ to $x_s(t)$ through modulation of the periodic impulse train

$$s(t) = \sum_{n=-\infty}^{\infty} \delta(t - nT), \qquad (4.2)$$

where $\delta(t)$ is the unit impulse function, or Dirac delta function. We modulate $s(t)$ with $x_c(t)$, obtaining

$$x_s(t) = x_c(t)s(t)$$
$$= x_c(t) \sum_{n=-\infty}^{\infty} \delta(t - nT). \qquad (4.3)$$

Through the "sifting property" of the impulse function, $x_s(t)$ can be expressed as

$$x_s(t) = \sum_{n=-\infty}^{\infty} x_c(nT)\delta(t - nT).   \tag{4.4}$$

Let us now consider the Fourier transform of $x_s(t)$. Since, from Eq. (4.3), $x_s(t)$ is the product of $x_c(t)$ and $s(t)$, the Fourier transform of $x_s(t)$ is the convolution of the Fourier transforms $X_c(j\Omega)$ and $S(j\Omega)$. The Fourier transform of a periodic impulse train is a periodic impulse train (Oppenheim and Willsky, 1997). Specifically,

$$S(j\Omega) = \frac{2\pi}{T} \sum_{k=-\infty}^{\infty} \delta(\Omega - k\Omega_s),   \tag{4.5}$$

where $\Omega_s = 2\pi/T$ is the sampling frequency in radians/s. Since

$$X_s(j\Omega) = \frac{1}{2\pi} X_c(j\Omega) * S(j\Omega),$$

where $*$ denotes the operation of continuous-variable convolution, it follows that

$$X_s(j\Omega) = \frac{1}{T} \sum_{k=-\infty}^{\infty} X_c(j(\Omega - k\Omega_s)).   \tag{4.6}$$

Equation (4.6) provides the relationship between the Fourier transforms of the input and the output of the impulse train modulator in Figure 4.2(a). We see from Eq. (4.6) that the Fourier transform of $x_s(t)$ consists of periodically repeated copies of the Fourier transform of $x_c(t)$. The copies of $X_c(j\Omega)$ are shifted by integer multiples of the sampling frequency and then superimposed to produce the periodic Fourier transform of the impulse train of samples. Figure 4.3 depicts the frequency-domain representation of impulse train sampling. Figure 4.3(a) represents a bandlimited Fourier transform whose highest nonzero frequency component in $X_c(j\Omega)$ is at Ω_N. Figure 4.3(b) shows the periodic impulse train $S(j\Omega)$, and Figure 4.3(c) shows $X_s(j\Omega)$, the result of convolving $X_c(j\Omega)$ with $S(j\Omega)$. From Figure 4.3(c), it is evident that when

$$\Omega_s - \Omega_N > \Omega_N, \quad \text{or} \quad \Omega_s > 2\Omega_N,   \tag{4.7}$$

the replicas of $X_c(j\Omega)$ do not overlap, and therefore, when they are added together in Eq. (4.6), there remains (to within a scale factor of $1/T$) a replica of $X_c(j\Omega)$ at each integer multiple of Ω_s. Consequently, $x_c(t)$ can be recovered from $x_s(t)$ with an ideal lowpass filter. This is depicted in Figure 4.4(a), which shows the impulse train modulator followed by a linear time-invariant system with frequency response $H_r(j\Omega)$. For $X_c(j\Omega)$ as in Figure 4.4(b), $X_s(j\Omega)$ would be as shown in Figure 4.4(c), where it is assumed that $\Omega_s > 2\Omega_N$. Since

$$X_r(j\Omega) = H_r(j\Omega)X_s(j\Omega),   \tag{4.8}$$

it follows that if $H_r(j\Omega)$ is an ideal lowpass filter with gain T and cutoff frequency Ω_c such that

$$\Omega_N < \Omega_c < (\Omega_s - \Omega_N),   \tag{4.9}$$

then

$$X_r(j\Omega) = X_c(j\Omega),   \tag{4.10}$$

as depicted in Figure 4.4(e).

Figure 4.3 Effect in the frequency domain of sampling in the time domain. (a) Spectrum of the original signal. (b) Spectrum of the sampling function. (c) Spectrum of the sampled signal with $\Omega_s > 2\Omega_N$. (d) Spectrum of the sampled signal with $\Omega_s < 2\Omega_N$.

If the inequality of Eq. (4.7) does not hold, i.e., if $\Omega_s \leq 2\Omega_N$, the copies of $X_c(j\Omega)$ overlap, so that when they are added together, $X_c(j\Omega)$ is no longer recoverable by lowpass filtering. This is illustrated in Figure 4.3(d). In this case, the reconstructed output $x_r(t)$ in Figure 4.4(a) is related to the original continuous-time input through a distortion referred to as *aliasing distortion,* or simply, *aliasing.* Figure 4.5 illustrates aliasing in the frequency domain for the simple case of a cosine signal. Figure 4.5(a) shows the Fourier transform of the signal

$$x_c(t) = \cos \Omega_0 t. \tag{4.11}$$

Figure 4.5(b) shows the Fourier transform of $x_s(t)$ with $\Omega_0 < \Omega_s/2$, and Figure 4.5(c) shows the Fourier transform of $x_s(t)$ with $\Omega_0 > \Omega_s/2$. Figures 4.5(d) and (e) correspond to the Fourier transform of the lowpass filter output for $\Omega_0 < \Omega_s/2 = \pi/T$ and $\Omega_0 > \pi/T$, respectively, with $\Omega_c = \Omega_s/2$. Figures 4.5(c) and (e) correspond to the case of

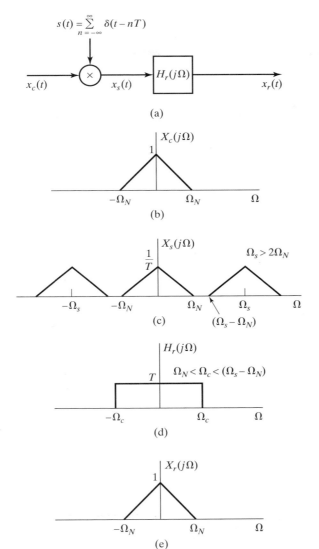

$$s(t) = \sum_{n=-\infty}^{\infty} \delta(t - nT)$$

(a)

(b)

(c)

(d)

(e)

Figure 4.4   Exact recovery of a continuous-time signal from its samples using an ideal lowpass filter.

aliasing. With no aliasing (Figures 4.5(b) and (d)), the reconstructed output is

$$x_r(t) = \cos \Omega_0 t. \tag{4.12}$$

With aliasing, the reconstructed output is

$$x_r(t) = \cos(\Omega_s - \Omega_0)t; \tag{4.13}$$

i.e., the higher frequency signal $\cos \Omega_0 t$ has taken on the identity (alias) of the lower frequency signal $\cos(\Omega_s - \Omega_0)t$ as a consequence of the sampling and reconstruction. This discussion is the basis for the Nyquist sampling theorem (Nyquist 1928; Shannon, 1949), stated as follows.

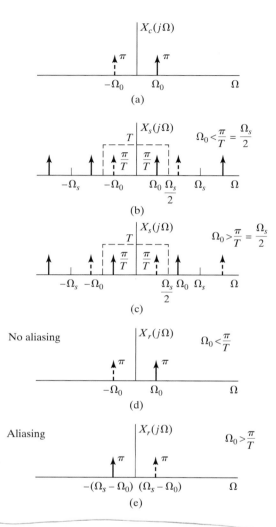

Figure 4.5   The effect of aliasing in the sampling of a cosine signal.

Nyquist Sampling Theorem: Let $x_c(t)$ be a bandlimited signal with

$$X_c(j\Omega) = 0 \qquad \text{for } |\Omega| \geq \Omega_N. \tag{4.14a}$$

Then $x_c(t)$ is uniquely determined by its samples $x[n] = x_c(nT)$, $n = 0, \pm1, \pm2, \ldots$, if

$$\Omega_s = \frac{2\pi}{T} \geq 2\Omega_N. \tag{4.14b}$$

The frequency Ω_N is commonly referred to as the *Nyquist frequency,* and the frequency $2\Omega_N$ that must be exceeded by the sampling frequency is called the *Nyquist rate*.

Thus far, we have considered only the impulse train modulator in Figure 4.2(a). Our eventual objective is to express $X(e^{j\omega})$, the discrete-time Fourier transform of the sequence $x[n]$, in terms of $X_s(j\Omega)$ and $X_c(j\Omega)$. To this end, let us consider an alternative expression for $X_s(j\Omega)$. Applying the continuous-time Fourier transform to Eq. (4.4), we obtain

$$X_s(j\Omega) = \sum_{n=-\infty}^{\infty} x_c(nT)e^{-j\Omega Tn}. \tag{4.15}$$

Since

$$x[n] = x_c(nT) \tag{4.16}$$

and

$$X(e^{j\omega}) = \sum_{n=-\infty}^{\infty} x[n]e^{-j\omega n}, \tag{4.17}$$

it follows that

$$X_s(j\Omega) = X(e^{j\omega})|_{\omega=\Omega T} = X(e^{j\Omega T}). \tag{4.18}$$

Consequently, from Eqs. (4.6) and (4.18),

$$X(e^{j\Omega T}) = \frac{1}{T} \sum_{k=-\infty}^{\infty} X_c(j(\Omega - k\Omega_s)), \tag{4.19}$$

or equivalently,

$$X(e^{j\omega}) = \frac{1}{T} \sum_{k=-\infty}^{\infty} X_c\left(j\left(\frac{\omega}{T} - \frac{2\pi k}{T}\right)\right). \tag{4.20}$$

From Eqs. (4.18)–(4.20), we see that $X(e^{j\omega})$ is simply a frequency-scaled version of $X_s(j\Omega)$ with the frequency scaling specified by $\omega = \Omega T$. This scaling can alternatively be thought of as a normalization of the frequency axis so that the frequency $\Omega = \Omega_s$ in $X_s(j\Omega)$ is normalized to $\omega = 2\pi$ for $X(e^{j\omega})$. The fact that there is a frequency scaling or normalization in the transformation from $X_s(j\Omega)$ to $X(e^{j\omega})$ is directly associated with the fact that there is a time normalization in the transformation from $x_s(t)$ to $x[n]$. Specifically, as we see in Figure 4.2, $x_s(t)$ retains a spacing between samples equal to the sampling period T. In contrast, the "spacing" of sequence values $x[n]$ is always unity; i.e., the time axis is normalized by a factor of T. Correspondingly, in the frequency domain the frequency axis is normalized by a factor of $f_s = 1/T$.

Example 4.1   Sampling and Reconstruction of a Sinusoidal Signal

If we sample the continuous-time signal $x_c(t) = \cos(4000\pi t)$ with sampling period $T = 1/6000$, we obtain $x[n] = x_c(nT) = \cos(4000\pi Tn) = \cos(\omega_0 n)$, where $\omega_0 = 4000\pi T = 2\pi/3$. In this case, $\Omega_s = 2\pi/T = 12000\pi$, and the highest frequency of the signal is $\Omega_0 = 4000\pi$, so the conditions of the Nyquist sampling theorem are satisfied and there is no aliasing. The Fourier transform of $x_c(t)$ is

$$X_c(j\Omega) = \pi\delta(\Omega - 4000\pi) + \pi\delta(\Omega + 4000\pi),$$

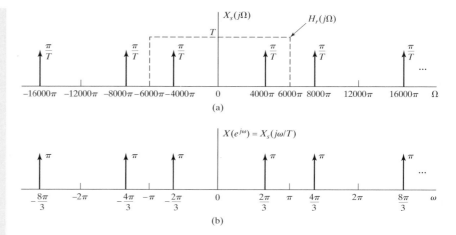

Figure 4.6   Continuous-time (a) and discrete-time (b) Fourier transforms for sampled cosine signal with frequency $\Omega_0 = 4000\pi$ and sampling period $T = 1/6000$.

and Figure 4.6(a) shows

$$X_s(j\Omega) = \frac{1}{T} \sum_{k=-\infty}^{\infty} X_c(j(\Omega - k\Omega_s)) \tag{4.21}$$

for $\Omega_s = 12{,}000\pi$. Note that $X_c(j\Omega)$ is a pair of impulses at $\Omega = \pm 4000\pi$, and we see shifted copies of this Fourier transform centered on $\pm\Omega_s, \pm 2\Omega_s$, etc. Plotting $X(e^{j\omega}) = X_s(j\omega/T)$ as a function of the normalized frequency $\omega = \Omega T$ gives Figure 4.6(b), where we have used the fact that scaling the independent variable of an impulse also scales its area, i.e., $\delta(\omega/T) = T\delta(\omega)$. Note that the original frequency $\Omega_0 = 4000\pi$ corresponds to the normalized frequency $\omega_0 = 4000\pi T = 2\pi/3$, which satisfies the inequality $\omega_0 < \pi$, corresponding to the fact that $\Omega_0 = 4000\pi < \pi/T = 6000\pi$. Figure 4.6(a) also shows the frequency response of the ideal reconstruction filter $H_r(j\Omega)$ for the given sampling rate of $\Omega_s = 12{,}000\pi$. It is clear from this figure that the signal that would be reconstructed would have frequency $\Omega_0 = 4000\pi$, which is the frequency of the original signal $x_c(t)$.

Example 4.2   Aliasing in the Reconstruction of an Undersampled Sinusoidal Signal

Now suppose that the continuous-time signal is $x_c(t) = \cos(16{,}000\pi t)$, but the sampling period is $T = 1/6000$, as it was in Example 4.1. This sampling period fails to satisfy the Nyquist criterion, since $\Omega_s = 2\pi/T = 12{,}000\pi < 2\Omega_0 = 32{,}000\pi$. Consequently, we expect to see aliasing. Now we see an interesting result. The Fourier transform $X_s(j\Omega)$ for this case is identical to that of Figure 4.6(a). However, now the impulse located at $\Omega = -4000\pi$ is from $X_c(j(\Omega - \Omega_s))$ in Eq. (4.21) rather than from $X_c(j\Omega_s)$ and the impulse at $\Omega = 4000\pi$ is from $X_c(j(\Omega + \Omega_s))$. Plotting $X(e^{j\omega}) = X_s(j\omega/T)$ as a function of ω yields the same graph as shown in Figure 4.6(b), since we are normalizing by the same sampling period. The fundamental reason for this is that the sequence of samples is the same in both cases; i.e.,

$$\cos(16{,}000\pi n/6000) = \cos(2\pi n + 4000\pi n/6000) = \cos(2\pi n/3).$$

(Note that we can add any integer multiple of 2π to the argument of the cosine without changing its value.) Thus, we have obtained the same sequence of samples, $x[n] = \cos(2\pi n/3)$, by sampling two different continuous-time signals with the same sampling frequency. In one case the sampling frequency satisfied the Nyquist criterion, and in the other case it did not. As before, Figure 4.6(a) shows the frequency response of the ideal reconstruction filter $H_r(j\Omega)$ for the given sampling rate of $\Omega_s = 12{,}000\pi$. It is clear from this figure that the signal that would be reconstructed would have frequency $\Omega_0 = 4000\pi$, which is *not* the frequency of the original signal $x_c(t)$.

Example 4.3   A Second Example of Aliasing

As a final example, suppose that the signal is again $x_c(t) = \cos(4000\pi t)$, as it was in Example 4.1; i.e., the frequency is again $\Omega_0 = 4000\pi$. However, now the sampling period is increased to $T = 1/1500$. Once more, this sampling period fails to satisfy the Nyquist criterion, since $\Omega_s = 2\pi/T = 3000\pi < 2\Omega_0 = 8000\pi$. Consequently, we expect to see aliasing again. Figure 4.7(a) shows the plot of $X_s(j\Omega)$ in this case. This time the impulse located at $\Omega = -1000\pi$ is from $X_c(j(\Omega - \Omega_s))$, and the impulse at $\Omega = 1000\pi$ is from $X_c(j(\Omega + \Omega_s))$. Plotting $X(e^{j\omega}) = X_s(j\omega/T)$ as a function of ω yields the graph shown in Figure 4.7(b), which we see is identical to Figure 4.6(b). Again, this Fourier transform corresponds to the sequence $x[n] = \cos(2\pi n/3)$. Accordingly, we see that the same discrete-time signal may result from sampling the same continuous-time signal at two different sampling rates if one of those sampling rates fails to satisfy the sampling theorem. The frequency response of the ideal reconstruction filter $H_r(j\Omega)$ for the given sampling rate of $\Omega_s = 3000\pi$ is shown Figure 4.7(a). It is clear from this figure that the signal that would be reconstructed using sampling period $T = 1/1500$ would have frequency $\Omega_0 = 1000\pi$ and not 4000π.

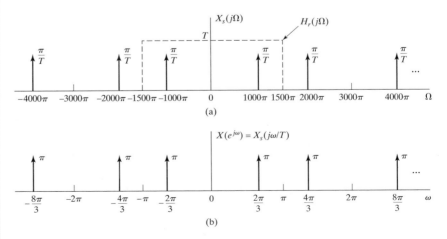

Figure 4.7   Continuous-time (a) and discrete-time (b) Fourier transforms for sampled cosine signal with frequency $\Omega_0 = 4000\pi$ and sampling period $T = 1/1500$.

4.3 RECONSTRUCTION OF A BANDLIMITED SIGNAL FROM ITS SAMPLES

According to the sampling theorem, samples of a continuous-time bandlimited signal taken frequently enough are sufficient to represent the signal exactly, in the sense that the signal can be recovered from the samples and with knowledge of the sampling period. Impulse train modulation provides a convenient means for understanding the process of reconstructing the continuous-time bandlimited signal from its samples.

In Section 4.2 we saw that if the conditions of the sampling theorem are met and if the modulated impulse train is filtered by an appropriate lowpass filter, then the Fourier transform of the filter output will be identical to the Fourier transform of the original continuous-time signal $x_c(t)$, and thus, the output of the filter will be $x_c(t)$. If we are given a sequence of samples, $x[n]$, we can form an impulse train $x_s(t)$ in which successive impulses are assigned an area equal to successive sequence values, i.e.,

$$x_s(t) = \sum_{n=-\infty}^{\infty} x[n]\delta(t - nT). \tag{4.22}$$

The nth sample is associated with the impulse at $t = nT$, where T is the sampling period associated with the sequence $x[n]$. If this impulse train is the input to an ideal lowpass continuous-time filter with frequency response $H_r(j\Omega)$ and impulse response $h_r(t)$, then the output of the filter will be

$$x_r(t) = \sum_{n=-\infty}^{\infty} x[n]h_r(t - nT). \tag{4.23}$$

A block diagram representation of this signal reconstruction process is shown in Figure 4.8(a). Recall that the ideal reconstruction filter has a gain of T (to compensate for the factor of $1/T$ in Eq. (4.19) or (4.20)) and a cutoff frequency Ω_c between Ω_N and $\Omega_s - \Omega_N$. A convenient and commonly used choice of the cutoff frequency is $\Omega_c = \Omega_s/2 = \pi/T$. This choice is appropriate for any relationship between Ω_s and Ω_N that avoids aliasing (i.e., so long as $\Omega_s > 2\Omega_N$). Figure 4.8(b) shows the frequency response of the ideal reconstruction filter. The corresponding impulse response, $h_r(t)$, is the inverse Fourier transform of $H_r(j\Omega)$, and for cutoff frequency π/T it is given by

$$h_r(t) = \frac{\sin(\pi t/T)}{\pi t/T}. \tag{4.24}$$

This impulse response is shown in Figure 4.8(c). From substituting Eq. (4.24) into Eq. (4.23), it follows that

$$x_r(t) = \sum_{n=-\infty}^{\infty} x[n]\frac{\sin[\pi(t - nT)/T]}{\pi(t - nT)/T}. \tag{4.25}$$

From the frequency-domain argument of Section 4.2, we saw that if $x[n] = x_c(nT)$, where $X_c(j\Omega) = 0$ for $|\Omega| \geq \pi/T$, then $x_r(t)$ is equal to $x_c(t)$. It is not immediately obvious that this is true by considering Eq. (4.25) alone. However, useful insight is gained by looking at that equation more closely. First let us consider the function $h_r(t)$ given by Eq. (4.24). We note that

$$h_r(0) = 1. \tag{4.26a}$$

Ideal reconstruction system

(a)

(b)

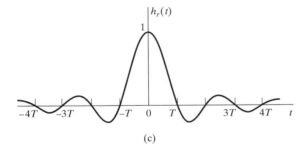

(c)

Figure 4.8 (a) Block diagram of an ideal bandlimited signal reconstruction system. (b) Frequency response of an ideal reconstruction filter. (c) Impulse response of an ideal reconstruction filter.

This follows from l'Hôpital's rule. In addition,

$$h_r(nT) = 0 \qquad \text{for } n = \pm1, \pm2, \ldots. \tag{4.26b}$$

It follows from Eqs. (4.26a) and (4.26b) and Eq. (4.23) that if $x[n] = x_c(nT)$, then

$$x_r(mT) = x_c(mT) \tag{4.27}$$

for all integer values of m. That is, the signal that is reconstructed by Eq. (4.25) has the same values at the sampling times as the original continuous-time signal, independently of the sampling period T.

In Figure 4.9, we show a continuous-time signal $x_c(t)$ and the corresponding modulated impulse train. Figure 4.9(c) shows several of the terms

$$x[n]\frac{\sin[\pi(t-nT)/T]}{\pi(t-nT)/T}$$

and the resulting reconstructed signal $x_r(t)$. As suggested by this figure, the ideal lowpass filter *interpolates* between the impulses of $x_s(t)$ to construct a continuous-time signal $x_r(t)$. From Eq. (4.27), the resulting signal is an exact reconstruction of $x_c(t)$ at the sampling times. The fact that, if there is no aliasing, the lowpass filter interpolates the

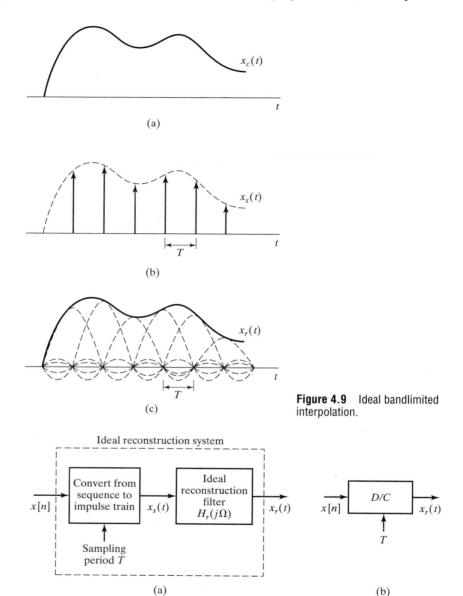

Figure 4.9   Ideal bandlimited interpolation.

Figure 4.10   (a) Ideal bandlimited signal reconstruction. (b) Equivalent represen-tation as an ideal D/C converter.

correct reconstruction between the samples follows from our frequency-domain analysis of the sampling and reconstruction process.

It is useful to formalize the preceding discussion by defining an ideal system for reconstructing a bandlimited signal from a sequence of samples. We will call this system the *ideal discrete-to-continuous-time (D/C) converter*. The desired system is depicted in Figure 4.10. As we have seen, the ideal reconstruction process can be represented as the conversion of the sequence to an impulse train, as in Eq. (4.22), followed by filtering with an ideal lowpass filter, resulting in the output given by Eq. (4.25). The intermediate step of conversion to an impulse train is a mathematical convenience in deriving Eq. (4.25) and in understanding the signal reconstruction process. However, once we are familiar with this process, it is useful to define a more compact representation, as depicted in Figure 4.10(b), where the input is the sequence $x[n]$ and the output is the continuous-time signal $x_r(t)$ given by Eq. (4.25).

The properties of the ideal D/C converter are most easily seen in the frequency domain. To derive an input/output relation in this domain, consider the Fourier transform of Eq. (4.23) or Eq. (4.25), which is

$$X_r(j\Omega) = \sum_{n=-\infty}^{\infty} x[n]H_r(j\Omega)e^{-j\Omega Tn}.$$

By factoring $H_r(j\Omega)$ out of the sum, we can write

$$X_r(j\Omega) = H_r(j\Omega)X(e^{j\Omega T}). \tag{4.28}$$

Equation (4.28) provides a frequency-domain description of the ideal D/C converter. According to Eq. (4.28), $X(e^{j\omega})$ is frequency scaled (i.e., ω is replaced by ΩT). The ideal lowpass filter $H_r(j\Omega)$ selects the base period of the resulting periodic Fourier transform $X(e^{j\Omega T})$ and compensates for the $1/T$ scaling inherent in sampling. Thus, if the sequence $x[n]$ has been obtained by sampling a bandlimited signal at the Nyquist rate or higher, then the reconstructed signal $x_r(t)$ will be equal to the original bandlimited signal. In any case, it is also clear from Eq. (4.28) that the output of the ideal D/C converter is always bandlimited to at most the cutoff frequency of the lowpass filter, which is typically taken to be one-half the sampling frequency.

4.4 DISCRETE-TIME PROCESSING OF CONTINUOUS-TIME SIGNALS

A major application of discrete-time systems is in the processing of continuous-time signals. This is accomplished by a system of the general form depicted in Figure 4.11. The system is a cascade of a C/D converter, followed by a discrete-time system, followed by a D/C converter. The block diagram of Figure 4.11 represents a large class of systems, since the sampling rate and the discrete-time system can be chosen as we wish. Note that the overall system is equivalent to a continuous-time system, since it transforms the continuous-time input signal $x_c(t)$ into the continuous-time output signal $y_r(t)$. The properties of the overall system are dependent on the choice of the discrete-time system and the sampling rate. We assume in Figure 4.11 that the C/D and D/C converters have the same sampling rate. This is not essential, and later sections of this chapter and some of the problems at the end of the chapter consider systems in which the input and output sampling rates are not the same.

Figure 4.11 Discrete-time processing of continuous-time signals.

The previous sections of the chapter have been devoted to understanding the C/D and D/C conversion operations in Figure 4.11. For convenience, and as a first step in understanding the overall system of Figure 4.11, we summarize the mathematical representations of these operations.

The C/D converter produces a discrete-time signal

$$x[n] = x_c(nT), \tag{4.29}$$

i.e., a sequence of samples of the continuous-time input signal $x_c(t)$. The discrete-time Fourier transform of this sequence is related to the continuous-time Fourier transform of the continuous-time input signal by

$$X(e^{j\omega}) = \frac{1}{T} \sum_{k=-\infty}^{\infty} X_c\left(j\left(\frac{\omega}{T} - \frac{2\pi k}{T}\right)\right). \tag{4.30}$$

The D/C converter creates a continuous-time output signal of the form

$$y_r(t) = \sum_{n=-\infty}^{\infty} y[n]\frac{\sin[\pi(t - nT)/T]}{\pi(t - nT)/T}, \tag{4.31}$$

where the sequence $y[n]$ is the output of the discrete-time system when the input to the system is $x[n]$. From Eq. (4.28), $Y_r(j\Omega)$, the continuous-time Fourier transform of $y_r(t)$, and $Y(e^{j\omega})$, the discrete-time Fourier transform of $y[n]$, are related by

$$Y_r(j\Omega) = H_r(j\Omega)Y(e^{j\Omega T})$$

$$= \begin{cases} TY(e^{j\Omega T}), & |\Omega| < \pi/T, \\ 0, & \text{otherwise.} \end{cases} \tag{4.32}$$

Next, let us relate the output sequence $y[n]$ to the input sequence $x[n]$, or equivalently, $Y(e^{j\omega})$ to $X(e^{j\omega})$. A simple example is the identity system, i.e., $y[n] = x[n]$. This is the case that we have studied in detail already. We know that if $x_c(t)$ has a bandlimited Fourier transform such that $X_c(j\Omega) = 0$ for $|\Omega| \geq \pi/T$ and if the discrete-time system in Figure 4.11 is the identity system such that $y[n] = x[n] = x_c(nT)$, then the output will be $y_r(t) = x_c(t)$. Recall that, in proving this result, we utilized the frequency-domain representations of the continuous-time and discrete-time signals, since the key concept of aliasing is most easily understood in the frequency domain. Likewise, when we deal with systems more complicated than the identity system, we generally carry out the analysis in the frequency domain. If the discrete-time system is nonlinear or time varying, it is usually difficult to obtain a general relationship between the Fourier transforms of the input and the output of the system. (In Problem 4.33, we consider an example of the system of Figure 4.11 in which the discrete-time system is nonlinear.) However, the linear time-invariant case leads to a rather simple and very useful result.

4.4.1 Linear Time-Invariant Discrete-Time Systems

If the discrete-time system in Figure 4.11 is linear and time invariant, we then have

$$Y(e^{j\omega}) = H(e^{j\omega})X(e^{j\omega}), \tag{4.33}$$

where $H(e^{j\omega})$ is the frequency response of the system or, equivalently, the Fourier transform of the unit sample response, and $X(e^{j\omega})$ and $Y(e^{j\omega})$ are the Fourier transforms of the input and output, respectively. Combining Eqs. (4.32) and (4.33), we obtain

$$Y_r(j\Omega) = H_r(j\Omega)H(e^{j\Omega T})X(e^{j\Omega T}). \tag{4.34}$$

Next, using Eq. (4.30) with $\omega = \Omega T$, we have

$$Y_r(j\Omega) = H_r(j\Omega)H(e^{j\Omega T})\frac{1}{T}\sum_{k=-\infty}^{\infty} X_c\left(j\left(\Omega - \frac{2\pi k}{T}\right)\right). \tag{4.35}$$

If $X_c(j\Omega) = 0$ for $|\Omega| \geq \pi/T$, then the ideal lowpass reconstruction filter $H_r(j\Omega)$ cancels the factor $1/T$ and selects only the term in Eq. (4.35) for $k = 0$; i.e.,

$$Y_r(j\Omega) = \begin{cases} H(e^{j\Omega T})X_c(j\Omega), & |\Omega| < \pi/T, \\ 0, & |\Omega| \geq \pi/T. \end{cases} \tag{4.36}$$

Thus, if $X_c(j\Omega)$ is bandlimited and the sampling rate is above the Nyquist rate, the output is related to the input through an equation of the form

$$Y_r(j\Omega) = H_{\text{eff}}(j\Omega)X_c(j\Omega), \tag{4.37}$$

where

$$H_{\text{eff}}(j\Omega) = \begin{cases} H(e^{j\Omega T}), & |\Omega| < \pi/T, \\ 0, & |\Omega| \geq \pi/T. \end{cases} \tag{4.38}$$

That is, the overall continuous-time system is equivalent to a linear time-invariant system whose *effective* frequency response is given by Eq. (4.38).

It is important to emphasize that the linear and time-invariant behavior of the system of Figure 4.11 depends on two factors. First, the discrete-time system must be linear and time invariant. Second, the input signal must be bandlimited, and the sampling rate must be high enough so that any aliased components are removed by the discrete-time system. As a simple illustration of this second condition being violated, consider the case when $x_c(t)$ is a single unit-amplitude pulse whose duration is less than the sampling period. If the pulse is unity at $t = 0$, then $x[n] = \delta[n]$. However, it is clearly possible to shift the pulse so that it is not aligned with any of the sampling times, i.e., $x[n] = 0$ for all n. Obviously, such a pulse, being limited in time, is not bandlimited. Even if the discrete-time system is the identity system, such that $y[n] = x[n]$, the overall system will not be time invariant. In general, if the discrete-time system in Figure 4.11 is linear and time invariant, and if the sampling frequency is above the Nyquist rate associated with the bandwidth of the input $x_c(t)$, then the overall system will be equivalent to a linear time-invariant continuous-time system with an effective frequency response given by Eq. (4.38). Furthermore, Eq. (4.38) is valid even if some aliasing occurs in the C/D converter, as long as $H(e^{j\omega})$ does not pass the aliased components. Example 4.4 is a simple illustration of this.

Example 4.4   Ideal Continuous-Time Lowpass Filtering Using a Discrete-Time Lowpass Filter

Consider Figure 4.11, with the linear time-invariant discrete-time system having frequency response

$$H(e^{j\omega}) = \begin{cases} 1, & |\omega| < \omega_c, \\ 0, & \omega_c < |\omega| \leq \pi. \end{cases} \tag{4.39}$$

(a)

(b)

Figure 4.12   (a) Frequency response of discrete-time system in Figure 4.11. (b) Corresponding effective continuous-time frequency response for bandlimited inputs.

This frequency response is, of course, periodic with period 2π, as shown in Figure 4.12(a). For bandlimited inputs sampled above the Nyquist rate, it follows from Eq. (4.38) that the overall system of Figure 4.11 will behave as a linear time-invariant continuous-time system with frequency response

$$
H_{\text{eff}}(j\Omega) = \begin{cases} 1, & |\Omega T| < \omega_c \text{ or } |\Omega| < \omega_c / T, \\ 0, & |\Omega T| > \omega_c \text{ or } |\Omega| > \omega_c / T. \end{cases} \tag{4.40}
$$

As shown in Figure 4.12(b), this effective frequency response is that of an ideal lowpass filter with cutoff frequency $\Omega_c = \omega_c / T$

As an interpretation of this result, consider the graphical illustration given in Figure 4.13. Figure 4.13(a) indicates the Fourier transform of a bandlimited signal. Figure 4.13(b) shows the Fourier transform of the intermediate modulated impulse train, which is identical to $X(e^{j\Omega T})$, the discrete-time Fourier transform of the sequence of samples evaluated for $\omega = \Omega T$. In Figure 4.13(c), the discrete-time Fourier transform of the sequence of samples and the frequency response of the discrete-time system are both plotted as a function of the normalized discrete-time frequency variable ω. Figure 4.13(d) shows $Y(e^{j\omega}) = H(e^{j\omega})X(e^{j\omega})$, the Fourier transform of the output of the discrete-time system. Figure 4.13(e) illustrates the Fourier transform of the output of the discrete-time system as a function of the continuous-time frequency Ω, together with the frequency response of the ideal reconstruction filter $H_r(j\Omega)$ of the D/C converter. Finally, Figure 4.13(f) shows the resulting Fourier transform of the output of the D/C converter. By comparing Figures 4.13(a) and 4.13(f), we see that the system behaves as a linear time-invariant system with frequency response given by Eq. (4.40) and plotted in Figure 4.12(b).

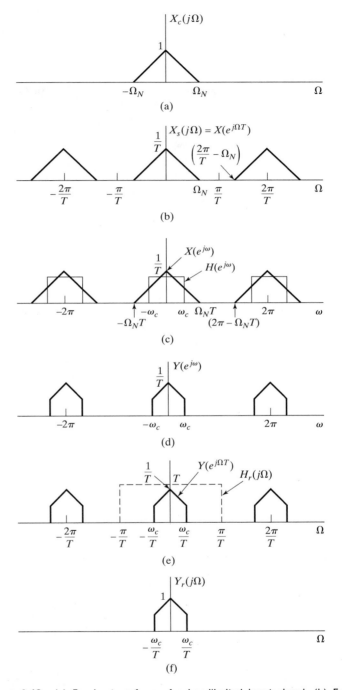

Figure 4.13 (a) Fourier transform of a bandlimited input signal. (b) Fourier transform of sampled input plotted as a function of continuous-time frequency Ω. (c) Fourier transform $X(e^{j\omega})$ of sequence of samples and frequency response $H(e^{j\omega})$ of discrete-time system plotted vs. ω. (d) Fourier transform of output of discrete-time system. (e) Fourier transform of output of discrete-time system and frequency response of ideal reconstruction filter plotted vs. Ω. (f) Fourier transform of output.

Several important points are illustrated in Example 4.4. First, note that the ideal lowpass discrete-time filter with discrete-time cutoff frequency ω_c has the effect of an ideal lowpass filter with cutoff frequency $\Omega_c = \omega_c/T$ when used in the configuration of Figure 4.11. This cutoff frequency depends on both ω_c and T. In particular, by using a fixed discrete-time lowpass filter, but varying the sampling period T, an equivalent continuous-time lowpass filter with a variable cutoff frequency can be implemented. For example, if T were chosen so that $\Omega_N T < \omega_c$, then the output of the system of Figure 4.11 would be $y_r(t) = x_c(t)$. Also, as illustrated in Problem 4.25, Eq. (4.40) will be valid even if some aliasing is present in Figures 4.13(b) and (c), as long as these distorted (aliased) components are eliminated by the filter $H(e^{j\omega})$. In particular, from Figure 4.13(c), we see that for no aliasing to be present in the output, we require that

$$(2\pi - \Omega_N T) > \omega_c, \tag{4.41}$$

compared with the Nyquist requirement that

$$(2\pi - \Omega_N T) > \Omega_N T. \tag{4.42}$$

As another example of continuous-time processing using a discrete-time system, let us consider the implementation of an ideal differentiator for bandlimited signals.

Example 4.5   Discrete-Time Implementation of an Ideal Continuous-Time Bandlimited Differentiator

The ideal continuous-time differentiator system is defined by

$$y_c(t) = \frac{d}{dt}[x_c(t)], \tag{4.43}$$

with corresponding frequency response

$$H_c(j\Omega) = j\Omega. \tag{4.44}$$

Since we are considering a realization in the form of Figure 4.11, the inputs are restricted to be bandlimited. For processing bandlimited signals, it is sufficient that

$$H_{\text{eff}}(j\Omega) = \begin{cases} j\Omega, & |\Omega| < \pi/T, \\ 0, & |\Omega| \geq \pi/T, \end{cases} \tag{4.45}$$

as depicted in Figure 4.14(a). The corresponding discrete-time system has frequency response

$$H(e^{j\omega}) = \frac{j\omega}{T}, \qquad |\omega| < \pi, \tag{4.46}$$

and is periodic with period 2π. This frequency response is plotted in Figure 4.14(b). The corresponding impulse response can be shown to be

$$h[n] = \frac{\pi n \cos \pi n - \sin \pi n}{\pi n^2 T}, \qquad -\infty < n < \infty,$$

or equivalently,

$$h[n] = \begin{cases} 0, & n = 0, \\ \dfrac{\cos \pi n}{nT}, & n \neq 0. \end{cases} \tag{4.47}$$

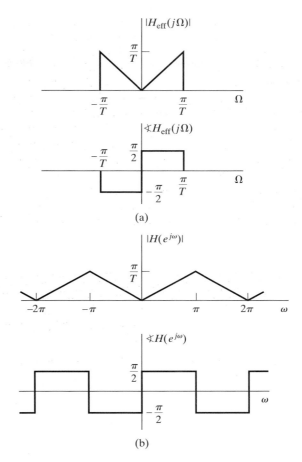

Figure 4.14   (a) Frequency response of a continuous-time ideal bandlimited differentiator $H_c(j\Omega) = j\Omega,\ |\Omega| < \pi/T$. (b) Frequency response of a discrete-time filter to implement a continuous-time bandlimited differentiator.

Thus, if a discrete-time system with this impulse response was used in the configuration of Figure 4.11, the output for every bandlimited input would be the derivative of the input.

Example 4.6   Illustration of Example 4.5 with a Sinusoidal Input

Suppose that the bandlimited differentiator of Example 4.5 has input $x_c(t) = \cos(\Omega_0 t)$ with $\Omega_0 < \pi/T$. The sampled input will be $x[n] = \cos(\omega_0 n)$, where $\omega_0 = \Omega_0 T < \pi$, and the discrete-time Fourier transform, expressed as a function of Ω, is

$$X(e^{j\Omega T}) = \frac{1}{T} \sum_{k=-\infty}^{\infty} [\pi\delta(\Omega - \Omega_0 - k\Omega_s) + \pi\delta(\Omega + \Omega_0 - k\Omega_s)].$$

If we focus on the base band of frequencies $-\pi/T < \Omega < \pi/T$, we obtain

$$X(e^{j\Omega T}) = \frac{\pi}{T}\delta(\Omega - \Omega_0) + \frac{\pi}{T}\delta(\Omega + \Omega_0) \qquad \text{for } |\Omega| \le \pi/T. \qquad (4.48)$$

To express the discrete-time Fourier transform in terms of ω, we substitute $\Omega = \omega/T$ into Eq. (4.48) and use the fact that $\delta(\omega/T) = T\delta(\omega)$. The result is

$$X(e^{j\omega}) = \pi\delta(\omega - \omega_0) + \pi\delta(\omega + \omega_0), \qquad |\omega| \le \pi.$$

The discrete-time Fourier transform $X(e^{j\omega})$ repeats periodically, of course, with period 2π in the variable ω, and $X(e^{j\Omega T})$ repeats periodically with period $2\pi/T$. Now, from Eq. (4.46), the discrete-time Fourier transform of the output is

$$Y(e^{j\omega}) = H(e^{j\omega})X(e^{j\omega})$$

$$= \frac{j\omega}{T}[\pi\delta(\omega - \omega_0) + \pi\delta(\omega + \omega_0)]$$

$$= \frac{j\omega_0\pi}{T}\delta(\omega - \omega_0) - \frac{j\omega_0\pi}{T}\delta(\omega + \omega_0), \qquad |\omega| \le \pi.$$

From Eq. (4.32), the continuous-time Fourier transform of the output of the D/C converter is, for $|\Omega| \le \pi/T$,

$$Y_r(j\Omega) = H_r(j\Omega)Y(e^{j\Omega T}) = TY(e^{j\Omega T})$$

$$= T\left[\frac{j\omega_0\pi}{T}\delta(\Omega T - \Omega_0 T) - \frac{j\omega_0\pi}{T}\delta(\Omega T + \Omega_0 T)\right]$$

$$= T\left[\frac{j\omega_0\pi}{T}\frac{1}{T}\delta(\Omega - \Omega_0) - \frac{j\omega_0\pi}{T}\frac{1}{T}\delta(\Omega + \Omega_0)\right]$$

$$= j\Omega_0\pi\delta(\Omega - \Omega_0) - j\Omega_0\pi\delta(\Omega + \Omega_0).$$

Thus, the reconstruction filter selects the two impulses at $\pm\Omega_0$, so it follows that

$$y_r(t) = j\Omega_0\frac{1}{2}e^{j\Omega_0 t} - j\Omega_0\frac{1}{2}e^{-j\Omega_0 t} = -\Omega_0\sin(\Omega_0 t),$$

and we obtain the expected result that

$$y_r(t) = \frac{d}{dt}[x_c(t)].$$

4.4.2 Impulse Invariance

We have shown that the cascade system of Figure 4.11 can be equivalent to a linear time-invariant system for bandlimited input signals. Let us now assume that, as depicted in Figure 4.15, we are given a desired continuous-time system that we wish to implement in the form of Figure 4.11. With $H_c(j\Omega)$ bandlimited, Eq. (4.38) specifies how to choose $H(e^{j\omega})$ so that $H_{\text{eff}}(j\Omega) = H_c(j\Omega)$. Specifically,

$$H(e^{j\omega}) = H_c(j\omega/T), \qquad |\omega| < \pi, \qquad (4.49)$$

(a)

$H_{\text{eff}}(j\Omega) = H_c(j\Omega)$

(b)

Figure 4.15   (a) Continuous-time LTI system. (b) Equivalent system for bandlimited inputs.

with the further requirement that T be chosen such that

$$H_c(j\Omega) = 0, \qquad |\Omega| \geq \pi/T. \tag{4.50}$$

Under the constraints of Eqs. (4.49) and (4.50), there is also a straightforward and useful relationship between the continuous-time impulse response $h_c(t)$ and the discrete-time impulse response $h[n]$. In particular, as we shall verify shortly,

$$h[n] = Th_c(nT); \tag{4.51}$$

i.e., the impulse response of the discrete-time system is a scaled, sampled version of $h_c(t)$. When $h[n]$ and $h_c(t)$ are related through Eq. (4.51), the discrete-time system is said to be an *impulse-invariant* version of the continuous-time system.

Equation (4.51) is a direct consequence of the discussion in Section 4.2. Specifically, with $x[n]$ and $x_c(t)$ respectively replaced by $h[n]$ and $h_c(t)$ in Eq. (4.16), i.e.,

$$h[n] = h_c(nT), \tag{4.52}$$

Eq. (4.20) becomes

$$H(e^{j\omega}) = \frac{1}{T} \sum_{k=-\infty}^{\infty} H_c\left(j\left(\frac{\omega}{T} - \frac{2\pi k}{T}\right)\right), \tag{4.53}$$

or, if Eq. (4.50) is satisfied,

$$H(e^{j\omega}) = \frac{1}{T} H_c\left(j\frac{\omega}{T}\right), \qquad |\omega| \leq \pi. \tag{4.54}$$

Modifying Eqs. (4.52) and (4.54) to account for the scale factor of T in Eq. (4.51), we have

$$h[n] = Th_c(nT), \tag{4.55}$$

$$H(e^{j\omega}) = H_c\left(j\frac{\omega}{T}\right), \qquad |\omega| \leq \pi. \tag{4.56}$$

Example 4.7   A Discrete-Time Lowpass Filter Obtained By Impulse Invariance

Suppose that we wish to obtain an ideal lowpass discrete-time filter with cutoff frequency $\omega_c < \pi$. We can do this by sampling a continuous-time ideal lowpass filter with cutoff frequency $\Omega_c = \omega_c/T < \pi/T$ defined by

$$H_c(j\Omega) = \begin{cases} 1, & |\Omega| < \Omega_c, \\ 0, & |\Omega| \le \Omega_c. \end{cases}$$

The impulse response of this continuous-time system is

$$h_c(t) = \frac{\sin(\Omega_c t)}{\pi t},$$

so we define the impulse response of the discrete-time system to be

$$h[n] = Th_c(nT) = T\frac{\sin(\Omega_c nT)}{\pi nT} = \frac{\sin(\omega_c n)}{\pi n},$$

where $\omega_c = \Omega_c T$. We have already shown that this sequence corresponds to the discrete-time Fourier transform

$$H(e^{j\omega}) = \begin{cases} 1, & |\omega| < \omega_c, \\ 0, & \omega_c < |\omega| \le \pi, \end{cases}$$

which is identical to $H_c(j\omega/T)$, as predicted by Eq. (4.56).

Example 4.8   Impulse Invariance Applied to Continuous-Time Systems with Rational System Functions

Many continuous-time systems have impulse responses composed of a sum of exponential sequences of the form

$$h_c(t) = Ae^{s_0 t}u(t).$$

Such time functions have Laplace transforms

$$H_c(s) = \frac{A}{s - s_0}.$$

If we apply the impulse invariance concept to such a continuous-time system, we obtain the impulse response

$$h[n] = Th_c(nT) = Ae^{s_0 Tn}u[n],$$

which has z-transform system function

$$H(z) = \frac{AT}{1 - e^{s_0 T}z^{-1}}$$

and frequency response

$$H(e^{j\omega}) = \frac{AT}{1 - e^{s_0 T}e^{-j\omega}}.$$

In this case, Eq. (4.56) does not hold exactly, because the original continuous-time system did not have a strictly bandlimited frequency response, and therefore, the resulting discrete-time frequency response is an *aliased* version of $H_c(j\Omega)$. Even though aliasing occurs in such a case as this, the effect may be small. Higher order systems whose impulse responses are sums of complex exponentials may in fact have frequency responses that fall off rapidly at high frequencies, so that aliasing is minimal if the sampling rate is high enough. Thus, one approach to the discrete-time simulation of continuous-time systems and also to the design of digital filters is through sampling of the impulse response of a corresponding analog filter.

4.5 CONTINUOUS-TIME PROCESSING OF DISCRETE-TIME SIGNALS

In Section 4.4, we discussed and analyzed the use of discrete-time systems for processing continuous-time signals in the configuration of Figure 4.11. In this section we consider the complementary situation depicted in Figure 4.16, which is appropriately referred to as continuous-time processing of discrete-time signals. While the system of Figure 4.16 is not typically used to implement discrete-time systems, it provides a useful interpretation of certain discrete-time systems.

From the definition of the ideal D/C converter, $X_c(j\Omega)$ and therefore also $Y_c(j\Omega)$, will necessarily be zero for $|\Omega| \geq \pi/T$. Thus, the C/D converter samples $y_c(t)$ without aliasing, and we can express $x_c(t)$ and $y_c(t)$ respectively as

$$x_c(t) = \sum_{n=-\infty}^{\infty} x[n]\frac{\sin[\pi(t-nT)/T]}{\pi(t-nT)/T} \tag{4.57}$$

and

$$y_c(t) = \sum_{n=-\infty}^{\infty} y[n]\frac{\sin[\pi(t-nT)/T]}{\pi(t-nT)/T}, \tag{4.58}$$

where $x[n] = x_c(nT)$ and $y[n] = y_c(nT)$. The frequency-domain relationships for Figure 4.16 are

$$X_c(j\Omega) = TX(e^{j\Omega T}), \qquad |\Omega| < \pi/T, \tag{4.59a}$$

$$Y_c(j\Omega) = H_c(j\Omega)X_c(j\Omega), \qquad |\Omega| < \pi/T, \tag{4.59b}$$

$$Y(e^{j\omega}) = \frac{1}{T}Y_c\left(j\frac{\omega}{T}\right), \qquad |\omega| < \pi. \tag{4.59c}$$

Therefore, by substituting Eqs. (4.59a) and (4.59b) into Eq. (4.59c), it follows that the overall system behaves as a discrete-time system whose frequency response is

$$H(e^{j\omega}) = H_c\left(j\frac{\omega}{T}\right), \qquad |\omega| < \pi, \tag{4.60}$$

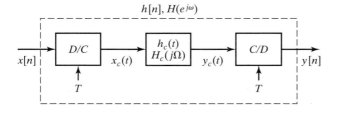

Figure 4.16   Continuous-time processing of discrete-time signals.

or equivalently, the overall frequency response of the system in Figure 4.16 will be equal to a given $H(e^{j\omega})$ if the frequency response of the continuous-time system is

$$H_c(j\Omega) = H(e^{j\Omega T}), \qquad |\Omega| < \pi/T. \tag{4.61}$$

Since $X_c(j\Omega) = 0$ for $|\Omega| \geq \pi/T$, $H_c(j\Omega)$ may be chosen arbitrarily above π/T. A convenient, but arbitrary, choice is $H_c(j\Omega) = 0$ for $|\Omega| \geq \pi/T$.

With this representation of a discrete-time system, we can focus on the equivalent effect of the continuous-time system on the bandlimited continuous-time signal $x_c(t)$. This is illustrated in Examples 4.9 and 4.10.

Example 4.9   Noninteger Delay

Let us consider a discrete-time system with frequency response

$$H(e^{j\omega}) = e^{-j\omega\Delta}, \qquad |\omega| < \pi. \tag{4.62}$$

When Δ is an integer, this system has a straightforward interpretation as a delay of Δ, i.e.,

$$y[n] = x[n - \Delta]. \tag{4.63}$$

When Δ is not an integer, Eq. (4.63) has no formal meaning, because we cannot shift the sequence $x[n]$ by anything but an integer. However, with the use of the system of Figure 4.16, a useful time-domain interpretation can be applied to the system specified by Eq. (4.62). Let $H_c(j\Omega)$ in Figure 4.16 be chosen to be

$$H_c(j\Omega) = H(e^{j\Omega T}) = e^{-j\Omega\Delta T}. \tag{4.64}$$

Then, from Eq. (4.61), the overall discrete-time system in Figure 4.16 will have the frequency response given by Eq. (4.62), whether or not Δ is an integer. To interpret the system of Eq. (4.62), we note that Eq. (4.64) represents a time delay of ΔT seconds. Therefore,

$$y_c(t) = x_c(t - \Delta T). \tag{4.65}$$

Furthermore, $x_c(t)$ is the bandlimited interpolation of $x[n]$, and $y[n]$ is obtained by sampling $y_c(t)$. For example, if $\Delta = \frac{1}{2}$, $y[n]$ would be the values of the bandlimited

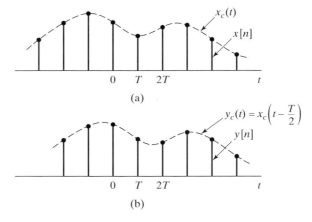

Figure 4.17   Continuous-time processing of the discrete-time sequence in part (a) can produce a new sequence with a "half-sample" delay, as in part (b).

interpolation halfway between the input sequence values. This is illustrated in Figure 4.17. We can also obtain a direct convolution representation for the system defined by Eq. (4.62). From Eqs. (4.65) and (4.57), we obtain

$$
\begin{aligned}
y[n] = y_c(nT) &= x_c(nT - \Delta T) \\
&= \sum_{k=-\infty}^{\infty} x[k] \left. \frac{\sin[\pi(t - \Delta T - kT)/T]}{\pi(t - \Delta T - kT)/T} \right|_{t=nT} \\
&= \sum_{k=-\infty}^{\infty} x[k] \frac{\sin \pi(n - k - \Delta)}{\pi(n - k - \Delta)},
\end{aligned}
\tag{4.66}
$$

which is, by definition, the convolution of $x[n]$ with

$$
h[n] = \frac{\sin \pi(n - \Delta)}{\pi(n - \Delta)}, \qquad -\infty < n < \infty.
$$

When Δ is not an integer, $h[n]$ has infinite extent. However, when $\Delta = n_0$ is an integer, it is easily shown that $h[n] = \delta[n - n_0]$, which is the impulse response of the ideal integer delay system.

The noninteger delay represented by Eq. (4.66) has considerable practical significance, since such a factor often arises in the frequency-domain representation of systems. When this kind of term is found in the frequency response of a causal discrete-time system, it can be interpreted in the light of this example. This interpretation is illustrated in Example 4.10.

Example 4.10   Moving-Average System with Noninteger Delay

In Example 2.20, we considered the general moving-average system and obtained its frequency response. For the case of the causal $(M + 1)$-point moving-average system, $M_1 = 0$ and $M_2 = M$, and the frequency response is

$$
H(e^{j\omega}) = \frac{1}{(M + 1)} \frac{\sin[\omega(M + 1)/2]}{\sin(\omega/2)} e^{-j\omega M/2}, \qquad |\omega| < \pi.
\tag{4.67}
$$

Figure 4.18 The moving-average system represented as a cascade of two systems.

This representation of the frequency response suggests the interpretation of the $(M + 1)$-point moving-average system as the cascade of two systems, as indicated in Figure 4.18. The first system imposes a frequency-domain amplitude weighting. The second system represents the linear-phase term in Eq. (4.67). If M is an even integer (meaning the moving average of an odd number of samples), then the linear-phase term corresponds to an integer delay, i.e.,

$$y[n] = w[n - M/2]. \tag{4.68}$$

However, if M is odd, the linear-phase term corresponds to a noninteger delay, specifically, an integer-plus-one-half sample interval. This noninteger delay can be interpreted in terms of the discussion in Example 4.9; i.e., $y[n]$ is equivalent to bandlimited interpolation of $w[n]$, followed by a continuous-time delay of $MT/2$ (where T is the assumed, but arbitrary, sampling period associated with the D/C interpolation of $w[n]$), followed by C/D conversion again with sampling period T. This fractional delay is illustrated in Figure 4.19. Figure 4.19(a) shows a discrete-time sequence $x[n] = \cos(0.25\pi n)$. This

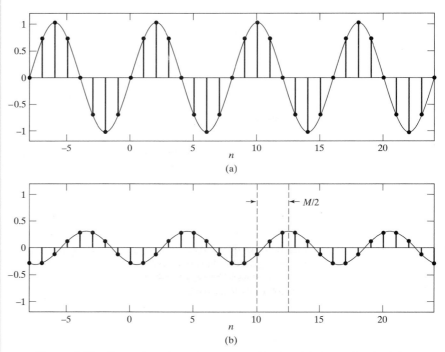

Figure 4.19 Illustration of moving-average filtering. (a) Input signal $x[n] = \cos(0.25\pi n)$. (b) Corresponding output of six-point moving-average filter.

sequence is the input to a six-point ($M = 5$) moving-average filter. In this example, the input is "turned on" far enough in the past so that the output consists only of the steady-state response for the time interval shown. Figure 4.19(b) shows the corresponding output sequence, which is given by

$$y[n] = H(e^{j0.25\pi})\frac{1}{2}e^{j0.25\pi n} + H(e^{-j0.25\pi})\frac{1}{2}e^{-j0.25\pi n}$$

$$= \frac{1}{2}\frac{\sin[3(0.25\pi)]}{6\sin(0.125\pi)}e^{-j(0.25\pi)5/2}e^{j0.25\pi n} + \frac{1}{2}\frac{\sin[3(-0.25\pi)]}{6\sin(-0.125\pi)}e^{j(0.25\pi)5/2}e^{-j0.25\pi n}$$

$$= 0.308\cos[0.25\pi(n - 2.5)].$$

Thus, the six-point moving-average filter reduces the amplitude of the cosine signal and introduces a phase shift that corresponds to 2.5 samples of delay. This is readily apparent in Figure 4.19, where we have plotted the continuous-time cosines that would be interpolated by the ideal D/C converter for both the input and the output sequence. Note in Figure 4.19(b) that the six-point moving-average filtering gives a sampled cosine signal such that the sample points have been shifted by 2.5 samples with respect to the sample points of the input. This can be seen from Figure 4.19(b) by comparing the positive peak at 10 in the interpolated cosine for the input to the positive peak at 12.5 in the interpolated cosine for the output. Thus, the six-point moving-average filter is seen to have a delay of $5/2 = 2.5$ samples.

4.6 CHANGING THE SAMPLING RATE USING DISCRETE-TIME PROCESSING

We have seen that a continuous-time signal $x_c(t)$ can be represented by a discrete-time signal consisting of a sequence of samples

$$x[n] = x_c(nT). \tag{4.69}$$

Alternatively, our previous discussion has shown that, even if $x[n]$ was not obtained originally by sampling, we can always use the bandlimited interpolation formula of Eq. (4.25) to find a continuous-time bandlimited signal $x_r(t)$ whose samples are $x[n] = x_c(nT)$.

It is often necessary to change the sampling rate of a discrete-time signal, i.e., to obtain a new discrete-time representation of the underlying continuous-time signal of the form

$$x'[n] = x_c(nT'), \tag{4.70}$$

where $T' \neq T$. One approach to obtaining the sequences $x'[n]$ from $x[n]$ is to reconstruct $x_c(t)$ from $x[n]$ using Eq. (4.25) and then resample $x_c(t)$ with period T' to obtain $x'[n]$. Often, however, this is not a desirable approach, because of the nonideal analog reconstruction filter, D/A converter, and A/D converter that would be used in a practical implementation. Thus, it is of interest to consider methods of changing the sampling rate that involve only discrete-time operations.

4.6.1 Sampling Rate Reduction by an Integer Factor

The sampling rate of a sequence can be reduced by "sampling" it, i.e., by defining a new sequence

$$x_d[n] = x[nM] = x_c(nMT). \tag{4.71}$$

Equation (4.71) defines the system depicted in Figure 4.20, which is called a *sampling rate compressor* (see Crochiere and Rabiner, 1983) or simply a *compressor*. From Eq. (4.71), it is clear that $x_d[n]$ is identical to the sequence that would be obtained from $x_c(t)$ by

$x[n]$

Sampling period T

$x_d[n] = x[nM]$

Sampling period $T' = MT$

Figure 4.20   Representation of a compressor or discrete-time sampler.

sampling with period $T' = MT$. Furthermore, if $X_c(j\Omega) = 0$ for $|\Omega| \geq \Omega_N$, then $x_d[n]$ is an exact representation of $x_c(t)$ if $\pi/T' = \pi/(MT) \geq \Omega_N$. That is, the sampling rate can be reduced by a factor of M without aliasing if the original sampling rate was at least M times the Nyquist rate or if the bandwidth of the sequence is first reduced by a factor of M by discrete-time filtering. In general, the operation of reducing the sampling rate (including any prefiltering) will be called *downsampling*.

As in the case of sampling a continuous-time signal, it is useful to obtain a frequency-domain relation between the input and output of the compressor. This time, however, it will be a relationship between discrete-time Fourier transforms. Although several methods can be used to derive the desired result, we will base our derivation on the results already obtained for sampling continuous-time signals. First recall that the discrete-time Fourier transform of $x[n] = x_c(nT)$ is

$$X(e^{j\omega}) = \frac{1}{T} \sum_{k=-\infty}^{\infty} X_c\left(j\left(\frac{\omega}{T} - \frac{2\pi k}{T}\right)\right). \tag{4.72}$$

Similarly, the discrete-time Fourier transform of $x_d[n] = x[nM] = x_c(nT')$ with $T' = MT$ is

$$X_d(e^{j\omega}) = \frac{1}{T'} \sum_{r=-\infty}^{\infty} X_c\left(j\left(\frac{\omega}{T'} - \frac{2\pi r}{T'}\right)\right). \tag{4.73}$$

Now, since $T' = MT$, we can write Eq. (4.73) as

$$X_d(e^{j\omega}) = \frac{1}{MT} \sum_{r=-\infty}^{\infty} X_c\left(j\left(\frac{\omega}{MT} - \frac{2\pi r}{MT}\right)\right). \tag{4.74}$$

To see the relationship between Eqs. (4.74) and (4.72), note that the summation index r in Eq. (4.74) can be expressed as

$$r = i + kM, \tag{4.75}$$

where k and i are integers such that $-\infty < k < \infty$ and $0 \leq i \leq M - 1$. Clearly, r is still an integer ranging from $-\infty$ to ∞, but now Eq. (4.74) can be expressed as

$$X_d(e^{j\omega}) = \frac{1}{M} \sum_{i=0}^{M-1} \left[\frac{1}{T} \sum_{k=-\infty}^{\infty} X_c\left(j\left(\frac{\omega}{MT} - \frac{2\pi k}{T} - \frac{2\pi i}{MT}\right)\right) \right]. \tag{4.76}$$

The term inside the square brackets in Eq. (4.76) is recognized from Eq. (4.72) as

$$X(e^{j(\omega - 2\pi i)/M}) = \frac{1}{T} \sum_{k=-\infty}^{\infty} X_c\left(j\left(\frac{\omega - 2\pi i}{MT} - \frac{2\pi k}{T}\right)\right). \tag{4.77}$$

Thus, we can express Eq. (4.76) as

$$X_d(e^{j\omega}) = \frac{1}{M} \sum_{i=0}^{M-1} X(e^{j(\omega/M - 2\pi i/M)}). \tag{4.78}$$

There is a strong analogy between Eqs. (4.72) and (4.78): Equation (4.72) expresses the Fourier transform of the sequence of samples, $x[n]$ (with period T), in terms of the Fourier transform of the continuous-time signal $x_c(t)$; Equation (4.78) expresses the Fourier transform of the discrete-time sampled sequence $x_d[n]$ (with sampling period M) in terms of the Fourier transform of the sequence $x[n]$. If we compare Eqs. (4.73) and (4.78), we see that $X_d(e^{j\omega})$ can be thought of as being composed of either an infinite set of copies of $X_c(j\Omega)$, frequency scaled through $\omega = \Omega T'$ and shifted by integer multiples of $2\pi/T'$ (Eq. (4.73)), or M copies of the periodic Fourier transform $X(e^{j\omega})$, frequency scaled by M and shifted by integer multiples of 2π (Eq. (4.78)). Either interpretation makes it clear that $X_d(e^{j\omega})$ is periodic with period 2π (as are all discrete-time Fourier transforms) and that aliasing can be avoided by ensuring that $X(e^{j\omega})$ is bandlimited, i.e.,

$$X(e^{j\omega}) = 0, \qquad \omega_N \le |\omega| \le \pi, \tag{4.79}$$

and $2\pi/M \ge 2\omega_N$.

Downsampling is illustrated in Figure 4.21. Figure 4.21(a) shows the Fourier transform of a bandlimited continuous-time signal, and Figure 4.21(b) shows the Fourier transform of the impulse train of samples when the sampling period is T. Figure 4.21(c) shows $X(e^{j\omega})$ and is related to Figure 4.21(b) through Eq. (4.18). As we have already seen, Figures 4.21(b) and (c) differ only in a scaling of the frequency variable. Figure 4.21(d) shows the discrete-time Fourier transform of the downsampled sequence when $M = 2$. We have plotted this Fourier transform as a function of the normalized frequency $\omega = \Omega T'$. Finally, Figure 4.21(e) shows the discrete-time Fourier transform of the downsampled sequence plotted as a function of the continuous-time frequency variable Ω. Figure 4.21(e) is identical to Figure 4.21(d), except for the scaling of the frequency axis through the relation $\Omega = \omega/T'$.

In this example, $2\pi/T = 4\Omega_N$; i.e., the original sampling rate is exactly twice the minimum rate to avoid aliasing. Thus, when the original sampled sequence is downsampled by a factor of $M = 2$, no aliasing results. If the downsampling factor is more than 2 in this case, aliasing will result, as illustrated in Figure 4.22.

Figure 4.22(a) shows the continuous-time Fourier transform of $x_c(t)$, and Figure 4.22(b) shows the discrete-time Fourier transform of the sequence $x[n] = x_c(nT)$, when $2\pi/T = 4\Omega_N$. Thus, $\omega_N = \Omega_N T = \pi/2$. Now, if we downsample by a factor of $M = 3$, we obtain the sequence $x_d[n] = x[3n] = x_c(n3T)$ whose discrete-time Fourier transform is plotted in Figure 4.22(c) with normalized frequency $\omega = \Omega T'$. Note that because $M\omega_N = 3\pi/2$, which is greater than π, aliasing occurs. In general, to avoid aliasing in downsampling by a factor of M requires that

$$\omega_N M < \pi \qquad \text{or} \qquad \omega_N < \pi/M. \tag{4.80}$$

If this condition does not hold, aliasing occurs, but it may be tolerable for some applications. In other cases, downsampling can be done without aliasing if we are willing to reduce the bandwidth of the signal $x[n]$ before downsampling. Thus, if $x[n]$ is filtered by an ideal lowpass filter with cutoff frequency π/M, then the output $\tilde{x}[n]$ can be downsampled without aliasing, as illustrated in Figures 4.22(d), (e), and (f). Note that the sequence $\tilde{x}_d[n] = \tilde{x}[nM]$ no longer represents the original underlying continuous-time signal $x_c(t)$. Rather, $\tilde{x}_d[n] = \tilde{x}_c(nT')$, where $T' = MT$, and $\tilde{x}_c(t)$ is obtained from $x_c(t)$ by lowpass filtering with cutoff frequency $\Omega_c = \pi/T' = \pi/(MT)$.

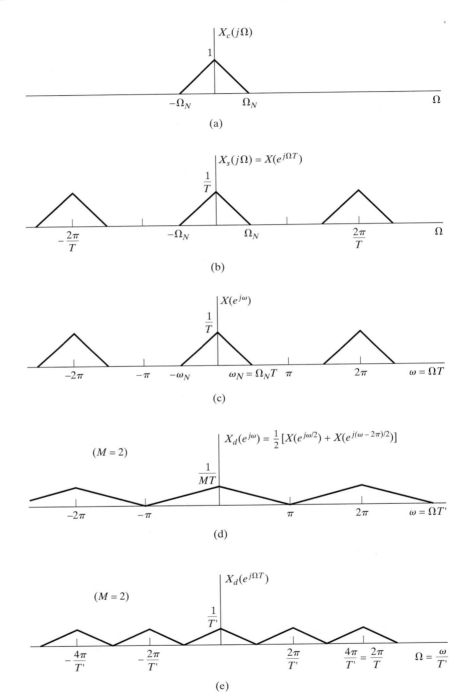

Figure 4.21   Frequency-domain illustration of downsampling.

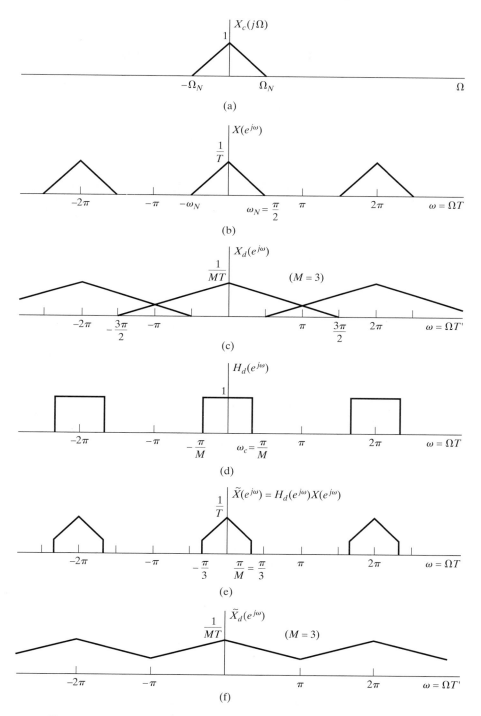

Figure 4.22 (a)–(c) Downsampling with aliasing. (d)–(f) Downsampling with prefiltering to avoid aliasing.

$x[n]$

Sampling
period T

$\tilde{x}[n]$

Sampling
period T

$\tilde{x}_d[n] = \tilde{x}[nM]$

Sampling
period $T' = MT$

Figure 4.23 General system for
sampling rate reduction by M.

From the preceding discussion, we see that a general system for downsampling by
a factor of M is the one shown in Figure 4.23. Such a system is called a *decimator*, and
downsampling by lowpass filtering followed by compression has been termed *decimation*
(Crochiere and Rabiner, 1983).

4.6.2 Increasing the Sampling Rate by an Integer Factor

We have seen that the reduction of the sampling rate of a discrete-time signal by an
integer factor involves sampling the sequence in a manner analogous to sampling a
continuous-time signal. Not surprisingly, increasing the sampling rate involves operations analogous to D/C conversion. To see this, consider a signal $x[n]$ whose sampling
rate we wish to increase by a factor of L. If we consider the underlying continuous-time
signal $x_c(t)$, the objective is to obtain samples

$$x_i[n] = x_c(nT'), \tag{4.81}$$

where $T' = T/L$, from the sequence of samples

$$x[n] = x_c(nT). \tag{4.82}$$

We will refer to the operation of increasing the sampling rate as *upsampling*.

From Eqs. (4.81) and (4.82) it follows that

$$x_i[n] = x[n/L] = x_c(nT/L), \qquad n = 0, \pm L, \pm 2L, \ldots. \tag{4.83}$$

Figure 4.24 shows a system for obtaining $x_i[n]$ from $x[n]$ using only discrete-time processing. The system on the left is called a *sampling rate expander* (see Crochiere and
Rabiner, 1983) or simply an *expander*. Its output is

$$x_e[n] = \begin{cases} x[n/L], & n = 0, \pm L, \pm 2L, \ldots, \\ 0, & \text{otherwise,} \end{cases} \tag{4.84}$$

or equivalently,

$$x_e[n] = \sum_{k=-\infty}^{\infty} x[k]\delta[n - kL]. \tag{4.85}$$

The system on the right is a lowpass discrete-time filter with cutoff frequency π/L and
gain L. This system plays a role similar to the ideal D/C converter in Figure 4.10(b).
First we create a discrete-time impulse train $x_e[n]$, and then we use a lowpass filter to
reconstruct the sequence.

The operation of the system in Figure 4.24 is most easily understood in the frequency domain. The Fourier transform of $x_e[n]$ can be expressed as

$$X_e(e^{j\omega}) = \sum_{n=-\infty}^{\infty} \left(\sum_{k=-\infty}^{\infty} x[k]\delta[n - kL] \right) e^{-j\omega n}$$

$$= \sum_{k=-\infty}^{\infty} x[k]e^{-j\omega Lk} = X(e^{j\omega L}). \tag{4.86}$$

Figure 4.24   General system for sampling rate increase by L.

Thus, the Fourier transform of the output of the expander is a frequency-scaled version of the Fourier transform of the input; i.e., ω is replaced by ωL so that ω is now normalized by

$$\omega = \Omega T'. \tag{4.87}$$

This effect is illustrated in Figure 4.25. Figure 4.25(a) shows a bandlimited continuous-time Fourier transform, and Figure 4.25(b) shows the discrete-time Fourier transform of the sequence $x[n] = x_c(nT)$, where $\pi/T = \Omega_N$. Figure 4.25(c) shows $X_e(e^{j\omega})$ according to Eq. (4.86), with $L = 2$, and Figure 4.25(e) shows the Fourier transform of the desired signal $x_i[n]$. We see that $X_i(e^{j\omega})$ can be obtained from $X_e(e^{j\omega})$ by correcting the amplitude scale from $1/T$ to $1/T'$ and by removing all the frequency-scaled images of $X_c(j\Omega)$ except at integer multiples of 2π. For the case depicted in Figure 4.25, this requires a lowpass filter with a gain of 2 and cutoff frequency $\pi/2$, as shown in Figure 4.25(d). In general, the required gain would be L, since $L(1/T) = [1/(T/L)] = 1/T'$, and the cutoff frequency would be π/L.

This example shows that the system of Figure 4.24 does indeed give an output satisfying Eq. (4.81) if the input sequence $x[n] = x_c(nT)$ was obtained by sampling without aliasing. That system is therefore called an *interpolator,* since it fills in the missing samples, and the operation of upsampling is therefore considered to be synonymous with *interpolation.*

As in the case of the D/C converter, it is possible to obtain an interpolation formula for $x_i[n]$ in terms of $x[n]$. First note that the impulse response of the lowpass filter in Figure 4.24 is

$$h_i[n] = \frac{\sin(\pi n/L)}{\pi n/L}. \tag{4.88}$$

Using Eq. (4.85), we obtain

$$x_i[n] = \sum_{k=-\infty}^{\infty} x[k] \frac{\sin[\pi(n - kL)/L]}{\pi(n - kL)/L}. \tag{4.89}$$

The impulse response $h_i[n]$ has the properties

$$\begin{aligned} h_i[0] &= 1, \\ h_i[n] &= 0, \qquad n = \pm L, \pm 2L, \ldots. \end{aligned} \tag{4.90}$$

Thus, for the ideal lowpass interpolation filter, we have

$$x_i[n] = x[n/L] = x_c(nT/L) = x_c(nT'), \qquad n = 0, \pm L, \pm 2L, \ldots, \tag{4.91}$$

as desired. The fact that $x_i[n] = x_c(nT')$ for all n follows from our frequency-domain argument.

In practice, ideal lowpass filters cannot be implemented exactly, but we will see in Chapter 7 that very good approximations can be designed. (Also, see Schafer and

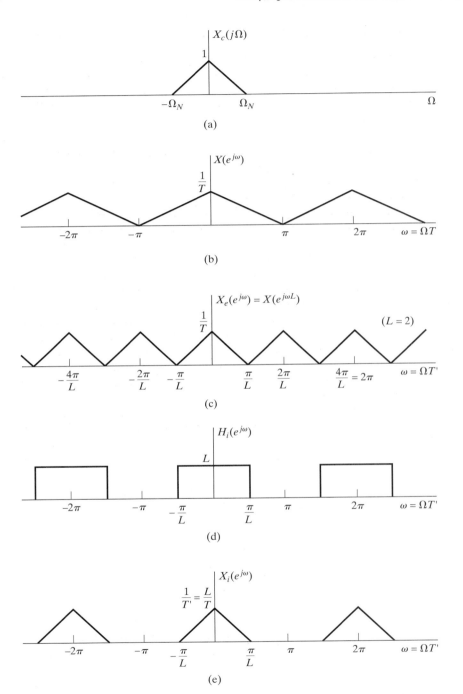

Figure 4.25   Frequency-domain illustration of interpolation.

Rabiner, 1973, and Oetken et al., 1975.) In some cases, very simple interpolation proce-dures are adequate. Since linear interpolation is often used (even though it is generally not very accurate), it is worthwhile to examine linear interpolation within the general framework that we have just developed.

Linear interpolation can be accomplished by the system of Figure 4.24 if the filter has impulse response

$$h_{\mathrm{lin}}[n] = \begin{cases} 1 - |n|/L, & |n| \le L, \\ 0, & \text{otherwise,} \end{cases} \tag{4.92}$$

as shown in Figure 4.26 for $L = 5$. With this filter, the interpolated output will be

$$x_{\mathrm{lin}}[n] = \sum_{k=-\infty}^{\infty} x_e[k] h_{\mathrm{lin}}[n-k] = \sum_{k=-\infty}^{\infty} x[k] h_{\mathrm{lin}}[n-kL]. \tag{4.93}$$

Figure 4.27(a) depicts $x_e[n]$ and $x_{\mathrm{lin}}[n]$ for the case $L = 5$. From this figure, we see that

Figure 4.26   Impulse response for linear interpolation.

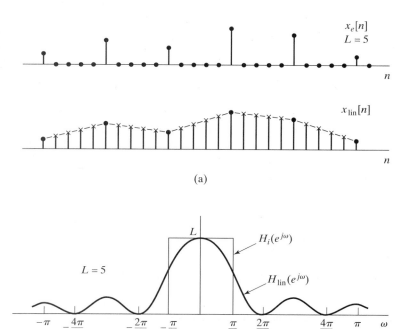

(a)

(b)

Figure 4.27   (a) Illustration of linear interpolation by filtering. (b) Frequency response of linear interpolator compared with ideal lowpass interpolation filter.

$x_{\text{lin}}[n]$ is identical to the sequence obtained by linear interpolation between the samples. Note that

$$h_{\text{lin}}[0] = 1,$$
$$h_{\text{lin}}[n] = 0, \qquad n = \pm L, \pm 2L, \ldots, \tag{4.94}$$

so that

$$x_{\text{lin}}[n] = x[n/L] \quad \text{at} \quad n = 0, \pm L, \pm 2L, \ldots . \tag{4.95}$$

The amount of distortion in the intervening samples can be gauged by comparing the frequency response of the linear interpolator with that of the ideal lowpass interpolator for a factor-of-L interpolation. It can be shown (see Problem 4.50) that

$$H_{\text{lin}}(e^{j\omega}) = \frac{1}{L} \left[\frac{\sin(\omega L/2)}{\sin(\omega/2)} \right]^2 . \tag{4.96}$$

This function is plotted in Figure 4.27(b) for $L = 5$, together with the ideal lowpass interpolation filter. From the figure we see that if the original signal is sampled at the Nyquist rate, linear interpolation will not be very good, since the output of the filter will contain considerable energy in the band $\pi/L < |\omega| \leq \pi$. However, if the original sampling rate is much higher than the Nyquist rate, then the linear interpolator will be more successful in removing the frequency-scaled images of $X_c(j\Omega)$ at multiples of $2\pi/L$. This is because $H_{\text{lin}}(e^{j\omega})$ is small at these normalized frequencies and at higher sampling rates the shifted copies of $X_c(j\Omega)$ are more localized at these frequencies. This is intuitively reasonable, since, if the original sampling rate greatly exceeds the Nyquist rate, the signal will not vary significantly between samples, and thus, linear interpolation should be more accurate for oversampled signals.

4.6.3 Changing the Sampling Rate by a Noninteger Factor

We have shown how to increase or decrease the sampling rate of a sequence by an integer factor. By combining decimation and interpolation, it is possible to change the sampling rate by a noninteger factor. Specifically, consider Figure 4.28(a), which shows an interpolator that decreases the sampling period from T to T/L, followed by a decimator that increases the sampling period by M, producing an output sequence $\tilde{x}_d[n]$ that has an effective sampling period of $T' = TM/L$. By choosing L and M appropriately, we can approach arbitrarily close to any desired ratio of sampling periods. For example, if $L = 100$ and $M = 101$, then $T' = 1.01T$.

If $M > L$, there is a net increase in the sampling period (a decrease in the sampling rate), and if $M < L$, the opposite is true. Since the interpolation and decimation filters in Figure 4.28(a) are in cascade, they can be combined as shown in Figure 4.28(b) into one lowpass filter with gain L and cutoff equal to the minimum of π/L and π/M. If $M > L$, then π/M is the dominant cutoff frequency, and there is a net reduction in sampling rate. As pointed out in Section 4.6.1, if $x[n]$ was obtained by sampling at the Nyquist rate, the sequence $\tilde{x}_d[n]$ will be a lowpass-filtered version of the original underlying bandlimited signal if we are to avoid aliasing. On the other hand, if $M < L$, then π/L is the dominant cutoff frequency, and there will be no need to further limit the bandwidth of the signal below the original Nyquist frequency.

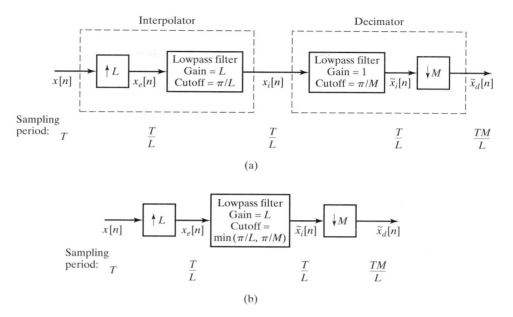

Figure 4.28   (a) System for changing the sampling rate by a noninteger factor. (b) Simplified system in which the decimation and interpolation filters are combined.

Example 4.11   Sampling Rate Conversion by a Noninteger Rational Factor

Figure 4.29 illustrates sampling rate conversion by a rational factor. Suppose that a bandlimited signal with $X_c(j\Omega)$ as given in Figure 4.29(a) is sampled at the Nyquist rate; i.e., $2\pi/T = 2\Omega_N$. The resulting discrete-time Fourier transform

$$X(e^{j\omega}) = \frac{1}{T} \sum_{k=-\infty}^{\infty} X_c \left(j \left(\frac{\omega}{T} - \frac{2\pi k}{T} \right) \right)$$

is plotted in Figure 4.29(b). If we wish to change the sampling period to $T' = (3/2)T$, we must first interpolate by a factor $L = 2$ and then decimate by a factor of $M = 3$. Since this implies a net decrease in sampling rate, and the original signal was sampled at the Nyquist rate, we must incorporate additional lowpass filtering in order to avoid aliasing.

Figure 4.29(c) shows the discrete-time Fourier transform of the output of the $L = 2$ upsampler. If we were interested only in interpolating by a factor of 2, we could choose the lowpass filter to have a cutoff frequency of $\omega_c = \pi/2$ and a gain of $L = 2$. However, since the output of the filter will be decimated by $M = 3$, we must use a cutoff frequency of $\omega_c = \pi/3$, but the gain of the filter should still be 2 as in Figure 4.29(d). The Fourier transform $\tilde{X}_i(e^{j\omega})$ of the output of the lowpass filter is shown in Figure 4.29(e). The shaded regions indicate the part of the signal spectrum that is removed due to the lower cutoff frequency for the interpolation filter. Finally, Figure 4.29(f) shows the discrete-time Fourier transform of the output of the downsampler by $M = 3$. Note that the shaded regions show the aliasing that would have occurred if the cutoff frequency of the interpolation lowpass filter had been $\pi/2$ instead of $\pi/3$.

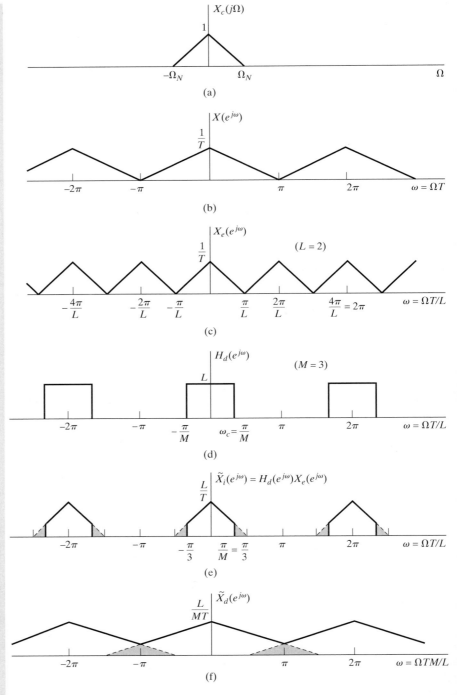

Figure 4.29   Illustration of changing the sampling rate by a noninteger factor.

4.7 MULTIRATE SIGNAL PROCESSING

As we have seen, it is possible to change the sampling rate of a discrete-time signal by a combination of interpolation and decimation. For example, if we want a new sampling period of $T' = 1.01T$, we can first interpolate by $L = 100$ using a lowpass filter that cuts off at $\omega_c = \pi/101$ and then decimate by $M = 101$. These large intermediate changes in sampling rate would require large amounts of computation for each output sample if we implement the filtering in a straightforward manner at the high intermediate sampling rate that is required. Fortunately, it is possible to greatly reduce the amount of computation required by taking advantage of some basic techniques in the area of *multirate signal processing*. Multirate techniques refer in general to utilizing upsampling, down-sampling, compressors, and expanders in a variety of ways to increase the efficiency of signal-processing systems. Besides their use in sampling rate conversion, they are exceedingly useful in A/D and D/A systems that exploit oversampling and noise shaping. Another important class of signal-processing algorithms that relies increasingly on multirate techniques is filter banks for the analysis and/or processing of signals.

Because of their widespread applicability, there is a large body of results on multirate signal processing. In this section, we will focus on two basic results and show how a combination of these results can greatly improve the efficiency of sampling rate conversion. The first result is concerned with the interchange of filtering and downsampling or upsampling operations. The second is the polyphase decomposition.

4.7.1 Interchange of Filtering and Downsampling/Upsampling

We will first derive two identities that aid in manipulating and understanding the operation of multirate systems. It is straightforward to show that the two systems in Figure 4.30 are equivalent. To see the equivalence, note that in Figure 4.30(b),

$$X_b(e^{j\omega}) = H(e^{j\omega M})X(e^{j\omega}), \tag{4.97}$$

and from Eq. (4.78),

$$Y(e^{j\omega}) = \frac{1}{M} \sum_{i=0}^{M-1} X_b(e^{j(\omega/M - 2\pi i/M)}). \tag{4.98}$$

Substituting Eq. (4.97) into Eq. (4.98) gives

$$Y(e^{j\omega}) = \frac{1}{M} \sum_{i=0}^{M-1} X(e^{j(\omega/M - 2\pi i/M)})H(e^{j(\omega - 2\pi i)}). \tag{4.99}$$

Figure 4.30 Two equivalent systems based on downsampling identities.

Since $H(e^{j(\omega-2\pi i)}) = H(e^{j\omega})$, Eq. (4.99) reduces to

$$Y(e^{j\omega}) = H(e^{j\omega})\frac{1}{M}\sum_{i=0}^{M-1} X(e^{j(\omega/M-2\pi i/M)}) \tag{4.100}$$
$$= H(e^{j\omega})X_a(e^{j\omega}),$$

which corresponds to Figure 4.30(a).

A similar identity applies to upsampling. Specifically, using Eq. (4.86) in Section 4.6.2, it is also straightforward to show the equivalence of the two systems in Figure 4.31. We have, from Eq. (4.86) and Figure 4.31(a),

$$Y(e^{j\omega}) = X_a(e^{j\omega L}) \tag{4.101}$$
$$= X(e^{j\omega L})H(e^{j\omega L}).$$

Since, from Eq. (4.86),

$$X_b(e^{j\omega}) = X(e^{j\omega L}),$$

it follows that Eq. (4.101) is, equivalently,

$$Y(e^{j\omega}) = H(e^{j\omega L})X_b(e^{j\omega}),$$

which corresponds to Figure 4.31(b).

In summary, then, we have shown that the operations of linear filtering and downsampling or upsampling can be interchanged if we modify the linear filter.

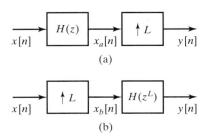

(a)

(b)

Figure 4.31   Two equivalent systems based on upsampling identities.

4.7.2 Polyphase Decompositions

The polyphase decomposition of a sequence is obtained by representing it as a superposition of M subsequences, each consisting of every Mth value of successively delayed versions of the sequence. When this decomposition is applied to a filter impulse response, it can lead to efficient implementation structures for linear filters in several contexts. Specifically, consider an impulse response $h[n]$ that we decompose into M subsequences $h_k[n]$ as follows:

$$h_k[n] = \begin{cases} h[n+k], & n = \text{integer multiple of } M, \\ 0, & \text{otherwise.} \end{cases} \tag{4.102}$$

By successively delaying these subsequences, we can reconstruct the original impulse response $h[n]$; i.e.,

$$h[n] = \sum_{k=0}^{M-1} h_k[n-k] \tag{4.103}$$

This decomposition can be represented with the block diagram in Figure 4.32. If we create a chain of advance elements at the input and a chain of delay elements at the output, the block diagram in Figure 4.33 is equivalent to that of Figure 4.32. In the decomposition in Figures 4.32 and 4.33, the sequences $e_k[n]$ are

$$e_k[n] = h[nM+k] = h_k[nM] \tag{4.104}$$

and are referred to in general as the polyphase components of $h[n]$. There are several

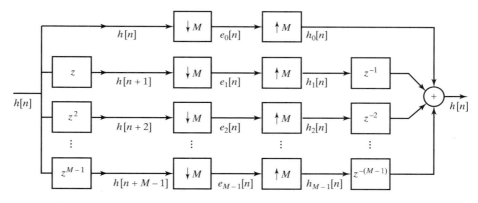

Figure 4.32   Polyphase decomposition of filter $h[n]$ using components $e_k[n]$.

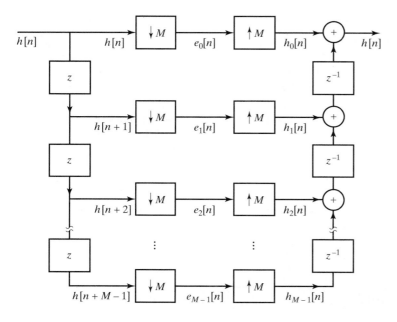

Figure 4.33   Polyphase decomposition of filter $h[n]$ using components $e_k[n]$ with chained delays.

other ways to derive the polyphase components, and there are other ways to index them for notational convenience (Vaidyanathan, 1993), but the definition in Eq. (4.104) is adequate for our purpose in this section.

Figures 4.32 and 4.33 are not realizations of the filter, but they show how the filter can be decomposed into M parallel filters. We see this by noting that Figures 4.32 and 4.33 show that, in the frequency or z-transform domain, the polyphase representation corresponds to expressing $H(z)$ as

$$H(z) = \sum_{k=0}^{M-1} E_k(z^M)z^{-k}. \tag{4.105}$$

Equation (4.105) expresses the system function $H(z)$ as a sum of delayed polyphase component filters. For example, from Eq. (4.105), we obtain the filter structure shown in Figure 4.34.

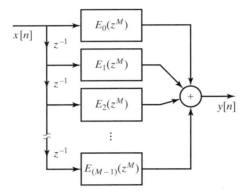

Figure 4.34   Realization structure based on polyphase decomposition of $h[n]$.

4.7.3 Polyphase Implementation of Decimation Filters

One of the important applications of the polyphase decomposition is in the implementation of filters whose output is then downsampled as indicated in Figure 4.35.

In the most straightforward implementation of Figure 4.35, the filter computes an output sample at each value of n, but then only one of every M output points is retained. Intuitively, we might expect that it should be possible to obtain a more efficient implementation which does not compute the samples that are thrown away.

To obtain a more efficient implementation, we can exploit a polyphase decomposition of the filter. Specifically, suppose we express $h[n]$ in polyphase form with polyphase components

$$e_k[n] = h[nM + k]. \tag{4.106}$$

From Eq. (4.105),

$$H(z) = \sum_{k=0}^{M-1} E_k(z^M)z^{-k}. \tag{4.107}$$

Figure 4.35   Decimation system.

With this decomposition and the fact that downsampling commutes with addition, Figure 4.35 can be redrawn as shown in Figure 4.36. Applying the identity in Figure 4.30 to the system in Figure 4.36, we see that the latter then becomes the system shown in Figure 4.37.

To illustrate the advantage of Figure 4.37 compared with Figure 4.35, suppose that the input $x[n]$ is clocked at a rate of 1 sample per unit time and that $H(z)$ is an N-point FIR filter. In the straightforward implementation of Figure 4.35, we require N multiplications and $(N - 1)$ additions per unit time. In the system of Figure 4.37, each of the filters $E_k(z)$ is of length N/M, and their inputs are clocked at a rate of 1 per M units of time. Consequently, each filter requires $\frac{1}{M}\left(\frac{N}{M}\right)$ multiplications per unit time and $\frac{1}{M}\left(\frac{N}{M} - 1\right)$ additions per unit time, and the entire system then requires (N/M) multiplications and $\left(\frac{N}{M} - 1\right) + (M - 1)$ additions per unit time. Thus, we can achieve a significant savings for some values of M and N.

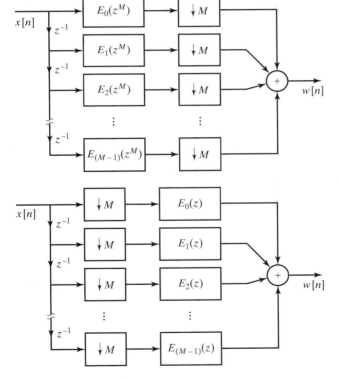

Figure 4.36   Implementation of decimation filter using polyphase decomposition.

Figure 4.37   Implementation of decimation filter after applying the downsampling identity to the polyphase decomposition.

4.7.4 Polyphase Implementation of Interpolation Filters

A savings similar to that just discussed for decimation can be achieved by applying the polyphase decomposition to systems in which a filter is preceded by an upsampler as shown in Figure 4.38. Since only every Lth sample of $w[n]$ is nonzero, the most straightforward implementation of Figure 4.38 would involve applying filter coefficients to sequence values that are known to be zero. Intuitively, here again we would expect that a more efficient implementation was possible.

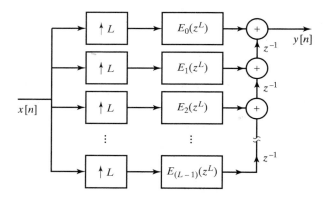

Figure 4.38   Interpolation system.

Figure 4.39   Implementation of interpolation filter using polyphase decomposition.

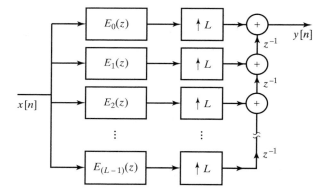

Figure 4.40   Implementation of interpolation filter after applying the upsampling identity to the polyphase decomposition.

To implement the system in Figure 4.38 more efficiently, we again utilize the polyphase decomposition of $H(z)$. For example, we can express $H(z)$ as in the form of Eq. (4.107) and represent Figure 4.38 as shown in Figure 4.39. Applying the identity in Figure 4.31, we can rearrange Figure 4.39 as shown in Figure 4.40.

To illustrate the advantage of Figure 4.40 compared with Figure 4.38, we note that in Figure 4.38 if $x[n]$ is clocked at a rate of 1 sample per unit time, then $w[n]$ is clocked at a rate of L samples per unit time. If $H(z)$ is an FIR filter of length N, we then require NL multiplications and $(NL - 1)$ additions per unit time. Figure 4.40, on the other hand, requires $L(N/L)$ multiplications and $L\left(\frac{N}{L} - 1\right)$ additions per unit time for the set of polyphase filters, plus $(L - 1)$ additions, to obtain $y[n]$. Thus, we again have the possibility of significant savings in computation for some values of L and N.

For both decimation and interpolation, gains in computational efficiency result from rearranging the operations so that the filtering is done at the low sampling rate. Combinations of interpolation and decimation systems for noninteger rate changes lead to significant savings when high intermediate rates are required.

4.8 DIGITAL PROCESSING OF ANALOG SIGNALS

So far, our discussions of the representation of continuous-time signals by discrete-time signals have focused on idealized models of periodic sampling and bandlimited interpolation. We have formalized those discussions in terms of an idealized sampling system that we have called the *ideal continuous-to-discrete* (C/D) *converter* and an idealized bandlimited interpolator system called the *ideal discrete-to-continuous* (D/C) *converter*. These idealized conversion systems allow us to concentrate on the essential mathematical details of the relationship between a bandlimited signal and its samples. For example, in Section 4.4 we used the idealized C/D and D/C conversion systems to show that linear time-invariant discrete-time systems can be used in the configuration of Figure 4.41(a) to implement linear time-invariant continuous-time systems if the input is bandlimited and the sampling rate exceeds the Nyquist rate. In a practical setting, continuous-time signals are not precisely bandlimited, ideal filters cannot be realized, and the ideal C/D and D/C converters can only be approximated by devices that are called analog-to-digital (A/D) and digital-to-analog (D/A) converters, respectively. The block diagram of Figure 4.41(b) shows a more realistic model for digital processing of continuous-time (analog) signals. In this section we will examine some of the considerations introduced by each of the components of the system in Figure 4.41(b).

4.8.1 Prefiltering to Avoid Aliasing

In processing analog signals using discrete-time systems, it is generally desirable to minimize the sampling rate. This is because the amount of arithmetic processing required to implement the system is proportional to the number of samples to be processed. If the input is not bandlimited or if the Nyquist frequency of the input is too high, prefiltering may be necessary. An example of such a situation occurs in processing speech signals, where often only the low-frequency band up to about 3–4 kHz is required for intelligibility, even though the speech signal may have significant frequency content in the 4 kHz to 20 kHz range. Also, even if the signal is naturally bandlimited, wideband

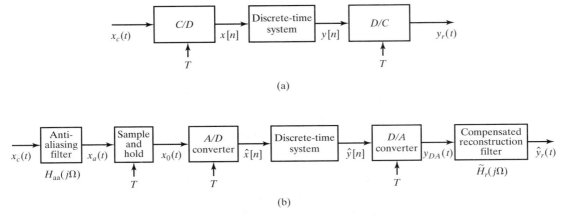

Figure 4.41   (a) Discrete-time filtering of continuous-time signals. (b) Digital processing of analog signals.

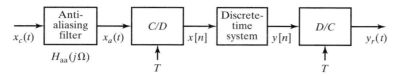

Figure 4.42   Use of prefiltering to avoid aliasing.

additive noise may fill in the higher frequency range, and as a result of sampling, these noise components would be aliased into the low-frequency band. If we wish to avoid aliasing, the input signal must be forced to be bandlimited to frequencies below one-half the desired sampling rate. This can be accomplished by lowpass filtering the continuous-time signal prior to C/D conversion, as shown in Figure 4.42. In this context, the lowpass filter that precedes the C/D converter is called an *antialiasing filter*. Ideally, the frequency response of the antialiasing filter would be

$$H_{aa}(j\Omega) = \begin{cases} 1, & |\Omega| < \Omega_c < \pi/T, \\ 0, & |\Omega| > \Omega_c. \end{cases} \tag{4.108}$$

From the discussion of Section 4.4.1, it follows that the overall system, from the output of the antialiasing filter $x_a(t)$ to the output $y_r(t)$, will always behave as a linear time-invariant system, since the input to the C/D converter, $x_a(t)$, is forced by the antialiasing filter to be bandlimited to frequencies below π/T radians/s. Thus the overall effective frequency response of Figure 4.42 will be the product of $H_{aa}(j\Omega)$ and the effective frequency response from $x_a(t)$ to $y_r(t)$. Combining Eqs. (4.108) and (4.38) gives

$$H_{eff}(j\Omega) = \begin{cases} H(e^{j\Omega T}), & |\Omega| < \Omega_c, \\ 0, & |\Omega| > \Omega_c. \end{cases} \tag{4.109}$$

Thus, for an ideal lowpass antialiasing filter, the system of Figure 4.42 behaves as a linear time-invariant system with frequency response given by Eq. (4.109), even when $X_c(j\Omega)$ is not bandlimited. In practice, the frequency response $H_{aa}(j\Omega)$ cannot be ideally bandlimited, but $H_{aa}(j\Omega)$ can be made small for $|\Omega| > \pi/T$ so that aliasing is minimized. In this case, the overall frequency response of the system in Figure 4.42 should be approximately

$$H_{eff}(j\Omega) \approx H_{aa}(j\Omega)H(e^{j\Omega T}). \tag{4.110}$$

To achieve a negligibly small frequency response above π/T, it would be necessary for $H_{aa}(j\Omega)$ to begin to "roll off," i.e., begin to introduce attenuation, at frequencies below π/T, Equation (4.110) suggests that the roll-off of the antialiasing filter (and other linear time-invariant distortions to be discussed later) could be at least partially compensated for by taking them into account in the design of the discrete-time system. This is illustrated in Problem 4.56.

The preceding discussion requires sharp-cutoff antialiasing filters. Such sharp-cutoff analog filters can be realized using active networks and integrated circuits. However, in applications involving powerful, but inexpensive, digital processors, these continuous-time filters may account for a major part of the cost of a system for discrete-time processing of analog signals. Sharp-cutoff filters are difficult and expensive to implement, and if the system is to operate with a variable sampling rate, adjustable filters would be required. Furthermore, sharp-cutoff analog filters generally have a highly non-linear phase response, particularly at the passband edge. Thus, it is desirable for several reasons to eliminate the continuous-time filters or simplify the requirements on them.

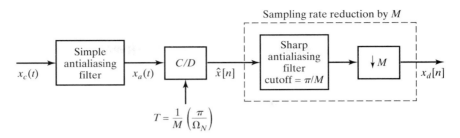

Figure 4.43   Using oversampled A/D conversion to simplify a continuous-time antialiasing filter.

One approach is depicted in Figure 4.43. With Ω_N denoting the highest frequency component to eventually be retained after the antialiasing filtering is completed, we first apply a very simple antialiasing filter that has a gradual cutoff with significant attenuation at $M\Omega_N$. Next, implement the C/D conversion at a sampling rate much higher than $2\Omega_N$, e.g., at $2M\Omega_N$. After that, sampling rate reduction by a factor of M that includes sharp antialiasing filtering is implemented in the discrete-time domain. Subsequent discrete-time processing can then be done at the low sampling rate to minimize computation.

This use of oversampling followed by sampling rate conversion is illustrated in Figure 4.44. Figure 4.44(a) shows the Fourier transform of a signal that occupies the band $|\Omega| < \Omega_N$, plus the Fourier transform of what might correspond to high-frequency "noise" or unwanted components that we eventually want to eliminate with the antialiasing filter. Also shown (dotted line) is the frequency response of an antialiasing filter that does not cut off sharply, but gradually falls to zero at frequencies above a frequency Ω_c. Figure 4.44(b) shows the Fourier transform of the output of this filter. If the signal $x_a(t)$ is sampled with period T such that $(2\pi/T - \Omega_c) > \Omega_N$, then the discrete-time Fourier transform of the sequence $\hat{x}[n]$ will be as shown in Figure 4.44(c). Note that the "noise" will be aliased, but aliasing will not affect the signal band $|\omega| < \omega_N = \Omega_N T$. Now, if T and T' are chosen so that $T' = MT$ and $\pi/T' = \Omega_N$, then $\hat{x}[n]$ can be filtered by a sharp-cutoff discrete-time filter (shown idealized in Figure 4.44(c)) with unity gain and cutoff frequency π/M. The output of the discrete-time filter can be downsampled by M to obtain the sampled sequence $x_d[n]$ whose Fourier transform is shown in Figure 4.44(d). Thus, all the sharp-cutoff filtering has been done by a discrete-time system, and only nominal continuous-time filtering is required. Since discrete-time FIR filters can have an exactly linear phase, it is possible using this oversampling approach to implement antialiasing filtering with virtually no phase distortion. This can be a significant advantage in situations where it is critical to preserve not only the frequency spectrum, but the waveshape as well.

4.8.2 Analog-to-Digital (A/D) Conversion

An ideal C/D converter converts a continuous-time signal into a discrete-time signal, where each sample is known with infinite precision. As an approximation to this for digital signal processing, the system of Figure 4.45 converts a continuous-time (analog) signal into a digital signal, i.e., a sequence of finite-precision or quantized samples. The two systems in Figure 4.45 are available as physical devices. The A/D converter

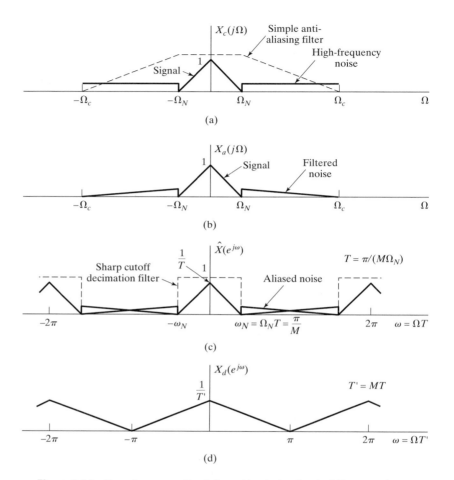

Figure 4.44 Use of oversampling followed by decimation in C/D conversion.

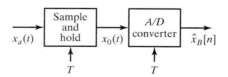

Figure 4.45 Physical configuration for analog-to-digital conversion.

is a physical device that converts a voltage or current amplitude at its input into a binary code representing a quantized amplitude value closest to the amplitude of the input. Under the control of an external clock, the A/D converter can be caused to start and complete an A/D conversion every T seconds. However, the conversion is not instantaneous, and for this reason, a high-performance A/D system typically includes a sample-and-hold, as in Figure 4.45. The ideal sample-and-hold system is the system whose output is

$$x_0(t) = \sum_{n=-\infty}^{\infty} x[n]h_0(t - nT), \tag{4.111}$$

Sample and hold

(a)

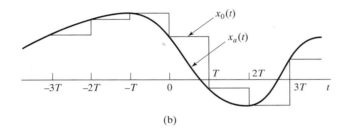

(b)

Figure 4.46 (a) Representation of an ideal sample-and-hold. (b) Representative input and output signals for the sample-and-hold.

where $x[n] = x_a(nT)$ are the ideal samples of $x_a(t)$ and $h_0(t)$ is the impulse response of the zero-order-hold system, i.e.,

$$h_0(t) = \begin{cases} 1, & 0 < t < T, \\ 0, & \text{otherwise.} \end{cases} \tag{4.112}$$

If we note that Eq. (4.111) has the equivalent form

$$x_0(t) = h_0(t) * \sum_{n=-\infty}^{\infty} x_a(nT)\delta(t - nT), \tag{4.113}$$

we see that the ideal sample-and-hold is equivalent to impulse train modulation followed by linear filtering with the zero-order-hold system, as depicted in Figure 4.46(a). The relationship between the Fourier transform of $x_0(t)$ and the Fourier transform of $x_a(t)$ can be worked out following the style of analysis of Section 4.2, and we will do a similar analysis when we discuss the D/A converter. However, the analysis is unnecessary at this point, since everything we need to know about the behavior of the system can be seen from the time-domain expression. Specifically, the output of the zero-order hold is a staircase waveform where the sample values are held constant during the sampling period of T seconds. This is illustrated in Figure 4.46(b). Physical sample-and-hold circuits are designed to sample $x_a(t)$ as nearly instantaneously as possible and to hold the sample value as nearly constant as possible until the next sample is taken. The purpose of this is to provide the constant input voltage (or current) required by the A/D converter. The details of the wide variety of A/D conversion processes and the details of sample-and-hold and A/D circuit implementations are outside the scope of this book. Many practical issues arise in obtaining a sample-and-hold that samples quickly and holds the sample value constant with no decay or "glitches." Likewise, many practical concerns dictate the speed and accuracy of conversion of A/D converter circuits. Such questions are considered in Hnatek (1988) and Schmid (1976), and details of the performance of specific products are available in manufacturers' specification and data sheets. Our concern in this section is the analysis of the quantization effects in A/D conversion.

Since the purpose of the sample-and-hold in Figure 4.45 is to implement ideal sampling and to hold the sample value for quantization by the A/D converter, we can represent the system of Figure 4.45 by the system of Figure 4.47, where the ideal C/D converter represents the sampling performed by the sample-and-hold and, as we will describe later, the quantizer and coder together represent the operation of the A/D converter.

The quantizer is a nonlinear system whose purpose is to transform the input sample $x[n]$ into one of a finite set of prescribed values. We represent this operation as

$$\hat{x}[n] = Q(x[n]) \tag{4.114}$$

and refer to $\hat{x}[n]$ as the quantized sample. Quantizers can be defined with either uniformly or nonuniformly spaced quantization levels; however, when numerical calculations are to be done on the samples, the quantization steps usually are uniform. Figure 4.48

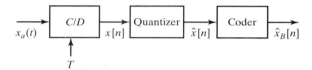

Figure 4.47   Conceptual representation of the system in Figure 4.45.

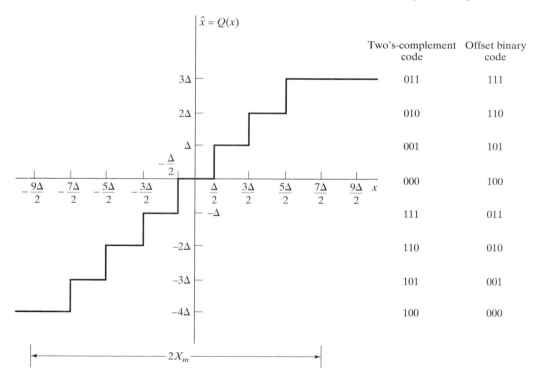

Figure 4.48   Typical quantizer for A/D conversion.

shows a typical uniform quantizer characteristic[1] in which the sample values are *rounded* to the nearest quantization level.

Several features of Figure 4.48 should be emphasized. First, note that this quantizer would be appropriate for a signal whose samples are both positive and negative (bipolar). If it is known that the input samples are always positive (or negative), then a different distribution of the quantization levels would be appropriate. Next, observe that the quantizer of Figure 4.48 has an even number of quantization levels. With an even number of levels, it is not possible to have a quantization level at zero amplitude and also have an equal number of positive and negative quantization levels. Generally, the number of quantization levels will be a power of 2, but the number will be much greater than eight, so this difference is usually inconsequential.

Figure 4.48 also depicts coding of the quantization levels. Since there are eight quantization levels, we can label them by a binary code of 3 bits. (In general, 2^{B+1} levels can be coded with a $(B + 1)$-bit binary code.) In principle, any assignment of symbols can be used, and many binary coding schemes exist, each with its own advantages and disadvantages, depending on the application. For example, the right-hand column of binary numbers in Figure 4.48 illustrates the *offset binary* coding scheme, in which the binary symbols are assigned in numeric order, starting with the most negative quantization level. However, in digital signal processing, we generally wish to use a binary code that permits us to do arithmetic directly with the code words as scaled representations of the quantized samples.

The left-hand column in Figure 4.48 shows an assignment according to the two's complement binary number system. This system for representing signed numbers is used in most computers and microprocessors; thus, it is perhaps the most convenient labeling of the quantization levels. Note, incidentally, that the offset binary code can be converted to two's complement code simply by complementing the most significant bit.

In the two's-complement system, the leftmost, or most significant, bit is considered as the sign bit, and we take the remaining bits as representing either binary integers or fractions. We will assume the latter; i.e., we assume a binary fraction point between the two most significant bits. Then, for the two's-complement interpretation, the binary symbols have the following meaning for $B = 2$:

Binary symbol	Numeric value, $\hat{x}_B$
$0_\diamond 1\ 1$	$3/4$
$0_\diamond 1\ 0$	$1/2$
$0_\diamond 0\ 1$	$1/4$
$0_\diamond 0\ 0$	0
$1_\diamond 1\ 1$	$-1/4$
$1_\diamond 1\ 0$	$-1/2$
$1_\diamond 0\ 1$	$-3/4$
$1_\diamond 0\ 0$	-1

In general, if we have a $(B + 1)$-bit binary two's-complement fraction of the form

$$a_0{}_\diamond a_1 a_2 \ldots a_B,$$

then its value is

$$-a_0 2^0 + a_1 2^{-1} + a_2 2^{-2} + \cdots + a_B 2^{-B}.$$

[1] Such quantizers are also called *linear* quantizers because of the linear progression of quantization steps.

Note that the symbol $\diamond$ denotes the "binary point" of the number. The relationship between the code words and the quantized signal levels depends on the parameter X_m in Figure 4.48. In general, this parameter is called the full-scale level of the A/D converter. Typical values are 10, 5, or 1 volt. From Figure 4.48, we see that the step size of the quantizer would in general be

$$\Delta = \frac{2X_m}{2^{B+1}} = \frac{X_m}{2^B}. \tag{4.115}$$

The smallest quantization levels ($\pm\Delta$) correspond to the least significant bit of the binary code word. Furthermore, the numeric relationship between the code words and the quantized samples is

$$\hat{x}[n] = X_m \hat{x}_B[n], \tag{4.116}$$

since we have assumed that $\hat{x}_B[n]$ is a binary number such that $-1 \le \hat{x}_B[n] < 1$ (for two's complement). In this scheme the binary coded samples $\hat{x}_B[n]$ are directly proportional to the quantized samples (in two's complement binary), and therefore, they can be used as a numeric representation of the amplitude of the samples. Indeed, it is generally appropriate to assume that the input signal is normalized, so that the numeric values of $\hat{x}[n]$ and $\hat{x}_B[n]$ are identical and there is no need to distinguish between the quantized samples and the binary coded samples.

Figure 4.49 shows a simple example of quantization and coding of the samples of a sine wave using a 3-bit quantizer. The unquantized samples $x[n]$ are illustrated with

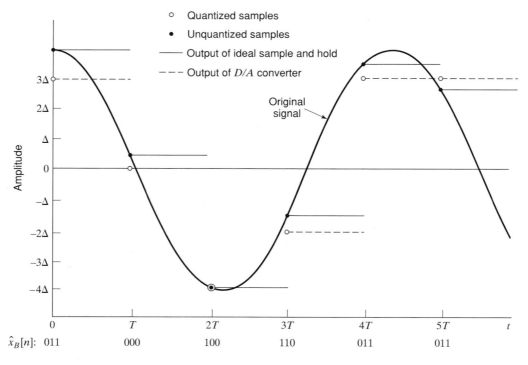

Figure 4.49 Sampling, quantization, coding, and D/A conversion with a 3-bit quantizer.

solid dots, and the quantized samples $\hat{x}[n]$ are illustrated with open circles. Also shown is the output of an ideal sample-and-hold. The dotted lines labeled "output of D/A converter" will be discussed later. Figure 4.49 shows, in addition, the 3-bit code words that represent each sample. Note that, since the analog input $x_a(t)$ exceeds the full-scale value of the quantizer, some of the positive samples are "clipped."

Although much of the preceding discussion pertains to two's-complement coding of the quantization levels, the basic principles of quantization and coding in A/D conversion are the same regardless of the binary code used to represent the samples. A more detailed discussion of the binary arithmetic systems used in digital computing can be found in texts on computer arithmetic. (See, for example, Knuth, 1997.) We now turn to an analysis of the effects of quantization. Since this analysis does not depend on the assignment of binary code words, it will lead to rather general conclusions.

4.8.3 Analysis of Quantization Errors

From Figures 4.48 and 4.49, we see that the quantized sample $\hat{x}[n]$ will generally be different from the true sample value $x[n]$. The difference between them is the *quantization error,* defined as

$$e[n] = \hat{x}[n] - x[n]. \tag{4.117}$$

For example, for the 3-bit quantizer of Figure 4.48, if $\Delta/2 < x[n] \le 3\Delta/2$, then $\hat{x}[n] = \Delta$, and it follows that

$$-\Delta/2 < e[n] \le \Delta/2. \tag{4.118}$$

In the case of Figure 4.48, Eq. (4.118) holds whenever

$$-9\Delta/2 < x[n] \le 7\Delta/2. \tag{4.119}$$

In the general case of a $(B+1)$-bit quantizer with Δ given by Eq. (4.115), the quantization error satisfies Eq. (4.118) whenever

$$(-X_m - \Delta/2) < x[n] \le (X_m - \Delta/2). \tag{4.120}$$

If $x[n]$ is outside this range, as it is for the sample at $t = 0$ in Figure 4.49, then the quantization error is larger in magnitude than $\Delta/2$, and such samples are said to be *clipped.*

A simplified, but useful, model of the quantizer is depicted in Figure 4.50. In this model, the quantization error samples are thought of as an additive noise signal. The model is exactly equivalent to the quantizer if we know $e[n]$. In most cases, however,

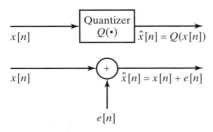

Figure 4.50   Additive noise model for quantizer.

$e[n]$ is not known, and a statistical model based on Figure 4.50 is then often useful in representing the effects of quantization. We will also use such a model in Chapters 6 and 9 to describe the effects of quantization in signal-processing algorithms. The statistical representation of quantization errors is based on the following assumptions:

1. The error sequence $e[n]$ is a sample sequence of a stationary random process.
2. The error sequence is uncorrelated with the sequence $x[n]$.
3. The random variables of the error process are uncorrelated; i.e., the error is a white-noise process.
4. The probability distribution of the error process is uniform over the range of quantization error.

As we will see, the preceding assumptions lead to a rather simple analysis of quantization effects. It is easy to find situations where these assumptions are not valid. For example, if $x_a(t)$ is a step function, the assumptions would not be justified. However, when the signal is a complicated signal, such as speech or music, where the signal fluctuates rapidly in a somewhat unpredictable manner, the assumptions are more realistic. Experiments have shown that, as the signal becomes more complicated, the measured correlation between the signal and the quantization error decreases, and the error also becomes uncorrelated. (See Bennett, 1948; Widrow, 1956, 1961.) In a heuristic sense, the assumptions of the statistical model appear to be valid if the signal is sufficiently complex and the quantization steps are sufficiently small so that the amplitude of the signal is likely to traverse many quantization steps from sample to sample.

Example 4.12   Quantization Error For a Sinusoidal Signal

As an illustration, Figure 4.51(a) shows the sequence of unquantized samples of the cosine signal $x[n] = 0.99 \cos(n/10)$. Figure 4.51(b) shows the quantized sample sequence $\hat{x}[n] = Q\{x[n]\}$ for a 3-bit quantizer ($B + 1 = 3$), assuming that $X_m = 1$. The dashed lines in this figure show the eight possible quantization levels. Figures 4.51(c) and 4.51(d) show the quantization error $e[n] = \hat{x}[n] - x[n]$ for 3-bit and 8-bit quantization, respectively. In each case, the scale of the quantization error is adjusted so that the range $\pm\Delta/2$ is indicated by the dashed lines.

Notice that in the 3-bit case, the error signal is highly correlated with the unquantized signal. For example, around the positive and negative peaks of the cosine, the quantized signal remains constant over many consecutive samples, so that the error has the shape of the input sequence during these intervals. Also, note that during the intervals around the positive peaks, the error is greater than $\Delta/2$ in magnitude because the signal level is too large for this setting of the quantizer parameters.

On the other hand, the quantization error for 8-bit quantization has no apparent patterns.[2] Visual inspection of these figures tends to confirm the preceding assertions about the quantization-noise properties in the finely quantized (8-bit) case; i.e., the error samples appear to vary randomly, with no correlation with the unquantized signal,

[2]For periodic cosine signals, the quantization error would, of course, be periodic, too. We used the frequency $\omega_0 = 1/10$ to avoid this case in the example.

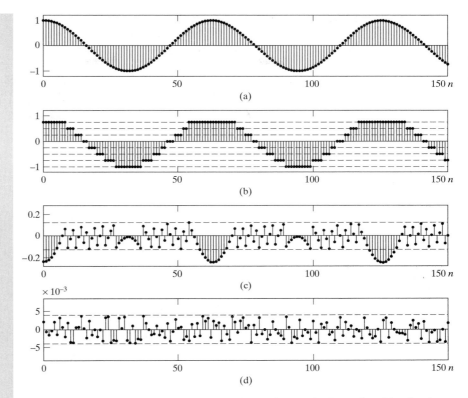

Figure 4.51   Example of quantization noise. (a) Unquantized samples of the signal $x[n] = 0.99\cos(n/10)$. (b) Quantized samples of the cosine waveform in part (a) with a 3-bit quantizer. (c) Quantization error sequence for 3-bit quantization of the signal in (a). (d) Quantization error sequence for 8-bit quantization of the signal in (a).

and they range between $-\Delta/2$ and $+\Delta/2$. In Chapter 10 this will be demonstrated more quantitatively when we calculate the power density spectrum and autocorrelation of a quantization-noise sequence.

For quantizers that round the sample value to the nearest quantization level, as shown in Figure 4.48, the amplitude of the quantization noise is in the range

$$-\Delta/2 < e[n] \le \Delta/2. \tag{4.121}$$

For small Δ, it is reasonable to assume that $e[n]$ is a random variable uniformly distributed from $-\Delta/2$ to $\Delta/2$. Therefore, the first-order probability density for the quantization noise is as shown in Figure 4.52. (If truncation rather than rounding is used in implementing quantization, then the error would always be negative, and we would assume a uniform probability density from $-\Delta$ to 0.) To complete the statistical model for the quantization noise, we assume that successive noise samples are uncorrelated with each other and that $e[n]$ is uncorrelated with $x[n]$. Thus, $e[n]$ is assumed to be a uniformly distributed white-noise sequence. The mean value of $e[n]$ is zero, and its

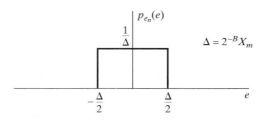

Figure 4.52 Probability density function of quantization error for a rounding quantizer such as that of Figure 4.48.

variance is

$$\sigma_e^2 = \int_{-\Delta/2}^{\Delta/2} e^2 \frac{1}{\Delta} de = \frac{\Delta^2}{12}. \tag{4.122}$$

For a $(B+1)$-bit quantizer with full-scale value X_m, the noise variance, or power, is

$$\sigma_e^2 = \frac{2^{-2B} X_m^2}{12}. \tag{4.123}$$

A common measure of the amount of degradation of a signal by additive noise is the signal-to-noise ratio, defined as the ratio of signal variance (power) to noise variance. Expressed in decibels (dB), the signal-to-noise ratio of a $(B+1)$-bit quantizer is

$$\begin{aligned}
\text{SNR} &= 10 \log_{10} \left(\frac{\sigma_x^2}{\sigma_e^2} \right) = 10 \log_{10} \left(\frac{12 \cdot 2^{2B} \sigma_x^2}{X_m^2} \right) \\
&= 6.02 B + 10.8 - 20 \log_{10} \left(\frac{X_m}{\sigma_x} \right).
\end{aligned} \tag{4.124}$$

From Eq. (4.124), we see that the signal-to-noise ratio increases approximately 6 dB for each bit added to the word length of the quantized samples, i.e., for each doubling of the number of quantization levels. It is particularly instructive to consider the term

$$-20 \log_{10} \left(\frac{X_m}{\sigma_x} \right) \tag{4.125}$$

in Eq. (4.124). First recall that X_m is a parameter of the quantizer, and it would usually be fixed in a practical system. The quantity σ_x is the rms value of the signal amplitude, and it would necessarily be less than the peak amplitude of the signal. For example, if $x_a(t)$ is a sine wave of peak amplitude X_p, then $\sigma_x = X_p/\sqrt{2}$. If σ_x is too large, the peak signal amplitude will exceed the full-scale amplitude X_m of the A/D converter. In this case Eq. (4.124) is no longer valid, and severe distortion results. If, on the other hand, σ_x is too small, then the term in Eq. (4.125) will become large and negative, thereby decreasing the signal-to-noise ratio in Eq. (4.124). In fact, it is easily seen that when σ_x is halved, SNR decreases by 6 dB. Thus, it is very important that the signal amplitude be carefully matched to the full-scale amplitude of the A/D converter.

For analog signals such as speech or music, the distribution of amplitudes tends to be concentrated about zero and falls off rapidly with increasing amplitude. In such cases, the probability that the magnitude of a sample will exceed three or four times the rms value is very low. For example, if the signal amplitude has a Gaussian distribution, only 0.064 percent of the samples would have an amplitude greater than $4\sigma_x$. Thus, to avoid clipping the peaks of the signal (as is assumed in our statistical model), we might set the gain of filters and amplifiers preceding the A/D converter so that $\sigma_x = X_m/4$. Using this value of σ_x in Eq. (4.124) gives

$$\text{SNR} \approx 6B - 1.25 \text{ dB.} \tag{4.126}$$

For example, obtaining a signal-to-noise ratio of about 90–96 dB for use in high-quality music recording and playback requires 16-bit quantization, but it should be remembered that such performance is obtained only if the input signal is carefully matched to the full-scale range of the A/D converter.

This trade-off between peak signal amplitude and absolute size of the quantization noise is fundamental to any quantization process. We will see its importance again in Chapter 6 when we discuss round-off noise in implementing discrete-time linear systems.

4.8.4 D/A Conversion

In Section 4.3, we discussed how a bandlimited signal can be reconstructed from a sequence of samples using ideal lowpass filtering. In terms of Fourier transforms, the reconstruction is represented as

$$X_r(j\Omega) = X(e^{j\Omega T})H_r(j\Omega), \tag{4.127}$$

where $X(e^{j\omega})$ is the discrete-time Fourier transform of the sequence of samples and $X_r(j\Omega)$ is the Fourier transform of the reconstructed continuous-time signal. The ideal reconstruction filter is

$$H_r(j\Omega) = \begin{cases} T, & |\Omega| < \pi/T, \\ 0, & |\Omega| > \pi/T. \end{cases} \tag{4.128}$$

For this choice of $H_r(j\Omega)$, the corresponding relation between $x_r(t)$ and $x[n]$ is

$$x_r(t) = \sum_{n=-\infty}^{\infty} x[n]\frac{\sin[\pi(t-nT)/T]}{\pi(t-nT)/T}. \tag{4.129}$$

The system that takes the sequence $x[n]$ as input and produces $x_r(t)$ as output is called the *ideal D/C converter*. A physically realizable counterpart to the ideal D/C converter is a *digital-to-analog converter* (D/A converter) followed by an approximate lowpass filter. As depicted in Figure 4.53(a), a D/A converter takes a sequence of binary code

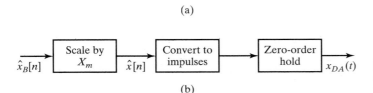

(a)

(b)

Figure 4.53   (a) Block diagram of D/A converter. (b) Representation in terms of a zero-order hold.

words as its input and produces a continuous-time output of the form

$$
\begin{aligned}
x_{DA}(t) &= \sum_{n=-\infty}^{\infty} X_m \hat{x}_B[n] h_0(t - nT) \\
&= \sum_{n=-\infty}^{\infty} \hat{x}[n] h_0(t - nT),
\end{aligned}
\tag{4.130}
$$

where $h_0(t)$ is the impulse response of the zero-order hold given by Eq. (4.112). The dotted lines in Figure 4.49 show the output of a D/A converter for the quantized examples of the sine wave. Note that the D/A converter holds the quantized sample for one sample period in the same way that the sample-and-hold holds the unquantized input sample. If we use the additive-noise model to represent the effects of quantization, Eq. (4.130) becomes

$$
x_{DA}(t) = \sum_{n=-\infty}^{\infty} x[n] h_0(t - nT) + \sum_{n=-\infty}^{\infty} e[n] h_0(t - nT).
\tag{4.131}
$$

To simplify our discussion, we define

$$
x_0(t) = \sum_{n=-\infty}^{\infty} x[n] h_0(t - nT),
\tag{4.132}
$$

$$
e_0(t) = \sum_{n=-\infty}^{\infty} e[n] h_0(t - nT),
\tag{4.133}
$$

so that Eq. (4.131) can be written as

$$
x_{DA}(t) = x_0(t) + e_0(t).
\tag{4.134}
$$

The signal component $x_0(t)$ is related to the input signal $x_a(t)$, since $x[n] = x_a(nT)$. The noise signal $e_0(t)$ depends on the quantization-noise samples $e[n]$ in the same way that $x_0(t)$ depends on the unquantized signal samples. The Fourier transform of Eq. (4.132) is

$$
\begin{aligned}
X_0(j\Omega) &= \sum_{n=-\infty}^{\infty} x[n] H_0(j\Omega) e^{-j\Omega nT} \\
&= \left(\sum_{n=-\infty}^{\infty} x[n] e^{-j\Omega Tn} \right) H_0(j\Omega) \\
&= X(e^{j\Omega T}) H_0(j\Omega).
\end{aligned}
\tag{4.135}
$$

Now, since

$$
X(e^{j\Omega T}) = \frac{1}{T} \sum_{k=-\infty}^{\infty} X_a \left(j \left(\Omega - \frac{2\pi k}{T} \right) \right),
\tag{4.136}
$$

it follows that

$$
X_0(j\Omega) = \left[\frac{1}{T} \sum_{k=-\infty}^{\infty} X_a \left(j \left(\Omega - \frac{2\pi k}{T} \right) \right) \right] H_0(j\Omega).
\tag{4.137}
$$

If $X_a(j\Omega)$ is bandlimited to frequencies below π/T, the shifted copies of $X_a(j\Omega)$ do not overlap in Eq. (4.137), and if we define a compensated reconstruction filter as

$$\tilde{H}_r(j\Omega) = \frac{H_r(j\Omega)}{H_0(j\Omega)}, \tag{4.138}$$

then the output of the filter will be $x_a(t)$ if the input is $x_0(t)$. The frequency response of the zero-order-hold filter is easily shown to be

$$H_0(j\Omega) = \frac{2\sin(\Omega T/2)}{\Omega}e^{-j\Omega T/2}. \tag{4.139}$$

Therefore, the compensated reconstruction filter is

$$\tilde{H}_r(j\Omega) = \begin{cases} \dfrac{\Omega T/2}{\sin(\Omega T/2)}e^{j\Omega T/2}, & |\Omega| < \pi/T, \\ 0, & |\Omega| > \pi/T. \end{cases} \tag{4.140}$$

Figure 4.54(a) shows $|H_0(j\Omega)|$ as given by Eq. (4.139), compared with the magnitude of the ideal interpolation filter $|H_r(j\Omega)|$ as given by Eq. (4.128). Both filters have a gain of T at $\Omega = 0$, but the zero-order-hold, although lowpass in nature, does not cut off sharply at $\Omega = \pi/T$. Figure 4.54(b) shows the magnitude of the frequency response of the ideal compensated reconstruction filter to be used following a zero-order-hold reconstruction system such as a D/A converter. The phase response would ideally correspond to an advance time shift of $T/2$ seconds to compensate for the delay of that amount introduced by the zero-order hold. Since such a time advance cannot be

(a)

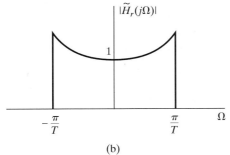

(b)

Figure 4.54   (a) Frequency response of zero-order hold compared with ideal interpolating filter. (b) Ideal compensated reconstruction filter for use with a zero-order-hold output.

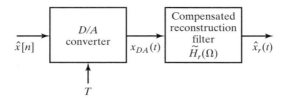

Figure 4.55   Physical configuration for digital-to-analog conversion.

realized in practical real-time approximations to the ideal compensated reconstruction filter, only the magnitude response would normally be compensated, and often even this compensation is neglected, since the gain of the zero-order hold drops only to $2/\pi$ (or -4 dB) at $\Omega = \pi/T$.

Figure 4.55 shows a D/A converter followed by an ideal compensated reconstruction filter. As can be seen from the preceding discussion, with the ideal compensated reconstruction filter following the D/A converter, the reconstructed output signal would be

$$
\begin{aligned}
\hat{x}_r(t) &= \sum_{n=-\infty}^{\infty} \hat{x}[n] \frac{\sin[\pi(t-nT)/T]}{\pi(t-nT)/T} \\
&= \sum_{n=-\infty}^{\infty} x[n] \frac{\sin[\pi(t-nT)/T]}{\pi(t-nT)/T} + \sum_{n=-\infty}^{\infty} e[n] \frac{\sin[\pi(t-nT)/T]}{\pi(t-nT)/T}.
\end{aligned}
\tag{4.141}
$$

In other words, the output would be

$$
\hat{x}_r(t) = x_a(t) + e_a(t),
\tag{4.142}
$$

where $e_a(t)$ would be a bandlimited white-noise signal.

Returning to a consideration of Figure 4.41(b), we are now in a position to understand the behavior of systems for digital processing of analog signals. If we assume that the output of the antialiasing filter is bandlimited to frequencies below π/T, that $\tilde{H}_r(j\Omega)$ is similarly bandlimited, and that the discrete-time system is linear and time invariant, then the output of the overall system will be of the form

$$
\hat{y}_r(t) = y_a(t) + e_a(t),
\tag{4.143}
$$

where

$$
Y_a(j\Omega) = \tilde{H}_r(j\Omega)H_0(j\Omega)H(e^{j\Omega T})H_{aa}(j\Omega)X_c(j\Omega),
\tag{4.144}
$$

in which $H_{aa}(j\Omega)$, $H_0(j\Omega)$, and $\tilde{H}_r(j\Omega)$ are the frequency responses of the antialiasing filter, the zero-order hold of the D/A converter, and the reconstruction lowpass filter, respectively. $H(e^{j\Omega T})$ is the frequency response of the discrete-time system. Similarly, assuming that the quantization noise introduced by the A/D converter is white noise with variance $\sigma_e^2 = \Delta^2/12$, it can be shown that the power spectrum of the output noise is

$$
P_{e_a}(j\Omega) = |\tilde{H}_r(j\Omega)H_0(j\Omega)H(e^{j\Omega T})|^2 \sigma_e^2,
\tag{4.145}
$$

i.e., the input quantization noise is changed by the successive stages of discrete- and continuous-time filtering. From Eq. (4.144), it follows that, under the assumption of negligible aliasing, the overall effective frequency response from $x_c(t)$ to $\hat{y}_r(t)$ is

$$
H_{\text{eff}}(j\Omega) = \tilde{H}_r(j\Omega)H_0(j\Omega)H(e^{j\Omega T})H_{aa}(j\Omega).
\tag{4.146}
$$

If the antialiasing filter is ideal, as in Eq. (4.108), and if the compensation of the reconstruction filter is ideal, as in Eq. (4.140), then the effective frequency response is as given in Eq. (4.109). Otherwise Eq. (4.146) provides a reasonable model for the effective response. Note that Eq. (4.146) suggests that compensation for imperfections in any of the four terms can, in principle, be included in any of the other terms; e.g., the discrete-time system can include appropriate compensation for the antialiasing filter or the zero-order hold or the reconstruction filter or all of these.

In addition to the filtering supplied by Eq. (4.146), Eq. (4.143) reminds us that the output will also be contaminated by the filtered quantization noise. In Chapter 6 we will see that noise can be introduced as well in the implementation of the discrete-time linear system. This internal noise will, in general, be filtered by parts of the discrete-time system implementation, by the zero-order hold of the D/A converter, and by the reconstruction filter.

4.9 OVERSAMPLING AND NOISE SHAPING IN A/D AND D/A CONVERSION

In Section 4.8.1, we showed that oversampling can make it possible to implement sharp-cutoff antialiasing filtering by incorporating digital filtering and decimation. As we discuss in Section 4.9.1, oversampling and subsequent discrete-time filtering and down-sampling also permit an increase in the step size Δ of the quantizer or, equivalently, a reduction in the number of bits required in the analog-to-digital conversion. In Section 4.9.2 we show how the step size can be reduced even further by using oversampling together with quantization-noise feedback, and in Section 4.9.3 we show how the oversampling principle can be applied in D/A conversion.

4.9.1 Oversampled A/D Conversion with Direct Quantization

To explore the relation between oversampling and the quantization step size, we consider the system in Figure 4.56. To analyze the effect of oversampling in this system, we consider $x_a(t)$ to be a zero-mean, wide-sense-stationary, random process with power-spectral density denoted by $\Phi_{x_a x_a}(j\Omega)$ and autocorrelation function denoted by $\phi_{x_a x_a}(\tau)$. To simplify our discussion, we assume initially that $x_a(t)$ is already bandlimited to Ω_N, i.e.,

$$\Phi_{x_a x_a}(j\Omega) = 0, \qquad |\Omega| \geq \Omega_N, \qquad (4.147)$$

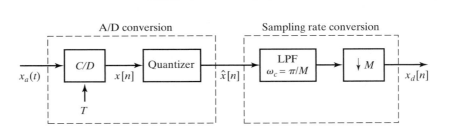

Figure 4.56   Oversampled A/D conversion with simple quantization and down-sampling.

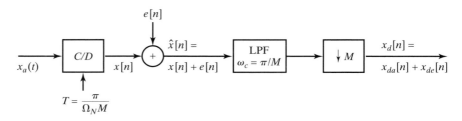

Figure 4.57   System of Figure 4.56 with quantizer replaced by linear noise model.

and we assume that $2\pi/T = 2M\Omega_N$. The constant M, which is assumed to be an integer, is called the *oversampling ratio*. Using the additive noise model discussed in detail in Section 4.8.3, we can replace Figure 4.56 by Figure 4.57. The decimation filter in Figure 4.57 is an ideal lowpass filter with unity gain and cutoff frequency $\omega_c = \pi/M$. Because the entire system of Figure 4.57 is linear, its output $x_d[n]$ has two components, one due to the signal input $x_a(t)$ and one due to the quantization noise input $e[n]$. We denote these components by $x_{da}[n]$ and $x_{de}[n]$, respectively.

Our goal is to determine the ratio of signal power $\mathcal{E}\{x_{da}^2[n]\}$ to quantization-noise power $\mathcal{E}\{x_{de}^2[n]\}$ in the output $x_d[n]$ as a function of the quantizer step size Δ and the oversampling ratio M. Since the system of Figure 4.57 is linear, and since the noise is assumed to be independent of the signal, we can treat the two sources separately in computing the respective powers of the signal and noise components at the output.

First we will consider the signal component of the output. We begin by relating the power spectral density, autocorrelation function, and signal power of the sampled signal $x[n]$ to the corresponding functions for the continuous-time analog signal $x_a(t)$. Let $\phi_{xx}[m]$ and $\Phi_{xx}(e^{j\omega})$ respectively denote the autocorrelation and power spectral density of $x[n]$. Then, by definition, $\phi_{xx}[m] = \mathcal{E}\{x[n+m]x[n]\}$, and since $x[n] = x_a(nT)$ and $x[n+m] = x_a(nT+mT)$,

$$\mathcal{E}\{x[n+m]x[n]\} = \mathcal{E}\{x_a((n+m)T)x_a(nT)\}. \tag{4.148}$$

Therefore,

$$\phi_{xx}[m] = \phi_{x_a x_a}(mT); \tag{4.149}$$

i.e., the autocorrelation function of the sequence of samples is a sampled version of the autocorrelation function of the corresponding continuous-time signal. In particular, Eqs. (4.148) and (4.149), together with the wide-sense-stationarity assumption, imply that $\mathcal{E}\{x_a^2(t)\}$ is a constant independent of t. It then follows that

$$\mathcal{E}\{x^2[n]\} = \mathcal{E}\{x_a^2(nT)\} = \mathcal{E}\{x_a^2(t)\} \qquad \text{for all } n \text{ or } t. \tag{4.150}$$

Since the power spectral densities are the Fourier transforms of the autocorrelation functions, as a consequence of Eq. (4.149),

$$\Phi_{xx}(e^{j\Omega T}) = \frac{1}{T} \sum_{k=-\infty}^{\infty} \Phi_{x_a x_a}\left(j\left(\Omega - \frac{2\pi k}{T}\right)\right). \tag{4.151}$$

Assuming that the input is bandlimited as in Eq. (4.147), and assuming oversampling

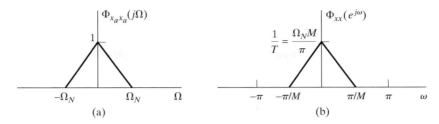

Figure 4.58   Illustration of frequency and amplitude scaling between $\Phi_{x_a x_a}(j\Omega)$ and $\Phi_{xx}(e^{j\omega})$.

by a factor of M so that $2\pi/T = 2M\Omega_N$, we obtain, by substituting $\Omega = \omega/T$ into Eq. (4.151)

$$\Phi_{xx}(e^{j\omega}) = \begin{cases} \dfrac{1}{T}\Phi_{x_a x_a}\left(j\dfrac{\omega}{T}\right), & |\omega| < \pi/M, \\[2ex] 0, & \pi/M < \omega \le \pi. \end{cases} \tag{4.152}$$

For example, if $\Phi_{x_a x_a}(j\Omega)$ is as depicted in Figure 4.58(a), and if we choose the sampling rate to be $2\pi/T = 2M\Omega_N$, then $\Phi_{xx}(e^{j\omega})$ will be as depicted in Figure 4.58(b).

It is instructive to demonstrate that Eq. (4.150) is true by utilizing the power spectrum. The total power of the original analog signal is given by

$$\mathcal{E}\{x_a^2(t)\} = \frac{1}{2\pi}\int_{-\Omega_N}^{\Omega_N} \Phi_{x_a x_a}(j\Omega)d\Omega.$$

From Eq. (4.152), the total power of the sampled signal is

$$\mathcal{E}\{x^2[n]\} = \frac{1}{2\pi}\int_{-\pi}^{\pi} \Phi_{xx}(e^{j\omega})d\omega \tag{4.153}$$

$$= \frac{1}{2\pi}\int_{-\pi/M}^{\pi/M} \frac{1}{T}\Phi_{x_a x_a}\left(j\frac{\omega}{T}\right)d\omega. \tag{4.154}$$

Using the fact that $\Omega_N T = \pi/M$ and making the substitution $\Omega = \omega/T$ in Eq. (4.154) gives

$$\mathcal{E}\{x^2[n]\} = \frac{1}{2\pi}\int_{-\Omega_N}^{\Omega_N} \Phi_{x_a x_a}(j\Omega)d\Omega = \mathcal{E}\{x_a^2(t)\}.$$

Thus, the total power of the sampled signal and the total power of the original analog signal are exactly the same. Since the decimation filter is an ideal lowpass filter with cutoff $\omega_c = \pi/M$, the signal $x[n]$ passes unaltered through the filter. Therefore, the downsampled signal component at the output, $x_{da}[n] = x[nM] = x_a(nMT)$, also has the same total power. This can be seen from the power spectrum by noting that, since $\Phi_{xx}(e^{j\omega})$ is bandlimited to $|\omega| < \pi/M$,

$$\Phi_{x_{da} x_{da}}(e^{j\omega}) = \frac{1}{M}\sum_{k=0}^{M-1} \Phi_{xx}(e^{j(\omega-2\pi k)/M})$$

$$= \frac{1}{M}\Phi_{xx}(e^{j\omega/M}) \qquad |\omega| < \pi. \tag{4.155}$$

Using Eq. (4.155), we obtain

$$\mathcal{E}\{x_{da}^2[n]\} = \frac{1}{2\pi} \int_{-\pi}^{\pi} \Phi_{x_{da}x_{da}}(e^{j\omega})d\omega$$

$$= \frac{1}{2\pi} \int_{-\pi}^{\pi} \frac{1}{M} \Phi_{xx}(e^{j\omega/M})d\omega$$

$$= \frac{1}{2\pi} \int_{-\pi/M}^{\pi/M} \Phi_{xx}(e^{j\omega})\, d\omega = \mathcal{E}\{x^2[n]\},$$

which shows that the power of the signal component stays the same as it traverses the entire system from the input $x_a(t)$ to the corresponding output component $x_{da}[n]$. In terms of the power spectrum, this occurs because, for each scaling of the frequency axis that results from sampling, we have a counterbalancing inverse scaling of the amplitude, so that the area under the power spectrum remains the same as we go from $\Phi_{x_a x_a}(j\Omega)$ to $\Phi_{xx}(e^{j\omega})$ to $\Phi_{x_{da}x_{da}}(e^{j\omega})$ by sampling.

Now let us consider the noise component that is generated by quantization. According to the model in Section 4.8.3, we assume that $e[n]$ is a wide-sense-stationary white-noise process with zero mean and variance[3]

$$\sigma_e^2 = \frac{\Delta^2}{12}.$$

Consequently, the autocorrelation function and power density spectrum for $e[n]$ are, respectively,

$$\phi_{ee}[m] = \sigma_e^2 \delta[m] \tag{4.156}$$

and

$$\Phi_{ee}(e^{j\omega}) = \sigma_e^2 \qquad |\omega| < \pi. \tag{4.157}$$

In Figure 4.59, we show the power density spectrum of $e[n]$ and of $x[n]$. The power density spectrum of the quantized signal $\hat{x}[n]$ is the sum of these, since the signal and quantization-noise samples are assumed to be independent in our model.

Although we have shown that the power in either $x[n]$ or $e[n]$ does not depend on M, we note that as the oversampling ratio M increases, less of the quantization-noise spectrum overlaps with the signal spectrum. It is this effect of the oversampling that lets us improve the signal-to-quantization-noise ratio by sampling-rate reduction. Specifically, the ideal lowpass filter removes the quantization noise in the band $\pi/M < |\omega| \leq \pi$, while it leaves the signal component unaltered. The noise power at the output of the ideal lowpass filter is

$$\mathcal{E}\{e^2[n]\} = \frac{1}{2\pi} \int_{-\pi/M}^{\pi/M} \sigma_e^2 d\omega = \frac{\sigma_e^2}{M}.$$

Next, the lowpass filtered signal is downsampled, and, as we have seen, the signal power in the downsampled output remains the same. In Figure 4.60, we show the resulting

---

[3]Since the random process has zero mean, the average power and the variance are the same.

Figure 4.59 Power spectral density of signal and quantization noise with an oversampling factor of M.

Figure 4.60 Power spectral density of signal and quantization noise after downsampling.

power density spectrum of both $x_{da}[n]$ and $x_{de}[n]$. Comparing Figures 4.58(b), 4.59, and 4.60, we can see that the area under the power density spectrum for the signal has not changed, since the frequency axis and amplitude axis scaling have been inverses of each other. On the other hand, the noise power in the decimated output is the same as at the output of the lowpass filter; i.e.,

$$\mathcal{E}\{x_{de}^2\} = \frac{1}{2\pi} \int_{-\pi}^{\pi} \frac{\sigma_e^2}{M} d\omega = \frac{\sigma_e^2}{M} = \frac{\Delta^2}{12M}. \tag{4.158}$$

Thus, the quantization-noise power $\mathcal{E}\{x_{de}^2[n]\}$ has been reduced by a factor of M through the filtering and downsampling, while the signal power has remained the same.

From Eq. (4.158), we see that for a given quantization noise power, there is a clear trade-off between the oversampling factor M and the quantizer step size Δ. Equation (4.115) states that for a quantizer with $(B+1)$ bits and maximum input signal level between plus and minus X_m, the step size is

$$\Delta = X_m/2^B,$$

and therefore,

$$\mathcal{E}\{x_{de}^2[n]\} = \frac{1}{12M} \left(\frac{X_m}{2^B} \right)^2. \tag{4.159}$$

Equation (4.159) shows that for a fixed quantizer, the noise power can be decreased by increasing the oversampling ratio M. Since the signal power is independent of M, increasing M will increase the signal-to-quantization-noise ratio. Alternatively, for a fixed quantization noise power $P_{de} = \mathcal{E}\{x_{de}^2[n]\}$, the required value for B is

$$B = -\frac{1}{2} \log_2 M - \frac{1}{2} \log_2 12 - \frac{1}{2} \log_2 P_{de} + \log_2 X_m. \tag{4.160}$$

From Eq. (4.160), we see that for every doubling of the oversampling ratio M, we need 1/2 bit less to achieve a given signal-to-quantization-noise ratio, or, in other words, if we oversample by a factor $M = 4$, we need one less bit to achieve a desired accuracy in representing the signal.

4.9.2 Oversampled A/D Conversion with Noise Shaping

In the previous section, we showed that oversampling and decimation can improve the signal-to-quantization-noise ratio. This seems to be a somewhat remarkable result. It implies that we can, in principle, use very crude quantization in our initial sampling of the signal, and if the oversampling ratio is high enough, we can still obtain an accurate representation of the original samples by doing digital computation on the noisy samples. The problem with what we have seen so far is that, to make a significant reduction in the required number of bits, we need very large oversampling ratios. For example, to reduce the number of bits from 16 to 12 would require $M = 4^4 = 256$. This seems to be a rather high cost. However, the basic oversampling principle can lead to much higher gains if we combine it with the concept of noise spectrum shaping by feedback.

As was indicated in Figure 4.59, with direct quantization the power density spectrum of the quantization noise is constant over the entire frequency band. The basic concept in noise shaping is to modify the A/D conversion procedure so that the power density spectrum of the quantization noise is no longer uniform, but rather, is shaped such that most of the noise power is outside the band $|\omega| < \pi/M$. In that way, the subsequent filtering and downsampling removes more of the quantization-noise power.

The noise-shaping quantizer, generally referred to as a sampled-data Delta-Sigma modulator, is shown in Figure 4.61. (See Candy and Temes, 1992 for a collection of papers on this topic.) Figure 4.61(a) shows a block diagram of how the system is implemented with integrated circuits. The integrator is a switched-capacitor discrete-time integrator. The A/D converter can be implemented in many ways, but generally, it is a simple 1-bit quantizer or comparator. The D/A converter takes the digital output and converts it back to an analog pulse that is subtracted from the input signal at the input to the integrator. This system can be represented by the discrete-time equivalent system shown in Figure 4.61(b). The switched-capacitor integrator is represented by an accumulator system, and the delay in the feedback path represents the delay introduced by the D/A converter.

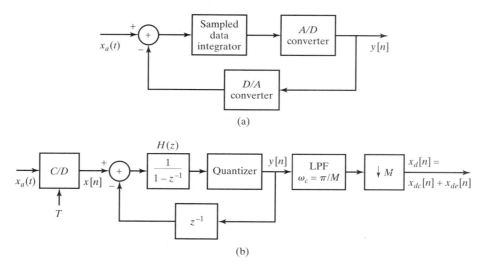

(a)

(b)

Figure 4.61   Oversampled quantizer with noise shaping.

As before, we model the quantization error as an additive noise source so that the system in Figure 4.61 can be replaced by the linear model in Figure 4.62. In this system, the output $y[n]$ is the sum of two components: $y_x[n]$ due to the input $x[n]$ alone and $\hat{e}[n]$ due to the noise $e[n]$ alone.

We denote the transfer function from $x[n]$ to $y[n]$ as $H_x(z)$ and from $e[n]$ to $y[n]$ as $H_e(z)$. These transfer functions can both be calculated in a straightforward manner and are

$$H_x(z) = 1, \tag{4.161a}$$

$$H_e(z) = (1 - z^{-1}). \tag{4.161b}$$

Consequently,

$$y_x[n] = x[n], \tag{4.162a}$$

and

$$\hat{e}[n] = e[n] - e[n-1]. \tag{4.162b}$$

Therefore, the output $y[n]$ can be represented equivalently as $y[n] = x[n] + \hat{e}[n]$, where $x[n]$ appears unmodified at the output and the quantization noise $e[n]$ is modified by the first-difference operator $H_e(z)$. This is depicted in the block diagram in Figure 4.63. With the power density spectrum for $e[n]$ given by Eq. (4.157), the power density spectrum of the quantization noise $\hat{e}[n]$ that is present in $y[n]$ is

$$\begin{aligned} \Phi_{\hat{e}\hat{e}}(e^{j\omega}) &= \sigma_e^2 |H_e(e^{j\omega})|^2 \\ &= \sigma_e^2 [2\sin(\omega/2)]^2. \end{aligned} \tag{4.163}$$

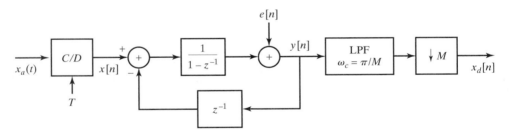

Figure 4.62   System of Figure 4.60 from $x_a(t)$ to $y[n]$ with quantizer replaced by a linear noise model.

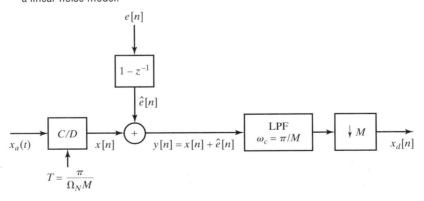

Figure 4.63   Equivalent representation of Figure 4.62.

In Figure 4.64, we show the power density spectrum of $\hat{e}[n]$, the power spectrum of $e[n]$, and the same signal power spectrum that was shown in Figure 4.58(b) and Figure 4.59. It is interesting to observe that the *total* noise power is increased from $\mathcal{E}\{e^2[n]\} = \sigma_e^2$ at the quantizer to $\mathcal{E}\{\hat{e}^2[n]\} = 2\sigma_e^2$ at the output of the noise-shaping system. However, note that in comparison with Figure 4.59, the quantization noise has been shaped in such a way that more of the noise power is outside the signal band $|\omega| < \pi/M$ than in the direct oversampled case, where the noise spectrum is flat.

In the system of Figure 4.61, this out-of-band noise power is removed by the low-pass filter. Specifically, in Figure 4.65 we show the power density spectrum of $\Phi_{x_{da}x_{da}}(e^{j\omega})$ superimposed on the power density spectrum of $\Phi_{x_{de}x_{de}}(e^{j\omega})$. Since the downsampler does not remove any of the signal power, the signal power in $x_{da}[n]$ is

$$P_{da} = \mathcal{E}\{x_{da}^2[n]\} = \mathcal{E}\{x^2[n]\} = \mathcal{E}\{x_a^2(t)\}.$$

The quantization-noise power in the final output is

$$P_{de} = \frac{1}{2\pi} \int_{-\pi}^{\pi} \Phi_{x_{de}x_{de}}(e^{j\omega})d\omega = \frac{1}{2\pi}\frac{\Delta^2}{12M}\int_{-\pi}^{\pi}\left(2\sin\left(\frac{\omega}{2M}\right)\right)^2 d\omega. \quad (4.164)$$

To compare this approximately with the results in Section 4.9.1, assume that M is sufficiently large so that

$$\sin\left(\frac{\omega}{2M}\right) \approx \frac{\omega}{2M}.$$

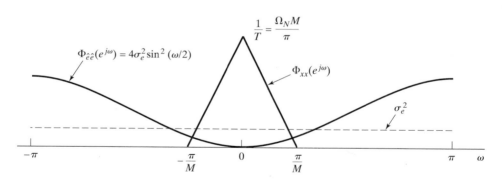

Figure 4.64   The power spectral density of the quantization noise and the signal.

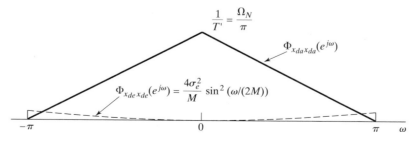

Figure 4.65   Power spectral density of the signal and quantization noise after downsampling.

With this approximation, Eq. (4.164) is easily evaluated to obtain

$$P_{de} = \frac{1}{36} \frac{\Delta^2 \pi^2}{M^3}. \tag{4.165}$$

From Eq. (4.165), we see again a trade-off between the oversampling ratio M and the quantizer step size Δ. For a $(B+1)$-bit quantizer and maximum input signal level between plus and minus X_m, $\Delta = X_m/2^B$. Therefore, to achieve a given quantization-noise power P_{de}, we must have

$$B = -\frac{3}{2} \log_2 M + \log_2(\pi/6) - \frac{1}{2} \log_2 P_{de} + \log_2 X_m. \tag{4.166}$$

Comparing Eq. (4.166) with Eq. (4.160), we see that, whereas with direct quantization a doubling of the oversampling ratio M gained $1/2$ bit in quantization, the use of noise shaping results in a gain of 1.5 bits.

Table 4.1 gives the equivalent savings in quantizer bits over direct quantization with no oversampling ($M = 1$) for (a) direct quantization with oversampling, as discussed in Section 4.9.1, and (b) oversampling with noise shaping, as examined in this section.

TABLE 4.1   EQUIVALENT SAVINGS IN QUANTIZER BITS RELATIVE TO $M = 1$ FOR DIRECT QUANTIZATION AND FIRST-ORDER NOISE SHAPING

M	Direct quantization	Noise shaping
4	1	2.2
8	1.5	3.7
16	2	5.1
32	2.5	6.6
64	3	8.1

The noise-shaping strategy in Figure 4.61 can be extended by incorporating a second stage of accumulation as shown in Figure 4.66. In this case, with the quantizer again modeled as an additive noise source $e[n]$, it can be shown that

$$y[n] = x[n] + \hat{e}[n]$$

where, in the two-stage case, $\hat{e}[n]$ is the result of processing the quantization noise $e[n]$

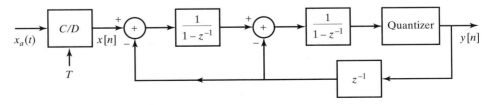

Figure 4.66   Oversampled quantizer with second-order noise shaping.

TABLE 4.2   REDUCTION IN QUANTIZER
BITS AS ORDER p OF NOISE SHAPING

Quantizer order p	Oversampling factor M				
	4	8	16	32	64
0	1.0	1.5	2.0	2.5	3.0
1	2.2	3.7	5.1	6.6	8.1
2	2.9	5.4	7.9	10.4	12.9
3	3.5	7.0	10.5	14.0	17.5
4	4.1	8.5	13.0	17.5	22.0
5	4.6	10.0	15.5	21.0	26.5

through the transfer function

$$H_e(z) = (1 - z^{-1})^2. \tag{4.167}$$

The corresponding power density spectrum of the quantization noise now present in $y[n]$ is

$$\Phi_{\hat{e}\hat{e}}(e^{j\omega}) = \sigma_e^2 [2\sin(\omega/2)]^4, \tag{4.168}$$

with the result that, although the total noise power at the output of the two-stage noise-shaping system is greater than for the one-stage case, even more of the noise lies outside the signal band. More generally, p stages of accumulation and feedback can be used, with corresponding noise shaping given by

$$\Phi_{\hat{e}\hat{e}}(e^{j\omega}) = \sigma_e^2 [2\sin(\omega/2)]^{2p}. \tag{4.169}$$

In Table 4.2, we show the equivalent reduction in quantizer bits as a function of the order p of the noise shaping and the oversampling ratio M. Note that with $p = 2$ and $M = 64$, we obtain almost 13 bits of increase in accuracy, suggesting that a 1-bit quantizer could achieve about 14-bit accuracy at the output of the decimator.

Although multiple feedback loops such as the one shown in Figure 4.66 promise greatly increased noise reduction, they are not without problems. Specifically, for large values of p, there is an increased potential for instability and oscillations to occur. An alternative structure known as multistage noise shaping (MASH) is considered in Problem 4.62.

4.9.3 Oversampling and Noise Shaping in D/A Conversion

In Sections 4.9.1 and 4.9.2, we discussed the use of oversampling to simplify the process of analog-to-digital conversion. As we mentioned, the signal is initially oversampled to simplify antialias filtering and improve accuracy, but the final output $x_d[n]$ of the A/D converter is sampled at the Nyquist rate for $x_a(t)$. The minimum sampling rate is, of course, highly desirable for digital processing or for simply representing the analog signal in digital form, as in the CD audio recording system. It is natural to apply the same principles in reverse to achieve improvements in the D/A conversion process.

The basic system, which is the counterpart to Figure 4.56, is shown in Figure 4.67. The sequence $y_d[n]$, which is to be converted to a continuous-time signal, is first

Figure 4.67   Oversampled D/A conversion.

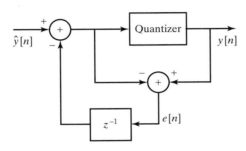

Figure 4.68   First-order noise-shaping system for oversampled D/A quantization.

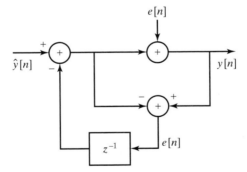

Figure 4.69   System of Figure 4.68 with quantizer replaced by linear noise model.

upsampled to produce the sequence $\hat{y}[n]$, which is then requantized before sending it to a D/A converter that accepts binary samples with the number of bits produced by the requantization process. We can use a simple D/A converter with few bits if we can be assured that the quantization noise does not occupy the signal band. Then the noise can be removed by inexpensive analog filtering.

In Figure 4.68, we show a structure for the quantizer that shapes the quantization noise in a similar manner to the first-order noise shaping provided by the system in Figure 4.61. To analyze the system in Figures 4.67 and 4.68, we replace the quantizer in Figure 4.68 by an additive white-noise source $e[n]$, so that Figure 4.68 is replaced by Figure 4.69. The transfer function from $\hat{y}[n]$ to $y[n]$ is unity, i.e. the upsampled signal $\hat{y}[n]$ appears at the output unaltered. The transfer function $H_e(z)$ from $e[n]$ to $y[n]$ is

$$H_e(z) = 1 - z^{-1}.$$

Therefore, the quantization noise component $\hat{e}[n]$ that appears at the output of the noise-shaping system in Figure 4.69 has the power density spectrum

$$\Phi_{\hat{e}\hat{e}}(e^{j\omega}) = \sigma_e^2 (2\sin\omega/2)^2, \tag{4.170}$$

where, again, $\sigma_e^2 = \Delta^2/12$.

An illustration of this approach to D/A conversion is given in Figure 4.70. Figure 4.70(a) shows the power spectrum $\Phi_{y_d y_d}(e^{j\omega})$ of the input $y_d[n]$ in Figure 4.67. Note that we assume that the signal $y_d[n]$ is sampled at the Nyquist rate. Figure 4.70(b) shows the corresponding power spectrum at the output of the upsampler (by M), and Figure 4.70(c) shows the quantization noise spectrum at the output of the quantizer/noise-shaper system. Finally, Figure 4.70(d) shows the power spectrum of the signal component

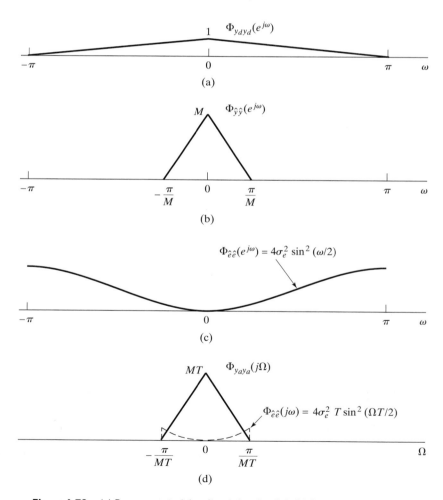

Figure 4.70   (a) Power spectral density of signal $y_d[n]$. (b) Power spectral density of signal $\hat{y}[n]$. (c) Power spectral density of quantization noise. (d) Power spectral density of the continuous-time signal and the quantization noise.

superimposed on the power spectrum of the noise component at the analog output of the D/C converter of Figure 4.67. In this case, we assume that the D/C converter has an ideal lowpass reconstruction filter with cutoff frequency $\pi/(MT)$, which will remove as much of the quantization noise as possible.

In a practical setting, we would like to avoid sharp-cutoff analog reconstruction filters. From Figure 4.70(d), it is clear that if we can tolerate somewhat more quantization noise, then the D/C reconstruction filter need not roll off so sharply. Furthermore, if we use multistage techniques in the noise shaping, we can obtain an output noise spectrum of the form

$$\Phi_{\hat{e}\hat{e}}(e^{j\omega}) = \sigma_e^2 (2\sin\omega/2)^{2p},$$

which would push more of the noise to higher frequencies. In this case, the analog reconstruction filter specifications could be relaxed even further.

4.10 SUMMARY

In this chapter, we developed and explored the relationship between continuous-time signals and the discrete-time sequences obtained by periodic sampling. The fundamental theorem that allows the continuous-time signal to be represented by a sequence of samples is the Nyquist theorem, which states that, for a bandlimited signal, periodic samples are a sufficient representation, as long as the sampling rate is sufficiently high relative to the highest frequency in the continuous-time signal. Under this condition, the continuous-time signal can be reconstructed from the samples by lowpass filtering, corresponding to bandlimited interpolation. If the sampling rate is too low relative to the bandwidth of the signal, then aliasing distortion occurs.

The ability to represent signals by sampling permits the discrete-time processing of continuous-time signals. This is accomplished by first sampling, then applying discrete-time processing, and, finally, reconstructing a continuous-time signal from the result. Examples given were lowpass filtering and differentiation.

A particularly important class of processing is the class corresponding to sampling rate changes. Downsampling a discrete-time signal corresponds in the frequency domain to a replication of the discrete-time spectrum and rescaling of the frequency axis, which may require additional bandlimiting to avoid aliasing. Upsampling corresponds to effectively increasing the sampling rate and is also represented in the frequency domain by a rescaling of the frequency axis. By combining upsampling and downsampling by integer amounts, noninteger sampling rate conversion can be achieved. We also showed how this can be efficiently done using multirate techniques.

In the final sections of the chapter, we explored a number of practical considerations associated with the discrete-time processing of continuous-time signals, including the use of prefiltering to avoid aliasing, quantization error in analog-to-digital conversion, and some issues associated with the filtering used in sampling and reconstructing the continuous-time signals. Finally, we showed how discrete-time decimation and interpolation and noise shaping can be used to simplify the analog side of A/D and D/A conversion.

PROBLEMS

Basic Problems with Answers

4.1. The signal

$$x_c(t) = \sin(2\pi(100)t)$$

was sampled with sampling period $T = 1/400$ second to obtain a discrete-time signal $x[n]$. What is the resulting signal $x[n]$?

4.2. The sequence

$$x[n] = \cos\left(\frac{\pi}{4}n\right), \qquad -\infty < n < \infty,$$

was obtained by sampling a continuous-time signal

$$x_c(t) = \cos(\Omega_0 t), \qquad -\infty < t < \infty,$$

at a sampling rate of 1000 samples/s. What are two possible positive values of Ω_0 that could have resulted in the sequence $x[n]$?

4.3. The continuous-time signal

$$x_c(t) = \cos(4000\pi t)$$

is sampled with a sampling period T to obtain a discrete-time signal

$$x[n] = \cos\left(\frac{\pi n}{3}\right).$$

(a) Determine a choice for T consistent with this information.
(b) Is your choice for T in Part (a) unique? If so, explain why. If not, specify another choice of T consistent with the information given.

4.4. The continuous-time signal

$$x_c(t) = \sin(20\pi t) + \cos(40\pi t)$$

is sampled with a sampling period T to obtain the discrete-time signal

$$x[n] = \sin\left(\frac{\pi n}{5}\right) + \cos\left(\frac{2\pi n}{5}\right).$$

(a) Determine a choice for T consistent with this information.
(b) Is your choice for T in Part (a) unique? If so, explain why. If not, specify another choice of T consistent with the information given.

4.5. Consider the system of Figure 4.11, with the discrete-time system an ideal lowpass filter with cutoff frequency $\pi/8$ radians/s.
(a) If $x_c(t)$ is bandlimited to 5 kHz, what is the maximum value of T that will avoid aliasing in the C/D converter?
(b) If $1/T = 10$ kHz, what will the cutoff frequency of the effective continuous-time filter be?
(c) Repeat Part (b) for $1/T = 20$ kHz.

4.6. Let $h_c(t)$ denote the impulse response of a linear time-invariant continuous-time filter and $h_d[n]$ the impulse response of a linear time-invariant discrete-time filter.

(a) If

$$h_c(t) = \begin{cases} e^{-at}, & t \ge 0, \\ 0, & t < 0, \end{cases}$$

where a is a positive real constant, determine the continuous-time filter frequency response and sketch its magnitude.

(b) If $h_d[n] = Th_c(nT)$ with $h_c(t)$ as in part (a), determine the discrete-time filter frequency response and sketch its magnitude.

(c) For a given value of a, determine, as a function of T, the minimum magnitude of the discrete-time filter frequency response.

4.7. A simple model of a multipath communication channel is indicated in Figure P4.7-1. Assume that $s_c(t)$ is bandlimited such that $S_c(j\Omega) = 0$ for $|\Omega| \ge \pi/T$ and that $x_c(t)$ is sampled with a sampling period T to obtain the sequence

$$x[n] = x_c(nT).$$

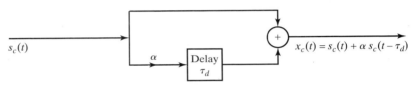

Figure P4.7-1

(a) Determine the Fourier transform of $x_c(t)$ and the Fourier transform of $x[n]$ in terms of $S_c(j\Omega)$.

(b) We want to simulate the multipath system with a discrete-time system by choosing $H(e^{j\omega})$ in Figure P4.7-2 so that the output $r[n] = x_c(nT)$ when the input is $s[n] = s_c(nT)$. Determine $H(e^{j\omega})$ in terms of T and τ_d.

(c) Determine the impulse response $h[n]$ in Figure P4.7-2 when (i) $\tau_d = T$ and (ii) $\tau_d = T/2$.

$$\xrightarrow{\quad s[n] = s_c(nT) \quad} \boxed{H(e^{j\omega})} \xrightarrow{\quad r[n] = x_c(nT) \quad} \qquad \textbf{Figure P4.7-2}$$

4.8. Consider the system in Figure P4.8-1 with the following relations:

$$X_c(j\Omega) = 0, \qquad |\Omega| \ge 2\pi \times 10^4,$$

$$x[n] = x_c(nT),$$

$$y[n] = T \sum_{k=-\infty}^{n} x[k].$$

$$\xrightarrow{\quad x_c(t) \quad} \boxed{C/D} \xrightarrow{\quad x[n] \quad} \boxed{\begin{array}{c} h[n] \\ H(e^{j\omega}) \end{array}} \xrightarrow{\quad y[n] \quad}$$

$$\uparrow T$$

Figure P4.8-1

(a) For this system, what is the maximum allowable value of T if aliasing is to be avoided, i.e., so that $x_c(t)$ can be recovered from $x[n]$.

(b) Determine $h[n]$.

(c) In terms of $X(e^{j\omega})$, what is the value of $y[n]$ for $n = \infty$?

(d) Determine whether there is any value of T for which

$$y[n]\Big|_{n=\infty} = \int_{-\infty}^{\infty} x_c(t)dt. \tag{P4.8-1}$$

If there is such a value for T, determine the maximum value. If there is not, explain and specify how T would be chosen so that the equality in Eq. (P4.8-1) is best approximated.

4.9. Consider a stable discrete-time signal $x[n]$ whose discrete-time Fourier transform $X(e^{j\omega})$ satisfies the equation

$$X(e^{j\omega}) = X\left(e^{j(\omega-\pi)}\right)$$

and has even symmetry, i.e., $x[n] = x[-n]$.

(a) Show that $X(e^{j\omega})$ is periodic with a period π.

(b) Find the value of $x[3]$. (*Hint:* Find values for all odd-indexed points.)

(c) Let $y[n]$ be the decimated version of $x[n]$, i.e., $y[n] = x[2n]$. Can you reconstruct $x[n]$ from $y[n]$ for all n. If yes, how? If no, justify your answer.

4.10. Each of the following continuous-time signals is used as the input $x_c(t)$ for an ideal C/D converter as shown in Figure 4.1 with the sampling period T specified. In each case, find the resulting discrete-time signal $x[n]$.

(a) $x_c(t) = \cos(2\pi(1000)t), \quad T = (1/3000)$ sec

(b) $x_c(t) = \sin(2\pi(1000)t), \quad T = (1/1500)$ sec

(c) $x_c(t) = \sin(2\pi(1000)t) / (\pi t), \quad T = (1/5000)$ sec

4.11. The following continuous-time input signals $x_c(t)$ and corresponding discrete-time output signals $x[n]$ are those of an ideal C/D as shown in Figure 4.1. Specify a choice for the sampling period T that is consistent with each pair of $x_c(t)$ and $x[n]$. In addition, indicate whether your choice of T is unique. If not, specify a second possible choice of T consistent with the information given.

(a) $x_c(t) = \sin(10\pi t), \quad x[n] = \sin(\pi n/4)$

(b) $x_c(t) = \sin(10\pi t)/(10\pi t), \quad x[n] = \sin(\pi n/2)/(\pi n/2)$

4.12. In the system of Figure 4.11, assume that

$$H(e^{j\omega}) = j\omega/T, \quad -\pi \leq \omega < \pi,$$

and $T = 1/10$ sec.

(a) For each of the following inputs $x_c(t)$, find the corresponding output $y_c(t)$.

    (i) $x_c(t) = \cos(6\pi t)$

    (ii) $x_c(t) = \cos(14\pi t)$

(b) Are the outputs $y_c(t)$ those you would expect from a differentiator?

4.13. In the system shown in Figure 4.16, $h_c(t) = \delta(t - T/2)$.

(a) Suppose the input $x[n] = \sin(\pi n/2)$ and $T = 10$. Find $y[n]$.

(b) Suppose you use the same $x[n]$ as in Part (a), but halve T to be 5. Find the resulting $y[n]$.

(c) In general, how does the continuous-time LTI system $h_c(t)$ limit the range of the sampling period T that can be used without changing $y[n]$?

4.14. Which of the following signals can be downsampled by a factor of 2 using the system in Figure 4.20 without any loss of information?

(a) $x[n] = \delta[n - n_0]$, for n_0 some unknown integer

(b) $x[n] = \cos(\pi n/4)$

(c) $x[n] = \cos(\pi n/4) + \cos(3\pi n/4)$

(d) $x[n] = \sin(\pi n/3)/(\pi n/3)$
(e) $x[n] = (-1)^n \sin(\pi n/3)/(\pi n/3)$

4.15. Consider the system shown in Figure P4.15-1. For each of the following input signals $x[n]$, indicate whether the output $x_r[n] = x[n]$.
  (a) $x[n] = \cos(\pi n/4)$
  (b) $x[n] = \cos(\pi n/2)$
  (c)

$$x[n] = \left[\frac{\sin(\pi n/8)}{\pi n}\right]^2$$

  Hint: Use the modulation property of the Fourier transform to find $X(e^{j\omega})$.

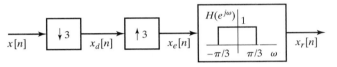

Figure P4.15-1

4.16. Consider the system in Figure 4.28. The input $x[n]$ and corresponding output $\tilde{x}_d[n]$ are given for a specific choice of M/L in each of the following parts. Determine a choice for M/L based on the information given, and specify whether your choice is unique.
  (a) $x[n] = \sin(\pi n/3)/(\pi n/3)$, $\tilde{x}_d[n] = \sin(5\pi n/6)/(5\pi n/6)$
  (b) $x[n] = \cos(3\pi n/4)$, $\tilde{x}_d[n] = \cos(\pi n/2)$

4.17. Each of the following parts lists an input signal $x[n]$ and the upsampling and downsampling rates L and M for the system in Figure 4.28. Determine the corresponding output $\tilde{x}_d[n]$.
  (a) $x[n] = \sin(2\pi n/3)/\pi n$, $L = 4$, $M = 3$
  (b) $x[n] = \sin(3\pi n/4)$, $L = 3$, $M = 5$

4.18. For the system shown in Figure 4.28, $X(e^{j\omega})$, the Fourier transform of the input signal $x[n]$, is shown in Figure P4.18-1. For each of the following choices of L and M, specify the maximum possible value of ω_0 such that $\tilde{X}_d(e^{j\omega}) = aX(e^{jM\omega/L})$ for some constant a.

Figure P4.18-1

  (a) $M = 3$, $L = 2$
  (b) $M = 5$, $L = 3$
  (c) $M = 2$, $L = 3$

4.19. The continuous-time signal $x_c(t)$ with the Fourier transform $X_c(j\Omega)$ shown in Figure P4.19-1 is passed through the system shown in Figure P4.19-2. Determine the range of values for T for which $x_r(t) = x_c(t)$.

Figure P4.19-1

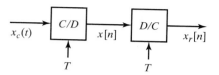

Figure P4.19-2

4.20. Consider the system in Figure 4.11. The input signal $x_c(t)$ has the Fourier transform shown in Figure P4.20-1 with $\Omega_0 = 2\pi(1000)$ radians/second. The discrete-time system is an ideal lowpass filter with frequency response

$$H(e^{j\omega}) = \begin{cases} 1, & |\omega| < \omega_c, \\ 0, & \text{otherwise.} \end{cases}$$

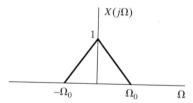

Figure P4.20-1

(a) What is the minimum sampling rate $F_s = 1/T$ such that no aliasing occurs in sampling the input?

(b) If $\omega_c = \pi/2$, what is the minimum sampling rate such that $y_c(t) = x_c(t)$?

Basic Problems

4.21. A complex-valued continuous-time signal $x_c(t)$ has the Fourier transform shown in Figure P4.21-1, where $(\Omega_2 - \Omega_1) = \Delta\Omega$. This signal is sampled to produce the sequence $x[n] = x_c(nT)$.

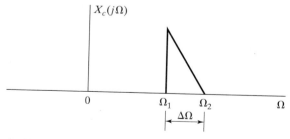

Figure P4.21-1

(a) Sketch the Fourier transform $X(e^{j\omega})$ of the sequence $x[n]$ for $T = \pi/\Omega_2$.

(b) What is the *lowest* sampling frequency that can be used without incurring any aliasing distortion, i.e., so that $x_c(t)$ can be recovered from $x[n]$?

(c) Draw the block diagram of a system that can be used to recover $x_c(t)$ from $x[n]$ if the sampling rate is greater than or equal to the rate determined in Part (b). Assume that (complex) ideal filters are available.

4.22. A continuous-time signal $x_c(t)$, with Fourier transform $X_c(j\Omega)$ shown in Figure P4.22-1, is sampled with sampling period $T = 2\pi/\Omega_0$ to form the sequence $x[n] = x_c(nT)$.

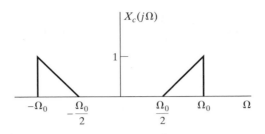

Figure P4.22-1

(a) Sketch the Fourier transform $X(e^{j\omega})$ for $|\omega| < \pi$.

(b) The signal $x[n]$ is to be transmitted across a digital channel. At the receiver, the original signal $x_c(t)$ must be recovered. Draw a block diagram of the recovery system and specify its characteristics. Assume that ideal filters are available.

(c) In terms of Ω_0, for what range of values of T can $x_c(t)$ be recovered from $x[n]$?

4.23. In Figure P4.23-1, assume that $X_c(j\Omega) = 0$, $|\Omega| \geq \pi/T_1$. For the general case in which $T_1 \neq T_2$ in the system, express $y_c(t)$ in terms of $x_c(t)$. Is the basic relationship different for $T_1 > T_2$ and $T_1 < T_2$?

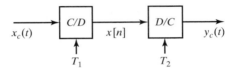

Figure P4.23-1

4.24. In the system of Figure P4.24-1, $X_c(j\Omega)$ and $H(e^{j\omega})$ are as shown. Sketch and label the Fourier transform of $y_c(t)$ for each of the following cases:

(a) $1/T_1 = 1/T_2 = 10^4$

(b) $1/T_1 = 1/T_2 = 2 \times 10^4$

(c) $1/T_1 = 2 \times 10^4, \quad 1/T_2 = 10^4$

(d) $1/T_1 = 10^4, \quad 1/T_2 = 2 \times 10^4$

Figure P4.24-1

4.25. Figure P4.25-1 shows the overall system for filtering a continuous-time signal using a discrete-time filter. The frequency responses of the reconstruction filter $H_r(j\Omega)$ and the discrete-time filter $H(e^{j\omega})$ are shown in Figure P4.25-2.

Figure P4.25-1

Figure P4.25-2

(a) For $X_c(j\Omega)$ as shown in Figure P4.25-3 and $1/T = 20\,\text{kHz}$, sketch $X_s(j\Omega)$ and $X(e^{j\omega})$.

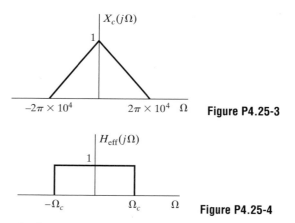

Figure P4.25-3

Figure P4.25-4

For a certain range of values of T, the overall system, with input $x_c(t)$ and output $y_c(t)$, is equivalent to a continuous-time lowpass filter with frequency response $H_{eff}(j\Omega)$ sketched in Figure P4.25-4.

(b) Determine the range of values of T for which the information presented in (a) is true when $X_c(j\Omega)$ is bandlimited to $|\Omega| \le 2\pi \times 10^4$ as shown in Fig. P4.25-3.

(c) For the range of values determined in (b), sketch Ω_c as a function of $1/T$.

Note: This is one way of implementing a variable-cutoff continuous-time filter using fixed continuous-time and discrete-time filters and a variable sampling rate.

4.26. Consider the sequence $x[n]$ whose Fourier transform $X(e^{j\omega})$ is shown in Figure P4.26-1. Define

$$x_s[n] = \begin{cases} x[n], & n = Mk, \quad k = 0, \pm 1, \pm 2, \ldots, \\ 0, & \text{otherwise,} \end{cases}$$

and

$$x_d[n] = x_s[Mn] = x[Mn].$$

(a) Sketch $X_s(e^{j\omega})$ and $X_d(e^{j\omega})$ for each of the following cases:
  (i) $M = 3, \omega_H = \pi/2$
  (ii) $M = 3, \omega_H = \pi/4$
(b) What is the maximum value of ω_H that will avoid aliasing when $M = 3$?

Figure P4.26-1

4.27. Using Parseval's theorem, briefly explain why the amplitude of the Fourier transform changes during downsampling but not during upsampling.

4.28. **(a)** Is the system in Figure 4.11 linear for a given choice of T ? If so, provide a brief argument demonstrating that it satisfies linearity. If not, provide a counterexample.
  (b) Is the system in Figure 4.11 time invariant for a given choice of T? If so, provide a brief argument demonstrating that it satisfies time invariance. If not, provide a counterexample.

Advanced Problems

4.29. Consider the systems shown in Figure P4.29-1. Suppose that $H_1(e^{j\omega})$ is fixed and known. Find $H_2(e^{j\omega})$, the frequency response of an LTI system, such that $y_2[n] = y_1[n]$ if the inputs to the systems are the same.

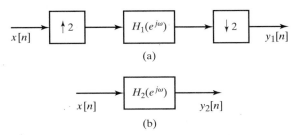

Figure P4.29-1

4.30. In the system of Figure 4.11, assume that the discrete-time system is linear and time invariant and that $X_c(j\Omega) = 0$ for $|\Omega| \geq 4000\pi$. Determine the largest possible value for T and the corresponding frequency response $H(e^{j\omega})$ for the discrete-time system such that

$$Y_c(j\Omega) = \begin{cases} |\Omega| X_c(j\Omega), & 1000\pi < |\Omega| < 2000\pi, \\ 0, & \text{otherwise.} \end{cases}$$

4.31. In the system of Figure 4.11, assume that $X_c(j\Omega) = 0$ for $|\Omega| > \pi/T$. Determine and plot the magnitude and phase of the frequency response of the discrete-time LTI system such that the output $y_r(t)$ is the running integral of the input, i.e.,

$$y_r(t) = \int_{-\infty}^{t} x_c(\tau) d\tau.$$

4.32. A bandlimited continuous-time signal is known to contain a 60-Hz component, which we
want to remove by processing with the system of Figure 4.11, where $T = 10^{-4}$.
   (a) What is the highest frequency that the continuous-time signal can contain if aliasing is
   to be avoided?
   (b) The discrete-time system to be used has frequency response

$$H(e^{j\omega}) = \frac{[1 - e^{-j(\omega-\omega_0)}][1 - e^{-j(\omega+\omega_0)}]}{[1 - 0.9e^{-j(\omega-\omega_0)}][1 - 0.9e^{-j(\omega+\omega_0)}]}.$$

   Sketch the magnitude and phase of $H(e^{j\omega})$.
   (c) What value should be chosen for ω_0 to eliminate the 60-Hz component?

4.33. Consider the system in Figure 4.11 with $X_c(j\Omega) = 0$ for $|\Omega| \geq 2\pi(1000)$ and the discrete-
time system a squarer, i.e., $y[n] = x^2[n]$. What is the largest value of T such that $y_c(t) = x_c^2(t)$?

4.34. For the LTI system in Figure P4.34-1,

$$H(e^{j\omega}) = e^{-j\omega/2}, \qquad |\omega| \leq \pi \text{ (half-sample delay)}.$$

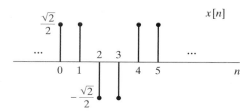

   Figure P4.34-1

   (a) Determine a choice for T and $h_c(t)$ in the system of Figure 4.16 so that the system in
   Figure P4.34-1 with $H(e^{j\omega})$ as specified is equivalent to the system in Figure 4.16.
   (b) Determine and sketch $y[n]$ when the input sequence is

$$x[n] = \cos\left(\frac{5\pi}{2}n - \frac{\pi}{4}\right),$$

   as sketched in Figure P4.34-2.

   Figure P4.34-2

4.35. Consider the system of Figure 4.16 with the continuous-time LTI system causal and characterized by the linear constant-coefficient differential equation

$$\frac{d^2 y_c(t)}{dt^2} + 4\frac{dy_c(t)}{dt} + 3y_c(t) = x_c(t).$$

The overall system is equivalent to a causal discrete-time LTI system. Determine the frequency response $H(e^{j\omega})$ of the equivalent discrete-time system when $T = 0.1$s.

4.36. In Figure P4.36-1, $x[n] = x_c(nT)$ and $y[n] = x[2n]$.
   (a) Assume that $x_c(t)$ has a Fourier transform such that $X_c(j\Omega) = 0, |\Omega| > 2\pi(100)$. What
   value of T is required so that

$$X(e^{j\omega}) = 0, \qquad \frac{\pi}{2} < |\omega| \leq \pi?$$

(b) How should T' be chosen so that $y_c(t) = x_c(t)$?

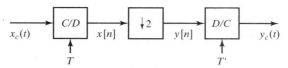

Figure P4.36-1

4.37. Suppose that you obtained a sequence $s[n]$ by filtering a speech signal $s_c(t)$ with a continuous-time lowpass filter with a cutoff frequency of 5 kHz and then sampling the resulting output at a 10-kHz rate, as shown in Figure P4.37-1. Unfortunately, the speech signal $s_c(t)$ was destroyed once the sequence $s[n]$ was stored on magnetic tape. Later, you find that what you should have done is followed the process shown in Figure P4.37-2. Develop a method to obtain $s_1[n]$ from $s[n]$ using discrete-time processing. Your method may require a very large amount of computation, but should *not* require a C/D or D/C converter. If your method uses a discrete-time filter, you should specify the frequency response of the filter.

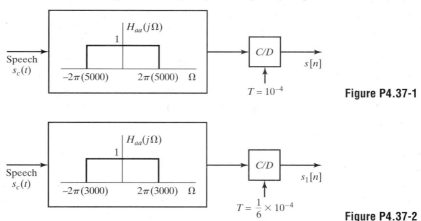

Figure P4.37-1

Figure P4.37-2

4.38. Consider the system shown in Figure P4.38-1, where

$$H(e^{j\omega}) = \begin{cases} 1, & |\omega| < \pi/L, \\ 0, & \pi/L < |\omega| \le \pi. \end{cases}$$

Sketch $Y_c(j\Omega)$ if $X_c(j\Omega)$ is as shown in Figure P4.38-2.

Figure P4.38-1

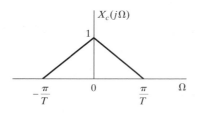

Figure P4.38-2

4.39. The system shown in Figure P4.39-1 approximately interpolates the sequence $x[n]$ by a factor L. Suppose that the linear filter has impulse response $h[n]$ such that $h[n] = h[-n]$ and $h[n] = 0$ for $|n| > (RL - 1)$, where R and L are integers; i.e., the impulse response is symmetric and of length $(2RL - 1)$ samples.

$x[n]$ $\uparrow L$ $v[n]$ $H(e^{j\omega})$ $y[n]$ **Figure P4.39-1**

(a) In answering the following, do not be concerned about the causality of the system; it can be made causal by including some delay. Specifically, how much delay must be inserted to make the system causal?
(b) What conditions must be satisfied by $h[n]$ in order that $y[n] = x[n/L]$ for $n = 0, \pm L$, $\pm 2L, \pm 3L, \ldots$?
(c) By exploiting the symmetry of the impulse response, show that each sample of $y[n]$ can be computed with no more than RL multiplications.
(d) By taking advantage of the fact that multiplications by zero need not be done, show that only $2R$ multiplications per output sample are required.

4.40. In the system of Figure P4.40-1,

$$X_c(j\Omega) = 0, \qquad |\Omega| \geq \pi/T,$$

and

$$H(e^{j\omega}) = \begin{cases} e^{-j\omega}, & |\omega| < \pi/L, \\ 0, & \pi/L < |\omega| \leq \pi. \end{cases}$$

How is $y[n]$ related to the input signal $x_c(t)$?

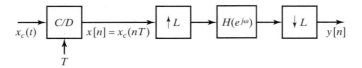

$x_c(t)$ C/D $x[n] = x_c(nT)$ $\uparrow L$ $H(e^{j\omega})$ $\downarrow L$ $y[n]$

T

Figure P4.40-1

4.41. Consider the system shown in Figure P4.41-1. The input to this system is the bandlimited signal whose Fourier transform is shown in Figure P4.20-1 with $\Omega_0 = \pi/T$. The discrete-time LTI system in Figure P4.41-1 has the frequency response shown in Figure P4.41-2.

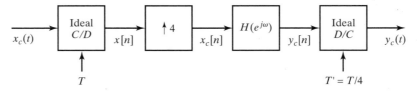

$x_c(t)$ Ideal C/D $x[n]$ $\uparrow 4$ $x_c[n]$ $H(e^{j\omega})$ $y_c[n]$ Ideal D/C $y_c(t)$

T $T' = T/4$

Figure P4.41-1

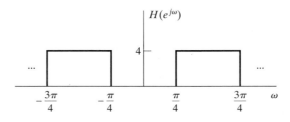

$H(e^{j\omega})$

4

$-\dfrac{3\pi}{4}$ $-\dfrac{\pi}{4}$ $\dfrac{\pi}{4}$ $\dfrac{3\pi}{4}$ ω

Figure P4.41-2

(a) Sketch the Fourier transforms $X(e^{j\omega})$, $X_e(e^{j\omega})$, $Y_e(e^{j\omega})$, and $Y_c(j\Omega)$.

(b) For the general case when $X_c(j\Omega) = 0$ for $|\Omega| \geq \pi/T$, express $Y_c(j\Omega)$ in terms of $X_c(j\Omega)$. Also, give a general expression for $y_c(t)$ in terms of $x_c(t)$ when $x_c(t)$ is band-limited in this manner.

4.42. Let $x_c(t)$ be a real-valued continuous-time signal with highest frequency $2\pi(250)$ radians/second. Furthermore, let $y_c(t) = x_c(t - 1/1000)$.

(a) If $x[n] = x_c(n/500)$, is it theoretically possible to recover $x_c(t)$ from $x[n]$? Justify your answer.

(b) If $y[n] = y_c(n/500)$, is it theoretically possible to recover $y_c(t)$ from $y[n]$? Justify your answer.

(c) Is it possible to obtain $y[n]$ from $x[n]$ using the system in Figure P4.42-1? If so, determine $H_1(e^{j\omega})$.

(d) It is also possible to obtain $y[n]$ from $x[n]$ without any upsampling or downsampling using a single LTI system with frequency response $H_2(e^{j\omega})$. Determine $H_2(e^{j\omega})$.

Figure P4.42-1

4.43. Consider the system shown in Figure P4.43-1 for discrete-time processing of the continuous-time input signal $g_c(t)$. The input signal $g_c(t)$ is of the form $g_c(t) = f_c(t) + e_c(t)$, where the Fourier transforms of $f_c(t)$ and $e_c(t)$ are shown in Figure P4.43-2. Since the input signal is not bandlimited, a continuous-time antialiasing filter $H_{aa}(j\Omega)$ is used. The magnitude of the frequency response for $H_{aa}(j\Omega)$ is shown in Figure P4.43-3, and the phase response of the antialiasing filter is $\angle H_{aa}(j\Omega) = -\Omega^3$.

Figure P4.43-1

Figure P4.43-2

Figure P4.43-3

(a) If the sampling rate is $2\pi/T = 1600\pi$, determine the magnitude and phase of $H(e^{j\omega})$, the frequency response of the discrete-time system, so that the output is $y_c(t) = f_c(t)$.

(b) Is it possible that $y_c(t) = f_c(t)$ if $2\pi/T < 1600\pi$? If so, what is the *minimum* value of $2\pi/T$? Determine $H(e^{j\omega})$ for this choice of $2\pi/T$.

4.44. Consider the system given in Figure P4.44-1. You may assume that $R_c(j\Omega)$ is bandlimited; i.e., $R_c(j\Omega) = 0$, $|\Omega| \geq 2\pi(1000)$, as shown in the figure.

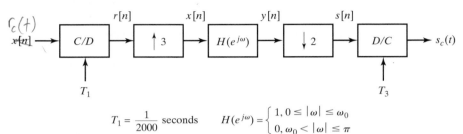

$$T_1 = \frac{1}{2000} \text{ seconds} \qquad H(e^{j\omega}) = \begin{cases} 1, & 0 \leq |\omega| \leq \omega_0 \\ 0, & \omega_0 < |\omega| \leq \pi \end{cases}$$

Figure P4.44-1

(a) Sketch $R(e^{j\omega})$ and $X(e^{j\omega})$.

(b) Choose nonzero values for ω_0 and T_2 such that

$$y[n] = \alpha r_c(nT_2)$$

for some nonzero constant α. (You do not have to determine the value of α.)

(c) Using the value of ω_0 you obtained in Part (b), determine a choice for T_3 such that

$$s_c(t) = \beta r_c(t)$$

for some nonzero constant β. (You do not have to determine the value of β.)

4.45. Assume that the continuous-time signal $x_c(t)$ in Figure P4.45-1 is exactly bandlimited and exactly time limited so that

$$x_c(t) = 0 \text{ for } t < 0 \text{ and } t > 10 \text{ seconds}$$

and

$$X_c(j\Omega) = 0 \text{ for } |\Omega| \geq 2\pi \times 10^4.$$

While no continuous-time signal can be exactly bandlimited and time limited, the assumption that a signal satisfies both constraints often an excellent approximation and one that we typically rely on in discrete-time processing of continuous-time signals. The continuous-time signal $x_c(t)$ is sampled as indicated in Figure P4.45-1 to obtain the sequence $x[n]$, which we want to process to estimate the total area A under $x_c(t)$ as precisely possible.

Figure P4.45-1

Specifically, we define

$$A = \int_0^{10} x_c(t)\, dt.$$

Specify a choice for the impulse response $h[n]$ for the discrete-time system and the largest possible value of T to obtain as accurate an estimate as possible of A. State specifically whether your estimate will be exact or approximate.

4.46. Consider the system in Figure P4.46-1 with $H_0(z)$, $H_1(z)$, and $H_2(z)$ as the system functions of LTI systems. Assume that $x[n]$ is an arbitrary stable complex signal without any symmetry properties.

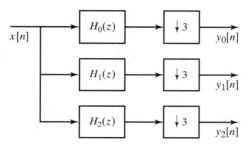

Figure P4.46-1

(a) Let $H_0(z) = 1$, $H_1(z) = z^{-1}$, and $H_2(z) = z^{-2}$. Can you reconstruct $x[n]$ from $y_0[n]$, $y_1[n]$, and $y_2[n]$? If so, how? If not, justify your answer.

(b) Assume that $H_0(e^{j\omega})$, $H_1(e^{j\omega})$, and $H_2(e^{j\omega})$ are as follows:

$$H_0(e^{j\omega}) = \begin{cases} 1, & |\omega| \le \pi/3, \\ 0, & \text{otherwise,} \end{cases}$$

$$H_1(e^{j\omega}) = \begin{cases} 1, & \pi/3 < |\omega| \le 2\pi/3, \\ 0, & \text{otherwise,} \end{cases}$$

$$H_2(e^{j\omega}) = \begin{cases} 1, & 2\pi/3 < |\omega| \le \pi, \\ 0, & \text{otherwise.} \end{cases}$$

Can you reconstruct $x[n]$ from $y_0[n]$, $y_1[n]$, and $y_2[n]$? If so, how? If not, justify your answer.

Now consider the system in Figure P4.46-2. Let $H_3(e^{j\omega})$ and $H_4(e^{j\omega})$ be the frequency responses of the LTI systems in this figure. Again, assume that $x[n]$ is an arbitrary stable complex signal with no symmetry properties.

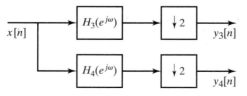

Figure P4.46-2

(c) Suppose that $H_3(e^{j\omega}) = 1$ and

$$H_4(e^{j\omega}) = \begin{cases} 1, & 0 \le \omega < \pi, \\ -1, & -\pi \le \omega < 0. \end{cases}$$

Can you reconstruct $x[n]$ from $y_3[n]$ and $y_4[n]$? If so, how? If not, justify your answer.

Extension Problems

4.47. In many applications, discrete-time random signals arise through periodic sampling of continuous-time random signals. We are concerned in this problem with a derivation of the sampling theorem for random signals. Consider a continuous-time, stationary, random process defined by the random variables $\{x_a(t)\}$, where t is a continuous variable. The autocorrelation function is defined as

$$\phi_{x_c x_c}(\tau) = \mathcal{E}\{x(t)x^*(t+\tau)\},$$

and the power density spectrum is

$$P_{x_c x_c}(\Omega) = \int_{-\infty}^{\infty} \phi_{x_c x_c}(\tau)e^{-j\Omega\tau} d\tau.$$

A discrete-time random process obtained by periodic sampling is defined by the set of random variables $\{x[n]\}$, where $x[n] = x_a(nT)$ and T is the sampling period.
  (a) What is the relationship between $\phi_{xx}[n]$ and $\phi_{x_c x_c}(\tau)$?
  (b) Express the power density spectrum of the discrete-time process in terms of the power density spectrum of the continuous-time process.
  (c) Under what condition is the discrete-time power density spectrum a faithful representation of the continuous-time power density spectrum?

4.48. Consider a continuous-time random process $x_c(t)$ with a bandlimited power density spectrum $P_{x_c x_c}(\Omega)$ as depicted in Figure P4.48-1. Suppose that we sample $x_c(t)$ to obtain the discrete-time random process $x[n] = x_c(nT)$.

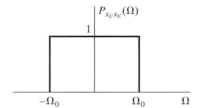

Figure P4.48-1

  (a) What is the autocorrelation sequence of the discrete-time random process?
  (b) For the continuous-time power density spectrum in Figure P4.48-1, how should T be chosen so that the discrete-time process is white, i.e., so that the power spectrum is constant for all ω?
  (c) If the continuous-time power density spectrum is as shown in Figure P4.48-2, how should T be chosen so that the discrete-time process is white?

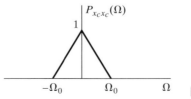

Figure P4.48-2

  (d) What is the general requirement on the continuous-time process and the sampling period such that the discrete-time process is white?

4.49. This problem explores the effect of interchanging the order of two operations on a signal, namely, sampling and performing a memoryless nonlinear operation.

(a) Consider the two signal-processing systems in Figure P4.49-1, where the C/D and D/C converters are ideal. The mapping $g[x] = x^2$ represents a memoryless nonlinear device. For the two systems in the figure, sketch the signal spectra at points 1, 2, and 3 when the sampling rate is selected to be $1/T = 2 f_m$ Hz and $x_c(t)$ has the Fourier transform shown in Figure P4.49-2. Is $y_1(t) = y_2(t)$? If not, why not? Is $y_1(t) = x^2(t)$? Explain your answer.

System 1:

System 2:

Figure P4.49-1

$-2\pi f_m$ 0 $2\pi f_m$ Ω **Figure P4.49-2**

(b) Consider System 1, and let $x(t) = A \cos(30\pi t)$. Let the sampling rate be $1/T = 40$ Hz. Is $y_1(t) = x_c^2(t)$? Explain why or why not.

(c) Consider the signal-processing system shown in Figure P4.49-3, where $g[x] = x^3$ and $g^{-1}[v]$ is the (unique) inverse, i.e., $g^{-1}[g(x)] = x$. Let $x(t) = A \cos(30\pi t)$ and $1/T = 40$ Hz. Express $v[n]$ in terms of $x[n]$. Is there spectral aliasing? Express $y[n]$ in terms of $x[n]$. What conclusion can you reach from this example? You may find the following identity helpful:

$$\cos^3 \Omega_0 t = \tfrac{3}{4} \cos \Omega_0 t + \tfrac{1}{4} \cos 3\Omega_0 t.$$

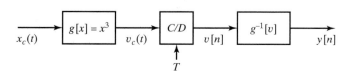

Figure P4.49-3

(d) One practical problem is that of digitizing a signal having a large dynamic range. Suppose we compress the dynamic range by passing the signal through a memoryless nonlinear device prior to A/D conversion and then expand it back after A/D conversion. What is the impact of the nonlinear operation prior to the A/D converter in our choice of the sampling rate?

4.50. Figure 4.24 depicts a system for interpolating a signal by a factor of L, where

$$x_e[n] = \begin{cases} x[n/L], & n = 0,\ \pm L,\ \pm 2L,\ \text{etc} \ldots, \\ 0, & \text{otherwise}, \end{cases}$$

and the lowpass filter interpolates between the nonzero values of $x_e[n]$ to generate the upsampled or interpolated signal $x_i[n]$. When the lowpass filter is ideal, the interpolation is referred to as bandlimited interpolation. As indicated in Section 4.6.2, simple interpolation procedures are adequate in many applications. Two simple procedures often used are zero-order-hold and linear interpolation. For zero-order-hold interpolation, each value of $x[n]$ is simply repeated L times; i.e.,

$$x_i[n] = \begin{cases} x_e[0], & n = 0,\ 1,\ldots, L-1, \\ x_e[L], & n = L,\ L+1,\ldots, 2L-1, \\ x_e[2L], & n = 2L,\ 2L+1,\ldots, \\ \vdots \end{cases}$$

Linear interpolation is described in Section 4.6.2.

(a) Determine an appropriate choice for the impulse response of the lowpass filter in Figure 4.24 to implement zero-order-hold interpolation. Also, determine the corresponding frequency response.

(b) Equation (4.92) specifies the impulse response for linear interpolation. Determine the corresponding frequency response. (You may find it helpful to use the fact that $h_{\text{lin}}[n]$ is triangular and consequently corresponds to the convolution of two rectangular sequences.)

(c) Sketch the magnitude of the filter frequency response for zero-order-hold and linear interpolation. Which is a better approximation to ideal bandlimited interpolation?

4.51. We wish to compute the autocorrelation function of an upsampled signal, as indicated in Figure P4.51-1. It is suggested that this can equivalently be accomplished with the system of Figure P4.51-2. Can $H_2(e^{j\omega})$ be chosen so that $\phi_3[n] = \phi_1[n]$? If not, why not? If so, specify $H_2(e^{j\omega})$.

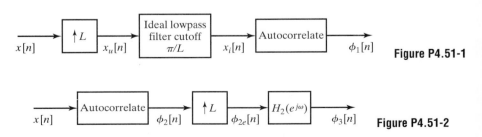

Figure P4.51-1

Figure P4.51-2

4.52. We are interested in upsampling a sequence by a factor of 2, using a system of the form of Figure 4.24. However, the lowpass filter in that figure is to be approximated by a five-point filter with impulse response $h[n]$ indicated in Figure P4.52-1. In this system, the output $y_1[n]$ is obtained by direct convolution of $h[n]$ with $w[n]$.

Figure P4.52-1

(a) A proposed implementation of the system with the preceding choice of $h[n]$ is shown in Figure P4.52-2. The three impulse responses $h_1[n]$, $h_2[n]$, and $h_3[n]$ are all restricted to be zero outside the range $0 \leq n \leq 2$. Determine and clearly justify a choice for $h_1[n]$, $h_2[n]$, and $h_3[n]$ so that $y_1[n] = y_2[n]$ for any $x[n]$, i.e., so that the two systems are identical.

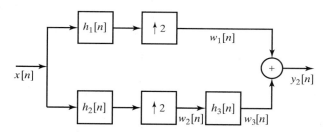

Figure P4.52-2

(b) Determine the number of multiplications per output point required in the system of Figure P4.52-1 and in the system of Figure P4.52-2. You should find that the system of Figure P4.52-2 is more efficient.

4.53. Consider the analysis–synthesis system shown in Figure P4.53-1. The lowpass filter $h_0[n]$ is identical in the analyzer and synthesizer, and the highpass filter $h_1[n]$ is identical in the analyzer and synthesizer. The Fourier transforms of $h_0[n]$ and $h_1[n]$ are related by

$$H_1(e^{j\omega}) = H_0(e^{j(\omega+\pi)}).$$

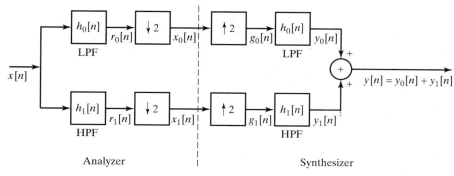

Figure P4.53-1

(a) If $X(e^{j\omega})$ and $H_0(e^{j\omega})$ are as shown in Figure P4.53-2, sketch (to within a scale factor) $X_0(e^{j\omega})$, $G_0(e^{j\omega})$, and $Y_0(e^{j\omega})$.

(b) Write a general expression for $G_0(e^{j\omega})$ in terms of $X(e^{j\omega})$ and $H_0(e^{j\omega})$. Do *not* assume that $X(e^{j\omega})$ and $H_0(e^{j\omega})$ are as shown in Figure 4.53-2.

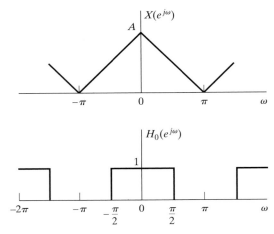

Figure P4.53-2

(c) Determine a set of conditions on $H_0(e^{j\omega})$ that is as general as possible and that will guarantee that $y[n]$ is proportional to $x[n - n_d]$ for any stable input $x[n]$.

Note: Analyzer–synthesizer filter banks of the form developed in this problem are very similar to quadrature mirror filter banks. For further reading, see Crochiere and Rabiner (1983), pp. 378–392.

4.54. Consider a real-valued sequence $x[n]$ for which

$$X(e^{j\omega}) = 0, \qquad \frac{\pi}{3} \le |\omega| \le \pi.$$

One value of $x[n]$ may have been corrupted, and we would like to approximately or exactly recover it. With $\hat{x}[n]$ denoting the corrupted signal,

$$\hat{x}[n] = x[n] \text{ for } n \ne n_0,$$

and $\hat{x}[n_0]$ is real but not related to $x[n_0]$. In each of the following three cases, specify a practical algorithm for exactly or approximately recovering $x[n]$ from $\hat{x}[n]$:

(a) The value of n_0 is known.

(b) The exact value of n_0 is *not* known, but we know that n_0 is an even number.

(c) Nothing about n_0 is known.

4.55. Communication systems often require conversion from time-division multiplexing (TDM) to frequency-division multiplexing (FDM). In this problem, we examine a simple example of such a system. The block diagram of the system to be studied is shown in Figure P4.55-1. The TDM input is assumed to be the sequence of interleaved samples

$$w[n] = \begin{cases} x_1[n/2] & \text{for } n \text{ an even integer,} \\ x_2[(n-1)/2] & \text{for } n \text{ an odd integer.} \end{cases}$$

Assume that the sequences $x_1[n] = x_{c1}(nT)$ and $x_2[n] = x_{c2}(nT)$ have been obtained by sampling the continuous-time signals $x_{c1}(t)$ and $x_{c2}(t)$, respectively, without aliasing.

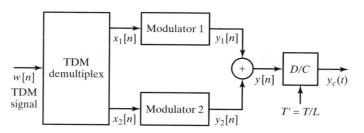

Figure P4.55-1

Assume also that these two signals have the same highest frequency, Ω_N, and that the sampling period is $T = \pi/\Omega_N$.

(a) Draw a block diagram of a system that produces $x_1[n]$ and $x_2[n]$ as outputs; i.e., obtain a system for demultiplexing a TDM signal using simple operations. State whether or not your system is linear, time invariant, causal, and stable.

The kth modulator system ($k = 1$ or 2) is defined by the block diagram in Figure P4.55-2. The lowpass filter $H_i(e^{j\omega})$, which is the same for both channels, has gain L and cutoff frequency π/L, and the highpass filters $H_k(e^{j\omega})$ have unity gain and cutoff frequency ω_k. The modulator frequencies are such that

$$\omega_2 = \omega_1 + \pi/L \qquad \text{and} \qquad \omega_2 + \pi/L \leq \pi \qquad (\text{assume } \omega_1 > \pi/2).$$

Figure P4.55-2

(b) Assume that $\Omega_N = 2\pi \times 5 \times 10^3$. Find ω_1 and L so that, after ideal D/C conversion with sampling period T/L, the Fourier transform of $y_c(t)$ is zero, except in the band of frequencies

$$2\pi \times 10^5 \leq |\omega| \leq 2\pi \times 10^5 + 2\Omega_N.$$

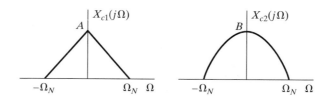

Figure P4.55-3

(c) Assume that the continuous-time Fourier transforms of the two original input signals are as sketched in Figure P4.55-3. Sketch the Fourier transforms at each point in the system.

(d) Based on your solution to Parts (a)–(c), discuss how the system could be generalized to handle M equal-bandwidth channels.

4.56. In Section 4.8.1, we considered the use of prefiltering to avoid aliasing. In practice, the antialiasing filter cannot be ideal. However, the nonideal characteristics can be at least partially compensated for with a discrete-time system applied to the sequence $x[n]$ that is the output of the C/D converter.

Consider the two systems in Figure P4.56-1. The antialiasing filters $H_{ideal}(j\Omega)$ and $H_{aa}(j\Omega)$ are shown in Figure P4.56-2. $H(e^{j\omega})$ in Figure P4.56-1 is to be specified to compensate for the nonideal characteristics of $H_{aa}(j\Omega)$.

Sketch $H(e^{j\omega})$ so that the two sequences $x[n]$ and $w[n]$ are identical.

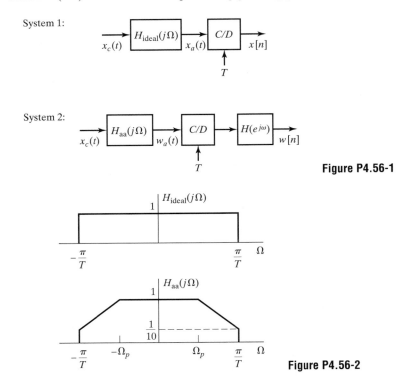

Figure P4.56-1

Figure P4.56-2

4.57. As discussed in Section 4.8.2, to process sequences on a digital computer, we must quantize the amplitude of the sequence to a set of discrete levels. This quantization can be expressed in terms of passing the input sequence $x[n]$ through a quantizer $Q(x)$ that has an input–output relation as shown in Figure 4.48.

As discussed in Section 4.8.3, if the quantization interval Δ is small compared with changes in the level of the input sequence, we can assume that the output of the quantizer is of the form

$$y[n] = x[n] + e[n],$$

where $e[n] = Q(x[n]) - x[n]$ and $e[n]$ is a stationary random process with a first-order probability density uniformly distributed between $-\Delta/2$ and $\Delta/2$, uncorrelated from sample to sample and uncorrelated with $x[n]$, so that $\mathcal{E}\{e[n]x[m]\} = 0$ for all m and n.

Let $x[n]$ be a stationary white-noise process with zero mean and variance σ_x^2.

(a) Find the mean, variance, and autocorrelation sequence of $e[n]$.

(b) What is the signal-to-quantizing-noise ratio σ_x^2/σ_e^2?

(c) The quantized signal $y[n]$ is to be filtered by a digital filter with impulse response $h[n] = \frac{1}{2}[a^n + (-a)^n]u[n]$. Determine the variance of the noise produced at the output due to the input quantization noise, and determine the signal-to-noise ratio at the output.

In some cases we may want to use nonlinear quantization steps, for example, logarithmically spaced quantization steps. This can be accomplished by applying uniform quantization to the logarithm of the input as depicted in Figure P4.57-1, where $Q[\cdot]$ is a uniform quantizer as specified in Figure 4.48. In this case, if we assume that Δ is small compared with changes in the sequence $\ln(x[n])$, then we can assume that the output of the quantizer is

$$\ln(y[n]) = \ln(x[n]) + e[n].$$

Thus,

$$y[n] = x[n] \cdot \exp(e[n]).$$

For small e, we can approximate $\exp(e[n])$ by $(1 + e[n])$, so that

$$y[n] \approx x[n](1 + e[n]) = x[n] + f[n]. \tag{P4.57}$$

This equation will be used to describe the effect of logarithmic quantization. We assume $e[n]$ to be a stationary random process, uncorrelated from sample to sample, independent of the signal $x[n]$, and with first-order probability density uniformly distributed between $\pm \Delta/2$.

Figure P4.57-1

(d) Determine the mean, variance, and autocorrelation sequence of the *additive* noise $f[n]$ defined in Eq. (P4.57).

(e) What is the signal-to-quantizing-noise ratio σ_x^2/σ_f^2? Note that in this case σ_x^2/σ_f^2 is independent of σ_x^2. Within the limits of our assumption, therefore, the signal-to-quantizing-noise ratio is independent of the input signal level, whereas, for linear quantization, the ratio σ_x^2/σ_e^2 depends directly on σ_x^2.

(f) The quantized signal $y[n]$ is to be filtered by means of a digital filter with impulse response $h[n] = \frac{1}{2}[a^n + (-a)^n]u[n]$. Determine the variance of the noise produced at the output due to the input quantization noise, and determine the signal-to-noise ratio at the output.

4.58. Figure P4.58-1 shows a system in which two continuous-time signals are multiplied and a discrete-time signal is then obtained from the product by sampling the product at the Nyquist rate; i.e., $y_1[n]$ is samples of $y_c(t)$ taken at the Nyquist rate. The signal $x_1(t)$ is bandlimited to 25 kHz $(X_1(j\Omega) = 0$ for $|\Omega| \geq 5\pi \times 10^4)$, and $x_2(t)$ is limited to 2.5 kHz $(X_2(j\Omega) = 0$ for $|\Omega| \geq (\pi/2) \times 10^4)$.

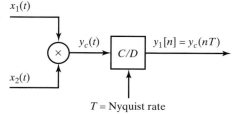

$T =$ Nyquist rate   **Figure P4.58-1**

In some situations (digital transmission, for example), the continuous-time signals have already been sampled at their individual Nyquist rates, and the multiplication is to be carried out in the discrete-time domain, perhaps with some additional processing before and after multiplication, as indicated in Figure P4.58-2. Each of the systems A, B, and C

either is an identity or can be implemented using one or more of the modules shown in Figure P4.58-3.

Figure P4.58-2

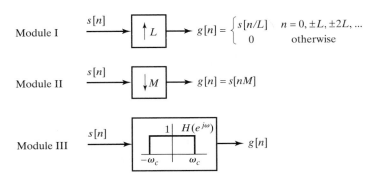

Figure P4.58-3

For each of the three systems A, B, and C, either specify that the system is an identity system or specify an appropriate interconnection of one or more of the modules shown in Figure P4.58-3. Also, specify all relevant parameters L, M, and ω_c. The systems A, B, and C should be constructed such that $y_2[n]$ is proportional to $y_1[n]$, i.e.,

$$y_2[n] = ky_1[n] = ky_c(nT) = kx_1(nT) \times x_2(nT),$$

and these samples are at the Nyquist rate, i.e., $y_2[n]$ does not represent oversampling or undersampling of $y_c(t)$.

4.59. Suppose $s_c(t)$ is a speech signal with the continuous-time Fourier transform $S_c(j\Omega)$ shown in Figure P4.59-1. We obtain a discrete-time sequence $s_r[n]$ from the system shown in Figure P4.59-2, where $H(e^{j\omega})$ is an ideal discrete-time lowpass filter with cutoff frequency ω_c and a gain of unity throughout the passband, as shown in Figure 2.17. The signal $s_r[n]$ will be used as an input to a speech coder, which operates correctly only on discrete-time samples representing speech sampled at an 8-kHz rate. Choose values of L, M, and ω_c that produce the correct input signal $s_r[n]$ for the speech coder.

Figure P4.59-1

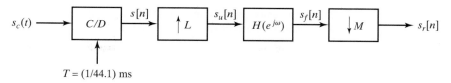

Figure P4.59-2

4.60. In many audio applications, it is necessary to sample a continuous-time signal $x_c(t)$ at a sampling rate $1/T = 44$ kHz. Figure P4.60-1 shows a straightforward system, including a continuous-time antialias filter $H_{a0}(j\Omega)$, to acquire the desired samples. In many applications, the "4x oversampling" system shown in Figure P4.60-2 is used instead of the conventional system shown in Figure P4.60-1. In the system in Figure P4.60-2,

$$H(e^{j\omega}) = \begin{cases} 1, & |\omega| \leq \pi/4, \\ 0, & \text{otherwise,} \end{cases}$$

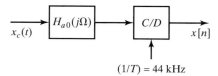

$(1/T) = 44$ kHz   **Figure P4.60-1**

an ideal lowpass filter, and

$$H_{a1}(j\Omega) = \begin{cases} 1, & |\Omega| \leq \Omega_p, \\ 0, & |\Omega| > \Omega_s, \end{cases}$$

for some $0 \leq \Omega_p \leq \Omega_s \leq \infty$.

$(1/T) = 4 \times 44$ kHz $= 176$ kHz   **Figure P4.60-2**

Assuming that $H(e^{j\omega})$ is ideal, find the minimal set of specifications on the antialias filter $H_{a1}(j\Omega)$, i.e., the smallest Ω_p and the largest Ω_s, such that the overall system of Figure P4.60-2 is equivalent to the system in Figure P4.60-1.

4.61. In this problem, we will consider the "double integration" system for quantization with noise shaping shown in Figure 4.61-1. In this system,

$$H_1(z) = \frac{1}{1 - z^{-1}} \quad \text{and} \quad H_2(z) = \frac{z^{-1}}{1 - z^{-1}},$$

and the frequency response of the decimation filter is

$$H_3(e^{j\omega}) = \begin{cases} 1, & |\omega| < \pi/M, \\ 0, & \pi/M \leq |\omega| \leq \pi. \end{cases}$$

The noise source $e[n]$, which represents a quantizer, is assumed to be a zero-mean white-noise (constant power spectrum) signal that is uniformly distributed in amplitude and has noise power $\sigma_e^2 = \Delta^2/12$.

Figure P4.61-1

(a) Determine an equation for $Y(z)$ in terms of $X(z)$ and $E(z)$. Assume for this part that $E(z)$ exists. From the z-transform relation, show that $y[n]$ can be expressed in the form $y[n] = x[n-1] + f[n]$, where $f[n]$ is the output due to the noise source $e[n]$. What is the time-domain relation between $f[n]$ and $e[n]$?

(b) Now assume that $e[n]$ is a white-noise signal as described prior to Part (a). Use the result from Part (a) to show that the power spectrum of the noise $f[n]$ is

$$P_{ff}(e^{j\omega}) = 16\sigma_e^2 \sin^4(\omega/2).$$

What is the *total* noise power (σ_f^2) in the noise component of the signal $y[n]$? On the same set of axes, sketch the power spectra $P_{ee}(e^{j\omega})$ and $P_{ff}(e^{j\omega})$ for $0 \le \omega \le \pi$.

(c) Now assume that $X(e^{j\omega}) = 0$ for $\pi/M < \omega \le \pi$. Argue that the output of $H_3(z)$ is $w[n] = x[n-1] + g[n]$. State in words what $g[n]$ is.

(d) Determine an expression for the noise power σ_g^2 at the output of the decimation filter. Assume that $\pi/M \ll \pi$, i.e., M is large, so that you can use a small-angle approximation to simplify the evaluation of the integral.

(e) After the decimator, the output is $v[n] = w[Mn] = x[Mn-1] + q[n]$, where $q[n] = g[Mn]$. Now suppose that $x[n] = x_c(nT)$ (i.e., $x[n]$ was obtained by sampling a continuous-time signal). What condition must be satisfied by $X_c(j\Omega)$ so that $x[n-1]$ will pass through the filter unchanged? Express the "signal component" of the output $v[n]$ in terms of $x_c(t)$. What is the total power σ_q^2 of the noise at the output? Give an expression for the power spectrum of the noise at the output, and, on the same set of axes, sketch the power spectra $P_{ee}(e^{j\omega})$ and $P_{qq}(e^{j\omega})$ for $0 \le \omega \le \pi$.

4.62. For sigma-delta oversampled A/D converters with high-order feedback loops, stability becomes a significant consideration. An alternative approach referred to as MASH achieves high-order noise shaping with only first-order feedback. The structure for second-order MASH noise shaping is shown in Figure P4.62-2 and analyzed in this problem.

Figure P4.62-1 is a first-order sigma-delta $(\Sigma - \Delta)$ noise shaping system, where the effect of the quantizer is represented by the additive noise signal $e[n]$. The noise $e[n]$ is explicitly shown in the diagram as a second output of the system. Assume that the input $x[n]$ is a zero-mean wide-sense stationary random process. Assume also that $e[n]$ is zero-mean, white, wide-sense stationary, and has variance σ_e^2. $e[n]$ is uncorrelated with $x[n]$.

(a) For the system in Figure P4.62-1, the output $y[n]$ has a component $y_x[n]$ due only to $x[n]$ and a component $y_e[n]$ due only to $e[n]$, i.e., $y[n] = y_x[n] + y_e[n]$.

(i) Determine $y_x[n]$ in terms of $x[n]$.

(ii) Determine $P_{y_e}(\omega)$, the power spectral density of $y_e[n]$.

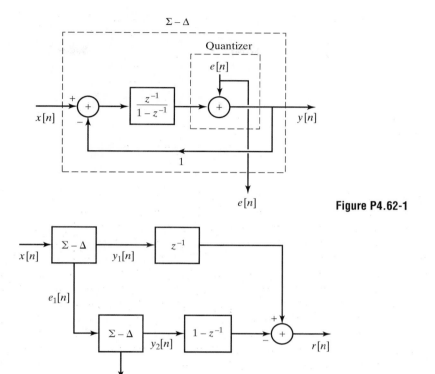

Figure P4.62-1

Figure P4.62-2

(b) The system of Figure P4.62-1 is now connected in the configuration shown in Figure P4.62-2, which shows the structure of the MASH system. Notice that $e_1[n]$ and $e_2[n]$ are the noise signals resulting from the quantizers in the sigma-delta noise shaping systems. The output of the system $r[n]$ has a component $r_x[n]$ due only to $x[n]$, and a component $r_e[n]$ due only to the quantization noise, i.e., $r[n] = r_x[n] + r_e[n]$. Assume that $e_1[n]$ and $e_2[n]$ are zero-mean, white, wide-sense stationary, each with variance σ_e^2. Also assume that $e_1[n]$ is uncorrelated with $e_2[n]$.

 (i) Determine $r_x[n]$ in terms of $x[n]$.

(ii) Determine $P_{r_e}(\omega)$, the power spectral density of $r_e[n]$.

5

TRANSFORM ANALYSIS OF LINEAR TIME–INVARIANT SYSTEMS

5.0 INTRODUCTION

In Chapter 2 we developed the Fourier transform representation of discrete-time signals and systems, and in Chapter 3 we extended that representation to the z-transform. In both chapters, the emphasis was on the transforms and their properties, with only a brief preview of the details of their use in the analysis of linear time-invariant (LTI) systems. In this chapter, we develop in more detail the representation and analysis of LTI systems using the Fourier and z-transforms. The material is essential background for our discussion in Chapter 6 of the implementation of LTI systems and in Chapter 7 of the design of such systems.

As developed in Chapter 2, an LTI system can be completely characterized in the time domain by its impulse response $h[n]$, with the output $y[n]$ due to a given input $x[n]$ specified through the convolution sum

$$y[n] = x[n] * h[n] = \sum_{k=-\infty}^{\infty} x[k]h[n-k]. \tag{5.1}$$

Alternatively, as discussed in Section 2.7, since the frequency response and impulse response are directly related through the Fourier transform, the frequency response, assuming it exists (i.e., converges), provides an equally complete characterization of LTI systems. In Chapter 3 we developed the z-transform as a generalization of the Fourier transform, and we showed that $Y(z)$, the z-transform of the output of an LTI system, is related to $X(z)$, the z-transform of the input, and $H(z)$, the z-transform of the system impulse response, by

$$Y(z) = H(z)X(z), \tag{5.2}$$

with an appropriate region of convergence. $H(z)$ is referred to as the *system function.* Since the z-transform and a sequence form a unique pair, it follows that any LTI system is completely characterized by its system function, again assuming convergence.

As we will see in this chapter, both the frequency response and the system function are extremely useful in the analysis and representation of LTI systems, because we can readily infer many properties of the system response from them.

5.1 THE FREQUENCY RESPONSE OF LTI SYSTEMS

The frequency response $H(e^{j\omega})$ of an LTI system was defined in Section 2.6 as the complex gain (eigenvalue) that the system applies to the complex exponential input (eigenfunction) $e^{j\omega n}$. Furthermore, in Section 2.9.6 we developed the fact that, since the Fourier transform of a sequence represents a decomposition as a linear combination of complex exponentials, the Fourier transforms of the system input and output are related by

$$Y(e^{j\omega}) = H(e^{j\omega})X(e^{j\omega}), \tag{5.3}$$

where $X(e^{j\omega})$ and $Y(e^{j\omega})$ are the Fourier transforms of the system input and output, respectively. With the frequency response expressed in polar form, the magnitude and phase of the Fourier transforms of the system input and output are related by

$$|Y(e^{j\omega})| = |H(e^{j\omega})| \cdot |X(e^{j\omega})|, \tag{5.4a}$$

$$\sphericalangle Y(e^{j\omega}) = \sphericalangle H(e^{j\omega}) + \sphericalangle X(e^{j\omega}). \tag{5.4b}$$

$|H(e^{j\omega})|$ is referred to as the *magnitude* response or the *gain* of the system, and $\sphericalangle H(e^{j\omega})$ is referred to as the *phase response* or *phase shift* of the system.

The magnitude and phase effects represented by Eqs. (5.4a) and (5.4b) can be either desirable, if the input signal is modified in a useful way, or undesirable, if the input signal is changed in a deleterious manner. In the latter case, we often refer to the effects of an LTI system on a signal, as represented by Eqs. (5.4a) and (5.4b), as *magnitude* and *phase distortions,* respectively.

5.1.1 Ideal Frequency-Selective Filters

An important implication of Eq. (5.4a) is that frequency components of the input are suppressed in the output if $|H(e^{j\omega})|$ is small at those frequencies. Whether this suppression of Fourier components is viewed as desirable or undesirable depends on the specific problem. Example 2.19 formalized the general notion of frequency-selective filters through the definition of certain ideal frequency responses. For example, the ideal lowpass filter was defined as the discrete-time linear time-invariant system whose frequency response is

$$H_{\text{lp}}(e^{j\omega}) = \begin{cases} 1, & |\omega| < \omega_c, \\ 0, & \omega_c < |\omega| \leq \pi, \end{cases} \tag{5.5}$$

and, of course, $H_{\mathrm{lp}}(e^{j\omega})$ is also periodic with period 2π. The ideal lowpass filter selects the low-frequency components of the signal and rejects the high-frequency components. The corresponding impulse response was shown in Example 2.22 to be

$$h_{\mathrm{lp}}[n] = \frac{\sin \omega_c n}{\pi n}, \qquad -\infty < n < \infty. \tag{5.6}$$

Analogously, the *ideal highpass filter* is defined as

$$H_{\mathrm{hp}}(e^{j\omega}) = \begin{cases} 0, & |\omega| < \omega_c, \\ 1, & \omega_c < |\omega| \leq \pi, \end{cases} \tag{5.7}$$

and since $H_{\mathrm{hp}}(e^{j\omega}) = 1 - H_{\mathrm{lp}}(e^{j\omega})$, its impulse response is

$$h_{\mathrm{hp}}[n] = \delta[n] - h_{\mathrm{lp}}[n] = \delta[n] - \frac{\sin \omega_c n}{\pi n}. \tag{5.8}$$

The ideal highpass filter passes the frequency band $\omega_c < \omega \leq \pi$ undistorted and rejects frequencies below ω_c. Other ideal frequency-selective filters were defined in Example 2.19.

     The ideal lowpass filters are noncausal, and their impulse responses extend from $-\infty$ to $+\infty$. Therefore, it is not possible to compute the output of either the ideal lowpass or the ideal highpass filter either recursively or nonrecursively; i.e., the systems are not *computationally realizable.*

     Another important property of the ideal lowpass filter as defined in Eq. (5.5) is that the phase response is specified to be zero. If it were not zero, the low-frequency band selected by the filter would also have phase distortion. It will become clear later in this chapter that causal approximations to ideal frequency-selective filters must have a nonzero phase response.

5.1.2 Phase Distortion and Delay

To understand the effect of the phase of a linear system, let us first consider the ideal delay system. The impulse response is

$$h_{\mathrm{id}}[n] = \delta[n - n_d], \tag{5.9}$$

and the frequency response is

$$H_{\mathrm{id}}(e^{j\omega}) = e^{-j\omega n_d}, \tag{5.10}$$

or

$$|H_{\mathrm{id}}(e^{j\omega})| = 1, \tag{5.11a}$$

$$\sphericalangle H_{\mathrm{id}}(e^{j\omega}) = -\omega n_d, \qquad |\omega| < \pi, \tag{5.11b}$$

with periodicity 2π in ω assumed. For now, we will assume that n_d is an integer.

     In many applications, delay distortion would be considered a rather mild form of phase distortion, since its effect is only to shift the sequence in time. Often this would be inconsequential, or it could easily be compensated for by introducing delay in other parts of a larger system. Thus, in designing approximations to ideal filters and other linear time-invariant systems, we frequently are willing to accept a linear phase response rather than a zero phase response as our ideal. For example, an ideal lowpass filter with linear phase would be defined as

$$H_{lp}(e^{j\omega}) = \begin{cases} e^{-j\omega n_d}, & |\omega| < \omega_c, \\ 0, & \omega_c < |\omega| \le \pi. \end{cases} \tag{5.12}$$

Its impulse response is

$$h_{lp}[n] = \frac{\sin \omega_c(n - n_d)}{\pi(n - n_d)}, \qquad -\infty < n < \infty. \tag{5.13}$$

In a similar manner, we could define other ideal frequency-selective filters with linear phase. These filters would have the desired effect of isolating a band of frequencies in the input signal, as well as the additional effect of delaying the output by n_d. Note, however, that no matter how large we make n_d, the ideal lowpass filter is always noncausal.

A convenient measure of the linearity of the phase is the *group delay*. The basic concept of group delay relates to the effect of the phase on a narrowband signal. Specifically, consider the output of a system with frequency response $H(e^{j\omega})$ for a narrowband input of the form $x[n] = s[n]\cos(\omega_0 n)$. Since it is assumed that $X(e^{j\omega})$ is nonzero only around $\omega = \omega_0$, the effect of the phase of the system can be approximated around $\omega = \omega_0$ as the linear approximation

$$\sphericalangle H(e^{j\omega}) \simeq -\phi_0 - \omega n_d. \tag{5.14}$$

With this approximation, it can be shown (see Problem 5.57) that the response $y[n]$ to $x[n] = s[n]\cos(\omega_0 n)$ is approximately $y[n] = |H(e^{j\omega_0})|s[n - n_d]\cos(\omega_0 n - \phi_0 - \omega_0 n_d)$. Consequently, the time delay of the envelope $s[n]$ of the narrowband signal $x[n]$ with Fourier transform centered at ω_0 is given by the negative of the slope of the phase at ω_0. In considering the linear approximation to $\sphericalangle H(e^{j\omega})$ around $\omega = \omega_0$, as given in Eq. (5.14), we must consider the phase response as a continuous function of ω. The phase response specified in this way will be denoted as $\arg[H(e^{j\omega})]$ and is referred to as the *continuous phase of $H(e^{j\omega})$*.

With phase specified as a continuous function of ω, the group delay of a system is defined as

$$\tau(\omega) = \text{grd}[H(e^{j\omega})] = -\frac{d}{d\omega}\{\arg[H(e^{j\omega})]\}. \tag{5.15}$$

The deviation of the group delay from a constant indicates the degree of nonlinearity of the phase.

Example 5.1   Effects of Attenuation and Group Delay

As an illustration of the effect of group delay, consider a filter with frequency response magnitude and group delay shown in Figure 5.1. In Figure 5.2, we show an input signal and its spectrum. In Figure 5.3 is the resulting output signal. Note that the input signal consists of three consecutive narrowband pulses, at frequencies $\omega = 0.85\pi$, $\omega = 0.25\pi$, and $\omega = 0.5\pi$.

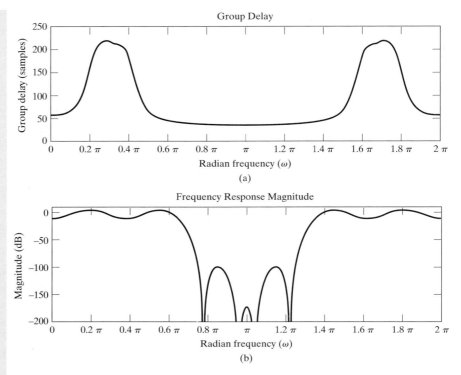

Figure 5.1 Frequency response magnitude and group delay for the filter in Example 5.1.

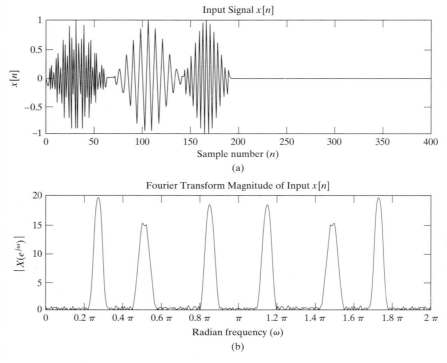

Figure 5.2 Input signal and associated Fourier transform magnitude for Example 5.1.

Figure 5.3   Output signal for Example 5.1.

Since the filter has considerable attenuation at $\omega = 0.85\pi$, the pulse at that frequency is not clearly present in the output. Also, since the group delay at $\omega = 0.25\pi$ is approximately 200 samples and at $\omega = 0.5\pi$ is approximately 50 samples, the second pulse in $x[n]$ will be delayed by about 200 samples and the third pulse by 50 samples, as we see is the case in Figure 5.3.

5.2 SYSTEM FUNCTIONS FOR SYSTEMS CHARACTERIZED BY LINEAR CONSTANT-COEFFICIENT DIFFERENCE EQUATIONS

While ideal frequency-selective filters are useful conceptually, they cannot be implemented with finite computation. Therefore, it is of interest to consider a class of systems that can be implemented as approximations to ideal frequency-selective filters.

In Section 2.5, we considered the class of systems whose input and output satisfy a linear constant-coefficient difference equation of the form

$$\sum_{k=0}^{N} a_k y[n-k] = \sum_{k=0}^{M} b_k x[n-k]. \tag{5.16}$$

We showed that if we further assume that the system is causal, the difference equation can be used to compute the output recursively. If the auxiliary conditions correspond to initial rest, the system will be causal, linear, and time invariant.

The properties and characteristics of LTI systems for which the input and output satisfy a linear constant-coefficient difference equation are best developed through the z-transform. Applying the z-transform to both sides of Eq. (5.16) and using the linearity property (Section 3.4.1) and the time-shifting property (Section 3.4.2), we obtain

$$\sum_{k=0}^{N} a_k z^{-k} Y(z) = \sum_{k=0}^{M} b_k z^{-k} X(z),$$

or equivalently,

$$\left(\sum_{k=0}^{N} a_k z^{-k}\right) Y(z) = \left(\sum_{k=0}^{M} b_k z^{-k}\right) X(z). \tag{5.17}$$

From Eq. (5.2) and Eq. (5.17), it follows that, for a system whose input and output satisfy a difference equation of the form of Eq. (5.16), the system function has the algebraic form

$$H(z) = \frac{Y(z)}{X(z)} = \frac{\displaystyle\sum_{k=0}^{M} b_k z^{-k}}{\displaystyle\sum_{k=0}^{N} a_k z^{-k}}. \tag{5.18}$$

$H(z)$ in Eq. (5.18) is a ratio of polynomials in z^{-1}, because Eq. (5.16) consists of a linear combination of delay terms. Although Eq. (5.18) can, of course, be rewritten so that the polynomials are expressed as powers of z rather than of z^{-1}, it is common practice not to do so. Also, it is often convenient to express Eq. (5.18) in factored form as

$$H(z) = \left(\frac{b_0}{a_0}\right) \frac{\displaystyle\prod_{k=1}^{M}(1 - c_k z^{-1})}{\displaystyle\prod_{k=1}^{N}(1 - d_k z^{-1})}. \tag{5.19}$$

Each of the factors $(1 - c_k z^{-1})$ in the numerator contributes a zero at $z = c_k$ and a pole at $z = 0$. Similarly, each of the factors $(1 - d_k z^{-1})$ in the denominator contributes a zero at $z = 0$ and a pole at $z = d_k$.

There is a straightforward relationship between the difference equation and the corresponding algebraic expression for the system function. Specifically, the numerator polynomial in Eq. (5.18) has the same coefficients and algebraic structure as the right-hand side of Eq. (5.16) (the terms of the form $b_k z^{-k}$ correspond to $b_k x[n-k]$), while the denominator polynomial in Eq. (5.18) has the same coefficients and algebraic structure as the left-hand side of Eq. (5.16) (the terms of the form $a_k z^{-k}$ correspond to $a_k y[n-k]$). Thus, given either the system function in the form of Eq. (5.18) or the difference equation in the form of Eq. (5.16), it is straightforward to obtain the other.

Example 5.2   Second-Order System

Suppose that the system function of a linear time-invariant system is

$$H(z) = \frac{(1 + z^{-1})^2}{\left(1 - \frac{1}{2}z^{-1}\right)\left(1 + \frac{3}{4}z^{-1}\right)}. \tag{5.20}$$

To find the difference equation that is satisfied by the input and output of this system, we express $H(z)$ in the form of Eq. (5.18) by multiplying the numerator and denominator factors to obtain the ratio of polynomials

$$H(z) = \frac{1 + 2z^{-1} + z^{-2}}{1 + \frac{1}{4}z^{-1} - \frac{3}{8}z^{-2}} = \frac{Y(z)}{X(z)}. \tag{5.21}$$

Thus,

$$\left(1 + \tfrac{1}{4}z^{-1} - \tfrac{3}{8}z^{-2}\right) Y(z) = (1 + 2z^{-1} + z^{-2})X(z),$$

and the difference equation is

$$y[n] + \tfrac{1}{4}y[n-1] - \tfrac{3}{8}y[n-2] = x[n] + 2x[n-1] + x[n-2]. \qquad (5.22)$$

Note that once the correspondence is well understood, it is possible to proceed directly from Eq. (5.21) to Eq. (5.22) without the intervening algebra (and vice versa).

5.2.1 Stability and Causality

To obtain Eq. (5.18) from Eq. (5.16), we assumed that the system was linear and time invariant, so that Eq. (5.2) applied, but we made no further assumption about stability or causality. Correspondingly, from the difference equation, we can obtain the algebraic expression for the system function, but not the region of convergence. Specifically, the region of convergence of $H(z)$ is not determined from the derivation leading to Eq. (5.18), since all that is required for Eq. (5.17) to hold is that $X(z)$ and $Y(x)$ have overlapping regions of convergence. This is consistent with the fact that, as we saw in Chapter 2, the difference equation does not uniquely specify the impulse response of a linear time-invariant system. For the system function of Eq. (5.18) or (5.19), there are a number of choices for the region of convergence. For a given ratio of polynomials, each possible choice for the region of convergence will lead to a different impulse response, but they will all correspond to the same difference equation. However, if we assume that the system is causal, it follows that $h[n]$ must be a right-sided sequence, and therefore, the region of convergence of $H(z)$ must be outside the outermost pole. Alternatively, if we assume that the system is stable, then, from the discussion in Section 2.4, the impulse response must be absolutely summable, i.e.,

$$\sum_{n=-\infty}^{\infty} |h[n]| < \infty. \qquad (5.23)$$

Since Eq. (5.23) is identical to the condition that

$$\sum_{n=-\infty}^{\infty} |h[n]z^{-n}| < \infty \qquad (5.24)$$

for $|z| = 1$, the condition for stability is equivalent to the condition that the ROC of $H(z)$ include the unit circle.

Example 5.3   Determining the ROC

Consider the LTI system with input and output related through the difference equation

$$y[n] - \tfrac{5}{2}y[n-1] + y[n-2] = x[n]. \qquad (5.25)$$

From the previous discussions, $H(z)$ is given by

$$H(z) = \frac{1}{1 - \tfrac{5}{2}z^{-1} + z^{-2}} = \frac{1}{\left(1 - \tfrac{1}{2}z^{-1}\right)\left(1 - 2z^{-1}\right)}. \qquad (5.26)$$

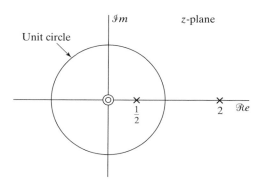

Figure 5.4   Pole–zero plot for Example 5.3.

The pole–zero plot for $H(z)$ is indicated in Figure 5.4. There are three possible choices for the ROC. If the system is assumed to be causal, then the ROC is outside the outermost pole, i.e., $|z| > 2$. In this case the system will not be stable, since the ROC does not include the unit circle. If we assume that the system is stable, then the ROC will be $\frac{1}{2} < |z| < 2$. For the third possible choice of ROC, $|z| < \frac{1}{2}$, the system will be neither stable nor causal.

As Example 5.3 suggests, causality and stability are not necessarily compatible require-ments. In order for a linear time-invariant system whose input and output satisfy a difference equation of the form of Eq. (5.16) to be both causal and stable, the ROC of the corresponding system function must be outside the outermost pole *and* include the unit circle. Clearly, this requires that all the poles of the system function be inside the unit circle.

5.2.2 Inverse Systems

For a given linear time-invariant system with system function $H(z)$, the corresponding inverse system is defined to be the system with system function $H_i(z)$ such that if it is cascaded with $H(z)$, the overall effective system function is unity; i.e.,

$$G(z) = H(z)H_i(z) = 1. \tag{5.27}$$

This implies that

$$H_i(z) = \frac{1}{H(z)}. \tag{5.28}$$

The time-domain condition equivalent to Eq. (5.27) is

$$g[n] = h[n] * h_i[n] = \delta[n]. \tag{5.29}$$

From Eq. (5.28), the frequency response of the inverse system, if it exists, is

$$H_i(e^{j\omega}) = \frac{1}{H(e^{j\omega})}; \tag{5.30}$$

i.e., $H_i(e^{j\omega})$ is the reciprocal of $H(e^{j\omega})$. Equivalently, the log magnitude, phase, and group delay of the inverse system are negatives of the corresponding functions for the original system. Not all systems have an inverse. For example, the ideal lowpass filter does not. There is no way to recover the frequency components above the cutoff frequency that are set to zero by such a filter.

Many systems do have inverses, and the class of systems with rational system functions provides a very useful and interesting example. Specifically, consider

$$H(z) = \left(\frac{b_0}{a_0}\right) \frac{\displaystyle\prod_{k=1}^{M}(1 - c_k z^{-1})}{\displaystyle\prod_{k=1}^{N}(1 - d_k z^{-1})}, \tag{5.31}$$

with zeros at $z = c_k$ and poles at $z = d_k$, in addition to possible zeros and/or poles at $z = 0$ and $z = \infty$. Then

$$H_i(z) = \left(\frac{a_0}{b_0}\right) \frac{\displaystyle\prod_{k=1}^{N}(1 - d_k z^{-1})}{\displaystyle\prod_{k=1}^{M}(1 - c_k z^{-1})}; \tag{5.32}$$

i.e., the poles of $H_i(z)$ are the zeros of $H(z)$ and vice versa. The question arises as to what region of convergence to associate with $H_i(z)$. The answer is provided by the convolution theorem, expressed in this case by Eq. (5.29). For Eq. (5.29) to hold, the regions of convergence of $H(z)$ and $H_i(z)$ must overlap. If $H(z)$ is causal, its region of convergence is

$$|z| > \max_k |d_k|. \tag{5.33}$$

Thus, any appropriate region of convergence for $H_i(z)$ that overlaps with the region specified by Eq. (5.33) is a valid region of convergence for $H_i(z)$. Some simple examples will illustrate some of the possibilities.

Example 5.4   Inverse System for First-Order System

Let $H(z)$ be

$$H(z) = \frac{1 - 0.5z^{-1}}{1 - 0.9z^{-1}}$$

with ROC $|z| > 0.9$. Then $H_i(z)$ is

$$H_i(z) = \frac{1 - 0.9z^{-1}}{1 - 0.5z^{-1}}.$$

Since $H_i(z)$ has only one pole, there are only two possibilities for its ROC, and the only choice for the ROC of $H_i(z)$ that overlaps with $|z| > 0.9$ is $|z| > 0.5$. Therefore, the impulse response of the inverse system is

$$h_i[n] = (0.5)^n u[n] - 0.9(0.5)^{n-1} u[n-1].$$

In this case, the inverse system is both causal and stable.

Example 5.5   Inverse for System with a Zero in the ROC

Suppose that $H(z)$ is

$$H(z) = \frac{z^{-1} - 0.5}{1 - 0.9z^{-1}}, \qquad |z| > 0.9.$$

The inverse system function is

$$H_i(z) = \frac{1 - 0.9z^{-1}}{z^{-1} - 0.5} = \frac{-2 + 1.8z^{-1}}{1 - 2z^{-1}}.$$

As before, there are two possible regions of convergence: $|z| < 2$ and $|z| > 2$. In this case, however, *both* regions overlap with $|z| > 0.9$, so both are valid inverse systems. The corresponding impulse response for an ROC $|z| < 2$ is

$$h_{i1}[n] = 2(2)^n u[-n-1] - 1.8(2)^{n-1} u[-n]$$

and, for an ROC $|z| > 2$, is

$$h_{i2}[n] = -2(2)^n u[n] + 1.8(2)^{n-1} u[n-1].$$

We see that $h_{i1}[n]$ is stable and noncausal, while $h_{i2}[n]$ is unstable and causal.

A generalization from Examples 5.4 and 5.5 is that if $H(z)$ is a causal system with zeros at $c_k, k = 1, \ldots, M$, then its inverse system will be causal if and only if we associate the region of convergence,

$$|z| > \max_k |c_k|,$$

with $H_i(z)$. If we also require that the inverse system be stable, then the region of convergence of $H_i(z)$ must include the unit circle. Therefore, it must be true that

$$\max_k |c_k| < 1;$$

i.e., all the zeros of $H(z)$ must be inside the unit circle. Thus, a linear time-invariant system is stable and causal and also has a stable and causal inverse if and only if both the poles and the zeros of $H(z)$ are inside the unit circle. Such systems are referred to as *minimum-phase* systems and will be discussed in more detail in Section 5.6.

5.2.3 Impulse Response for Rational System Functions

The discussion of the partial fraction expansion technique for finding inverse z-transforms (Section 3.3.2) can be applied to the system function $H(z)$ to obtain a general expression for the impulse response of a system that has a rational system function as in Eq. (5.19). Recall that any rational function of z^{-1} with only first-order poles can be expressed in the form

$$H(z) = \sum_{r=0}^{M-N} B_r z^{-r} + \sum_{k=1}^{N} \frac{A_k}{1 - d_k z^{-1}}, \tag{5.34}$$

where the terms in the first summation would be obtained by long division of the denominator into the numerator and would be present only if $M \geq N$. The coefficients A_k in the second set of terms are obtained using Eq. (3.41). If $H(z)$ has a multiple-order pole, its partial fraction expansion would have the form of Eq. (3.44). If the system is assumed to be causal, then the ROC is outside all of the poles in Eq. (5.34), and it follows that

$$h[n] = \sum_{r=0}^{M-N} B_r \delta[n - r] + \sum_{k=1}^{N} A_k d_k^n u[n], \tag{5.35}$$

where the first summation is included only if $M \geq N$.

In discussing LTI systems, it is useful to identify two classes. In the first class, at least one nonzero pole of $H(z)$ is not canceled by a zero. In this case there will be at least one term of the form $A_k(d_k)^n u[n]$, and $h[n]$ will not be of finite length, i.e., will not be zero outside a finite interval. Systems of this class are therefore called *infinite impulse response* (IIR) systems. A simple IIR system is discussed in the following example.

Example 5.6   A First-Order IIR System

Consider a causal system whose input and output satisfy the difference equation

$$y[n] - ay[n-1] = x[n]. \tag{5.36}$$

The system function is (by inspection)

$$H(z) = \frac{1}{1 - az^{-1}}. \tag{5.37}$$

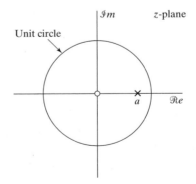

Figure 5.5   Pole–zero plot for Example 5.6.

Figure 5.5 shows the pole–zero plot of $H(z)$. The region of convergence is $|z| > |a|$, and the condition for stability is $|a| < 1$. The inverse z-transform of $H(z)$ is

$$h[n] = a^n u[n]. \tag{5.38}$$

For the second class of systems, $H(z)$ has no poles except at $z = 0$; i.e., $N = 0$ in Eqs. (5.16) and (5.18). Thus, a partial fraction expansion is not possible, and $H(z)$ is simply a polynomial in z^{-1} of the form

$$H(z) = \sum_{k=0}^{M} b_k z^{-k}. \tag{5.39}$$

(We assume, without loss of generality, that $a_0 = 1$.) In this case, $H(z)$ is determined to within a constant multiplier by its zeros. From Eq. (5.39), $h[n]$ is seen by inspection to be

$$h[n] = \sum_{k=0}^{M} b_k \delta[n - k] = \begin{cases} b_n, & 0 \le n \le M, \\ 0, & \text{otherwise.} \end{cases} \tag{5.40}$$

In this case, the impulse response is finite in length; i.e., it is zero outside a finite interval. Consequently, these systems are called *finite impulse response* (FIR) systems. Note that for FIR systems, the difference equation of Eq. (5.16) is identical to the convolution sum, i.e.,

$$y[n] = \sum_{k=0}^{M} b_k x[n - k]. \tag{5.41}$$

Example 5.7 gives a simple example of an FIR system.

Example 5.7   A Simple FIR System

Consider an impulse response that is a truncation of the impulse response of Example 5.6:

$$h[n] = \begin{cases} a^n, & 0 \le n \le M, \\ 0 & \text{otherwise.} \end{cases}$$

Then the system function is

$$H(z) = \sum_{n=0}^{M} a^n z^{-n} = \frac{1 - a^{M+1} z^{-M-1}}{1 - az^{-1}}. \tag{5.42}$$

Since the zeros of the numerator are at

$$z_k = ae^{j2\pi k/(M+1)}, \qquad k = 0, 1, \ldots, M, \tag{5.43}$$

where a is assumed real and positive, the pole at $z = a$ is canceled by a zero. The pole–zero plot for the case $M = 7$ is shown in Figure 5.6.

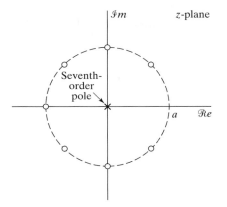

Figure 5.6   Pole–zero plot for Example 5.7.

The difference equation satisfied by the input and output of the linear time-invariant system is the discrete convolution

$$y[n] = \sum_{k=0}^{M} a^k x[n - k].$$

(5.44)

However, Eq. (5.42) suggests that the input and output also satisfy the difference equation

$$y[n] - ay[n - 1] = x[n] - a^{M+1} x[n - M - 1].$$

(5.45)

These two equivalent difference equations result from the two equivalent forms of $H(z)$ in Eq. (5.42).

5.3 FREQUENCY RESPONSE FOR RATIONAL SYSTEM FUNCTIONS

If a stable linear time-invariant system has a rational system function (i.e., if its input and output satisfy a difference equation of the form of Eq. (5.16), then its frequency response (the system function of Eq. (5.18) evaluated on the unit circle) has the form

$$H(e^{j\omega}) = \frac{\displaystyle\sum_{k=0}^{M} b_k e^{-j\omega k}}{\displaystyle\sum_{k=0}^{N} a_k e^{-j\omega k}}.$$

(5.46)

That is, $H(e^{j\omega})$ is a ratio of polynomials in the variable $e^{-j\omega}$. To determine the magnitude, phase, and group delay associated with the frequency response of such systems, it is useful to express $H(e^{j\omega})$ in terms of the poles and zeros of $H(z)$. Such an expression results from substituting $z = e^{j\omega}$ into Eq. (5.19):

$$H(e^{j\omega}) = \left(\frac{b_0}{a_0}\right) \frac{\prod\limits_{k=1}^{M}(1 - c_k e^{-j\omega})}{\prod\limits_{k=1}^{N}(1 - d_k e^{-j\omega})}. \tag{5.47}$$

From Eq. (5.47), it follows that

$$|H(e^{j\omega})| = \left|\frac{b_0}{a_0}\right| \frac{\prod\limits_{k=1}^{M}|1 - c_k e^{-j\omega}|}{\prod\limits_{k=1}^{N}|1 - d_k e^{-j\omega}|}. \tag{5.48}$$

Sometimes it is convenient to consider the magnitude squared, rather than the magnitude, of the system function. The magnitude-squared function is

$$|H(e^{j\omega})|^2 = H(e^{j\omega})H^*(e^{j\omega}),$$

where * denotes complex conjugation, and for $H(e^{j\omega})$ as in Eq. (5.47),

$$|H(e^{j\omega})|^2 = \left(\frac{b_0}{a_0}\right)^2 \frac{\prod\limits_{k=1}^{M}(1 - c_k e^{-j\omega})(1 - c_k^* e^{j\omega})}{\prod\limits_{k=1}^{N}(1 - d_k e^{-j\omega})(1 - d_k^* e^{j\omega})}. \tag{5.49}$$

From Eq. (5.48), we see that $|H(e^{j\omega})|$ is the product of the magnitudes of all the zero factors of $H(z)$ evaluated on the unit circle, divided by the product of the magnitudes of all the pole factors evaluated on the unit circle. It is common practice to transform these products into a corresponding sum of terms by considering $20 \log_{10} |H(e^{j\omega})|$ instead of $|H(e^{j\omega})|$. The logarithm of Eq. (5.48) is

$$20 \log_{10} |H(e^{j\omega})| = 20 \log_{10} \left|\frac{b_0}{a_0}\right| + \sum_{k=1}^{M} 20 \log_{10} |1 - c_k e^{-j\omega}|$$
$$- \sum_{k=1}^{N} 20 \log_{10} |1 - d_k e^{-j\omega}|. \tag{5.50}$$

The function $20 \log_{10} |H(e^{j\omega})|$ is referred to as the *log magnitude* of $H(e^{j\omega})$ and is expressed in *decibels* (dB). Sometimes this quantity is called the *gain in dB;* i.e.,

$$\text{Gain in dB} = 20 \log_{10} |H(e^{j\omega})|. \tag{5.51}$$

Note that zero dB corresponds to a value of $|H(e^{j\omega})| = 1$, while $|H(e^{j\omega})| = 10^m$ is $20m$ dB. Also, $|H(e^{j\omega})| = 2^m$ is approximately $6m$ dB. When $|H(e^{j\omega})| < 1$, the quantity $20\log_{10}|H(e^{j\omega})|$ is negative. This would be the case, for example, in the stopband of a frequency-selective filter. It is common practice to define

$$\text{Attenuation in dB} = -20\log_{10}|H(e^{j\omega})|$$
$$= -\text{Gain in dB}. \tag{5.52}$$

The attenuation is therefore a positive number when the magnitude response is less than unity. For example, a 60-dB attenuation at a given frequency ω means that at that frequency $|H(e^{j\omega})| = 0.001$.

Another advantage to expressing the magnitude in decibels stems from Eq. (5.4a), which, after taking logarithms of both sides, becomes

$$20\log_{10}|Y(e^{j\omega})| = 20\log_{10}|H(e^{j\omega})| + 20\log_{10}|X(e^{j\omega})|, \tag{5.53}$$

so the frequency response in dB is added to the log magnitude of the input Fourier transform to find the log magnitude of the output Fourier transform. If Eq. (5.53) replaces Eq. (5.4a) in Eqs. (5.4), then the effects of both magnitude and phase are additive.

The phase response for a rational system function has the form

$$\sphericalangle H(e^{j\omega}) = \sphericalangle \left[\frac{b_0}{a_0}\right] + \sum_{k=1}^{M} \sphericalangle[1 - c_k e^{-j\omega}] - \sum_{k=1}^{N} \sphericalangle[1 - d_k e^{-j\omega}]. \tag{5.54}$$

As in Eq. (5.50), the zero factors contribute with a plus sign and the pole factors contribute with a minus sign.

The corresponding group delay for a rational system function is

$$\text{grd}[H(e^{j\omega})] = \sum_{k=1}^{N} \frac{d}{d\omega}(\arg[1 - d_k e^{-j\omega}]) - \sum_{k=1}^{M} \frac{d}{d\omega}(\arg[1 - c_k e^{-j\omega}]), \tag{5.55}$$

where arg[ ] represents the continuous phase. An equivalent expression is

$$\text{grd}[H(e^{j\omega})] = \sum_{k=1}^{N} \frac{|d_k|^2 - \mathcal{R}e\{d_k e^{-j\omega}\}}{1 + |d_k|^2 - 2\mathcal{R}e\{d_k e^{-j\omega}\}} - \sum_{k=1}^{M} \frac{|c_k|^2 - \mathcal{R}e\{c_k e^{-j\omega}\}}{1 + |c_k|^2 - 2\mathcal{R}e\{c_k e^{-j\omega}\}}. \tag{5.56}$$

In Eq. (5.54), as written, the phase of each of the terms is ambiguous; i.e., any integer multiple of 2π can be added to each term at each value of ω without changing the value of the complex number. The expression for the group delay, on the other hand, involves differentiating the continuous phase.

When the angle of a complex number is computed, with the use of an arctangent subroutine on a calculator or with a computer system subroutine, the principal value is obtained. The principal value of the phase of $H(e^{j\omega})$ is denoted as $\text{ARG}[H(e^{j\omega})]$, where

$$-\pi < \text{ARG}[H(e^{j\omega})] \leq \pi. \tag{5.57}$$

Any other angle that gives the correct complex value of the function $H(e^{j\omega})$ can be represented in terms of the principal value as

$$\sphericalangle H(e^{j\omega}) = \text{ARG}[H(e^{j\omega})] + 2\pi r(\omega), \tag{5.58}$$

where $r(\omega)$ is a positive or negative integer that can be different at each value of ω. Similarly, in calculating any of the individual terms in Eq. (5.54), we would typically obtain the principal value.

If the principal value is used to compute the phase response as a function of ω, then $\text{ARG}[H(e^{j\omega})]$ may be a discontinuous function. The discontinuities introduced by taking the principal value will be jumps of 2π radians. This is illustrated in Figure 5.7(a), which shows a continuous-phase function $\arg[H(e^{j\omega})]$ and its principal value $\text{ARG}[H(e^{j\omega})]$ plotted over the range $0 \le \omega \le \pi$. The phase function plotted in Figure 5.7(a) exceeds the range $-\pi$ to $+\pi$. The principal value, shown in Figure 5.7(b),

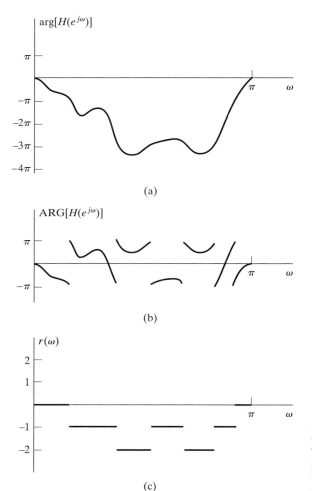

Figure 5.7 (a) Continuous-phase curve for a system function evaluated on the unit circle. (b) Principal value of the phase curve in part (a). (c) Integer multiples of 2π to be added to $\text{ARG}[H(e^{j\omega})]$ to obtain $\arg[H(e^{j\omega})]$.

has jumps of 2π due to the integer multiples of 2π that must be subtracted in certain regions to bring the phase curve within the range of the principal value. Figure 5.7(c) shows the corresponding value of $r(\omega)$ in Eq. (5.58).

Let us consider Eq. (5.54) when the principal value is used to compute the individual contributions to the phase. It is not difficult to see that

$$
\begin{aligned}
\text{ARG}[H(e^{j\omega})] = \text{ARG}\left[\frac{b_0}{a_0}\right] + \sum_{k=1}^{M} \text{ARG}[1 - c_k e^{-j\omega}] \\
- \sum_{k=1}^{N} \text{ARG}[1 - d_k e^{-j\omega}] + 2\pi r(\omega),
\end{aligned}
\tag{5.59}
$$

where $r(\omega)$ is an integer that can be different at each value of ω. The last term, $+2\pi r$, is required because, in general, the principal value of a sum of angles is *not* equal to the sum of the principal values of the individual angles. This is of considerable importance in the theory of cepstral analysis and homomorphic systems. (See Oppenheim, Schafer, and Stockham, 1968 and Tribolet, 1977.) However, it presents no problem in plotting phase functions, since the principal value can be used to compute the phase function for each pole and zero, and an appropriate multiple of 2π can then be added or subtracted as in Eq. (5.59) to obtain the principal value of the total phase function.

The principal value of the phase function can be computed using Eq. (5.59). Alternatively, we can use the relation

$$
\text{ARG}[H(e^{j\omega})] = \arctan\left[\frac{H_I(e^{j\omega})}{H_R(e^{j\omega})}\right],
\tag{5.60}
$$

where $H_R(e^{j\omega})$ and $H_I(e^{j\omega})$ are the real and imaginary parts, respectively, of $H(e^{j\omega})$. However, in computing the group delay function of Eq. (5.15), it is the derivative of the continuous phase function, i.e., $\arg[H(e^{j\omega})]$, in which we are interested:

$$
\text{grd}[H(e^{j\omega})] = -\frac{d}{d\omega}\{\arg[H(e^{j\omega})]\}.
\tag{5.61}
$$

Except at the discontinuities of $\text{ARG}[H(e^{j\omega})]$ corresponding to jumps between $+\pi$ and $-\pi$,

$$
\frac{d}{d\omega}\{\arg[H(e^{j\omega})]\} = \frac{d}{d\omega}\{\text{ARG}[H(e^{j\omega})]\}.
\tag{5.62}
$$

Consequently, the group delay can be obtained from the principal value by differentiating, except at the discontinuities. Similarly, we can express the group delay in terms of the ambiguous phase $\sphericalangle H(e^{j\omega})$ as

$$
\text{grd}[H(e^{j\omega})] = -\frac{d}{d\omega}[\sphericalangle H(e^{j\omega})],
\tag{5.63}
$$

with the interpretation that impulses caused by discontinuities of size 2π in $\sphericalangle H(e^{j\omega})$ are ignored.

5.3.1 Frequency Response of a Single Zero or Pole

Equations (5.50), (5.54), and (5.56) represent the magnitude in dB, the phase, and the group delay, respectively, as a sum of contributions from each of the poles and zeros of the system function. To obtain further insight into the properties of frequency responses of stable linear time-invariant systems with rational system functions, it is worthwhile to first examine the properties of a single factor of the form $(1 - re^{j\theta}e^{-j\omega})$, where r is the radius and θ is the angle of the pole or zero in the z-plane. This factor is typical of either a pole or a zero at a radius r and angle θ in the z-plane.

The square of the magnitude of such a factor is

$$|1 - re^{j\theta}e^{-j\omega}|^2 = (1 - re^{j\theta}e^{-j\omega})(1 - re^{-j\theta}e^{j\omega}) = 1 + r^2 - 2r\cos(\omega - \theta). \quad (5.64)$$

Since, for any complex quantity C,

$$10\log_{10}|C|^2 = 20\log_{10}|C|,$$

the log magnitude in dB is

$$20\log_{10}|1 - re^{j\theta}e^{-j\omega}| = 10\log_{10}[1 + r^2 - 2r\cos(\omega - \theta)]. \quad (5.65)$$

The principal value phase for such a factor is

$$\text{ARG}[1 - re^{j\theta}e^{-j\omega}] = \arctan\left[\frac{r\sin(\omega - \theta)}{1 - r\cos(\omega - \theta)}\right]. \quad (5.66)$$

Differentiating the right-hand side of Eq. (5.66) (except at discontinuities) gives the group delay of the factor as

$$\text{grd}[1 - re^{j\theta}e^{-j\omega}] = \frac{r^2 - r\cos(\omega - \theta)}{1 + r^2 - 2r\cos(\omega - \theta)} = \frac{r^2 - r\cos(\omega - \theta)}{|1 - re^{j\theta}e^{-j\omega}|^2}. \quad (5.67)$$

The functions in Eqs. (5.64)–(5.67) are, of course, periodic in ω with period 2π. Figure 5.8(a) shows a plot of Eq. (5.65) as a function of ω over one period ($0 \le \omega < 2\pi$) for several values of θ with $r = 0.9$. Note that the function dips sharply in the vicinity of $\omega = \theta$. Note also, from Eq. (5.65), that when r is fixed, the log magnitude is a function of $(\omega - \theta)$, so as θ changes, the dip is shifted in frequency. In general, the maximum value of Eq. (5.65) occurs at $(\omega - \theta) = \pi$ and is

$$10\log_{10}(1 + r^2 + 2r) = 20\log_{10}(1 + r),$$

which, for $r = 0.9$, is equal to 5.57 dB. Similarly, the minimum value of Eq. (5.65), occurring at $\omega = \theta$, is

$$10\log_{10}(1 + r^2 - 2r) = 20\log_{10}|1 - r|,$$

which is equal to −20 dB for $r = 0.9$. Note that the plot of the magnitude-squared function in Eq. (5.64) would look similar to Figure 5.8(a), except that it would have a much wider relative range of values. Hence, its plot would appear much sharper for the same value of r.

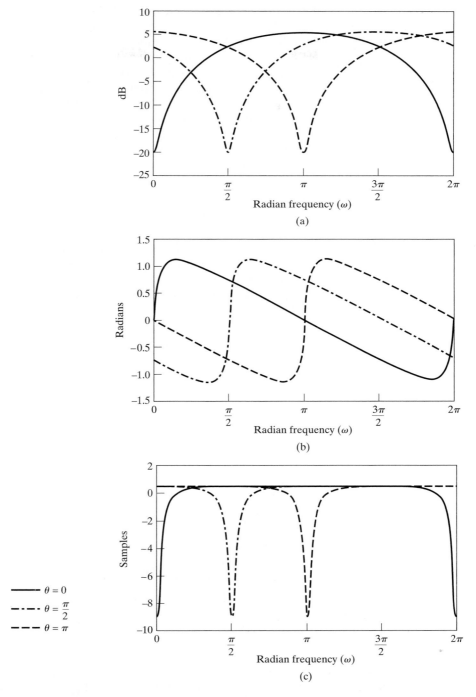

Figure 5.8 Frequency response for a single zero, with $r = 0.9$ and the three values of θ shown. (a) Log magnitude. (b) Phase. (c) Group delay.

Figure 5.8(b) shows the phase function in Eq. (5.66) as a function of ω for $r = 0.9$ and several values of θ. Note that the phase is zero at $\omega = \theta$ and that, for fixed r, the function simply shifts with θ. Figure 5.8(c) shows the group delay function in Eq. (5.67) for the same conditions on r and θ. Note that the high positive slope of the phase around $\omega = \theta$ corresponds to a large negative peak in the group delay function at $\omega = \theta$.

A simple geometric construction is often very useful in approximate sketching of frequency-response functions directly from the pole–zero plot. The procedure is based on the facts that the frequency response corresponds to the system function evaluated on the unit circle in the z-plane and that the complex value of each pole and zero factor can be represented by a vector in the z-plane from the pole or zero to a point on the unit circle. Let us first illustrate the procedure for a first-order system function of the form

$$H(z) = (1 - re^{j\theta}z^{-1}) = \frac{(z - re^{j\theta})}{z}, \qquad r < 1. \tag{5.68}$$

In Section 5.3.2, we will consider higher order examples. Such a factor has a pole at $z = 0$ and a zero at $z = re^{j\theta}$, as illustrated in Figure 5.9. Also indicated in this figure are the vectors v_1, v_2, and $v_3 = v_1 - v_2$, representing the complex numbers $e^{j\omega}$, $re^{j\theta}$, and $(e^{j\omega} - re^{j\theta})$, respectively. In terms of these vectors, the magnitude of the complex number

$$\frac{e^{j\omega} - re^{j\theta}}{e^{j\omega}}$$

is the ratio of the magnitudes of the vectors v_3 and v_1, i.e.,

$$|1 - re^{j\theta}e^{-j\omega}| = \left|\frac{e^{j\omega} - re^{j\theta}}{e^{j\omega}}\right| = \frac{|v_3|}{|v_1|}, \tag{5.69}$$

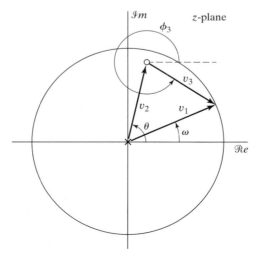

Figure 5.9 z-plane vectors for a first-order system function evaluated on the unit circle, with $r < 1$.

or, since $|v_1| = 1$, Eq. (5.69) is just equal to $|v_3|$. The corresponding phase is

$$\sphericalangle(1 - re^{j\theta}e^{-j\omega}) = \sphericalangle(e^{j\omega} - re^{j\theta}) - \sphericalangle(e^{j\omega}) = \sphericalangle(v_3) - \sphericalangle(v_1)$$
$$= \phi_3 - \phi_1 = \phi_3 - \omega. \tag{5.70}$$

Typically, a vector such as v_3 from a zero to the unit circle is referred to as a zero vector, and a vector from a pole to the unit circle is called a pole vector. Thus, the contribution of a single zero factor $(1 - re^{j\theta}z^{-1})$ to the magnitude function at frequency ω is the length of the zero vector v_3 from the zero to the point $z = e^{j\omega}$ on the unit circle. The vector has minimum length when $\omega = \theta$. This accounts for the sharp dip in the magnitude function at $\omega = \theta$ in Figure 5.8(a). Note that the pole vector v_1 from the pole at $z = 0$ to $z = e^{j\omega}$ always has unit length. Thus, it does not have any effect on the magnitude response. Equation (5.70) states that the phase function is equal to the difference between the angle of the zero vector from the zero at $re^{j\theta}$ to the point $z = e^{j\omega}$ and the angle of the pole vector from the pole at $z = 0$ to the point $z = e^{j\omega}$.

The pole–zero plot for the case $\theta = \pi$ is shown in Figure 5.10, and the pole and zero vectors are shown for two different values of ω. Clearly, as ω increases from zero, the magnitude of the vector v_3 decreases until it reaches a minimum at $\omega = \pi$, thereby accounting for the shape of the curve corresponding to $\theta = \pi$ in Figure 5.8(a). The angle of vector v_3 in Figure 5.10 increases more slowly than ω at first, so that the phase curve starts out negative; then, when ω is close to π, the angle of vector v_3 increases more rapidly than ω, thereby accounting for the steep positive slope of the phase function around $\omega = \pi$. Note that when $\omega = \pi$, the angles of vectors v_3 and v_1 are equal, so the net phase is zero.

The dependence of the frequency-response contributions of a single factor $(1 - re^{j\theta}e^{-j\omega})$ on the radius r is shown in Figure 5.11 for $\theta = \pi$ and several values of r. Note that the log magnitude function plotted in Figure 5.11(a) dips more sharply as r becomes closer to 1; indeed, the magnitude in dB approaches $-\infty$ at $\omega = \theta$ as r approaches 1. The phase function plotted in Figure 5.11(b) has positive slope around $\omega = \theta$, which becomes infinite as r approaches 1. Thus, for $r = 1$, the phase function is discontinuous, with a jump of π radians at $\omega = \theta$. Away from $\omega = \theta$, the slope of the phase function is negative. Since the group delay is the negative of the slope of the phase curve, the group

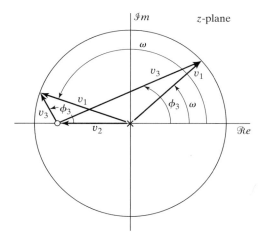

Figure 5.10   z-plane vectors for a first-order system function evaluated on the unit circle, with $\theta = \pi$, $r < 1$. The pole vector v_1 and the zero vector v_3 are shown for two different values of ω.

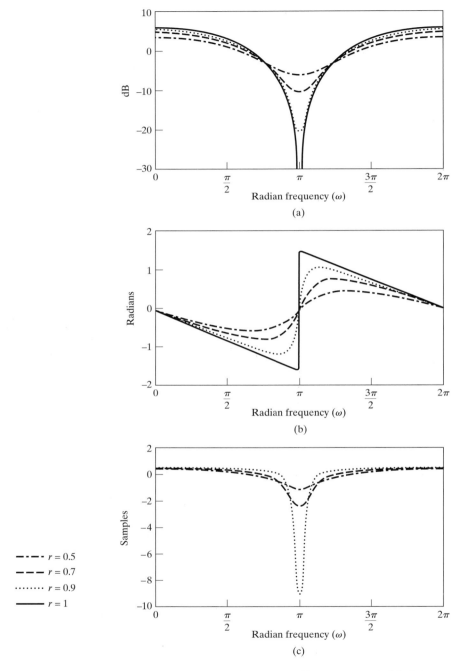

Figure 5.11 Frequency response for a single zero, with $\theta = \pi$, $r = 1$, 0.9, 0.7, and 0.5. (a) Log magnitude. (b) Phase. (c) Group delay for $r = 0.9$, 0.7, and 0.5.

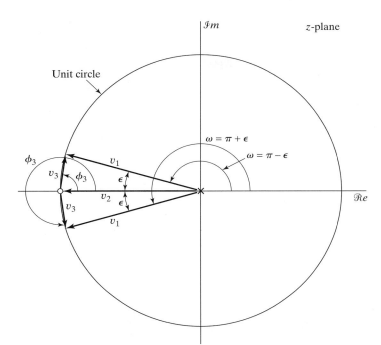

$\mathscr{I}m$ z-plane

Unit circle

$\omega = \pi + \epsilon$

$\omega = \pi - \epsilon$

ϕ_3

v_3

ϕ_3

v_1

ϵ

v_2

ϵ

$\mathscr{R}e$

v_3

v_1

Figure 5.12   z-plane vectors for a zero at $z = -1$ for two different frequencies close to π ($\omega = \pi - \varepsilon$ and $\pi + \varepsilon$).

delay is negative around $\omega = \theta$, and it dips sharply as r approaches 1. Figure 5.11(c) shows that as we move away from $\omega = \theta$, the group delay becomes positive and relatively flat. When $r = 1$, the group delay is equal to $\frac{1}{2}$ everywhere, except at $\omega = \theta$, where it is undefined.

The geometric construction for a zero on the unit circle at $z = -1$ is shown in Figure 5.12. Indicated are vectors for two different frequencies, $\omega = (\pi - \varepsilon)$ and $\omega = (\pi + \varepsilon)$, where ε is small. Two observations can be made. First, the length of the vector v_3 approaches zero as ω approaches the angle of the zero vector ($\varepsilon \to 0$). Therefore, the multiplicative contribution to the frequency response is zero ($-\infty$ dB). Second, the vector v_3 changes its angle discontinuously by π radians as ω goes from $(\pi - \varepsilon)$ to $(\pi + \varepsilon)$.

Figures 5.8 and 5.11 were restricted to $r \leq 1$. If $r > 1$, the log magnitude function behaves similarly to the case $r < 1$; i.e., it dips more sharply as $r \to 1$, as shown in Figure 5.13(a). The phase function in Figure 5.13(b) shows a discontinuity of 2π radians at $\omega = \theta$ for all values of $r > 1$. The source of this discontinuity can be seen from Figure 5.14, which shows vectors for $\omega = (\pi - \varepsilon)$ and $\omega = (\pi + \varepsilon)$. Note that the pole vector v_1 has an angle of ω, which varies continuously from $\omega = 0$ to $\omega = 2\pi$. The angle of the zero vector v_3 is labeled ϕ_3 in the figure. If this angle is measured positively in the counterclockwise direction, the figure shows that ϕ_3 jumps from zero to 2π radians as ω goes from $(\pi - \varepsilon)$ to $(\pi + \varepsilon)$. This jump of 2π radians is evident in Figure 5.13(b). The discontinuity of 2π radians can be interpreted as being due to computing the principal-value phase function. The angle ϕ_3 can also be seen to be positive for $\omega = (\pi - \varepsilon)$ and negative for $\omega = (\pi + \varepsilon)$. With this interpretation, the angle is continuous at $\omega = \theta$. However, since the total angle of the factor $(1 - re^{j\theta}e^{-j\omega})$ is less than $-\pi$ radians at $\omega = (\pi + \varepsilon)$, the principal value would appear as in Figure 5.13(b).

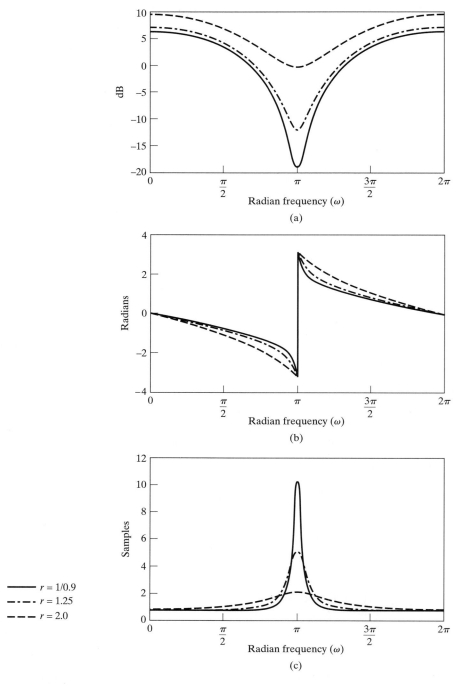

Figure 5.13 Frequency response for a single real zero outside the unit circle, with $\theta = \pi$, $r = 1/0.9$, 1.25, 2. (a) Log magnitude. (b) Phase (principal value). (c) Group delay.

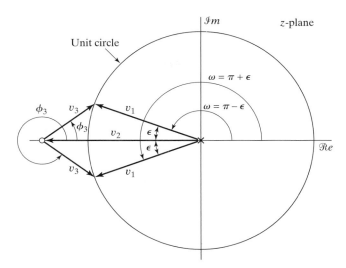

Figure 5.14   z-plane vectors for a single zero evaluated on the unit circle, with $\theta = \pi, r > 1$.

The phase curves in Figure 5.13(b) all have negative slope. Therefore, the group delay function for $r > 1$ is positive for all ω. This is also easily seen by considering Eq. (5.67) for $r > 1$.

The preceding discussion and Figures 5.8, 5.11, and 5.13 all pertain to a single factor of the form $(1 - re^{j\theta}e^{-j\omega})$. If the factor represents a zero of $H(z)$, then the curves of Figures 5.8, 5.11, and 5.13 will contribute to the frequency-response functions with positive algebraic sign. If the factor represents a pole of $H(z)$, then all the contributions will enter with opposite sign. Thus, the contribution of a pole $z = re^{j\theta}$ would be the negative of the curves in Figures 5.8 and 5.11. Instead of dipping toward zero ($-\infty$ dB), the magnitude function would peak around $\omega = \theta$. The dependence on r would be the same as for a zero; i.e., the closer r is to 1, the more peaked will be the contribution to the magnitude function. For stable and causal systems, there will, of course, be no poles outside the unit circle; i.e., r will always be less than 1.

5.3.2 Examples with Multiple Poles and Zeros

In this section, we illustrate the use of the results of Section 5.3.1 to determine the frequency response of systems with rational system functions.

Example 5.8   Second-Order IIR System

Consider the second-order system

$$H(z) = \frac{1}{(1 - re^{j\theta}z^{-1})(1 - re^{-j\theta}z^{-1})} = \frac{1}{1 - 2r\cos\theta z^{-1} + r^2 z^{-2}}. \tag{5.71}$$

The difference equation satisfied by the input and output of the system is

$$y[n] - 2r\cos\theta\, y[n-1] + r^2 y[n-2] = x[n].$$

Using the partial fraction expansion technique, we can show that the impulse response

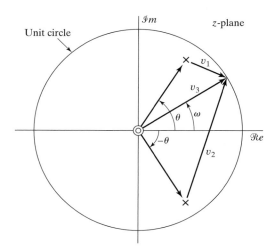

Figure 5.15   Pole–zero plot for Example 5.8.

of a causal system with this system function is

$$h[n] = \frac{r^n \sin[\theta(n+1)]}{\sin\theta} u[n].\tag{5.72}$$

The system function in Eq. (5.71) has a pole at $z = re^{j\theta}$ and at the conjugate location, $z = re^{-j\theta}$, and two zeros at $z = 0$. The pole–zero plot is shown in Figure 5.15. From our discussion in Section 5.3.1,

$$20\log_{10}|H(e^{j\omega})| = -10\log_{10}[1 + r^2 - 2r\cos(\omega - \theta)] \\ -10\log_{10}[1 + r^2 - 2r\cos(\omega + \theta)],\tag{5.73a}$$

$$\sphericalangle H(e^{j\omega}) = -\arctan\left[\frac{r\sin(\omega - \theta)}{1 - r\cos(\omega - \theta)}\right] - \arctan\left[\frac{r\sin(\omega + \theta)}{1 - r\cos(\omega + \theta)}\right],\tag{5.73b}$$

and

$$\text{grd}[H(e^{j\omega})] = -\frac{r^2 - r\cos(\omega - \theta)}{1 + r^2 - 2r\cos(\omega - \theta)} - \frac{r^2 - r\cos(\omega + \theta)}{1 + r^2 - 2r\cos(\omega + \theta)}.\tag{5.73c}$$

These functions are plotted in Figure 5.16 for $r = 0.9$ and $\theta = \pi/4$.

Figure 5.15 shows the pole and zero vectors v_1, v_2, and v_3. The magnitude response is the product of the lengths of the zero vectors (which in this case are always unity), divided by the product of the lengths of the pole vectors. That is,

$$|H(e^{j\omega})| = \frac{|v_3|^2}{|v_1| \cdot |v_2|} = \frac{1}{|v_1| \cdot |v_2|}.\tag{5.74}$$

When $\omega \approx \theta$, the length of the vector $v_1 = e^{j\omega} - re^{j\theta}$ becomes small and changes significantly as ω varies about θ, while the length of the vector $v_2 = e^{j\omega} - re^{-j\theta}$ changes only slightly as ω varies around $\omega = \theta$. Thus, the pole at angle θ dominates the frequency response around $\omega = \theta$, as is evident from Figure 5.16. By symmetry, the pole at angle $-\theta$ dominates the frequency response around $\omega = -\theta$.

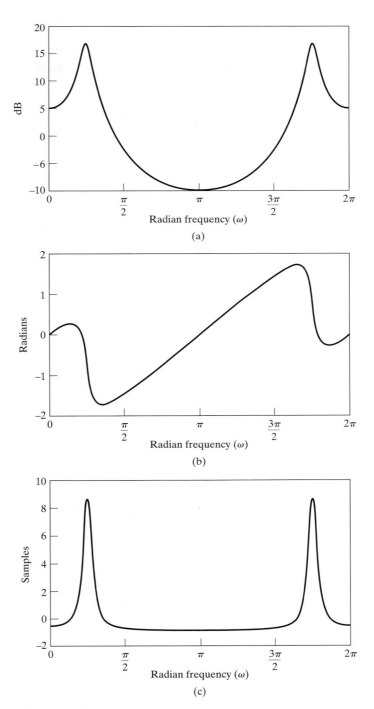

Figure 5.16 Frequency response for a complex-conjugate pair of poles as in Example 5.8, with $r = 0.9$, $\pi/4$. (a) Log magnitude. (b) Phase. (c) Group delay.

Example 5.9   Second-Order FIR System

Consider an FIR system whose impulse response is

$$h[n] = \delta[n] - 2r\cos\theta\delta[n-1] + r^2\delta[n-2]. \tag{5.75}$$

The corresponding system function is

$$H(z) = 1 - 2r\cos\theta z^{-1} + r^2 z^{-2}, \tag{5.76}$$

which is the reciprocal of the system function in Example 5.8. Therefore, the frequency-response plots for this FIR system are simply the negative of the plots in Figure 5.16. Note that the pole and zero locations are interchanged in the reciprocal.

Example 5.10   Third-Order IIR System

In this example, we consider a lowpass filter designed using one of the approximation methods to be described in Chapter 7. The system function to be considered is

$$H(z) = \frac{0.05634(1 + z^{-1})(1 - 1.0166z^{-1} + z^{-2})}{(1 - 0.683z^{-1})(1 - 1.4461z^{-1} + 0.7957z^{-2})}, \tag{5.77}$$

and the system is specified to be stable. The zeros of this system function are at the following locations:

Radius	Angle
1	π rad
1	± 1.0376 rad (59.45°)

The poles are at the following locations:

Radius	Angle
0.683	0
0.892	± 0.6257 rad (35.85°)

The pole–zero plot for this system is shown in Figure 5.17. Figure 5.18 shows the

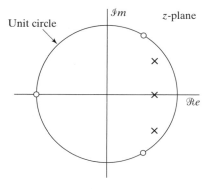

Figure 5.17   Pole–zero plot for the lowpass filter of Example 5.10.

Figure 5.18   Frequency response for the lowpass filter of Example 5.10. (a) Log magnitude. (b) Phase. (c) Group delay.

log magnitude, phase, and group delay of the system. The effect of the zeros that are on the unit circle at $\omega = \pm 1.0376$ and π is clearly evident. However, the poles are placed so that, rather than peaking for frequencies close to their angles, the total log magnitude remains close to 0 dB over a band from $\omega = 0$ to $\omega = 0.2\pi$ (and, by symmetry, from $\omega = 1.8\pi$ to $\omega = 2\pi$), and then it drops abruptly and remains below -25 dB from about $\omega = 0.3\pi$ to 1.7π. As suggested by this example, useful approximations to frequency-selective filter responses can be achieved using poles to build up the magnitude response and zeros to suppress it.

In this example, we see both types of discontinuities in the plotted phase curve. At $\omega \approx 0.22\pi$, there is a discontinuity of 2π due to the use of the principal value in plotting. At $\omega = \pm 1.0376$ and $\omega = \pi$, the discontinuities of π are due to the zeros on the unit circle.

5.4 RELATIONSHIP BETWEEN MAGNITUDE AND PHASE

The frequency response of a linear time-invariant system is the Fourier transform of the impulse response. In general, knowledge about the magnitude provides no information about the phase, and vice versa. However, for systems described by linear constant-coefficient difference equations, i.e., rational system functions, there is some constraint between magnitude and phase. In particular, as we discuss in this section, if the magnitude of the frequency response and the number of poles and zeros are known, then there are only a finite number of choices for the associated phase. Similarly, if the number of poles and zeros and the phase are known, then, to within a scale factor, there are only a finite number of choices for the magnitude. Furthermore, under a constraint referred to as minimum phase, the frequency-response magnitude specifies the phase uniquely, and the frequency-response phase specifies the magnitude to within a scale factor.

To explore the possible choices of system function, given the square of the magnitude of the system frequency response, we consider $|H(e^{j\omega})|^2$ expressed as

$$
\begin{aligned}
|H(e^{j\omega})|^2 &= H(e^{j\omega})H^*(e^{j\omega}) \\
&= H(z)H^*(1/z^*)|_{z=e^{j\omega}}.
\end{aligned}
\tag{5.78}
$$

Restricting the system function $H(z)$ to be rational in the form of Eq. (5.19), i.e.,

$$
H(z) = \left(\frac{b_0}{a_0}\right) \frac{\displaystyle\prod_{k=1}^{M}(1 - c_k z^{-1})}{\displaystyle\prod_{k=1}^{N}(1 - d_k z^{-1})},
\tag{5.79}
$$

we see that $H^*(1/z^*)$ in Eq. (5.78) is

$$
H^*\left(\frac{1}{z^*}\right) = \left(\frac{b_0}{a_0}\right) \frac{\displaystyle\prod_{k=1}^{M}(1 - c_k^* z)}{\displaystyle\prod_{k=1}^{N}(1 - d_k^* z)},
\tag{5.80}
$$

where we have assumed that a_0 and b_0 are real. Therefore, Eq. (5.78) states that the square of the magnitude of the frequency response is the evaluation on the unit circle of the z-transform

$$C(z) = H(z)H^*(1/z^*)$$ (5.81)

$$= \left(\frac{b_0}{a_0}\right)^2 \frac{\displaystyle\prod_{k=1}^{M}(1 - c_k z^{-1})(1 - c_k^* z)}{\displaystyle\prod_{k=1}^{N}(1 - d_k z^{-1})(1 - d_k^* z)}.$$ (5.82)

If we are given $|H(e^{j\omega})|^2$, then by replacing $e^{j\omega}$ by z, we can construct $C(z)$. From $C(z)$, we would like to infer as much as possible about $H(z)$. We first note that for each pole d_k of $H(z)$, there is a pole of $C(z)$ at d_k and $(d_k^*)^{-1}$. Similarly, for each zero c_k of $H(z)$, there is a zero of $C(z)$ at c_k and $(c_k^*)^{-1}$. Consequently, the poles and zeros of $C(z)$ occur in conjugate reciprocal pairs, with one element of each pair associated with $H(z)$ and one element of each pair associated with $H^*(1/z^*)$. Furthermore, if one element of each pair is inside the unit circle, then the other (i.e., the conjugate reciprocal) will be outside the unit circle. The only other alternative is for both to be on the unit circle in the same location.

If $H(z)$ is assumed to correspond to a causal, stable system, then all its poles must lie inside the unit circle. With this constraint, the poles of $H(z)$ can be identified from the poles of $C(z)$. However, with this constraint alone, the zeros of $H(z)$ cannot be uniquely identified from the zeros of $C(z)$. This can be seen from the following example.

Example 5.11   Systems with the Same C(z)

Consider two stable systems with system functions

$$H_1(z) = \frac{2(1 - z^{-1})(1 + 0.5z^{-1})}{(1 - 0.8e^{j\pi/4}z^{-1})(1 - 0.8e^{-j\pi/4}z^{-1})}$$ (5.83)

and

$$H_2(z) = \frac{(1 - z^{-1})(1 + 2z^{-1})}{(1 - 0.8e^{j\pi/4}z^{-1})(1 - 0.8e^{-j\pi/4}z^{-1})}.$$ (5.84)

The pole–zero plots for these systems are shown in Figures 5.19(a) and 5.19(b), respectively.

Now,

$$C_1(z) = H_1(z)H_1^*(1/z^*)$$

$$= \frac{2(1 - z^{-1})(1 + 0.5z^{-1})2(1 - z)(1 + 0.5z)}{(1 - 0.8e^{j\pi/4}z^{-1})(1 - 0.8e^{-j\pi/4}z^{-1})(1 - 0.8e^{-j\pi/4}z)(1 - 0.8e^{j\pi/4}z)}$$ (5.85)

and

$$C_2(z) = H_2(z)H_2^*(1/z^*)$$

$$= \frac{(1 - z^{-1})(1 + 2z^{-1})(1 - z)(1 + 2z)}{(1 - 0.8e^{j\pi/4}z^{-1})(1 - 0.8e^{-j\pi/4}z^{-1})(1 - 0.8e^{-j\pi/4}z)(1 - 0.8e^{j\pi/4}z)}.$$ (5.86)

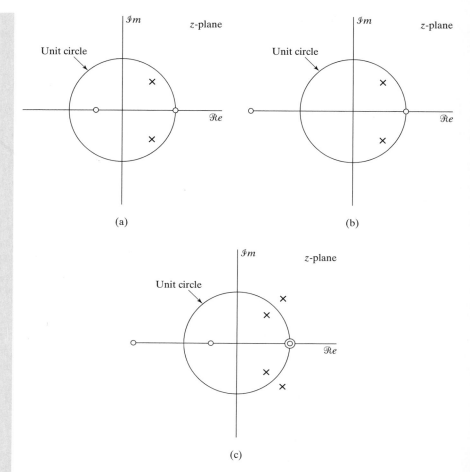

Figure 5.19 Pole–zero plots for two system functions and their common magnitude-squared function. (a) $H_1(z)$. (b) $H_2(z)$. (c) $C_1(z)$, $C_2(z)$.

Using the fact that

$$4(1 + 0.5z^{-1})(1 + 0.5z) = (1 + 2z^{-1})(1 + 2z), \tag{5.87}$$

we see that $C_1(z) = C_2(z)$. The pole–zero plot for $C_1(z)$ and $C_2(z)$ is shown in Figure 5.19(c).

The system functions $H_1(z)$ and $H_2(z)$ in Example 5.11 differ only in the location of the zeros. In the example, the factor $2(1 + 0.5z^{-1}) = (z^{-1} + 2)$ contributes the same to the square of the magnitude of the frequency response as the factor $(1 + 2z^{-1})$, and consequently, $|H_1(e^{j\omega})|$ and $|H_2(e^{j\omega})|$ are equal. However, the phase functions for these two frequency responses are different.

Example 5.12

Suppose we are given the pole–zero plot for $C(z)$ in Figure 5.20 and want to determine the poles and zeros to associate with $H(z)$. The conjugate reciprocal pairs of poles and zeros for which one element of each is associated with $H(z)$ and one with $H^*(1/z^*)$ are as follows:

$$\text{Pole pair 1:} \quad (P_1, P_4)$$
$$\text{Pole pair 2:} \quad (P_2, P_5)$$
$$\text{Pole pair 3:} \quad (P_3, P_6)$$
$$\text{Zero pair 1:} \quad (Z_1, Z_4)$$
$$\text{Zero pair 2:} \quad (Z_2, Z_5)$$
$$\text{Zero pair 3:} \quad (Z_3, Z_6)$$

Knowing that $H(z)$ corresponds to a stable, causal system, we must choose the poles from each pair that are inside the unit circle, i.e., P_1, P_2, and P_3. No such constraint is imposed on the zeros. However, if we assume that the coefficients a_k and b_k are real in Eqs. (5.16) and (5.18), the zeros (and poles) either are real or occur in complex conjugate pairs. Consequently, the zeros to associate with $H(z)$ are

$$Z_3 \quad \text{or} \quad Z_6$$

and

$$(Z_1, Z_2) \quad \text{or} \quad (Z_4, Z_5).$$

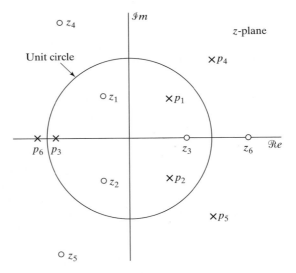

Figure 5.20   Pole–zero plot for the magnitude-squared function in Example 5.12.

Therefore, there are a total of four different stable, causal systems with three poles and three zeros for which the pole–zero plot of $C(z)$ is that shown in Figure 5.20 and, equivalently, for which the frequency-response magnitude is the same. If we had not assumed that the coefficients a_k and b_k were real, the number of choices would be greater. Furthermore, if the number of poles and zeros of $H(z)$ were not restricted, the number of choices for $H(z)$ would be unlimited. To see this, assume that $H(z)$ has a factor of the form

$$\frac{z^{-1} - a^*}{1 - az^{-1}},$$

i.e.,

$$H(z) = H_1(z)\frac{z^{-1} - a^*}{1 - az^{-1}}. \tag{5.88}$$

Factors of this form are referred to as *all-pass factors,* since they have unity magnitude on the unit circle; they are discussed in more detail in Section 5.5. It is easily verified that

$$C(z) = H(z)H^*(1/z^*) = H_1(z)H_1^*(1/z^*); \tag{5.89}$$

i.e., all-pass factors cancel in $C(z)$ and therefore would not be identifiable from the pole–zero plot of $C(z)$. Consequently, if the number of poles and zeros of $H(z)$ is unspecified, then, given $C(z)$, any choice for $H(z)$ can be cascaded with an arbitrary number of all-pass factors with poles inside the unit circle (i.e., $|a| < 1$).

5.5 ALL-PASS SYSTEMS

As indicated in the discussion of Example 5.12, a stable system function of the form

$$H_{ap}(z) = \frac{z^{-1} - a^*}{1 - az^{-1}} \tag{5.90}$$

has a frequency-response magnitude that is independent of ω. This can be seen by writing $H_{ap}(e^{j\omega})$ in the form

$$H_{ap}(e^{j\omega}) = \frac{e^{-j\omega} - a^*}{1 - ae^{-j\omega}}$$

$$= e^{-j\omega}\frac{1 - a^*e^{j\omega}}{1 - ae^{-j\omega}}. \tag{5.91}$$

In Eq. (5.91), the term $e^{-j\omega}$ has unity magnitude, and the remaining numerator and denominator factors are complex conjugates of each other and therefore have the same magnitude. Consequently, $|H_{ap}(e^{j\omega})| = 1$. A system for which the frequency-response magnitude is a constant is called an *all-pass system,* since the system passes all of the frequency components of its input with constant gain or attenuation. The most general form for the system function of an all-pass system with a real-valued impulse response is a product of factors like Eq. (5.90), with complex poles being paired with their con-jugates; i.e.,

$$H_{ap}(z) = A\prod_{k=1}^{M_r}\frac{z^{-1} - d_k}{1 - d_kz^{-1}}\prod_{k=1}^{M_c}\frac{(z^{-1} - e_k^*)(z^{-1} - e_k)}{(1 - e_kz^{-1})(1 - e_k^*z^{-1})}, \tag{5.92}$$

where A is a positive constant and the d_k's are the real poles, and the e_k's the complex poles, of $H_{ap}(z)$. For causal and stable all-pass systems, $|d_k| < 1$ and $|e_k| < 1$. In terms

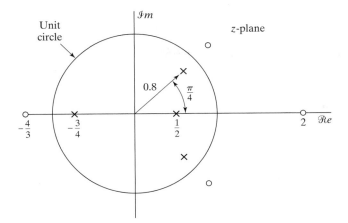

Figure 5.21 Typical pole–zero plot for an all-pass system.

of our general notation for system functions, all-pass systems have $M = N = 2M_c + M_r$ poles and zeros. Figure 5.21 shows a typical pole–zero plot for an all-pass system. In this case $M_r = 2$ and $M_c = 1$. Note that each pole of $H_{ap}(z)$ is paired with a conjugate reciprocal zero.

The frequency response for a general all-pass system can be expressed in terms of the frequency responses of first-order all-pass systems like that specified in Eq. (5.90). For a causal all-pass system, each of these terms consists of a single pole inside the unit circle and a zero at the conjugate reciprocal location. The magnitude response for such a term is, as we have shown, unity. Thus, the log magnitude in dB is zero. With a expressed in polar form as $a = re^{j\theta}$, the phase function for Eq. (5.90) is

$$\sphericalangle\left[\frac{e^{-j\omega} - re^{-j\theta}}{1 - re^{j\theta}e^{-j\omega}}\right] = -\omega - 2\arctan\left[\frac{r\sin(\omega - \theta)}{1 - r\cos(\omega - \theta)}\right]. \tag{5.93}$$

Likewise, the phase of a second-order all-pass system with poles at $z = re^{j\theta}$ and $z = re^{-j\theta}$ is

$$\sphericalangle\left[\frac{(e^{-j\omega} - re^{-j\theta})(e^{-j\omega} - re^{j\theta})}{(1 - re^{j\theta}e^{-j\omega})(1 - re^{-j\theta}e^{-j\omega})}\right] = -2\omega - 2\arctan\left[\frac{r\sin(\omega - \theta)}{1 - r\cos(\omega - \theta)}\right]$$

$$-2\arctan\left[\frac{r\sin(\omega + \theta)}{1 - r\cos(\omega + \theta)}\right]. \tag{5.94}$$

Example 5.13   First- and Second-Order All-Pass Systems

Figure 5.22 shows plots of the log magnitude, phase, and group delay for two first-order all-pass systems, one with a real pole at $z = 0.9$ ($\theta = 0, r = 0.9$) and another with a pole at $z = -0.9$ ($\theta = \pi, r = 0.9$). For both systems, the radii of the poles are $r = 0.9$. Likewise, Figure 5.23 shows the same functions for a second-order all-pass system with poles at $z = 0.9e^{j\pi/4}$ and $z = 0.9e^{-j\pi/4}$.

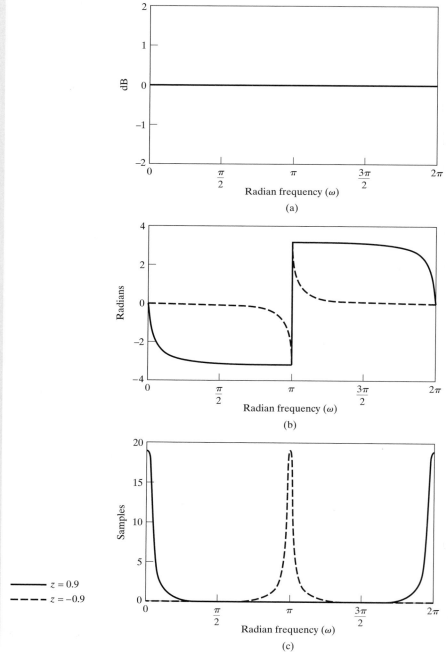

Figure 5.22 Frequency response for all-pass filters with real poles at $z = 0.9$ (solid line) and $z = -0.9$ (dashed line). (a) Log magnitude. (b) Phase (principal value). (c) Group delay.

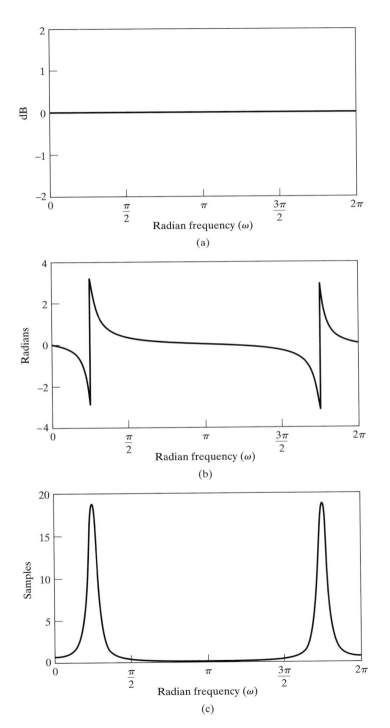

Figure 5.23   Frequency response of second-order all-pass system with poles at $z = 0.9e^{\pm j\pi/4}$. (a) Log magnitude. (b) Phase (principal value). (c) Group delay.

Example 5.13 illustrates a general property of causal all-pass systems. In Figure 5.22(b), we see that the phase is nonpositive for $0 < \omega < \pi$. Similarly, in Figure 5.23(b), if the discontinuity of 2π resulting from the computation of the principal value is removed, the resulting continuous-phase curve is nonpositive for $0 < \omega < \pi$. Since the more general all-pass system given by Eq. (5.92) is just a product of such first- and second-order factors, we can conclude that the (continuous) phase, $\arg[H_{ap}(e^{j\omega})]$, of a causal all-pass system is always nonpositive for $0 < \omega < \pi$. This may not appear to be true if the principal value is plotted, as is illustrated in Figure 5.24, which shows the log magnitude, phase, and group delay for an all-pass system with poles and zeros as in Figure 5.21. However, we can establish this result by first considering the group delay.

The group delay of the simple one-pole all-pass system of Eq. (5.90) is the negative derivative of the phase given by Eq. (5.93). With a small amount of algebra, it can be shown that

$$\mathrm{grd}\left[\frac{e^{-j\omega} - re^{-j\theta}}{1 - re^{j\theta}e^{-j\omega}}\right] = \frac{1 - r^2}{1 + r^2 - 2r\cos(\omega - \theta)} = \frac{1 - r^2}{|1 - re^{j\theta}e^{-j\omega}|^2}. \tag{5.95}$$

Since $r < 1$ for a stable and causal all-pass system, from Eq. (5.95) the group delay contributed by a single causal all-pass factor is always positive. Since the group delay of a higher order all-pass system will be a sum of positive terms, as in Eq. (5.95), it is true in general that the group delay of a causal rational all-pass system is always positive. This is confirmed by Figures 5.22(c), 5.23(c), and 5.24(c), which show the group delay for first-order, second-order, and third-order all-pass systems, respectively.

The positivity of the group delay of a causal all-pass system is the basis for a simple proof of the negativity of the phase of such a system. First, note that

$$\arg[H_{ap}(e^{j\omega})] = -\int_0^\omega \mathrm{grd}[H_{ap}(e^{j\phi})]d\phi + \arg[H_{ap}(e^{j0})] \tag{5.96}$$

for $0 \le \omega \le \pi$. From Eq. (5.92), it follows that

$$H_{ap}(e^{j0}) = A\prod_{k=1}^{M_r}\frac{1 - d_k}{1 - d_k}\prod_{k=1}^{M_c}\frac{|1 - e_k|^2}{|1 - e_k|^2} = A. \tag{5.97}$$

Therefore, $\arg[H_{ap}(e^{j0})] = 0$, and since

$$\mathrm{grd}[H_{ap}(e^{j\omega})] \ge 0, \tag{5.98}$$

it follows from Eq. (5.96) that

$$\arg[H_{ap}(e^{j\omega})] \le 0 \qquad \text{for } 0 \le \omega < \pi. \tag{5.99}$$

The positivity of the group delay and the nonpositivity of the continuous phase are important properties of causal all-pass systems.

All-pass systems have many uses. They can be used as compensators for phase (or group delay) distortion, as we will see in Chapter 7, and they are useful in the theory of minimum-phase systems, as we will see in Section 5.6. They are also useful in transforming frequency-selective lowpass filters into other frequency-selective forms and in obtaining variable-cutoff frequency-selective filters. These applications are discussed in Chapter 7 and applied in the problems in that chapter.

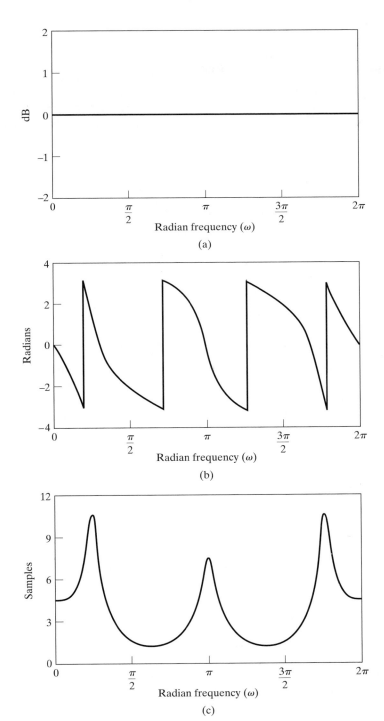

Figure 5.24 Frequency response for an all-pass system with the pole–zero plot in Figure 5.21. (a) Log magnitude. (b) Phase (principal value). (c) Group delay.

5.6 MINIMUM-PHASE SYSTEMS

In Section 5.4, we showed that the frequency-response magnitude for an LTI system with rational system function does not uniquely characterize the system. If the system is stable and causal, the poles must be inside the unit circle, but stability and causality place no such restriction on the zeros. For certain classes of problems, it is useful to impose the additional restriction that the inverse system (one with system function $1/H(z)$) also be stable and causal. As discussed in Section 5.2.2, this then restricts the zeros, as well as the poles, to be inside the unit circle, since the poles of $1/H(z)$ are the zeros of $H(z)$. Such systems are commonly referred to as *minimum-phase* systems. The name *minimum-phase* comes from a property of the phase response, which is not obvious from the preceding definition. This and other fundamental properties that we discuss are unique to this class of systems, and therefore, any one of them could be taken as the definition of the class. These properties are developed in Section 5.6.3.

If we are given a magnitude-squared function in the form of Eq. (5.82) and we know that the system is a minimum-phase system, then $H(z)$ is uniquely determined and will consist of all the poles and zeros of $C(z) = H(z)H^*(1/z^*)$ that lie inside the unit circle.[1] This approach is often followed in filter design when only the magnitude response is determined by the design method used. (See Chapter 7.)

5.6.1 Minimum-Phase and All-Pass Decomposition

In Section 5.4 we showed that, from the square of the magnitude of the frequency response alone, we could not uniquely determine the system function $H(z)$, since any choice that had the given frequency-response magnitude could be cascaded with arbitrary all-pass factors without affecting the magnitude. A related observation is that any rational system function[2] can be expressed as

$$H(z) = H_{\min}(z)H_{\rm ap}(z), \tag{5.100}$$

where $H_{\min}(z)$ is a minimum-phase system and $H_{\rm ap}(z)$ is an all-pass system.

To show this, suppose that $H(z)$ has one zero outside the unit circle at $z = 1/c^*$, where $|c| < 1$, and the remaining poles and zeros are inside the unit circle. Then $H(z)$ can be expressed as

$$H(z) = H_1(z)(z^{-1} - c^*), \tag{5.101}$$

where, by definition, $H_1(z)$ is minimum phase. An equivalent expression for $H(z)$ is

$$H(z) = H_1(z)(1 - cz^{-1})\frac{z^{-1} - c^*}{1 - cz^{-1}}. \tag{5.102}$$

Since $|c| < 1$, the factor $H_1(z)(1 - cz^{-1})$ also is minimum phase, and it differs from $H(z)$ only in that the zero of $H(z)$ that was outside the unit circle at $z = 1/c^*$ is reflected inside

---

[1] We have assumed that $C(z)$ has no poles or zeros on the unit circle. Strictly speaking, systems with poles on the unit circle are unstable and are generally to be avoided in practice. Zeros on the unit circle, however, often occur in practical filter designs. By our definition, such systems are nonminimum phase, but many of the properties of minimum-phase systems hold even in this case.

[2] Somewhat for convenience, we will restrict the discussion to stable, causal systems, although the observation applies more generally.

the unit circle to the conjugate reciprocal location $z = c$. The term $(z^{-1} - c^*)/(1 - cz^{-1})$ is all-pass. This example can be generalized in a straightforward way to include more zeros outside the unit circle, thereby showing that, in general, any system function can be expressed as

$$H(z) = H_{\min}(z)H_{\text{ap}}(z), \tag{5.103}$$

where $H_{\min}(z)$ contains the poles and zeros of $H(z)$ that lie inside the unit circle, plus zeros that are the conjugate reciprocals of the zeros of $H(z)$ that lie outside the unit circle. $H_{\text{ap}}(z)$ is comprised of all the zeros of $H(z)$ that lie outside the unit circle, together with poles to cancel the reflected conjugate reciprocal zeros in $H_{\min}(z)$.

Using Eq. (5.103), we can form a nonminimum-phase system from a minimum-phase system by reflecting one or more zeros lying inside the unit circle to their conjugate reciprocal locations outside the unit circle, or, conversely, we can form a minimum-phase system from a nonminimum-phase system by reflecting all the zeros lying outside the unit circle to their conjugate reciprocal locations inside. In either case, both the minimum-phase and the nonminimum-phase systems will have the same frequency-response magnitude.

Example 5.14   Minimum-Phase/All-Pass Decomposition

To illustrate the decomposition of a stable, causal system into the cascade of a minimum-phase and an all-pass system, consider the two stable, causal systems specified by the system functions

$$H_1(z) = \frac{(1 + 3z^{-1})}{1 + \frac{1}{2}z^{-1}}$$

and

$$H_2(z) = \frac{\left(1 + \frac{3}{2}e^{+j\pi/4}z^{-1}\right)\left(1 + \frac{3}{2}e^{-j\pi/4}z^{-1}\right)}{\left(1 - \frac{1}{3}z^{-1}\right)}.$$

The first system function, $H_1(z)$, has a pole inside the unit circle at $z = -\frac{1}{2}$, but a zero outside at $z = -3$. We will need to choose the appropriate all-pass system to reflect this zero inside the unit circle. From Eq. (5.101), we have $c = -\frac{1}{3}$. Therefore, from Eqs. (5.102) and (5.103), the all-pass component will be

$$H_{\text{ap}}(z) = \frac{z^{-1} + \frac{1}{3}}{1 + \frac{1}{3}z^{-1}},$$

and the minimum-phase component will be

$$H_{\min}(z) = 3\frac{1 + \frac{1}{3}z^{-1}}{1 + \frac{1}{2}z^{-1}};$$

i.e.,

$$H_1(z) = \left(3\frac{1 + \frac{1}{3}z^{-1}}{1 + \frac{1}{2}z^{-1}}\right)\left(\frac{z^{-1} + \frac{1}{3}}{1 + \frac{1}{3}z^{-1}}\right).$$

The second system function, $H_2(z)$, has two complex zeros outside the unit circle and a real pole inside. We can express $H_2(z)$ in the form of Eq. (5.101) by factoring $\frac{3}{2}e^{j\pi/4}$ and $\frac{3}{2}e^{-j\pi/4}$ out of the numerator terms to get

$$H_2(z) = \frac{9}{4}\frac{\left(z^{-1} + \frac{2}{3}e^{-j\pi/4}\right)\left(z^{-1} + \frac{2}{3}e^{j\pi/4}\right)}{1 - \frac{1}{3}z^{-1}}.$$

Factoring as in Eq. (5.102) yields

$$H_2(z) = \left[\frac{9}{4}\frac{\left(1 + \frac{2}{3}e^{-j\pi/4}z^{-1}\right)\left(1 + \frac{2}{3}e^{j\pi/4}z^{-1}\right)}{1 - \frac{1}{3}z^{-1}}\right]$$

$$\times \left[\frac{\left(z^{-1} + \frac{2}{3}e^{-j\pi/4}\right)\left(z^{-1} + \frac{2}{3}e^{j\pi/4}\right)}{\left(1 + \frac{2}{3}e^{-j\pi/4}z^{-1}\right)\left(1 + \frac{2}{3}e^{j\pi/4}z^{-1}\right)}\right].$$

The first term in square brackets is a minimum-phase system, while the second term is an all-pass system.

5.6.2 Frequency-Response Compensation

In many signal-processing contexts, a signal has been distorted by an LTI system with an undesirable frequency response. It may then be of interest to process the distorted signal with a compensating system, as indicated in Figure 5.25. This situation may arise, for example, in transmitting signals over a communication channel. If perfect compensation is achieved, then $s_c[n] = s[n]$, i.e., $H_c(z)$ is the inverse of $H_d(z)$. However, if we assume that the distorting system is stable and causal and require the compensating system to be stable and causal, then perfect compensation is possible only if $H_d(z)$ is a minimum-phase system, so that it has a stable, causal inverse.

Based on the previous discussions, assuming that $H_d(z)$ is known or approximated as a rational system function, we can form a minimum-phase system $H_{d\min}(z)$ by reflecting all the zeros of $H_d(z)$ that are outside the unit circle to their conjugate reciprocal locations inside the unit circle. $H_d(z)$ and $H_{d\min}(z)$ have the same frequency-response magnitude and are related through an all-pass system $H_{ap}(z)$, i.e.,

$$H_d(z) = H_{d\min}(z)H_{ap}(z). \tag{5.104}$$

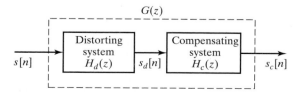

Figure 5.25 Illustration of distortion compensation by linear filtering.

Choosing the compensating filter to be

$$H_c(z) = \frac{1}{H_{d\min}(z)}, \tag{5.105}$$

we find that the overall system function relating $s[n]$ and $s_c[n]$ is

$$G(z) = H_d(z)H_c(z) = H_{\mathrm{ap}}(z); \tag{5.106}$$

i.e., $G(z)$ corresponds to an all-pass system. Consequently, the frequency-response magnitude is exactly compensated for, while the phase response is modified to $\sphericalangle H_{\mathrm{ap}}(e^{j\omega})$.

The following example illustrates compensation of the frequency response magnitude when the system to be compensated for is a nonminimum-phase FIR system.

Example 5.15   Compensation of an FIR System

Consider the distorting system function to be

$$H_d(z) = (1 - 0.9e^{j0.6\pi}z^{-1})(1 - 0.9e^{-j0.6\pi}z^{-1})$$
$$\times (1 - 1.25e^{j0.8\pi}z^{-1})(1 - 1.25e^{-j0.8\pi}z^{-1}). \tag{5.107}$$

The pole–zero plot is shown in Figure 5.26. Since $H_d(z)$ has only zeros (all poles are at $z = 0$), it follows that the system has a finite-duration impulse response. Therefore the system is stable; and since $H_d(z)$ is a polynomial with only negative powers of z, the system is causal. However, since two of the zeros are outside the unit circle, the system is nonminimum phase. Figure 5.27 shows the log magnitude, phase, and group delay for $H_d(e^{j\omega})$.

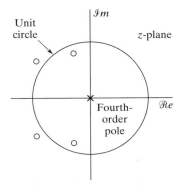

Figure 5.26   Pole–zero plot of FIR system in Example 5.15.

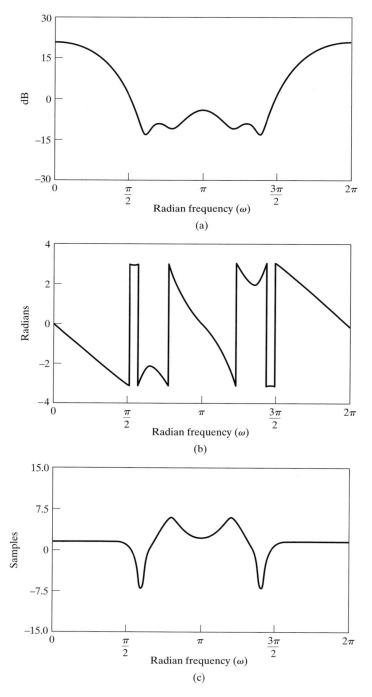

Figure 5.27 Frequency response for FIR system with pole–zero plot in Figure 5.26. (a) Log magnitude. (b) Phase (principal value). (c) Group delay.

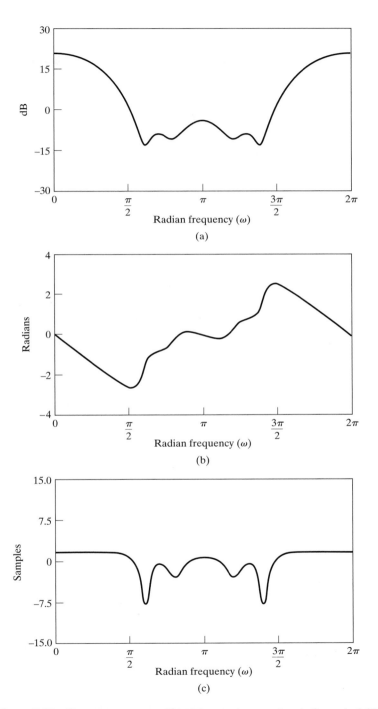

Figure 5.28   Frequency response for minimum-phase system in Example 5.15. (a) Log magnitude. (b) Phase. (c) Group delay.

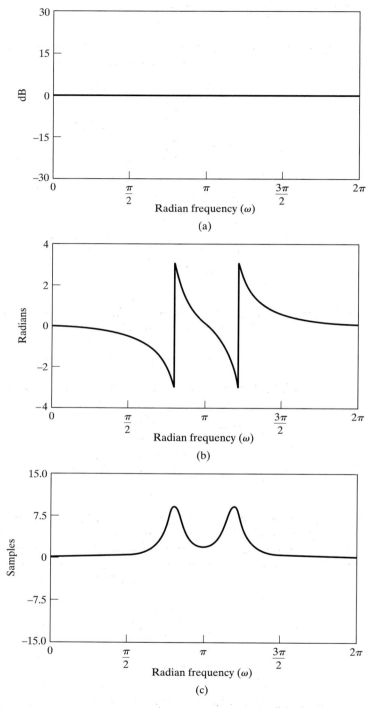

Figure 5.29 Frequency response of all-pass system of Example 5.15. (The sum of corresponding curves in Figures 5.28 and 5.29 equals the corresponding curve in Figure 5.27 with the sum of the phase curves taken modulo 2π.) (a) Log magnitude. (b) Phase (principal value). (c) Group delay.

The corresponding minimum-phase system is obtained by reflecting the zeros that occur at $z = 1.25e^{\pm j0.8\pi}$ to their conjugate reciprocal locations inside the unit circle. If we express $H_d(z)$ as

$$H_d(z) = (1 - 0.9e^{j0.6\pi}z^{-1})(1 - 0.9e^{-j0.6\pi}z^{-1})(1.25)^2$$
$$\times (z^{-1} - 0.8e^{-j0.8\pi})(z^{-1} - 0.8e^{j0.8\pi}), \tag{5.108}$$

then

$$H_{\min}(z) = (1.25)^2(1 - 0.9e^{j0.6\pi}z^{-1})(1 - 0.9e^{-j0.6\pi}z^{-1})$$
$$\times (1 - 0.8e^{-j0.8\pi}z^{-1})(1 - 0.8e^{j0.8\pi}z^{-1}), \tag{5.109}$$

and the all-pass system that relates $H_{\min}(z)$ and $H_d(z)$ is

$$H_{ap}(z) = \frac{(z^{-1} - 0.8e^{-j0.8\pi})(z^{-1} - 0.8e^{j0.8\pi})}{(1 - 0.8e^{j0.8\pi}z^{-1})(1 - 0.8e^{-j0.8\pi}z^{-1})}. \tag{5.110}$$

The log magnitude, phase, and group delay of $H_{\min}(e^{j\omega})$ are shown in Figure 5.28. Figures 5.27(a) and 5.28(a) are, of course, identical. The log magnitude, phase, and group delay for $H_{ap}(e^{j\omega})$ are plotted in Figure 5.29.

Note that the inverse system for $H_d(z)$ would have poles at $z = 1.25e^{\pm j0.8\pi}$ and at $z = 0.9e^{\pm j0.6\pi}$, and thus, the causal inverse would be unstable. The minimum-phase inverse would be the reciprocal of $H_{\min}(z)$, as given by Eq. (5.109), and if this inverse were used in the cascade system of Figure 5.25, the overall effective system function would be $H_{ap}(z)$, as given in Eq. (5.110).

5.6.3 Properties of Minimum-Phase Systems

We have been using the term "minimum phase" to refer to systems that are causal and stable and that have a causal and stable inverse. This choice of name is motivated by a property of the phase function that, while not obvious, follows from our chosen definition. In this section, we develop a number of interesting and important properties of minimum-phase systems relative to all other systems that have the same frequency-response magnitude.

The Minimum Phase-Lag Property

The use of the terminology "minimum phase" as a descriptive name for a system having all its poles and zeros inside the unit circle is suggested by Example 5.15. Recall that, as a consequence of Eq. (5.100), the continuous phase, i.e., $\arg[H(e^{j\omega})]$, of any nonminimum-phase system can be expressed as

$$\arg[H(e^{j\omega})] = \arg[H_{\min}(e^{j\omega})] + \arg[H_{ap}(e^{j\omega})]. \tag{5.111}$$

Therefore, the continuous phase that would correspond to the principal-value phase of Figure 5.27(b) is the sum of the continuous phase associated with the minimum-phase function of Figure 5.28(b) and the continuous phase of the all-pass system associated with the principal-value phase shown in Figure 5.29(b). As was shown in Section 5.5, and as indicated by the principal-value phase curves of Figures 5.22(b), 5.23(b), 5.24(b), and 5.29(b), the continuous-phase curve of an all-pass system is negative for

$0 \leq \omega \leq \pi$. Thus, the reflection of zeros of $H_{\min}(z)$ from inside the unit circle to conjugate reciprocal locations outside always *decreases* the (continuous) phase or *increases* the negative of the phase, which is called the *phase-lag* function. Hence, the causal, stable system that has $|H_{\min}(e^{j\omega})|$ as its magnitude response and also has all its zeros (and, of course, poles) inside the unit circle has the minimum phase-lag function (for $0 \leq \omega < \pi$) of all the systems having that same magnitude response. Therefore, a more precise terminology is *minimum phase-lag* system, but *minimum phase* is historically the established terminology.

To make the interpretation of minimum phase-lag systems more precise, it is necessary to impose the additional constraint that $H(e^{j\omega})$ be positive at $\omega = 0$, i.e.,

$$H(e^{j0}) = \sum_{n=-\infty}^{\infty} h[n] > 0. \tag{5.112}$$

Note that $H(e^{j0})$ will be real if we restrict $h[n]$ to be real. The condition of Eq. (5.112) is necessary because a system with impulse response $-h[n]$ has the same poles and zeros for its system function as a system with impulse response $h[n]$. However, multiplying by -1 would alter the phase by π radians. Thus, to remove this ambiguity, we must impose the condition of Eq. (5.112) to ensure that a system with all its poles and zeros inside the unit circle also has the minimum phase-lag property. However, this constraint is often of little significance, and our definition at the beginning of Section 5.6, which does not include it, is the generally accepted definition of the class of minimum-phase systems.

The Minimum Group-Delay Property

Example 5.15 illustrates another property of systems whose poles and zeros are all inside the unit circle. First note that the group delay for the systems that have the same magnitude response is

$$\text{grd}[H(e^{j\omega})] = \text{grd}[H_{\min}(e^{j\omega})] + \text{grd}[H_{\text{ap}}(e^{j\omega})]. \tag{5.113}$$

The group delay for the minimum-phase system shown in Figure 5.28(c) is always less than the group delay for the nonminimum-phase system shown in Figure 5.27(c). This is because, as Figure 5.29(c) shows, the all-pass system that converts the minimum-phase system into the nonminimum-phase system has a positive group delay. In Section 5.5, we showed this to be a general property of all-pass systems; they always have positive group delay for all ω. Thus, if we again consider all the systems that have a given magnitude response $|H_{\min}(e^{j\omega})|$, the one that has all its poles and zeros inside the unit circle has the minimum group delay. An equally appropriate name for such systems would therefore be *minimum group-delay* systems, but this terminology is not generally used.

The Minimum Energy-Delay Property

In Example 5.15, there are a total of four causal FIR systems with real impulse responses that have the same frequency-response magnitude as the system in Eq. (5.107). The associated pole–zero plots are shown in Figure 5.30, where Figure 5.30(d) corresponds to Eq. (5.107) and Figure 5.30(a) to the minimum-phase system of Eq. (5.109). The impulse responses for these four cases are plotted in Figure 5.31. If we compare the

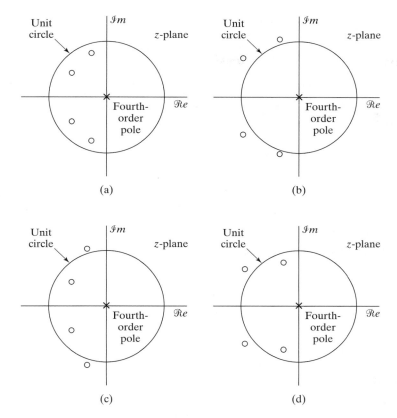

Figure 5.30 Four systems, all having the same frequency-response magnitude. Zeros are at all combinations of $0.9e^{\pm j0.6\pi}$ and $0.8e^{\pm j0.8\pi}$ and their reciprocals.

four sequences in this figure, we observe that the minimum-phase sequence appears to have larger samples at its left-hand end than do all the other sequences. Indeed, it is true for this example and, in general, that

$$|h[0]| \leq |h_{\min}[0]| \qquad\qquad (5.114)$$

for any causal, stable sequence $h[n]$ for which

$$|H(e^{j\omega})| = |H_{\min}(e^{j\omega})|. \qquad\qquad (5.115)$$

A proof of this property is suggested in Problem 5.65.

All the impulse responses whose frequency-response magnitude is equal to $|H_{\min}(e^{j\omega})|$ have the same total energy as $h_{\min}[n]$, since, by Parseval's theorem,

$$\sum_{n=0}^{\infty} |h[n]|^2 = \frac{1}{2\pi} \int_{-\pi}^{\pi} |H(e^{j\omega})|^2 d\omega = \frac{1}{2\pi} \int_{-\pi}^{\pi} |H_{\min}(e^{j\omega})|^2 d\omega$$

$$= \sum_{n=0}^{\infty} |h_{\min}[n]|^2. \qquad\qquad (5.116)$$

(a)

(b)

(c)

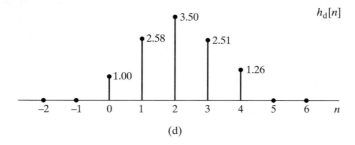

(d)

Figure 5.31 Sequences corresponding to the pole–zero plots of Figure 5.30.

If we define the *partial energy* of the impulse response as

$$E[n] = \sum_{m=0}^{n} |h[m]|^2,$$ (5.117)

then it can be shown that (see Problem 5.66)

$$\sum_{m=0}^{n} |h[m]|^2 \leq \sum_{m=0}^{n} |h_{\min}[m]|^2$$ (5.118)

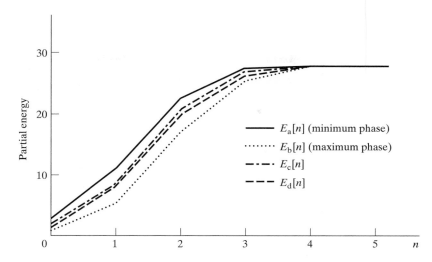

Figure 5.32   Partial energies for the four sequences of Figure 5.31. (Note that $E_a[n]$ is for the minimum-phase sequence $h_a[n]$ and $E_b[n]$ is for the maximum-phase sequence $h_b[n]$.)

for all impulse responses $h[n]$ belonging to the family of systems that have magnitude response given by Eq. (5.115). According to Eq. (5.118), the partial energy of the minimum-phase system is most concentrated around $n = 0$; i.e., the energy of the minimum-phase system is delayed the least of all systems having the same magnitude response function. For this reason, minimum-phase (lag) systems are also called *minimum energy-delay systems,* or simply, *minimum-delay systems.* This delay property is illustrated by Figure 5.32, which shows plots of the partial energy for the four sequences in Figure 5.31. We note for this example, and it is true in general, that the minimum energy delay occurs for the system that has all its zeros *inside* the unit circle (i.e., the minimum-phase system) and the maximum energy delay occurs for the system that has all its zeros *outside* the unit circle. Maximum energy-delay systems are also often called *maximum-phase systems.*

5.7 LINEAR SYSTEMS WITH GENERALIZED LINEAR PHASE

In designing filters and other signal-processing systems that pass some portion of the frequency band undistorted, it is desirable to have approximately constant frequency-response magnitude and zero phase in that band. For causal systems, zero phase is not attainable, and consequently, some phase distortion must be allowed. As we saw in Section 5.1.2, the effect of linear phase with integer slope is a simple time shift. A nonlinear phase, on the other hand, can have a major effect on the shape of a signal, even when the frequency-response magnitude is constant. Thus, in many situations it is particularly desirable to design systems to have exactly or approximately linear phase. In this section, we consider a formalization and generalization of the notions of linear phase and ideal time delay by considering the class of systems that have constant group delay. We begin by reconsidering the concept of delay in a discrete-time system.

5.7.1 Systems with Linear Phase

Consider an LTI system whose frequency response over one period is

$$H_{\mathrm{id}}(e^{j\omega}) = e^{-j\omega\alpha}, \qquad |\omega| < \pi, \tag{5.119}$$

where α is a real number, not necessarily an integer. Such a system is an "ideal delay" system, where α is the delay introduced by the system. Note that this system has constant magnitude response, linear phase, and constant group delay; i.e.,

$$|H_{\mathrm{id}}(e^{j\omega})| = 1, \tag{5.120a}$$

$$\sphericalangle H_{\mathrm{id}}(e^{j\omega}) = -\omega\alpha, \tag{5.120b}$$

$$\mathrm{grd}[H_{\mathrm{id}}(e^{j\omega})] = \alpha. \tag{5.120c}$$

The inverse Fourier transform of $H_{\mathrm{id}}(e^{j\omega})$ is the impulse response

$$h_{\mathrm{id}}[n] = \frac{\sin\pi(n-\alpha)}{\pi(n-\alpha)}, \qquad -\infty < n < \infty. \tag{5.121}$$

The output of this system for an input $x[n]$ is

$$y[n] = x[n] * \frac{\sin\pi(n-\alpha)}{\pi(n-\alpha)} = \sum_{k=-\infty}^{\infty} x[k]\frac{\sin\pi(n-k-\alpha)}{\pi(n-k-\alpha)}. \tag{5.122}$$

If $\alpha = n_d$, where n_d is an integer, then, as mentioned in Section 5.1.2,

$$h_{\mathrm{id}}[n] = \delta[n-n_d] \tag{5.123}$$

and

$$y[n] = x[n] * \delta[n-n_d] = x[n-n_d]. \tag{5.124}$$

That is, if $\alpha = n_d$ is an integer, the system with linear phase and unity gain in Eq. (5.119) simply shifts the input sequence by n_d samples. If α is not an integer, the most straightforward interpretation is the one developed in Example 4.9 in Chapter 4. Specifically, a representation of the system of Eq. (5.119) is that shown in Figure 5.33, with $h_c(t) = \delta(t-\alpha T)$ and $H_c(j\Omega) = e^{-j\Omega\alpha T}$, so that

$$H(e^{j\omega}) = e^{-j\omega\alpha}, \qquad |\omega| < \pi. \tag{5.125}$$

In this representation, the choice of T is irrelevant and could simply be normalized to unity. It is important to stress again that the representation is valid whether or not $x[n]$ was originally obtained by sampling a continuous-time signal. According to the representation in Figure 5.33, $y[n]$ is the sequence of samples of the time-shifted, band-limited interpolation of the input sequence $x[n]$; i.e., $y[n] = x_c(nT - \alpha T)$. The system of Eq. (5.119) is said to have a time shift of α samples, even if α is not an integer. If the group delay α is positive, the time shift is a time delay. If α is negative, the time shift is a time advance.

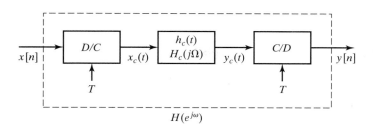

Figure 5.33 Interpretation of noninteger delay in discrete-time systems.

Figure 5.34 Representation of a linear-phase LTI system as a cascade of a magnitude filter and a time shift.

This discussion also provides a useful interpretation of linear phase when it is associated with a nonconstant magnitude response. For example, consider a more general frequency response with linear phase, i.e.,

$$H(e^{j\omega}) = |H(e^{j\omega})|e^{-j\omega\alpha}, \qquad |\omega| < \pi. \tag{5.126}$$

Equation (5.126) suggests the interpretation of Figure 5.34. The signal $x[n]$ is filtered by the zero-phase frequency response $|H(e^{j\omega})|$, and the filtered output is then "time shifted" by the (integer or noninteger) amount α. Suppose, for example, that $H(e^{j\omega})$ is the linear-phase ideal lowpass filter

$$H_{lp}(e^{j\omega}) = \begin{cases} e^{-j\omega\alpha}, & |\omega| < \omega_c, \\ 0, & \omega_c < |\omega| \le \pi. \end{cases} \tag{5.127}$$

The corresponding impulse response is

$$h_{lp}[n] = \frac{\sin \omega_c(n - \alpha)}{\pi(n - \alpha)}. \tag{5.128}$$

Note that Eq. (5.121) is obtained if $\omega_c = \pi$.

Example 5.16   Ideal Lowpass with Linear Phase

The impulse response of the ideal lowpass filter illustrates some interesting properties of linear-phase systems. Figure 5.35(a) shows $h_{lp}[n]$ for $\omega_c = 0.4\pi$ and $\alpha = n_d = 5$. Note that when α is an integer, the impulse response is symmetric about $n = n_d$; i.e.,

$$h_{lp}[2n_d - n] = \frac{\sin \omega_c(2n_d - n - n_d)}{\pi(2n_d - n - n_d)}$$

$$= \frac{\sin \omega_c(n_d - n)}{\pi(n_d - n)} \tag{5.129}$$

$$= h_{lp}[n].$$

In this case we could define a *zero-phase system*

$$\hat{H}_{lp}(e^{j\omega}) = H_{lp}(e^{j\omega})e^{j\omega n_d} = |H_{lp}(e^{j\omega})|, \tag{5.130}$$

where the impulse response is shifted to the left by n_d samples, yielding an even sequence

$$\hat{h}_{lp}[n] = \frac{\sin \omega_c n}{\pi n} = \hat{h}_{lp}[-n]. \tag{5.131}$$

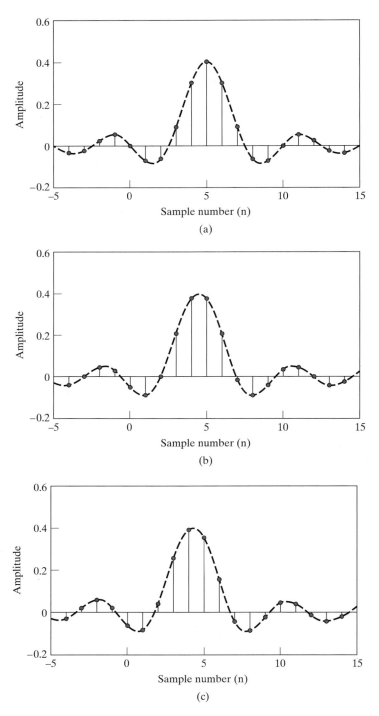

Figure 5.35 Ideal lowpass filter impulse responses, with $\omega_c = 0.4\pi$. (a) Delay $= \alpha = 5$. (b) Delay $= \alpha = 4.5$. (c) Delay $= \alpha = 4.3$.

Figure 5.35(b) shows $h_{lp}[n]$ for $\omega_c = 0.4\pi$ and $\alpha = 4.5$. This is typical of the case when the linear phase corresponds to an integer plus one-half. As in the case of the integer delay, it is easily shown that if α is an integer plus one-half (or 2α is an integer), then

$$h_{lp}[2\alpha - n] = h_{lp}[n]. \tag{5.132}$$

In this case, the point of symmetry is α, which is not an integer. Therefore, since the symmetry is not about a point of the sequence, it is not possible to shift the sequence to obtain an even sequence that has zero phase. This is similar to the case of Example 4.10 with M odd.

Figure 5.35(c) represents a third case, in which there is no symmetry at all. In this case, $\omega_c = 0.4\pi$ and $\alpha = 4.3$.

In general a linear-phase system has frequency response

$$H(e^{j\omega}) = |H(e^{j\omega})|e^{-j\omega\alpha}. \tag{5.133}$$

As illustrated in Example 5.16, if 2α is an integer (i.e., if α is an integer or an integer plus one-half), the corresponding impulse response has even symmetry about α; i.e.,

$$h[2\alpha - n] = h[n]. \tag{5.134}$$

If 2α is not an integer, then the impulse response will not have symmetry. This is illustrated in Figure 5.35(c), which shows an impulse response that is not symmetric, but that has linear phase, or equivalently, constant group delay.

5.7.2 Generalized Linear Phase

In the discussion in Section 5.7.1, we considered a class of systems whose frequency response is of the form of Eq. (5.126), i.e., a real-valued nonnegative function of ω multiplied by a linear phase term $e^{-j\omega\alpha}$. For a frequency response of this form, the phase of $H(e^{j\omega})$ is entirely associated with the linear phase factor $e^{-j\omega\alpha}$, i.e., $\sphericalangle H(e^{j\omega}) = -\omega\alpha$, and consequently, systems in this class are referred to as linear-phase systems. In the moving average of Example 4.10, the frequency response in Eq. (4.67) is a real-valued function of ω multiplied by a linear-phase term, but the system is not, strictly speaking, a linear-phase system, since, at frequencies for which the factor

$$\frac{1}{M+1} \frac{\sin[\omega(M+1)/2]}{\sin(\omega/2)}$$

is negative, this term contributes an additional phase of π radians to the total phase.

Many of the advantages of linear-phase systems apply to systems with frequency response having the form of Eq. (4.67) as well, and consequently, it is useful to generalize somewhat the definition and concept of linear phase. Specifically, a system is referred to as a *generalized linear-phase system* if its frequency response can be expressed in the form

$$H(e^{j\omega}) = A(e^{j\omega})e^{-j\alpha\omega+j\beta}, \tag{5.135}$$

where α and β are constants and $A(e^{j\omega})$ is a real (possibly bipolar) function of ω. For the linear-phase system of Eq. (5.127) and the moving-average filter of Example 4.10, $\beta = 0$. We see, however, that the bandlimited differentiator of Example 4.5 has the form of Eq. (5.135) with $\alpha = 0$, $\beta = \pi/2$, and $A(e^{j\omega}) = \omega/T$.

A system whose frequency response has the form of Eq. (5.135) is called a generalized linear-phase system because the phase of such a system consists of constant terms added to the linear function $-\omega\alpha$; i.e. $-\omega\alpha + \beta$ is the equation of a straight line. However, if we ignore any discontinuities that result from the addition of constant phase over all or part of the band $|\omega| < \pi$, then such a system can be characterized by constant group delay. That is, the class of systems such that

$$\tau(\omega) = \text{grd}[H(e^{j\omega})] = -\frac{d}{d\omega}\{\arg[H(e^{j\omega})]\} = \alpha \tag{5.136}$$

have linear phase of the more general form

$$\arg[H(e^{j\omega})] = \beta - \omega\alpha, \qquad 0 < \omega < \pi, \tag{5.137}$$

where β and α are both real constants.

Recall that we showed in Section 5.7.1 that the impulse responses of linear-phase systems may have symmetry about α if 2α is an integer. To see the implication of this for generalized linear-phase systems, it is useful to derive an equation that must be satisfied by $h[n]$, α, and β for constant group-delay systems. This equation is derived by noting that, for such systems, the frequency response can be expressed as

$$\begin{aligned} H(e^{j\omega}) &= A(e^{j\omega})e^{j(\beta - \alpha\omega)} \\ &= A(e^{j\omega})\cos(\beta - \omega\alpha) + jA(e^{j\omega})\sin(\beta - \omega\alpha), \end{aligned} \tag{5.138}$$

or equivalently, as

$$\begin{aligned} H(e^{j\omega}) &= \sum_{n=-\infty}^{\infty} h[n]e^{-j\omega n} \\ &= \sum_{n=-\infty}^{\infty} h[n]\cos\omega n - j\sum_{n=-\infty}^{\infty} h[n]\sin\omega n, \end{aligned} \tag{5.139}$$

where we have assumed that $h[n]$ is real. The tangent of the phase angle of $H(e^{j\omega})$ can be expressed as

$$\tan(\beta - \omega\alpha) = \frac{\sin(\beta - \omega\alpha)}{\cos(\beta - \omega\alpha)} = \frac{-\sum\limits_{n=-\infty}^{\infty} h[n]\sin\omega n}{\sum\limits_{n=-\infty}^{\infty} h[n]\cos\omega n}.$$

Cross multiplying and combining terms with a trigonometric identity leads to the equation

$$\sum_{n=-\infty}^{\infty} h[n]\sin[\omega(n - \alpha) + \beta] = 0 \quad \text{for all } \omega. \tag{5.140}$$

This equation is a necessary condition on $h[n]$, α, and β for the system to have constant group delay. It is not a sufficient condition, however, and, due to its implicit nature, it does not tell us how to find a linear-phase system. For example, it can be shown that one set of conditions that satisfies Eq. (5.140) is

$$\beta = 0 \quad \text{or} \quad \pi, \tag{5.141a}$$

$$2\alpha = M = \text{an integer}, \tag{5.141b}$$

$$h[2\alpha - n] = h[n]. \tag{5.141c}$$

With $\beta = 0$ or π, Eq. (5.140) becomes

$$\sum_{n=-\infty}^{\infty} h[n] \sin[\omega(n - \alpha)] = 0, \tag{5.142}$$

from which it can be shown that if 2α is an integer, terms in Eq. (5.142) can be paired so that each pair of terms is identically zero for all ω. These conditions in turn imply that the corresponding frequency response has the form of Eq. (5.135) with $\beta = 0$ or π and $A(e^{j\omega})$ an even (and, of course, real) function of ω.

Alternatively, if $\beta = \pi/2$ or $3\pi/2$, then Eq. (5.140) becomes

$$\sum_{n=-\infty}^{\infty} h[n] \cos[\omega(n - \alpha)] = 0, \tag{5.143}$$

and it can be shown that

$$\beta = \pi/2 \quad \text{or} \quad 3\pi/2, \tag{5.144a}$$

$$2\alpha = M = \text{an integer,} \tag{5.144b}$$

and

$$h[2\alpha - n] = -h[n] \tag{5.144c}$$

satisfy Eq. (5.143) for all ω. Equations (5.144) imply that the frequency response has the form of Eq. (5.135) with $\beta = \pi/2$ and $A(e^{j\omega})$ an odd function of ω.

Note that Eqs. (5.141) and (5.144) give two sets of conditions that guarantee generalized linear phase or constant group delay, but as we have already seen in Figure 5.35(c), there are other systems that satisfy Eq. (5.135) without these symmetry conditions.

5.7.3 Causal Generalized Linear-Phase Systems

If the system is causal, then Eq. (5.140) becomes

$$\sum_{n=0}^{\infty} h[n] \sin[\omega(n - \alpha) + \beta] = 0 \qquad \text{for all } \omega. \tag{5.145}$$

Causality and the conditions in Eqs. (5.141) and (5.144) imply that

$$h[n] = 0, \qquad n < 0 \quad \text{and} \quad n > M;$$

i.e., causal FIR systems have generalized linear phase if they have impulse response length $(M + 1)$ and satisfy either Eq. (5.141c) or (5.144c). Specifically, it can be shown that if

$$h[n] = \begin{cases} h[M - n], & 0 \le n \le M, \\ 0, & \text{otherwise,} \end{cases} \tag{5.146a}$$

then

$$H(e^{j\omega}) = A_e(e^{j\omega})e^{-j\omega M/2}, \tag{5.146b}$$

where $A_e(e^{j\omega})$ is a real, even, periodic function of ω. Similarly, if

$$h[n] = \begin{cases} -h[M-n], & 0 \le n \le M, \\ 0, & \text{otherwise,} \end{cases} \tag{5.147a}$$

then it follows that

$$H(e^{j\omega}) = j A_o(e^{j\omega})e^{-j\omega M/2} = A_o(e^{j\omega})e^{-j\omega M/2+j\pi/2}, \tag{5.147b}$$

where $A_o(e^{j\omega})$ is a real, odd, periodic function of ω. Note that in both cases the length of the impulse response is $(M+1)$ samples.

The conditions in Eqs. (5.146a) and (5.147a) are sufficient to guarantee a causal system with generalized linear phase. However, they are not necessary conditions. Clements and Pease (1989) have shown that causal infinite-duration impulse responses can also have Fourier transforms with generalized linear phase. The corresponding system functions, however, are not rational, and thus, the systems cannot be implemented with difference equations.

Expressions for the frequency response of FIR linear-phase systems are useful in filter design and in understanding some of the properties of such systems. In deriving these expressions, it turns out that significantly different expressions result, depending on the type of symmetry and whether M is an even or odd integer. For this reason, it is generally useful to define four types of FIR generalized linear-phase systems.

Type I FIR Linear-Phase Systems

A type I system is defined as a system that has a symmetric impulse response

$$h[n] = h[M-n], \qquad 0 \le n \le M, \tag{5.148}$$

with M an even integer. The delay $M/2$ is an integer. The frequency response is

$$H(e^{j\omega}) = \sum_{n=0}^{M} h[n]e^{-j\omega n}. \tag{5.149}$$

By applying the symmetry condition, Eq. (5.148), the sum in Eq. (5.149) can be rewritten in the form

$$H(e^{j\omega}) = e^{-j\omega M/2}\left(\sum_{k=0}^{M/2} a[k] \cos \omega k \right), \tag{5.150a}$$

where

$$a[0] = h[M/2], \tag{5.150b}$$

$$a[k] = 2h[(M/2)-k], \qquad k = 1, 2, \ldots, M/2. \tag{5.150c}$$

Thus, from Eq. (5.150a), we see that $H(e^{j\omega})$ has the form of Eq. (5.146b), and in particular, β in Eq. (5.135) is either 0 or π.

Type II FIR Linear-Phase Systems

A type II system has a symmetric impulse response as in Eq. (5.148), with M an odd integer. $H(e^{j\omega})$ for this case can be expressed as

$$H(e^{j\omega}) = e^{-j\omega M/2} \left\{ \sum_{k=1}^{(M+1)/2} b[k] \cos\left[\omega\left(k - \tfrac{1}{2}\right)\right] \right\}, \qquad (5.151a)$$

where

$$b[k] = 2h[(M+1)/2 - k], \qquad k = 1, 2, \ldots, (M+1)/2. \qquad (5.151b)$$

Again, $H(e^{j\omega})$ has the form of Eq. (5.146b) with a time delay of $M/2$, which in this case is an integer plus one-half, and β in Eq. (5.135) is either 0 or π.

Type III FIR Linear-Phase Systems

If the system has an antisymmetric impulse response

$$h[n] = -h[M - n], \qquad 0 \le n \le M, \qquad (5.152)$$

with M an even integer, then $H(e^{j\omega})$ has the form

$$H(e^{j\omega}) = je^{-j\omega M/2} \left[\sum_{k=1}^{M/2} c[k] \sin\omega k \right], \qquad (5.153a)$$

where

$$c[k] = 2h[(M/2) - k], \qquad k = 1, 2, \ldots, M/2. \qquad (5.153b)$$

In this case, $H(e^{j\omega})$ has the form of Eq. (5.147b) with a delay of $M/2$, which is an integer, and β in Eq. (5.135) is $\pi/2$ or $3\pi/2$.

Type IV FIR Linear-Phase Systems

If the impulse response is antisymmetric as in Eq. (5.152) and M is odd, then

$$H(e^{j\omega}) = je^{-j\omega M/2} \left[\sum_{k=1}^{(M+1)/2} d[k] \sin\left[\omega\left(k - \tfrac{1}{2}\right)\right] \right], \qquad (5.154a)$$

where

$$d[k] = 2h[(M+1)/2 - k], \qquad k = 1, 2, \ldots, (M+1)/2. \qquad (5.154b)$$

As in the case of type III systems, $H(e^{j\omega})$ has the form of Eq. (5.147b) with delay $M/2$, which is an integer plus one-half, and β in Eq. (5.135) is $\pi/2$ or $3\pi/2$.

(a)

(b)

(c)

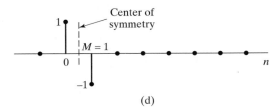

(d)

Figure 5.36 Examples of FIR linear-phase systems. (a) Type I, M even, $h[n] = h[M - n]$. (b) Type II, M odd, $h[n] = h[M - n]$. (c) Type III, M even, $h[n] = -h[M - n]$. (d) Type IV, M odd, $h[n] = -h[M - n]$.

Examples of FIR Linear-Phase Systems

Figure 5.36 shows an example of each of the four types of FIR linear-phase systems. The associated frequency responses are given in Examples 5.17–5.20.

Example 5.17   Type I Linear-Phase System

If the impulse response is

$$h[n] = \begin{cases} 1, & 0 \le n \le 4, \\ 0, & \text{otherwise,} \end{cases} \qquad (5.155)$$

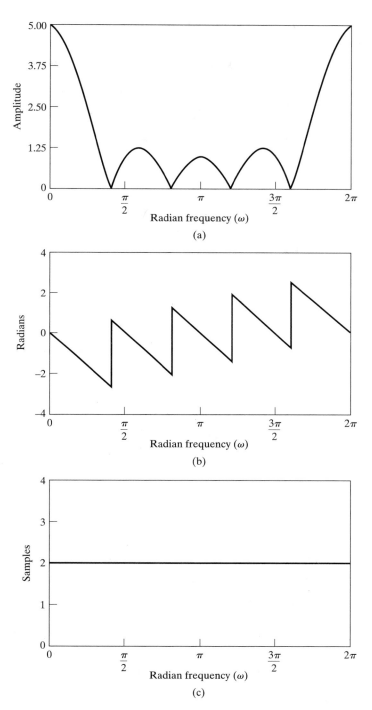

Figure 5.37   Frequency response of type I system of Example 5.17. (a) Magnitude. (b) Phase. (c) Group delay.

as shown in Figure 5.36(a), the system satisfies the condition of Eq. (5.148). The frequency response is

$$H(e^{j\omega}) = \sum_{n=0}^{4} e^{-j\omega n} = \frac{1 - e^{-j\omega 5}}{1 - e^{-j\omega}}$$

$$= e^{-j\omega 2} \frac{\sin(5\omega/2)}{\sin(\omega/2)}. \tag{5.156}$$

The magnitude, phase, and group delay of the system are shown in Figure 5.37. Since $M = 4$ is even, the group delay is an integer, i.e., $\alpha = 2$.

Example 5.18 Type II Linear-Phase System

If the length of the impulse response of the previous example is extended by one sample, we obtain the impulse response of Figure 5.36(b), which has frequency response

$$H(e^{j\omega}) = e^{-j\omega 5/2} \frac{\sin(3\omega)}{\sin(\omega/2)}. \tag{5.157}$$

The frequency-response functions for this system are shown in Figure 5.38. Note that the group delay in this case is constant with $\alpha = 5/2$.

Example 5.19 Type III Linear-Phase System

If the impulse response is

$$h[n] = \delta[n] - \delta[n - 2], \tag{5.158}$$

as in Figure 5.36(c), then

$$H(e^{j\omega}) = 1 - e^{-j2\omega}$$

$$= j[2 \sin(\omega)]e^{-j\omega}. \tag{5.159}$$

The frequency-response plots for this example are given in Figure 5.39. Note that the group delay in this case is constant with $\alpha = 1$.

Example 5.20 Type IV Linear-Phase System

In this case (Figure 5.36(d)), the impulse response is

$$h[n] = \delta[n] - \delta[n - 1], \tag{5.160}$$

for which the frequency response is

$$H(e^{j\omega}) = 1 - e^{-j\omega}$$

$$= j[2 \sin(\omega/2)]e^{-j\omega/2}. \tag{5.161}$$

The frequency response for this system is shown in Figure 5.40. Note that the group delay is equal to $\frac{1}{2}$ for all ω.

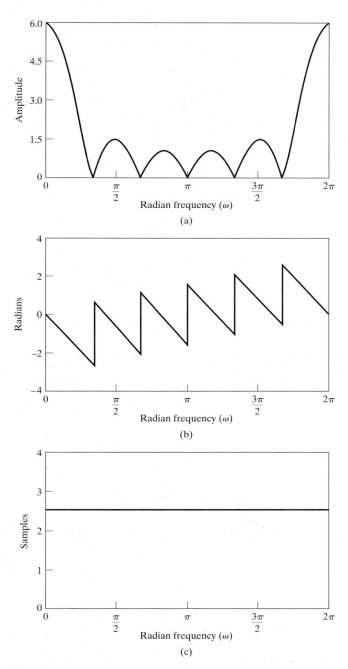

Figure 5.38 Frequency response of type II system of Example 5.18. (a) Magnitude. (b) Phase. (c) Group delay.

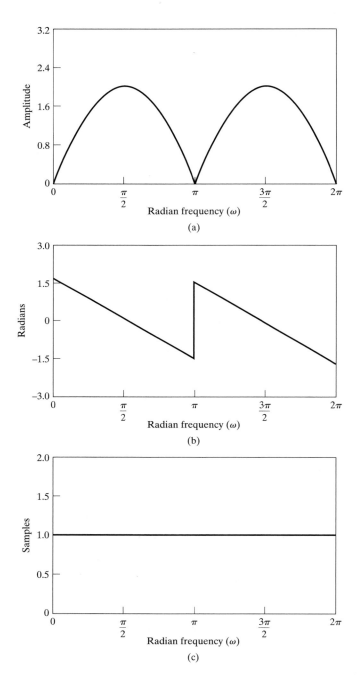

Figure 5.39   Frequency response of type III system of Example 5.19. (a) Magnitude. (b) Phase. (c) Group delay.

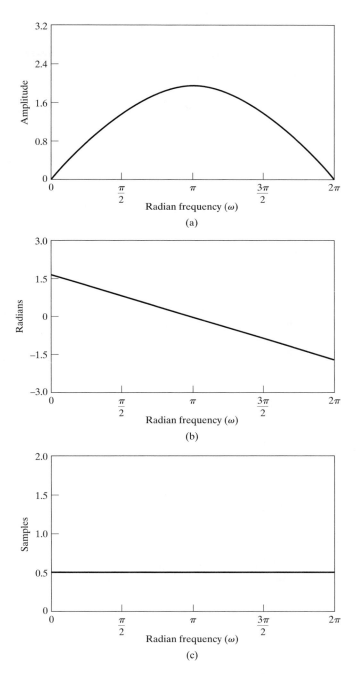

Figure 5.40 Frequency response of type IV system of Example 5.20. (a) Magnitude. (b) Phase. (c) Group delay.

Locations of Zeros for FIR Linear-Phase Systems

The preceding examples illustrate the properties of the impulse response and the frequency response for all four types of FIR linear-phase systems. It is also instructive to consider the locations of the zeros of the system function for FIR linear-phase systems. The system function is

$$H(z) = \sum_{n=0}^{M} h[n]z^{-n}. \tag{5.162}$$

In the symmetric cases (types I and II), we can use Eq. (5.148) to express $H(z)$ as

$$H(z) = \sum_{n=0}^{M} h[M-n]z^{-n} = \sum_{k=M}^{0} h[k]z^k z^{-M} \tag{5.163}$$

$$= z^{-M} H(z^{-1}).$$

From Eq. (5.163), we conclude that if z_0 is a zero of $H(z)$, then

$$H(z_0) = z_0^{-M} H(z_0^{-1}) = 0. \tag{5.164}$$

This implies that if $z_0 = re^{j\theta}$ is a zero of $H(z)$, then $z_0^{-1} = r^{-1}e^{-j\theta}$ is also a zero of $H(z)$. When $h[n]$ is real and z_0 is a zero of $H(z)$, $z_0^* = re^{-j\theta}$ will also be a zero of $H(z)$, and by the preceding argument, so will $(z_0^*)^{-1} = r^{-1}e^{j\theta}$. Therefore, when $h[n]$ is real, each complex zero not on the unit circle will be part of a set of four conjugate reciprocal zeros of the form

$$(1 - re^{j\theta}z^{-1})(1 - re^{-j\theta}z^{-1})(1 - r^{-1}e^{j\theta}z^{-1})(1 - r^{-1}e^{-j\theta}z^{-1}).$$

If a zero of $H(z)$ is on the unit circle, i.e., $z_0 = e^{j\theta}$, then $z_0^{-1} = e^{-j\theta} = z_0^*$, so zeros on the unit circle come in pairs of the form

$$(1 - e^{j\theta}z^{-1})(1 - e^{-j\theta}z^{-1}).$$

If a zero of $H(z)$ is real and not on the unit circle, the reciprocal will also be a zero of $H(z)$, and $H(z)$ will have factors of the form

$$(1 \pm rz^{-1})(1 \pm r^{-1}z^{-1}).$$

Finally, a zero of $H(z)$ at $z = \pm 1$ can appear by itself, since ± 1 is its own reciprocal and its own conjugate. Thus, we may also have factors of $H(z)$ of the form

$$(1 \pm z^{-1}).$$

The case of a zero at $z = -1$ is particularly important. From Eq. (5.163),

$$H(-1) = (-1)^M H(-1).$$

If M is even, we have a simple identity, but if M is odd, $H(-1) = -H(-1)$, so $H(-1)$ must be zero. Thus, for symmetric impulse responses with M odd, the system function

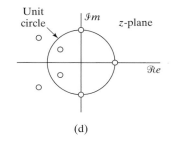

Figure 5.41   Typical plots of zeros for linear-phase systems. (a) Type I. (b) Type II. (c) Type III. (d) Type IV.

must have a zero at $z = -1$. Figures 5.41(a) and 5.41(b) show typical locations of zeros for type I (M even) and type II (M odd) systems, respectively.

If the impulse response is antisymmetric (types III and IV), then, following the approach used to obtain Eq. (5.163), we can show that

$$H(z) = -z^{-M}H(z^{-1}). \tag{5.165}$$

This equation can be used to show that the zeros of $H(z)$ for the antisymmetric case are constrained in the same way as the zeros for the symmetric case. In the antisymmetric case, however, both $z = 1$ and $z = -1$ are of special interest. If $z = 1$, Eq. (5.165) becomes

$$H(1) = -H(1). \tag{5.166}$$

Thus, $H(z)$ *must* have a zero at $z = 1$ for both M even and M odd. If $z = -1$, Eq. (5.165) gives

$$H(-1) = (-1)^{-M+1}H(-1). \tag{5.167}$$

In this case, if $(M - 1)$ is odd (i.e., if M is even), $H(-1) = -H(-1)$, so $z = -1$ *must* be a zero of $H(z)$ if M is even. Figures 5.41(c) and 5.41(d) show typical zero locations for type III and IV systems, respectively.

These constraints on the zeros are important in designing FIR linear-phase systems, since they impose limitations on the types of frequency responses that can be achieved. For example, we note that, in approximating a highpass filter using a symmetric impulse response, M should not be odd, since the frequency response is constrained to be zero at $\omega = \pi (z = -1)$.

5.7.4 Relation of FIR Linear-Phase Systems to Minimum-Phase Systems

The previous discussion shows that all FIR linear-phase systems with real impulse response have zeros either on the unit circle or at conjugate reciprocal locations. Thus, it is easily shown that the system function of any FIR linear-phase system can be factored into a minimum-phase term $H_{\min}(z)$, a maximum-phase term $H_{\max}(z)$, and a term $H_{uc}(z)$ containing only zeros on the unit circle; i.e.,

$$H(z) = H_{\min}(z) H_{uc}(z) H_{\max}(z), \tag{5.168a}$$

where

$$H_{\max}(z) = H_{\min}(z^{-1}) z^{-M_i} \tag{5.168b}$$

and M_i is the number of zeros of $H_{\min}(z)$. In Eq. (5.168a), $H_{\min}(z)$ has all M_i of its zeros *inside* the unit circle, and $H_{uc}(z)$ has all M_o of its zeros *on* the unit circle. $H_{\max}(z)$ has all M_i of its zeros *outside* the unit circle, and, from Eq. (5.168b), its zeros are the reciprocals of the zeros of $H_{\min}(z)$. The order of the system function $H(z)$ is therefore $M = 2M_i + M_o$.

Example 5.21   Decomposition of a Linear-Phase System

As a simple example of the use of Eqs. (5.168), consider the minimum-phase system function of Eq. (5.109), for which the frequency response is plotted in Figure 5.28. The system obtained by applying Eq. (5.168b) to $H_{\min}(z)$ in Eq. (5.109) is

$$H_{\max}(z) = (0.9)^2 (1 - 1.1111 e^{j0.6\pi} z^{-1})(1 - 1.1111 e^{-j0.6\pi} z^{-1})$$

$$\times (1 - 1.25 e^{-j0.8\pi} z^{-1})(1 - 1.25 e^{j0.8\pi} z^{-1}).$$

$H_{\max}(z)$ has the frequency response shown in Figure 5.42. Now, if these two systems are cascaded, it follows from Eq. (5.168) that the overall system

$$H(z) = H_{\min}(z) H_{\max}(z)$$

has linear phase. The frequency response of the composite system would be obtained by adding the respective log magnitude, phase, and group-delay functions. Therefore,

$$20 \log_{10} |H(e^{j\omega})| = 20 \log_{10} |H_{\min}(e^{j\omega})| + 20 \log_{10} |H_{\max}(e^{j\omega})|$$

$$= 40 \log_{10} |H_{\min}(e^{j\omega})|. \tag{5.169}$$

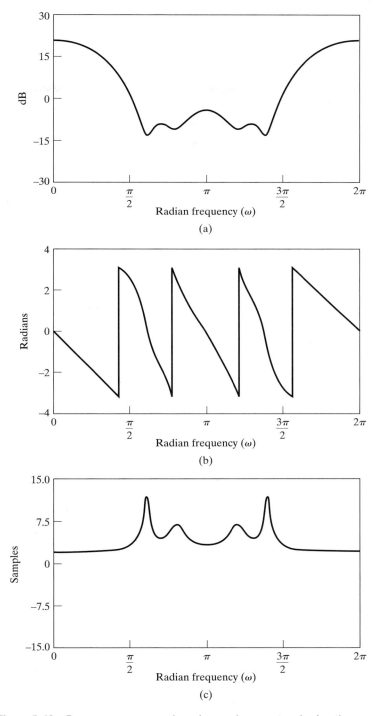

Figure 5.42 Frequency response of maximum-phase system having the same magnitude as the system in Figure 5.28. (a) Log magnitude. (b) Phase (principal value). (c) Group delay.

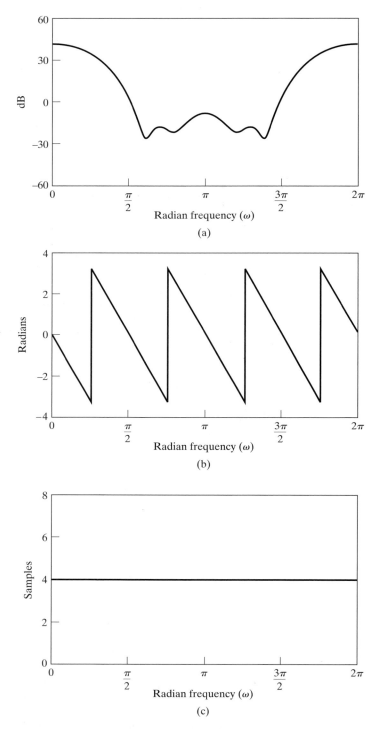

Figure 5.43 Frequency response of cascade of maximum-phase and minimum-phase systems, yielding a linear-phase system. (a) Log magnitude. (b) Phase (principal value). (c) Group delay.

Similarly,

$$\sphericalangle H(e^{j\omega}) = \sphericalangle H_{\min}(e^{j\omega}) + \sphericalangle H_{\max}(e^{j\omega}). \tag{5.170}$$

From Eq. (5.168b), it follows that

$$\sphericalangle H_{\max}(e^{j\omega}) = -\omega M_i - \sphericalangle H_{\min}(e^{j\omega}). \tag{5.171}$$

and

$$\sphericalangle H(e^{j\omega}) = -\omega M_i,$$

where $M_i = 4$ is the number of zeros of $H_{\min}(z)$. In like manner, the group-delay functions of $H_{\min}(e^{j\omega})$ and $H_{\max}(e^{j\omega})$ combine to give

$$\mathrm{grd}[H(e^{j\omega})] = M_i = 4.$$

The frequency-response plots for the composite system are given in Figure 5.43. Note that the curves are sums of the corresponding functions in Figures 5.28 and 5.42.

5.8 SUMMARY

In this chapter, we developed and explored the representation and analysis of LTI systems using the Fourier and z-transforms. The importance of transform analysis for LTI systems stems directly from the fact that complex exponentials are eigenfunctions of such systems and the associated eigenvalues correspond to the system function or frequency response.

A particularly important class of LTI systems is that characterized by linear constant-coefficient difference equations. Transform analysis is particularly useful for analyzing these systems, since the Fourier transform or z-transform converts a difference equation to an algebraic equation. In particular, the system function is a ratio of polynomials, the coefficients of which correspond directly to the coefficients in the difference equation. The roots of these polynomials provide a useful system representation in terms of the pole–zero plot. Systems characterized by difference equations may have an impulse response that is infinite in duration (IIR) or finite in duration (FIR).

The frequency response of LTI systems is often characterized in terms of magnitude and phase or group delay, which is the negative of the derivative of the phase. Linear phase is often a desirable characteristic of a system frequency response, since it is a relatively mild form of phase distortion, corresponding to a time shift. The importance of FIR systems lies in part in the fact that such systems can be easily designed to have exactly linear phase (or generalized linear phase), while, for a given set of frequency response magnitude specifications, IIR systems are more efficient. These and other trade-offs will be discussed in detail in Chapter 7.

While, in general, for LTI systems, the frequency-response magnitude and phase are independent, for minimum-phase systems the magnitude uniquely specifies the phase and the phase uniquely specifies the magnitude to within a scale factor. Nonminimum-phase systems can be represented as the cascade combination of a minimum-phase system and an all-pass system. Relations between Fourier transform magnitude and phase will be discussed in considerably more detail in Chapter 11.

PROBLEMS

Basic Problems with Answers

5.1. In the system shown in Figure P5.1-1, $H(e^{j\omega})$ is an ideal lowpass filter. Determine whether for some choice of input $x[n]$ and cutoff frequency ω_c, the output can be the pulse

$$y[n] = \begin{cases} 1, & 0 \le n \le 10, \\ 0, & \text{otherwise,} \end{cases}$$

shown in Figure P5.1-2.

Figure P5.1-1

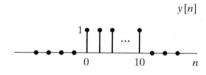

Figure P5.1-2

5.2. Consider a stable linear time-invariant system with input $x[n]$ and output $y[n]$. The input and output satisfy the difference equation

$$y[n-1] - \tfrac{10}{3}y[n] + y[n+1] = x[n].$$

(a) Plot the poles and zeros in the z-plane.
(b) Find the impulse response $h[n]$.

5.3. Consider a linear time-invariant discrete-time system for which the input $x[n]$ and output $y[n]$ are related by the second-order difference equation

$$y[n-1] + \tfrac{1}{3}y[n-2] = x[n].$$

From the following list, choose *two* possible impulse responses for the system:

(a) $\left(-\tfrac{1}{3}\right)^{n+1} u[n+1]$

(b) $3^{n+1}u[n+1]$

(c) $3(-3)^{n+2}u[-n-2]$

(d) $\tfrac{1}{3}\left(-\tfrac{1}{3}\right)^{n} u[-n-2]$

(e) $\left(-\frac{1}{3}\right)^{n+1} u[-n-2]$

(f) $\left(\frac{1}{3}\right)^{n+1} u[n+1]$

(g) $(-3)^{n+1}u[n]$

(h) $n^{1/3}u[n]$

5.4. When the input to a linear time-invariant system is

$$x[n] = \left(\frac{1}{2}\right)^n u[n] + (2)^n u[-n-1],$$

the output is

$$y[n] = 6 \left(\frac{1}{2}\right)^n u[n] - 6 \left(\frac{3}{4}\right)^n u[n].$$

(a) Find the system function $H(z)$ of the system. Plot the poles and zeros of $H(z)$, and indicate the region of convergence.

(b) Find the impulse response $h[n]$ of the system for all values of n.

(c) Write the difference equation that characterizes the system.

(d) Is the system stable? Is it causal?

5.5. Consider a system described by a linear constant-coefficient difference equation with initial-rest conditions. The step response of the system is given by

$$y[n] = \left(\frac{1}{3}\right)^n u[n] + \left(\frac{1}{4}\right)^n u[n] + u[n].$$

(a) Determine the difference equation.

(b) Determine the impulse response of the system.

(c) Determine whether or not the system is stable.

5.6. The following information is known about a linear time-invariant system:

(a) The system is causal.

(b) When the input is

$$x[n] = -\frac{1}{3} \left(\frac{1}{2}\right)^n u[n] - \frac{4}{3}(2)^n u[-n-1],$$

the z-transform of the output is

$$Y(z) = \frac{1 - z^{-2}}{\left(1 - \frac{1}{2}z^{-1}\right)(1 - 2z^{-1})}.$$

(c) Find the z-transform of $x[n]$.

(d) What are the possible choices for the region of convergence of $Y(z)$?

(e) What are the possible choices for the impulse response of the system?

5.7. When the input to a linear time-invariant system is

$$x[n] = 5u[n],$$

the output is

$$y[n] = \left[2 \left(\frac{1}{2}\right)^n + 3 \left(-\frac{3}{4}\right)^n\right] u[n].$$

(a) Find the system function $H(z)$ of the system. Plot the poles and zeros of $H(z)$, and indicate the region of convergence.

(b) Find the impulse response of the system for all values of n.

(c) Write the difference equation that characterizes the system.

5.8. A causal linear time-invariant system is described by the difference equation

$$y[n] = \tfrac{3}{2}y[n-1] + y[n-2] + x[n-1].$$

(a) Find the system function $H(z) = Y(z)/X(z)$ for this system. Plot the poles and zeros of $H(z)$, and indicate the region of convergence.

(b) Find the impulse response of the system.

(c) You should have found the system to be unstable. Find a stable (noncausal) impulse response that satisfies the difference equation.

5.9. Consider a linear time-invariant system with input $x[n]$ and output $y[n]$ for which

$$y[n-1] - \tfrac{5}{2}y[n] + y[n+1] = x[n].$$

The system may or may not be stable or causal.

By considering the pole–zero pattern associated with the preceding difference equation, determine three possible choices for the impulse response of the system. Show that each choice satisfies the difference equation. Indicate which choice corresponds to a stable system and which choice corresponds to a causal system.

5.10. If the system function $H(z)$ of a linear time-invariant system has a pole–zero diagram as shown in Figure P5.10-1 and the system is causal, can the inverse system $H_i(z)$, where $H(z)H_i(z) = 1$, be both causal and stable? Clearly justify your answer.

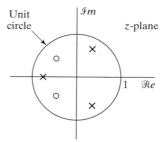

Figure P5.10-1

5.11. The system function of a linear time-invariant system has the pole–zero plot shown in Figure P5.11-1 Specify whether each of the following statements is true, is false, or cannot be determined from the information given.

(a) The system is stable.

(b) The system is causal.

(c) If the system is causal, then it must be stable.

(d) If the system is stable, then it must have a two-sided impulse response.

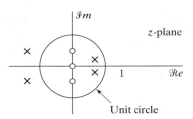

z-plane

Unit circle   **Figure P5.11-1**

5.12. A discrete-time causal LTI system has the system function

$$H(z) = \frac{(1 + 0.2z^{-1})(1 - 9z^{-2})}{(1 + 0.81z^{-2})}.$$

(a) Is the system stable?

(b) Find expressions for a minimum-phase system $H_1(z)$ and an all-pass system $H_{ap}(z)$ such that

$$H(z) = H_1(z)H_{ap}(z).$$

5.13. Figure P5.13-1 shows the pole–zero plots for four different LTI systems. Based on these plots, state whether or not each system is an all-pass system.

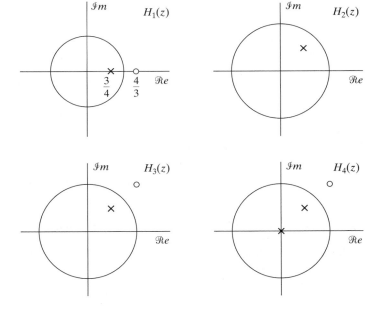

Figure P5.13-1

5.14. Determine the group delay for $0 < \omega < \pi$ for each of the following sequences:
(a)

$$x_1[n] = \begin{cases} n-1, & 1 \le n \le 5, \\ 9-n, & 5 < n \le 9, \\ 0, & \text{otherwise.} \end{cases}$$

(b)

$$x_2[n] = \left(\frac{1}{2}\right)^{|n-1|} + \left(\frac{1}{2}\right)^{|n|}.$$

5.15. Consider the class of discrete-time filters whose frequency response has the form

$$H(e^{j\omega}) = |H(e^{j\omega})|e^{-j\alpha\omega},$$

where $|H(e^{j\omega})|$ is a real and nonnegative function of ω and α is a real constant. As discussed in Section 5.7.1, this class of filters is referred to as *linear-phase* filters.

Consider also the class of discrete-time filters whose frequency response has the form

$$H(e^{j\omega}) = A(e^{j\omega})e^{-j\alpha\omega+j\beta},$$

where $A(e^{j\omega})$ is a real function of ω, α is a real constant, and β is a real constant. As discussed in Section 5.7.2, filters in this class are referred to as *generalized linear-phase* filters.

For each of the filters in Figure P5.15-1, determine whether it is a generalized linear-phase filter. If it is, then find $A(e^{j\omega})$, α, and β. In addition, for each filter you determine to be a generalized linear-phase filter, indicate whether it also meets the more stringent criterion for being a linear-phase filter.

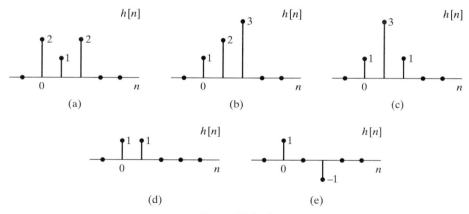

Figure P5.15-1

5.16. Figure P5.16-1 plots the continuous-phase $\arg[H(e^{j\omega})]$ for the frequency response of a specific LTI system, where

$$\arg[H(e^{j\omega})] = -\alpha\omega$$

for $|\omega| < \pi$ and α is a positive integer.

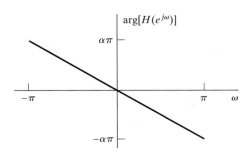

Figure P5.16-1

Is the impulse response $h[n]$ of this system a causal sequence? If the system is definitely causal, or if it is definitely not causal, give a proof. If the causality of the system cannot be determined from Figure P5.16-1, give examples of a noncausal sequence and a causal sequence that both have the foregoing phase response $\arg[H(e^{j\omega})]$.

5.17. For each of the following system functions, state whether or not it is a minimum-phase system. Justify your answers:

$$H_1(z) = \frac{\left(1 - 2z^{-1}\right)\left(1 + \frac{1}{2}z^{-1}\right)}{\left(1 - \frac{1}{3}z^{-1}\right)\left(1 + \frac{1}{3}z^{-1}\right)},$$

$$H_2(z) = \frac{\left(1 + \frac{1}{4}z^{-1}\right)\left(1 - \frac{1}{4}z^{-1}\right)}{\left(1 - \frac{2}{3}z^{-1}\right)\left(1 + \frac{2}{3}z^{-1}\right)},$$

$$H_3(z) = \frac{1 - \frac{1}{3}z^{-1}}{\left(1 - \frac{j}{2}z^{-1}\right)\left(1 + \frac{j}{2}z^{-1}\right)},$$

$$H_4(z) = \frac{z^{-1}\left(1 - \frac{1}{3}z^{-1}\right)}{\left(1 - \frac{j}{2}z^{-1}\right)\left(1 + \frac{j}{2}z^{-1}\right)}.$$

5.18. For each of the following system functions $H_k(z)$, specify a minimum-phase system function $H_{\min}(z)$ such that the frequency-response magnitudes of the two systems are equal, i.e., $|H_k(e^{j\omega})| = |H_{\min}(e^{j\omega})|$.

(a)

$$H_1(z) = \frac{1 - 2z^{-1}}{1 + \frac{1}{3}z^{-1}}$$

(b)

$$H_2(z) = \frac{\left(1 + 3z^{-1}\right)\left(1 - \frac{1}{2}z^{-1}\right)}{z^{-1}\left(1 + \frac{1}{3}z^{-1}\right)}$$

(c)

$$H_3(z) = \frac{\left(1 - 3z^{-1}\right)\left(1 - \frac{1}{4}z^{-1}\right)}{\left(1 - \frac{3}{4}z^{-1}\right)\left(1 - \frac{4}{3}z^{-1}\right)}$$

5.19. Figure P5.19-1 shows the impulse responses for several different LTI systems. Find the group delay associated with each system.

Figure P5.19-1

5.20. Figure P5.20-1 shows just the zero locations for several different system functions. For each plot, state whether the system function could be a generalized linear-phase system implemented by a linear constant-coefficient difference equation with real coefficients.

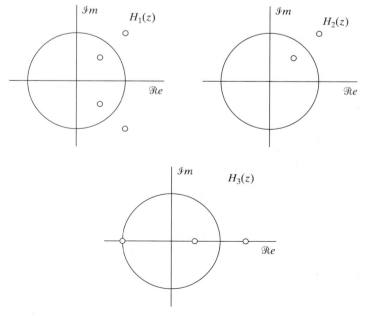

Figure P5.20-1

Basic Problems

5.21. Let $h_{lp}[n]$ denote the impulse response of an ideal lowpass filter with unity passband gain and cutoff frequency $\omega_c = \pi/4$. Figure P5.21-1 shows five systems, each of which is equivalent to an ideal LTI frequency-selective filter. For each system shown, sketch the equivalent frequency response, indicating explicitly the band-edge frequencies in terms of ω_c. In each case, specify whether the system is a lowpass, highpass, bandpass, bandstop, or multiband filter.

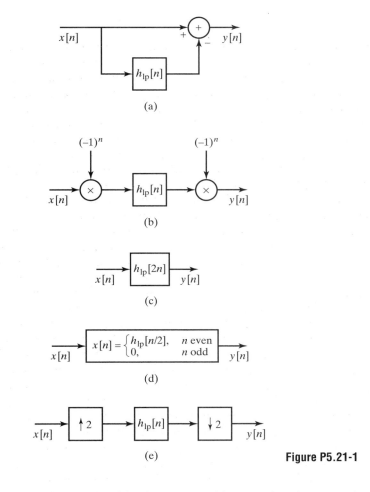

(a)

(b)

(c)

(d)

(e)

Figure P5.21-1

5.22. Consider a causal linear time-invariant system with system function

$$H(z) = \frac{1 - a^{-1}z^{-1}}{1 - az^{-1}},$$

where a is real.

(a) Write the difference equation that relates the input and the output of this system.

(b) For what range of values of a is the system stable?

(c) For $a = \frac{1}{2}$, plot the pole–zero diagram and shade the region of convergence.

(d) Find the impulse response $h[n]$ for the system.
(e) Show that the system is an all-pass system, i.e., that the magnitude of the frequency response is a constant. Also, specify the value of the constant.

5.23. (a) For each of the four types of causal linear phase FIR filters discussed in Section 5.7.3, determine whether the associated symmetry imposes any constraint on the frequency response at $\omega = 0$ and/or $\omega = \pi$.

(b) For each of the following types of desired filter, indicate which of the four FIR filter types would be useful to consider in approximating the desired filter:

> Lowpass
> Bandpass
> Highpass
> Bandstop
> Differentiator

5.24. Let $x[n]$ be a causal, N-point sequence that is zero outside the range $0 \le n \le N-1$. When $x[n]$ is the input to the causal LTI system represented by the difference equation

$$y[n] - \tfrac{1}{4}y[n-2] = x[n-2] - \tfrac{1}{4}x[n],$$

the output is $y[n]$, also a causal, N-point sequence.

(a) Show that the causal LTI system described by this difference equation represents an all-pass filter.
(b) Given that

$$\sum_{n=0}^{N-1} |x[n]|^2 = 5,$$

determine the value of

$$\sum_{n=0}^{N-1} |y[n]|^2.$$

5.25. Is the following statement true or false?

> **Statement:** It is not possible for a noncausal system to have a positive constant group delay; i.e., $\mathrm{grd}\big[H(e^{j\omega})\big] = \tau_0 > 0$.

If the statement is true, give a brief argument justifying it. If the statement is false, provide a counterexample.

5.26. Consider the z-transform

$$H(z) = \frac{rz^{-1}}{1 - (2r\cos\omega_0)z^{-1} + r^2 z^{-2}}, \qquad |z| > r.$$

Assume first that $\omega_0 \ne 0$.
(a) Draw a labeled pole–zero diagram and determine $h[n]$.
(b) Repeat Part (a) when $\omega_0 = 0$. This is known as a critically damped system.

5.27. An LTI system with impulse response $h_1[n]$ is an ideal lowpass filter with cutoff frequency $\omega_c = \pi/2$. The frequency response of the system is $H_1(e^{j\omega})$. Suppose a new LTI system with impulse response $h_2[n]$ is obtained from $h_1[n]$ by

$$h_2[n] = (-1)^n h_1[n].$$

Sketch the frequency response $H_2(e^{j\omega})$.

Advanced Problems

5.28. The system function $H(z)$ of a causal linear time-invariant system has the pole-zero configuration shown in Figure P5.28-1. It is also known that $H(z) = 6$ when $z = 1$.

Figure P5.28-1

(a) Determine $H(z)$.
(b) Determine the impulse response $h[n]$ of the system.
(c) Determine the response of the system to the following input signals:
   (i) $x[n] = u[n] - \frac{1}{2}u[n-1]$
   (ii) The sequence $x[n]$ obtained from sampling the continuous-time signal

$$x(t) = 50 + 10\cos 20\pi t + 30\cos 40\pi t$$

   at a sampling frequency $\Omega_s = 2\pi(40)$ rad/s

5.29. The system function of a linear time-invariant system is given by

$$H(z) = \frac{21}{\left(1 - \frac{1}{2}z^{-1}\right)(1 - 2z^{-1})(1 - 4z^{-1})}.$$

It is known that the system is not stable and that the impulse response is two sided.
(a) Determine the impulse response $h[n]$ of the system.
(b) The impulse response found in Part (a) can be expressed as the sum of a causal impulse response $h_1[n]$ and an anticausal impulse response $h_2[n]$. Determine the corresponding system functions $H_1(z)$ and $H_2(z)$.

5.30. A signal $x[n]$ is processed by a linear time-invariant system $H(z)$ and then downsampled by a factor of 2 to yield $y[n]$, as shown in Figure P5.30-1. The pole–zero plot for $H(z)$ is shown in Figure P5.30-2.
(a) Determine and sketch $h[n]$, the impulse response of the system $H(z)$.

(b) A second system is shown in Figure P5.30-3, in which the signal $x[n]$ is first time compressed by a factor of 2 and then passed through an LTI system $G(z)$ to obtain $r[n]$.

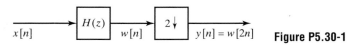

$x[n] \quad\boxed{H(z)}\quad w[n] \quad\boxed{2\downarrow}\quad y[n] = w[2n]$ **Figure P5.30-1**

Unit circle

$\mathcal{I}m$

Mth-order pole at $z = 0$, $M \geq 2$ and M is an integer

z-plane

$-\dfrac{1}{2} \qquad \dfrac{1}{2} \qquad \mathcal{R}e$

Figure P5.30-2

$x[n] \quad\boxed{2\downarrow}\quad\quad\boxed{G(z)}\quad r[n]$ **Figure P5.30-3**

Determine whether $G(z)$ can be chosen so that $y[n] = r[n]$ for any input $x[n]$. If your answer is no, clearly explain. If your answer is yes, specify $G(z)$. If your answer depends on the value of M, clearly explain how. (M is constrained to be an integer greater than or equal to 2.)

5.31. Consider a linear time-invariant system whose system function is

$$H(z) = \frac{z^{-2}}{\left(1 - \frac{1}{2}z^{-1}\right)\left(1 - 3z^{-1}\right)}.$$

(a) Suppose the system is known to be stable. Determine the output $y[n]$ when the input $x[n]$ is the unit step sequence.

(b) Suppose the region of convergence of $H(z)$ includes $z = \infty$. Determine $y[n]$ evaluated at $n = 2$ when $x[n]$ is as shown in Figure P5.31-1.

$x[n]$

$$\cdots \qquad 0 \quad -1 \qquad \cdots \qquad n$$

Figure P5.31-1

(c) Suppose we wish to recover $x[n]$ from $y[n]$ by processing $y[n]$ with an LTI system whose impulse response is given by $h_i[n]$. Determine $h_i[n]$. Does $h_i[n]$ depend on the region of convergence of $H(z)$?

5.32. The Fourier transform of a stable linear time-invariant system is purely real and is shown in Figure P5.32-1. Determine whether this system has a stable inverse system.

Figure P5.32-1

5.33. A sequence $x[n]$ is the output of a linear time-invariant system whose input is $s[n]$. This system is described by the difference equation

$$x[n] = s[n] - e^{-8\alpha}s[n-8],\qquad\qquad (\text{P5.33-1})$$

where $0 < \alpha$.
(a) Find the system function

$$H_1(z) = \frac{X(z)}{S(z)},$$

and plot its poles and zeros in the z-plane. Indicate the region of convergence.
(b) We wish to recover $s[n]$ from $x[n]$ with a linear time-invariant system. Find the system function

$$H_2(z) = \frac{Y(z)}{X(z)}$$

such that $y[n] = s[n]$. Find all possible regions of convergence for $H_2(z)$, and for each, tell whether or not the system is causal and/or stable.
(c) Find all possible choices for the impulse response $h_2[n]$ such that

$$y[n] = h_2[n] * x[n] = s[n].\qquad\qquad (\text{P5.33-2})$$

(d) For all choices determined in Part (c), demonstrate, by explicitly evaluating the convolution in Eq. P5.33-2, that when $s[n] = \delta[n]$, $y[n] = \delta[n]$.
Note: As discussed in Problem 4.7, Eq. P5.33-1 represents a simple model for a multipath channel. The systems determined in Parts (b) and (c), then, correspond to compensation systems to correct for the multipath distortion.

5.34. Consider a linear time-invariant system whose impulse response is

$$h[n] = \left(\tfrac{1}{2}\right)^n u[n] + \left(\tfrac{1}{3}\right)^n u[n].$$

The input $x[n]$ is zero for $n < 0$, but in general, may be nonzero for $0 \le n \le \infty$. We would like to compute the output $y[n]$ for $0 \le n \le 10^9$, and in particular, we want to compare the use of an FIR filter with that of an IIR filter for obtaining $y[n]$ over this interval.
(a) Determine the linear constant-coefficient difference equation for the IIR system relating $x[n]$ and $y[n]$.
(b) Determine the impulse response $h_1[n]$ of the minimum-length LTI FIR filter whose output $y_1[n]$ is identical to the output $y[n]$ for $0 \le n \le 10^9$.
(c) Specify the linear constant-coefficient difference equation associated with the FIR filter in Part (b).

(d) Compare the number of arithmetic operations (multiplications and additions) required to obtain $y[n]$ for $0 \le n \le 10^9$ using the linear constant-coefficient difference equations in Part (a) and in Part (c).

5.35. Consider a causal linear time-invariant system with system function $H(z)$ and real impulse response. $H(z)$ evaluated for $z = e^{j\omega}$ is shown in Figure P5.35-1.

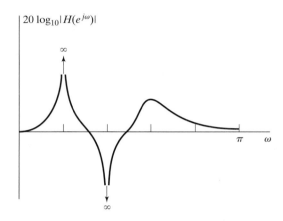

Figure P5.35-1

(a) Carefully sketch a pole–zero plot for $H(z)$ showing all information about the pole and zero locations that can be inferred from the figure.
(b) What can be said about the length of the impulse response?
(c) Specify whether $\sphericalangle H(e^{j\omega})$ is linear.
(d) Specify whether the system is stable.

5.36. A causal linear time-invariant system has the system function

$$H(z) = \frac{(1 - 1.5z^{-1} - z^{-2})(1 + 0.9z^{-1})}{(1 - z^{-1})(1 + 0.7jz^{-1})(1 - 0.7jz^{-1})}.$$

(a) Write the difference equation that is satisfied by the input and the output of the system.
(b) Plot the pole–zero diagram and indicate the region of convergence for the system function.
(c) Sketch $|H(e^{j\omega})|$.
(d) State whether the following are true or false about the system:
   (i) The system is stable.
   (ii) The impulse response approaches a constant for large n.
   (iii) The magnitude of the frequency response has a peak at approximately $\omega = \pm\pi/4$.
   (iv) The system has a stable and causal inverse.

5.37. Consider a causal sequence $x[n]$ with the z-transform

$$X(z) = \frac{\left(1 - \frac{1}{2}z^{-1}\right)\left(1 - \frac{1}{4}z^{-1}\right)\left(1 - \frac{1}{5}z\right)}{\left(1 - \frac{1}{6}z\right)}.$$

For what values of α is $\alpha^n x[n]$ a real, minimum-phase sequence?

5.38. Consider the linear time-invariant system whose system function is

$$H(z) = (1 - 0.9e^{j0.6\pi}z^{-1})(1 - 0.9e^{-j0.6\pi}z^{-1})(1 - 1.25e^{j0.8\pi}z^{-1})(1 - 1.25e^{-j0.8\pi}z^{-1}).$$

(a) Find all causal system functions that result in the same frequency-response magnitude as $H(z)$ and for which the impulse responses are real valued and of the same length as the impulse response associated with $H(z)$. (There are four different such system functions.) Identify which system function is minimum phase and which, to within a time shift, is maximum phase.

(b) Find the impulse responses for the system functions in Part (a).

(c) For each of the sequences in Part (b), compute and plot the quantity

$$E[n] = \sum_{m=0}^{n}(h[m])^2$$

for $0 \leq n \leq 5$. Indicate explicitly which plot corresponds to the minimum-phase system.

5.39. Shown in Figure P5.39-1 are eight different finite-duration sequences. Each sequence is four points long. The magnitude of the Fourier transform is the same for all sequences. Which of the sequences has all the zeros of its z-transform *inside* the unit circle?

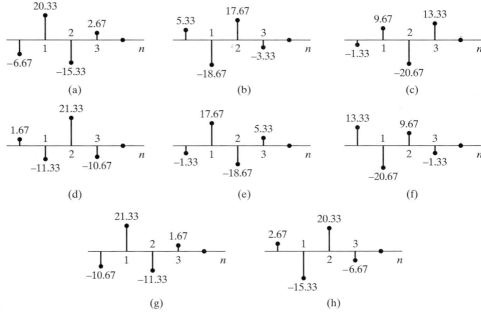

Figure P5.39-1

5.40. Each of the pole–zero plots in Figure P5.40-1, together with the specification of the region of convergence, describes a linear time-invariant system with system function $H(z)$. In each case, determine whether any of the following statements are true. Justify your answer with a brief statement or a counterexample.

(a) The system is a zero-phase or a generalized linear-phase system.

(b) The system has a stable inverse $H_i(z)$.

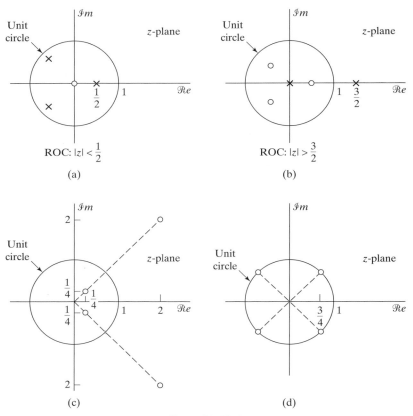

ROC: $|z| < \frac{1}{2}$

(a)

ROC: $|z| > \frac{3}{2}$

(b)

(c)

(d)

Figure P5.40-1

5.41. Figure P5.41-1 shows two different interconnections of three systems. The impulse responses $h_1[n]$, $h_2[n]$, and $h_3[n]$ are as shown in Figure P5.41-2. Determine whether system A and/or system B is a generalized linear-phase system.

Figure P5.41-1

$h_1[n]$ $h_2[n]$ $h_3[n]$

Figure P5.41-2

5.42. The overall system of Figure P5.42-1 is a discrete-time linear time-invariant system with frequency response $H(e^{j\omega})$ and impulse response $h[n]$.

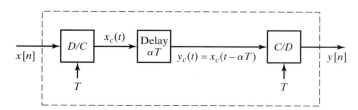

Figure P5.42-1

(a) $H(e^{j\omega})$ can be expressed in the form

$$H(e^{j\omega}) = A(e^{j\omega})e^{j\phi(\omega)},$$

with $A(e^{j\omega})$ real. Determine and sketch $A(e^{j\omega})$ and $\phi(\omega)$ for $|\omega| < \pi$.

(b) Sketch $h[n]$ for the following:
    (i) $\alpha = 3$
    (ii) $\alpha = 3\frac{1}{2}$
    (iii) $\alpha = 3\frac{1}{4}$

(c) Consider a discrete-time linear time-invariant system for which

$$H(e^{j\omega}) = A(e^{j\omega})e^{j\alpha\omega}, \qquad |\omega| < \pi,$$

with $A(e^{j\omega})$ real. What can be said about the symmetry of $h[n]$ for the following?
    (i) $\alpha = $ integer
    (ii) $\alpha = M/2$, where M is an odd integer
    (iii) General α

5.43. Consider the class of FIR filters that have $h[n]$ real, $h[n] = 0$ for $n < 0$ and $n > M$, and one of the following symmetry properties:

$$\text{Symmetric: } h[n] = h[M - n]$$

$$\text{Antisymmetric: } h[n] = -h[M - n]$$

All filters in this class have generalized linear phase, i.e., have frequency response of the form

$$H(e^{j\omega}) = A(e^{j\omega})e^{-j\alpha\omega + j\beta},$$

where $A(e^{j\omega})$ is a real function of ω, α is a real constant, and β is a real constant.

For the following table, show that $A(e^{j\omega})$ has the indicated form, and find the values of α and β.

Type	Symmetry	Filter length $(M+1)$	Form of $A(e^{j\omega})$	α	β
I	Symmetric	Odd	$\displaystyle\sum_{n=0}^{M/2} a[n]\cos\omega n$		
II	Symmetric	Even	$\displaystyle\sum_{n=1}^{(M+1)/2} b[n]\cos\omega(n-1/2)$		
III	Antisymmetric	Odd	$\displaystyle\sum_{n=1}^{M/2} c[n]\sin\omega n$		
IV	Antisymmetric	Even	$\displaystyle\sum_{n=1}^{(M+1)/2} d[n]\sin\omega(n-1/2)$		

Here are several helpful suggestions.

- For type I filters, first show that $H(e^{j\omega})$ can be written in the form

$$H(e^{j\omega}) = \sum_{n=0}^{(M-2)/2} h[n]e^{-j\omega n} + \sum_{n=0}^{(M-2)/2} h[M-n]e^{-j\omega[M-n]} + h[M/2]e^{-j\omega(M/2)}.$$

- The analysis for type III filters is very similar to that for type I, with the exception of a sign change and removal of one of the preceding terms.
- For type II filters, first write $H(e^{j\omega})$ in the form

$$H(e^{j\omega}) = \sum_{n=0}^{(M-1)/2} h[n]e^{-j\omega n} + \sum_{n=0}^{(M-1)/2} h[M-n]e^{-j\omega[M-n]},$$

and then pull out a common factor of $e^{-j\omega(M/2)}$ from both sums.
- The analysis for type IV filters is very similar to that for type II filters.

5.44. Let $h_{\text{lp}}[n]$ denote the impulse response of an FIR generalized linear-phase lowpass filter. The impulse response $h_{\text{hp}}[n]$ of an FIR generalized linear-phase highpass filter can be obtained by the transformation

$$h_{\text{hp}}[n] = (-1)^n h_{\text{lp}}[n].$$

If we decide to design a highpass filter using this transformation and we wish the resulting highpass filter to be symmetric, which of the four types of generalized linear-phase FIR filters can we use for the design of the lowpass filter? Your answer should consider *all* the possible types.

5.45. A causal linear time-invariant discrete-time system has system function

$$H(z) = \frac{(1 - 0.5z^{-1})(1 + 4z^{-2})}{(1 - 0.64z^{-2})}.$$

(a) Find expressions for a minimum-phase system $H_1(z)$ and an all-pass system $H_{\text{ap}}(z)$ such that

$$H(z) = H_1(z)H_{\text{ap}}(z).$$

(b) Find expressions for a different minimum-phase system $H_2(z)$ and a generalized linear-phase FIR system $H_{\text{lin}}(z)$ such that

$$H(z) = H_2(z)H_{\text{lin}}(z).$$

5.46. (a) A minimum-phase system has system function $H_{\text{min}}(z)$ is such that

$$H_{\text{min}}(z)H_{\text{ap}}(z) = H_{\text{lin}}(z),$$

where $H_{\text{ap}}(z)$ is an all-pass system function and $H_{\text{lin}}(z)$ is a causal generalized linear-phase system. What does this information tell you about the poles and zeros of $H_{\text{min}}(z)$?

(b) A generalized linear-phase FIR system has an impulse response with real values and $h[n] = 0$ for $n < 0$ and for $n \geq 8$, and $h[n] = -h[7 - n]$. The system function of this system has a zero at $z = 0.8e^{j\pi/4}$ and another zero at $z = -2$. What is $H(z)$?

5.47. Consider an LTI system with input $x[n]$ and output $y[n]$. When the input to the system is

$$x[n] = 5\frac{\sin(0.4\pi n)}{\pi n} + 10\cos(0.5\pi n),$$

the corresponding output is

$$y[n] = 10\frac{\sin[0.3\pi(n - 10)]}{\pi(n - 10)}.$$

Determine the frequency response $H(e^{j\omega})$ and the impulse response $h[n]$ for the LTI system.

5.48. Figure P5.48-1 shows the pole–zero plots for three different causal LTI systems with real impulse responses. Indicate which of the following properties apply to each of the systems pictured: stable, IIR, FIR, minimum phase, all-pass, generalized linear phase, positive group delay at all ω.

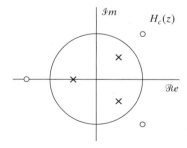

Figure P5.48-1

5.49. Let S_1 be an LTI system with system function:

$$H_1(z) = \frac{1 - z^{-5}}{1 - z^{-1}}, \qquad |z| > 0,$$

and impulse response $h_1[n]$.

(a) Is S_1 causal? Explain.

(b) Let $g[n] = h_1[n] * h_2[n]$. Specify an $h_2[n]$ such that $g[n]$ has at least nine nonzero samples and $g[n]$ can be considered the impulse response of a causal LTI system with strictly linear phase; i.e., $G(e^{j\omega}) = |G(e^{j\omega})|e^{-j\omega n_0}$ for some integer n_0.

(c) Let $q[n] = h_1[n] * h_3[n]$. Specify an $h_3[n]$ such that

$$q[n] = \delta[n] \qquad \text{for } 0 \le n \le 19.$$

5.50. The LTI systems $H_1(e^{j\omega})$ and $H_2(e^{j\omega})$ are generalized linear-phase systems. Which, if any, of the following systems also must be generalized linear-phase systems?

(a)

$$G_1(e^{j\omega}) = H_1(e^{j\omega}) + H_2(e^{j\omega})$$

(b)

$$G_2(e^{j\omega}) = H_1(e^{j\omega})H_2(e^{j\omega})$$

(c)

$$G_3(e^{j\omega}) = \frac{1}{2\pi} \int_{-\pi}^{\pi} H_1(e^{j\theta})H_2(e^{j(\omega-\theta)})d\theta$$

5.51. This problem concerns a discrete-time filter with a real-valued impulse response $h[n]$. Determine whether the following statement is true or false:

> **Statement:** If the group delay of the filter is a constant for $0 < \omega < \pi$, then the impulse response must have the property that either
>
> $$h[n] = h[M - n]$$
>
> or
>
> $$h[n] = -h[M - n],$$
>
> where M is an integer.

If you believe that the statement is true, clearly show your reasoning. If you believe that it is false, provide a counterexample.

5.52. The system function $H_{II}(z)$ represents a type II FIR generalized linear-phase system with impulse response $h_{II}[n]$. This system is cascaded with an LTI system whose system function is $(1 - z^{-1})$ to produce a third system with system function $H(z)$ and impulse response $h[n]$. Prove that the overall system is a generalized linear-phase system, and determine what type of linear phase system it is.

5.53. In this problem, you will consider three different LTI systems. All three are causal and have real impulse responses. You will be given additional information about each system. Using this information, state as much as possible about the poles and zeros of each system function and about the length of the impulse response of the system.

(a) $H_1(z)$ has a pole at $z = 0.9e^{j\pi/3}$, and when $x[n] = u[n]$, $\lim_{n\to\infty} y[n] = 0$.

(b) $H_2(z)$ has a zero at $z = 0.8e^{j\pi/4}$, $H_2(e^{j\omega})$ has linear phase with $\sphericalangle H_2(e^{j\omega}) = -2.5\omega$, and $20\log_{10}|H_2(e^{j0})| = -\infty$.

(c) $H_3(z)$ has a pole at $z = 0.8e^{j\pi/4}$ and $|H_3(e^{j\omega})| = 1$ for all ω.

5.54. The following three things are known about a signal $x[n]$ with z-transform $X(z)$:

   (i) $x[n]$ is real valued and minimum phase,

   (ii) $x[n]$ is zero outside the interval $0 \le n \le 4$,

   (iii) $X(z)$ has a zero at $z = \frac{1}{2}e^{j\pi/4}$ and a zero at $z = \frac{1}{2}e^{j3\pi/4}$.

Based on this information, answer the following questions:

(a) Is $X(z)$ rational? Justify your answer.

(b) Sketch the complete pole–zero plot for $X(z)$ and specify its ROC.

(c) If $y[n]*x[n] = \delta[n]$ and $y[n]$ is rightsided, sketch the pole–zero plot for $Y(z)$ and specify its ROC.

5.55. Consider a real sequence $x[n]$ and its DTFT $X(e^{j\omega})$. Given the following information, determine and plot the sequence $x[n]$:

1. $x[n]$ is a finite-length sequence.
2. At $z = 0$, $X(z)$ has exactly five poles and no zeros. $X(z)$ *may* have poles or zeros at other locations.
3. The unwrapped phase function is

$$\arg\left[X(e^{j\omega})\right] = \begin{cases} -\alpha\omega + \frac{\pi}{2}, & 0 < \omega < \pi, \\ -\alpha\omega - \frac{\pi}{2}, & -\pi < \omega < 0, \end{cases}$$

for some real constant α.

4. The group delay of the sequence evaluated at $\omega = \frac{\pi}{2}$ is 2; i.e.,

$$\text{grd}\left[X(e^{j\omega})\right]_{\omega=\pi/2} = 2.$$

5.

$$\frac{1}{2\pi} \int_{-\pi}^{\pi} |X(e^{j\omega})|^2 d\omega = 28.$$

6. If $y[n] = x[n] * u[n]$, then

$$\frac{1}{2\pi} \int_{-\pi}^{\pi} Y(e^{j\omega}) d\omega = 4,$$

$$\frac{1}{2\pi} \int_{-\pi}^{\pi} Y(e^{j\omega}) e^{j\omega} d\omega = 6.$$

7. $X(e^{j\omega})|_{\omega=\pi} = 0$.
8. The sequence $v[n]$ whose DTFT is $V(e^{j\omega}) = \mathcal{R}e\{X(e^{j\omega})\}$ satisfies $v[5] = -\frac{3}{2}$.

5.56. Let S_1 be a causal and stable LTI system with impulse response $h_1[n]$ and frequency response $H_1(e^{j\omega})$. The input $x[n]$ and output $y[n]$ for S_1 are related by the difference equation

$$y[n] - y[n-1] + \tfrac{1}{4}y[n-2] = x[n].$$

(a) If an LTI system S_2 has a frequency response given by $H_2(e^{j\omega}) = H_1(-e^{j\omega})$, would you characterize S_2 as being a lowpass filter, a bandpass filter, or a highpass filter? Justify your answer.

(b) Let S_3 be a causal LTI system whose frequency response $H_3(e^{j\omega})$ has the property that

$$H_3(e^{j\omega}) H_1(e^{j\omega}) = 1.$$

Is S_3 a minimum-phase filter? Could S_3 be classified as one of the four types of FIR filters with generalized linear phase? Justify your answers.

(c) Let S_4 be a stable and *noncausal* LTI system whose frequency response is $H_4(e^{j\omega})$ and whose input $x[n]$ and output $y[n]$ are related by the difference equation:

$$y[n] + \alpha_1 y[n-1] + \alpha_2 y[n-2] = \beta_0 x[n],$$

where α_1, α_2, and β_0 are all real and nonzero constants. Specify a value for α_1, a value for α_2, and a value for β_0 such that $|H_4(e^{j\omega})| = |H_1(e^{j\omega})|$.

(d) Let S_5 be an FIR filter whose impulse response is $h_5[n]$ and whose frequency response, $H_5(e^{j\omega})$, has the property that $H_5(e^{j\omega}) = |A(e^{j\omega})|^2$ for some DTFT $A(e^{j\omega})$ (i.e., S_5 is a *zero-phase* filter). Find $h_5[n]$ such that $h_5[n] * h_1[n]$ is the impulse response of a noncausal FIR filter.

Extension Problems

5.57. In the system shown in Figure P5.57-1, assume that the input can be expressed in the form
$$x[n] = s[n]\cos(\omega_0 n).$$

Assume also that $s[n]$ is lowpass and relatively narrowband; i.e., $S(e^{j\omega}) = 0$ for $|\omega| > \Delta$, with Δ very small and $\Delta \ll \omega_0$, so that $X(e^{j\omega})$ is narrowband around $\omega = \pm\omega_0$.

$x[n]$ $y[n]$ **Figure P5.57-1**

(a) If $|H(e^{j\omega})| = 1$ and $\sphericalangle H(e^{j\omega})$ is as illustrated in Figure P5.57-2, show that $y[n] = s[n]\cos(\omega_0 n - \phi_0)$.

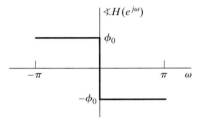

Figure P5.57-2

(b) If $|H(e^{j\omega})| = 1$ and $\sphericalangle H(e^{j\omega})$ is as illustrated in Figure P5.57-3, show that $y[n]$ can be expressed in the form
$$y[n] = s[n - n_d]\cos(\omega_0 n - \phi_0 - \omega_0 n_d).$$

Show also that $y[n]$ can be equivalently expressed as
$$y[n] = s[n - n_d]\cos(\omega_0 n - \phi_1),$$

where $-\phi_1$ is the phase of $H(e^{j\omega})$ at $\omega = \omega_0$.

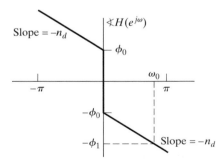

Figure P5.57-3

(c) The group delay associated with $H(e^{j\omega})$ is defined as
$$\tau_{\text{gr}}(\omega) = -\frac{d}{d\omega}\arg[H(e^{j\omega})],$$

and the phase delay is defined as $\tau_{\text{ph}}(\omega) = -(1/\omega)\sphericalangle H(e^{j\omega})$. Assume that $|H(e^{j\omega})|$ is unity over the bandwidth of $x[n]$. Based on your results in Parts (a) and (b) and on the assumption that $x[n]$ is narrowband, show that if $\tau_{\text{gr}}(\omega_0)$ and $\tau_{\text{ph}}(\omega_0)$ are both integers, then
$$y[n] = s[n - \tau_{\text{gr}}(\omega_0)]\cos\{\omega_0[n - \tau_{\text{ph}}(\omega_0)]\}.$$

This equation shows that, for a narrowband signal $x[n]$, $\sphericalangle H(e^{j\omega})$ effectively applies a delay of $\tau_{\text{gr}}(\omega_0)$ to the envelope $s[n]$ of $x[n]$ and a delay of $\tau_{\text{ph}}(\omega_0)$ to the carrier $\cos \omega_0 n$.

(d) Referring to the discussion in Section 4.5 associated with noninteger delays of a sequence, how would you interpret the effect of group delay and phase delay if $\tau_{\text{gr}}(\omega_0)$ or $\tau_{\text{ph}}(\omega_0)$ (or both) is not an integer?

5.58. The signal $y[n]$ is the output of a linear time-invariant system with input $x[n]$, which is zero-mean white noise. The system is described by the difference equation

$$y[n] = \sum_{k=1}^{N} a_k y[n-k] + \sum_{k=0}^{M} b_k x[n-k], \qquad b_0 = 1.$$

(a) What is the z-transform $\Phi_{yy}(z)$ of the autocorrelation function $\phi_{yy}[n]$?

Sometimes it is of interest to process $y[n]$ with a linear filter such that the power spectrum of the linear filter's output will be flat when the input to the linear filter is $y[n]$. This procedure is known as "whitening" $y[n]$, and the linear filter that accomplishes the task is said to be the "whitening filter" for the signal $y[n]$. Suppose that we know the autocorrelation function $\phi_{yy}[n]$ and its z-transform $\Phi_{yy}(z)$, but not the a_k's and the b_k's.

(b) Discuss a procedure for finding a system function $H_w(z)$ of the whitening filter.

(c) Is the whitening filter unique?

5.59. In many practical situations, we are faced with the problem of recovering a signal that has been "blurred" by a convolution process. We can model this blurring process as a linear filtering operation, as depicted in Figure P5.59-1, where the blurring impulse response is as shown in Figure P5.59-2. This problem will consider ways to recover $x[n]$ from $y[n]$.

$\begin{array}{ccc} x[n] & & y[n] \\ \text{Desired signal} & & \text{Blurred signal} \end{array}$ **Figure P5.59-1**

$$h[n] = \begin{cases} 1, & 0 \le n \le M-1 \\ 0, & \text{otherwise} \end{cases}$$

Figure P5.59-2

(a) One approach to recovering $x[n]$ from $y[n]$ is to use an inverse filter; i.e., $y[n]$ is filtered by a system whose frequency response is

$$H_i(e^{j\omega}) = \frac{1}{H(e^{j\omega})},$$

where $H(e^{j\omega})$ is the Fourier transform of $h[n]$. For the impulse response $h[n]$ shown in Figure P5.59-2, discuss the practical problems involved in implementing the inverse filtering approach. Be complete, but also be brief and to the point.

(b) Because of the difficulties involved in inverse filtering, the following approach is suggested for recovering $x[n]$ from $y[n]$: The blurred signal $y[n]$ is processed by the system shown in Figure P5.59-3, which produces an output $w[n]$ from which we can extract an improved replica of $x[n]$. The impulse responses $h_1[n]$ and $h_2[n]$ are shown in Figure P5.59-4. Explain in detail the working of this system. In particular, state precisely the conditions under which we can recover $x[n]$ exactly from $w[n]$. *Hint:* Consider the impulse response of the overall system from $x[n]$ to $w[n]$.

Figure P5.59-3

Figure P5.59-4

(c) Let us now attempt to generalize this approach to arbitrary finite-length blurring impulse responses $h[n]$; i.e., assume only that $h[n] = 0$ for $n < 0$ or $n \geq M$. Further, assume that $h_1[n]$ is the same as in Figure P5.59-4. How must $H_2(z)$ and $H(z)$ be related for the system to work as in Part (b)? What condition must $H(z)$ satisfy in order that it be possible to implement $H_2(z)$ as a causal system?

5.60. In this problem, we demonstrate that, for a rational z-transform, a factor of the form $(z - z_0)$ and a factor of the form $z/(z - z_0^*)$ contribute the same phase.

(a) Let $H(z) = z - 1/a$, where a is real and $0 < a < 1$. Sketch the poles and zeros of the system, including an indication of those at $z = \infty$. Determine $\sphericalangle H(e^{j\omega})$, the phase of the system.

(b) Let $G(z)$ be specified such that it has poles at the conjugate-reciprocal locations of zeros of $H(z)$ and zeros at the conjugate-reciprocal locations of poles of $H(z)$, including those at zero and ∞. Sketch the pole–zero diagram of $G(z)$. Determine $\sphericalangle G(e^{j\omega})$, the phase of the system, and show that it is identical to $\sphericalangle H(e^{j\omega})$.

5.61. Prove the validity of the following two statements:

(a) The convolution of two minimum-phase sequences is also a minimum-phase sequence.

(b) The sum of two minimum-phase sequences is not necessarily a minimum-phase sequence. Specifically, give an example of both a minimum-phase and a nonminimum-phase sequence that can be formed as the sum of two minimum-phase sequences.

5.62. A sequence is defined by the relationship

$$r[n] = \sum_{m=-\infty}^{\infty} h[m]h[n+m] = h[n] * h[-n],$$

where $h[n]$ is a minimum-phase sequence and

$$r[n] = \tfrac{4}{3}\left(\tfrac{1}{2}\right)^n u[n] + \tfrac{4}{3}2^n u[-n-1].$$

(a) Find $R(z)$ and sketch the pole–zero diagram.

(b) Determine the minimum-phase sequence $h[n]$ to within a scale factor of ± 1. Also, determine the z-transform $H(z)$ of $h[n]$.

5.63. A *maximum-phase* sequence is a stable sequence whose z-transform has all its poles and zeros *outside* the unit circle.

(a) Show that maximum-phase sequences are anticausal, i.e., that they are zero for $n > 0$.

  FIR maximum-phase sequences can be made causal by including a finite amount of delay. A finite-duration causal maximum-phase sequence having a Fourier transform of a given magnitude can be obtained by reflecting all the zeros of the z-transform of a minimum-phase sequence to conjugate-reciprocal positions outside the unit circle. That is, we can express the z-transform of a maximum-phase causal finite-duration sequence as

$$H_{\max}(z) = H_{\min}(z) H_{\mathrm{ap}}(z).$$

Obviously, this process ensures that $|H_{\max}(e^{j\omega})| = |H_{\min}(e^{j\omega})|$. Now, the z-transform of a finite-duration minimum-phase sequence can be expressed as

$$H_{\min}(z) = h_{\min}[0] \prod_{k=1}^{M}(1 - c_k z^{-1}), \qquad |c_k| < 1.$$

(b) Obtain an expression for the all-pass system function required to reflect all the zeros of $H_{\min}(z)$ to positions outside the unit circle.

(c) Show that $H_{\max}(z)$ can be expressed as

$$H_{\max}(z) = z^{-M} H_{\min}(z^{-1}).$$

(d) Using the result of Part (c), express the maximum-phase sequence $h_{\max}[n]$ in terms of $h_{\min}[n]$.

5.64. It is not possible to obtain a causal and stable inverse system (a perfect compensator) for a nonminimum-phase system. In this problem, we study an approach to compensating for only the magnitude of the frequency response of a nonminimum-phase system.

  Suppose that a stable nonminimum-phase linear time-invariant discrete-time system with a rational system function $H(z)$ is cascaded with a compensating system $H_c(z)$ as shown in Figure P5.64-1.

Figure P5.64-1

(a) How should $H_c(z)$ be chosen so that it is stable and causal and so that the magnitude of the overall effective frequency response is unity? (Recall that $H(z)$ can always be represented as $H(z) = H_{\mathrm{ap}}(z) H_{\min}(z)$.)

(b) What are the corresponding system functions $H_c(z)$ and $G(z)$?

(c) Assume that

$$H(z) = (1 - 0.8e^{j0.3\pi}z^{-1})(1 - 0.8e^{-j0.3\pi}z^{-1})(1 - 1.2e^{j0.7\pi}z^{-1})(1 - 1.2e^{-j0.7\pi}z^{-1}).$$

Find $H_{\min}(z)$, $H_{\mathrm{ap}}(z)$, $H_c(z)$, and $G(z)$ for this case, and construct the pole–zero plots for each system function.

5.65. Let $h_{\min}[n]$ denote a minimum-phase sequence with z-transform $H_{\min}(z)$. If $h[n]$ is a causal nonminimum-phase sequence whose Fourier transform magnitude is equal to $|H_{\min}(e^{j\omega})|$, show that

$$|h[0]| < |h_{\min}[0]|.$$

(Use the initial-value theorem together with Eq. (5.103).)

5.66. One of the interesting and important properties of minimum-phase sequences is the minimum-energy delay property; i.e., of all the causal sequences having the same Fourier transform magnitude function $|H(e^{j\omega})|$, the quantity

$$E[n] = \sum_{m=0}^{n} |h[m]|^2$$

is maximum for all $n \geq 0$ when $h[n]$ is the minimum-phase sequence. This result is proved as follows: Let $h_{\min}[n]$ be a minimum-phase sequence with z-transform $H_{\min}(z)$. Furthermore, let z_k be a zero of $H_{\min}(z)$ so that we can express $H_{\min}(z)$ as

$$H_{\min}(z) = Q(z)(1 - z_k z^{-1}), \qquad |z_k| < 1,$$

where $Q(z)$ is again minimum phase. Now consider another sequence $h[n]$ with z-transform $H(z)$ such that

$$|H(e^{j\omega})| = |H_{\min}(e^{j\omega})|$$

and such that $H(z)$ has a zero at $z = 1/z_k^*$ instead of at z_k.
(a) Express $H(z)$ in terms of $Q(z)$.
(b) Express $h[n]$ and $h_{\min}[n]$ in terms of the minimum-phase sequence $q[n]$ that has z-transform $Q(z)$.
(c) To compare the distribution of energy of the two sequences, show that

$$\varepsilon = \sum_{m=0}^{n} |h_{\min}[m]|^2 - \sum_{m=0}^{n} |h[m]|^2 = (1 - |z_k|^2)|q[n]|^2.$$

(d) Using the result of Part (c), argue that

$$\sum_{m=0}^{n} |h[m]|^2 \leq \sum_{m=0}^{n} |h_{\min}[m]|^2 \qquad \text{for all } n.$$

5.67. A causal all-pass system $H_{ap}(z)$ has input $x[n]$ and output $y[n]$.
(a) If $x[n]$ is a real minimum-phase sequence (which also implies that $x[n] = 0$ for $n < 0$), using Eq. (5.118), show that

$$\sum_{k=0}^{n} |x[k]|^2 \geq \sum_{k=0}^{n} |y[k]|^2. \qquad (P5.67\text{-}1)$$

(b) Show that Eq. (P5.67-1) holds even if $x[n]$ is not minimum phase, but is zero for $n < 0$.

5.68. In the design of either continuous-time or discrete-time filters, we often approximate a specified magnitude characteristic without particular regard to the phase. For example, standard design techniques for lowpass and bandpass filters are derived from a consideration of the magnitude characteristics only.

In many filtering problems, we would prefer that the phase characteristics be zero or linear. For causal filters, it is impossible to have zero phase. However, for many filtering applications, it is not necessary that the impulse response of the filter be zero for $n < 0$ if the processing is not to be carried out in real time.

One technique commonly used in discrete-time filtering when the data to be filtered are of finite duration and are stored, for example, in computer memory is to process the data forward and then backward through the same filter.

Let $h[n]$ be the impulse response of a causal filter with an arbitrary phase characteristic. Assume that $h[n]$ is real, and denote its Fourier transform by $H(e^{j\omega})$. Let $x[n]$ be the data that we want to filter.

(a) *Method A:* The filtering operation is performed as shown in Figure P5.68-1.

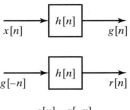

$$s[n] = r[-n]$$ **Figure P5.68-1**

1. Determine the overall impulse response $h_1[n]$ that relates $x[n]$ to $s[n]$, and show that it has a zero-phase characteristic.
2. Determine $|H_1(e^{j\omega})|$, and express it in terms of $|H(e^{j\omega})|$ and $\sphericalangle H(e^{j\omega})$.

(b) *Method B:* As depicted in Figure P5.68b-2, process $x[n]$ through the filter $h[n]$ to get $g[n]$. Also, process $x[n]$ backward through $h[n]$ to get $r[n]$. The output $y[n]$ is then taken as the sum of $g[n]$ and $r[-n]$. This composite set of operations can be represented by a filter with input $x[n]$, output $y[n]$, and impulse response $h_2[n]$.

$$y[n] = g[n] + r[-n]$$ **Figure P5.68-2**

1. Show that the composite filter $h_2[n]$ has a zero-phase characteristic.
2. Determine $|H_2(e^{j\omega})|$, and express it in terms of $|H(e^{j\omega})|$ and $\sphericalangle H(e^{j\omega})$.

(c) Suppose that we are given a sequence of finite duration on which we would like to perform a bandpass zero-phase filtering operation. Furthermore, assume that we are given the bandpass filter $h[n]$, with frequency response as specified in Figure P5.68-3, which has the magnitude characteristic that we desire, but has linear phase. To achieve zero phase, we could use either method A or B. Determine and sketch $|H_1(e^{j\omega})|$ and $|H_2(e^{j\omega})|$. From these results, which method would you use to achieve the desired bandpass filtering operation? Explain why. More generally, if $h[n]$ has the desired magnitude, but a nonlinear phase characteristic, which method is preferable to achieve a zero-phase characteristic?

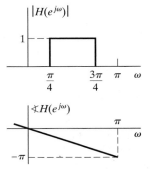

Figure P5.68-3

5.69. Determine whether the following statement is true or false. If it is true, concisely state your reasoning. If it is false, give a counterexample.

Statement: If the system function $H(z)$ has poles anywhere other than at the origin or infinity, then the system cannot be a zero-phase or a generalized linear-phase system.

5.70. Figure P5.70-1 shows the zeros of the system function $H(z)$ for a real causal linear-phase FIR filter. All of the indicated zeros represent factors of the form $(1 - az^{-1})$. The corresponding poles at $z = 0$ for these factors are not shown in the figure. The filter has approximately unity gain in its passband.

    (a) One of the zeros has magnitude 0.5 and angle 153 degrees. Determine the exact location of as many other zeros as you can from this information.

    (b) The system function $H(z)$ is used in the system for discrete-time processing of continuous-time signals shown in Figure 4.11, with the sampling period $T = 0.5$ msec. Assume that the continuous-time input $X_c(j\Omega)$ is bandlimited and that the sampling rate is high enough to avoid aliasing. What is the time delay (in msec) through the entire system, assuming that both C/D and D/C conversion require negligible amounts of time?

    (c) For the system in Part (b), sketch the overall effective continuous-time frequency response $20 \log_{10} |H_{\text{eff}}(j\Omega)|$ for $0 \leq \Omega \leq \pi/T$ as accurately as possible using the given information. Estimate the frequencies at which $H_{\text{eff}}(j\Omega) = 0$, and mark them on your plot.

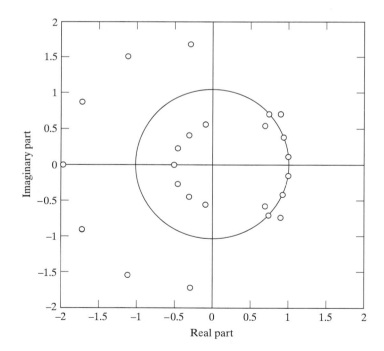

Figure P5.70-1

5.71. A signal $x[n]$ is processed through an LTI system $H(z)$ and then downsampled by a factor of 2 to yield $y[n]$ as indicated in Figure P5.71-1. Also, as shown in the same figure, $x[n]$ is first downsampled and then processed through an LTI system $G(z)$ to obtain $r[n]$.

(a) Specify a choice for $H(z)$ (other than a constant) and $G(z)$ so that $r[n] = y[n]$ for an arbitrary $x[n]$.

(b) Specify a choice for $H(z)$ so that there is no choice for $G(z)$ that will result in $r[n] = y[n]$ for an arbitrary $x[n]$.

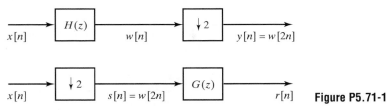

Figure P5.71-1

(c) Determine as general a set of conditions as you can on $H(z)$ such that $G(z)$ can be chosen so that $r[n] = y[n]$ for an arbitrary $x[n]$. The conditions should not depend on $x[n]$. If you first develop the conditions in terms of $h[n]$, restate them in terms of $H(z)$.

(d) For the conditions determined in Part (c), what is $g[n]$ in terms of $h[n]$ so that $r[n] = y[n]$.

5.72. Consider a discrete-time LTI system with a real-valued impulse response $h[n]$. We want to find $h[n]$, or equivalently, the system function $H(z)$ from the autocorrelation $c_{hh}[\ell]$ of the impulse response. The definition of the autocorrelation is

$$c_{hh}[\ell] = \sum_{k=-\infty}^{\infty} h[k]h[k + \ell].$$

(a) If the system $h[n]$ is causal and stable, can you uniquely recover $h[n]$ from $c_{hh}[\ell]$? Justify your answer.

(b) Assume that $h[n]$ is causal and stable and that, in addition, you know that the system function has the form

$$H(z) = \frac{1}{1 - \sum_{k=1}^{N} a_k z^{-k}}$$

for some finite a_k. Can you uniquely recover $h[n]$ from $c_{hh}[\ell]$? Clearly justify your answer.

5.73. Let $h[n]$ and $H(z)$ denote the impulse response and system function of a stable all-pass LTI system. Let $h_i[n]$ denote the impulse response of the (stable) LTI inverse system. Assume that $h[n]$ is real. Show that $h_i[n] = h[-n]$.

5.74. Consider a real-valued sequence $x[n]$ for which $X(e^{j\omega}) = 0$ for $\frac{\pi}{4} \le |\omega| \le \pi$. One sequence value of $x[n]$ may have been corrupted, and we would like to recover it approximately or exactly. With $g[n]$ denoting the corrupted signal,

$$g[n] = x[n] \qquad \text{for } n \ne n_0,$$

and $g[n_0]$ is real but not related to $x[n_0]$. In each of the following two cases, specify a practical algorithm for recovering $x[n]$ from $g[n]$ exactly or approximately.

(a) The exact value of n_0 is not known, but we know that n_0 is an odd number.

(b) Nothing about n_0 is known.

5.75. Show that if $h[n]$ is an N-point FIR filter such that $h[n] = h[N - 1 - n]$ and $H(z_0) = 0$, then $H(1/z_0) = 0$. This shows that even symmetric linear-phase FIR filters have zeros that are reciprocal images. (If $h[n]$ is real, the zeros also will be real or will occur in complex conjugates.)

6

STRUCTURES FOR DISCRETE-TIME SYSTEMS

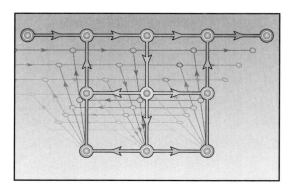

6.0 INTRODUCTION

As we saw in Chapter 5, a linear time-invariant system with a rational system function has the property that the input and output sequences satisfy a linear constant-coefficient difference equation. Since the system function is the z-transform of the impulse response, and since the difference equation satisfied by the input and output can be determined by inspection of the system function, it follows that the difference equation, the impulse response, and the system function are equivalent characterizations of the input–output relation of a linear time-invariant discrete-time system. When such systems are implemented with discrete-time analog or digital hardware, the difference equation or the system function representation must be converted to an algorithm or structure that can be realized in the desired technology. As we will see in this chapter, systems described by linear constant-coefficient difference equations can be represented by structures consisting of an interconnection of the basic operations of addition, multiplication by a constant, and delay, the exact implementation of which is dictated by the technology to be used.

As an illustration of the computation associated with a difference equation, consider the system described by the system function

$$H(z) = \frac{b_0 + b_1 z^{-1}}{1 - a z^{-1}}, \qquad |z| > |a|. \tag{6.1}$$

The impulse response of this system is

$$h[n] = b_0 a^n u[n] + b_1 a^{n-1} u[n-1], \tag{6.2}$$

and the first-order difference equation that is satisfied by the input and output sequences is

$$y[n] - ay[n-1] = b_0 x[n] + b_1 x[n-1]. \tag{6.3}$$

Since the system has an infinite-duration impulse response, it is not possible to implement the system by discrete convolution. However, rewriting Eq. (6.3) in the form

$$y[n] = ay[n-1] + b_0 x[n] + b_1 x[n-1] \tag{6.4}$$

provides the basis for an algorithm for recursive computation of the output at any time n in terms of the previous output $y[n-1]$, the current input sample $x[n]$, and the previous input sample $x[n-1]$. As discussed in Section 2.5, if we further assume initial-rest conditions (i.e., if $x[n] = 0$ for $n < 0$, then $y[n] = 0$ for $n < 0$), and if we use Eq. (6.4) as a recurrence formula for computing the output from past values of the output and present and past values of the input, the system will be linear and time invariant. A similar procedure can be applied to the more general case of an Nth-order difference equation. However, the algorithm suggested by Eq. (6.4) and its generalization for higher order difference equations is not the only computational algorithm for implementing a particular system, and often it is not the most preferable. As we will see, an unlimited variety of computational structures result in the same relation between the input sequence $x[n]$ and the output sequence $y[n]$.

In the remainder of this chapter, we consider the important issues in the implementation of linear time-invariant discrete-time systems. We first present the block diagram and signal flow graph descriptions of computational structures or networks for linear constant-coefficient difference equations representing linear time-invariant causal systems. Using a combination of algebraic manipulations and manipulations of block diagram representations, we derive a number of basic equivalent structures for implementing a causal linear time-invariant system. Although two structures may be equivalent with regard to their input–output characteristics for infinite-precision representations of coefficients and variables, they may have vastly different behavior when the numerical precision is limited. This is the major reason that it is of interest to study different implementation structures. The effects of finite-precision representation of the system coefficients and the effects of truncation or rounding of intermediate computations are considered in the latter sections of the chapter.

6.1 BLOCK DIAGRAM REPRESENTATION OF LINEAR CONSTANT-COEFFICIENT DIFFERENCE EQUATIONS

The implementation of a linear time-invariant discrete-time system by iteratively evaluating a recurrence formula obtained from a difference equation requires that delayed values of the output, input, and intermediate sequences be available. The delay of sequence values implies the need for storage of past sequence values. Also, we must provide means for multiplication of the delayed sequence values by the coefficients, as well as means for adding the resulting products. Therefore, the basic elements required for the implementation of a linear time-invariant discrete-time system are adders,

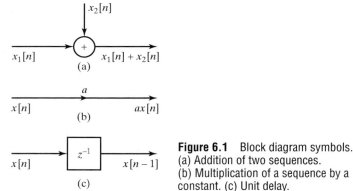

Figure 6.1 Block diagram symbols.
(a) Addition of two sequences.
(b) Multiplication of a sequence by a
constant. (c) Unit delay.

multipliers, and memory for storing delayed sequence values. The interconnection of
these basic elements is conveniently depicted by block diagrams composed of the basic pictorial symbols shown in Figure 6.1. Figure 6.1(a) represents the addition of two
sequences. In general block diagram notation, an adder may have any number of inputs. However, in almost all practical implementations, adders have only two inputs.
In all the diagrams of this chapter, we indicate this explicitly by limiting the number of inputs as in Figure 6.1(a). Figure 6.1(b) depicts multiplication of a sequence
by a constant, and Figure 6.1(c) depicts delaying a sequence by one sample. In digital implementations, the delay operation can be implemented by providing a storage
register for each unit delay that is required. In analog discrete-time implementations,
such as switched-capacitor filters, the delays are implemented by charge storage devices. The unit delay system is represented in Figure 6.1(c) by its system function, z^{-1}.
Delays of more than one sample can be denoted as in Figure 6.1(c), with a system
function of z^{-M}, where M is the number of samples of delay; however, the actual implementation of M samples of delay would generally be done by cascading M unit
delays. In an integrated-circuit implementation, these unit delays might form a shift
register that is clocked at the sampling rate of the input signal. In a software implementation, M cascaded unit delays would be implemented as M consecutive memory
registers.

Example 6.1 Block Diagram Representation of a Difference Equation

As an example of the representation of a difference equation in terms of the elements
in Figure 6.1, consider the second-order difference equation

$$y[n] = a_1 y[n-1] + a_2 y[n-2] + b_0 x[n]. \tag{6.5}$$

The corresponding system function is

$$H(z) = \frac{b_0}{1 - a_1 z^{-1} - a_2 z^{-2}}. \tag{6.6}$$

The block diagram representation of the system realization based on Eq. (6.5) is
shown in Figure 6.2. Such diagrams give a pictorial representation of a computational

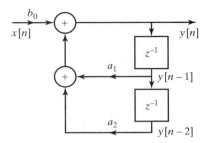

Figure 6.2   Example of a block diagram representation of a difference equation.

algorithm for implementing the system. When the system is implemented on either a general-purpose computer or a digital signal processing (DSP) chip, network structures such as the one shown in Figure 6.2 serve as the basis for a program that implements the system. If the system is implemented with discrete components or as a complete system with VLSI technology, the block diagram is the basis for determining a hardware architecture for the system. In both cases, diagrams such as Figure 6.2 show explicitly that we must provide storage for the delayed variables (in this case, $y[n-1]$ and $y[n-2]$) and also the coefficients of the difference equation (in this case, a_1, a_2, and b_0). Furthermore, we see from Figure 6.2 that an output sequence value $y[n]$ is computed by first forming the products $a_1 y[n-1]$ and $a_2 y[n-2]$, then adding them, and, finally, adding the result to $b_0 x[n]$. Thus, Figure 6.2 conveniently depicts the complexity of the associated computational algorithm, the steps of the algorithm, and the amount of hardware required to realize the system.

Example 6.1 can be generalized to higher order difference equations of the form[1]

$$y[n] - \sum_{k=1}^{N} a_k y[n-k] = \sum_{k=0}^{M} b_k x[n-k], \tag{6.7}$$

with the corresponding system function

$$H(z) = \frac{\displaystyle\sum_{k=0}^{M} b_k z^{-k}}{1 - \displaystyle\sum_{k=1}^{N} a_k z^{-k}}. \tag{6.8}$$

---

[1]The form used in previous chapters for a general Nth-order difference equation was

$$\sum_{k=0}^{N} a_k y[n-k] = \sum_{k=0}^{M} b_k x[n-k].$$

In the remainder of the book, it will be more convenient to use the form in Eq. (6.7), where the coefficient of $y[n]$ is normalized to unity and the coefficients associated with the delayed output appear with a positive sign after they have been moved to the right-hand side of the equation. (See Eq. (6.9).)

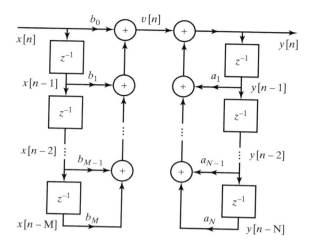

Figure 6.3 Block diagram representation for a general Nth-order difference equation.

Rewriting Eq. (6.7) as a recurrence formula for $y[n]$ in terms of a linear combination of past values of the output sequence and current and past values of the input sequence leads to the relation

$$y[n] = \sum_{k=1}^{N} a_k y[n-k] + \sum_{k=0}^{M} b_k x[n-k]. \tag{6.9}$$

The block diagram of Figure 6.3 is an explicit pictorial representation of Eq. (6.9). More precisely, it represents the pair of difference equations

$$v[n] = \sum_{k=0}^{M} b_k x[n-k], \tag{6.10a}$$

$$y[n] = \sum_{k=1}^{N} a_k y[n-k] + v[n]. \tag{6.10b}$$

The assumption of a two-input adder implies that the additions are done in a specified order. That is, Figure 6.3 shows that the products $a_N y[n-N]$ and $a_{N-1} y[n-N+1]$ must be computed, then added, and the resulting sum added to $a_{N-2} y[n-N+2]$, and so on. After $y[n]$ has been computed, the delay variables must be updated by moving $y[n-N+1]$ into the register holding $y[n-N]$, and so on.

A block diagram can be rearranged or modified in a variety of ways without changing the overall system function. Each appropriate rearrangement represents a *different* computational algorithm for implementing the *same* system. For example, the block diagram of Figure 6.3 can be viewed as a cascade of two systems, the first representing the computation of $v[n]$ from $x[n]$ and the second representing the computation of $y[n]$ from $v[n]$. Since each of the two systems is a linear time-invariant system (assuming initial-rest conditions for the delay registers), the order in which the two systems are cascaded can be reversed, as shown in Figure 6.4, without affecting the overall system

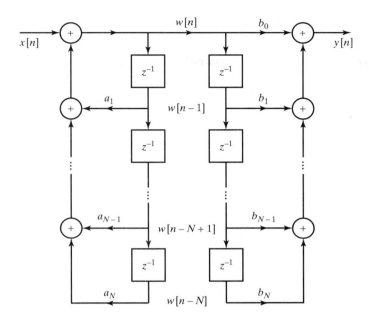

Figure 6.4   Rearrangement of block diagram of Figure 6.3. We assume for convenience that $N = M$. If $N \neq M$, some of the coefficients will be zero.

function. In Figure 6.4, for convenience, we have assumed that $M = N$. Clearly, there is no loss of generality, since if $M \neq N$, some of the coefficients a_k or b_k in the figure would be zero, and the diagram could be simplified accordingly.

In terms of the system function $H(z)$ in Eq. (6.8), Figure 6.3 can be viewed as an implementation of $H(z)$ through the decomposition

$$H(z) = H_2(z)H_1(z) = \left(\frac{1}{1 - \displaystyle\sum_{k=1}^{N} a_k z^{-k}} \right) \left(\sum_{k=0}^{M} b_k z^{-k} \right) \tag{6.11}$$

or, equivalently, through the pair of equations

$$V(z) = H_1(z)X(z) = \left(\sum_{k=0}^{M} b_k z^{-k} \right) X(z), \tag{6.12a}$$

$$Y(z) = H_2(z)V(z) = \left(\frac{1}{1 - \displaystyle\sum_{k=1}^{N} a_k z^{-k}} \right) V(z). \tag{6.12b}$$

Figure 6.4, on the other hand, represents $H(z)$ as

$$H(z) = H_1(z)H_2(z) = \left(\sum_{k=0}^{M} b_k z^{-k}\right)\left(\frac{1}{1 - \sum_{k=1}^{N} a_k z^{-k}}\right) \tag{6.13}$$

or, equivalently, through the equations

$$W(z) = H_2(z)X(z) = \left(\frac{1}{1 - \sum_{k=1}^{N} a_k z^{-k}}\right)X(z), \tag{6.14a}$$

$$Y(z) = H_1(z)W(z) = \left(\sum_{k=0}^{M} b_k z^{-k}\right)W(z). \tag{6.14b}$$

In the time domain, Figure 6.4 and, equivalently, Eqs. (6.14a) and (6.14b) can be represented by the pair of difference equations

$$w[n] = \sum_{k=1}^{N} a_k w[n-k] + x[n], \tag{6.15a}$$

$$y[n] = \sum_{k=0}^{M} b_k w[n-k]. \tag{6.15b}$$

The block diagrams of Figures 6.3 and 6.4 have several important differences. In Figure 6.3, the zeros of $H(z)$, represented by $H_1(z)$, are implemented first, followed by the poles, represented by $H_2(z)$. In Figure 6.4, the poles are implemented first, followed by the zeros. Theoretically, the order of implementation does not affect the overall system function. However, as we will see, when a difference equation is implemented with finite-precision arithmetic, there can be a significant difference between two systems that are theoretically equivalent. Another important point concerns the number of delay elements in the two systems. As drawn, the systems in Figures 6.3 and 6.4 each have a total of $(N+M)$ delay elements. However, the block diagram of Figure 6.4 can be redrawn by noting that exactly the same signal, $w[n]$, is stored in the two chains of delay elements in the figure. Consequently, the two can be collapsed into one chain, as indicated in Figure 6.5.

The total number of delay elements in Figure 6.5 is less than in either Figure 6.3 or Figure 6.4, and in fact it is the minimum number required to implement a system with system function given by Eq. (6.8). Specifically, the minimum number of delays required is, in general, $\max(N, M)$. An implementation with the minimum number of delay elements is commonly referred to as a *canonic form* implementation. The noncanonic block diagram in Figure 6.3 is referred to as the *direct form I* implementation

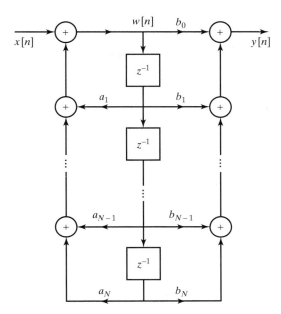

Figure 6.5   Combination of delays in Figure 6.4.

of the general Nth-order system because it is a direct realization of the difference equation satisfied by the input $x[n]$ and the output $y[n]$, which in turn can be written directly from the system function by inspection. Figure 6.5 is often referred to as the *direct form II* or *canonic direct form* implementation. Knowing that Figure 6.5 is an appropriate realization structure for $H(z)$ given by Eq. (6.8), we can go directly back and forth in a straightforward manner between the system function and the block diagram (or the equivalent difference equation).

Example 6.2   Direct Form I and Direct Form II Implementation of an LTI System

Consider the LTI system with system function

$$H(z) = \frac{1 + 2z^{-1}}{1 - 1.5z^{-1} + 0.9z^{-2}}.$$ (6.16)

Comparing this system function with Eq. (6.8), we find $b_0 = 1$, $b_1 = 2$, $a_1 = +1.5$, and $a_2 = -0.9$, so it follows from Figure 6.3 that we can implement the system in a direct form I block diagram as shown in Figure 6.6. Referring to Figure 6.5, we can also implement the system function in direct form II, as shown in Figure 6.7. In both cases, note that the coefficients in the feedback branches in the block diagram have opposite signs from the corresponding coefficients of z^{-1} and z^{-2} in Eq. (6.16). Although this change of sign is sometimes confusing, it is essential to remember that the feedback coefficients $\{a_k\}$ always have the opposite sign in the difference equation from their sign in the system function. Note also that the direct form II requires only two delays to implement $H(z)$, one less than the direct form I implementation.

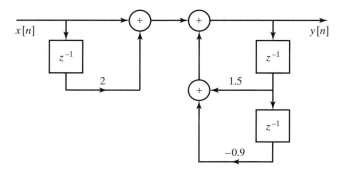

Figure 6.6   Direct form I implementation of Eq. (6.16).

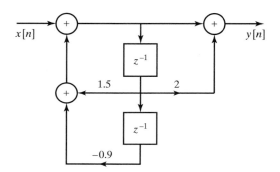

Figure 6.7   Direct form II implementation of Eq. (6.16).

In the preceding discussion, we developed two equivalent block diagrams for implementing a linear time-invariant system with system function given by Eq. (6.8). These block diagrams, which represent different computational algorithms for implementing the system, were obtained by manipulations based on the linearity of the system and the algebraic properties of the system function. Indeed, since the basic difference equations that represent a linear time-invariant system are linear, equivalent sets of difference equations can be obtained simply by linear transformations of the variables of the difference equations. Thus, there are an unlimited number of equivalent realizations of any given system. In Section 6.3, using an approach similar to that employed in this section, we will develop a number of other important and useful equivalent structures for implementing a system with system function as in Eq. (6.8). Before discussing these other forms, however, it is convenient to introduce signal flow graphs as an alternative to block diagrams for representing difference equations.

6.2 SIGNAL FLOW GRAPH REPRESENTATION OF LINEAR CONSTANT-COEFFICIENT DIFFERENCE EQUATIONS

A signal flow graph representation of a difference equation is essentially the same as a block diagram representation, except for a few notational differences. Formally, a signal flow graph is a network of directed branches that connect at nodes. Associated

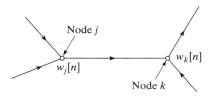

Figure 6.8 Example of nodes and branches in a signal flow graph.

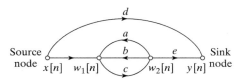

Figure 6.9 Example of a signal flow graph showing source and sink nodes.

with each node is a variable or node value. The value associated with node k might be denoted w_k, or, since node variables for digital filters are generally sequences, we often indicate this explicitly with the notation $w_k[n]$. Branch (j, k) denotes a branch originating at node j and terminating at node k, with the direction from j to k being indicated by an arrowhead on the branch. This is shown in Figure 6.8. Each branch has an input signal and an output signal. The input signal from node j to branch (j, k) is the node value $w_j[n]$. In a linear signal flow graph, which is the only class we will consider, the output of a branch is a linear transformation of the input to the branch. The simplest example is a constant gain, i.e., when the output of the branch is simply a constant multiple of the input to the branch. The linear operation represented by the branch is typically indicated next to the arrowhead showing the direction of the branch. For the case of a constant multiplier, the constant is simply shown next to the arrowhead. When an explicit indication of the branch operation is omitted, this indicates a branch transmittance of unity, or the identity transformation. By definition, the value at each node in a graph is the sum of the outputs of all the branches entering the node.

To complete the definition of signal flow graph notation, we define two special types of nodes. *Source nodes* are nodes that have no entering branches. Source nodes are used to represent the injection of external inputs or signal sources into a graph. *Sink nodes* are nodes that have only entering branches. Sink nodes are used to extract outputs from a graph. Source nodes, sink nodes, and simple branch gains are illustrated in the signal flow graph of Figure 6.9. The linear equations represented by the figure are as follows:

$$w_1[n] = x[n] + aw_2[n] + bw_2[n],$$

$$w_2[n] = cw_1[n], \tag{6.17}$$

$$y[n] = dx[n] + ew_2[n].$$

Addition, multiplication by a constant, and delay are the basic operations required to implement a linear constant-coefficient difference equation. Since these are all linear operations, it is possible to use signal flow graph notation to depict algorithms for implementing linear time-invariant discrete-time systems. As an example of how the flow graph concepts just discussed can be applied to the representation of a difference equation, consider the block diagram in Figure 6.10(a), which is the direct form II realization of the system whose system function is given by Eq. (6.1). A signal flow graph corresponding to this system is shown in Figure 6.10(b). In the representation of difference equations, the node variables are sequences. In Figure 6.10(b), node 0 is a source node whose value is determined by the input sequence $x[n]$, and node 5 is a sink node whose value is denoted $y[n]$. Notice that the source and sink nodes are connected to the rest of the graph by unity-gain branches to clearly denote the input and output of the system. Obviously, nodes 3 and 5 have identical values. In Figure 6.10(b), all branches except one (the delay branch $(2, 4)$) can be represented by a simple branch gain; i.e., the output signal is a constant multiple of the branch input. A delay cannot be represented in the time domain by a branch gain. However, the z-transform representation of a unit delay is multiplication by the factor z^{-1}. If we represented the difference equations by their corresponding z-transform equations, all the branches would be characterized by their system functions. In this case, each branch gain would be a function of z; e.g., a unit delay branch would have a gain of z^{-1}. By convention, we represent the variables in a signal flow graph as sequences rather than as z-transforms of sequences. However, to simplify the notation, we normally indicate a delay branch by showing its branch gain as z^{-1}, but it is understood that the output of such a branch is the branch input delayed by one sequence value. The graph of Figure 6.10(b) is shown in Figure 6.11 with this convention. The equations represented by Figure 6.11 are

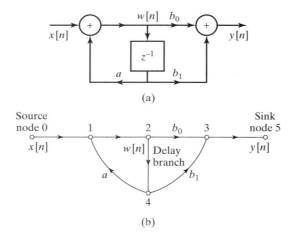

Figure 6.10   (a) Block diagram representation of a first-order digital filter. (b) Structure of the signal flow graph corresponding to the block diagram in (a).

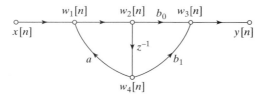

Figure 6.11   Signal flow graph of Figure 6.10(b) with the delay branch indicated by z^{-1}.

as follows:

$$w_1[n] = aw_4[n] + x[n], \tag{6.18a}$$

$$w_2[n] = w_1[n], \tag{6.18b}$$

$$w_3[n] = b_0w_2[n] + b_1w_4[n], \tag{6.18c}$$

$$w_4[n] = w_2[n-1], \tag{6.18d}$$

$$y[n] = w_3[n], \tag{6.18e}$$

A comparison of Figure 6.10(a) and Figure 6.11 shows that there is a direct correspondence between branches in the block diagram and branches in the flow graph. In fact, the important difference between the two is that nodes in the flow graph represent both branching points and adders, whereas in the block diagram a special symbol is used for adders. A branching point in the block diagram is represented in the flow graph by a node that has only one incoming branch and one or more outgoing branches. An adder in the block diagram is represented in the signal flow graph by a node that has two (or more) incoming branches. Signal flow graphs are therefore totally equivalent to block diagrams as pictorial representations of difference equations, but they are simpler to draw. Like block diagrams, they can be manipulated graphically to gain insight into the properties of a given system. A large body of signal flow graph theory exists that can be directly applied to discrete-time systems when they are represented in this form. (See Mason and Zimmermann, 1960; Chow and Cassignol, 1962; and Phillips and Nagle, 1995.) Although we will use flow graphs primarily for their pictorial value, we will utilize certain theorems relating to signal flow graphs in examining alternative structures for implementing linear systems.

Equations (6.18a)–(6.18e) define a multistep algorithm for computing the output of the linear time-invariant system from the input sequence $x[n]$. This example illustrates the kind of data precedence relations that generally arise in the implementation of IIR systems. Equations (6.18a)–(6.18e) cannot be computed in arbitrary order. Equations (6.18a) and (6.18c) require multiplications and additions, but Eqs. (6.18b) and (6.18e) simply rename variables. Equation (6.18d) represents the "updating" of the memory of the system. It would be implemented simply by replacing the contents of the memory register representing $w_4[n]$ by the value of $w_2[n]$, but this would have to be done consistently either *before* or *after* the evaluation of all the other equations. Initial-rest conditions would be imposed in this case by defining $w_2[-1] = 0$ or $w_4[0] = 0$. Clearly, Eqs. (6.18a)–(6.18e) must be computed in the order given, except that the last two could be interchanged or Eq. (6.18d) could be consistently evaluated first.

The flow graph represents a set of difference equations, with one equation being written at each node of the network. In the case of the flow graph of Figure 6.11, we can eliminate some of the variables rather easily to obtain the pair of equations

$$w_2[n] = aw_2[n-1] + x[n], \tag{6.19a}$$

$$y[n] = b_0w_2[n] + b_1w_2[n-1], \tag{6.19b}$$

which are in the form of Eqs. (6.15a) and (6.15b); i.e., in direct form II. Often, the manipulation of the difference equations of a flow graph is difficult when dealing with the time-domain variables, due to feedback of delayed variables. In such cases, it is always possible to work with the z-transform representation, wherein all branches are simple gains. Problems 6.1–6.26 illustrate the utility of z-transform analysis of flow graphs for obtaining equivalent sets of difference equations.

Example 6.3   Determination of the System Function from a Flow Graph

To illustrate the use of the z-transform in determining the system function from a flow graph, consider Figure 6.12. The flow graph in this figure is not in direct form. Therefore, the system function cannot be written down by inspection of the graph. However, the set of difference equations represented by the graph can be written down by writing an equation for the value of each node variable in terms of the other node variables. The five equations are

$$w_1[n] = w_4[n] - x[n], \tag{6.20a}$$

$$w_2[n] = \alpha w_1[n], \tag{6.20b}$$

$$w_3[n] = w_2[n] + x[n], \tag{6.20c}$$

$$w_4[n] = w_3[n - 1], \tag{6.20d}$$

$$y[n] = w_2[n] + w_4[n]. \tag{6.20e}$$

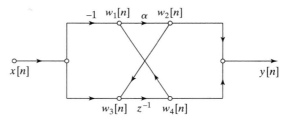

Figure 6.12   Flow graph not in standard direct form.

These are the equations that would be used to implement the system in the form described by the flow graph. Equations (6.20a)–(6.20e) can be represented by the z-transform equations

$$W_1(z) = W_4(z) - X(z), \tag{6.21a}$$

$$W_2(z) = \alpha W_1(z), \tag{6.21b}$$

$$W_3(z) = W_2(z) + X(z), \tag{6.21c}$$

$$W_4(z) = z^{-1} W_3(z), \tag{6.21d}$$

$$Y(z) = W_2(z) + W_4(z). \tag{6.21e}$$

We can eliminate $W_1(z)$ and $W_3(z)$ from this set of equations by substituting Eq. (6.21a) into Eq. (6.21b) and Eq. (6.21c) into Eq. (6.21d), obtaining

$$W_2(z) = \alpha(W_4(z) - X(z)), \tag{6.22a}$$

$$W_4(z) = z^{-1}(W_2(z) + X(z)), \tag{6.22b}$$

$$Y(z) = W_2(z) + W_4(z). \tag{6.22c}$$

Equations (6.22a) and (6.22b) can be solved for $W_2(z)$ and $W_4(z)$, yielding

$$W_2(z) = \frac{\alpha(z^{-1} - 1)}{1 - \alpha z^{-1}} X(z), \tag{6.23a}$$

$$W_4(z) = \frac{z^{-1}(1 - \alpha)}{1 - \alpha z^{-1}} X(z), \tag{6.23b}$$

and substituting Eqs. (6.23a) and (6.23b) into Eq. (6.22c) leads to

$$Y(z) = \left(\frac{\alpha(z^{-1} - 1) + z^{-1}(1 - \alpha)}{1 - \alpha z^{-1}} \right) X(z) = \left(\frac{z^{-1} - \alpha}{1 - \alpha z^{-1}} \right) X(z). \tag{6.24}$$

Therefore, the system function of the flow graph of Figure 6.12 is

$$H(z) = \frac{z^{-1} - \alpha}{1 - \alpha z^{-1}}, \tag{6.25}$$

from which it follows that the impulse response of the system is

$$h[n] = \alpha^{n-1} u[n - 1] - \alpha^{n+1} u[n]$$

and the direct form I flow graph is as shown in Figure 6.13.

Figure 6.13   Direct form I equivalent of Figure 6.12.

Example 6.3 shows how the z-transform converts the time-domain expressions, which involve feedback and thus are difficult to solve, into linear equations that can be solved by algebraic techniques. The example also illustrates that different flow graph representations define computational algorithms that require different amounts of computational resources. By comparing Figures 6.12 and 6.13, we see that the original implementation requires only one multiplication and one delay (memory) element, while the direct form I implementation would require two multiplications and two delay elements. The direct form II implementation would require one less delay, but it still would require two multiplications.

6.3 BASIC STRUCTURES FOR IIR SYSTEMS

In Section 6.1, we introduced two alternative structures for implementing a linear time-invariant system with system function as in Eq. (6.8). In this section we present the signal flow graph representations of those systems, and we also develop several other commonly used equivalent flow graph network structures. Our discussion will make it clear that, for any given rational system function, a wide variety of equivalent sets of difference equations or network structures exists. One consideration in the choice among these different structures is computational complexity. For example, in some digital implementations, structures with the fewest constant multipliers and the fewest delay branches are often most desirable. This is because multiplication is generally a time-consuming and costly operation in digital hardware and because each delay element corresponds to a memory register. Consequently, a reduction in the number of constant multipliers means an increase in speed, and a reduction in the number of delay elements means a reduction in memory requirements.

Other, more subtle, trade-offs arise in VLSI implementations, in which the area of a chip is often an important measure of efficiency. Modularity and simplicity of data transfer on the chip are also frequently very desirable in such implementations. In multiprocessor implementations, the most important considerations are often related to partitioning of the algorithm and communication requirements between processors. Another major consideration is the effects of a finite register length and finite-precision arithmetic. These effects depend on the way in which the computations are organized, i.e., on the structure of the signal flow graph. Sometimes it is desirable to use a structure that does not have the minimum number of multipliers and delay elements if that structure is less sensitive to finite register length effects.

In this section, we develop several of the most commonly used forms for implementing a linear time-invariant IIR system and obtain their flow graph representations.

6.3.1 Direct Forms

In Section 6.1, we obtained block diagram representations of the direct form I (Figure 6.3) and direct form II, or canonic direct form (Figure 6.5), structures for a linear time-invariant system whose input and output satisfy a difference equation of the form

$$y[n] - \sum_{k=1}^{N} a_k y[n-k] = \sum_{k=0}^{M} b_k x[n-k], \tag{6.26}$$

with the corresponding rational system function

$$H(z) = \frac{\displaystyle\sum_{k=0}^{M} b_k z^{-k}}{1 - \displaystyle\sum_{k=1}^{N} a_k z^{-k}}. \tag{6.27}$$

In Figure 6.14, the direct form I structure of Figure 6.3 is shown using signal flow graph conventions, and Figure 6.15 shows the signal flow graph representation of the direct form II structure of Figure 6.5. Again, we have assumed for convenience that $N = M$.

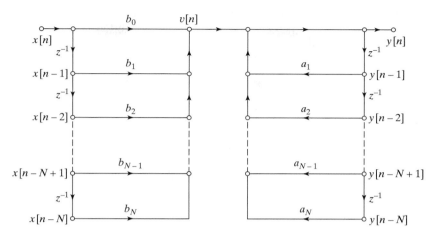

Figure 6.14   Signal flow graph of direct form I structure for an Nth-order system.

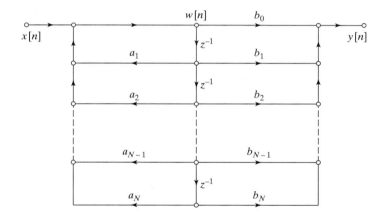

Figure 6.15   Signal flow graph of direct form II structure for an Nth-order system.

Note that we have drawn the flow graph so that each node has no more than two inputs. A node in a signal flow graph may have any number of inputs, but, as indicated earlier, this two-input convention results in a graph that is more closely related to programs and architectures for implementing the computation of the difference equations represented by the graph.

Example 6.4   Illustration of Direct Form I and Direct Form II Structures

Consider the system function

$$H(z) = \frac{1 + 2z^{-1} + z^{-2}}{1 - 0.75z^{-1} + 0.125z^{-2}}. \tag{6.28}$$

Since the coefficients in the direct form structures correspond directly to the coefficients of the numerator and denominator polynomials (taking into account the minus sign in the denominator of Eq. (6.27)), we can draw these structures by inspection with reference to Figures 6.14 and 6.15. The direct form I and direct form II structures for this example are shown in Figures 6.16 and 6.17, respectively.

Figure 6.16   Direct form I structure for Example 6.4.

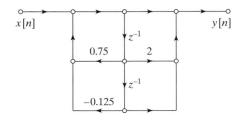

Figure 6.17   Direct form II structure for Example 6.4.

6.3.2 Cascade Form

The direct form structures were obtained directly from the system function $H(z)$, written as a ratio of polynomials in the variable z^{-1} as in Eq. (6.27). If we factor the numerator and denominator polynomials, we can express $H(z)$ in the form

$$H(z) = A \frac{\displaystyle\prod_{k=1}^{M_1}(1 - f_k z^{-1})\prod_{k=1}^{M_2}(1 - g_k z^{-1})(1 - g_k^* z^{-1})}{\displaystyle\prod_{k=1}^{N_1}(1 - c_k z^{-1})\prod_{k=1}^{N_2}(1 - d_k z^{-1})(1 - d_k^* z^{-1})}, \qquad (6.29)$$

where $M = M_1 + 2M_2$ and $N = N_1 + 2N_2$. In this expression, the first-order factors represent real zeros at f_k and real poles at c_k, and the second-order factors represent complex conjugate pairs of zeros at g_k and g_k^* and complex conjugate pairs of poles at d_k and d_k^*. This represents the most general distribution of poles and zeros when all the coefficients in Eq. (6.27) are real. Equation (6.29) suggests a class of structures consisting of a cascade of first- and second-order systems. There is considerable freedom in the choice of composition of the subsystems and in the order in which the subsystems are cascaded. In practice, however, it is often desirable to implement the cascade realization using a minimum of storage and computation. A modular structure that is advantageous

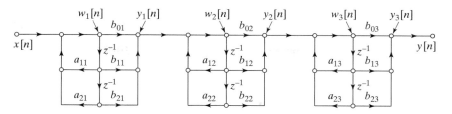

Figure 6.18 Cascade structure for a sixth-order system with a direct form II realization of each second-order subsystem.

for many types of implementations is obtained by combining pairs of real factors and complex conjugate pairs into second-order factors so that Eq. (6.29) can be expressed as

$$H(z) = \prod_{k=1}^{N_s} \frac{b_{0k} + b_{1k}z^{-1} + b_{2k}z^{-2}}{1 - a_{1k}z^{-1} - a_{2k}z^{-2}}, \tag{6.30}$$

where $N_s = \lfloor (N+1)/2 \rfloor$ is the largest integer contained in $(N+1)/2$. In writing $H(z)$ in this form, we have assumed that $M \leq N$ and that the real poles and zeros have been combined in pairs. If there are an odd number of real zeros, one of the coefficients b_{2k} will be zero. Likewise, if there are an odd number of real poles, one of the coefficients a_{2k} will be zero. The individual second-order sections can be implemented using either of the direct form structures; however, the previous discussion shows that we can implement a cascade structure with a minimum number of multiplications and a minimum number of delay elements if we use the direct form II structure for each second-order section. A cascade structure for a sixth-order system using three direct form II second-order sections is shown in Figure 6.18. The difference equations represented by a general cascade of direct form II second-order sections are of the form

$$y_0[n] = x[n], \tag{6.31a}$$

$$w_k[n] = a_{1k}w_k[n-1] + a_{2k}w_k[n-2] + y_{k-1}[n], \quad k = 1, 2, \ldots, N_s, \tag{6.31b}$$

$$y_k[n] = b_{0k}w_k[n] + b_{1k}w_k[n-1] + b_{2k}w_k[n-2], \quad k = 1, 2, \ldots, N_s, \tag{6.31c}$$

$$y[n] = y_{N_s}[n]. \tag{6.31d}$$

It is easy to see that a variety of theoretically equivalent systems can be obtained by simply pairing the poles and zeros in different ways and by ordering the second-order sections in different ways. Indeed, if there are N_s second-order sections, there are $N_s!$ (N_s factorial) pairings of the poles with zeros and $N_s!$ orderings of the resulting second-order sections, or a total of $(N_s!)^2$ different pairings and orderings. Although these all have the same overall system function and corresponding input–output relation when infinite-precision arithmetic is used, their behavior with finite-precision arithmetic can be quite different, as we will see in Section 6.8.

Example 6.5   Illustration of Cascade Structures

Let us again consider the system function of Eq. (6.28). Since this is a second-order system, a cascade structure with direct form II second-order sections reduces to the structure of Figure 6.17. Alternatively, to illustrate the cascade structure, we can use first-order systems by expressing $H(z)$ as a product of first-order factors, as in

$$H(z) = \frac{1 + 2z^{-1} + z^{-2}}{1 - 0.75z^{-1} + 0.125z^{-2}} = \frac{(1 + z^{-1})(1 + z^{-1})}{(1 - 0.5z^{-1})(1 - 0.25z^{-1})}. \tag{6.32}$$

Since all of the poles and zeros are real, a cascade structure with first-order sections has real coefficients. If the poles and/or zeros were complex, only a second-order section would have real coefficients. Figure 6.19 shows two equivalent cascade structures, each of which has the system function in Eq. (6.32). The difference equations represented by the flow graphs in the figure can be written down easily. Problem 6.22 is concerned with finding other, equivalent system configurations.

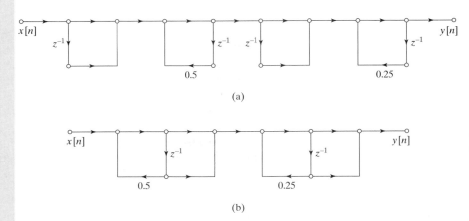

Figure 6.19   Cascade structures for Example 6.5. (a) Direct form I subsections. (b) Direct form II subsections.

A final comment should be made about our definition of the system function for the cascade form. As defined in Eq. (6.30), each second-order section has five constant multipliers. For comparison, let us assume that $M = N$ in $H(z)$ as given by Eq. (6.27), and furthermore, assume that N is an even integer, so that $N_s = N/2$. Then the direct form I and II structures have $2N + 1$ constant multipliers, while the cascade form structure suggested by Eq. (6.30) has $5N/2$ constant multipliers. For the sixth-order system in Figure 6.18, we require a total of 15 multipliers, while the equivalent direct forms would require a total of 13 multipliers. Another definition of the cascade form is

$$H(z) = b_0 \prod_{k=1}^{N_s} \frac{1 + \tilde{b}_{1k} z^{-1} + \tilde{b}_{2k} z^{-2}}{1 - a_{1k} z^{-1} - a_{2k} z^{-2}}, \tag{6.33}$$

where b_0 is the leading coefficient in the numerator polynomial of Eq. (6.27) and

$\tilde{b}_{ik} = b_{ik}/b_{0k}$ for $i = 1, 2$ and $k = 1, 2, \ldots, N_s$. This form for $H(z)$ suggests a cascade of four-multiplier second-order sections, with a single overall gain constant b_0. This cascade form has the same number of constant multipliers as the direct form structures. As discussed in Section 6.8, the five-multiplier second-order sections are commonly used when implemented with fixed-point arithmetic, because they make it possible to distribute the gain of the system and thereby control the size of signals at various critical points in the system. When floating-point arithmetic is used and dynamic range is not a problem, the four-multiplier second-order sections can be used to decrease the amount of computation. Further simplification results for zeros on the unit circle. In this case, $\tilde{b}_{2k} = 1$, and we require only three multipliers per second-order section.

6.3.3 Parallel Form

As an alternative to factoring the numerator and denominator polynomials of $H(z)$, we can express a rational system function as given by Eq. (6.27) or (6.29) as a partial fraction expansion in the form

$$H(z) = \sum_{k=0}^{N_p} C_k z^{-1} + \sum_{k=1}^{N_1} \frac{A_k}{1 - c_k z^{-1}} + \sum_{k=1}^{N_2} \frac{B_k(1 - e_k z^{-1})}{(1 - d_k z^{-1})(1 - d_k^* z^{-1})}, \qquad (6.34)$$

where $N = N_1 + 2N_2$. If $M \geq N$, then $N_p = M - N$; otherwise, the first summation in Eq. (6.34) is not included. If the coefficients a_k and b_k are real in Eq. (6.27), then the quantities A_k, B_k, C_k, c_k, and e_k are all real. In this form, the system function can be interpreted as representing a parallel combination of first- and second-order IIR systems, with possibly N_p simple scaled delay paths. Alternatively, we may group the real poles in pairs, so that $H(z)$ can be expressed as

$$H(z) = \sum_{k=0}^{N_p} C_k z^{-k} + \sum_{k=1}^{N_s} \frac{e_{0k} + e_{1k} z^{-1}}{1 - a_{1k} z^{-1} - a_{2k} z^{-2}}, \qquad (6.35)$$

where, as in the cascade form, $N_s = \lfloor (N + 1)/2 \rfloor$ is the largest integer contained in $(N+1)/2$, and if $N_p = M - N$ is negative, the first sum is not present. A typical example for $N = M = 6$ is shown in Figure 6.20. The general difference equations for the parallel form with second-order direct form II sections are

$$w_k[n] = a_{1k} w_k[n-1] + a_{2k} w_k[n-2] + x[n], \quad k = 1, 2, \ldots, N_s, \qquad (6.36a)$$

$$y_k[n] = e_{0k} w_k[n] + e_{1k} w_k[n-1], \qquad\qquad k = 1, 2, \ldots, N_s, \qquad (6.36b)$$

$$y[n] = \sum_{k=0}^{N_p} C_k x[n-k] + \sum_{k=1}^{N_s} y_k[n]. \qquad (6.36c)$$

If $M < N$, then the first summation in Eq. (6.36c) is not included.

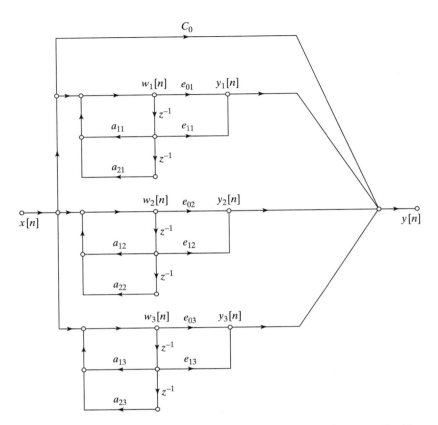

Figure 6.20 Parallel-form structure for sixth-order system ($M = N = 6$) with the real and complex poles grouped in pairs.

Example 6.6   Illustration of Parallel-Form Structures

Consider again the system function used in Examples 6.4 and 6.5. For the parallel form, we must express $H(z)$ in the form of either Eq. (6.34) or Eq. (6.35). If we use second-order sections,

$$H(z) = \frac{1 + 2z^{-1} + z^{-2}}{1 - 0.75z^{-1} + 0.125z^{-2}} = 8 + \frac{-7 + 8z^{-1}}{1 - 0.75z^{-1} + 0.125z^{-2}}. \qquad (6.37)$$

The parallel-form realization for this example with a second-order section is shown in Figure 6.21.

Since all the poles are real, we can obtain an alternative parallel form realization by expanding $H(z)$ as

$$H(z) = 8 + \frac{18}{1 - 0.5z^{-1}} - \frac{25}{1 - 0.25z^{-1}}. \qquad (6.38)$$

The resulting parallel form with first-order sections is shown in Figure 6.22. As in the general case, the difference equations represented by both Figures 6.21 and 6.22 can be written down by inspection.

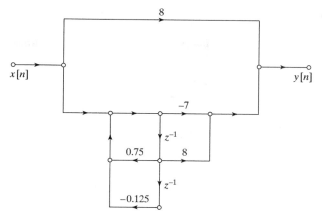

Figure 6.21 Parallel-form structure for Example 6.6 using a second-order system.

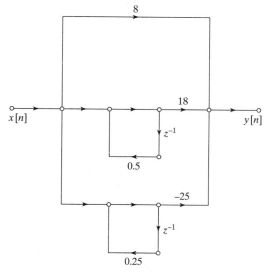

Figure 6.22 Parallel-form structure for Example 6.6 using first-order systems.

6.3.4 Feedback in IIR Systems

All the flow graphs of this section have feedback loops; i.e., they have closed paths that begin at a node and return to that node by traversing branches only in the direction of their arrowheads. Such a structure in the flow graph implies that a node variable in a loop depends directly or indirectly on itself. A simple example is shown in Figure 6.23(a), which represents the difference equation

$$y[n] = ay[n-1] + x[n]. \tag{6.39}$$

Such loops are necessary (but not sufficient) to generate infinitely long impulse responses. This can be seen if we consider a network with no feedback loops. In such a

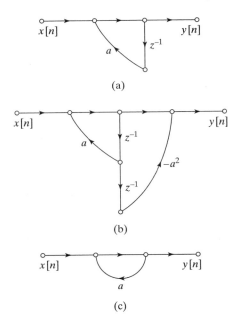

Figure 6.23   (a) System with feedback loop. (b) FIR system with feedback loop. (c) Noncomputable system.

case, any path from the input to the output can pass through each delay element only once. Therefore, the longest delay between the input and output would occur for a path that passes through all of the delay elements in the network. Thus, for a network with no loops, the impulse response is no longer than the total number of delay elements in the network. From this, we conclude that if a network has no loops, then the system function has only zeros (except for poles at $z = 0$), and the number of zeros can be no more than the number of delay elements in the network.

Returning to the simple example of Figure 6.23(a), we see that when the input is the impulse sequence, the single-input sample continually recirculates in the feedback loop with either increasing (if $|a| > 1$) or decreasing (if $|a| < 1$) amplitude due to multiplication by the constant a, so that the impulse response is $h[n] = a^n u[n]$. This is the way that feedback can create an infinitely long impulse response.

If a system function has poles, a corresponding block diagram or signal flow graph will have feedback loops. On the other hand, neither poles in the system function nor loops in the network are sufficient for the impulse response to be infinitely long. Figure 6.23(b) shows a network with a loop, but with an impulse response of finite length. This is because the pole of the system function cancels with a zero; i.e., for Figure 6.23(b),

$$H(z) = \frac{1 - a^2 z^{-2}}{1 - az^{-1}} = \frac{(1 - az^{-1})(1 + az^{-1})}{1 - az^{-1}} = 1 + az^{-1}. \qquad (6.40)$$

The impulse response of this system is $h[n] = \delta[n] + a\delta[n - 1]$. The system is a simple example of a general class of FIR systems called *frequency-sampling systems*. This class of systems is considered in more detail in Problems 6.30 and 6.37.

Loops in a network pose special problems in implementing the computations implied by the network. As we have discussed, it must be possible to compute the node variables in a network in sequence such that all necessary values are available when needed. In some cases, there is no way to order the computations so that the node variables of a flow graph can be computed in sequence. Such a network is called *noncomputable* (Crochiere and Oppenheim, 1975). A simple noncomputable network is shown in Figure 6.23(c). The difference equation for this network is

$$y[n] = ay[n] + x[n]. \tag{6.41}$$

In this form, we cannot compute $y[n]$ because the right-hand side of the equation involves the quantity we wish to compute. The fact that a flow graph is noncomputable does *not* mean the equations represented by the flow graph cannot be solved; indeed, the solution to Eq. (6.41) is $y[n] = x[n]/(1 - a)$. It simply means that the flow graph does not represent a set of difference equations that can be solved successively for the node variables. The key to the computability of a flow graph is that all loops must contain at least one unit delay element. Thus, in manipulating flow graphs representing implementations of linear time-invariant systems, we must be careful not to create delay-free loops. Problem 6.28 deals with a system having a delay-free loop. Problem 7.45 shows how a delay-free loop can be introduced.

6.4 TRANSPOSED FORMS

The theory of linear signal flow graphs provides a variety of procedures for transforming such graphs into different forms while leaving the overall system function between input and output unchanged. One of these procedures, called *flow graph reversal* or *transposition,* leads to a set of transposed system structures that provide some useful alternatives to the structures discussed in the previous section.

Transposition of a flow graph is accomplished by reversing the directions of all branches in the network while keeping the branch transmittances as they were and reversing the roles of the input and output so that source nodes become sink nodes and vice versa. For single-input, single-output systems, the resulting flow graph has the same system function as the original graph if the input and output nodes are interchanged. Although we will not formally prove this result here,[2] we will demonstrate that it is valid with two examples.

Example 6.7 Transposed Form for a First-Order System with No Zeroes

The first-order system corresponding to the flow graph in Figure 6.24 has system function

$$H(z) = \frac{1}{1 - az^{-1}}. \tag{6.42}$$

To obtain the transposed form for this system, we reverse the directions of all the branch arrows, taking the output where the input was and injecting the input where the output was. The result is shown in Figure 6.25. It is usually convenient to draw the transposed network with the input on the left and the output on the right, as shown in Figure 6.26. Comparing Figures 6.24 and 6.26 we note that the only difference is that in Figure 6.24 we multiply the *delayed* output sequence $y[n - 1]$ by the coefficient a,

---

[2]The theorem follows directly from Mason's gain formula of signal flow graph theory. (See Mason and Zimmermann, 1960; Chow and Cassignol, 1962; or Phillips and Nagle, 1995.)

Figure 6.24   Flow graph of simple first-order system.

Figure 6.25   Transposed form of Figure 6.24.

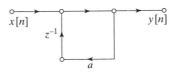

Figure 6.26   Structure of Figure 6.25 redrawn with input on left.

whereas in Figure 6.26 we multiply the output $y[n]$ by the coefficient a and then delay the resulting product. Since the two operations can be interchanged, we can conclude by inspection that the original system in Figure 6.24 and the corresponding transposed system in Figure 6.26 have the same system function.

In Example 6.7, it is straightforward to see that the original system and its transpose have the same system function. However, for more complicated graphs, the result is often not so obvious. This is illustrated by the next example.

Example 6.8   Transposed Form for a Basic Second-Order Section

Consider the basic second-order section depicted in Figure 6.27. The corresponding difference equations for this system are

$$w[n] = a_1 w[n-1] + a_2 w[n-2] + x[n], \tag{6.43a}$$

$$y[n] = b_0 w[n] + b_1 w[n-1] + b_2 w[n-2]. \tag{6.43b}$$

The transposed flow graph is shown in Figure 6.28; its corresponding difference equations are

$$v_0[n] = b_0 x[n] + v_1[n-1], \tag{6.44a}$$

$$y[n] = v_0[n], \tag{6.44b}$$

$$v_1[n] = a_1 y[n] + b_1 x[n] + v_2[n-1], \tag{6.44c}$$

$$v_2[n] = a_2 y[n] + b_2 x[n]. \tag{6.44d}$$

Equations (6.43a)–(6.43b) and Eqs. (6.44a)–(6.44d) are different ways to orga- nize the computation of the output samples $y[n]$ from the input samples $x[n]$, and

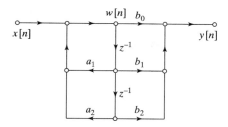

Figure 6.27   Direct form II structure for Example 6.8.

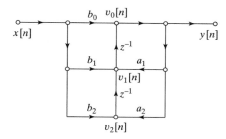

Figure 6.28   Transposed direct form II structure for Example 6.8.

it is not immediately clear that the two sets of difference equations are equivalent. One way to show this equivalence is to use the z-transform representations of both sets of equations, solve for the ratio $Y(z)/X(z) = H(z)$ in both cases, and compare the results. Another way is to substitute Eq. (6.44d) into Eq. (6.44c), substitute the result into Eq. (6.44a), and finally, substitute that result into Eq. (6.44b). The final result is

$$y[n] = a_1 y[n-1] + a_2 y[n-2] + b_0 x[n] + b_1 x[n-1] + b_2 x[n-2]. \qquad (6.45)$$

Since the network of Figure 6.27 is a direct form II structure, it is easily seen that the input and output of the system in Figure 6.27 also satisfies the difference equation (6.45). Therefore, for initial-rest conditions, the systems in Figures 6.27 and 6.28 are equivalent.

The transposition theorem can be applied to any of the structures that we have discussed so far. For example, the result of applying the theorem to the direct form I structure of Figure 6.14 is shown in Figure 6.29, and similarly, the structure obtained by transposing the direct form II structure of Figure 6.15 is shown in Figure 6.30. Clearly, if a signal flow graph configuration is transposed, the number of delay branches and the number of coefficients remain the same. Thus, the transposed direct form II structure is also a canonic structure.

An important point becomes evident through a comparison of Figures 6.15 and 6.30. Whereas the direct form II structure implements the poles first and then the zeros, the transposed direct form II structure implements the zeros first and then the poles. These differences can become important in the presence of quantization in finite-precision digital implementations or in the presence of noise in discrete-time analog implementations.

When the transposition theorem is applied to cascade or parallel structures, the individual second-order systems are replaced by transposed structures. For example,

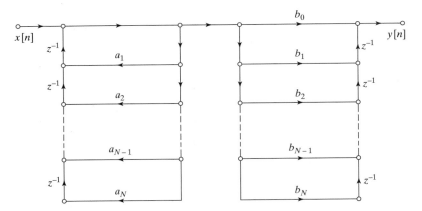

Figure 6.29 General flow graph resulting from applying the transposition theorem to the direct form I structure of Figure 6.14.

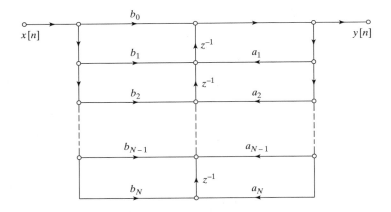

Figure 6.30 General flow graph resulting from applying the transposition theorem to the direct form II structure of Figure 6.15.

applying the transposition theorem to Figure 6.18 results in a cascade of three transposed direct form II sections (like the one in Example 6.8) with the same coefficients as in Figure 6.18, but with the order of the cascade reversed. A similar statement can be made about the transposition of Figure 6.20.

The transposition theorem further emphasizes that an infinite variety of implementation structures exists for any given rational system function. The transposition theorem provides a simple procedure for generating new structures. The problems of implementing systems with finite-precision arithmetic have motivated the development of many more classes of equivalent structures than we can discuss here. Thus, we concentrate only on the most commonly used structures.

6.5 BASIC NETWORK STRUCTURES FOR FIR SYSTEMS

The direct, cascade, and parallel-form structures discussed in Sections 6.3 and 6.4 are the most common basic structures for IIR systems. These structures were developed under the assumption that the system function had both poles and zeros. Although the

Figure 6.31   Direct-form realization of an FIR system.

Figure 6.32   Transposition of the network of Figure 6.31.

direct and cascade forms for IIR systems include FIR systems as a special case, there are additional specific forms for FIR systems.

6.5.1 Direct Form

For causal FIR systems, the system function has only zeros (except for poles at $z = 0$), and since the coefficients a_k are all zero, the difference equation of Eq. (6.9) reduces to

$$y[n] = \sum_{k=0}^{M} b_k x[n - k]. \tag{6.46}$$

This can be recognized as the discrete convolution of $x[n]$ with the impulse response

$$h[n] = \begin{cases} b_n & n = 0, 1, \ldots, M, \\ 0 & \text{otherwise.} \end{cases} \tag{6.47}$$

In this case, the direct form I and direct form II structures in Figures 6.14 and 6.15 both reduce to the direct-form FIR structure shown in Figure 6.31. Because of the chain of delay elements across the top of the diagram, this structure is also referred to as a *tapped delay line* structure or a *transversal filter* structure. As seen from Figure 6.31, the signal at each tap along this chain is weighted by the appropriate coefficient (impulse-response value), and the resulting products are summed to form the output $y[n]$.

The transposed direct form for the FIR case is obtained by applying the transposition theorem to Figure 6.31 or, equivalently, by setting the coefficients a_k to zero in Figure 6.29 or Figure 6.30. The result is shown in Figure 6.32.

6.5.2 Cascade Form

The cascade form for FIR systems is obtained by factoring the polynomial system function. That is, we represent $H(z)$ as

$$H(z) = \sum_{n=0}^{M} h[n] z^{-n} = \prod_{k=1}^{M_s} (b_{0k} + b_{1k} z^{-1} + b_{2k} z^{-2}), \tag{6.48}$$

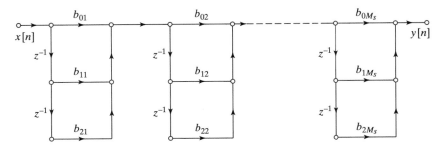

Figure 6.33 Cascade-form realization of an FIR system.

where $M_s = \lfloor (M+1)/2 \rfloor$ is the largest integer contained in $(M+1)/2$. If M is odd, one of the coefficients b_{2k} will be zero, since $H(z)$ in that case would have an odd number of real zeros. The flow graph representing Eq. (6.48) is shown in Figure 6.33, which is identical in form to Figure 6.18 with the coefficients a_{1k} and a_{2k} all zero. Each of the second-order sections in Figure 6.33 uses the direct form shown in Figure 6.31. Another alternative is to use transposed direct-form second-order sections or, equivalently, to apply the transposition theorem to Figure 6.33.

6.5.3 Structures for Linear-Phase FIR Systems

In Chapter 5, we showed that causal FIR systems have generalized linear phase if the impulse response satisfies the symmetry condition

$$h[M-n] = h[n] \qquad n = 0, 1, \dots, M \tag{6.49a}$$

or

$$h[M-n] = -h[n] \qquad n = 0, 1, \dots, M. \tag{6.49b}$$

With either of these conditions, the number of coefficient multipliers can be essentially halved. To see this, consider the following manipulations of the discrete convolution equation, assuming that M is an even integer corresponding to type I or type III systems:

$$y[n] = \sum_{k=0}^{M} h[k]x[n-k]$$

$$= \sum_{k=0}^{M/2-1} h[k]x[n-k] + h[M/2]x[n-M/2] + \sum_{k=M/2+1}^{M} h[k]x[n-k]$$

$$= \sum_{k=0}^{M/2-1} h[k]x[n-k] + h[M/2]x[n-M/2] + \sum_{k=0}^{M/2-1} h[M-k]x[n-M+k].$$

For type I systems, we use Eq. (6.49a) to obtain

$$y[n] = \sum_{k=0}^{M/2-1} h[k](x[n-k] + x[n-M+k]) + h[M/2]x[n-M/2]. \tag{6.50}$$

For type III systems, we use Eq. (6.49b) to obtain

$$y[n] = \sum_{k=0}^{M/2-1} h[k](x[n-k] - x[n-M+k]). \tag{6.51}$$

For the case of M an odd integer, the corresponding equations are, for type II systems,

$$y[n] = \sum_{k=0}^{(M-1)/2} h[k](x[n-k] + x[n-M+k]) \tag{6.52}$$

and, for type IV systems,

$$y[n] = \sum_{k=0}^{(M-1)/2} h[k](x[n-k] - x[n-M+k]). \tag{6.53}$$

Equations (6.50)–(6.53) imply structures with either $M/2 + 1$, $M/2$, or $(M+1)/2$ coefficient multipliers, rather than the M coefficient multipliers of the general direct-form structure of Figure 6.31. Figure 6.34 shows the structure implied by Eq. (6.50), and Figure 6.35 shows the structure implied by Eq. (6.52).

In our discussion of linear-phase systems in Section 5.7.3, we showed that the symmetry conditions of Eqs. (6.49a) and (6.49b) cause the zeros of $H(z)$ to occur in mirror-image pairs. That is, if z_0 is a zero of $H(z)$, then $1/z_0$ is also a zero of $H(z)$. Furthermore, if $h[n]$ is real, then the zeros of $H(z)$ occur in complex-conjugate pairs.

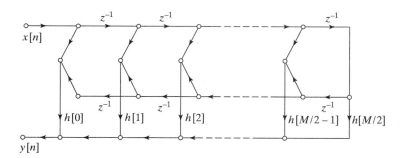

Figure 6.34 Direct-form structure for an FIR linear-phase system when M is an even integer.

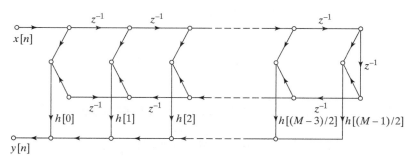

Figure 6.35 Direct-form structure for an FIR linear-phase system when M is an odd integer.

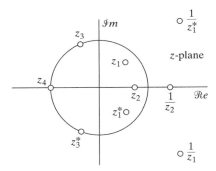

Figure 6.36 Symmetry of zeros for a linear-phase FIR filter.

As a consequence, real zeros not on the unit circle occur in reciprocal pairs. Complex zeros not on the unit circle occur in groups of four, corresponding to the complex conjugates and reciprocals. If a zero is on the unit circle, its reciprocal is also its conjugate. Consequently, complex zeros on the unit circle are conveniently grouped into pairs. Zeros at $z = \pm 1$ are their own reciprocal and complex conjugate. The four cases are summarized in Figure 6.36, where the zeros at z_1, z_1^*, $1/z_1$, and $1/z_1^*$ are considered as a group of four. The zeros at z_2 and $1/z_2$ are considered as a group of two, as are the zeros at z_3 and z_3^*. The zero at z_4 is considered singly. If $H(z)$ has the zeros shown in Figure 6.36, it can be factored into a product of first-, second-, and fourth-order factors. Each of these factors is a polynomial whose coefficients have the same symmetry as the coefficients of $H(z)$; i.e., each factor is a linear-phase polynomial in z^{-1}. Therefore, the system can be implemented as a cascade of first-, second-, and fourth-order systems. For example, the system function corresponding to the zeros of Figure 6.36 can be expressed as

$$H(z) = h[0](1 + z^{-1})(1 + az^{-1} + z^{-2})(1 + bz^{-1} + z^{-2})$$
$$\times (1 + cz^{-1} + dz^{-2} + cz^{-3} + z^{-4}), \tag{6.54}$$

where

$$a = (z_2 + 1/z_2), \qquad b = 2\mathcal{R}e\{z_3\}, \qquad c = -2\mathcal{R}e\{z_1 + 1/z_1\}, \qquad d = 2 + |z_1 + 1/z_1|^2.$$

This representation suggests a cascade structure consisting of linear-phase elements. It can be seen that the order of the system function polynomial is $M = 9$ and the number of different coefficient multipliers is five. This is the same number $((M + 1)/2 = 5)$ of constant multipliers required for implementing the system in the linear-phase direct form of Figure 6.34. Thus, with no additional multiplications, we obtain a modular structure in terms of a cascade of short linear-phase FIR systems.

6.6 OVERVIEW OF FINITE-PRECISION NUMERICAL EFFECTS

We have seen that a particular linear time-invariant discrete-time system can be implemented by a variety of computational structures. One motivation for considering alternatives to the simple direct-form structures is that different structures that are theoretically equivalent may behave differently when implemented with finite numerical precision. In this section, we give a brief introduction to the major numerical problems that arise in implementing discrete-time systems. A more detailed analysis of these finite-word-length effects is given in Sections 6.7–6.9.

6.6.1 Number Representations

In theoretical analyses of discrete-time systems, we generally assume that signal values and system coefficients are represented in the real-number system. However, with analog discrete-time systems, the limited precision of the components of a circuit makes it difficult to realize coefficients exactly. Similarly, when implementing digital signal-processing systems, we must represent signals and coefficients in some digital number system that must always be of finite precision. Most general-purpose digital computers, DSP chips, or special-purpose hardware use a binary number system.

The problem of finite numerical precision has already been discussed in Section 4.8.2 in the context of A/D conversion. We showed there that the output samples from an A/D converter are quantized and thus can be represented by fixed-point binary numbers. For compactness and simplicity in implementing arithmetic, one of the bits of the binary number is assumed to indicate the algebraic sign of the number. Formats such as *sign and magnitude, one's complement,* and *two's complement* are possible, but two's complement is most common.[3] A real number can be represented with infinite precision in two's-complement form as

$$x = X_m \left(-b_0 + \sum_{i=1}^{\infty} b_i 2^{-i} \right),$$ (6.55)

where X_m is an arbitrary scale factor and the b_i's are either 0 or 1. The quantity b_0 is referred to as the *sign bit*. If $b_0 = 0$, then $0 \le x \le X_m$, and if $b_0 = 1$, then $-X_m \le x < 0$. Thus, any real number whose magnitude is less than or equal to X_m can be represented by Eq. (6.55). An arbitrary real number x would require an infinite number of bits for its exact binary representation. As we saw in the case of A/D conversion, if we use only a finite number of bits ($B + 1$), then the representation of Eq. (6.55) must be modified to

$$\hat{x} = Q_B[x] = X_m \left(-b_0 + \sum_{i=1}^{B} b_i 2^{-i} \right) = X_m \hat{x}_B.$$ (6.56)

The resulting binary representation is quantized, so that the smallest difference between numbers is

$$\Delta = X_m 2^{-B}.$$ (6.57)

In this case, the quantized numbers are in the range $-X_m \le \hat{x} < X_m$. The fractional part of $\hat{x}$ can be represented with the positional notation

$$\hat{x}_B = b_0{}_\diamond b_1 b_2 b_3 \cdots b_B,$$ (6.58)

where $\diamond$ represents the binary point.

---

[3] A detailed description of binary number systems and corresponding arithmetic is given by Knuth (1997).

(a)

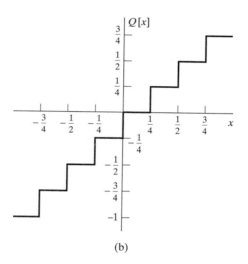

(b)

Figure 6.37 Nonlinear relationships representing two's-complement (a) rounding and (b) truncation for $B = 2$.

The operation of quantizing a number to $(B + 1)$ bits can be implemented by rounding or by truncation, but in either case quantization is a nonlinear memoryless operation. Figures 6.37(a) and 6.37(b) show the input–output relation for two's-complement rounding and truncation, respectively, for the case $B = 2$. In considering the effects of quantization, we often define the *quantization error* as

$$e = Q_B[x] - x. \tag{6.59}$$

For the case of two's-complement rounding, $-\Delta/2 < e \le \Delta/2$, and for two's-complement truncation, $-\Delta < e \le 0$.[4]

---

[4]Note that Eq. (6.56) also represents the result of rounding or truncating any $(B_1 + 1)$-bit binary representation, where $B_1 > B$. In this case Δ would be replaced by $(\Delta - X_m 2^{-B_1})$ in the bounds on the size of the quantization error.

If a number is larger than X_m (a situation called an overflow), we must implement some method of determining the quantized result. In the two's-complement arithmetic system, this need arises when we add two numbers whose sum is greater than X_m. For example, consider the 4-bit two's-complement number 0111, which in decimal form is 7. If we add the number 0001, the carry propagates all the way to the sign bit, so that the result is 1000, which in decimal form is -8. Thus, the resulting error can be very large when overflow occurs. Figure 6.38(a) shows the two's-complement quantizer, including the

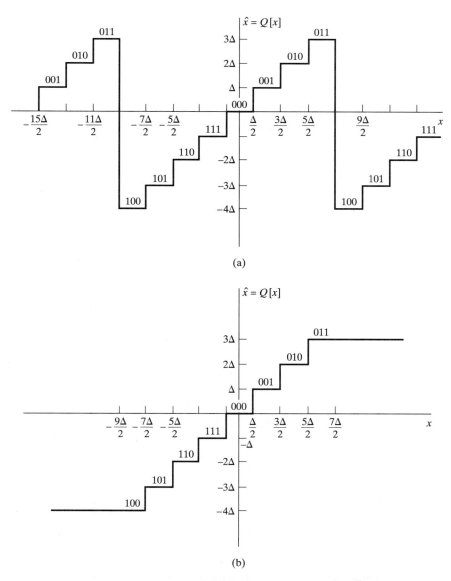

(a)

(b)

Figure 6.38   Two's-complement rounding. (a) Natural overflow. (b) Saturation.

effect of regular two's-complement arithmetic overflow. An alternative, which is called *saturation overflow* or *clipping,* is shown in Figure 6.38(b). This method of handling overflow is generally implemented for A/D conversion, and it sometimes is implemented in specialized DSP microprocessors for the addition of two's-complement numbers. With saturation overflow, the size of the error does not increase abruptly when overflow occurs; however, a disadvantage of the method is that it voids the following interesting and useful property of two's-complement arithmetic: *If several two's-complement numbers whose sum would not overflow are added, then the result of two's-complement accumulation of these numbers is correct, even though intermediate sums might overflow.*

Both quantization and overflow introduce errors in digital representations of numbers. Unfortunately, to minimize overflow while keeping the number of bits the same, we must increase X_m and thus increase the size of quantization errors proportionately. Hence, to simultaneously achieve wider dynamic range and lower quantization error, we must increase the number of bits in the binary representation.

So far, we have simply stated that the quantity X_m is an arbitrary scale factor; however, this factor has several useful interpretations. In A/D conversion, we considered X_m to be the full-scale amplitude of the A/D converter. In this case, X_m would probably represent a voltage or current in the analog part of the system. Thus, X_m serves as a calibration constant for relating binary numbers in the range $-1 \leq \hat{x}_B < 1$ to analog signal amplitudes.

In digital signal-processing implementations, it is common to assume that all signal variables and all coefficients are binary fractions. Thus, if we multiply a $(B+1)$-bit signal variable by a $(B + 1)$-bit coefficient, the result is a $(2B + 1)$-bit fraction that can be conveniently reduced to $(B+1)$ bits by rounding or truncating the least significant bits. With this convention, the quantity X_m can be thought of as a scale factor that allows the representation of numbers that are greater than unity in magnitude. For example, in fixed-point computations, it is common to assume that each binary number has a scale factor of the form $X_m = 2^c$. Accordingly, a value $c = 2$ implies that the binary point is actually located between b_2 and b_3 of the binary word in Eq. (6.58). Often, this scale factor is not explicitly represented; instead, it is implicit in the implementation program or hardware architecture.

Still another way of thinking about the scale factor X_m leads to the *floating-point representations,* in which the exponent c of the scale factor is called the *characteristic* and the fractional part $\hat{x}_B$ is called the *mantissa.* The characteristic and the mantissa are each represented explicitly as binary numbers in floating-point arithmetic systems. Floating-point representations provide a convenient means for maintaining both a wide dynamic range and a small quantization noise; however, quantization error manifests itself in a somewhat different way.

6.6.2 Quantization in Implementing Systems

Numerical quantization affects the implementation of linear time-invariant discrete-time systems in several ways. As a simple illustration, consider Figure 6.39(a), which shows a block diagram for a system in which a bandlimited continuous-time signal $x_c(t)$ is sampled to obtain the sequence $x[n]$, which is the input to a linear time-invariant

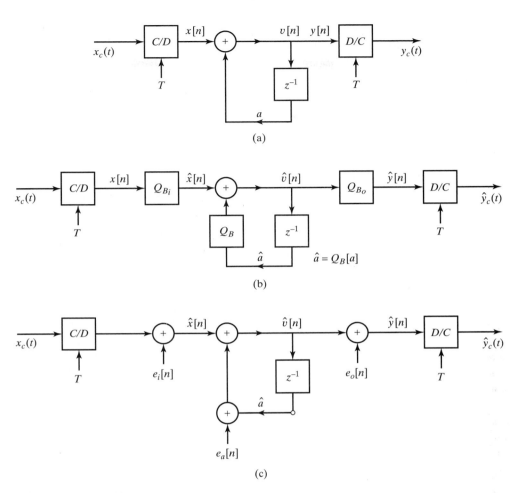

Figure 6.39 Implementation of discrete-time filtering of an analog signal. (a) Ideal system. (b) Nonlinear model. (c) Linearized model.

system whose system function is

$$H(z) = \frac{1}{1 - az^{-1}}. \tag{6.60}$$

The output of this system, $y[n]$, is converted by ideal bandlimited interpolation to the bandlimited signal $y_c(t)$.

A more realistic model is shown in Figure 6.39(b). In a practical setting, sampling would be done with an A/D converter with finite precision of $(B_i + 1)$ bits. The system would be implemented with binary arithmetic of $(B + 1)$ bits. The coefficient a in Figure 6.39(a) would be represented with $(B + 1)$ bits of precision. Also, the delayed variable $\hat{v}[n - 1]$ would be stored in a $(B + 1)$-bit register, and when the $(B + 1)$-bit number $\hat{v}[n - 1]$ is multiplied by the $(B + 1)$-bit number $\hat{a}$, the resulting product would be $(2B + 1)$ bits in length. If we assume that a $(B + 1)$-bit adder is used, the product $\hat{a}\hat{v}[n - 1]$ must be quantized (i.e., rounded or truncated) to $(B + 1)$ bits before it can be added to the $(B_i + 1)$-bit input sample $\hat{x}[n]$. When $B_i < B$, the $(B_i + 1)$ bits of the input samples can be placed anywhere in the $(B + 1)$-bit binary word with appropriate extension of the sign. Different choices correspond to different scalings of the input. The coefficient a has been quantized, so leaving aside the other quantization errors, the system response cannot in general be the same as in Figure 6.39(a). Finally, the $(B + 1)$-bit samples $\hat{v}[n]$, computed by iterating the difference equation represented by the block diagram, would be converted to an analog signal by a $(B_o + 1)$-bit D/A converter. When $B_o < B$, the output samples must be quantized before D/A conversion.

Although the model of Figure 6.39(b) could be an accurate representation of a real system, it would be difficult to analyze. The system is nonlinear due to the quantizers and the possibility of overflow at the adder. Also, quantization errors are introduced at many points in the system. The effects of these errors are impossible to analyze precisely, since they depend on the input signal, which we generally consider to be unknown. Thus, we are forced to adopt several different approximation approaches to simplify the analysis of such systems.

The effect of quantizing the system parameters, such as the coefficient a in Figure 6.39(a), is generally determined separately from the effect of quantization in data conversion or in implementing difference equations. That is, the ideal coefficients of a system function are replaced by their quantized values, and the resulting response functions are tested to see whether quantization has degraded the performance of the system to unacceptable levels. For the example of Figure 6.39, if the real number a is quantized to $(B + 1)$ bits, we must consider whether the resulting system with system function

$$\hat{H}(z) = \frac{1}{1 - \hat{a}z^{-1}} \tag{6.61}$$

is close enough to the desired system function $H(z)$ given by Eq. (6.60). Since there are only 2^{B+1} different $(B + 1)$-bit binary numbers, the pole of $H(z)$ can occur only at 2^{B+1} locations on the real axis of the z-plane. This type of analysis is discussed in more general terms in Section 6.7.

The nonlinearity of the system of Figure 6.39(b) causes behavior that cannot occur in a linear system. Specifically, systems such as this can exhibit *zero-input limit cycles*, whereby the output oscillates periodically when the input becomes zero after having been non-zero. Limit cycles are caused both by quantization and by overflow. Although the analysis of such phenomena is difficult, some useful approximate results have been developed. Limit cycles are discussed briefly in Section 6.9.

If care is taken in the design of a digital implementation, we can ensure that overflow occurs only rarely and quantization errors are small. Under these conditions, the system of Figure 6.39(b) behaves very much like a linear system (with quantized coefficients) in which quantization errors are injected at the input and output and at internal points in the network where rounding or truncation occurs. Therefore, we can replace the model of Figure 6.39(b) by the linearized model of Figure 6.39(c), where the quantizers are replaced by additive noise sources (Gold and Rader, 1969; Jackson, 1970a, 1970b). Figure 6.39(c) is equivalent to Figure 6.39(b) if we know each of the noise sources exactly. However, as discussed in Section 4.8.3, useful results are obtained if we assume a random noise model for the quantization noise in A/D conversion. This same approach can be used in analyzing the effects of arithmetic quantization in digital implementations of linear systems. As seen in Figure 6.39(c), each noise source injects a random signal that is processed by a different part of the system, but since we assume that all parts of the system are linear, we can compute the overall effect by superposition. In Section 6.8, we illustrate this style of analysis for several important systems.

In the simple example of Figure 6.39, there is little flexibility in the choice of structure. However, for higher order systems, we have seen that there is a wide variety of choices. Some of the structures are less sensitive to coefficient quantization than others. Similarly, because different structures have different quantization noise sources and because these noise sources are filtered in different ways by the system, we will find that structures which are theoretically equivalent sometimes perform quite differently when finite-precision arithmetic is used to implement them.

6.7 THE EFFECTS OF COEFFICIENT QUANTIZATION

Linear time-invariant discrete-time systems are generally used to perform a filtering operation. Methods for designing FIR and IIR filters are discussed in Chapter 7. These design methods typically assume a particular form for the system function. The result of the filter design process is a system function for which we must choose an implementation structure (a set of difference equations) from infinitely many theoretically equivalent implementations. Although we are almost always interested in implementations that require the least hardware or software complexity, it is not always possible to base the choice of implementation structure on this criterion alone. As we will see in Section 6.8, the implementation structure determines the quantization noise generated internally in the system. Also, some structures are more sensitive than others to perturbations of the coefficients. As we pointed out in Section 6.6, the standard approach to the study of these issues is to treat them independently. In this section, we consider the effects of quantizing the system parameters.

6.7.1 Effects of Coefficient Quantization in IIR Systems

When the parameters of a rational system function or corresponding difference equation are quantized, the poles and zeros of the system function move to new positions in the z-plane. Equivalently, the frequency response is perturbed from its original value. If the system implementation structure is highly sensitive to perturbations of the coefficients, the resulting system may no longer meet the original design specifications, or an IIR system might even become unstable.

A detailed sensitivity analysis for the general case is complex and usually of limited value in specific cases of digital filter implementation. Using powerful simulation tools, it is usually easiest to simply quantize the coefficients of the difference equations employed in implementing the system and then compute the frequency response and compare it with the desired frequency-response function. Even though simulation of the system is generally necessary in specific cases, it is still worthwhile to consider, in general, how the system function is affected by quantization of the coefficients of the difference equations. For example, the system function representation corresponding to both direct forms is the ratio of polynomials

$$H(z) = \frac{\displaystyle\sum_{k=0}^{M} b_k z^{-k}}{1 - \displaystyle\sum_{k=1}^{N} a_k z^{-k}}. \tag{6.62}$$

The sets of coefficients $\{a_k\}$ and $\{b_k\}$ are the ideal infinite-precision coefficients in both direct-form implementation structures. If we quantize these coefficients, we obtain the system function

$$\hat{H}(z) = \frac{\displaystyle\sum_{k=0}^{M} \hat{b}_k z^{-k}}{1 - \displaystyle\sum_{k=1}^{N} \hat{a}_k z^{-k}}, \tag{6.63}$$

where $\hat{a}_k = a_k + \Delta a_k$ and $\hat{b}_k = b_k + \Delta b_k$ are the quantized coefficients that differ from the original coefficients by the quantization errors Δa_k and Δb_k.

Now consider how the roots of the denominator and numerator polynomials (the poles and zeros of $H(z)$) are affected by the errors in the coefficients. Clearly, each polynomial root is affected by *all* of the errors in the coefficients of the polynomial. Thus, each pole and zero will be affected by all of the quantization errors in the denominator and numerator polynomials, respectively. More specifically, Kaiser (1966) showed that if the poles (or zeros) are tightly clustered, it is possible that small errors in the denominator (numerator) coefficients can cause large shifts of the poles (zeros) for the direct-form structures. Thus, if the poles (zeros) are tightly clustered, corresponding to a narrow-bandpass filter or a narrow-bandwidth lowpass filter, then we can expect the poles of the direct-form structure to be quite sensitive to quantization errors in the coefficients. Furthermore, Kaiser's analysis showed that the larger the number of clustered poles (zeros), the greater is the sensitivity.

The cascade- and parallel-form system functions, which are given by Eqs. (6.30) and (6.35), respectively, consist of combinations of second-order direct-form systems. However, in both cases, each pair of complex-conjugate poles is realized independently of all the other poles. Thus, the error in a particular pole pair is independent of its distance from the other poles of the system function. For the cascade form, the same argument holds for the zeros, since they are realized as independent second-order factors. Thus, the cascade form is generally much less sensitive to coefficient quantization than the equivalent direct-form realization.

As seen in Eq. (6.35), the zeros of the parallel-form system function are realized implicitly, through combining the quantized second-order sections to obtain a common denominator. Thus, a particular zero is affected by quantization errors in the numerator and denominator coefficients of *all* the second-order sections. However, for most practical filter designs, the parallel form is also found to be much less sensitive to coefficient quantization than the equivalent direct forms because the second-order subsystems are not extremely sensitive to quantization. In many practical filters, the zeros are often widely distributed around the unit circle, or in some cases they may all be located at $z = \pm1$. In the latter situation, the zeros mainly provide much higher attenuation around frequencies $\omega = 0$ and $\omega = \pi$ than is specified, and thus, movements of zeros away from $z = \pm1$ do not significantly degrade the performance of the system.

6.7.2 Example of Coefficient Quantization in an Elliptic Filter

To illustrate the effect of coefficient quantization, consider the example of a bandpass IIR elliptic filter. The filter was designed to meet the following specifications:

$$0.99 \leq |H(e^{j\omega})| \leq 1.01, \qquad 0.3\pi \leq |\omega| \leq 0.4\pi,$$

$$|H(e^{j\omega})| \leq 0.01 (\text{i.e.,} -40\,\text{dB}), \qquad |\omega| \leq 0.29\pi,$$

$$|H(e^{j\omega})| \leq 0.01 (\text{i.e.,} -40\,\text{dB}), \qquad 0.41\pi \leq |\omega| \leq \pi.$$

A filter of order 12 is required to meet these sharp-cutoff narrowband specifications.

In the filter design process, the coefficients of the cascade form were computed with 32-bit floating-point accuracy. These are coefficients defined as "unquantized." Table 6.1 gives the coefficients of the six second-order sections (as defined in Eq. (6.30)). Figure 6.40(a) shows the log magnitude in dB, and Figure 6.40(b) shows the magnitude response in the passband (i.e., $0.3\pi \leq |\omega| \leq 0.4\pi$) for the unquantized system.

To illustrate how coefficients are quantized and represented as fixed-point numbers, the coefficients in Table 6.1 were quantized to 16-bit accuracy. The resulting coefficients are presented in Table 6.2. The fixed-point coefficients are shown as a decimal integer times a power-of-2 scale factor. The binary representation would be obtained by converting the decimal integer to a binary number. In a fixed-point implementation, the scale factor would be represented only implicitly in the data shifts that would be necessary to line up the binary points of products prior to their addition to other products. Notice that binary points of the coefficients are not all in the same location. For example, all the coefficients with scale factor 2^{-15} have their binary points between the

TABLE 6.1   UNQUANTIZED CASCADE-FORM COEFFICIENTS FOR A 12TH-ORDER ELLIPTIC FILTER

k	a_{1k}	a_{2k}	b_{0k}	b_{1k}	b_{2k}
1	0.738409	−0.850835	0.135843	0.026265	0.135843
2	0.960374	−0.860000	0.278901	−0.444500	0.278901
3	0.629449	−0.931460	0.535773	−0.249249	0.535773
4	1.116458	−0.940429	0.697447	−0.891543	0.697447
5	0.605182	−0.983693	0.773093	−0.425920	0.773093
6	1.173078	−0.986166	0.917937	−1.122226	0.917937

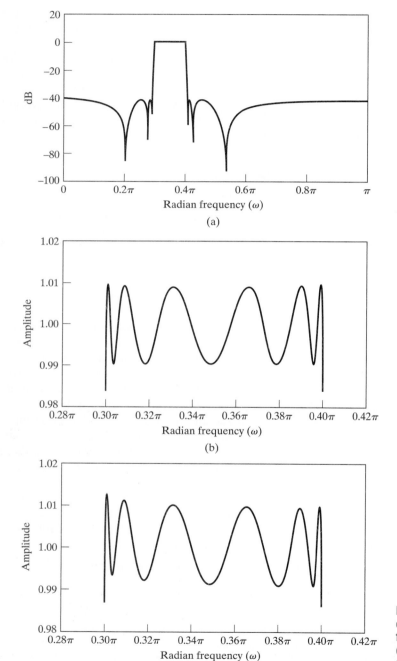

Figure 6.40 IIR coefficient quantization example. (a) Log magnitude for unquantized elliptic bandpass filter. (b) Passband for unquantized cascade case. (c) Passband for cascade structure with 16-bit coefficients.

Figure 6.40 (*continued*) (d) Passband for parallel structure with 16-bit coefficients. (e) Log magnitude for direct form with 16-bit coefficients.

TABLE 6.2 SIXTEEN-BIT QUANTIZED CASCADE-FORM COEFFICIENTS FOR A 12TH-ORDER ELLIPTIC FILTER

k	a_{1k}	a_{2k}	b_{0k}	b_{1k}	b_{2k}
1	24196×2^{-15}	-27880×2^{-15}	17805×2^{-17}	3443×2^{-17}	17805×2^{-17}
2	31470×2^{-15}	-28180×2^{-15}	18278×2^{-16}	-29131×2^{-16}	18278×2^{-16}
3	20626×2^{-15}	-30522×2^{-15}	17556×2^{-15}	-8167×2^{-15}	17556×2^{-15}
4	18292×2^{-14}	-30816×2^{-15}	22854×2^{-15}	-29214×2^{-15}	22854×2^{-15}
5	19831×2^{-15}	-32234×2^{-15}	25333×2^{-15}	-13957×2^{-15}	25333×2^{-15}
6	19220×2^{-14}	-32315×2^{-15}	15039×2^{-14}	-18387×2^{-14}	15039×2^{-14}

sign bit, b_0, and the highest fractional bit, b_1, as shown in Eq. (6.58). However, numbers whose magnitudes do not exceed 0.5, such as the coefficient b_{02}, can be shifted left by one or more bit positions.[5] Thus, the binary point for b_{02} is actually to the left of the

[5]The use of different binary point locations retains greater accuracy in the coefficients, but it complicates the programming or system architecture.

sign bit. On the other hand, numbers whose magnitudes exceed 1 but are less than 2, such as a_{16}, must have their binary points moved one position to the right, i.e., between b_1 and b_2 in Eq. (6.58).

Figure 6.40(c) shows the magnitude response in the passband for the quantized cascade-form implementation. The frequency response is only slightly degraded in the passband region.

To obtain other equivalent structures, the cascade-form system function must be rearranged into a different form. For example, if a parallel-form structure is determined (by partial fraction expansion of the unquantized system function), and the resulting coefficients are quantized to 16 bits as before, the frequency response in the passband is as shown in Figure 6.40(d).

If the second-order factors in the numerator and denominator of the unquantized cascade form are multiplied together to form 12th-order numerator and denominator polynomials, and if the resulting coefficients are quantized to 16 bits and used in a direct-form structure, the log magnitude of the frequency response is as shown in Figure 6.40(e). In this case, the passband response is virtually destroyed by the quantization. Comparing Figure 6.40(e) with Figure 6.40(a) shows that the widely separated zeros on either side of the passband (at approximately $\omega = .2\pi$ and $.54\pi$) have not moved off the unit circle. However, the unquantized system has a pair of zeros on the unit circle just at both stopband edges. These zeros, being clustered, move away from the unit circle with quantization and thus contribute to the overall degradation of the frequency response.

6.7.3 Poles of Quantized Second-Order Sections

The example just discussed illustrates the robustness of the cascade and parallel forms to the effects of coefficient quantization, and it also illustrates the extreme sensitivity of the direct forms for high-order filters. Because of this sensitivity, the direct forms are rarely used for implementing anything other than second-order systems.[6] Since the cascade and parallel forms can be configured to require the same amount of memory and the same or only slightly more computation as the canonic direct form, these modular structures are the most commonly used. More complex structures may be more robust for very short word lengths, but they require significantly more computation for systems of the same order.

Even for the second-order systems that are used to implement the cascade and parallel forms, there remains some flexibility to improve the robustness to coefficient quantization. Consider a complex-conjugate pole pair implemented using the direct form, as in Figure 6.41. With infinite-precision coefficients, this network has poles at $z = re^{j\theta}$ and $z = re^{-j\theta}$. However, if the coefficients $2r\cos\theta$ and $-r^2$ are quantized, only a finite number of different pole locations is possible. The poles must lie on a grid in the z-plane defined by the intersection of concentric circles (corresponding to the quantization of r^2) and vertical lines (corresponding to the quantization of $2r\cos\theta$). Such a grid is illustrated in Figure 6.42(a) for 4-bit quantization (3 bits plus 1 bit for the sign); i.e., r^2 is restricted to seven positive values and zero, while $2r\cos\theta$ is restricted to

---

[6]An exception is in speech synthesis, where systems of 10th order and higher are routinely implemented using the direct form. This is possible because in speech synthesis the poles of the system function are widely separated (Rabiner and Schafer, 1978).

Figure 6.41   Direct-form implementation of a complex-conjugate pole pair.

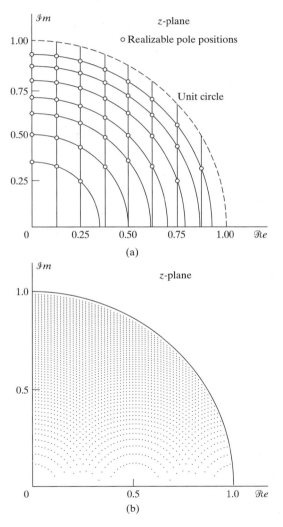

Figure 6.42   Pole-locations for the second-order IIR direct-form system of Figure 6.41. (a) Four-bit quantization of coefficients. (b) Seven-bit quantization.

seven positive values, eight negative values, and zero. Figure 6.42(b) shows a denser grid obtained with 7-bit quantization (6 bits plus 1 bit for the sign). The plots of Figure 6.42 are, of course, symmetrically mirrored into each of the other quadrants of the z-plane. Notice that for the direct form, the grid is rather sparse around the real axis. Thus, poles located around $\theta = 0$ or $\theta = \pi$ may be shifted more than those around $\theta = \pi/2$. Of

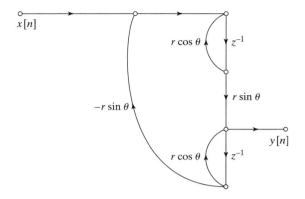

Figure 6.43 Coupled-form implementation of a complex-conjugate pole pair.

course, it is always possible that the infinite-precision pole location is very close to one of the allowed quantized poles. In this case, quantization causes no problem whatsoever, but in general, quantization can be expected to degrade performance.

An alternative second-order structure for realizing poles at $z = re^{j\theta}$ and $z = re^{-j\theta}$ is shown in Figure 6.43. This structure is referred to as the *coupled form* for the second-order system (Rader and Gold, 1967). It is easily verified that the systems of Figures 6.41 and 6.43 have the same poles for infinite-precision coefficients. To implement the system of Figure 6.43, we must quantize $r\cos\theta$ and $r\sin\theta$. Since these quantities are the real and imaginary parts, respectively, of the pole locations, the quantized pole locations are at intersections of evenly spaced horizontal and vertical lines in the z-plane. Figures 6.44(a) and 6.44(b) show the possible pole locations for 4-bit and 7-bit quantization, respectively. In this case, the density of pole locations is uniform throughout the interior of the unit circle. Twice as many constant multipliers are required to achieve this more uniform density. In some situations, the extra computation might be justified to achieve more accurate pole location with reduced word length.

6.7.4 Effects of Coefficient Quantization in FIR Systems

For FIR systems, we must be concerned only with the locations of the zeros of the system function, since, for causal FIR systems, all the poles are at $z = 0$. Although we have just seen that the direct-form structure should be avoided for high-order IIR systems, it turns out that the direct-form structure is commonly used for FIR systems. To understand why this is so, we express the system function for a direct-form FIR system in the form

$$H(z) = \sum_{n=0}^{M} h[n]z^{-n}. \tag{6.64}$$

Now, suppose that the coefficients $\{h[n]\}$ are quantized, resulting in a new set of coefficients $\{\hat{h}[n] = h[n] + \Delta h[n]\}$. The system function for the quantized system is then

$$\hat{H}(z) = \sum_{n=0}^{M} \hat{h}[n]z^{-n} = H(z) + \Delta H(z), \tag{6.65}$$

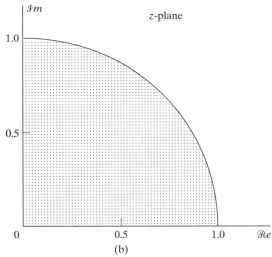

Figure 6.44   Pole locations for coupled-form second-order IIR system of Figure 6.43. (a) Four-bit quantization of coefficients. (b) Seven-bit quantization.

where

$$\Delta H(z) = \sum_{n=0}^{M} \Delta h[n] z^{-n}. \qquad (6.66)$$

Thus, the system function (and therefore, also the frequency response) of the quantized system is linearly related to the quantization errors in the impulse-response coefficients. For this reason, the quantized system can be represented as in Figure 6.45, which shows the unquantized system in parallel with an error system whose impulse response is the sequence of quantization error samples $\{\Delta h[n]\}$ and whose system function is the corresponding z-transform, $\Delta H(z)$.

Another approach to studying the sensitivity of the direct-form FIR structure would be to examine the sensitivity of the zeros to quantization errors in the impulse-

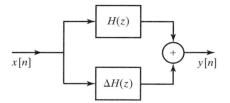

Figure 6.45 Representation of coefficient quantization in FIR systems.

response coefficients, which are, of course the coefficients of the polynomial $H(z)$. If the zeros of $H(z)$ are tightly clustered, then their locations will be highly sensitive to quantization errors in the impulse-response coefficients. The reason that the direct form FIR system is widely used is that, for most linear phase FIR filters, the zeros are more or less uniformly spread in the z-plane. We demonstrate this by the following example.

6.7.5 Example of Quantization of an Optimum FIR Filter

As an example of the effect of coefficient quantization in the FIR case, consider a linear-phase lowpass filter designed to meet the following specifications:

$$0.99 \leq |H(e^{j\omega})| \leq 1.01, \qquad 0 \leq |\omega| \leq 0.4\pi,$$

$$|H(e^{j\omega})| \leq 0.001 (\text{i.e.}, -60\,\text{dB}), \qquad 0.6\pi \leq |\omega| \leq \pi.$$

This filter was designed using the Parks-McClellan design technique, which will be discussed in Section 7.4.3. The details of the design for this example are considered in Section 7.5.1.

Table 6.3 shows the unquantized impulse-response coefficients for the system, along with quantized coefficients for 16-, 14-, 13-, and 8-bit quantization. Figure 6.46 gives a comparison of the frequency responses of the various systems. Figure 6.46(a) shows the log magnitude in dB of the frequency response for unquantized coefficients.

TABLE 6.3   UNQUANTIZED AND QUANTIZED COEFFICIENTS FOR AN OPTIMUM FIR LOWPASS FILTER ($M = 27$)

Coefficient	Unquantized	16 bits	14 bits	13 bits	8 bits
$h[0] = h[27]$	1.359657×10^{-3}	45×2^{-15}	11×2^{-13}	6×2^{-12}	0×2^{-7}
$h[1] = h[26]$	-1.616993×10^{-3}	-53×2^{-15}	-13×2^{-13}	-7×2^{-12}	0×2^{-7}
$h[2] = h[25]$	-7.738032×10^{-3}	-254×2^{-15}	-63×2^{-13}	-32×2^{-12}	-1×2^{-7}
$h[3] = h[24]$	-2.686841×10^{-3}	-88×2^{-15}	-22×2^{-13}	-11×2^{-12}	0×2^{-7}
$h[4] = h[23]$	1.255246×10^{-2}	411×2^{-15}	103×2^{-13}	51×2^{-12}	2×2^{-7}
$h[5] = h[22]$	6.591530×10^{-3}	216×2^{-15}	54×2^{-13}	27×2^{-12}	1×2^{-7}
$h[6] = h[21]$	-2.217952×10^{-2}	-727×2^{-15}	-182×2^{-13}	-91×2^{-12}	-3×2^{-7}
$h[7] = h[20]$	-1.524663×10^{-2}	-500×2^{-15}	-125×2^{-13}	-62×2^{-12}	-2×2^{-7}
$h[8] = h[19]$	3.720668×10^{-2}	1219×2^{-15}	305×2^{-13}	152×2^{-12}	5×2^{-7}
$h[9] = h[18]$	3.233332×10^{-2}	1059×2^{-15}	265×2^{-13}	132×2^{-12}	4×2^{-7}
$h[10] = h[17]$	-6.537057×10^{-2}	-2142×2^{-15}	-536×2^{-13}	-268×2^{-12}	-8×2^{-7}
$h[11] = h[16]$	-7.528754×10^{-2}	-2467×2^{-15}	-617×2^{-13}	-308×2^{-12}	-10×2^{-7}
$h[12] = h[15]$	1.560970×10^{-1}	5115×2^{-15}	1279×2^{-13}	639×2^{-12}	20×2^{-7}
$h[13] = h[14]$	4.394094×10^{-1}	14399×2^{-15}	3600×2^{-13}	1800×2^{-12}	56×2^{-7}

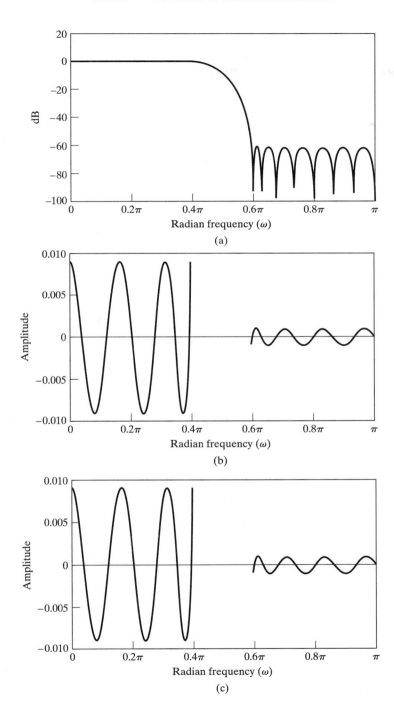

Figure 6.46   FIR quantization example. (a) Log magnitude for unquantized case. (b) Approximation error for unquantized case. (Error not defined in transition band.) (c) Approximation error for 16-bit quantization.

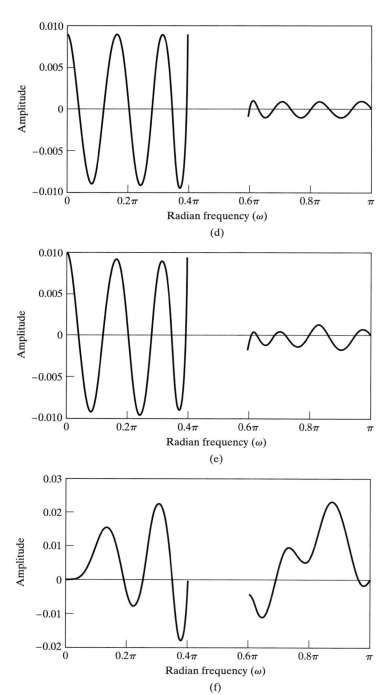

Figure 6.46 *(continued)* (d) Approximation error for 14-bit quantization. (e) Approximation error for 13-bit quantization. (f) Approximation error for 8-bit quantization.

Figures 6.46(b), (c), (d), (e), and (f) show the passband and stopband approximation errors (errors in approximating unity in the passband and zero in the stopband) for the unquantized, 16-, 14-, 13-, and 8-bit quantized cases, respectively. From Figure 6.46, we see that the system meets the specifications for the unquantized case and both the 16-bit and 14-bit quantized cases. However, with 13-bit quantization the stopband approximation error becomes greater than 0.001, and with 8-bit quantization the stopband approximation error is over 10 times as large as specified. Thus, we see that at least 14-bit coefficients are required for a direct-form implementation of the system. However, this is not a serious limitation, since 16-bit or 14-bit coefficients are well matched to many of the technologies that might be used to implement such a filter.

The effect of quantization of the filter coefficients on the locations of the zeros of the filter is shown in Figure 6.47. Note that in the unquantized case, shown in Figure 6.47(a), the zeros are spread around the z-plane, although there is some clustering

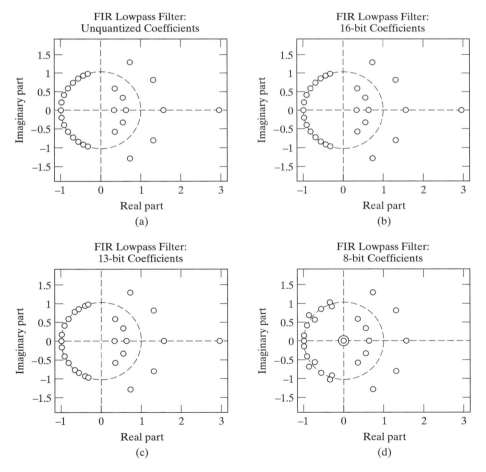

Figure 6.47 Effect of impulse response quantization on zeros of $H(z)$. (a) Unquantized. (b) Sixteen-bit quantization. (c) Thirteen-bit quantization. (d) Eight-bit quantization.

on the unit circle. The zeros on the unit circle are primarily responsible for developing the stopband attenuation, while those at conjugate reciprocal locations off the unit circle are primarily responsible for forming the passband. Note that little difference is observed in Figure 6.47(b) for 16-bit quantization, but in Figure 6.47(c), showing 13-bit quantization, the zeros on the unit circle have moved significantly. Finally, in Figure 6.47(d), we see that 8-bit quantization causes several of the zeros on the unit circle to pair up and move off the circle to conjugate reciprocal locations. This behavior of the zeros is entirely consistent with the behavior of the frequency response shown in Figure 6.46.

A final point about this example is worth mentioning. All of the unquantized coefficients have magnitudes less than 0.5. Consequently, if all of the coefficients (and therefore, the impulse response) are doubled prior to quantization, more efficient use of the available bits will result, corresponding in effect to increasing B by 1. In Table 6.3 and Figure 6.46, we did not take this potential for increased accuracy into account.

6.7.6 Maintaining Linear Phase

So far, we have not made any assumptions about the phase response of the FIR system. However, the possibility of generalized linear phase is one of the major advantages of an FIR system. Recall that a linear-phase FIR system has either a symmetric ($h[M - n] = h[n]$) or an antisymmetric ($h[M - n] = -h[n]$) impulse response. These linear-phase conditions are easily preserved for the direct-form quantized system. Thus, all the systems discussed in the example of the previous subsection have a precisely linear phase, regardless of the coarseness of the quantization. This can be seen in the way in which the conjugate reciprocal locations are preserved in Figure 6.47.

Figure 6.47(d) suggests that, in situations where quantization is very coarse or for high-order systems with closely spaced zeros, it may be worthwhile to realize smaller sets of zeros independently with a cascade-form FIR system. To maintain linear phase, each of the sections in the cascade must also have linear phase. Recall that the zeros of a linear-phase system must occur as illustrated in Figure 6.36. For example, if we use second-order sections of the form $(1 + az^{-1} + z^{-2})$ for each complex-conjugate pair of zeros on the unit circle, the zero can move only on the unit circle when the coefficient a is quantized. This prevents zeros from moving away from the unit circle, thereby lessening their attenuating effect. Similarly, real zeros inside the unit circle and at the reciprocal location outside the unit circle would remain real. Also, zeros at $z = \pm 1$ can be realized exactly by first-order systems. If a pair of complex-conjugate zeros inside the unit circle is realized by a second-order system rather than a fourth-order system, then we must ensure that, for each complex zero inside the unit circle, there is a conjugate reciprocal zero outside the unit circle. This can be done by expressing the fourth-order factor corresponding to zeros at $z = re^{j\theta}$ and $z = r^{-1}e^{-j\theta}$ as

$$
1 + cz^{-1} + dz^{-2} + cz^{-3} + z^{-4}
$$
$$
= (1 - 2r\cos\theta z^{-1} + r^2 z^{-2})\frac{1}{r^2}(r^2 - 2r\cos\theta z^{-1} + z^{-2}). \tag{6.67}
$$

This condition corresponds to the subsystem shown in Figure 6.48. This system uses the same coefficients, $-2r\cos\theta$ and r^2, to realize both the zeros inside the unit circle and the conjugate reciprocal zeros outside the unit circle. Thus, the linear-phase condition

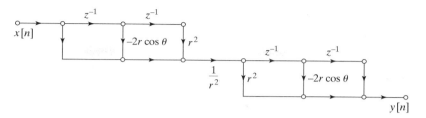

Figure 6.48   Subnetwork to implement fourth-order factors in a linear-phase FIR system such that linearity of the phase is maintained independently of parameter quantization.

is preserved under quantization. Notice that the factor $(1 - 2r\cos\theta z^{-1} + r^2 z^{-2})$ is identical to the denominator of the second-order direct-form IIR system of Figure 6.41. Therefore, the set of quantized zeros is as depicted in Figure 6.42. More details on cascade realizations of FIR systems are given by Herrmann and Schüssler (1970b).

6.8 EFFECTS OF ROUND-OFF NOISE IN DIGITAL FILTERS

Difference equations realized with finite-precision arithmetic are nonlinear systems. Although it is important in general to understand how this nonlinearity affects the performance of discrete-time systems, a precise analysis of arithmetic quantization effects is generally not required in practical applications, where we are typically concerned with the performance of a specific system. Indeed, just as with coefficient quantization, the most effective approach is often to simulate the system and measure its performance. For example, a common objective in quantization error analysis is to choose the digital word length such that the digital system is a sufficiently accurate realization of the desired linear system and at the same time requires a minimum of hardware or software complexity. The digital word length can, of course, be changed only in steps of 1 bit, and as we have already seen in Section 4.8.2, the addition of 1 bit to the word length reduces the size of quantization errors by a factor of 2. Thus, the choice of word length is insensitive to inaccuracies in the quantization error analysis; an analysis that is correct to within 30–40% is often adequate. For this reason, many of the important effects of quantization can be studied using linear models. We develop such models in this section and illustrate their use with several examples. An exception is the phenomenon of zero-input limit cycles, which are strictly nonlinear phenomena. We restrict our study of nonlinear models for digital filters to a brief introduction to zero-input limit cycles in Section 6.9.

6.8.1 Analysis of the Direct-Form IIR Structures

To introduce the basic ideas, let us consider the direct-form structure for a linear time-invariant discrete-time system. The flow graph of a direct form I second-order system is shown in Figure 6.49(a). The general Nth-order difference equation for the direct form I structure is

$$y[n] = \sum_{k=1}^{N} a_k y[n-k] + \sum_{k=0}^{M} b_k x[n-k], \qquad (6.68)$$

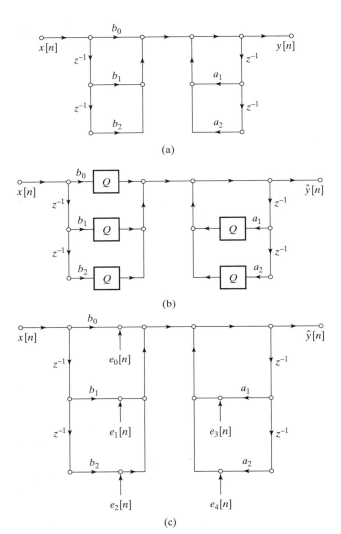

Figure 6.49 Models for direct form I system. (a) Infinite-precision model. (b) Nonlinear quantized model. (c) Linear-noise model.

and the system function is

$$H(z) = \frac{\displaystyle\sum_{k=0}^{M} b_k z^{-k}}{1 - \displaystyle\sum_{k=1}^{N} a_k z^{-k}} = \frac{B(z)}{A(z)}. \tag{6.69}$$

Let us assume that all signal values and coefficients are represented by $(B+1)$-bit fixed-point binary numbers. Then, in implementing Eq. (6.68) with a $(B+1)$-bit adder, it would be necessary to reduce the length of the $(2B+1)$-bit products to $(B+1)$ bits. Since all numbers are treated as fractions, we would discard the least significant B bits by either rounding or truncation. This is represented by replacing each constant multiplier branch in Figure 6.49(a) by a constant multiplier followed by a quantizer, as in the nonlinear model of Figure 6.49(b). The difference equation corresponding to Figure 6.49(b) is the nonlinear equation

$$\hat{y}[n] = \sum_{k=1}^{N} Q[a_k \hat{y}[n-k]] + \sum_{k=0}^{M} Q[b_k x[n-k]]. \tag{6.70}$$

Figure 6.49(c) shows an alternative representation in which the quantizers are replaced by noise sources that are equal to the quantization error at the output of each quantizer. For example, rounding or truncation of a product $bx[n]$ is represented by a noise source of the form

$$e[n] = Q[bx[n]] - bx[n]. \tag{6.71}$$

If the noise sources are known exactly, then Figure 6.49(c) is exactly equivalent to Figure 6.49(b). However, Figure 6.49(c) is most useful when we assume that each quantization noise source has the following properties:

1. Each quantization noise source $e[n]$ is a wide-sense-stationary white-noise process.
2. Each quantization noise source has a uniform distribution of amplitudes over one quantization interval.
3. Each quantization noise source is *uncorrelated* with the input to the corresponding quantizer, all other quantization noise sources, and the input to the system.

These assumptions are identical to those made in the analysis of A/D conversion in Section 4.8. Strictly speaking, our assumptions here cannot be valid, since the quantization error depends directly on the input to the quantizer. This is readily apparent for constant and sinusoidal signals. However, experimental and theoretical analyses have shown (see Bennett, 1948; Widrow, 1956, 1961) that in many situations the model just described leads to accurate predictions of statistical averages such as the mean, variance, and correlation function. This is true when the input signal is a complicated wideband signal such as speech, in which the signal fluctuates rapidly among all the quantization levels and traverses many of those levels in going from sample to sample (Gold and Rader, 1969). The simple linear-noise model presented here allows us to characterize the noise generated in the system by averages such as the mean and variance and to determine how these averages are modified by the system.

For $(B+1)$-bit quantization, we showed in Section 6.6 that, for rounding,

$$-\frac{1}{2}2^{-B} < e[n] \le \frac{1}{2}2^{-B}, \tag{6.72a}$$

and for two's-complement truncation,

$$-2^{-B} < e[n] \le 0. \tag{6.72b}$$

Thus, according to our second assumption, the probability density functions for the random variables representing quantization error are the uniform densities shown in Figure 6.50(a) for rounding and in Figure 6.50(b) for truncation. The mean and variance for rounding are, respectively,

$$m_e = 0, \tag{6.73a}$$

$$\sigma_e^2 = \frac{2^{-2B}}{12}. \tag{6.73b}$$

(a)

(b)

Figure 6.50 Probability density function for quantization errors. (a) Rounding. (b) Truncation.

For two's-complement truncation, the mean and variance are

$$m_e = -\frac{2^{-B}}{2}, \tag{6.74a}$$

$$\sigma_e^2 = \frac{2^{-2B}}{12}. \tag{6.74b}$$

In general, the autocorrelation sequence of a quantization noise source is, according to the first assumption,

$$\phi_{ee}[n] = \sigma_e^2 \, \delta[n] + m_e^2. \tag{6.75}$$

In the case of rounding, which we will assume for convenience henceforth, $m_e = 0$, so the autocorrelation function is $\phi_{ee}[n] = \sigma_e^2 \, \delta[n]$, and the power spectrum is $\Phi_{ee}(e^{j\omega}) = \sigma_e^2$ for $|\omega| \leq \pi$. In this case, the variance and the average power are identical. In the case of truncation, the mean is not zero, so average-power results derived for rounding must be corrected by computing the mean of the signal and adding its square to the average-power results for rounding.

With this model for each of the noise sources in Figure 6.49(c), we can now proceed to determine the effect of the quantization noise on the output of the system. To aid us in doing this, it is helpful to observe that all of the noise sources in that figure are effectively injected between the part of the system that implements the zeros and the part that implements the poles. Thus, Figure 6.51 is equivalent to Figure 6.49(c) if $e[n]$ in Figure 6.51 is

$$e[n] = e_0[n] + e_1[n] + e_2[n] + e_3[n] + e_4[n]. \tag{6.76}$$

Since we assume that all the noise sources are independent of the input and independent of each other, the variance of the combined noise sources for the second-order direct form I case is

$$\sigma_e^2 = \sigma_{e_0}^2 + \sigma_{e_1}^2 + \sigma_{e_2}^2 + \sigma_{e_3}^2 + \sigma_{e_4}^2 = 5 \cdot \frac{2^{-2B}}{12}, \tag{6.77}$$

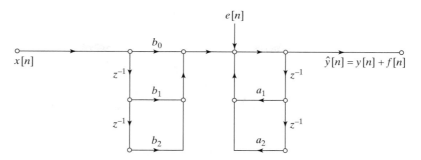

Figure 6.51   Linear-noise model for direct form I with noise sources combined.

and for the general direct form I case, it is

$$\sigma_e^2 = (M + 1 + N)\frac{2^{-2B}}{12}. \tag{6.78}$$

To obtain an expression for the output noise, we note from Figure 6.51 that the system has two inputs, $x[n]$ and $e[n]$, and since the system is now assumed to be linear, the output can be represented as $\hat{y}[n] = y[n] + f[n]$, where $y[n]$ is the response of the ideal unquantized system to the input sequence $x[n]$ and $f[n]$ is the response of the system to the input $e[n]$. The output $y[n]$ is given by the difference equation (6.68), but since $e[n]$ is injected after the zeros and before the poles, the output noise satisfies the difference equation

$$f[n] = \sum_{k=1}^{N} a_k f[n-k] + e[n]; \tag{6.79}$$

i.e., the properties of the output noise in the direct form I implementation depend only on the poles of the system.

To determine the mean and variance of the output noise sequence, we can use some results from Section 2.10. Consider a linear system with system function $H_{ef}(z)$ with a white-noise input $e[n]$ and corresponding output $f[n]$. Then, from Eqs. (2.189) and (2.190), the mean of the output is

$$m_f = m_e \sum_{n=-\infty}^{\infty} h_{ef}[n] = m_e H_{ef}(e^{j0}). \tag{6.80}$$

Since $m_e = 0$ for rounding, the mean of the output will be zero, and we need not be concerned with the mean value of the noise if we assume rounding. From Eqs. (6.75) and (2.195), it follows that, because, for rounding, $e[n]$ is a zero-mean white-noise sequence, the power density spectrum of the output noise is simply

$$P_{ff}(\omega) = \Phi_{ff}(e^{j\omega}) = \sigma_e^2 |H_{ef}(e^{j\omega})|^2. \tag{6.81}$$

From Eq. (2.197), the variance of the output noise can be shown to be

$$\sigma_f^2 = \frac{1}{2\pi} \int_{-\pi}^{\pi} P_{ff}(\omega) d\omega = \sigma_e^2 \frac{1}{2\pi} \int_{-\pi}^{\pi} |H_{ef}(e^{j\omega})|^2 d\omega. \tag{6.82}$$

Using Parseval's theorem in the form of Eq. (2.167), we can also express σ_f^2 as

$$\sigma_f^2 = \sigma_e^2 \sum_{n=-\infty}^{\infty} |h_{ef}[n]|^2. \tag{6.83}$$

When the system function corresponding to $h_{ef}[n]$ is a rational function, as it will always be for difference equations of the type considered in this chapter, we can use Eq. (A.66) in Appendix A for evaluating infinite sums of squares of the form of Eq. (6.83).

We will use the results summarized in Eqs. (6.80)–(6.83) often in our analysis of quantization noise in linear systems. For example, for the direct form I system of Figure 6.51, $H_{ef}(z) = 1/A(z)$; i.e., the system function from the point where all the noise sources are injected to the output consists only of the poles of the system function $H(z)$ in Eq. (6.69). Thus, we conclude that, in general, the total output variance due to internal round-off or truncation is

$$\sigma_f^2 = (M+1+N)\frac{2^{-2B}}{12}\frac{1}{2\pi}\int_{-\pi}^{\pi}\frac{d\omega}{|A(e^{j\omega})|^2}$$
$$= (M+1+N)\frac{2^{-2B}}{12}\sum_{n=-\infty}^{\infty}|h_{ef}[n]|^2, \tag{6.84}$$

where $h_{ef}[n]$ is the impulse response corresponding to $H_{ef}(z) = 1/A(z)$. The use of the preceding results is illustrated by the following examples.

Example 6.9   Round-off Noise in a First-Order System

Suppose that we wish to implement a stable system having the system function

$$H(z) = \frac{b}{1-az^{-1}}, \qquad |a| < 1. \tag{6.85}$$

Figure 6.52 shows the flow graph of the linear-noise model for the implementation in which products are quantized before addition. Each noise source is filtered by the system from $e[n]$ to the output, for which the impulse response is $h_{ef}[n] = a^n u[n]$. From Eq. (6.81), the power spectrum of the output noise is

$$P_{ff}(\omega) = 2\frac{2^{-2B}}{12}\left(\frac{1}{1+a^2-2a\cos\omega}\right), \tag{6.86}$$

The total noise variance at the output is

$$\sigma_f^2 = 2\frac{2^{-2B}}{12}\sum_{n=0}^{\infty}|a|^{2n} = 2\frac{2^{-2B}}{12}\left(\frac{1}{1-|a|^2}\right). \tag{6.87}$$

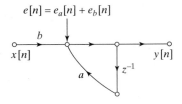

Figure 6.52   First-order linear noise model.

From Eq. (6.87), we see that the output noise variance increases as the pole at $z = a$ approaches the unit circle. Thus, to maintain the noise variance below a specified level as $|a|$ approaches unity, we must use longer word lengths. The following example also illustrates this point.

Example 6.10   Round-off Noise in a Second-Order System

Consider a stable second-order direct form I system with system function

$$H(z) = \frac{b_0 + b_1 z^{-1} + b_2 z^{-2}}{(1 - re^{j\theta} z^{-1})(1 - re^{-j\theta} z^{-1})}. \tag{6.88}$$

The linear-noise model for this system is shown in Figure 6.49(c) or, equivalently Figure 6.51, with $a_1 = 2r \cos \theta$ and $a_2 = -r^2$. In this case, the total output noise power can be expressed in the form

$$\sigma_f^2 = 5 \frac{2^{-2B}}{12} \frac{1}{2\pi} \int_{-\pi}^{\pi} \frac{d\omega}{|(1 - re^{j\theta} e^{-j\omega})(1 - re^{-j\theta} e^{-j\omega})|^2}. \tag{6.89}$$

Using Eq. (A.66) in Appendix A, the output noise power is found to be

$$\sigma_f^2 = 5 \frac{2^{-2B}}{12} \left(\frac{1+r^2}{1-r^2} \right) \frac{1}{r^4 + 1 - 2r^2 \cos 2\theta}. \tag{6.90}$$

As in Example 6.9, we see that as the complex conjugate poles approach the unit circle ($r \to 1$), the total output noise variance increases, thus requiring longer word lengths to maintain the variance below a prescribed level.

The techniques of analysis developed so far for the direct form I structure can also be applied to the direct form II structure. The nonlinear difference equations for the direct form II structure are of the form

$$\hat{w}[n] = \sum_{k=1}^{N} Q[a_k \hat{w}[n-k]] + x[n], \tag{6.91a}$$

$$\hat{y}[n] = \sum_{k=0}^{M} Q[b_k \hat{w}[n-k]]. \tag{6.91b}$$

Figure 6.53(a) shows the linear-noise model for a second-order direct form II system. A noise source has been introduced after each multiplication, indicating that the products are quantized to $(B+1)$ bits before addition. Figure 6.53(b) shows an equivalent linear model wherein we have moved the noise sources resulting from implementation of the poles and combined them into a single noise source $e_a[n] = e_3[n] + e_4[n]$ at the input. Likewise, the noise sources due to implementation of the zeros are combined into the single noise source $e_b[n] = e_0[n] + e_1[n] + e_2[n]$ that is added directly to the output. From this equivalent model, it follows that for M zeros and N poles and rounding ($m_e = 0$), the power spectrum of the output noise is

$$P_{ff}(\omega) = N \frac{2^{-2B}}{12} |H(e^{j\omega})|^2 + (M+1) \frac{2^{-2B}}{12}, \tag{6.92}$$

(a)

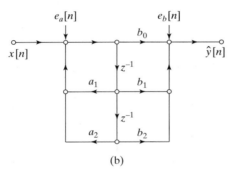

(b)

Figure 6.53   Linear-noise models for direct form II. (a) Showing quantization of individual products. (b) With noise sources combined.

and the output noise variance is

$$\sigma_f^2 = N \frac{2^{-2B}}{12} \frac{1}{2\pi} \int_{-\pi}^{\pi} |H(e^{j\omega})|^2 d\omega + (M+1)\frac{2^{-2B}}{12}$$

$$= N \frac{2^{-2B}}{12} \sum_{n=-\infty}^{\infty} |h[n]|^2 + (M+1)\frac{2^{-2B}}{12}. \qquad (6.93)$$

That is, the white noise produced in implementing the poles is filtered by the entire system, while the white noise produced in implementing the zeros is added directly to the output of the system. In writing Eq. (6.93), we have assumed that the N noise sources at the input are independent, so that their sum has N times the variance of a single quantization noise source. The same assumption was made about the $(M+1)$ noise sources at the output. These results are easily modified for two's-complement truncation. Recall from Eqs. (6.73a)–(6.73b) and Eqs. (6.74a)–(6.74b) that the variance of a truncation noise source is the same as that of a rounding noise source, but the mean of a truncation noise source is not zero. Consequently, the formulas in Eqs. (6.84) and (6.93) for the total output noise variance also hold for truncation. However, the output noise will have a nonzero average value that can be computed using Eq. (6.80).

A comparison of Eq. (6.84) with Eq. (6.93) shows that the direct form I and direct form II structures are affected differently by the quantization of products in implementing the corresponding difference equations. In general, other equivalent structures such as cascade, parallel, and transposed forms will have a total output noise variance different from that of either of the direct-form structures. However, even though Eqs. (6.84) and (6.93) are different, we cannot say which system will have the smaller output noise variance unless we know specific values for the coefficients of the system. In other words, it is not possible to state that a particular structural form will always produce the least output noise.

It is possible to improve the noise performance of the direct-form systems (and therefore cascade and parallel forms as well) by using a $(2B+1)$-bit adder to accumulate the sum of products required in both direct-form systems. For example, for the direct form I implementation, we could use a difference equation of the form

$$\hat{y}[n] = Q\left[\sum_{k=1}^{N} a_k \hat{y}[n-k] + \sum_{k=0}^{M} b_k x[n-k]\right]; \tag{6.94}$$

i.e., the sums of products are accumulated with $(2B+1)$- or $(2B+2)$-bit accuracy, and the result is quantized to $(B+1)$ bits for output and storage in the delay memory. In the direct form I case, this means that the quantization noise is still filtered by the poles, but the factor $(M+1+N)$ in Eq. (6.84) is replaced by unity. Similarly, for the direct form II realization, the difference equations (6.91a)–(6.91b) can respectively be replaced by

$$\hat{w}[n] = Q\left[\sum_{k=1}^{N} a_k \hat{w}[n-k] + x[n]\right] \tag{6.95a}$$

and

$$\hat{y}[n] = Q\left[\sum_{k=0}^{M} b_k \hat{w}[n-k]\right]. \tag{6.95b}$$

This implies a single noise source at both the input and output, so the factors N and $(M+1)$ in Eq. (6.93) are each replaced by unity. Thus, the double-length accumulator word provided in most DSP chips can be used to significantly reduce quantization noise in direct-form systems.

6.8.2 Scaling in Fixed-Point Implementations of IIR Systems

The possibility of overflow is another important consideration in the implementation of IIR systems using fixed-point arithmetic. If we follow the convention that each fixed-point number represents a fraction (possibly times a known scale factor), each node in the network must be constrained to have a magnitude less than unity to avoid overflow. If $w_k[n]$ denotes the value of the kth node variable and $h_k[n]$ denotes the impulse response from the input $x[n]$ to the node variable $w_k[n]$, then

$$|w_k[n]| = \left|\sum_{m=-\infty}^{\infty} x[n-m]h_k[m]\right|. \tag{6.96}$$

The bound

$$|w_k[n]| \leq x_{\max} \sum_{m=-\infty}^{\infty} |h_k[m]| \tag{6.97}$$

is obtained by replacing $x[n-m]$ by its maximum value $x_{\max}$ and using the fact that the magnitude of a sum is less than or equal to the sum of the magnitudes of the summands. Therefore, a sufficient condition for $|w_k[n]| < 1$ is

$$x_{\max} < \frac{1}{\displaystyle\sum_{m=-\infty}^{\infty} |h_k[m]|} \tag{6.98}$$

for all nodes in the network. If $x_{\max}$ does not satisfy Eq. (6.98), then we can multiply $x[n]$ by a scaling multiplier s at the input to the system so that $s x_{\max}$ satisfies Eq. (6.98) for all nodes in the network; i.e.,

$$s x_{\max} < \frac{1}{\displaystyle\max_k \left[\sum_{m=-\infty}^{\infty} |h_k[m]| \right]}. \tag{6.99}$$

Scaling the input in this way guarantees that overflow never occurs at any of the nodes in the network. Equation (6.98) is necessary as well as sufficient, since an input always exists such that Eq. (6.97) is satisfied with equality. (See Eq. (2.68) in the discussion of stability in Section 2.4.) However, Eq. (6.98) leads to a very conservative scaling of the input for most signals.

Another approach to scaling is to assume that the input is a narrowband signal, modeled as $x[n] = x_{\max} \cos \omega_0 n$. In this case, the node variables will be

$$w_k[n] = |H_k(e^{j\omega_0})| x_{\max} \cos(\omega_0 n + \sphericalangle H_k(e^{j\omega_0})). \tag{6.100}$$

Therefore, overflow is avoided for *all* sinusoidal signals if

$$\max_{k, |\omega| \leq \pi} |H_k(e^{j\omega})| x_{\max} < 1 \tag{6.101}$$

or if the input is scaled by the scale factor s such that

$$s x_{\max} < \frac{1}{\displaystyle\max_{k, |\omega| \leq \pi} |H_k(e^{j\omega})|}. \tag{6.102}$$

Still another approach is to scale the input so that the total energy of each node variable sequence is less than or equal to the total energy of the input sequence. We can derive the appropriate scale factor by applying the Schwarz inequality and Parseval's theorem. The total energy of the sequence $w_k[n]$ is bounded by

$$\begin{aligned}
\sum_{n=-\infty}^{\infty} |w_k[n]|^2 &= \frac{1}{2\pi} \int_{-\pi}^{\pi} |H_k(e^{j\omega}) X(e^{j\omega})|^2 d\omega \\
&\leq \sum_{n=-\infty}^{\infty} |x[n]|^2 \frac{1}{2\pi} \int_{-\pi}^{\pi} |H_k(e^{j\omega})|^2 d\omega.
\end{aligned} \tag{6.103}$$

Therefore, to ensure that $\sum |w_k[n]|^2 \le \sum |x[n]|^2$, we can multiply the sequence $x[n]$ by a scale factor s such that

$$s^2 \le \frac{1}{\dfrac{1}{2\pi} \displaystyle\int_{-\pi}^{\pi} |H_k(e^{j\omega})|^2 d\omega} = \frac{1}{\displaystyle\sum_{n=-\infty}^{\infty} |h_k[n]|^2}. \tag{6.104}$$

It can be shown that

$$\left\{ \sum_{n=-\infty}^{\infty} |h_k[n]|^2 \right\}^{1/2} \le \max_{k,\omega} |H_k(e^{j\omega})| \le \sum_{n=-\infty}^{\infty} |h_k[n]|. \tag{6.105}$$

Hence, Eqs. (6.99), (6.102), and (6.104) give three decreasingly conservative ways of obtaining scaling multipliers for the input to a digital filter. Of the three, Eq. (6.104) is generally the easiest to evaluate analytically, because the sum of squares can be evaluated by the partial fraction method described in Appendix A. On the other hand, Eq. (6.99) is difficult to evaluate analytically, except for the simplest systems. Of course, for specific systems with known coefficients, the scale factors can be estimated by computing the impulse response or frequency response numerically.

If the input must be scaled down ($s < 1$), the signal-to-noise ratio at the output of the system will be reduced. Figure 6.54 shows second-order direct form I and direct form II systems with scaling multipliers at the input. In determining the scaling multiplier for these systems, it is not necessary to examine each node in the network. Some nodes

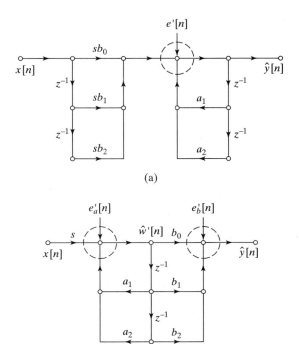

(a)

(b)

Figure 6.54 Scaling of direct form systems. (a) Direct form I. (b) Direct form II.

do not represent addition and thus cannot overflow. Other nodes represent partial sums. If we use nonsaturation two's-complement arithmetic, such nodes are permitted to overflow if certain key nodes do not. For example, in Figure 6.54(a), we can focus on the node enclosed by the dashed circle. In the figure, the scaling multiplier is shown combined with the b_k's, so that the noise source is the same as in Figure 6.51; i.e., it has five times the power of a single quantization noise source.[7] Since the noise source is again filtered only by the poles, the output noise power is the same in Figures 6.51 and 6.54(a). However, the overall system function of the system in Figure 6.54(a) is $s\,H(z)$ instead of $H(z)$, so the unquantized component of the output $\hat{y}[n]$ is $s\,y[n]$ instead of $y[n]$. Since the noise is injected after the scaling, the ratio of signal power to noise power in the scaled system is s^2 times the signal-to-noise ratio for Figure 6.51. Because $s < 1$ if scaling is required to avoid overflow, the signal-to-noise ratio is reduced by scaling.

The same is true for the direct form II system of Figure 6.54(b). In this case, we must determine the scaling multiplier to avoid overflow at both of the circled nodes. Again, the overall gain of the system is s times the gain of the system in Figure 6.53(b), but it may be necessary to implement the scaling multiplier explicitly in this case to avoid overflow at the node on the left. This scaling multiplier adds an additional noise component to $e_a[n]$, so the noise power at the input is, in general, $(N + 1)2^{-2B}/12$. Otherwise, the noise sources are filtered by the system in exactly the same way in both Figure 6.53(b) and Figure 6.54(b). Therefore, the signal power is multiplied by s^2, and the noise power at the output is again given by Eq. (6.93), with N replaced by $(N + 1)$. The signal-to-noise ratio is again reduced if scaling is required to avoid overflow.

Example 6.11   Interaction Between Scaling and Round-off Noise

To illustrate the interaction of scaling and round-off noise, consider the system of Example 6.9 with system function given by Eq. (6.85). If the scaling multiplier is combined with the coefficient b, we obtain the flow graph of Figure 6.55 for the scaled system. Suppose that the input is white noise with amplitudes uniformly distributed between -1 and $+1$. Then the total signal variance is $\sigma_x^2 = 1/3$. To guarantee no overflow in computing $\hat{y}[n]$, we use Eq. (6.99) to compute the scale factor

$$s = \frac{1}{\displaystyle\sum_{n=0}^{\infty} |b|\,|a|^n} = \frac{1 - |a|}{|b|}. \tag{6.106}$$

The output noise variance was determined in Example 6.9 to be

$$\sigma_{f'}^2 = 2\frac{2^{-2B}}{12}\frac{1}{1 - a^2} = \sigma_f^2 \tag{6.107}$$

and the variance of the output $y'[n]$ is

$$\sigma_{y'}^2 = \left(\frac{1}{3}\right)\frac{s^2 b^2}{1 - a^2} = s^2 \sigma_y^2. \tag{6.108}$$

---

[7]This eliminates a separate scaling multiplication and quantization noise source. However, scaling (and quantizing) the b_k's can change the frequency response of the system. If a separate input scaling multiplier precedes the implementation of the zeros in Figure 6.54(a), then an additional quantization noise source would contribute to the output noise after going through the entire system $H(z)$.

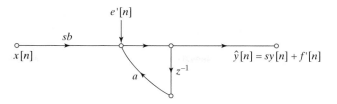

Figure 6.55   Scaled first-order system.

Therefore, the signal-to-noise ratio is

$$\frac{\sigma_{y'}^2}{\sigma_{f'}^2} = s^2 \frac{\sigma_y^2}{\sigma_f^2} = \left(\frac{1 - |a|}{|b|}\right)^2 \frac{\sigma_y^2}{\sigma_f^2}. \tag{6.109}$$

As the pole of the system approaches the unit circle, the signal-to-noise ratio decreases because the quantization noise is amplified by the system and because the high gain of the system forces the input to be scaled down to avoid overflow. Again, we see that overflow and quantization noise work in opposition to decrease the performance of the system.

6.8.3 Example of Analysis of a Cascade IIR Structure

The previous results of this section can be applied directly to the analysis of either parallel or cascade structures composed of second-order direct-form subsystems. The interaction of scaling and quantization is particularly interesting in the cascade form. Our general comments on cascade systems will be interwoven with a specific example.

An elliptic lowpass filter was designed to meet the following specifications:

$$0.99 \leq |H(e^{j\omega})| \leq 1.01, \qquad |\omega| \leq 0.5\pi,$$

$$|H(e^{j\omega})| \leq 0.01, \qquad 0.56\pi \leq |\omega| \leq \pi.$$

The system function of the resulting system is

$$H(z) = 0.079459 \prod_{k=1}^{3} \left(\frac{1 + b_{1k}z^{-1} + z^{-2}}{1 - a_{1k}z^{-1} - a_{2k}z^{-2}}\right) = 0.079459 \prod_{k=1}^{3} H_k(z), \tag{6.110}$$

where the coefficients are given in Table 6.4. Notice that all the zeros of $H(z)$ are on the unit circle in this example; however, that need not be the case in general.

TABLE 6.4   COEFFICIENTS FOR ELLIPTIC LOWPASS FILTER IN CASCADE FORM

k	a_{1k}	a_{2k}	b_{1k}
1	0.478882	−0.172150	1.719454
2	0.137787	−0.610077	0.781109
3	−0.054779	−0.902374	0.411452

Figure 6.56(a) shows a flow graph of a possible implementation of this system as a cascade of second-order transposed direct form II subsystems. The gain constant, 0.079459, is such that the overall gain of the system is approximately unity in the passband, and it is assumed that this guarantees no overflow at the output of the system. Figure 6.56(a) shows the gain constant placed at the input to the system. This approach reduces the amplitude of the signal immediately, with the result that the subsequent filter sections must have high gain to produce an overall gain of unity. Since the quantization noise sources are introduced after the gain of 0.079459, but are likewise amplified

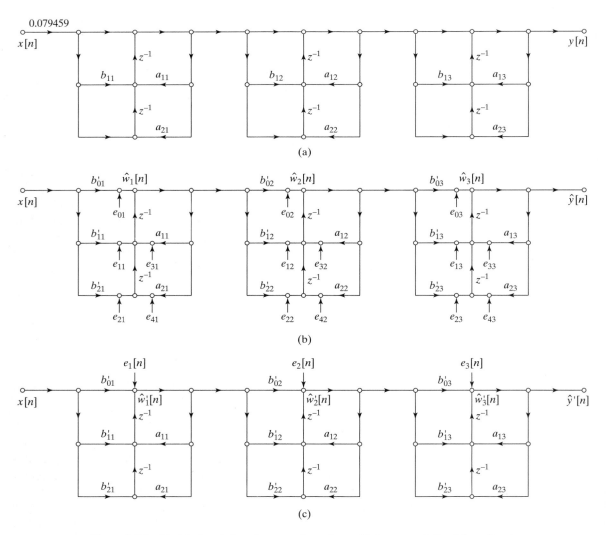

Figure 6.56 Models for sixth-order cascade system with transposed direct form II subsystems. (a) Infinite-precision model. (b) Linear-noise model for scaled system, showing quantization of individual multiplications. (c) Linear-noise model with noise sources combined.

by the rest of the system, this is not a good approach. Ideally, the overall gain constant, being less than unity, should be placed at the very end of the cascade, so that the signal and noise will be attenuated by the same amount. However, this creates the possibility of overflow along the cascade. Therefore, a better approach is to distribute the gain among the three stages of the system so that overflow is just avoided at each stage of the cascade. This distribution is represented by

$$H(z) = s_1 H_1(z) s_2 H_2(z) s_3 H_3(z), \tag{6.111}$$

where $s_1 s_2 s_3 = 0.079459$. The scaling multipliers can be incorporated into the coefficients of the numerators of the individual system functions $H'_k(z) = s_k H_k(z)$, as in

$$H(z) = \prod_{k=1}^{3} \left(\frac{b'_{0k} + b'_{1k} z^{-1} + b'_{2k} z^{-2}}{1 - a_{1k} z^{-1} - a_{2k} z^{-2}} \right) = \prod_{k=1}^{3} H'_k(z), \tag{6.112}$$

where $b'_{0k} = b'_{2k} = s_k$ and $b'_{1k} = s_k b_{1k}$. The resulting scaled system is depicted in Figure 6.56(b).

Also shown in Figure 6.56(b) are quantization noise sources representing the quantization of the products before addition. Figure 6.56(c) shows an equivalent noise model, for which it is recognized that all the noise sources in a particular section are filtered only by the poles of that section (and the subsequent subsystems). Figure 6.56(c) also uses the fact that delayed white-noise sources are still white noise and are independent of all the other noise sources, so that all five sources in a subsection can be combined into a single noise source having five times the variance of a single quantization noise source.[8] Since the noise sources are assumed independent, the variance of the output noise is the sum of the variances due to the three noise sources in Figure 6.56(c). Therefore, for rounding, the power spectrum of the output noise is

$$P_{f'f'}(\omega) = 5 \frac{2^{-2B}}{12} \left[\frac{s_2^2 |H_2(e^{j\omega})|^2 s_3^2 |H_3(e^{j\omega})|^2}{|A_1(e^{j\omega})|^2} + \frac{s_3^2 |H_3(e^{j\omega})|^2}{|A_2(e^{j\omega})|^2} + \frac{1}{|A_3(e^{j\omega})|^2} \right], \tag{6.113}$$

and the total output noise variance is

$$\begin{aligned}
\sigma_{f'}^2 = 5 \frac{2^{-2B}}{12} \Bigg[&\frac{1}{2\pi} \int_{-\pi}^{\pi} \frac{s_2^2 |H_2(e^{j\omega})|^2 s_3^2 |H_3(e^{j\omega})|^2}{|A_1(e^{j\omega})|^2} d\omega \\
&+ \frac{1}{2\pi} \int_{-\pi}^{\pi} \frac{s_3^2 |H_3(e^{j\omega})|^2}{|A_2(e^{j\omega})|^2} d\omega + \frac{1}{2\pi} \int_{-\pi}^{\pi} \frac{1}{|A_3(e^{j\omega})|^2} d\omega \Bigg].
\end{aligned} \tag{6.114}$$

If a double-length accumulator is available, it would be necessary to quantize only the sums that are the inputs to the delay elements in Figure 6.56(b). In this case the factor of 5 in Eqs. (6.113) and (6.114) would be changed to 3. Furthermore, if a double-length register were used to implement the delay elements, only the variables $\hat{w}_k[n]$ would have to be quantized, and there would be only one quantization noise source per subsystem. In that case, the factor of 5 in Eqs. (6.113) and (6.114) would be changed to unity.

---

[8]This discussion can be generalized to show that the transposed direct form II has the same noise behavior as the direct form I system.

The scale factors s_k are chosen to avoid overflow at points along the cascade system. We will use the scaling convention of Eq. (6.102). Therefore, the scaling constants are chosen to satisfy

$$s_1 \max_{|\omega| \le \pi} |H_1(e^{j\omega})| < 1, \tag{6.115a}$$

$$s_1 s_2 \max_{|\omega| \le \pi} |H_1(e^{j\omega}) H_2(e^{j\omega})| < 1, \tag{6.115b}$$

$$s_1 s_2 s_3 = 0.079459. \tag{6.115c}$$

The last condition ensures that there will be no overflow at the output of the system for unit-amplitude sinusoidal inputs, because the maximum overall gain of the filter is unity. For the coefficients of Table 6.4, the resulting scale factors are $s_1 = 0.186447$, $s_2 = 0.529236$, and $s_3 = 0.805267$.

Equations (6.113) and (6.114) show that the shape of the output noise power spectrum and the total output noise variance depends on the way that zeros and poles are paired to form the second-order sections and on the order of the second-order sections in the cascade-form realization. Indeed, it is easily seen that, for N sections, there are ($N!$) ways to pair the poles and zeros, and there are likewise ($N!$) ways to order the resulting second-order sections, a total of ($N!$)^2 different systems. In addition, we can choose either direct form I or direct form II (or their transposes) for the implementation of the second-order sections. In our example, this implies that there are 144 different cascade systems to consider if we wish to find the system with the lowest output noise variance. For five cascaded sections, there would be 57,600 different systems! Clearly, the complete analysis of even low-order systems is a tedious task, since an expression like Eq. (6.114) must be evaluated for each pairing and ordering. Hwang (1974) used dynamic programming and Liu and Peled (1975) used a heuristic approach to reduce the amount of computation. Dehner's program for the design and analysis of cascade-form IIR filters includes the optimization of both pairing and ordering (DSP Committee, 1979).

In spite of the difficulty of finding the optimum pairing and ordering, Jackson (1970a, 1970b, 1996) found that good results are almost always obtained by applying simple rules of the following form:

1. The pole that is closest to the unit circle should be paired with the zero that is closest to it in the z-plane.

2. Rule 1 should be repeatedly applied until all the poles and zeros have been paired.

3. The resulting second-order sections should be ordered according to either increasing closeness to the unit circle or decreasing closeness to the unit circle.

The pairing rules are based on the observation that subsystems with high peak gain are undesirable because they can cause overflow and because they can amplify quantization noise. Pairing a pole that is close to the unit circle with an adjacent zero tends to reduce the peak gain of that section.

One motivation for rule 3 is suggested by Eq. (6.113). We see that the frequency responses of some of the subsystems appear more than once in the equation for the power spectrum of the output noise. If we do not want the output noise variance spectrum to have a high peak around a pole that is close to the unit circle, then it is advantageous to have the frequency-response component due to that pole not appear frequently in Eq. (6.113). This suggests moving such "high Q" poles to the beginning of the cascade.

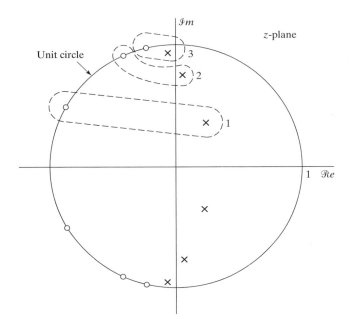

Figure 6.57 Pole–zero plot for sixth-order system of Figure 6.56, showing pairing of poles and zeros.

On the other hand, the frequency response from the input to a particular node in the network will involve a product of the frequency responses of the subsystems that precede the node. Thus, to avoid excessive reduction of the signal level in the early stages of the cascade, we should place the poles that are close to the unit circle last in order. Clearly, then, the question of ordering hinges on a variety of considerations, including total output noise variance and the shape of the output noise spectrum. Jackson (1970a, 1970b) used L_p norms to quantify the analysis of the pairing-and-ordering problem and gave a much more detailed set of "rules of thumb" for obtaining good results without having to evaluate all possibilities.

The pole–zero plot for the system in our example is shown in Figure 6.57. The paired poles and zeros are circled. In this case, we have chosen to order the sections from least peaked to most peaked frequency response. Figure 6.58 illustrates how the frequency responses of the individual sections combine to form the overall frequency response. Figures 6.58(a)–(c) show the frequency responses of the individual unscaled subsystems. Figures 6.58(d)–(f) show how the overall frequency response is built up. Notice that Figures 6.58(d)–(f) demonstrate that the scaling equations (6.115a)–(6.115c) ensure that the maximum gain from the input to the output of any subsystem is less than unity. The solid curve in Figure 6.59 shows the power spectrum of the output noise for the ordering 123 (least peaked to most peaked). We assume that $B + 1 = 16$ for the plot. Note that the spectrum peaks in the vicinity of the pole that is closest to the unit circle. The dotted curve shows the power spectrum of the output noise when the section order is reversed (i.e., 321). Since section 1 has high gain at low frequencies, the noise spectrum is appreciably larger at low frequencies and slightly lower around the peak. The high Q pole still filters the noise sources of the first section in the cascade, so it still tends to dominate the spectrum. The total noise power for the two orderings turns out to be almost the same in this case.

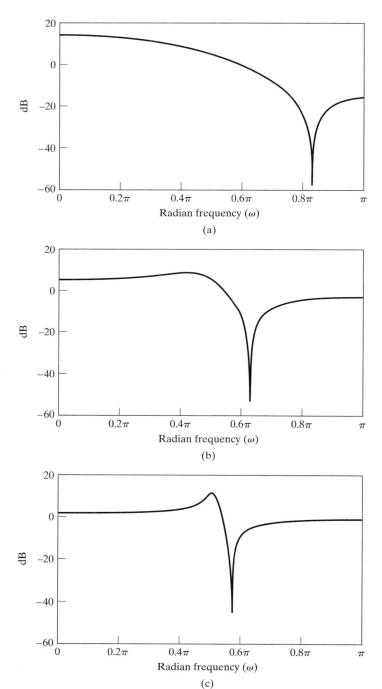

Figure 6.58 Frequency-response functions for example system.
(a) $20 \log_{10} |H_1(e^{j\omega})|$.
(b) $20 \log_{10} |H_2(e^{j\omega})|$.
(c) $20 \log_{10} |H_3(e^{j\omega})|$.

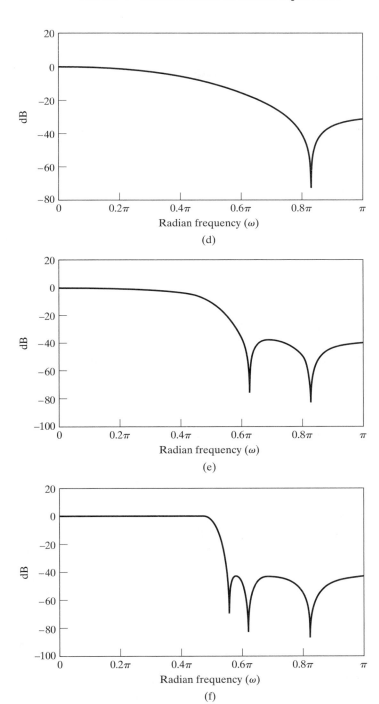

Figure 6.58 (*continued*)
(d) $20 \log_{10} |H_1'(e^{j\omega})|$.
(e) $20 \log_{10} |H_1'(e^{j\omega}) H_2'(e^{j\omega})|$.
(f) $20 \log_{10} |H_1'(e^{j\omega}) H_2'(e^{j\omega}) H_3'(e^{j\omega})|$
$= 20 \log_{10} |H'(e^{j\omega})|$.

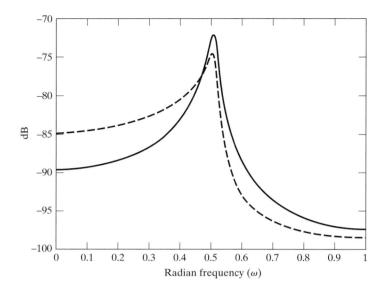

Figure 6.59   Output noise power spectrum for 123 ordering (solid line) and 321 ordering of second-order sections.

The example we have just presented shows the complexity of the issues that arise in fixed-point implementations of cascade IIR systems. The parallel form is somewhat simpler because the issue of pairing and ordering does not arise. However, scaling is still required to avoid overflow in individual second-order subsystems and when the outputs of the subsystems are summed to produce the overall output. The techniques that we have developed must therefore be applied for the parallel form as well. Jackson (1996) discusses the analysis of the parallel form in detail and concludes that its total output noise power is typically comparable to that of the best pairings and orderings of the cascade form. Even so, the cascade form is more common, because, for widely used IIR filters such that the zeros of the system function are on the unit circle, the cascade form can be implemented with fewer multipliers and with more control over the locations of the zeros.

6.8.4 Analysis of Direct-Form FIR Systems

Since the direct form I and direct form II IIR systems include the direct-form FIR system as a special case (i.e., the case where all a_k's in Figures 6.14 and 6.15 are zero), the results and analysis techniques of Sections 6.8.1 and 6.8.2 apply to FIR systems if we eliminate all reference to the poles of the system function and eliminate the feedback paths in all the signal flow graphs.

The direct-form FIR system is simply the discrete convolution

$$y[n] = \sum_{k=0}^{M} h[k]x[n-k]. \tag{6.116}$$

Figure 6.60(a) shows the ideal unquantized direct-form FIR system, and Figure 6.60(b) shows the linear-noise model for the system, assuming that all products are quantized before additions are performed. The effect is to inject $(M + 1)$ white-noise sources

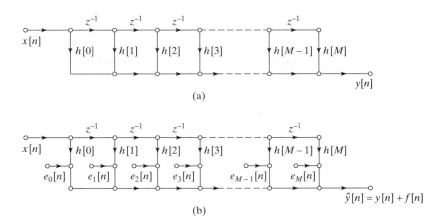

Figure 6.60   Direct-form realization of an FIR system. (a) Infinite-precision model. (b) Linear-noise model.

directly at the output of the system, so that the total output noise variance is

$$\sigma_f^2 = (M+1)\frac{2^{-2B}}{12}. \tag{6.117}$$

This is exactly the result we would obtain by setting $N = 0$ and $h_{ef}[n] = \delta[n]$ in Eqs. (6.84) and (6.93). When a double-length accumulator is available, we would need to quantize only the output. Therefore, the factor $(M + 1)$ in Eq. (6.117) would be replaced by unity. This makes the double-length accumulator a very attractive hardware feature for implementing FIR systems.

Overflow is also a problem for fixed-point realizations of FIR systems in direct form. For two's-complement arithmetic, we need to be concerned only about the size of the output, since all the other sums in Figure 6.60(b) are partial sums. Thus, the impulse-response coefficients can be scaled to reduce the possibility of overflow. Scaling multipliers can be determined using any of the alternatives discussed in Section 6.8.2. Of course, scaling the impulse response reduces the gain of the system, and therefore the signal-to-noise ratio at the output is reduced as discussed in that section.

Example 6.12   Scaling Considerations for the FIR System in Section 6.7.5

The impulse-response coefficients for the system in Section 6.7.5 are given in Table 6.3. A simple calculation shows, and from Figure 6.46(b) we see, that

$$\sum_{n=0}^{27} |h[n]| = 1.751352,$$

$$\left(\sum_{n=0}^{27} |h[n]|^2\right)^{1/2} = 0.679442,$$

$$\max_{|\omega| \le \pi} |H(e^{j\omega})| \approx 1.009.$$

> These numbers satisfy the relationship in Eq. (6.105). Thus, the system, as given, is scaled so that overflow is theoretically possible for a sinusoidal signal whose amplitude is greater than $1/1.009 = 0.9911$, but even so, overflow is unlikely for most signals. Indeed, since the filter has a linear phase, we can argue that, for wideband signals, since the gain in the passband is approximately unity and the gain elsewhere is less than unity, the output signal should be smaller than the input signal.

In Section 6.5.3, we showed that linear-phase systems like the one in Example 6.12 can be implemented with about half the number of multiplications of the general FIR system. This is evident from the signal flow graphs of Figures 6.34 and 6.35. In these cases, it should be clear that the output noise variance would be halved if products were quantized before addition. However, the utilization of such structures involves a more complicated indexing algorithm than the direct form. The architecture of most DSP chips combines a double-length accumulator with an efficient pipelined multiply–accumulate operation and simple looping control to optimize for the case of the direct-form FIR system. For this reason, direct-form FIR implementations are often most attractive, even compared with IIR filters that meet frequency-response specifications with fewer multiplications.

In Section 6.5.3, we discussed cascade realizations of FIR systems. The results and analysis techniques of Section 6.8.3 apply to these realizations; but for FIR systems with no poles, the pairing and ordering problem reduces to just an ordering problem. As in the case of IIR cascade systems, the analysis of all possible orderings can be very difficult if the system is composed of many subsystems. Chan and Rabiner (1973a, 1973b) studied this problem and found experimentally that the noise performance is relatively insensitive to the ordering. Their results suggest that a good ordering is an ordering for which the frequency response from each noise source to the output is relatively flat and for which the peak gain is small.

6.8.5 Floating-Point Realizations of Discrete-Time Systems

From the preceding discussion, it is clear that the limited dynamic range of fixed-point arithmetic makes it necessary to carefully scale the input and intermediate signal levels in fixed-point digital realizations of discrete-time systems. The need for such scaling can be essentially eliminated by using floating-point numerical representations and floating-point arithmetic.

In floating-point representations, a real number x is represented by the binary number $2^c \hat{x}_M$, where the exponent c of the scale factor is called the *characteristic* and $\hat{x}_M$ is a fractional part called the *mantissa*. Both the characteristic and the mantissa are represented explicitly as fixed-point binary numbers in floating-point arithmetic systems. Floating-point representations provide a convenient means for maintaining both a wide dynamic range and low quantization noise; however, quantization error manifests itself in a somewhat different way. Floating-point arithmetic generally maintains its high accuracy and wide dynamic range by adjusting the characteristics and normalizing the mantissa so that $0.5 \leq \hat{x}_M < 1$. When floating-point numbers are multiplied, their characteristics are added and their mantissas are multiplied. Thus, the mantissa must be quantized. When two floating-point numbers are added, their characteristics must be adjusted to be the same by moving the binary point of the mantissa of the smaller number. Hence, addition results in quantization, too. If we assume that the range of the characteristic is sufficient so that no numbers become larger than 2^c, then quantization affects only the mantissa, but the error in the mantissa is also scaled by 2^c. Thus, a quantized floating-point number is conveniently represented as

$$\hat{x} = x(1 + \varepsilon) = x + \varepsilon x. \tag{6.118}$$

By representing the quantization error as a fraction ε of x, we automatically represent the fact that the quantization error is scaled up and down with the signal level.

The aforementioned properties of floating-point arithmetic complicate the quantization error analysis of floating-point implementations of discrete-time systems. First, noise sources must be inserted both after each multiplication and after each addition. An important consequence is that, in contrast to fixed-point arithmetic, the *order* in which multiplications and additions are performed can sometimes make a big difference. More important for analysis, we can no longer justify the assumption that the quantization noise sources are white noise and are independent of the signal. In fact, in Eq. (6.118), the noise is expressed explicitly in terms of the signal. Therefore, we can no longer analyze the noise without making assumptions about the nature of the input signal. If the input is assumed to be known (e.g., white noise), a reasonable assumption is that the relative error ε is independent of x and is uniformly distributed white noise.

With these types of assumptions, useful results have been obtained by Sandberg (1967), Liu and Kaneko (1969), Weinstein and Oppenheim (1969), and Kan and Aggarwal (1971). In particular, Weinstein and Oppenheim, comparing floating-point and fixed-point realizations of first- and second-order IIR systems, showed that if the number of bits representing the floating-point mantissa is equal to the length of the fixed-point word, then floating-point arithmetic leads to higher signal-to-noise ratio at the output. Not surprisingly, the difference was found to be greater for poles close to the unit circle. However, additional bits are required to represent the characteristic, and the greater the desired dynamic range, the more bits are required for the characteristic. Also, the hardware to implement floating-point arithmetic is much more complex than that for fixed-point arithmetic. Therefore, the use of floating-point arithmetic entails an increased word length and increased complexity in the arithmetic unit. Its major advantage is that it essentially eliminates the problem of overflow, and if a sufficiently long mantissa is used, quantization also becomes much less of a problem. This translates into greater simplicity in system design and implementation.

6.9 ZERO-INPUT LIMIT CYCLES IN FIXED-POINT REALIZATIONS OF IIR DIGITAL FILTERS

For stable IIR discrete-time systems implemented with infinite-precision arithmetic, if the excitation becomes zero and remains zero for n greater than some value n_0, the output for $n > n_0$ will decay asymptotically toward zero. For the same system, implemented with finite-register-length arithmetic, the output may continue to oscillate indefinitely with a periodic pattern while the input remains equal to zero. This effect is often referred to as *zero-input limit cycle behavior* and is a consequence either of the nonlinear quantizers in the feedback loop of the system or of overflow of additions. The limit cycle behavior of a digital filter is complex and difficult to analyze, and we will not attempt to treat the topic in any general sense. To illustrate the point, however, we will give two simple examples that will show how such limit cycles can arise.

6.9.1 Limit Cycles due to Round-off and Truncation

Successive round-off or truncation of products in an iterated difference equation can create repeating patterns. This is illustrated in the following example.

Example 6.13   Limit Cycle Behavior in a First-Order System

Consider the first-order system characterized by the difference equation

$$y[n] = ay[n-1] + x[n], \qquad |a| < 1. \tag{6.119}$$

The signal flow graph of this system is shown in Figure 6.61(a). Let us assume that the register length for storing the coefficient a, the input $x[n]$, and the filter node variable $y[n-1]$ is 4 bits (i.e., a sign bit to the left of the binary point and 3 bits to the

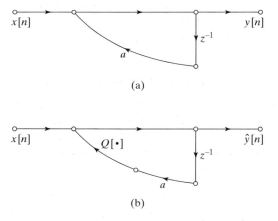

(a)

(b)

Figure 6.61   First-order IIR system. (a) Infinite-precision linear system. (b) Non-linear system due to quantization.

right of the binary point). Because of the finite-length registers, the product $ay[n-1]$ must be rounded or truncated to 4 bits before being added to $x[n]$. The flow graph representing the actual realization based on Eq. (6.119) is shown in Figure 6.61(b). Assuming rounding of the product, the actual output $\hat{y}[n]$ satisfies the nonlinear difference equation

$$\hat{y}[n] = Q[a\hat{y}[n-1]] + x[n], \tag{6.120}$$

where $Q[\cdot]$ represents the rounding operation. Let us assume that $a = 1/2 = 0_\diamond100$ and that the input is $x[n] = (7/8)\delta[n] = (0_\diamond111)\delta[n]$. Using Eq. (6.120), we see that for $n = 0$, $\hat{y}[0] = 7/8 = 0_\diamond111$. To obtain $\hat{y}[1]$, we multiply $\hat{y}[0]$ by a, obtaining the result $a\hat{y}[0] = 0_\diamond011100$, a 7-bit number that must be rounded to 4 bits. This number, 7/16, is exactly halfway between the two 4-bit quantization levels 4/8 and 3/8. If we choose always to round upward in such cases, then $0_\diamond011100$ rounded to 4 bits is $0_\diamond100 = 1/2$.

Since $x[1] = 0$, it follows that $\hat{y}[1] = 0{\scriptstyle\circ}100 = 1/2$. Continuing to iterate the difference equation gives $\hat{y}[2] = Q[a\hat{y}[1]] = 0{\scriptstyle\circ}010 = 1/4$ and $\hat{y}[3] = 0{\scriptstyle\circ}001 = 1/8$. In both these cases, no rounding is necessary. However, to obtain $\hat{y}[4]$, we must round the 7-bit number $a\hat{y}[3] = 0{\scriptstyle\circ}000100$ to $0{\scriptstyle\circ}001$. The same result is obtained for all values of $n \geq 3$. The output sequence for this example is shown in Figure 6.62(a). If $a = -1/2$, we can carry out the preceding computation again to demonstrate that the output is as shown in Figure 6.62(b). Thus, because of rounding of the product $a\hat{y}[n-1]$, the output reaches a constant value of $1/8$ when $a = 1/2$ and a periodic steady-state oscillation between $+1/8$ and $-1/8$ when $a = -1/2$. These are periodic outputs similar to those that would be obtained from a first-order pole at $z = \pm 1$ instead of at $z = \pm 1/2$.

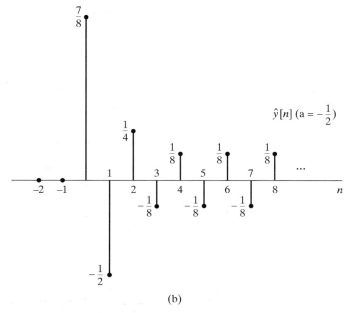

Figure 6.62 Response of the first-order system of Figure 6.61 to an impulse. (a) $a = \frac{1}{2}$. (b) $a = -\frac{1}{2}$.

When $a = +1/2$, the period of the oscillation is 1, and when $a = -1/2$, the period of oscillation is 2. Such steady-state periodic outputs are called *limit cycles,* and their existence was first noted by Blackman (1965), who referred to the amplitude intervals to which such limit cycles are confined as *dead bands.* In this case the dead band is $-2^{-B} \leq \hat{y}[n] \leq +2^{-B}$, where $B = 3$.

The foregoing example has illustrated that a zero-input limit cycle can result from rounding in a first-order IIR system. Similar results can be demonstrated for truncation. Second-order systems can also exhibit limit cycle behavior. In the case of parallel realizations of higher order systems, the outputs of the individual second-order systems are independent when the input is zero. In this case, one or more of the second-order sections could contribute a limit cycle to the output sum. In the case of cascade realizations, only the first section has zero input; succeeding sections may exhibit their own characteristic limit cycle behavior, or they may appear to be simply filtering the limit cycle output of a previous section. For high-order systems realized by other filter structures, the limit cycle behavior becomes more complex, as does its analysis.

In addition to giving an understanding of limit cycle effects in digital filters, the preceding results are useful when the zero-input limit cycle response of a system is the desired output. This is the case, for example, when one is concerned with digital sine wave oscillators for signal generation or for the generation of coefficients for calculation of the discrete Fourier transform.

6.9.2 Limit Cycles Due to Overflow

In addition to the classes of limit cycles discussed in the preceding section, a more severe type of limit cycle can occur due to overflow. The effect of overflow is to insert a gross error in the output, and in some cases the filter output thereafter oscillates between large-amplitude limits. Such limit cycles have been referred to as *overflow oscillation.* The problem of oscillations caused by overflow is discussed in detail by Ebert et al. (1969). Overflow oscillations are illustrated by the following example.

Example 6.14 Overflow Oscillations in a Second-Order System

Consider a second-order system realized by the difference equation

$$\hat{y}[n] = x[n] + Q[a_1 \hat{y}[n-1]] + Q[a_2 \hat{y}[n-2]], \tag{6.121}$$

where $Q[\cdot]$ represents two's-complement rounding with a word length of 3 bits plus 1 bit for the sign. Overflow can occur with two's-complement addition of the rounded products. Suppose that $a_1 = 3/4 = 0_\circ 110$ and $a_2 = -3/4 = 1_\circ 010$, and assume that $x[n]$ remains equal to zero for $n \geq 0$. Furthermore, assume that $\hat{y}[-1] = 3/4 = 0_\circ 110$ and $\hat{y}[-2] = -3/4 = 1_\circ 010$. Now the output at sample $n = 0$ is

$$\hat{y}[0] = 0_\circ 110 \times 0_\circ 110 + 1_\circ 010 \times 1_\circ 010.$$

If we evaluate the products using two's-complement arithmetic, we obtain

$$\hat{y}[0] = 0_\circ 100100 + 0_\circ 100100,$$

and if we choose to round upward when a number is halfway between two quantization levels, the result of two's-complement addition is

$$\hat{y}[0] = 0_\circ 101 + 0_\circ 101 = 1_\circ 010 = -\tfrac{3}{4}.$$

In this case the binary carry overflows into the sign bit, thus changing the positive sum into a negative number. Repeating the process gives

$$\hat{y}[1] = 1_\circ 011 + 1_\circ 011 = 0_\circ 110 = \tfrac{3}{4}.$$

The binary carry resulting from the sum of the sign bits is lost, and the negative sum is mapped into a positive number. Clearly, $\hat{y}[n]$ will continue to oscillate between $+3/4$ and $-3/4$ until an input is applied. Thus, $\hat{y}[n]$ has entered a periodic limit cycle with a period of 2 and an amplitude of almost the full-scale amplitude of the implementation.

The preceding example illustrates how overflow oscillations occur. Much more complex behavior can be exhibited by higher order systems, and other frequencies can occur. Some results are available for predicting when overflow oscillations can be supported by a difference equation (Ebert et al., 1969). Overflow oscillations can be avoided by using the saturation overflow characteristic of Figure 6.38(b) (Ebert et al., 1969).

6.9.3 Avoiding Limit Cycles

The possible existence of a zero-input limit cycle is important in applications where a digital filter is to be in continuous operation, since it is generally desired that the output approach zero when the input is zero. For example, suppose that a speech signal is sampled, filtered by a digital filter, and then converted back to an acoustic signal using a D/A converter. In such a situation it would be very undesirable for the filter to enter a periodic limit cycle whenever the input is zero, since the limit cycle would produce an audible tone.

One approach to the general problem of limit cycles is to seek structures that do not support limit cycle oscillations. Such structures have been derived by using state-space representations (Barnes and Fam, 1977; Mills et al., 1978) and concepts analogous to passivity in analog systems (Rao and Kailath, 1984; Fettweis, 1986). However, these structures generally require more computation than an equivalent cascade- or parallel-form implementation. By adding more bits to the computational word length, we can generally avoid overflow. Similarly, since round-off limit cycles usually are limited to the least significant bits of the binary word, additional bits can be used to reduce the effective amplitude of the limit cycle. Also, Claasen et al. (1973) showed that if a double-length accumulator is used so that quantization occurs after the accumulation of products, then limit cycles due to round-off are much less likely to occur in second-order systems. Thus, the trade-off between word length and computational algorithm complexity arises for limit cycles just as it does for coefficient quantization and round-off noise.

Finally, it is important to point out that zero-input limit cycles due to both overflow and round-off are a phenomenon unique to IIR systems: FIR systems cannot support zero-input limit cycles, because they have no feedback paths. The output of an

FIR system will be zero no later than $(M + 1)$ samples after the input goes to zero and remains there. This is a major advantage of FIR systems in applications wherein limit cycle oscillations cannot be tolerated.

6.10 SUMMARY

In this chapter, we have considered many aspects of the problem of implementing a linear time-invariant discrete-time system. The first half of the chapter was devoted to basic implementation structures. After introducing block diagram and signal flow graphs as pictorial representations of difference equations, we discussed a number of basic structures for IIR and FIR discrete-time systems. These included the direct form I, direct form II, cascade form, parallel form, and transposed form. We showed that these forms are all equivalent when implemented with infinite-precision arithmetic. However, the different structures are most significant in the context of finite-precision implementations. Therefore, the remainder of the chapter addressed problems associated with finite precision or quantization in digital implementations of the basic structures.

We began the discussion of finite precision effects with a brief review of digital number representation and an overview which showed that the quantization effects that are important in sampling (discussed in Chapter 4) are also important in representing the coefficients of a discrete-time system and in implementing systems using finite-precision arithmetic. We illustrated the effect of quantization of the coefficients of a difference equation through several examples. This issue was treated independently of the effects of finite-precision arithmetic, which we showed introduces nonlinearity into the system. We demonstrated that in some cases this nonlinearity was responsible for limit cycles that may persist after the input to a system has become zero. We also showed that quantization effects can be modeled in terms of independent random white-noise sources that are injected internally into the network. Such linear-noise models were developed for the direct-form structures and for the cascade structure. In all of our discussion of quantization effects, the underlying theme was the conflict between the desire for fine quantization and the need to maintain a wide range of signal amplitudes. We saw that in fixed-point implementations, one can be improved at the expense of the other, but to improve one while leaving the other unaffected requires that we increase the number of bits used to represent coefficients and signal amplitudes. This can be done either by increasing the fixed-point word length or by adopting a floating-point representation.

Our discussion of quantization effects serves two purposes. First, we developed several results that can be useful in guiding the design of practical implementations. We found that quantization effects depend greatly on the structure used and on the specific parameters of the system to be implemented, and even though simulation of the system is generally necessary to evaluate its performance, many of the results discussed are useful in making intelligent decisions in the design process. A second, equally important, purpose of this part of the chapter was to illustrate a style of analysis that can be applied in studying quantization effects in a variety of digital signal-processing algorithms. The examples of the chapter indicate the types of assumptions and approximations that are commonly made in studying quantization effects. In Chapter 9 we will apply the analysis techniques developed here to the study of quantization in the computation of the discrete Fourier transform.

PROBLEMS

Basic Problems with Answers

6.1. Determine the system function of the two networks in Figure P6.1-1, and show that they have the same poles.

Network 1

(a)

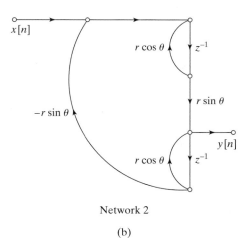

Network 2

(b)

Figure P6.1-1

6.2. The signal flow graph of Figure P6.2-1 represents a linear difference equation with constant coefficients. Determine the difference equation that relates the output $y[n]$ to the input $x[n]$.

Figure P6.2-1

6.3. Figure P6.3-1 shows six systems. Determine which one of the last five, (b)–(f), has the same system function as (a). You should be able to eliminate some of the possibilities by inspection.

(a)

(b)

(c)

(d)

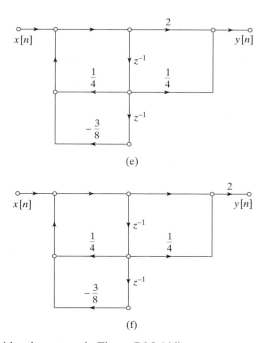

(e)

(f)   **Figure P6.3-1 (*Continued*)**

6.4. Consider the system in Figure P6.3-1(d).

   (a) Find the system function relating the z-transforms of the input and output.
   (b) Write the difference equation that is satisfied by the input sequence $x[n]$ and the output sequence $y[n]$.

6.5. A linear time-invariant system is realized by the flow graph shown in Figure P6.5-1.

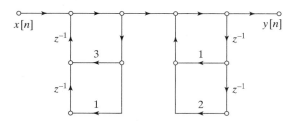

Figure P6.5-1

   (a) Write the difference equation relating $x[n]$ and $y[n]$ for this flow graph.
   (b) What is the system function of the system?
   (c) In the realization of Figure P6.5-1, how many real multiplications and real additions are required to compute each sample of the output? (Assume that $x[n]$ is real, and assume that multiplication by 1 does not count in the total.)
   (d) The realization of Figure P6.5-1 requires four storage registers (delay elements). Is it possible to reduce the number of storage registers by using a different structure? If so, draw the flow graph; if not, explain why the number of storage registers cannot be reduced.

6.6. Determine the impulse response of each of the systems in Figure P6.6-1.

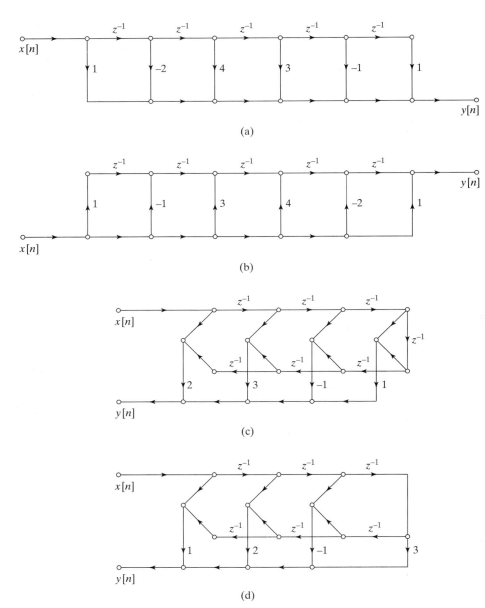

Figure P6.6-1

6.7. Let $x[n]$ and $y[n]$ be N-point sequences ($N > 3$) related by the following difference equation:

$$y[n] - \frac{1}{4}y[n-2] = x[n-2] - \frac{1}{4}x[n].$$

Draw a direct form II signal flow graph for the causal LTI system corresponding to this difference equation.

6.8. The signal flow graph in Figure P6.8-1 represents an LTI system. Determine a difference equation that gives a relationship between the input $x[n]$ and the output $y[n]$ of this system. As usual, all branches of the signal flow graph have unity gain unless specifically indicated otherwise.

Figure P6.8-1

6.9. Figure P6.9-1 shows the signal flow graph for a causal discrete-time LTI system. Branches without gains explicitly indicated have a gain of unity.
(a) Determine $h[1]$, the impulse response at $n = 1$.
(b) Determine the difference equation relating $x[n]$ and $y[n]$.

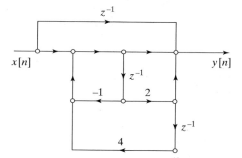

Figure P6.9-1

6.10. Consider the signal flow graph shown in Figure P6.10-1.
(a) Using the node variables indicated, write the set of difference equations represented by this network.
(b) Draw the flow graph of an equivalent system that is the cascade of two first-order systems.
(c) Is the system stable? Explain.

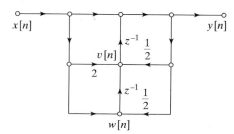

Figure P6.10-1

6.11. Consider a causal LTI system **S** with impulse response $h[n]$ and system function

$$H(z) = \frac{(1 - 2z^{-1})(1 - 4z^{-1})}{z\left(1 - \frac{1}{2}z^{-1}\right)}$$

(a) Draw a direct form II flow graph for the system **S**.
(b) Draw the transposed form of the flow graph in Part (a).

6.12. For the linear time-invariant system described by the flow graph in Figure P6.12-1, determine the difference equation relating the input $x[n]$ to the output $y[n]$.

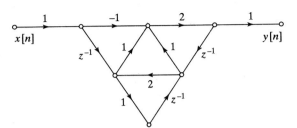

Figure P6.12-1

6.13. Draw the signal flow graph for the direct form I implementation of the LTI system with system function

$$H(z) = \frac{1 - \frac{1}{2}z^{-2}}{1 - \frac{1}{4}z^{-1} - \frac{1}{8}z^{-2}}.$$

6.14. Draw the signal flow graph for the direct form II implementation of the LTI system with system function

$$H(z) = \frac{1 + \frac{5}{6}z^{-1} + \frac{1}{6}z^{-2}}{1 - \frac{1}{2}z^{-1} - \frac{1}{2}z^{-2}}.$$

6.15. Draw the signal flow graph for the transposed direct form II implementation of the LTI system with system function

$$H(z) = \frac{1 - \frac{7}{6}z^{-1} + \frac{1}{6}z^{-2}}{1 + z^{-1} + \frac{1}{2}z^{-2}}.$$

6.16. Consider the signal flow graph shown in Figure P6.16-1.
  (a) Draw the signal flow graph that results from applying the transposition theorem to this signal flow graph.
  (b) Confirm that the transposed signal flow graph that you found in (a) has the same system function $H(z)$ as the original system in the figure.

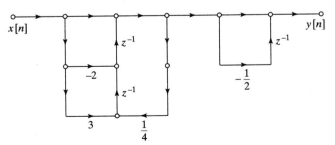

Figure P6.16-1

6.17. Consider the causal LTI system with system function

$$H(z) = 1 - \frac{1}{3}z^{-1} + \frac{1}{6}z^{-2} + z^{-3}.$$

  (a) Draw the signal flow graph for the direct form implementation of this system.
  (b) Draw the signal flow graph for the transposed direct form implementation of the system.

6.18. For some choices of the parameter a, the signal flow graph in Figure P6.18-1 can be replaced by a second-order direct form II signal flow graph implementing the same system function. Give one such choice for a and the system function $H(z)$ that results.

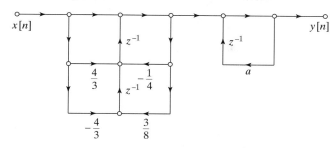

Figure P6.18-1

6.19. Consider the causal LTI system with the system function

$$H(z) = \frac{2 - \frac{8}{3}z^{-1} - 2z^{-2}}{\left(1 - \frac{1}{3}z^{-1}\right)\left(1 + \frac{2}{3}z^{-1}\right)}.$$

Draw a signal flow graph that implements this system as a parallel combination of first-order transposed direct form II sections.

6.20. Draw a signal flow graph implementing the system function

$$H(z) = \frac{(1 + (1 - j/2)z^{-1})(1 + (1 + j/2)z^{-1})}{(1 + (j/2)z^{-1})(1 - (j/2)z^{-1})(1 - (1/2)z^{-1})(1 - 2z^{-1})}$$

as a cascade of second-order transposed direct form II sections with real coefficients.

Basic Problems

6.21. For many applications, it is useful to have a system that will generate a sinusoidal sequence. One possible way to do this is with a system whose impulse response is $h[n] = e^{j\omega_0 n}u[n]$. The real and imaginary parts of $h[n]$ are therefore $h_r[n] = (\cos \omega_0 n)u[n]$ and $h_i[n] = (\sin \omega_0 n)u[n]$, respectively.

In implementing a system with a complex impulse response, the real and imaginary parts are distinguished as separate outputs. By first writing the complex difference equation required to produce the desired impulse response and then separating it into its real and imaginary parts, draw a flow graph that will implement this system. The flow graph that you draw can have only real coefficients. This implementation is sometimes called the *coupled form oscillator,* since, when the input is excited by an impulse, the outputs are sinusoidal.

6.22. For the system function

$$H(z) = \frac{1 + 2z^{-1} + z^{-2}}{1 - \frac{3}{4}z^{-1} + \frac{1}{8}z^{-2}},$$

draw the flow graphs of all possible realizations for this system as cascades of first-order systems.

6.23. Consider a causal linear time-invariant system whose system function is

$$H(z) = \frac{1 - \frac{1}{5}z^{-1}}{\left(1 - \frac{1}{2}z^{-1} + \frac{1}{3}z^{-2}\right)\left(1 + \frac{1}{4}z^{-1}\right)}.$$

(a) Draw the signal flow graphs for implementations of the system in each of the following forms:

    (i) Direct form I

    (ii) Direct form II

    (iii) Cascade form using first- and second-order direct form II sections

    (iv) Parallel form using first- and second-order direct form II sections

    (v) Transposed direct form II

(b) Write the difference equations for the flow graph of (v) in Part (a), and show that this system has the correct system function.

6.24. Several flow graphs are shown in Figure P6.24-1. Determine the transpose of each flow graph, and verify that in each case the original and transposed flow graphs have the same system function.

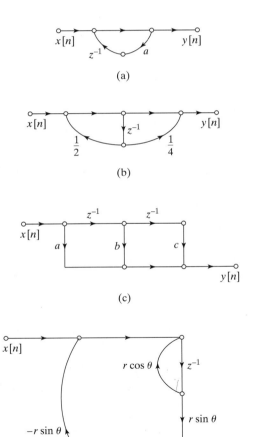

(a)

(b)

(c)

(d)

Figure P6.24-1

6.25. Consider the system in Figure P6.25-1.

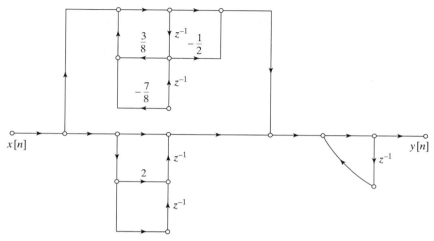

Figure P6.25-1

(a) Find the system function relating the z-transforms of the input and output.

(b) Write the difference equation that is satisfied by the input sequence $x[n]$ and the output sequence $y[n]$.

(c) Draw a signal flow graph that has the same input–output relationship as the system in Figure P6.25-1, but that has the smallest possible number of delay elements.

6.26. A linear time-invariant system with system function

$$H(z) = \frac{0.2(1 + z^{-1})^6}{\left(1 - 2z^{-1} + \frac{7}{8}z^{-2}\right)\left(1 + z^{-1} + \frac{1}{2}z^{-2}\right)\left(1 - \frac{1}{2}z^{-1} + z^{-2}\right)}$$

is to be implemented using a flow graph of the form shown in Figure P6.26-1.

Figure P6.26-1

(a) Fill in all the coefficients in the diagram of Figure P6.26-1. Is your solution unique?

(b) Define appropriate node variables in Figure P6.26-1, and write the set of difference equations that is represented by the flow graph.

Advanced Problems

6.27. Consider a general flow graph (denoted network A) consisting of coefficient multipliers and delay elements, as shown in Figure P6.27-1. If the system is initially at rest, its behavior is completely specified by its impulse response $h[n]$. We wish to modify the system to create a new flow graph (denoted network A_1) with impulse response $h_1[n] = (-1)^n h[n]$.

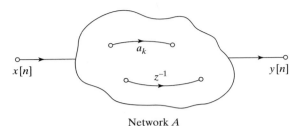

Network *A*   **Figure P6.27-1**

(a) If $H(e^{j\omega})$ is as given in Figure P6.27-2, sketch $H_1(e^{j\omega})$.

Figure P6.27-2

(b) Explain how to modify network *A* by simple modifications of its coefficient multipliers and/or the delay branches to form the new network A_1 whose impulse response is $h_1[n]$.

(c) If network *A* is as given in Figure P6.27-3, show how to modify it by simple modifications to *only the coefficient multipliers* so that the resulting network A_1 has impulse response $h_1[n]$.

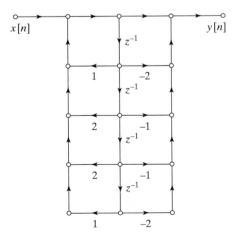

Figure P6.27-3

6.28. The flow graph shown in Figure P6.28-1 is noncomputable; i.e., it is not possible to compute the output using the difference equations represented by the flow graph because it contains a closed loop having no delay elements.

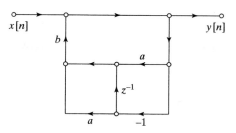

Figure P6.28-1

(a) Write the difference equations for the system of Figure P6.28-1, and from them, find the system function of the network.

(b) From the system function, obtain a flow graph that is computable.

6.29. The impulse response of a linear time-invariant system is

$$h[n] = \begin{cases} a^n, & 0 \le n \le 7, \\ 0, & \text{otherwise.} \end{cases}$$

(a) Draw the flow graph of a direct-form nonrecursive implementation of the system.

(b) Show that the corresponding system function can be expressed as

$$H(z) = \frac{1 - a^8 z^{-8}}{1 - az^{-1}}, \quad |z| > |a|.$$

(c) Draw the flow graph of an implementation of $H(z)$, as expressed in Part (b), corresponding to a cascade of an FIR system (numerator) with an IIR system (denominator).

(d) Is the implementation in Part (c) recursive or nonrecursive? Is the overall system FIR or IIR?

(e) Which implementation of the system requires
   (i) the most storage (delay elements)?
   (ii) the most arithmetic (multiplications and additions per output sample)?

6.30. Consider an FIR system whose impulse response is

$$h[n] = \begin{cases} \frac{1}{15}(1 + \cos[(2\pi/15)(n - n_0)]), & 0 \le n \le 14, \\ 0, & \text{otherwise.} \end{cases}$$

This system is an example of a class of filters known as frequency-sampling filters. Problem 6.37 discusses these filters in detail. In this problem, we consider just one specific case.

(a) Sketch the impulse response of the system for the cases $n_0 = 0$ and $n_0 = 15/2$.

(b) Show that the system function of the system can be expressed as

$$H(z) = (1 - z^{-15}) \cdot \frac{1}{15} \left[\frac{1}{1 - z^{-1}} + \frac{\frac{1}{2}e^{-j2\pi n_0/15}}{1 - e^{j2\pi/15}z^{-1}} + \frac{\frac{1}{2}e^{j2\pi n_0/15}}{1 - e^{-j2\pi/15}z^{-1}} \right].$$

(c) Show that if $n_0 = 15/2$, the frequency response of the system can be expressed as

$$H(e^{j\omega}) = \frac{1}{15}e^{-j\omega 7} \left\{ \frac{\sin(\omega 15/2)}{\sin(\omega/2)} + \frac{1}{2}\frac{\sin[(\omega - 2\pi/15)15/2]}{\sin[(\omega - 2\pi/15)/2]} \right.$$

$$\left. + \frac{1}{2}\frac{\sin[(\omega + 2\pi/15)15/2]}{\sin[(\omega + 2\pi/15)/2]} \right\}.$$

Use this expression to sketch the magnitude of the frequency response of the system for $n_0 = 15/2$. Obtain a similar expression for $n_0 = 0$. Sketch the magnitude response for $n_0 = 0$. For which choices of n_0 does the system have a generalized linear phase?

(d) Draw a signal flow graph of an implementation of the system as a cascade of an FIR system whose system function is $1 - z^{-15}$ and a parallel combination of a first- and second-order IIR system.

6.31. Consider the discrete-time system depicted in Figure P6.31-1.

Figure P6.31-1

(a) Write the set of difference equations represented by the flow graph of Figure P6.31-1.

(b) Determine the system function $H_1(z) = Y_1(z)/X(z)$ of the system in Figure P6.31-2, and determine the magnitudes and angles of the poles of $H_1(z)$ as a function of r for $-1 < r < 1$.

(c) Figure P6.31-2 shows a different flow graph obtained from the flow graph of Figure P6.31-1 by moving the delay elements to the opposite top branch. How is the system function $H_2(z) = Y_2(z)/X(z)$ related to $H_1(z)$?

Figure P6.31-2

6.32. The three networks in Figure P6.32-1 are all equivalent implementations of the same two-input, two-output linear time-invariant system.

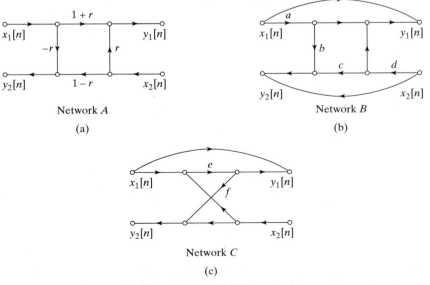

Figure P6.32-1

(a) Write the difference equations for network A.

(b) Find values of a, b, c, and d for network B in terms of r in network A such that the two systems are equivalent.

(c) Find values of e and f for network C in terms of r in network A such that the two systems are equivalent.

(d) Why might network B or C be preferred over network A? What possible advantage could network A have over network B or C?

6.33. Consider an all-pass system with system function

$$H(z) = \frac{z^{-1} - 0.54}{1 - 0.54z^{-1}}.$$

A flow graph for an implementation of this system is shown in Figure P6.33-1.

Figure P6.33-1

(a) Determine the coefficients b, c, and d such that the flow graph in Figure P6.33-1 is a realization of $H(z)$.

(b) In a practical implementation of the network in Figure P6.33-1, the coefficients b, c, and d might be quantized by rounding the exact value to the nearest tenth (e.g., 0.54 will be rounded to 0.5 and 1.8518...will be rounded to 1.9). Would the resulting system still be an all-pass system?

The difference equation relating the input and output of the all-pass system with system function $H(z)$ can be expressed as

$$y[n] = 0.54(y[n-1] - x[n]) + x[n-1].$$

(c) Draw the flow graph of a network that implements this difference equation using two delay elements, but that requires only one multiplication by a constant other than ± 1.

(d) With quantized coefficients, would the network of part (c) be an all-pass system?

The primary disadvantage of the implementation in part (c) compared with the implementation in part (a) is that it requires two delay elements. However, for higher order systems, it is necessary to implement a cascade of all-pass systems. For N all-pass sections in cascade, it is possible to use all-pass sections in the form determined in Part (c) while requiring only $(N + 1)$ delay elements. This is accomplished by sharing a delay element between sections.

(e) Consider the all-pass system with system function

$$H(z) = \left(\frac{z^{-1} - a}{1 - az^{-1}} \right) \left(\frac{z^{-1} - b}{1 - bz^{-1}} \right).$$

Draw the flow graph of a "cascade" realization composed of two sections of the form obtained in Part (c) with one delay element shared between the sections. The resulting network should have only three delay elements.

(f) With quantized coefficients a and b, would the network in Part (e) be an all-pass system?

6.34. All branches of the signal flow graphs in this problem have unity gain unless specifically indicated otherwise.

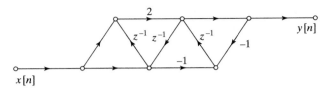

Figure P6.34-1

(a) The signal flow graph of System A, shown in Figure P6.34-1, represents a causal LTI system. Is it possible to implement the same input–output relationship using fewer delays? If it is possible, what is the minimum number of delays required to implement an equivalent system? If it is not possible, explain why not.

(b) Does the System B shown in Figure P6.34-2 represent the same input–output relationship as System A in Figure P6.34-1? Explain clearly.

Figure P6.34-2

6.35. Consider an all-pass system whose system function is

$$H(z) = \frac{z^{-1} - \frac{1}{3}}{1 - \frac{1}{3}z^{-1}}.$$

(a) Draw the direct form I signal flow graph for the system. How many delays and multipliers do you need? (Do not count multiplying by ± 1.)

(b) Draw a signal flow graph for the system that uses one multiplier. Minimize the number of delays.

(c) Now consider another all-pass system whose system function is

$$H(z) = \frac{(z^{-1} - \frac{1}{3})(z^{-1} - 2)}{(1 - \frac{1}{3}z^{-1})(1 - 2z^{-1})}.$$

Find and draw a signal flow graph for the system with two multipliers and three delays.

6.36. The flow graph shown in Figure P6.36-1 is an implementation of a causal, LTI system.

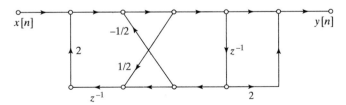

Figure P6.36-1

(a) Draw the transpose of the signal flow graph.

(b) For either the original system or its transpose, determine the difference equation relating the input $x[n]$ to the output $y[n]$. (*Note:* The difference equations will be the same for both structures.)

(c) Is the system BIBO stable?

(d) Determine $y[2]$ if $x[n] = (1/2)^n u[n]$.

Extension Problems

6.37. In this problem, we will develop some of the properties of a class of discrete-time systems called frequency-sampling filters. This class of filters has system functions of the form

$$H(z) = (1 - z^{-N}) \cdot \sum_{k=0}^{N-1} \frac{\tilde{H}[k]/N}{1 - z_k z^{-1}},$$

where $z_k = e^{j(2\pi/N)k}$ for $k = 0, 1, \ldots, N-1$.

(a) System functions such as $H(z)$ can be implemented as a cascade of an FIR system whose system function is $(1 - z^{-N})$ with a parallel combination of first-order IIR systems. Draw the signal flow graph of such an implementation.

(b) Show that $H(z)$ is an $(N-1)$st-degree polynomial in z^{-1}. To do this, it is necessary to show that $H(z)$ has no poles other than $z = 0$ and that it has no powers of z^{-1} higher than $(N-1)$. What do these conditions imply about the length of the impulse response of the system?

(c) Show that the impulse response is given by the expression

$$h[n] = \left(\frac{1}{N} \sum_{k=0}^{N-1} \tilde{H}[k] e^{j(2\pi/N)kn} \right) (u[n] - u[n - N]).$$

Hint: Find the impulse responses of the FIR and the IIR parts of the system, and convolve them to find the overall impulse response.

(d) Use l'Hôpital's rule to show that

$$H(z_m) = H(e^{j(2\pi/N)m}) = \tilde{H}[m], \qquad m = 0, 1, \ldots, N-1;$$

i.e., show that the constants $\tilde{H}[m]$ are samples of the frequency response of the system, $H(e^{j\omega})$, at equally spaced frequencies $\omega_m = (2\pi/N)m$ for $m = 0, 1, \ldots, N-1$. It is this property that accounts for the name of this class of FIR systems.

(e) In general, both the poles z_k of the IIR part and the samples of the frequency response $\tilde{H}[k]$ will be complex. However, if $h[n]$ is real, we can find an implementation involving only real quantities. Specifically, show that if $h[n]$ is real and N is an even integer, then $H(z)$ can be expressed as

$$H(z) = (1 - z^{-N}) \left\{ \frac{H(1)/N}{1 - z^{-1}} + \frac{H(-1)/N}{1 + z^{-1}} \right.$$

$$\left. + \sum_{k=1}^{(N/2)-1} \frac{2|H(e^{j(2\pi/N)k})|}{N} \cdot \frac{\cos[\theta(2\pi k/N)] - z^{-1}\cos[\theta(2\pi k/N) - 2\pi k/N]}{1 - 2\cos(2\pi k/N)z^{-1} + z^{-2}} \right\},$$

where $H(e^{j\omega}) = |H(e^{j\omega})| e^{j\theta(\omega)}$. Draw the signal flow graph representation of such a system when $N = 16$ and $H(e^{j\omega_k}) = 0$ for $k = 3, 4, \ldots, 14$.

6.38. In Chapter 4, we showed that, in general, the sampling rate of a discrete-time signal can be reduced by a combination of linear filtering and time compression. Figure P6.38-1 shows a block diagram of an M-to-1 decimator that can be used to reduce the sampling rate by an integer factor M. According to the model, the linear filter operates at the high sampling rate. However, if M is large, most of the output samples of the filter will be discarded by the compressor. In some cases, more efficient implementations are possible.

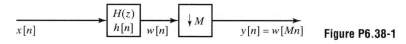

Figure P6.38-1

(a) Assume that the filter is an FIR system with impulse response such that $h[n] = 0$ for $n < 0$ and for $n > 10$. Draw the system in Figure P6.38-1, but replace the filter $h[n]$ with an equivalent signal flow graph based on the given information. Note that it is not possible to implement the M-to-1 compressor using a signal flow graph, so you must leave this as a box, as shown in Figure P6.38-1.

(b) Note that some of the branch operations can be commuted with the compression operation. Using this fact, draw the flow graph of a more efficient realization of the system of Part (a). By what factor has the total number of computations required in obtaining the output $y[n]$ been decreased?

(c) Now suppose that the filter in Figure P6.38-1 has system function

$$H(z) = \frac{1}{1 - \frac{1}{2}z^{-1}}, \qquad |z| > \frac{1}{2}.$$

Draw the flow graph of the direct-form realization of the complete system in the figure. With this system for the linear filter, can the total computation per output sample be reduced? If so, by what factor?

(d) Finally, suppose that the filter in Figure P6.38-1 has system function

$$H(z) = \frac{1 + \frac{7}{8}z^{-1}}{1 - \frac{1}{2}z^{-1}}, \qquad |z| > \frac{1}{2}.$$

Draw the flow graph for the complete system of the figure, using each of the following forms for the linear filter:

  (i)   direct form I
  (ii) direct form II
  (iii) transposed direct form I
  (iv) transposed direct form II

For which of the four forms can the system of Figure P6.38-1 be more efficiently implemented by commuting operations with the compressor?

6.39. Speech production can be modeled by a linear system representing the vocal cavity, which is excited by puffs of air released through the vibrating vocal cords. One approach to synthesizing speech involves representing the vocal cavity as a connection of cylindrical acoustic tubes of equal length, but with varying cross-sectional areas, as depicted in Figure P6.39-1. Let us assume that we want to simulate this system in terms of the volume velocity representing airflow. The input is coupled into the vocal tract through a small constriction, the vocal cords. We will assume that the input is represented by a change in volume velocity at the left end, but that the boundary condition for traveling waves at the left end is that the net volume velocity must be zero. This is analogous to an electrical transmission line driven by a current source at one end and with an open circuit at the far end. Current in the transmission line is then analogous to volume velocity in the acoustic tube, while voltage is analogous to acoustic pressure. The output of the acoustic tube is the volume velocity at the right end. We assume that each section is a lossless acoustic transmission line.

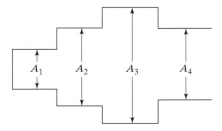

Figure P6.39-1

At each interface between sections, a forward-traveling wave f^+ is transmitted to the next section with one coefficient and reflected as a backward-traveling wave f^- with a different coefficient. Similarly, a backward-traveling wave f^- arriving at an interface is transmitted with one coefficient and reflected with a different coefficient. Specifically, if we consider a forward-traveling wave f^+ in a tube with cross-sectional area A_1 arriving at the interface with a tube of cross-sectional area A_2, then the forward-traveling wave transmitted is $(1+r)f^+$ and the reflected wave is rf^+, where

$$r = \frac{A_2 - A_1}{A_2 + A_1}.$$

Consider the length of each section to be 3.4 cm, with the velocity of sound in air equal to 34,000 cm/s. Draw a flow graph that will implement the four-section model in Figure P6.39-1, with the output sampled at 20,000 samples/s.

In spite of the lengthy introduction, this a reasonably straightforward problem. If you find it difficult to think in terms of acoustic tubes, think in terms of transmission-line sections with different characteristic impedances. Just as with transmission lines, it is difficult to express the impulse response in closed form. Therefore, draw the network directly from physical considerations, in terms of forward- and backward-traveling pulses in each section.

6.40. In modeling the effects of round-off and truncation in digital filter implementations, quantized variables are represented as

$$\hat{x}[n] = Q[x[n]] = x[n] + e[n],$$

where $Q[\cdot]$ denotes either rounding or truncation to $(B+1)$ bits and $e[n]$ is the *quantization error*. We assume that the quantization noise sequence is a stationary white-noise sequence such that

$$\mathcal{E}\{(e[n] - m_e)(e[n+m] - m_e)\} = \sigma_e^2 \, \delta[m]$$

and that the amplitudes of the noise sequence values are uniformly distributed over the quantization step $\Delta = 2^{-B}$. The first-order probability densities for rounding and truncation are shown in Figures P6.40-1(a) and (b), respectively.

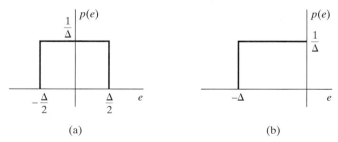

(a)   (b)   **Figure P6.40-1**

(a) Find the mean m_e and the variance σ_e^2 for the noise due to rounding.
(b) Find the mean m_e and the variance σ_e^2 for the noise due to truncation.

6.41. Consider a linear time-invariant system with two inputs, as depicted in Figure P6.41-1. Let $h_1[n]$ and $h_2[n]$ be the impulse responses from nodes 1 and 2, respectively, to the output, node 3. Show that if $x_1[n]$ and $x_2[n]$ are uncorrelated, then their corresponding outputs $y_1[n]$ and $y_2[n]$ are also uncorrelated.

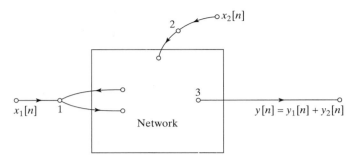

Figure P6.41-1

6.42. The networks in Figure P6.42-1 all have the same system function. Assume that the systems in the figure are implemented using $(B + 1)$-bit fixed-point arithmetic in all the computations. Assume also that all products are rounded to $(B + 1)$ bits *before* additions are performed.

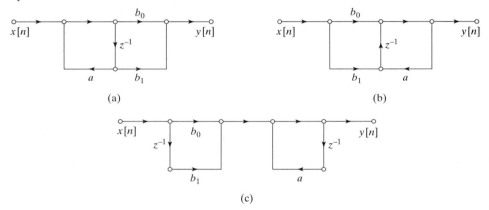

Figure P6.42-1

(a) Draw linear-noise models for each of the systems in Figure P6.42-1.
(b) Two of the networks in Figure P6.42-1 have the *same* total output noise power due to arithmetic round-off. Without explicitly computing the output noise power, determine which two have the same output noise power.
(c) Determine the output noise power for each of the networks in Figure P6.42-1. Express your answer in terms of σ_B^2, the power of a single source of round-off noise.

6.43. The flow graph of a first-order system is shown in Figure P6.43-1.

Figure P6.43-1

(a) Assuming infinite-precision arithmetic, find the response of the system to the input

$$x[n] = \begin{cases} \frac{1}{2}, & n \geq 0, \\ 0, & n < 0. \end{cases}$$

What is the response of the system for large n?

Now suppose that the system is implemented with fixed-point arithmetic. The coefficient and all variables in the network are represented in sign-and-magnitude notation with

5-bit registers. That is, all numbers are to be considered signed fractions represented as

$$b_0 b_1 b_2 b_3 b_4,$$

where b_0, b_1, b_2, b_3, and b_4 are either 0 or 1 and

$$|\text{Register value}| = b_1 2^{-1} + b_2 2^{-2} + b_3 2^{-3} + b_4 2^{-4}.$$

If $b_0 = 0$, the fraction is positive, and if $b_0 = 1$, the fraction is negative. The result of a multiplication of a sequence value by a coefficient is truncated before additions occur; i.e., only the sign bit and the most significant four bits are retained.

(b) Compute the response of the quantized system to the input of Part (a), and plot the responses of both the quantized and unquantized systems for $0 \le n \le 5$. How do the responses compare for large n?

(c) Now consider the system depicted in Figure P6.43-2, where

$$x[n] = \begin{cases} \frac{1}{2}(-1)^n, & n \ge 0, \\ 0, & n < 0. \end{cases}$$

Repeat Parts (a) and (b) for this system and input.

Figure P6.43-2

6.44. A causal LTI system has a system function

$$H(z) = \frac{1}{1 - 1.04z^{-1} + 0.98z^{-2}}.$$

(a) Is this system stable?

(b) If the coefficients are rounded to the nearest tenth, would the resulting system be stable?

6.45. When implemented with infinite-precision arithmetic, the flow graphs in Figure P6.45-1 have the same system function.

Network 1

Network 2

Figure P6.45-1

(a) Show that the two systems have the same overall system function from input $x[n]$ to output $y[n]$.

(b) Assume that the preceding systems are implemented with two's complement fixed-point arithmetic and that products are rounded *before* additions are performed. Draw signal flow graphs that insert round-off noise sources at appropriate locations in the signal flow graphs of Figure P6.45-1.

(c) Circle the nodes in your figure from Part (b) where overflow can occur.

(d) Determine the maximum size of the input samples such that overflow cannot occur in either of the two systems.

(e) Assume that $|a| < 1$. Find the total noise power at the output of each system, and determine the maximum value of $|a|$ such that Network 1 has lower output noise power than Network 2.

7
FILTER DESIGN TECHNIQUES

7.0 INTRODUCTION

Filters are a particularly important class of linear time-invariant systems. Strictly speaking, the term *frequency-selective filter* suggests a system that passes certain frequency components and totally rejects all others, but in a broader context any system that modifies certain frequencies relative to others is also called a filter. While the primary emphasis in this chapter is on the design of frequency-selective filters, some of the techniques are more broadly applicable. Also, we concentrate on the design of causal filters, although in many contexts filters need not be restricted to causal designs. Very often, noncausal filters are designed and implemented by modifying causal designs.

The design of filters involves the following stages: (1) the specification of the desired properties of the system, (2) the approximation of the specifications using a causal discrete-time system, and (3) the realization of the system. Although these three steps are certainly not independent, we focus our attention primarily on the second step, the first being highly dependent on the application and the third dependent on the technology to be used for the implementation. In a practical setting, the desired filter is generally implemented with digital computation and used to filter a signal that is derived from a continuous-time signal by means of periodic sampling followed by analog-to-digital conversion. For this reason, it has become common to refer to discrete-time filters as *digital filters,* even though the underlying design techniques most often relate only to the discrete-time nature of the signals and systems.

When a discrete-time filter is to be used for discrete-time processing of continuous-time signals in the configuration of Figure 7.1, the specifications for both the discrete-time filter and the effective continuous-time filter are typically (but not always) given in the frequency domain. This is especially common for frequency-selective filters such

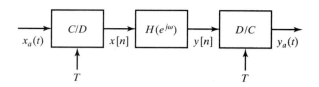

Figure 7.1 Basic system for discrete-time filtering of continuous-time signals.

as lowpass, bandpass, and highpass filters. As shown in Section 4.4, if a linear time-invariant discrete-time system is used as in Figure 7.1, and if the input is bandlimited and the sampling frequency is high enough to avoid aliasing, then the overall system behaves as a linear time-invariant continuous-time system with frequency response

$$H_{\text{eff}}(j\Omega) = \begin{cases} H(e^{j\Omega T}), & |\Omega| < \pi/T, \\ 0, & |\Omega| > \pi/T. \end{cases} \tag{7.1a}$$

In such cases, it is straightforward to convert from specifications on the effective continuous-time filter to specifications on the discrete-time filter through the relation $\omega = \Omega T$. That is, $H(e^{j\omega})$ is specified over one period by the equation

$$H(e^{j\omega}) = H_{\text{eff}}\left(j\frac{\omega}{T}\right), \qquad |\omega| < \pi. \tag{7.1b}$$

This type of conversion is illustrated in Example 7.1.

Example 7.1   Determining Specifications for a Discrete-Time Filter

Consider a discrete-time filter that is to be used to lowpass filter a continuous-time signal using the basic configuration of Figure 7.1. Specifically, we want the overall system of that figure to have the following properties when the sampling rate is 10^4 samples/s ($T = 10^{-4}$ s):

1. The gain $|H_{\text{eff}}(j\Omega)|$ should be within ± 0.01 of unity in the frequency band $0 \leq \Omega \leq 2\pi(2000)$.

2. The gain should be no greater than 0.001 in the frequency band $2\pi(3000) \leq \Omega$.

Such a set of lowpass specifications on $|H_{\text{eff}}(j\Omega)|$ can be depicted as in Figure 7.2(a), where the limits of tolerable approximation error are indicated by the shaded horizontal lines. For this specific example, the parameters would be

$$\delta_1 = 0.01,$$

$$\delta_2 = 0.001,$$

$$\Omega_p = 2\pi(2000),$$

$$\Omega_s = 2\pi(3000).$$

Therefore, in this case, the ideal passband gain is unity. The passband gain varies between $(1 + \delta_1)$, and $(1 - \delta_1)$, and the stopband gain varies betwen 0 and δ_2. It is common to express the maximum passband and stopband gains in units of decibels. For this example:

$$\begin{aligned} \text{ideal passband gain in decibels} &= 20\log_{10}(1) &&= 0\,\text{dB} \\ \text{maximum passband gain in decibels} &= 20\log_{10}(1.01) &&= 0.086\,\text{dB} \\ \text{maximum stopband gain in decibels} &= 20\log_{10}(0.001) &&= -60\,\text{dB} \end{aligned}$$

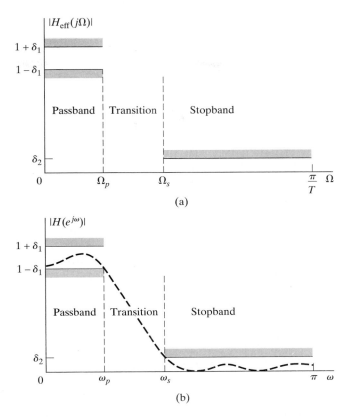

Figure 7.2 (a) Specifications for effective frequency response of overall system in Figure 7.1 for the case of a lowpass filter. (b) Corresponding specifications for the discrete-time system in Figure 7.1.

Since the sampling rate is 10^4 samples/s, the gain of the overall system is identically zero above $\Omega = 2\pi(5000)$, due to the ideal discrete-to-continuous (D/C) converter in Figure 7.1.

The tolerance scheme for the discrete-time filter is shown in Figure 7.2(b). It is the same as that in Figure 7.2(a), except that it is plotted as a function of normalized frequency ($\omega = \Omega T$), and it need only be plotted in the range $0 \le \omega \le \pi$, since the remainder can be inferred from symmetry properties (assuming that $h[n]$ is real) and the periodicity of $H(e^{j\omega})$. From Eq. (7.1b), it follows that in the *passband* the magnitude of the frequency response must approximate unity within an error of $\pm\delta_1$, i.e.,

$$(1 - \delta_1) \le |H(e^{j\omega})| \le (1 + \delta_1), \qquad |\omega| \le \omega_p, \tag{7.2}$$

where $\delta_1 = 0.01$ and $\omega_p = 2\pi(2000) \cdot 10^{-4} = 0.4\pi$ radians. The other approximation band is the *stopband*, in which the magnitude response must approximate zero with an error less than δ_2; i.e.,

$$|H(e^{j\omega})| \le \delta_2, \qquad \omega_s \le |\omega| \le \pi. \tag{7.3}$$

In this example, $\delta_2 = 0.001$ and $\omega_s = 2\pi(3000) \cdot 10^{-4} = 0.6\pi$ radians. The passband cutoff frequency ω_p and the stopband cutoff frequency ω_s are given in terms of normalized radian frequency or equivalently in terms of angle in the z-plane. To approximate the ideal lowpass filter in this way with a realizable system, we must provide a transition band of nonzero width $(\omega_s - \omega_p)$ in which the magnitude response changes smoothly from passband to stopband. The dashed curve in Figure 7.2(b) is the magnitude response of a system that meets the prescribed specification.

There are many applications in which a discrete-time signal to be filtered is not derived from a continuous-time signal, and there are a variety of means besides periodic sampling for representing continuous-time signals in terms of sequences. (See, for example, Steiglitz, 1965, and Oppenheim and Johnson, 1972.) Also, in most of the design techniques that we discuss, the sampling period plays no role whatsoever in the approximation procedure. For these reasons, we take the point of view that the filter design problem begins from a set of desired specifications in terms of the discrete-time frequency variable ω. Depending on the specific application or context, these specifications may or may not have been obtained from a consideration of filtering in the framework of Figure 7.1.

Many of the filters used in practice are specified by a tolerance scheme similar to that in Example 7.1, with no constraints on the phase response other than those imposed implicitly by stability and causality requirements. For example, the poles of the system function for a causal and stable infinite impulse response (IIR) filter must lie inside the unit circle. Similarly, in designing finite impulse response (FIR) filters, we often impose the constraint of a linear phase. This again removes the phase of the signal from consideration in the design process.

Given a set of specifications in the form of Figure 7.2(b), we must determine the system function of a discrete-time linear system whose frequency response falls within the prescribed tolerances. This is a problem in functional approximation. Designing IIR filters implies approximation by a rational function of z, while designing FIR filters implies polynomial approximation. Our discussion distinguishes between design techniques that are appropriate for IIR filters and those that are appropriate for FIR filters. We discuss a variety of design techniques for both types of filter, ranging from closed-form procedures, which involve only substitution of design specifications into design formulas, to algorithmic techniques, in which a solution is obtained by an iterative procedure.

7.1 DESIGN OF DISCRETE-TIME IIR FILTERS FROM CONTINUOUS-TIME FILTERS

The traditional approach to the design of discrete-time IIR filters involves the transformation of a continuous-time filter into a discrete-time filter meeting prescribed specifications. This is a reasonable approach for several reasons:

- The art of continuous-time IIR filter design is highly advanced, and since useful results can be achieved, it is advantageous to use the design procedures already developed for continuous-time filters.

- Many useful continuous-time IIR design methods have relatively simple closed-form design formulas. Therefore, discrete-time IIR filter design methods based on such standard continuous-time design formulas are rather simple to carry out.

- The standard approximation methods that work well for continuous-time IIR filters do not lead to simple closed-form design formulas when these methods are applied directly to the discrete-time IIR case.

The fact that continuous-time filter designs can be mapped to discrete-time filter designs is totally unrelated to, and independent of, whether the discrete-time filter is to be used in the configuration of Figure 7.1 for processing continuous-time signals. We emphasize again that the design procedure for the discrete-time system begins from a set of *discrete-time* specifications. Henceforth, we assume that these specifications have been determined by an analysis like that of Example 7.1 or by some other method. We will use continuous-time filter approximation methods only as a convenience in determining the discrete-time filter that meets the desired specifications. Indeed, the continuous-time filter on which the approximation is based may have a frequency response that is vastly different from the effective frequency response when the discrete-time filter is used in the configuration of Figure 7.1.

In designing a discrete-time filter by transforming a prototype continuous-time filter, the specifications for the continuous-time filter are obtained by a transformation of the specifications for the desired discrete-time filter. The system function $H_c(s)$ or impulse response $h_c(t)$ of the continuous-time filter is then obtained through one of the established approximation methods used for continuous-time filter design, examples of which are discussed in Appendix B. Next, the system function $H(z)$ or impulse response $h[n]$ for the discrete-time filter is obtained by applying to $H_c(s)$ or $h_c(t)$ a transformation of the type discussed in this section.

In such transformations, we generally require that the essential properties of the continuous-time frequency response be preserved in the frequency response of the resulting discrete-time filter. Specifically, this implies that we want the imaginary axis of the s-plane to map onto the unit circle of the z-plane. A second condition is that a stable continuous-time filter should be transformed to a stable discrete-time filter. This means that if the continuous-time system has poles only in the left half of the s-plane, then the discrete-time filter must have poles only inside the unit circle in the z-plane. These constraints are basic to all the techniques discussed in this section.

7.1.1 Filter Design by Impulse Invariance

In Section 4.4.2 we discussed the concept of *impulse invariance,* wherein a discrete-time system is defined by sampling the impulse response of a continuous-time system. We showed that impulse invariance provides a direct means of computing samples of the output of a bandlimited continuous-time system for bandlimited input signals. Alternatively, in the context of filter design, we can think of impulse invariance as a method for obtaining a discrete-time system whose frequency response is determined by the frequency response of a continuous-time system.

In the impulse invariance design procedure for transforming continuous-time filters into discrete-time filters, the impulse response of the discrete-time filter is chosen proportional to equally spaced samples of the impulse response of the continuous-time filter; i.e.,

$$h[n] = T_d h_c(nT_d), \tag{7.4}$$

where T_d represents a sampling interval. As we will see, because we begin the design problem with the discrete-time filter specifications, the parameter T_d in Eq. (7.4) in fact has no role whatsoever in the design process or the resulting discrete-time filter. However, since it is customary to specify this parameter in defining the procedure, we include it in the following discussion. As we will see, even if the filter is used in the basic configuration of Figure 7.1, the design sampling period T_d need not be the same as the sampling period T associated with the C/D and D/C conversion.

When impulse invariance is used as a means for designing a discrete-time filter with a specified frequency response, we are especially interested in the relationship between the frequency responses of the discrete-time and continuous-time filters. From the discussion of sampling in Chapter 4, it follows that the frequency response of the discrete-time filter obtained through Eq. (7.4) is related to the frequency response of the continuous-time filter by

$$H(e^{j\omega}) = \sum_{k=-\infty}^{\infty} H_c\left(j\frac{\omega}{T_d} + j\frac{2\pi}{T_d}k\right). \tag{7.5}$$

If the continuous-time filter is bandlimited, so that

$$H_c(j\Omega) = 0, \qquad |\Omega| \geq \pi/T_d, \tag{7.6}$$

then

$$H(e^{j\omega}) = H_c\left(j\frac{\omega}{T_d}\right), \qquad |\omega| \leq \pi; \tag{7.7}$$

i.e., the discrete-time and continuous-time frequency responses are related by a linear scaling of the frequency axis, namely, $\omega = \Omega T_d$ for $|\omega| < \pi$. Unfortunately, any practical continuous-time filter cannot be exactly bandlimited, and consequently, interference between successive terms in Eq. (7.5) occurs, causing aliasing, as illustrated in Figure 7.3. However, if the continuous-time filter approaches zero at high frequencies, the aliasing may be negligibly small, and a useful discrete-time filter can result from the sampling of the impulse response of a continuous-time filter.

In the impulse invariance design procedure, the discrete-time filter specifications are first transformed to continuous-time filter specifications through the use of Eq. (7.7). Assuming that the aliasing involved in the transformation from $H_c(j\Omega)$ to $H(e^{j\omega})$ will be negligible, we obtain the specifications on $H_c(j\Omega)$ by applying the relation

$$\Omega = \omega/T_d \tag{7.8}$$

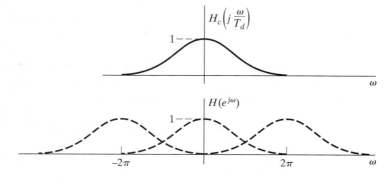

Figure 7.3 Illustration of aliasing in the impulse invariance design technique.

to obtain the continuous-time filter specifications from the specifications on $H(e^{j\omega})$. After obtaining a suitable continuous-time filter based on these specifications, the continuous-time filter with system function $H_c(s)$ is transformed to the desired discrete-time filter with system function $H(z)$. We develop the algebraic details of the transformation from $H_c(s)$ to $H(z)$ shortly. Note, however, that in the transformation back to discrete-time frequency, $H(e^{j\omega})$ will be related to $H_c(j\Omega)$ through Eq. (7.5), which again applies the transformation of Eq. (7.8) to the frequency axis. As a consequence, the "sampling" parameter T_d cannot be used to control aliasing. Since the basic specifications are in terms of discrete-time frequency, if the sampling rate is increased (i.e., if T_d is made smaller), then the cutoff frequency of the continuous-time filter must increase in proportion. In practice, to compensate for aliasing that might occur in the transformation from $H_c(s)$ to $H(z)$, the continuous-time filter may be somewhat overdesigned, i.e., designed to exceed the specifications, particularly in the stopband.

While the impulse invariance transformation from continuous time to discrete time is defined in terms of time-domain sampling, it is easy to carry out as a transformation on the system function. To develop this transformation, let us consider the system function of the continuous-time filter expressed in terms of a partial fraction expansion, so that[1]

$$H_c(s) = \sum_{k=1}^{N} \frac{A_k}{s - s_k}. \tag{7.9}$$

The corresponding impulse response is

$$h_c(t) = \begin{cases} \displaystyle\sum_{k=1}^{N} A_k e^{s_k t}, & t \geq 0, \\ 0, & t < 0. \end{cases} \tag{7.10}$$

The impulse response of the discrete-time filter obtained by sampling $T_d h_c(t)$ is

$$\begin{aligned} h[n] = T_d h_c(nT_d) &= \sum_{k=1}^{N} T_d A_k e^{s_k n T_d} u[n] \\ &= \sum_{k=1}^{N} T_d A_k (e^{s_k T_d})^n u[n]. \end{aligned} \tag{7.11}$$

The system function of the discrete-time filter is therefore given by

$$H(z) = \sum_{k=1}^{N} \frac{T_d A_k}{1 - e^{s_k T_d} z^{-1}}. \tag{7.12}$$

In comparing Eqs. (7.9) and (7.12), we observe that a pole at $s = s_k$ in the s-plane transforms to a pole at $z = e^{s_k T_d}$ in the z-plane and the coefficients in the partial fraction expansions of $H_c(s)$ and $H(z)$ are equal, except for the scaling multiplier T_d. If the continuous-time filter is stable, corresponding to the real part of s_k being less than zero, then the magnitude of $e^{s_k T_d}$ will be less than unity, so that the corresponding pole in the discrete-time filter is inside the unit circle. Therefore, the causal discrete-time filter is

---

[1] For simplicity, we assume in the discussion that all poles of $H(s)$ are single order. In Problem 7.24, we consider the modifications required for multiple-order poles.

also stable. While the poles in the s-plane map to poles in the z-plane according to the relationship $z_k = e^{s_k T_d}$, it is important to recognize that the impulse invariance design procedure does not correspond to a simple mapping of the s-plane to the z-plane by that relationship. In particular, the zeros in the discrete-time system function are a function of the poles and the coefficients $T_d A_k$ in the partial fraction expansion, and they will not in general be mapped in the same way the poles are mapped. We illustrate the impulse invariance design procedure with the following example.

Example 7.2   Impulse Invariance with a Butterworth Filter

Let us consider the design of a lowpass discrete-time filter by applying impulse invariance to an appropriate Butterworth continuous-time filter.[2] The specifications for the discrete-time filter are

$$0.89125 \leq |H(e^{j\omega})| \leq 1, \qquad 0 \leq |\omega| \leq 0.2\pi, \tag{7.13a}$$

$$|H(e^{j\omega})| \leq 0.17783, \qquad 0.3\pi \leq |\omega| \leq \pi. \tag{7.13b}$$

Since the parameter T_d cancels in the impulse invariance procedure, we can choose $T_d = 1$, so that $\omega = \Omega$. In Problem 7.2, this same example is considered, but with the parameter T_d explicitly included to illustrate how and where it cancels.

In designing the filter using impulse invariance on a continuous-time Butterworth filter, we must first transform the discrete-time specifications to specifications on the continuous-time filter. Recall that impulse invariance corresponds to a linear mapping between Ω and ω in the absence of aliasing. For this example, we will assume that the effect of aliasing is negligible. After the design is complete, we can evaluate the resulting frequency response against the specifications in Eqs. (7.13a) and (7.13b).

Because of the preceding considerations, we want to design a continuous-time Butterworth filter with magnitude function $|H_c(j\Omega)|$ for which

$$0.89125 \leq |H_c(j\Omega)| \leq 1, \qquad 0 \leq |\Omega| \leq 0.2\pi, \tag{7.14a}$$

$$|H_c(j\Omega)| \leq 0.17783, \qquad 0.3\pi \leq |\Omega| \leq \pi. \tag{7.14b}$$

Since the magnitude response of an analog Butterworth filter is a monotonic function of frequency, Eqs. (7.14a) and (7.14b) will be satisfied if

$$|H_c(j0.2\pi)| \geq 0.89125 \tag{7.15a}$$

and

$$|H_c(j0.3\pi)| \leq 0.17783. \tag{7.15b}$$

Specifically, the magnitude-squared function of a Butterworth filter is of the form

$$|H_c(j\Omega)|^2 = \frac{1}{1 + (\Omega/\Omega_c)^{2N}}, \tag{7.16}$$

so that the filter design process consists of determining the parameters N and Ω_c to meet the desired specifications. Using Eq. (7.16) in Eqs. (7.15) with equality leads

---

[2]Continuous-time Butterworth and Chebyshev filters are discussed in Appendix B.

$$2N\left(\log 0.2\pi\right) - \text{(}\log \Omega_c\text{)}$$

$$2N \log\left(\frac{0.2\pi}{\Omega_c}\right) = \log(\#)$$

to the equations

$$1 + \left(\frac{0.2\pi}{\Omega_c}\right)^{2N} = \left(\frac{1}{0.89125}\right)^2 \tag{7.17a}$$

and

$$1 + \left(\frac{0.3\pi}{\Omega_c}\right)^{2N} = \left(\frac{1}{0.17783}\right)^2 . \tag{7.17b}$$

The solution of these two equations is $N = 5.8858$ and $\Omega_c = 0.70474$. The parameter N, however, must be an integer. Therefore, so that the specifications are met or exceeded, we must round N up to the nearest integer, $N = 6$. Because we have rounded N up to the next highest integer, the filter will not exactly satisfy both Eqs. (7.17a) and (7.17b) simultaneously. With $N = 6$, the filter parameter Ω_c can be chosen to exceed the specified requirements in either the passband, the stopband, or both. Specifically, as the value of Ω_c varies, there is a trade-off in the amount by which the stopband and passband specifications are exceeded. If we substitute $N = 6$ into Eq. (7.17a), we obtain $\Omega_c = 0.7032$. With this value, the passband specifications (of the continuous-time filter) will be met exactly, and the stopband specifications (of the continuous-time filter) will be exceeded. This allows some margin for aliasing in the discrete-time filter. With $\Omega_c = 0.7032$ and with $N = 6$, the 12 poles of the magnitude-squared function $H_c(s)H_c(-s) = 1/[1 + (s/j\Omega_c)^{2N}]$ are uniformly distributed in angle on a circle of radius $\Omega_c = 0.7032$, as indicated in Figure 7.4. Consequently, the poles of $H_c(s)$ are the three pole pairs in the left half of the s-plane with the following coordinates:

Pole pair 1: $-0.182 \pm j(0.679),$

Pole pair 2: $-0.497 \pm j(0.497),$

Pole pair 3: $-0.679 \pm j(0.182).$

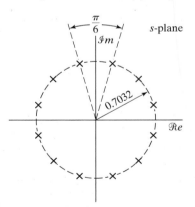

Figure 7.4 *s*-plane locations for poles of $H_c(s)H_c(-s)$ for sixth-order Butterworth filter in Example 7.2.

Therefore,

$$H_c(s) = \frac{0.12093}{(s^2 + 0.3640s + 0.4945)(s^2 + 0.9945s + 0.4945)(s^2 + 1.3585s + 0.4945)} \tag{7.18}$$

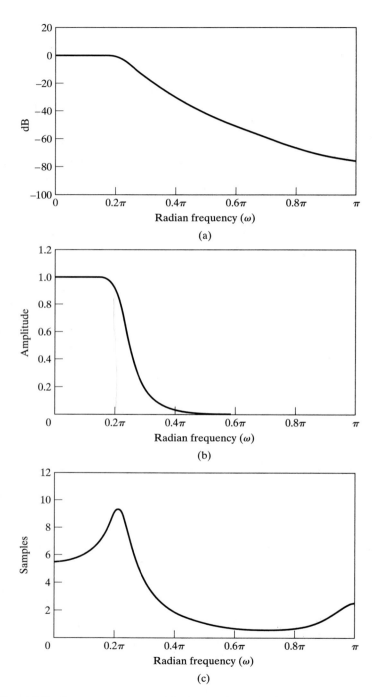

Figure 7.5 Frequency response of sixth-order Butterworth filter transformed by impulse invariance. (a) Log magnitude in dB. (b) Magnitude. (c) Group delay.

If we express $H_c(s)$ as a partial fraction expansion, perform the transformation of Eq. (7.12), and then combine complex-conjugate terms, the resulting system function of the discrete-time filter is

$$H(z) = \frac{0.2871 - 0.4466z^{-1}}{1 - 1.2971z^{-1} + 0.6949z^{-2}} + \frac{-2.1428 + 1.1455z^{-1}}{1 - 1.0691z^{-1} + 0.3699z^{-2}}$$

$$+ \frac{1.8557 - 0.6303z^{-1}}{1 - 0.9972z^{-1} + 0.2570z^{-2}}.$$

(7.19)

As is evident from Eq. (7.19), the system function resulting from the impulse invariance design procedure may be realized directly in parallel form. If cascade or direct form is desired, the separate second-order terms must be combined in an appropriate way.

The frequency-response functions of the discrete-time system are shown in Figure 7.5. Recall that the prototype continuous-time filter was designed to meet the specifications exactly at the passband edge and to exceed the specifications at the stopband edge, and this turns out to be true for the resulting discrete-time filter. This is an indication that the continuous-time filter was sufficiently bandlimited so that aliasing presented no problem. Indeed, the difference between $20 \log_{10} |H(e^{j\omega})|$ and $20 \log_{10} |H_c(j\Omega)|$ would not be visible on this plotting scale, except for a slight deviation around $\omega = \pi$. (Recall that $T_d = 1$, so $\Omega = \omega$.) Sometimes, aliasing is much more of a problem. If the resulting discrete-time filter fails to meet the specifications because of aliasing, there is no alternative with impulse invariance but to try again with a higher order filter or with different filter parameters, holding the order fixed.

The basis for impulse invariance is to choose an impulse response for the discrete-time filter that is similar in some sense to the impulse response of the continuous-time filter. The use of this procedure is often motivated not so much by a desire to maintain the shape of the impulse response as by the knowledge that if the continuous-time filter is bandlimited, then the discrete-time filter frequency response will closely approximate the continuous-time frequency response. However, in some filter design problems, a primary objective may be to control some aspect of the time response, such as the impulse response or the step response. In these cases, a natural approach might be to design the discrete-time filter by impulse invariance or by step invariance. In the latter case, the response of the filter to a sampled unit step function is defined to be the sequence obtained by sampling the continuous-time step response. If the continuous-time filter has good step response characteristics, such as a small rise time and low peak overshoot, these characteristics will be preserved in the discrete-time filter. Clearly, this concept of waveform invariance can be extended to the preservation of the output waveshape for a variety of inputs, as illustrated in Problem 7.1. The problem points up the fact that transforming the same continuous-time filter by impulse invariance and also by step invariance (or some other waveform invariance criterion) does not lead to the same discrete-time filter in the two cases.

In the impulse invariance design procedure, the relationship between continuous-time and discrete-time frequency is linear; consequently, except for aliasing, the shape of the frequency response is preserved. This is in contrast to the procedure discussed next, which is based on an algebraic transformation. We note, in concluding this subsection, that the impulse invariance technique is appropriate only for bandlimited filters; highpass or bandstop continuous-time filters, for example, would require additional bandlimiting to avoid severe aliasing distortion if impulse invariance design is used.

7.1.2 Bilinear Transformation

The technique discussed in this subsection avoids the problem of aliasing by using the bilinear transformation, an algebraic transformation between the variables s and z that maps the entire $j\Omega$-axis in the s-plane to one revolution of the unit circle in the z-plane. Since $-\infty \leq \Omega \leq \infty$ maps onto $-\pi \leq \omega \leq \pi$, the transformation between the continuous-time and discrete-time frequency variables must be nonlinear. Therefore, the use of this technique is restricted to situations in which the corresponding warping of the frequency axis is acceptable.

With $H_c(s)$ denoting the continuous-time system function and $H(z)$ the discrete-time system function, the bilinear transformation corresponds to replacing s by

$$s = \frac{2}{T_d}\left(\frac{1 - z^{-1}}{1 + z^{-1}}\right); \tag{7.20}$$

that is,

$$H(z) = H_c\left[\frac{2}{T_d}\left(\frac{1 - z^{-1}}{1 + z^{-1}}\right)\right]. \tag{7.21}$$

As in impulse invariance, a "sampling" parameter T_d is included in the definition of the bilinear transformation. Historically, this parameter has been included because the difference equation corresponding to $H(z)$ can be obtained by applying the trapezoidal integration rule to the differential equation corresponding to $H_c(s)$, with T_d representing the step size of the numerical integration. (See Kaiser, 1966, and Problem 7.43.) However, in filter design, our use of the bilinear transformation is based on the properties of the algebraic transformation given in Eq. (7.20). As with impulse invariance, the parameter T_d is of no consequence in the design procedure, since we assume that the design problem always begins with specifications on the discrete-time filter $H(e^{j\omega})$. When these specifications are mapped to continuous-time specifications and the continuous-time filter is then mapped back to a discrete-time filter, the effect of T_d will cancel. Although we will retain the parameter T_d in our discussion, in specific problems and examples any convenient value of T_d can be chosen.

To develop the properties of the algebraic transformation specified in Eq. (7.20), we solve for z to obtain

$$z = \frac{1 + (T_d/2)s}{1 - (T_d/2)s}, \tag{7.22}$$

and, substituting $s = \sigma + j\Omega$ into Eq. (7.22), we obtain

$$z = \frac{1 + \sigma T_d/2 + j\Omega T_d/2}{1 - \sigma T_d/2 - j\Omega T_d/2}. \tag{7.23}$$

If $\sigma < 0$, then, from Eq. (7.23), it follows that $|z| < 1$ for any value of Ω. Similarly, if $\sigma > 0$, then $|z| > 1$ for all Ω. That is, if a pole of $H_c(s)$ is in the left-half s-plane, its image in the z-plane will be inside the unit circle. Therefore, causal stable continuous-time filters map into causal stable discrete-time filters.

Next, to show that the $j\Omega$-axis of the s-plane maps onto the unit circle, we substitute $s = j\Omega$ into Eq. (7.22), obtaining

$$z = \frac{1 + j\Omega T_d/2}{1 - j\Omega T_d/2}.$$ (7.24)

From Eq. (7.24), it is clear that $|z| = 1$ for all values of s on the $j\Omega$-axis. That is, the $j\Omega$-axis maps onto the unit circle, so Eq. (7.24) takes the form

$$e^{j\omega} = \frac{1 + j\Omega T_d/2}{1 - j\Omega T_d/2}.$$ (7.25)

To derive a relationship between the variables ω and Ω, it is useful to return to Eq. (7.20) and substitute $z = e^{j\omega}$. We obtain

$$s = \frac{2}{T_d} \left(\frac{1 - e^{-j\omega}}{1 + e^{-j\omega}} \right),$$ (7.26)

or, equivalently,

$$s = \sigma + j\Omega = \frac{2}{T_d} \left[\frac{2e^{-j\omega/2}(j \sin \omega/2)}{2e^{-j\omega/2}(\cos \omega/2)} \right] = \frac{2j}{T_d} \tan(\omega/2).$$ (7.27)

Equating real and imaginary parts on both sides of Eq. (7.27) leads to the relations $\sigma = 0$ and

$$\Omega = \frac{2}{T_d} \tan(\omega/2),$$ (7.28)

or

$$\omega = 2 \arctan(\Omega T_d/2).$$ (7.29)

These properties of the bilinear transformation as a mapping from the s-plane to the z-plane are summarized in Figures 7.6 and 7.7. From Eq. (7.29) and Figure 7.7, we see that the range of frequencies $0 \le \Omega \le \infty$ maps to $0 \le \omega \le \pi$, while the range $-\infty \le \Omega \le 0$ maps to $-\pi \le \omega \le 0$. The bilinear transformation avoids the problem of aliasing encountered with the use of impulse invariance, because it maps the entire imaginary axis of the s-plane onto the unit circle in the z-plane. The price paid for this, however, is the nonlinear compression of the frequency axis depicted in Figure 7.7. Consequently, the design of discrete-time filters using the bilinear transformation is useful only when this compression can be tolerated or compensated for, as in the case of filters that approximate ideal piecewise-constant magnitude-response characteristics. This is illustrated in Figure 7.8, where we show how a continuous-time frequency response and tolerance scheme maps to a corresponding discrete-time frequency response and tolerance scheme through the frequency warping of Eqs. (7.28) and (7.29). If the critical frequencies (such as the passband and stopband edge frequencies) of the continuous-time filter are prewarped according to Eq. (7.28) then, when the continuous-time filter is transformed to the discrete-time filter using Eq. (7.21), the discrete-time filter will meet the desired specifications.

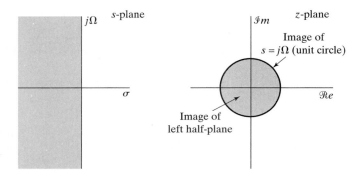

Figure 7.6 Mapping of the *s*-plane onto the *z*-plane using the bilinear transformation.

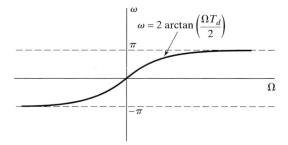

$$\omega = 2 \arctan\left(\frac{\Omega T_d}{2}\right)$$

Figure 7.7 Mapping of the continuous-time frequency axis onto the discrete-time frequency axis by bilinear transformation.

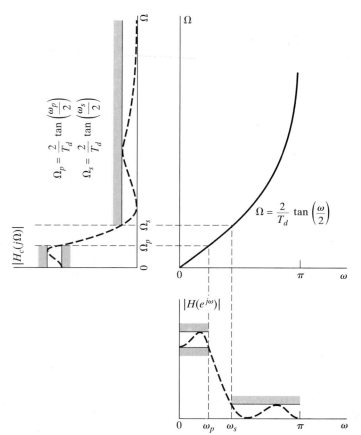

$$\Omega_p = \frac{2}{T_d} \tan\left(\frac{\omega_p}{2}\right)$$

$$\Omega_s = \frac{2}{T_d} \tan\left(\frac{\omega_s}{2}\right)$$

$$\Omega = \frac{2}{T_d} \tan\left(\frac{\omega}{2}\right)$$

Figure 7.8 Frequency warping inherent in the bilinear transformation of a continuous-time lowpass filter into a discrete-time lowpass filter. To achieve the desired discrete-time cutoff frequencies, the continuous-time cutoff frequencies must be prewarped as indicated.

Typical frequency-selective continuous-time approximations are Butterworth, Chebyshev, and elliptic filters. The closed-form design formulas of these continuous-time approximation methods make the design procedure rather straightforward. As discussed in Appendix B a Butterworth continuous-time filter is monotonic in the passband and in the stopband. A type I Chebyshev filter has an equiripple characteristic in the passband and varies monotonically in the stopband. A type II Chebyshev filter is monotonic in the passband and equiripple in the stopband. An elliptic filter is equiripple in both the passband and the stopband. Clearly, these properties will be preserved when the filter is mapped to a digital filter with the bilinear transformation. This is illustrated by the dashed approximation shown in Figure 7.8.

Although the bilinear transformation can be used effectively in mapping a piecewise-constant magnitude-response characteristic from the s-plane to the z-plane, the distortion in the frequency axis also manifests itself as a warping of the phase response of the filter. For example, Figure 7.9 shows the result of applying the bilinear transformation to an ideal linear phase factor $e^{-s\alpha}$. If we substitute Eq. (7.20) for s and evaluate the result on the unit circle, the phase angle is $-(2\alpha/T_d)\tan(\omega/2)$. In Figure 7.9, the solid curve shows the function $-(2\alpha/T_d)\tan(\omega/2)$, and the dotted curve is the periodic linear phase function $-(\omega\alpha/T_d)$, which is obtained by using the small angle approximation $\omega/2 \approx \tan(\omega/2)$. From this, it should be evident that if we were interested in a discrete-time lowpass filter with a linear phase characteristic, we could not obtain such a filter by applying the bilinear transformation to a continuous-time lowpass filter with a linear phase characteristic.

As mentioned previously, because of the frequency warping, the use of the bilinear transformation is restricted to the design of approximations to filters with

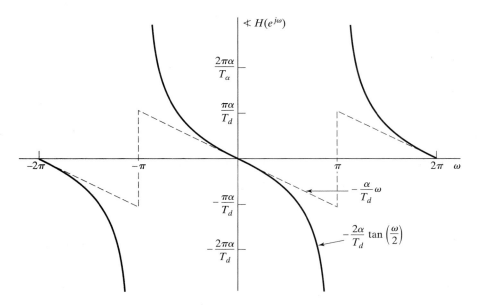

Figure 7.9   Illustration of the effect of the bilinear transformation on a linear phase characteristic. (Dashed line is linear phase and solid line is phase resulting from bilinear transformation.)

piecewise-constant frequency magnitude characteristics, such as highpass, lowpass and bandpass filters. As demonstrated in Example 7.2, impulse invariance can also be used to design lowpass filters. However, impulse invariance cannot be used to map highpass continuous-time designs to highpass discrete-time designs, since highpass continuous-time filters are not bandlimited.

In Example 4.5, we discussed a class of filters often referred to as discrete-time differentiators. A significant feature of the frequency response of this class of filters is that it is linear with frequency. The nonlinear warping of the frequency axis introduced by the bilinear transformation will not preserve that property. Consequently, the bilinear transformation applied to a continuous-time differentiator will not result in a discrete-time differentiator. However, impulse invariance applied to an appropriately bandlimited continuous-time differentiator will result in a discrete-time differentiator.

7.1.3 Examples of Bilinear Transformation Design

In the following discussion, we present a number of examples to illustrate IIR filter design using the bilinear transformation. Example 7.3 serves to illustrate the design procedure based on the bilinear transformation, in comparison with the use of impulse invariance. Examples 7.4, 7.5, and 7.6 illustrate a Butterworth, Chebyshev, and elliptic filter, respectively, each designed to the same specifications using the bilinear transformation.

Example 7.3 Bilinear Transformation of a Butterworth Filter

Consider the discrete-time filter specifications of Example 7.2, in which we illustrated the impulse invariance technique for the design of a discrete-time filter. The specifications on the discrete-time filter are

$$0.89125 \leq |H(e^{j\omega})| \leq 1, \qquad 0 \leq \omega \leq 0.2\pi, \qquad (7.30a)$$

$$|H(e^{j\omega})| \leq 0.17783, \qquad 0.3\pi \leq \omega \leq \pi. \qquad (7.30b)$$

In carrying out the design using the bilinear transformation, the critical frequencies of the discrete-time filter must be prewarped to the corresponding continuous-time frequencies using Eq. (7.28) so that the frequency distortion inherent in the bilinear transformation will map the continuous-time frequencies back to the correct discrete-time critical frequencies. For this specific filter, with $|H_c(j\Omega)|$ representing the magnitude-response function of the continuous-time filter, we require that

$$0.89125 \leq |H_c(j\Omega)| \leq 1, \qquad 0 \leq \Omega \leq \frac{2}{T_d} \tan\left(\frac{0.2\pi}{2}\right), \qquad (7.31a)$$

$$|H_c(j\Omega)| \leq 0.17783, \qquad \frac{2}{T_d} \tan\left(\frac{0.3\pi}{2}\right) \leq \Omega \leq \infty. \qquad (7.31b)$$

For convenience, we choose $T_d = 1$. Also, as with Example 7.2, since a continuous-time Butterworth filter has a monotonic magnitude response, we can equivalently require that

$$|H_c(j2\tan(0.1\pi))| \geq 0.89125 \qquad (7.32a)$$

and

$$|H_c(j2\tan(0.15\pi))| \le 0.17783. \tag{7.32b}$$

The form of the magnitude-squared function for the Butterworth filter is

$$|H_c(j\Omega)|^2 = \frac{1}{1 + (\Omega/\Omega_c)^{2N}}. \tag{7.33}$$

Solving for N and Ω_c with the equality sign in Eqs. (7.32a) and (7.32b), we obtain

$$1 + \left(\frac{2\tan(0.1\pi)}{\Omega_c}\right)^{2N} = \left(\frac{1}{0.89}\right)^2 \tag{7.34a}$$

and

$$1 + \left(\frac{2\tan(0.15\pi)}{\Omega_c}\right)^{2N} = \left(\frac{1}{0.178}\right)^2, \tag{7.34b}$$

and solving for N in Eqs. (7.34a) and (7.34b) gives

$$N = \frac{\log\left[\left(\left(\frac{1}{0.178}\right)^2 - 1\right) \Big/ \left(\left(\frac{1}{0.89}\right)^2 - 1\right)\right]}{2\log[\tan(0.15\pi)/\tan(0.1\pi)]} \tag{7.35}$$

$$= 5.305.$$

Since N must be an integer, we choose $N = 6$. Substituting $N = 6$ into Eq. (7.34b), we obtain $\Omega_c = 0.766$. For this value of Ω_c, the passband specifications are exceeded and the stopband specifications are met exactly. This is reasonable for the bilinear transformation, since we do not have to be concerned with aliasing. That is, with proper prewarping, we can be certain that the resulting discrete-time filter will meet the specifications exactly at the desired stopband edge.

In the s-plane, the 12 poles of the magnitude-squared function are uniformly distributed in angle on a circle of radius 0.766, as shown in Figure 7.10. The system function of the continuous-time filter obtained by selecting the left half-plane

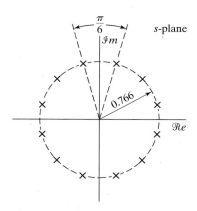

Figure 7.10 s-plane locations for poles of $H_c(s)H_c(-s)$ for sixth-order Butterworth filter in Example 7.3.

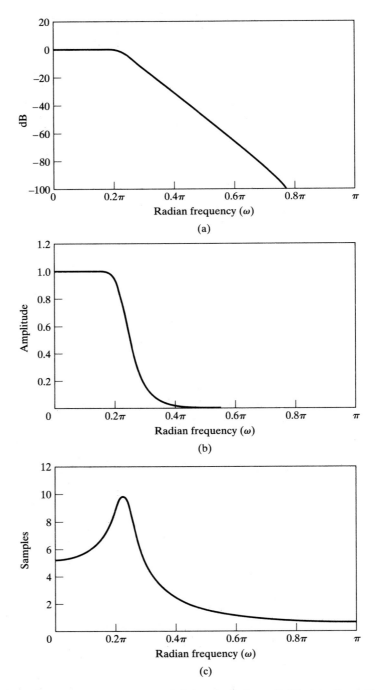

Figure 7.11   Frequency response of sixth-order Butterworth filter transformed by bilinear transform. (a) Log magnitude in dB. (b) Magnitude. (c) Group delay.

poles is

$$H_c(s) = \frac{0.20238}{(s^2 + 0.3996s + 0.5871)(s^2 + 1.0836s + 0.5871)(s^2 + 1.4802s + 0.5871)}.$$

(7.36)

The system function for the discrete-time filter is then obtained by applying the bilinear transformation to $H_c(s)$ with $T_d = 1$. The result is

$$H(z) = \frac{0.0007378(1 + z^{-1})^6}{(1 - 1.2686z^{-1} + 0.7051z^{-2})(1 - 1.0106z^{-1} + 0.3583z^{-2})}$$

(7.37)

$$\times \frac{1}{(1 - 0.9044z^{-1} + 0.2155z^{-2})}.$$

The magnitude, log magnitude, and group delay of the frequency response of the discrete-time filter are shown in Figure 7.11. At $\omega = 0.2\pi$ the log magnitude is -0.56 dB, and at $\omega = 0.3\pi$ the log magnitude is exactly -15 dB.

Since the bilinear transformation maps the entire $j\Omega$-axis of the s-plane onto the unit circle in the z-plane, the magnitude response of the discrete-time filter falls off much more rapidly than that of the original continuous-time filter. In particular, the behavior of $H(e^{j\omega})$ at $\omega = \pi$ corresponds to the behavior of $H_c(j\Omega)$ at $\Omega = \infty$. Therefore, since the continuous-time Butterworth filter has a sixth-order zero at $s = \infty$, the resulting discrete-time filter has a sixth-order zero at $z = -1$.

It is interesting to note that, since the general form of the Nth-order Butterworth continuous-time filter is as given by Eq. (7.33), and since ω and Ω are related by Eq. (7.28), it follows that the general Nth-order Butterworth discrete-time filter has magnitude-squared function

$$|H(e^{j\omega})|^2 = \frac{1}{1 + \left(\dfrac{\tan(\omega/2)}{\tan(\omega_c/2)}\right)^{2N}},$$

(7.38)

where $\tan(\omega_c/2) = \Omega_c T_d/2$.

The frequency-response function of Eq. (7.38) has the same properties as the continuous-time Butterworth response; i.e., it is maximally flat[3] and $|H(e^{j\omega_c})|^2 = 0.5$. However, the function in Eq. (7.38) is periodic with period 2π and falls off more sharply than the continuous-time Butterworth response.

We do not design discrete-time Butterworth filters directly by starting with Eq. (7.38), because it is not straightforward to determine the z-plane locations of the poles (all the zeros are at $z = -1$) associated with the magnitude-squared function of Eq. (7.38). It is necessary to determine the poles so as to factor the magnitude-squared

---

[3]The first $(2N - 1)$ derivatives of $|H(e^{j\omega})|^2$ are zero at $\omega = 0$.

function into $H(z)H(z^{-1})$ and thereby determine $H(z)$. It is much easier to find the s-plane pole locations (all the zeros are at infinity), factor the continuous-time system function, and then transform the left half-plane poles by the bilinear transformation as we did in Example 7.3.

Equations of the form of Eq. (7.38) may also be obtained for discrete-time Chebyshev filters, but the same difficulties arise in their use. Thus, the two-step approach just described has become the established method of designing IIR frequency-selective filters.

The major approximation methods for frequency-selective IIR analog filters are the Butterworth, Chebyshev, and elliptic function approximation methods. The details of these methods can be found in Guillemin (1957), Daniels (1974), Weinberg (1975), and Lam (1979). The methods are generally explained and developed in terms of lowpass filter approximations. This is the approach followed in Appendix B, where we summarize the essential features of some of the methods. In the next three examples, we illustrate the realization of a set of filter specifications for each of these classes of filters. The details of the design computations are not presented, since they are tedious and lengthy and are best carried out by computer programs that incorporate the appropriate closed-form design equations.

The lowpass discrete-time filter specifications for these examples are those used in Example 7.1, i.e.,

$$0.99 \leq |H(e^{j\omega})| \leq 1.01, \qquad |\omega| \leq 0.4\pi, \tag{7.39a}$$

and

$$|H(e^{j\omega})| \leq 0.001, \qquad 0.6\pi \leq |\omega| \leq \pi. \tag{7.39b}$$

In terms of the tolerance scheme of Figure 7.2(b), $\delta_1 = 0.01, \delta_2 = 0.001, \omega_p = 0.4\pi$, and $\omega_s = 0.6\pi$. These specifications are sufficient to determine the input parameters to the Butterworth, Chebyshev, and elliptic design formulas. Note that the specifications are only on the magnitudes of the frequency response. The phase is implicitly determined by the nature of the approximating functions.

Example 7.4   Butterworth Approximation

For the specification of Eqs. (7.39a) and (7.39b), the Butterworth approximation method requires a system of 14th order. The frequency response of the discrete-time filter that results from the bilinear transformation of the appropriate prewarped Butterworth filter is shown in Figure 7.12. Figure 7.12(a) shows the log magnitude in dB, Figure 7.12(b) shows the magnitude of $H(e^{j\omega})$ in the passband only, and Figure 7.12(c) shows the group delay of the filter. From these plots, we see that the Butterworth frequency response decreases monotonically with frequency and the gain of the filter becomes very small above about $\omega = 0.7\pi$. Note from Figure 7.12(b) that in this example the Butterworth frequency response has been normalized so that it has gain greater than unity in the passband as is allowed in the specifications in Eqs. (7.39).

Figure 7.12   Frequency response of 14th-order Butterworth filter in Example 7.4. (a) Log magnitude in dB. (b) Detailed plot of magnitude in passband. (c) Group delay.

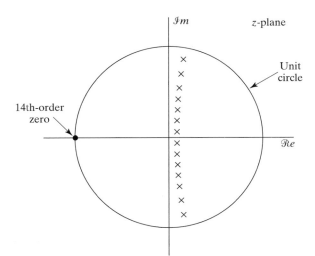

Figure 7.13   Pole–zero plot of 14th-order Butterworth filter in Example 7.4.

In the Butterworth example, the specifications are exceeded at the passband and stopband edges because of rounding the order up to the next integer. However, the specifications are *far* exceeded in the stopband. The reason for this is evident from Figure 7.13, which shows the pole–zero plot for the 14th-order Butterworth filter. Because the continuous-time Butterworth filter has 14 zeros at $s = \infty$, the bilinear transformation creates 14 zeros at $z = -1$ for the discrete-time filter. It is reasonable to expect that a lower order filter might still satisfy the specifications, even if it did not exceed them so greatly in the stopband. This expectation motivates the use of Chebyshev or equiripple approximation.

Example 7.5   Chebyshev Approximation

This method has two forms. Chebyshev type I approximations have equiripple behavior in the passband, and Chebyshev type II approximations have equiripple behavior in the stopband. Both methods lead to the same order for a given set of specifications. For the specifications of Eqs. (7.39a) and (7.39b), the required order is 8 rather than 14, as for the Butterworth approximation. Figure 7.14 shows the log magnitude, passband magnitude, and group delay for the type I approximation to the specifications of Eqs. (7.39a) and (7.39b). Note that the frequency response oscillates with equal maximum error on either side of the desired gain of unity in the passband.

Figure 7.15 shows the frequency-response functions for the Chebyshev type II approximation to the specifications of Eqs. (7.39a) and (7.39b). In this case, the equiripple approximation behavior is in the stopband. The pole–zero plots for the Chebyshev filters are shown in Figure 7.16. Note that the Chebyshev type I system is similar to the Butterworth system in that it has all eight of its zeros at $z = -1$. On the other hand, the type II system has its zeros arrayed on the unit circle. These zeros

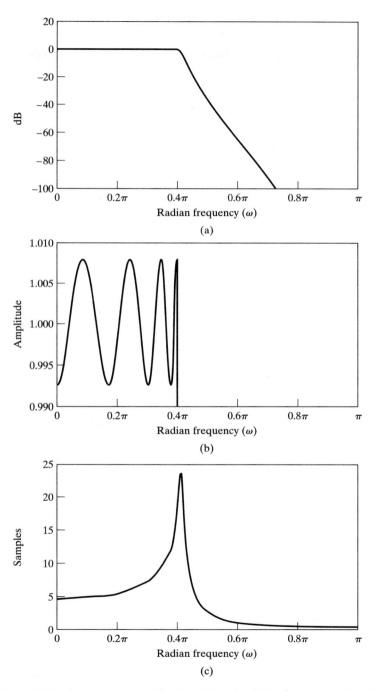

Figure 7.14 Frequency response of eighth-order Chebyshev type I filter in Example 7.5. (a) Log magnitude in dB. (b) Detailed plot of magnitude in pass-band. (c) Group delay.

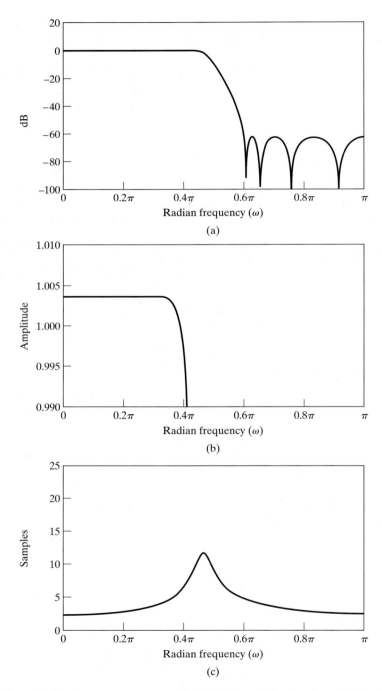

Figure 7.15 Frequency response of eighth-order Chebyshev type II filter in Example 7.5. (a) Log magnitude in dB. (b) Detailed plot of magnitude in passband. (c) Group delay.

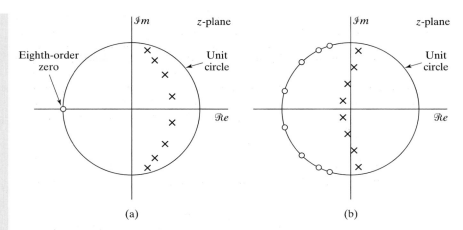

Figure 7.16 Pole–zero plot of eighth-order Chebyshev filters in Example 7.5. (a) Type I. (b) Type II.

are positioned by the design equations so as to achieve the equiripple behavior in the stopband.

In both cases of Chebyshev approximation, the monotonic behavior in either the stopband or the passband suggests that perhaps a lower order system might be obtained if equiripple approximation were used in both the passband and the stopband. Indeed, it can be shown (see Papoulis, 1957) that for fixed values of δ_1, δ_2, ω_p, and ω_s in the tolerance scheme of Figure 7.2(b), the lowest order filter is obtained when the approximation error ripples equally between the extremes of the two approximation bands. Since this equiripple behavior is achieved with a rational function that involves elliptic functions, such systems are generally called elliptic filters.

Example 7.6   Elliptic Approximation[4]

The specifications of Eqs. (7.39a) and (7.39b) are met by an elliptic filter of order six. This is the *lowest* order rational function approximation to the specifications. Figure 7.17 clearly shows the equiripple behavior in both approximation bands. Figure 7.18 shows that the elliptic filter, like the Chebyshev type II, has its zeros arrayed in the stopband region of the unit circle.

---

[4]The design equations for elliptic filters are too involved to be appropriately summarized in Appendix B. They can be found in Storer (1957), Weinberg (1975), and Parks and Burrus (1987). A program for elliptic filter design was given by Gray and Markel (1976), and extensive tables for elliptic filter designs are available in Zverev (1967). All three types of filters can be designed using functions in the Signal Processing Toolbox of MATLAB®.

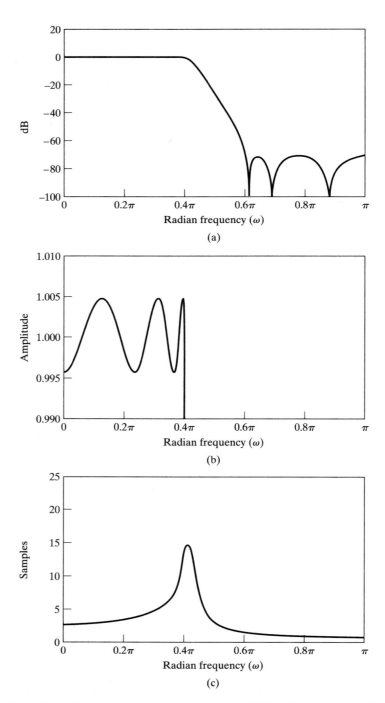

Figure 7.17 Frequency response of sixth-order elliptic filter in Example 7.6. (a) Log magnitude in dB. (b) Detailed plot of magnitude in passband. (c) Group delay.

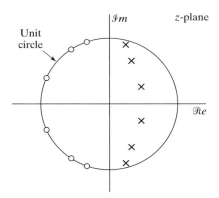

Figure 7.18   Pole–zero plot of sixth-order elliptic filter in Example 7.6.

Bilinear transformation of analog filters designed by Butterworth, Chebyshev, or elliptic approximation methods is a standard method of design of IIR discrete-time filters. The previous examples illustrate several important general features of such filters. In all cases, the resulting system function $H(z)$ has all its zeros on the unit circle and (for stability) all its poles inside the unit circle. As a result, all the approximation methods yield digital filters with nonconstant group delay or, equivalently, nonlinear phase. The greatest deviation from constant group delay occurs in all cases at the edge of the passband or in the transition band. In general, the Chebyshev type II approximation method yields the smallest delay in the passband and the widest region of the passband over which the group delay is approximately constant. However, if phase linearity is not an issue, then elliptic approximation yields the lowest order system function, and therefore, elliptic filters will generally require the least computation to implement a given filter specification.

7.2 DESIGN OF FIR FILTERS BY WINDOWING

As discussed in Section 7.1, commonly used techniques for the design of IIR filters are based on transformations of continuous-time IIR systems into discrete-time IIR systems. This is partly because continuous-time filter design was a highly advanced art before discrete-time filters were of interest and partly because of the difficulty of implementing a noniterative direct design method for IIR filters.

In contrast, FIR filters are almost entirely restricted to discrete-time implementations. Consequently, the design techniques for FIR filters are based on directly approximating the desired frequency response of the discrete-time system. Furthermore, most techniques for approximating the magnitude response of an FIR system assume a linear phase constraint, thereby avoiding the problem of spectrum factorization that complicates the direct design of IIR filters.

The simplest method of FIR filter design is called the *window method*. This method generally begins with an ideal desired frequency response that can be represented as

$$H_d(e^{j\omega}) = \sum_{n=-\infty}^{\infty} h_d[n]e^{-j\omega n}, \tag{7.40}$$

where $h_d[n]$ is the corresponding impulse response sequence, which can be expressed in terms of $H_d(e^{j\omega})$ as

$$h_d[n] = \frac{1}{2\pi} \int_{-\pi}^{\pi} H_d(e^{j\omega})e^{j\omega n} d\omega. \tag{7.41}$$

Many idealized systems are defined by piecewise-constant or piecewise-functional frequency responses with discontinuities at the boundaries between bands. As a result, these systems have impulse responses that are noncausal and infinitely long. The most straightforward approach to obtaining a causal FIR approximation to such systems is to truncate the ideal response. Equation (7.40) can be thought of as a Fourier series representation of the periodic frequency response $H_d(e^{j\omega})$, with the sequence $h_d[n]$ playing the role of the Fourier coefficients. Thus, the approximation of an ideal filter by truncation of the ideal impulse response is identical to the issue of the convergence of Fourier series, a subject that has received a great deal of study. A particularly important concept from this theory is the Gibbs phenomenon, which was discussed in Example 2.22. In the following discussion, we will see how this nonuniform convergence phenomenon manifests itself in the design of FIR filters.

The simplest way to obtain a causal FIR filter from $h_d[n]$ is to define a new system with impulse response $h[n]$ given by[5]

$$h[n] = \begin{cases} h_d[n], & 0 \leq n \leq M, \\ 0, & \text{otherwise.} \end{cases} \tag{7.42}$$

More generally, we can represent $h[n]$ as the product of the desired impulse response and a finite-duration "window" $w[n]$; i.e.,

$$h[n] = h_d[n]w[n], \tag{7.43}$$

where, for simple truncation as in Eq. (7.42), the window is the *rectangular window*

$$w[n] = \begin{cases} 1, & 0 \leq n \leq M, \\ 0, & \text{otherwise.} \end{cases} \tag{7.44}$$

It follows from the modulation, or windowing, theorem (Section 2.9.7) that

$$H(e^{j\omega}) = \frac{1}{2\pi} \int_{-\pi}^{\pi} H_d(e^{j\theta})W(e^{j(\omega-\theta)}) d\theta. \tag{7.45}$$

That is, $H(e^{j\omega})$ is the periodic convolution of the desired ideal frequency response with the Fourier transform of the window. Thus, the frequency response $H(e^{j\omega})$ will be a "smeared" version of the desired response $H_d(e^{j\omega})$. Figure 7.19(a) depicts typical functions $H_d(e^{j\theta})$ and $W(e^{j(\omega-\theta)})$, as required in Eq. (7.45).

If $w[n] = 1$ for all n (i.e., if we do not truncate at all), $W(e^{j\omega})$ is a periodic impulse train with period 2π, and therefore, $H(e^{j\omega}) = H_d(e^{j\omega})$. This interpretation suggests

---

[5]The notation for FIR systems was established in Chapter 5. That is, M is the order of the system function polynomial. Thus, $(M+1)$ is the length, or duration, of the impulse response. Often in the literature, N is used for the length of the impulse response of an FIR filter; however, we have used N to denote the order of the denominator polynomial in the system function of an IIR filter. Thus, to avoid confusion and maintain consistency throughout this book, we will always consider the length of the impulse response of an FIR filter to be $(M+1)$.

Figure 7.19   (a) Convolution process implied by truncation of the ideal impulse response. (b) Typical approximation resulting from windowing the ideal impulse response.

that if $w[n]$ is chosen so that $W(e^{j\omega})$ is concentrated in a narrow band of frequencies around $\omega = 0$, then $H(e^{j\omega})$ will "look like" $H_d(e^{j\omega})$, except where $H_d(e^{j\omega})$ changes very abruptly. Consequently, the choice of window is governed by the desire to have $w[n]$ as short as possible in duration, so as to minimize computation in the implementation of the filter, while having $W(e^{j\omega})$ approximate an impulse; that is, we want $W(e^{j\omega})$ to be highly concentrated in frequency so that the convolution of Eq. (7.45) faithfully reproduces the desired frequency response. These are conflicting requirements, as can be seen in the case of the rectangular window of Eq. (7.44), where

$$W(e^{j\omega}) = \sum_{n=0}^{M} e^{-j\omega n} = \frac{1 - e^{-j\omega(M+1)}}{1 - e^{-j\omega}} = e^{-j\omega M/2} \frac{\sin[\omega(M+1)/2]}{\sin(\omega/2)}. \qquad (7.46)$$

The magnitude of the function $\sin[\omega(M+1)/2]\sin(\omega/2)$ is plotted in Figure 7.20 for the case $M = 7$. Note that $W(e^{j\omega})$ for the rectangular window has a generalized linear phase. As M increases, the width of the "main lobe" decreases. The main lobe is usually defined as the region between the first zero-crossings on either side of the origin. For the rectangular window, the width of the main lobe is $\Delta_{\omega m} = 4\pi/(M + 1)$. However, for the rectangular window, the side lobes are large, and in fact, as M increases, the peak amplitudes of the main lobe and the side lobes grow in a manner such that the area under each lobe is a constant while the width of each lobe decreases with M. Consequently, as $W(e^{j(\omega-\theta)})$ "slides by" a discontinuity of $H_d(e^{j\theta})$ with increasing ω, the integral of $W(e^{j(\omega-\theta)})H_d(e^{j\theta})$ will oscillate as each side lobe of $W(e^{j(\omega-\theta)})$ moves past the discontinuity. This result is depicted in Figure 7.19 (b). Since the area under

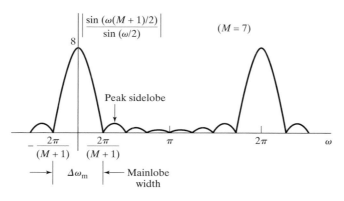

Figure 7.20 Magnitude of the Fourier transform of a rectangular window ($M = 7$).

each lobe remains constant with increasing M, the oscillations occur more rapidly, but do not decrease in amplitude as M increases.

In the theory of Fourier series, it is well known that this nonuniform convergence, the *Gibbs phenomenon,* can be moderated through the use of a less abrupt truncation of the Fourier series. By tapering the window smoothly to zero at each end, the height of the side lobes can be diminished; however, this is achieved at the expense of a wider main lobe and thus a wider transition at the discontinuity.

7.2.1 Properties of Commonly Used Windows

Some commonly used windows are shown in Figure 7.21.[6] These windows are defined by the following equations:

Rectangular

$$w[n] = \begin{cases} 1, & 0 \le n \le M, \\ 0, & \text{otherwise} \end{cases} \tag{7.47a}$$

Bartlett (triangular)

$$w[n] = \begin{cases} 2n/M, & 0 \le n \le M/2, \\ 2 - 2n/M, & M/2 < n \le M, \\ 0, & \text{otherwise} \end{cases} \tag{7.47b}$$

Hanning

$$w[n] = \begin{cases} 0.5 - 0.5\cos(2\pi n/M), & 0 \le n \le M, \\ 0, & \text{otherwise} \end{cases} \tag{7.47c}$$

Hamming

$$w[n] = \begin{cases} 0.54 - 0.46\cos(2\pi n/M), & 0 \le n \le M, \\ 0, & \text{otherwise} \end{cases} \tag{7.47d}$$

---

[6]The Bartlett, Hanning, Hamming, and Blackman windows are all named after their originators. The Hanning window is associated with Julius von Hann, an Austrian meteorologist, and is sometimes referred to as the Hann window. The term "hanning" was used by Blackman and Tukey (1958) to describe the operation of applying this window to a signal and has since become the most widely used name for the window, with varying preferences for the choice of "Hanning" or "hanning."

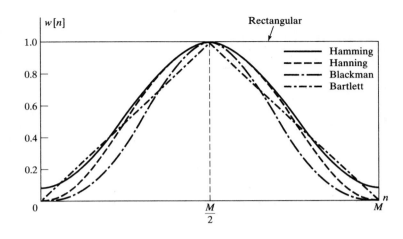

Figure 7.21   Commonly used windows.

Blackman

$$w[n] = \begin{cases} 0.42 - 0.5\cos(2\pi n/M) + 0.08\cos(4\pi n/M), & 0 \le n \le M, \\ 0, & \text{otherwise} \end{cases} \quad (7.47e)$$

(For convenience, Figure 7.21 shows these windows plotted as functions of a continuous variable; however, as specified in Eqs. (7.47), the window sequence is defined only at integer values of n.)

As will be discussed in Chapter 10, the windows defined in Eqs. (7.47) are commonly used for spectrum analysis as well as for FIR filter design. They have the desirable property that their Fourier transforms are concentrated around $\omega = 0$, and they have a simple functional form that allows them to be computed easily. The Fourier transform of the Bartlett window can be expressed as a product of Fourier transforms of rectangular windows, and the Fourier transforms of the other windows can be expressed as sums of frequency-shifted Fourier transforms of the rectangular window, as given by Eq. (7.46). (See Problem 7.34.)

The function $20\log_{10}|W(e^{j\omega})|$ is plotted in Figure 7.22 for each of these windows with $M = 50$. The rectangular window clearly has the narrowest main lobe, and thus, for a given length, it should yield the sharpest transitions of $H(e^{j\omega})$ at a discontinuity of $H_d(e^{j\omega})$. However, the first side lobe is only about 13 dB below the main peak, resulting in oscillations of $H(e^{j\omega})$ of considerable size around discontinuities of $H_d(e^{j\omega})$. Table 7.1, which compares the windows of Eqs. (7.47), shows that, by tapering the window smoothly to zero, as with the Hamming, Hanning, and Blackman windows, the side lobes (second column) are greatly reduced in amplitude; however, the price paid is a much wider main lobe (third column) and thus wider transitions at discontinuities of $H_d(e^{j\omega})$. The other columns of Table 7.1 will be discussed later.

7.2.2 Incorporation of Generalized Linear Phase

In designing many types of FIR filters, it is desirable to obtain causal systems with a generalized linear phase response. All the windows of Eqs. (7.47) have been defined in

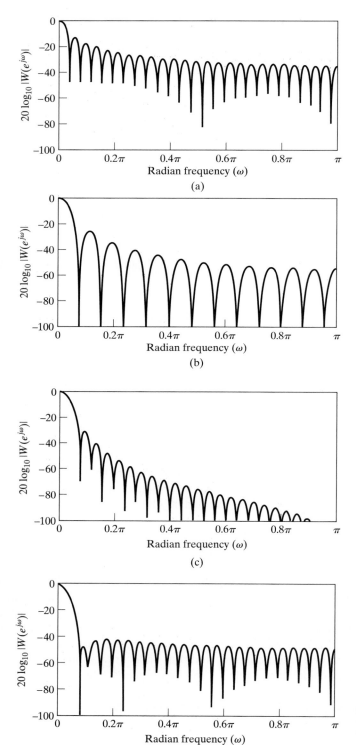

Figure 7.22 Fourier transforms (log magnitude) of windows of Figure 7.21. with $M = 50$. (a) Rectangular. (b) Bartlett. (c) Hanning. (d) Hamming. (e) Blackman.

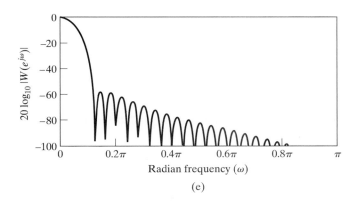

Figure 7.22   (*Continued*)
(e) Blackman.

TABLE 7.1   COMPARISON OF COMMONLY USED WINDOWS

Type of Window	Peak Side-Lobe Amplitude (Relative)	Approximate Width of Main Lobe	Peak Approximation Error, $20 \log_{10} \delta$ (dB)	Equivalent Kaiser Window, β	Transition Width of Equivalent Kaiser Window
Rectangular	-13	$4\pi/(M+1)$	-21	0	$1.81\pi/M$
Bartlett	-25	$8\pi/M$	-25	1.33	$2.37\pi/M$
Hanning	-31	$8\pi/M$	-44	3.86	$5.01\pi/M$
Hamming	-41	$8\pi/M$	-53	4.86	$6.27\pi/M$
Blackman	-57	$12\pi/M$	-74	7.04	$9.19\pi/M$

anticipation of this need. Specifically, note that all the windows have the property that

$$w[n] = \begin{cases} w[M-n], & 0 \le n \le M, \\ 0, & \text{otherwise;} \end{cases} \tag{7.48}$$

i.e., they are symmetric about the point $M/2$. As a result, their Fourier transforms are of the form

$$W(e^{j\omega}) = W_e(e^{j\omega})e^{-j\omega M/2}, \tag{7.49}$$

where $W_e(e^{j\omega})$ is a real, even function of ω. This is illustrated by Eq. (7.46). The convention of Eq. (7.48) leads to causal filters in general, and if the desired impulse response is also symmetric about $M/2$, i.e., if $h_d[M-n] = h_d[n]$, then the windowed impulse response will also have that symmetry, and the resulting frequency response will have a generalized linear phase; that is,

$$H(e^{j\omega}) = A_e(e^{j\omega})e^{-j\omega M/2}, \tag{7.50}$$

where $A_e(e^{j\omega})$ is real and is an even function of ω. Similarly, if the desired impulse response is antisymmetric about $M/2$, i.e., if $h_d[M-n] = -h_d[n]$, then the windowed impulse response will also be antisymmetric about $M/2$, and the resulting frequency response will have a generalized linear phase with a constant phase shift of ninety degrees; i.e.,

$$H(e^{j\omega}) = jA_o(e^{j\omega})e^{-j\omega M/2}, \tag{7.51}$$

where $A_o(e^{j\omega})$ is real and is an odd function of ω.

Although the preceding statements are straightforward if we consider the product of the symmetric window with the symmetric (or antisymmetric) desired impulse response, it is useful to consider the frequency-domain representation. Suppose $h_d[M - n] = h_d[n]$. Then

$$H_d(e^{j\omega}) = H_e(e^{j\omega})e^{-j\omega M/2}, \tag{7.52}$$

where $H_e(e^{j\omega})$ is real and even.

If the window is symmetric, we can substitute Eqs. (7.49) and (7.52) into Eq. (7.45) to obtain

$$H(e^{j\omega}) = \frac{1}{2\pi} \int_{-\pi}^{\pi} H_e(e^{j\theta})e^{-j\theta M/2} W_e(e^{j(\omega-\theta)})e^{-j(\omega-\theta)M/2} d\theta. \tag{7.53}$$

A simple manipulation of the phase factors leads to

$$H(e^{j\omega}) = A_e(e^{j\omega})e^{-j\omega M/2}, \tag{7.54}$$

where

$$A_e(e^{j\omega}) = \frac{1}{2\pi} \int_{-\pi}^{\pi} H_e(e^{j\theta}) W_e(e^{j(\omega-\theta)}) d\theta. \tag{7.55}$$

Thus, we see that the resulting system has a generalized linear phase and, moreover, the real function $A_e(e^{j\omega})$ is the result of the periodic convolution of the real functions $H_e(e^{j\omega})$ and $W_e(e^{j\omega})$.

The detailed behavior of the convolution of Eq. (7.55) determines the magnitude response of the filter that results from windowing. The following example illustrates this for a linear-phase lowpass filter.

Example 7.7   Linear-Phase Lowpass Filter

The desired frequency response is defined as

$$H_{lp}(e^{j\omega}) = \begin{cases} e^{-j\omega M/2}, & |\omega| < \omega_c, \\ 0, & \omega_c < |\omega| \le \pi, \end{cases} \tag{7.56}$$

where the generalized linear phase factor has been incorporated into the definition of the ideal lowpass filter. The corresponding ideal impulse response is

$$h_{lp}[n] = \frac{1}{2\pi} \int_{-\omega_c}^{\omega_c} e^{-j\omega M/2} e^{j\omega n} d\omega = \frac{\sin[\omega_c(n - M/2)]}{\pi(n - M/2)} \tag{7.57}$$

for $-\infty < n < \infty$. It is easily shown that $h_{lp}[M - n] = h_{lp}[n]$, so if we use a symmetric window in the equation

$$h[n] = \frac{\sin[\omega_c(n - M/2)]}{\pi(n - M/2)} w[n], \tag{7.58}$$

then a linear-phase system will result.

The upper part of Figure 7.23 depicts the character of the amplitude response that would result for all the windows of Eqs. (7.47), except the Bartlett window, which is rarely used for filter design. (The Bartlett window would produce a monotonic function $A_e(e^{j\omega})$, because $W_e(e^{j\omega})$ is a positive function.) The figure displays the important properties of window method approximations to desired frequency responses that have step discontinuities. It applies accurately when ω_c is not close to zero or to π and when the width of the main lobe is smaller than $2\omega_c$. At the bottom of the figure is a typical Fourier transform for a symmetric window (except for the linear phase). This function should be visualized in different positions as an aid in understanding the shape of the approximation $A_e(e^{j\omega})$ in the vicinity of ω_c.

When $\omega = \omega_c$, the symmetric function $W_e(e^{j(\omega-\theta)})$ is centered on the discontinuity, and about one-half its area contributes to $A_e(e^{j\omega})$. Similarly, we can see that the peak overshoot occurs when $W_e(e^{j(\omega-\theta)})$ is shifted such that the first negative side lobe on the right is just to the right of ω_c. Similarly, the peak negative undershoot occurs when the first negative side lobe on the left is just to the left of ω_c. This means that the distance between the peak ripples on either side of the discontinuity is approximately the main-lobe width $\Delta\omega_m$, as shown in Figure 7.23. The transition width $\Delta\omega$ as defined in the figure is therefore somewhat less than the main-lobe width. Finally, due to the symmetry of $W_e(e^{j(\omega-\theta)})$, the approximation tends to be symmetric around ω_c; i.e., the approximation overshoots by an amount δ in the passband and undershoots by the same amount in the stopband.

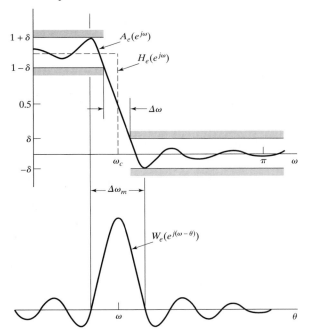

Figure 7.23   Illustration of type of approximation obtained at a discontinuity of the ideal frequency response.

The fourth column of Table 7.1 shows the peak approximation error (in dB) for the windows of Eqs. (7.47). Clearly, the windows with the smaller side lobes yield better approximations of the ideal response at a discontinuity . Also, the third column, which shows the width of the main lobe, suggests that narrower transition regions can be achieved by increasing M. Thus, through the choice of the shape and duration of the window, we can control the properties of the resulting FIR filter. However, trying different windows and adjusting lengths by trial and error is not a very satisfactory way to design filters. Fortunately, a simple formalization of the window method has been developed by Kaiser (1974).

7.2.3 The Kaiser Window Filter Design Method

The trade-off between the main-lobe width and side-lobe area can be quantified by seeking the window function that is maximally concentrated around $\omega = 0$ in the frequency domain. The issue was considered in depth in a series of classic papers by Slepian et al. (1961). The solution found in this work involves prolate spheroidal wave functions, which are difficult to compute and therefore unattractive for filter design. However, Kaiser (1966, 1974) found that a near-optimal window could be formed using the zeroth-order modified Bessel function of the first kind, a function that is much easier to compute. The Kaiser window is defined as

$$
w[n] = \begin{cases} \dfrac{I_0[\beta(1 - [(n - \alpha)/\alpha]^2)^{1/2}]}{I_0(\beta)}, & 0 \leq n \leq M, \\ 0, & \text{otherwise,} \end{cases} \tag{7.59}
$$

where $\alpha = M/2$, and $I_0(\cdot)$ represents the zeroth-order modified Bessel function of the first kind. In contrast to the other windows in Eqs. (7.47), the Kaiser window has two parameters: the length $(M + 1)$ and a shape parameter β. By varying $(M + 1)$ and β, the window length and shape can be adjusted to trade side-lobe amplitude for main-lobe width. Figure 7.24(a) shows continuous envelopes of Kaiser windows of length $M + 1 = 21$ for $\beta = 0, 3$, and 6. Notice from Eq. (7.59) that the case $\beta = 0$ reduces to the rectangular window. Figure 7.24(b) shows the corresponding Fourier transforms of the Kaiser windows in Figure 7.24(a). Figure 7.24(c) shows Fourier transforms of Kaiser windows with $\beta = 6$ and $M = 10, 20$, and 40. The plots in Figs. 7.24(b) and (c) clearly show that the desired trade-off can be achieved. If the window is tapered more, the side lobes of the Fourier transform become smaller, but the main lobe becomes wider. Figure 7.24(c) shows that increasing M while holding β constant causes the main lobe to decrease in width, but does not affect the amplitude of the side lobes. In fact, through extensive numerical experimentation, Kaiser obtained a pair of formulas that permit the filter designer to predict in advance the values of M and β needed to meet a given frequency-selective filter specification. Figure 7.23 is also typical of approximations obtained using the Kaiser window, and Kaiser (1974) found that, over a usefully wide range of conditions, the peak approximation error (δ in Figure 7.23) is determined by the choice of β. Given that δ is fixed, the passband cutoff frequency ω_p of the lowpass filter is defined to be the highest frequency such that $|H(e^{j\omega})| \geq 1 - \delta$. The stopband cutoff frequency ω_s is defined to be the lowest frequency such that $|H(e^{j\omega})| \leq \delta$. Therefore, the transition region has width

$$
\Delta\omega = \omega_s - \omega_p, \tag{7.60}
$$

for the lowpass filter approximation. Defining

$$
A = -20 \log_{10} \delta, \tag{7.61}
$$

Kaiser determined empirically that the value of β needed to achieve a specified value of A is given by

$$
\beta = \begin{cases} 0.1102(A - 8.7), & A > 50, \\ 0.5842(A - 21)^{0.4} + 0.07886(A - 21), & 21 \leq A \leq 50, \\ 0.0, & A < 21. \end{cases} \tag{7.62}
$$

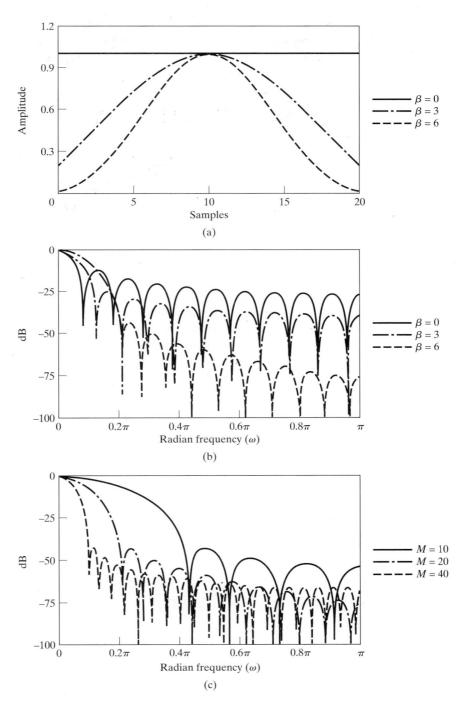

Figure 7.24   (a) Kaiser windows for $\beta = 0$, 3, and 6 and $M = 20$. (b) Fourier transforms corresponding to windows in (a). (c) Fourier transforms of Kaiser windows with $\beta = 6$ and $M = 10$, 20, and 40.

(Recall that the case $\beta = 0$ is the rectangular window for which $A = 21$.) Furthermore, Kaiser found that to achieve prescribed values of A and $\Delta\omega$, M must satisfy

$$M = \frac{A - 8}{2.285\Delta\omega}. \tag{7.63}$$

Equation (7.63) predicts M to within ± 2 over a wide range of values of $\Delta\omega$ and A. Thus, with these formulas, the Kaiser window design method requires almost no iteration or trial and error. Example 7.8 outlines and illustrates the procedure.

Example 7.8   Kaiser Window Design of a Lowpass Filter

With the use of the design formulas for the Kaiser window, it is straightforward to design an FIR lowpass filter to meet prescribed specifications. The procedure is as follows:

1. First the specifications must be established. This means selecting the desired ω_p and ω_s and the maximum tolerable approximation error. For window design, the resulting filter will have the same peak error δ in both the passband and the stopband. For this example, we use the same specifications as in Examples 7.4, 7.5, and 7.6, i.e., $\omega_p = 0.4\pi$, $\omega_s = 0.6\pi$, $\delta_1 = 0.01$, and $\delta_2 = 0.001$. Since filters designed by the window method inherently have $\delta_1 = \delta_2$, we must set $\delta = 0.001$.

2. The cutoff frequency of the underlying ideal lowpass filter must be found. Due to the symmetry of the approximation at the discontinuity of $H_d(e^{j\omega})$, we would set

$$\omega_c = \frac{\omega_p + \omega_s}{2} = 0.5\pi.$$

3. To determine the parameters of the Kaiser window, we first compute

$$\Delta\omega = \omega_s - \omega_p = 0.2\pi, \qquad A = -20\log_{10}\delta = 60.$$

We substitute these two quantities into Eqs. (7.62) and (7.63) to obtain the required values of β and M. For this example the formulas predict

$$\beta = 5.653, \qquad M = 37.$$

4. The impulse response of the filter is computed using Eqs. (7.58) and (7.59). We obtain

$$h[n] = \begin{cases} \dfrac{\sin \omega_c(n - \alpha)}{\pi(n - \alpha)} \cdot \dfrac{I_0[\beta(1 - [(n - \alpha)/\alpha]^2)^{1/2}]}{I_0(\beta)}, & 0 \le n \le M, \\ 0, & \text{otherwise}, \end{cases}$$

where $\alpha = M/2 = 37/2 = 18.5$. Since $M = 37$ is an odd integer, the resulting linear-phase system would be of type II. (See Section 5.7.3 for the definitions of the four types of FIR systems with generalized linear phase.) The response characteristics of the filter are shown in Figure 7.25. Figure 7.25 (a), which shows the impulse response, displays the characteristic symmetry of a type II system. Figure 7.25(b), which shows the log magnitude response in dB, indicates that

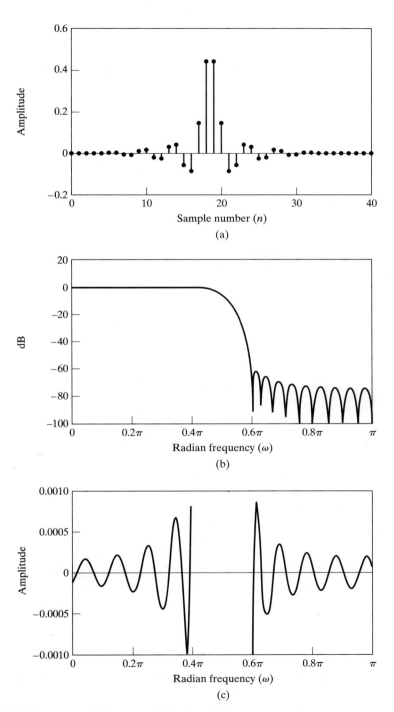

Figure 7.25 Response functions for Example 7.8. (a) Impulse response ($M = 37$). (b) Log magnitude. (c) Approximation error.

$H(e^{j\omega})$ is zero at $\omega = \pi$ or, equivalently, that $H(z)$ has a zero at $z = -1$, as required for a type II FIR system. Figure 7.25(c) shows the approximation error in the passband and stopbands. This error function is defined as

$$E_A(\omega) = \begin{cases} 1 - A_e(e^{j\omega}), & 0 \leq \omega \leq \omega_p, \\ 0 - A_e(e^{j\omega}), & \omega_s \leq \omega \leq \pi. \end{cases} \tag{7.64}$$

(The error is not defined in the transition region, $0.4\pi < \omega < 0.6\pi$.) Note the symmetry of the approximation error, and note also that the peak approximation error is slightly greater than $\delta = 0.001$. Increasing M to 38 results in a type I filter for which $\delta = 0.0008$.

Finally, observe that it is not necessary to plot either the phase or the group delay, since we know that the phase is precisely linear and the delay is $M/2 = 18.5$ samples.

7.2.4 Relationship of the Kaiser Window to Other Windows

The basic principle of the window design method is to truncate the ideal impulse response with a finite-length window such as one of those discussed in this section. The corresponding effect in the frequency domain is that the ideal frequency response is convolved with the Fourier transform of the window. If the ideal filter is a lowpass filter, the discontinuity in its frequency response is smeared as the main lobe of the Fourier transform of the window moves across the discontinuity in the convolution process. To a first approximation, the width of the resulting transition band is determined by the width of the main lobe of the Fourier transform of the window, and the passband and stopband ripples are determined by the side lobes of the Fourier transform of the window. Because the passband and stopband ripples are produced by integration of the symmetric window side lobes, the ripples in the passband and the stopband are approximately the same. Furthermore, to a very good approximation, the maximum passband and stopband deviations are not dependent on M and can be changed only by changing the shape of the window used. This is illustrated by Kaiser's formula, Eq. (7.62), for the window shape parameter, which is independent of M. The last two columns of Table 7.1 compare the Kaiser window with the windows of Eqs. (7.47). The fifth column gives the Kaiser window shape parameter that yields the same peak approximation error (δ) as the window indicated in the first column. The sixth column shows the corresponding transition width (from Eq. 7.63) for filters designed with the Kaiser window. This formula would be a much better predictor of the transition width for the other windows than would the main-lobe width given in the third column of the table.

7.3 EXAMPLES OF FIR FILTER DESIGN BY THE KAISER WINDOW METHOD

The use of the window method is, of course, not restricted to lowpass filters: Windows can be used to truncate *any* ideal impulse response to obtain a causal FIR approximation. In this section, we give several examples that illustrate the technique using the Kaiser window. These examples also serve to point out some important properties of FIR systems.

7.3.1 Highpass Filter

The ideal highpass filter with generalized linear phase has the frequency response

$$H_{hp}(e^{j\omega}) = \begin{cases} 0, & |\omega| < \omega_c, \\ e^{-j\omega M/2}, & \omega_c < |\omega| \leq \pi. \end{cases} \tag{7.65}$$

The corresponding impulse response can be found by evaluating the inverse transform of $H_{hp}(e^{j\omega})$, or we can observe that

$$H_{hp}(e^{j\omega}) = e^{-j\omega M/2} - H_{lp}(e^{j\omega}), \tag{7.66}$$

where $H_{lp}(e^{j\omega})$ is given by Eq. (7.56). Thus, $h_{hp}[n]$ is

$$h_{hp}[n] = \frac{\sin \pi(n - M/2)}{\pi(n - M/2)} - \frac{\sin \omega_c(n - M/2)}{\pi(n - M/2)}, \qquad -\infty < n < \infty. \tag{7.67}$$

To design an FIR approximation to the highpass filter, we can proceed in a manner similar to that of Example 7.8.

Example 7.9   Kaiser Window Design of a Highpass Filter

Suppose that we wish to design a filter to meet the highpass specifications

$$|H(e^{j\omega})| \leq \delta_2, \qquad |\omega| \leq \omega_s$$

$$1 - \delta_1 \leq |H(e^{j\omega})| \leq 1 + \delta_1, \qquad \omega_p \leq |\omega| \leq \pi$$

where $\omega_s = 0.35\pi$, $\omega_p = 0.5\pi$, and $\delta_1 = \delta_2 = \delta = 0.021$. Since the ideal response also has a discontinuity, we can apply Kaiser's formulas in Eqs. (7.62) and (7.63) to estimate the required values of $\beta = 2.6$ and $M = 24$. Figure 7.26 shows the response characteristics that result when a Kaiser window with these parameters is applied to $h_{hp}[n]$ with $\omega_c = (0.35\pi + 0.5\pi)/2$. Note that, since M is an even integer, the filter is a type I FIR system with linear phase, and the delay is precisely $M/2 = 12$ samples. In this case, the actual peak approximation error is $\delta = 0.0213$ rather than 0.021, as specified. Since the error is less than 0.021 everywhere except at the stopband edge, it is tempting to simply increase M to 25, keeping β the same, thereby narrowing the transition region. This type II filter, which is shown in Figure 7.27, is highly unsatisfactory, due to the zero of $H(z)$ that is forced by the linear phase constraint to be at $z = -1$, i.e., $\omega = \pi$. Although increasing the order by 1 leads to a worse result, increasing M to 26 would, of course, lead to a type I system that would exceed the specifications. Clearly, type II FIR linear-phase systems are generally not appropriate approximations for either highpass or bandstop filters.

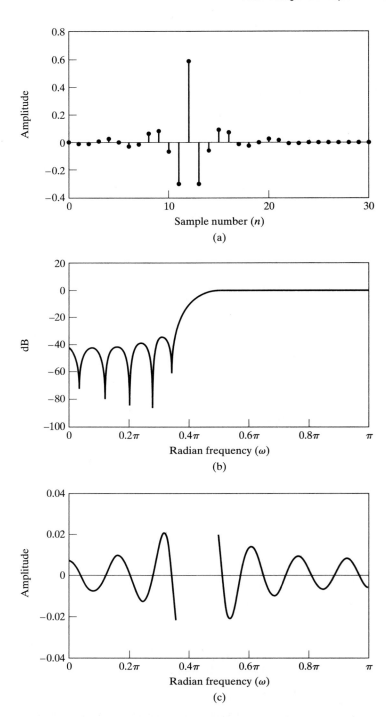

Figure 7.26 Response functions for type I FIR highpass filter. (a) Impulse response ($M = 24$). (b) Log magnitude. (c) Approximation error.

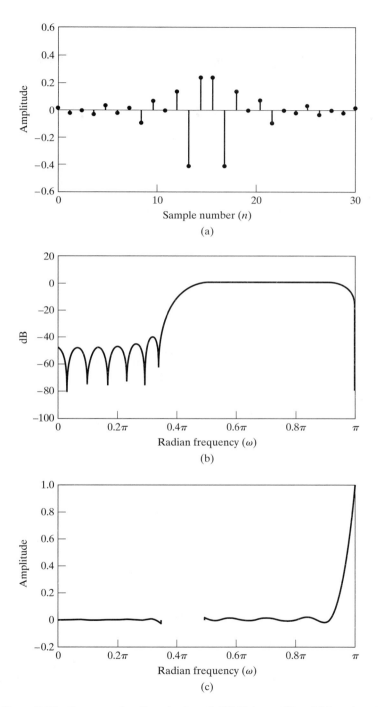

Figure 7.27   Response functions for type II FIR highpass filter. (a) Impulse response ($M = 25$). (b) Log magnitude of Fourier transform. (c) Approximation error.

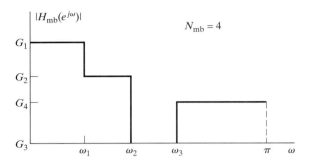

Figure 7.28 Ideal frequency response for multiband filter.

The previous discussion of highpass filter design can be generalized to the case of multiple passbands and stopbands. Figure 7.28 shows an ideal multiband frequency-selective frequency response. This generalized multiband filter includes lowpass, highpass, bandpass, and bandstop filters as special cases. If such a magnitude function is multiplied by a linear phase factor $e^{-j\omega M/2}$, the corresponding ideal impulse response is

$$h_{\text{mb}}[n] = \sum_{k=1}^{N_{\text{mb}}} (G_k - G_{k+1}) \frac{\sin \omega_k (n - M/2)}{\pi (n - M/2)}, \tag{7.68}$$

where N_{mb} is the number of bands and $G_{N_{\text{mb}}+1} = 0$. If $h_{\text{mb}}[n]$ is multiplied by a Kaiser window, the type of approximations that we have observed at the single discontinuity of the lowpass and highpass systems will occur at *each* of the discontinuities. The behavior will be the same at each discontinuity, provided that the discontinuities are far enough apart. Thus, Kaiser's formulas for the window parameters can be applied to this case to predict approximation errors and transition widths. Note that the approximation errors will be scaled by the size of the jump that produces them. That is, if a discontinuity of unity produces a peak error of δ, then a discontinuity of one-half will have a peak error of $\delta/2$.

7.3.2 Discrete-Time Differentiators

As illustrated in Example 4.5, sometimes it is of interest to obtain samples of the derivative of a bandlimited signal from samples of the signal itself. Since the Fourier transform of the derivative of a continuous-time signal is $j\Omega$ times the Fourier transform of the signal, it follows that, for bandlimited signals, a discrete-time system with frequency response $(j\omega/T)$ for $-\pi < \omega < \pi$ (and that is periodic, with period 2π) will yield output samples that are equal to samples of the derivative of the continuous-time signal. A system with this property is referred to as a discrete-time differentiator.

For an ideal discrete-time differentiator with a linear phase, the appropriate frequency response is

$$H_{\text{diff}}(e^{j\omega}) = (j\omega)e^{-j\omega M/2}, \qquad -\pi < \omega < \pi. \tag{7.69}$$

(We have omitted the factor $1/T$.) The corresponding ideal impulse response is

$$h_{\text{diff}}[n] = \frac{\cos \pi (n - M/2)}{(n - M/2)} - \frac{\sin \pi (n - M/2)}{\pi (n - M/2)^2}, \qquad -\infty < n < \infty. \qquad (7.70)$$

If $h_{\text{diff}}[n]$ is multiplied by a symmetric window of length $(M + 1)$, then it is easily shown that $h[n] = -h[M - n]$. Thus, the resulting system is either a type III or a type IV generalized linear-phase system.

Since Kaiser's formulas were developed for frequency responses with simple magnitude discontinuities, it is not straightforward to apply them to differentiators wherein the discontinuity in the ideal frequency response is introduced by the phase. Nevertheless, as we show in the next example, the window method is very effective in designing such systems.

Example 7.10   Kaiser Window Design of a Differentiator

To illustrate the window design of a differentiator, suppose $M = 10$ and $\beta = 2.4$. The resulting response characteristics are shown in Figure 7.29. Figure 7.29(a) shows the antisymmetric impulse response. Since M is even, the system is a type III linear-phase system, which implies that $H(z)$ has zeros at *both* $z = +1$ ($\omega = 0$) and $z = -1$ ($\omega = \pi$). This is clearly displayed in the magnitude response shown in Figure 7.29(b). The phase is exact, since type III systems have a $\pi/2$-radian constant phase shift plus a linear phase corresponding in this case to $M/2 = 5$ samples delay. Figure 7.29(c) shows the amplitude approximation error

$$E_{\text{diff}}(\omega) = \omega - A_o(e^{j\omega}), \qquad 0 \le \omega \le 0.8\pi, \qquad (7.71)$$

where $A_o(e^{j\omega})$ is the amplitude of the approximation. (Note that the error is large around $\omega = \pi$ and is not plotted for frequencies above $\omega = 0.8\pi$.) Clearly, the linearly increasing magnitude is not achieved over the whole band, and, obviously, the relative error (i.e., $E_{\text{diff}}(\omega)/\omega$) is very large for low frequencies or high frequencies (around $\omega = \pi$).

Type IV linear-phase systems do not constrain $H(z)$ to have a zero at $z = -1$. This type of system leads to much better approximations to the amplitude function, as shown in Figure 7.30, for $M = 5$ and $\beta = 2.4$. In this case, the amplitude approximation error is very small up to and beyond $\omega = 0.8\pi$. The phase for this system is again a $\pi/2$-radian constant phase shift plus a linear phase corresponding to a delay of $M/2 = 2.5$ samples. This noninteger delay is the price paid for the exceedingly good amplitude approximation. Instead of obtaining samples of the derivative of the continuous-time signal at the original sampling times $t = nT$, we obtain samples of the derivative at times $t = (n - 2.5)T$. However, in many applications this noninteger delay may not cause a problem, or it could be compensated for by other noninteger delays in a more complex system involving other linear-phase filters.

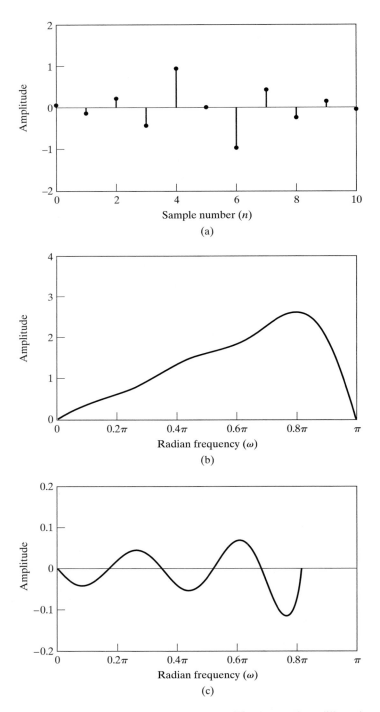

Figure 7.29   Response functions for type III FIR discrete-time differentiator.
(a) Impulse response ($M = 10$). (b) Magnitude. (c) Approximation error.

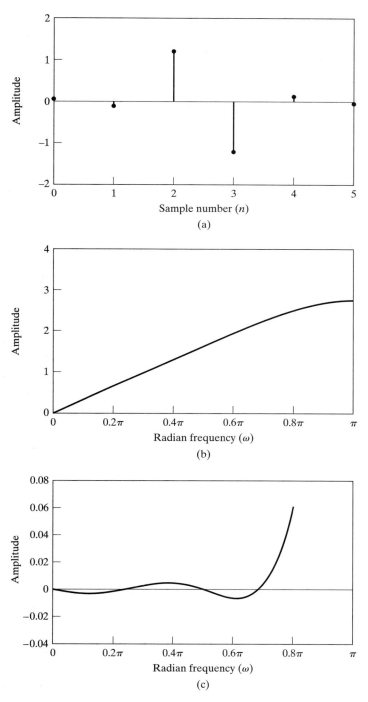

Figure 7.30   Response functions for type IV FIR discrete-time differentiator.
(a) Impulse response ($M = 5$). (b) Magnitude. (c) Approximation error.

7.4 OPTIMUM APPROXIMATIONS OF FIR FILTERS

The design of FIR filters by windowing is straightforward and is quite general, even though it has a number of limitations. However, we often wish to design a filter that is the "best" that can be achieved for a given value of M. It is meaningless to discuss this question in the absence of an approximation criterion. For example, in the case of the window design method, it follows from the theory of Fourier series that the rectangular window provides the best mean-square approximation to a desired frequency response for a given value of M. That is,

$$h[n] = \begin{cases} h_d[n], & 0 \le n \le M, \\ 0, & \text{otherwise,} \end{cases} \tag{7.72}$$

minimizes the expression

$$\varepsilon^2 = \frac{1}{2\pi} \int_{-\pi}^{\pi} |H_d(e^{j\omega}) - H(e^{j\omega})|^2 d\omega. \tag{7.73}$$

(See Problem 7.32.) However, as we have seen, this approximation criterion leads to adverse behavior at discontinuities of $H_d(e^{j\omega})$. Furthermore, the window method does not permit individual control over the approximation errors in different bands. For many applications, better filters result from a minimax strategy (minimization of the maximum errors) or a frequency-weighted error criterion. Such designs can be achieved using algorithmic techniques.

As the previous examples show, frequency-selective filters designed by windowing often have the property that the error is greatest on either side of a discontinuity of the ideal frequency response, and the error becomes smaller for frequencies away from the discontinuity. Furthermore, as suggested by Figure 7.23, such filters typically result in approximately equal errors in the passband and stopband. (See Figures 7.25(c) and 7.26(c), for example.) We have already seen that, for IIR filters, if the approximation error is spread out uniformly in frequency and if the passband and stopband ripples are adjusted separately, a given design specification can be met with a lower order filter than if the approximation just meets the specification at one frequency and far exceeds it at others. This intuitive notion is confirmed for FIR systems by a theorem to be discussed later in the section.

In the following discussion, we consider a particularly effective and widely used algorithmic procedure for the design of FIR filters with a generalized linear phase. Although we consider only type I filters in detail, where appropriate we indicate how the results apply to types II, III, and IV generalized linear-phase filters.

In designing a causal type I linear-phase FIR filter, it is convenient first to consider the design of a zero-phase filter, i.e., one for which

$$h_e[n] = h_e[-n], \tag{7.74}$$

and then to insert a delay sufficient to make it causal. Consequently, we consider $h_e[n]$ satisfying the condition of Eq. (7.74). The corresponding frequency response is given by

$$A_e(e^{j\omega}) = \sum_{n=-L}^{L} h_e[n]e^{-j\omega n}, \tag{7.75}$$

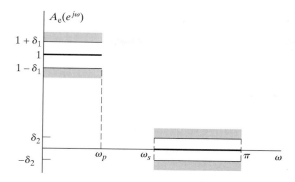

Figure 7.31   Tolerance scheme and ideal response for lowpass filter.

with $L = M/2$ an integer, or, because of Eq. (7.74),

$$A_e(e^{j\omega}) = h_e[0] + \sum_{n=1}^{L} 2h_e[n] \cos(\omega n). \tag{7.76}$$

Note that $A_e(e^{j\omega})$ is a real, even, and periodic function of ω. A causal system can be obtained from $h_e[n]$ by delaying it by $L = M/2$ samples. The resulting system has impulse response

$$h[n] = h_e[n - M/2] = h[M - n] \tag{7.77}$$

and frequency response

$$H(e^{j\omega}) = A_e(e^{j\omega})e^{-j\omega M/2}. \tag{7.78}$$

Figure 7.31 shows a tolerance scheme for an approximation to a lowpass filter with a real function such as $A_e(e^{j\omega})$. Unity is to be approximated in the band $0 \leq |\omega| \leq \omega_p$ with maximum absolute error δ_1, and zero is to be approximated in the band $\omega_s \leq |\omega| \leq \pi$ with maximum absolute error δ_2. An algorithmic technique for designing a filter to meet these specifications must, in effect, systematically vary the $(L + 1)$ unconstrained impulse response values $h_e[n]$, where $0 \leq n \leq L$. Design algorithms have been developed in which some of the parameters $L, \delta_1, \delta_2, \omega_p$, and ω_s are fixed and an iterative procedure is used to obtain optimum adjustments of the remaining parameters. Two distinct approaches have been developed. Herrmann (1970), Herrmann and Schüssler (1970a), and Hofstetter et al. (1971) developed procedures in which L, δ_1, and δ_2 are fixed and ω_p and ω_s are variable. Parks and McClellan (1972a, 1972b), McClellan and Parks (1973), and Rabiner (1972a, 1972b) developed procedures in which L, ω_p, ω_s and the ratio δ_1/δ_2 are fixed and δ_1 (or δ_2) is variable. Since the time when these different approaches were developed, the Parks–McClellan algorithm has become the dominant method for optimum design of FIR filters. This is because it is the most flexible and the most computationally efficient. Thus, we will discuss only that algorithm here.

The Parks–McClellan algorithm is based on reformulating the filter design problem as a problem in polynomial approximation. Specifically, the terms $\cos(\omega n)$ in Eq. (7.76) can be expressed as a sum of powers of $\cos \omega$ in the form

$$\cos(\omega n) = T_n(\cos \omega), \tag{7.79}$$

where $T_n(x)$ is an nth-order polynomial.[7] Consequently, Eq. (7.76) can be rewritten as an Lth-order polynomial in $\cos \omega$, namely,

$$A_e(e^{j\omega}) = \sum_{k=0}^{L} a_k(\cos \omega)^k, \tag{7.80}$$

where the a_k's are constants that are related to $h_e[n]$, the values of the impulse response. With the substitution $x = \cos \omega$, we can express Eq. (7.80) as

$$A_e(e^{j\omega}) = P(x)|_{x=\cos \omega}, \tag{7.81}$$

where $P(x)$ is the Lth-order polynomial

$$P(x) = \sum_{k=0}^{L} a_k x^k. \tag{7.82}$$

We will see that it is not necessary to know the relationship between the a_k's and $h_e[n]$ (although a formula can be obtained); it is enough to know that $A_e(e^{j\omega})$ can be expressed as the Lth-degree trigonometric polynomial of Eq. (7.80).

The key to gaining control over ω_p and ω_s is to fix them at their desired values and let δ_1 and δ_2 vary. Parks and McClellan (1972a, 1972b) showed that with L, ω_p, and ω_s fixed, the frequency-selective filter design problem becomes a problem in Chebyshev approximation over disjoint sets, an important problem in approximation theory and one for which several useful theorems and procedures have been developed. (See Cheney, 1982.) To formalize the approximation problem in this case, let us define an approximation error function

$$E(\omega) = W(\omega)[H_d(e^{j\omega}) - A_e(e^{j\omega})], \tag{7.83}$$

where the weighting function $W(\omega)$ incorporates the approximation error parameters into the design process. In this design method, the error function $E(\omega)$, the weighting function $W(\omega)$, and the desired frequency response $H_d(e^{j\omega})$ are defined only over closed subintervals of $0 \leq \omega \leq \pi$. For example, to approximate a lowpass filter, these functions are defined for $0 \leq \omega \leq \omega_p$ and $\omega_s \leq \omega \leq \pi$. The approximating function $A_e(e^{j\omega})$ is not constrained in the transition region(s) (e.g., $\omega_p < \omega < \omega_s$), and it may take any shape necessary to achieve the desired response in the other subintervals.

For example, suppose that we wish to obtain an approximation as in Figure 7.31, where L, ω_p, and ω_s are fixed design parameters. For this case,

$$H_d(e^{j\omega}) = \begin{cases} 1, & 0 \leq \omega \leq \omega_p, \\ 0, & \omega_s \leq \omega \leq \pi. \end{cases} \tag{7.84}$$

The weighting function $W(\omega)$ allows us to weight the approximation errors differently in the different approximation intervals. For the lowpass filter approximation problem, the weighting function is

$$W(\omega) = \begin{cases} \dfrac{1}{K} & 0 \leq \omega \leq \omega_p, \\ 1, & \omega_s \leq \omega \leq \pi, \end{cases} \tag{7.85}$$

where $K = \delta_1/\delta_2$. If $A_e(e^{j\omega})$ is as shown in Figure 7.32, the weighted approximation error, $E(\omega)$ in Eq. (7.83), would be as indicated in Figure 7.33. Note that with this

---

[7]More specifically, $T_n(x)$ is the nth-order Chebyshev polynomial, defined as $T_n(x) = \cos(n \cos^{-1} x)$.

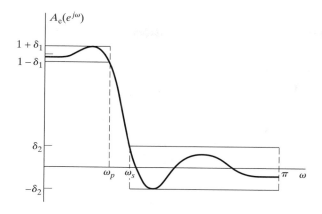

Figure 7.32 Typical frequency response meeting the specifications of Figure 7.31.

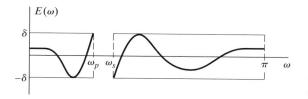

Figure 7.33 Weighted error for the approximation of Figure 7.32.

weighting, the maximum weighted absolute approximation error is $\delta = \delta_2$ in both bands.

The particular criterion used in this design procedure is the so-called minimax or Chebyshev criterion, where, within the frequency intervals of interest (the passband and stopband for a lowpass filter), we seek a frequency response $A_e(e^{j\omega})$ that *minimizes* the *maximum* weighted approximation error of Eq. (7.83). Stated more compactly, the best approximation is to be found in the sense of

$$\min_{\{h_e[n]:0\leq n\leq L\}} \left(\max_{\omega\in F} |E(\omega)| \right),$$

where F is the closed subset of $0 \leq \omega \leq \pi$ such that $0 \leq \omega \leq \omega_p$ or $\omega_s \leq \omega \leq \pi$. Thus, we seek the set of impulse response values that minimizes δ in Figure 7.33.

Parks and McClellan (1972a, 1972b) applied the following theorem of approximation theory to this filter design problem.

Alternation Theorem: Let F_P denote the closed subset consisting of the disjoint union of closed subsets of the real axis x. Then

$$P(x) = \sum_{k=0}^{r} a_k x^k$$

is an rth-order polynomial. Also, $D_P(x)$ denotes a given desired function of x that is continuous on F_P; $W_P(x)$ is a positive function, continuous on F_P, and

$$E_P(x) = W_P(x)[D_P(x) - P(x)].$$

is the weighted error. The maximum error is defined as

$$\|E\| = \max_{x \in F_P} |E_P(x)|.$$

A necessary and sufficient condition that $P(x)$ be the unique rth-order polynomial that minimizes $\|E\|$ is that $E_p(x)$ exhibit *at least* $(r + 2)$ alternations; i.e., there must exist at least $(r + 2)$ values x_i in F_p such that $x_1 < x_2 < \cdots < x_{r+2}$ and such that $E_p(x_i) = -E_P(x_{i+1}) = \pm \|E\|$ for $i = 1, 2, \ldots, (r + 1)$.

At first glance, it may seem to be difficult to relate this formal theorem to the problem of filter design. However, in the discussion that follows, all of the elements of the theorem will be shown to be important in developing the design algorithm. To aid in understanding the alternation theorem, in Section 7.4.1 we will interpret it specifically for the design of a type I lowpass filter. Before proceeding to apply the alternation theorem to filter design, however, we illustrate in Example 7.11 how the theorem is applied to polynomials.

Example 7.11   Alternation Theorem and Polynomials

The alternation theorem provides a necessary and sufficient condition for a polynomial to satisfy in order that it be the polynomial that minimizes the maximum weighted error for a given order. To illustrate how the theorem is applied, suppose we want to examine polynomials $P(x)$ that approximate unity for $-1 \le x \le -0.1$ and zero for $0.1 \le x \le 1$. Consider three such polynomials, as shown in Figure 7.34. Each of these polynomials is of fifth order, and we would like to determine which, if any, satisfy the alternation theorem. The closed subsets of the real axis x referred to in the theorem are the regions $-1 \le x \le -0.1$ and $0.1 \le x \le 1$. We will weight errors equally in both regions, i.e., $W_p(x) = 1$.

According to the alternation theorem, the optimal fifth-order polynomial must exhibit *at least* seven alternations of the error in the regions corresponding to the closed subset F_p. $P_1(x)$ has only five alternations—three in the region $-1 \le x \le -0.1$ and two in the region $0.1 \le x \le 1$. The point with zero slope that does not touch the dotted line is not an alternation, because it does not reach the positive extreme value. The alternation theorem specifies that adjacent alternations must alternate sign, so the extremal value at $x = 1$ cannot be an alternation either, since the previous alternation was a negative extremal value at the first point with zero slope in $0.1 \le x \le 1$.

$P_2(x)$ also has only five alternations and thus is not optimal. Specifically, $P_2(x)$ has three alternations in $-1 \le x \le -0.1$, but again, only two alternations in $0.1 \le x \le 1$. The difficulty occurs because $x = 0.1$ is not a positive extremal value. The previous alternation at $x = -0.1$ is a negative extremal value, so we need a positive extremal value for the next alternation. The first point with zero slope inside $0.1 \le x \le 1$ also cannot be counted, since it is a negative extremal value, like $x = -0.1$, and does not alternate sign. We can count the second point with zero slope in this region and $x = 1$, giving two alternations in $0.1 \le x \le 1$ and a total of 5.

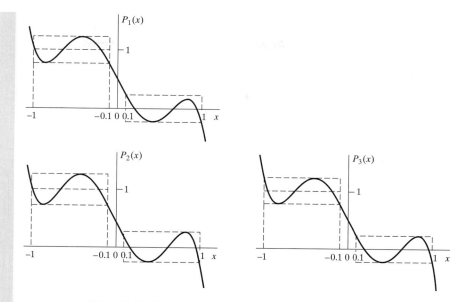

Figure 7.34 Fifth-order polynomials for Example 7.11.

$P_3(x)$ has eight alternations; all points of zero slope, $x = -1$, $x = -0.1$, $x = 0.1$ and $x = 1$. Since eight alternations satisfies the alternation theorem, which specifies a minimum of seven, $P_3(x)$ is the unique optimal fifth-order polynomial approximation for this region.

7.4.1 Optimal Type I Lowpass Filters

For type I filters, the polynomial $P(x)$ is the cosine polynomial $A_e(e^{j\omega})$ in Eq. (7.80), with the transformation of variable $x = \cos \omega$ and $r = L$:

$$P(\cos \omega) = \sum_{k=0}^{L} a_k (\cos \omega)^k. \qquad (7.86)$$

$D_P(x)$ is the desired lowpass filter frequency response in Eq. (7.84), with $x = \cos \omega$:

$$D_P(\cos \omega) = \begin{cases} 1, & \cos \omega_p \leq \cos \omega \leq 1, \\ 0, & -1 \leq \cos \omega \leq \cos \omega_s. \end{cases} \qquad (7.87)$$

$W_P(\cos \omega)$ is given by Eq. (7.85), rephrased in terms of $\cos \omega$:

$$W_P(\cos \omega) = \begin{cases} \dfrac{1}{K}, & \cos \omega_p \leq \cos \omega \leq 1, \\ 1, & -1 \leq \cos \omega \leq \cos \omega_s. \end{cases} \qquad (7.88)$$

And the weighted approximation error is

$$E_P(\cos \omega) = W_P(\cos \omega)[D_P(\cos \omega) - P(\cos \omega)]. \qquad (7.89)$$

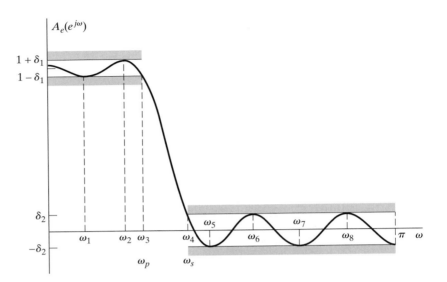

Figure 7.35 Typical example of a lowpass filter approximation that is optimal according to the alternation theorem for $L = 7$.

The closed subset F_P is made up of the union of the intervals $0 \leq \omega \leq \omega_p$ and $\omega_s \leq \omega \leq \pi$, or, in terms of $\cos \omega$, of the intervals $\cos \omega_p \leq \cos \omega \leq 1$ and $-1 \leq \cos \omega \leq \cos \omega_s$. The alternation theorem then states that a set of coefficients a_k in Eq. (7.86) will correspond to the filter representing the unique best approximation to the ideal lowpass filter with the ratio δ_1/δ_2 fixed at K and with passband and stopband edges ω_p and ω_s if and only if $E_P(\cos \omega)$ exhibits at least $(L + 2)$ alternations on F_P, i.e., if and only if $E_P(\cos \omega)$ alternately equals plus and minus its maximum value at least $(L + 2)$ times. Such approximations are called *equiripple approximations.*

Figure 7.35 shows a filter frequency response that is optimal according to the alternation theorem for $L = 7$. In this figure, $A_e(e^{j\omega})$ is plotted against ω. To formally test the alternation theorem, we should first redraw $A_e(e^{j\omega})$ as a function of $x = \cos \omega$. Furthermore, we want to explicitly examine the alternations of $E_P(x)$. Consequently, in Figure 7.36(a), (b), and (c), we show $P(x)$, $W_P(x)$, and $E_P(x)$, respectively, as a function of $x = \cos \omega$. In this example, where $L = 7$, we see that there are nine alternations of the error. Consequently, the alternation theorem is satisfied. An important point is that, in counting alternations, we include the points $\cos \omega_p$ and $\cos \omega_s$, since, according to the alternation theorem, the subsets (or subintervals) included in F_P are closed, i.e., the endpoints of the intervals are counted. While this might seem to be a small issue, it is in fact very significant, as we will see.

Comparing Figures 7.35 and 7.36 suggests that when the desired filter is a lowpass filter (or any piecewise-constant filter) we could actually count the alternations by direct examination of the frequency response, keeping in mind that the maximum error is different (in the ratio $K = \delta_1/\delta_2$) in the passband and stopband.

The alternation theorem states that the optimum filter must have a minimum of $(L + 2)$ alternations, but does not exclude the possibility of more than $(L + 2)$

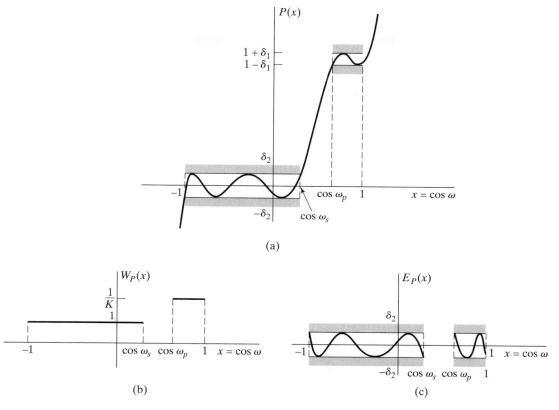

Figure 7.36 Equivalent polynomial approximation functions as a function of $x = \cos\omega$. (a) Approximating polynomial. (b) Weighting function. (c) Approximation error.

alternations. In fact, we will show that for a lowpass filter, the maximum possible number of alternations is $(L+3)$. First, however, we illustrate this in Figure 7.37 for $L = 7$. Figure 7.37(a) has $L + 3 = 10$ alternations, while Figures 7.37(b), (c), and (d) each have $L + 2 = 9$ alternations. The case of $L+3$ alternations (Figure 7.37a) is often referred to as the *extraripple case*. Note that for the extraripple filter, there are alternations at $\omega = 0$ and π, as well as at $\omega = \omega_p$ and $\omega = \omega_s$, i.e., at all the band edges. For Figures 7.37(b) and (c), there are again alternations at ω_p and ω_s, but not at both $\omega = 0$ and $\omega = \pi$. In Figure 7.37(d), there are alternations at 0, π, ω_p, and ω_s, but there is one less extremum (point of zero slope) inside the stopband. We also observe that all of these cases are equiripple inside the passband and stopband; i.e., all points of zero slope inside the interval $0 < \omega < \pi$ are frequencies at which the magnitude of the weighted error is maximal. Finally, because all of the filters in Figure 7.37 satisfy the alternation theorem for $L = 7$ and for the same value of $K = \delta_1/\delta_2$, it follows that ω_p and/or ω_s must be different for each, since the alternation theorem states that the optimum filter under the conditions of the theorem is unique.

(a)

(b)

(c)

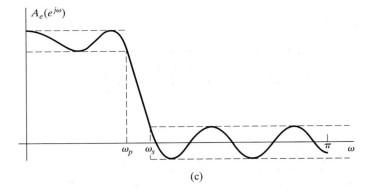

(d)

Figure 7.37 Possible optimum lowpass filter approximations for $L = 7$. (a) $L + 3$ alternations (extraripple case). (b) $L + 2$ alternations (extremum at $\omega = \pi$). (c) $L + 2$ alternations (extremum at $\omega = 0$). (d) $L + 2$ alternations (extremum at both $\omega = 0$ and $\omega = \pi$).

The properties referred to in the preceding paragraph for the filters in Figure 7.37 result from the alternation theorem. Specifically we will show that for type I lowpass filters:

- The maximum possible number of alternations of the error is $(L+3)$.
- Alternations will always occur at ω_p and ω_s.
- All points with zero slope inside the passband and all points with zero slope inside the stopband (for $0 < \omega < \omega_p$ and $\omega_s < \omega < \pi$) will correspond to alternations; i.e., the filter will be equiripple, except possibly at $\omega = 0$ and $\omega = \pi$.

The maximum possible number of alternations is $(L + 3)$

Reference to Figure 7.35 or Figure 7.37 suggests that the maximum possible number of locations for extrema, or alternations, are the four band edges ($\omega = 0, \pi, \omega_p, \omega_s$) and the local extrema, i.e., the frequencies at which $A_e(e^{j\omega})$ has zero slope. Since an Lth-degree polynomial can have at most $(L-1)$ points with zero slope in an open interval, the maximum possible number of locations for alternations are the $(L-1)$ extrema of the polynomial plus the four band edges, a total of $(L+3)$. In considering extrema and points with zero slope for trigonometric polynomials, it is important to observe that the trigonometric polynomial

$$P(\cos\omega) = \sum_{k=0}^{L} a_k(\cos\omega)^k, \tag{7.90}$$

when considered as a function of ω, will always have zero slope at $\omega = 0$ and $\omega = \pi$, even though $P(x)$ considered as a function of x may not have zero slope at the corresponding points $x = 1$ and $x = -1$. This is because

$$\frac{dP(\cos\omega)}{d\omega} = -\sin\omega \left(\sum_{k=0}^{L} k a_k(\cos\omega)^{k-1} \right)$$

$$= -\sin\omega \left(\sum_{k=0}^{L-1} (k+1)a_{k+1}(\cos\omega)^{k} \right), \tag{7.91}$$

which is always zero at $\omega = 0$ and $\omega = \pi$, as well as at the $(L-1)$ roots of the $(L-1)$st-order polynomial represented by the sum. This behavior at $\omega = 0$ and $\omega = \pi$ is evident in Figure 7.37. In Figure 7.37(d), it happens that both $\sin\omega$ and

$$\sum_{k=0}^{L} k a_k(\cos\omega)^{k-1}$$

are zero in Equation (7.91); i.e., the polynomial $P(x)$ also has zero slope at $x = -1 = \cos\pi$.

Alternations always occur at ω_p and ω_s

For all of the frequency responses in Figure 7.37, $A_e(e^{j\omega})$ is exactly equal to $1 - \delta_1$ at the passband edge ω_p and exactly equal to $+\delta_2$ at the stopband edge ω_s. To suggest why this must always be the case, let us consider whether the filter in Figure 7.37(a) could also be optimal if we decreased ω_p as indicated in Figure 7.38. The frequencies at which the magnitude of the maximum weighted error are equal are the frequencies $\omega = 0, \omega_1, \omega_2, \omega_s, \omega_3, \omega_4, \omega_5, \omega_6,$ and $\omega = \pi,$ for a total of $(L+2) = 9.$ However, not all of the frequencies are alternations, since, to be counted in the alternation theorem, the error must *alternate* between $\delta = \pm\|E\|$ at these frequencies. Therefore, because the error is negative at both ω_2 and ω_s, the frequencies counted in the alternation theorem are $\omega = 0, \omega_1, \omega_2, \omega_3, \omega_4, \omega_5, \omega_6,$ and $\pi,$ for a total of 8. Since $(L+2) = 9,$ the conditions of the alternation theorem are not satisfied, and the frequency response of Figure 7.38 is not optimal with ω_p and ω_s as indicated. In other words, the removal of ω_p as an alternation frequency removes two alternations. Since the maximum number is $(L+3)$, this leaves at most $(L+1)$, which is not a sufficient number. An identical argument would hold if ω_s were removed as an alternation frequency. A similar argument can be constructed for highpass filters, but this is not necessarily the case for bandpass or multiband filters. (See Problem 7.57.)

The filter will be equiripple except possibly at $\omega = 0$ or $\omega = \pi$

The argument here is very similar to the one used to show that both ω_p and ω_s must be alternations. Suppose, for example, that the filter in Figure 7.37(a) was modified as indicated in Figure 7.39, so that one point with zero slope did not achieve the maximum error. Then, while the maximum error occurs at nine frequencies, only eight of these can be counted as alternations. Consequently, eliminating one ripple as a point of max-

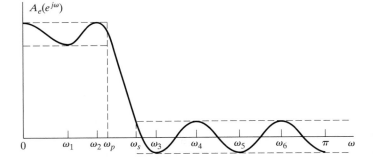

Figure 7.38   Illustration that the passband edge ω_p must be an alternation frequency.

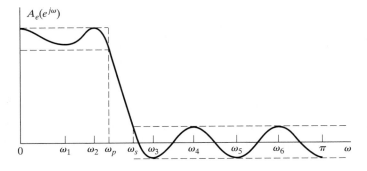

Figure 7.39   Illustration that the frequency response must be equiripple in the approximation bands.

imum error reduces the number of alternations by two, leaving $(L+1)$ as the maximum possible number.

The foregoing represent only a few of many properties that can be inferred from the alternation theorem. A variety of others are discussed in Rabiner and Gold (1975). Furthermore, we have considered only type I lowpass filters. While a much broader and detailed discussion of type II, III, and IV filters or filters with more general desired frequency responses is beyond the scope of this book, we briefly consider type II lowpass filters to further emphasize a number of aspects of the alternation theorem.

7.4.2 Optimal Type II Lowpass Filters

A type II causal filter is a filter for which $h[n] = 0$ outside the range $0 \le n \le M$, with the filter length $(M+1)$ even, i.e., M odd, and with the symmetry property

$$h[n] = h[M - n]. \tag{7.92}$$

Consequently, the frequency response $H(e^{j\omega})$ can be expressed in the form

$$H(e^{j\omega}) = e^{-j\omega M/2} \sum_{n=0}^{(M-1)/2} 2h[n] \cos\left[\omega\left(\frac{M}{2} - n\right)\right]. \tag{7.93}$$

Letting $b[n] = 2h[(M+1)/2 - n], n = 1, 2, \ldots, (M+1)/2$, we can rewrite Eq. (7.93) as

$$H(e^{j\omega}) = e^{-j\omega M/2} \left\{ \sum_{n=1}^{(M+1)/2} b[n] \cos\left[\omega\left(n - \frac{1}{2}\right)\right] \right\}. \tag{7.94}$$

To apply the alternation theorem to the design of type II filters, we must be able to identify the problem as one of polynomial approximation. To accomplish this, we express the summation in Eq. (7.94) in the form

$$\sum_{n=1}^{(M+1)/2} b[n] \cos\left[\omega\left(n - \frac{1}{2}\right)\right] = \cos(\omega/2) \left[\sum_{n=0}^{(M-1)/2} \tilde{b}[n] \cos(\omega n)\right]. \tag{7.95}$$

(See Problem 7.52.) The summation on the right-hand side of Eq. (7.95) can now be represented as a trigonometric polynomial $P(\cos \omega)$ so that

$$H(e^{j\omega}) = e^{-j\omega M/2} \cos(\omega/2) P(\cos \omega), \tag{7.96a}$$

where

$$P(\cos \omega) = \sum_{k=0}^{L} a_k (\cos \omega)^k \tag{7.96b}$$

and $L = (M-1)/2$. The coefficients a_k in Eq. (7.96b) are related to the coefficients $\tilde{b}[n]$ in Eq. (7.95), which in turn are related to the coefficients $b[n] = 2h[(M+1)/2 - n]$ in Eq. (7.94). As in the type I case, it is not necessary to obtain an explicit relationship between the impulse response and the a_k's. We now can apply the alternation theorem to the weighted error between $P(\cos\omega)$ and the desired frequency response. For a type I lowpass filter with a specified ratio K of passband to stopband ripple, the desired function is given by Eq. (7.84), and the weighting function for the error is given by Eq. (7.85). For type II lowpass filters, because of the presence of the factor $\cos(\omega/2)$ in Eq. (7.96a), the function to be approximated by the polynomial $P(\cos\omega)$ is defined as

$$H_d(e^{j\omega}) = D_P(\cos\omega) = \begin{cases} \dfrac{1}{\cos(\omega/2)}, & 0 \le \omega \le \omega_p, \\ 0, & \omega_s \le \omega \le \pi, \end{cases} \tag{7.97}$$

and the weighting function to be applied to the error is

$$W(\omega) = W_P(\cos\omega) = \begin{cases} \dfrac{\cos(\omega/2)}{K}, & 0 \le \omega \le \omega_p, \\ \cos(\omega/2), & \omega_s \le \omega \le \pi. \end{cases} \tag{7.98}$$

Consequently, type II filter design is a different polynomial approximation problem than type I filter design.

In this section we have only outlined the design of type II filters, principally to highlight the requirement that the design problem first be formulated as a polynomial approximation problem. A similar set of issues arises in the design of type III and type IV linear-phase FIR filters. Specifically, these classes also can be formulated as polynomial approximation problems, but in each class the weighting function applied to the error has a trigonometric form, just as it does for type II filters. (See Problem 7.52.) A detailed discussion of the design and properties of these classes of filters can be found in Rabiner and Gold (1975).

The details of the formulation of the problem for type I and type II linear-phase systems have been illustrated for the case of the lowpass filter. However, the discussion of type II systems in particular should suggest that there is great flexibility in the choice of both the desired response function $H_d(e^{j\omega})$ and the weighting function $W(\omega)$. For example, the weighting function can be defined in terms of the desired function so as to yield equiripple percentage error approximation. This approach is valuable in designing type III and type IV differentiator systems.

7.4.3 The Parks–McClellan Algorithm

The alternation theorem gives necessary and sufficient conditions on the error for optimality in the Chebyshev or minimax sense. Although the theorem does not state explicitly how to find the optimum filter, the conditions that are presented serve as the basis for an efficient algorithm for finding it. While our discussion is phrased in terms of type I lowpass filters, the algorithm easily generalizes.

From the alternation theorem, we know that the optimum filter $A_e(e^{j\omega})$ will satisfy the set of equations

$$W(\omega_i)[H_d(e^{j\omega_i}) - A_e(e^{j\omega_i})] = (-1)^{i+1}\delta, \qquad i = 1, 2, \ldots, (L+2), \qquad (7.99)$$

where δ is the optimum error and $A_e(e^{j\omega})$ is given by either Eq. (7.76) or Eq. (7.80). Using Eq. (7.80) for $A_e(e^{j\omega})$, we can write these equations as

$$
\begin{bmatrix}
1 & x_1 & x_1^2 & \cdots & x_1^L & \dfrac{1}{W(\omega_1)} \\
1 & x_2 & x_2^2 & \cdots & x_2^L & \dfrac{-1}{W(\omega_2)} \\
\vdots & \vdots & \vdots & & \vdots & \vdots \\
1 & x_{L+2} & x_{L+2}^2 & \cdots & x_{L+2}^L & \dfrac{(-1)^{L+2}}{W(\omega_{L+2})}
\end{bmatrix}
\begin{bmatrix}
a_0 \\
a_1 \\
\vdots \\
\delta
\end{bmatrix}
=
\begin{bmatrix}
H_d(e^{j\omega_1}) \\
H_d(e^{j\omega_2}) \\
\vdots \\
H_d(e^{j\omega_{L+2}})
\end{bmatrix},
\qquad (7.100)
$$

where $x_i = \cos\omega_i$. This set of equations serves as the basis for an iterative algorithm for finding the optimum $A_e(e^{j\omega})$. The procedure begins by guessing a set of alternation frequencies ω_i for $i = 1, 2, \ldots, (L+2)$. Note that ω_p and ω_s are fixed and, based on our discussion in Section 7.4.1, are necessarily members of the set of alternation frequencies. Specifically, if $\omega_\ell = \omega_p$, then $\omega_{\ell+1} = \omega_s$. The set of equations (7.100) could be solved for the set of coefficients a_k and δ. However, a more efficient alternative is to use polynomial interpolation. In particular, Parks and McClellan (1972a, 1972b) found that, for the given set of the extremal frequencies,

$$\delta = \dfrac{\displaystyle\sum_{k=1}^{L+2} b_k H_d(e^{j\omega_k})}{\displaystyle\sum_{k=1}^{L+2} \dfrac{b_k(-1)^{k+1}}{W(\omega_k)}}, \qquad (7.101)$$

where

$$b_k = \prod_{\substack{i=1 \\ i \neq k}}^{L+2} \dfrac{1}{(x_k - x_i)} \qquad (7.102)$$

and, as before, $x_i = \cos\omega_i$. That is, if $A_e(e^{j\omega})$ is determined by the set of coefficients a_k that satisfy Eq. (7.100), with δ given by Eq. (7.101), then the error function goes through $\pm\delta$ at the $(L+2)$ frequencies ω_i, or, equivalently, $A_e(e^{j\omega})$ has values $1 \pm K\delta$ if $0 \le \omega_i \le \omega_p$ and $\pm\delta$ if $\omega_s \le \omega_i \le \pi$. Now, since $A_e(e^{j\omega})$ is known to be an Lth-order trigonometric polynomial, we can interpolate a trigonometric polynomial through $(L+1)$ of the $(L+2)$ known values $E(\omega_i)$ (or equivalently, $A_e(e^{j\omega_i})$). Parks and McClellan used the Lagrange interpolation formula to obtain

$$A_e(e^{j\omega}) = P(\cos\omega) = \dfrac{\displaystyle\sum_{k=1}^{L+1} [d_k/(x - x_k)]C_k}{\displaystyle\sum_{k=1}^{L+1} [d_k/(x - x_k)]}, \qquad (7.103a)$$

where $x = \cos\omega$, $x_i = \cos\omega_i$,

$$C_k = H_d(e^{j\omega_k}) - \frac{(-1)^{k+1}\delta}{W(\omega_k)}, \tag{7.103b}$$

and

$$d_k = \prod_{\substack{i=1 \\ i \neq k}}^{L+1} \frac{1}{(x_k - x_i)} = b_k(x_k - x_{L+2}). \tag{7.103c}$$

Although only the frequencies $\omega_1, \omega_2, \ldots, \omega_{L+1}$ are used in fitting the Lth-degree polynomial, we can be assured that the polynomial also takes on the correct value at ω_{L+2} because Eqs. (7.100) are satisfied by the resulting $A_e(e^{j\omega})$.

Now $A_e(e^{j\omega})$ is available at any desired frequency, without the need to solve the set of equations (7.100) for the coefficients a_k. The polynomial of Eq. (7.103a) can be used to evaluate $A_e(e^{j\omega})$ and also $E(\omega)$ on a dense set of frequencies in the passband and stopband. If $|E(\omega)| \leq \delta$ for all ω in the passband and stopband, then the optimum approximation has been found. Otherwise we must find a new set of extremal frequencies.

Figure 7.40 shows a typical example for a type I lowpass filter before the optimum has been found. Clearly, the set of frequencies ω_i used to find δ (as represented by open circles in the figure) was such that δ was too small. Adopting the philosophy of the Remez method, we find that the extremal frequencies are exchanged for a completely new set defined by the $(L+2)$ largest peaks of the error curve. The points marked with $\times$ would be the new set of frequencies for the example shown in the figure. As before, ω_p and ω_s must be selected as extremal frequencies. Recall that there are at most $(L-1)$ local minima and maxima in the open intervals $0 < \omega < \omega_p$ and $\omega_s < \omega < \pi$. The remaining extrema can be at either $\omega = 0$ or $\omega = \pi$. If there is a maximum of the error function at both 0 and π, then the frequency at which the greatest error occurs is taken as the new estimate of the frequency of the remaining extremum. The cycle—computing the value of δ, fitting a polynomial to the assumed error peaks, and then locating the actual error peaks—is repeated until δ does not change from its previous value by more

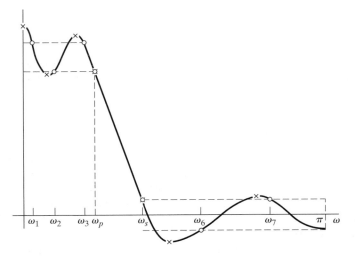

Figure 7.40 Illustration of the Parks–McClellan algorithm for equiripple approximation.

than a prescribed small amount. This value of δ is then the desired minimum maximum weighted approximation error.

A flowchart for the Parks–McClellan algorithm is shown in Figure 7.41. In this algorithm, all the impulse response values $h_e[n]$ are implicitly varied on each iteration to obtain the desired optimal approximation, but the values of $h_e[n]$ are never explicitly computed. After the algorithm has converged, the impulse response can be computed from samples of the polynomial representation using the discrete Fourier transform, as will be discussed in Chapter 8.

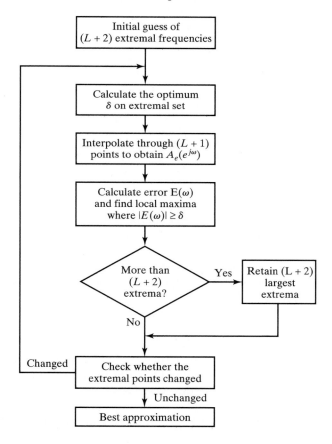

Figure 7.41   Flowchart of Parks–McClellan algorithm.

7.4.4 Characteristics of Optimum FIR Filters

Optimum lowpass FIR filters have the smallest maximum weighted approximation error δ for prescribed passband and stopband edge frequencies ω_p and ω_s. For the weighting function of Eq. (7.85), the resulting maximum stopband approximation error is $\delta_2 = \delta$, and the maximum passband approximation error is $\delta_1 = K\delta$. In Figure 7.42, we illustrate how δ varies with the order of the filter and the passband cutoff frequency. For this example, $K = 1$ and the transition width is fixed at $(\omega_s - \omega_p) = 0.2\pi$. The curves show that as ω_p increases, the error δ attains local minima. These minima on the curves correspond

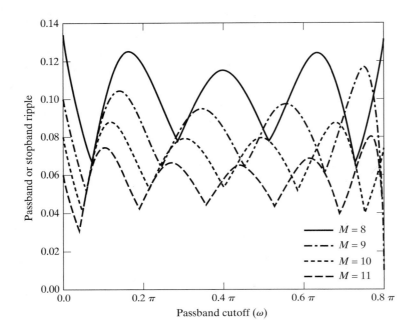

Figure 7.42 Illustration of the dependence of passband and stopband error on cutoff frequency for optimal approximations of a lowpass filter. For this example, $K = 1$ and $(\omega_s - \omega_p) = 0.2\pi$.

to the extraripple ($L + 3$ extrema) filters. All points between the minima correspond to filters that are optimal according to the alternation theorem. The filters for $M = 8$ and $M = 10$ are type I filters, while $M = 9$ and $M = 11$ correspond to a type II filter. It is interesting to note that, for some choices of parameters, a shorter filter ($M = 9$) may be better (i.e., it yields a smaller error) than a longer filter ($M = 10$). This may at first seem surprising and even contradictory. However, the cases $M = 9$ and $M = 10$ represent fundamentally different types of filters. Interpreted another way, filters for $M = 9$ cannot be considered to be special cases of $M = 10$ with one point set to zero, since this would violate the linear-phase symmetry requirement. On the other hand, $M = 8$ could always be thought of as a special case of $M = 10$ with the first and last samples set to zero. For that reason, an optimal filter for $M = 8$ cannot be better than one for $M = 10$. This restriction can be seen in Figure 7.42, where the curve for $M = 8$ is always above or equal to the one for $M = 10$. The points at which the two curves touch correspond to identical impulse responses, with the $M = 10$ filter having the first and last points equal to zero.

Herrmann et al. (1973) did an extensive computational study of the relationships among the parameters M, δ_1, δ_2, ω_p, and ω_s for equiripple lowpass approximations, and Kaiser (1974) subsequently obtained the simplified formula

$$M = \frac{-10 \log_{10}(\delta_1 \delta_2) - 13}{2.324 \Delta \omega}, \tag{7.104}$$

where $\Delta \omega = \omega_s - \omega_p$, as a fit to their data. By comparing Eq. (7.104) with the design formula of Eq. (7.63) for the Kaiser window method, we can see that, for the comparable case ($\delta_1 = \delta_2 = \delta$), the optimal approximations provide about 5 dB better approximation error for a given value of M. Another important advantage of the equiripple filters is that δ_1 and δ_2 need not be equal, as must be the case for the window method.

7.5 EXAMPLES OF FIR EQUIRIPPLE APPROXIMATION

The Parks–McClellan algorithm for optimum equiripple approximation of FIR filters can be used to design a wide variety of such filters. In this section, we give several examples that illustrate some of the properties of the optimum approximation and suggest the great flexibility that is afforded by the design method.

7.5.1 Lowpass Filter

For the lowpass filter case, we again approximate the set of specifications used in Examples 7.4, 7.5, 7.6, and 7.8 so that we can compare all the major design methods on the same lowpass filter specifications. These specifications call for $\omega_p = 0.4\pi$, $\omega_s = 0.6\pi$, $\delta_1 = 0.01$, and $\delta_2 = 0.001$. In contrast to the window method, the Parks–McClellan algorithm can accommodate the different approximation error in the passband versus that in the stopband by fixing the weighting function parameter at $K = \delta_1/\delta_2 = 10$.

Substituting the foregoing specifications into Eq. (7.104) and rounding up yields the estimate $M = 26$ for the value of M that is necessary to achieve the specifications. Figures 7.43(a), (b), and (c) show the impulse response, log magnitude, and approximation error, respectively, for the optimum filter with $M = 26$, $\omega_p = 0.4\pi$, and $\omega_s = 0.6\pi$. Figure 7.43(c) shows the *unweighted* approximation error

$$E_A(\omega) = \frac{E(\omega)}{W(\omega)} = \begin{cases} 1 - A_e(e^{j\omega}), & 0 \leq \omega \leq \omega_p, \\ 0 - A_e(e^{j\omega}), & \omega_s \leq \omega \leq \pi, \end{cases} \tag{7.105}$$

rather than the weighted error used in the formulation of the design algorithm. The weighted error would be identical to Figure 7.43(c), except that the error would be divided by 10 in the passband.[8] The alternations of the approximation error are clearly in evidence in Figure 7.43(c). There are seven alternations in the passband and eight in the stopband, for a total of fifteen alternations. Since $L = M/2$ for type I (M even) systems, and $M = 26$, the minimum number of alternations is $(L + 2) = (26/2 + 2) = 15$. Thus, the filter of Figure 7.43 is the optimum filter for $M = 26$, $\omega_p = 0.4\pi$, and $\omega_s = 0.6\pi$. However, Figure 7.43(c) shows that the filter fails to meet the original specifications on passband and stopband error. (The maximum errors in the passband and stopband are 0.0116 and 0.00116, respectively.) To meet or exceed the specifications, we must increase M.

The filter response functions for the case $M = 27$ are shown in Figure 7.44. Now the passband and stopband approximation errors are slightly less than the specified values. (The maximum errors in the passband and stopband are 0.0092 and 0.00092, respectively.) In this case there are again seven alternations in the passband and eight alternations in the stopband, for a total of fifteen. Note that, since $M = 27$, this is a type II system, and for type II systems, the order of the implicit approximating polynomial is $L = (M-1)/2 = (27-1)/2 = 13$. Thus, the minimum number of alternations is still 15. Note also that in the type II case, the system is constrained to have a zero of its system function at $z = -1$ or $\omega = \pi$. This is clearly shown in Figures 7.44(b) and (c).

---

[8]For frequency-selective filters, the unweighted approximation error also conveniently displays the passband and stopband behavior, since $A_e(e^{j\omega}) = 1 - E(\omega)$ in the passband and $A_e(e^{j\omega}) = -E(\omega)$ in the stopband.

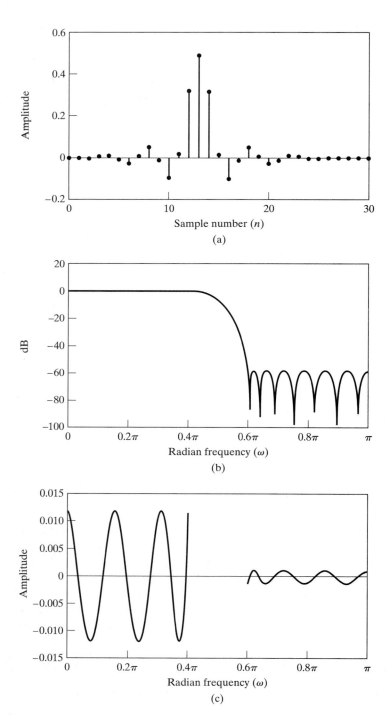

Figure 7.43 Optimum type I FIR lowpass filter for $\omega_p = 0.4\pi$, $\omega_s = 0.6\pi$, $K = 10$, and $M = 26$. (a) Impulse response. (b) Log magnitude of the frequency response. (c) Approximation error (unweighted).

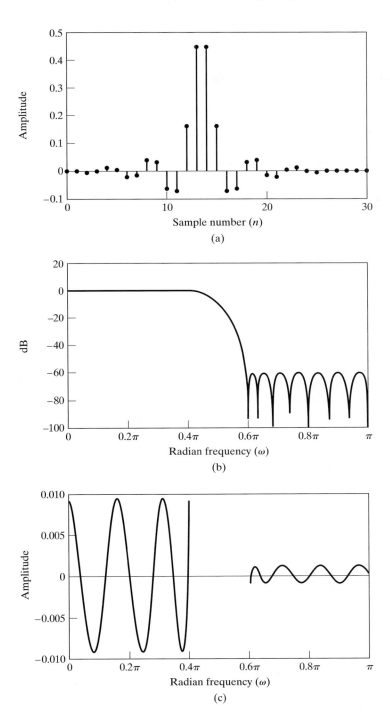

Figure 7.44 Optimum type II FIR lowpass filter for $\omega_p = 0.4\pi$, $\omega_s = 0.6\pi$, $K = 10$, and $M = 27$. (a) Impulse response. (b) Log magnitude of frequency response. (c) Approximation error (unweighted).

If we compare the results of this example with the results of Example 7.8, we find that the Kaiser window method requires a value $M = 38$ to meet or exceed the specifications, while the Parks–McClellan method requires $M = 27$. This disparity is accentuated because the window method produces approximately equal maximum errors in the passband and stopband, while the Parks–McClellan method can weight the errors differently.

7.5.2 Compensation for Zero-Order Hold

In many cases, a discrete-time filter is designed to be used in a system such as that depicted in Figure 7.45; i.e., the filter is used to process a sequence of samples $x[n]$ to obtain a sequence $y[n]$, which is then the input to a D/A converter and continuous-time lowpass filter (as an approximation to the ideal D/C converter) used for the reconstruction of a continuous-time signal $y_c(t)$. Such a system arises as part of a system for discrete-time filtering of a continuous-time signal, as discussed in Section 4.8. If the D/A converter holds its output constant for the entire sampling period T, the Fourier transform of the output $y_c(t)$ is

$$Y_c(j\Omega) = \tilde{H}_r(j\Omega)H_o(j\Omega)H(e^{j\Omega T})X(e^{j\Omega T}), \tag{7.106}$$

where $H_r(j\Omega)$ is the frequency response of an appropriate lowpass reconstruction filter and

$$H_o(j\Omega) = \frac{\sin(\Omega T/2)}{\Omega/2}e^{-j\Omega T/2} \tag{7.107}$$

is the frequency response of the zero-order hold of the D/A converter. (Note that $H_r(j\Omega)$ can have unity gain because $H_o(j\Omega)$ has a gain of T at $\Omega = 0$.) In Section 4.8.4, we suggested that compensation for $H_o(j\Omega)$ could be incorporated into the continuous-time reconstruction filter; i.e., $\tilde{H}_r(j\Omega)$ could be chosen as

$$\tilde{H}_r(j\Omega) = \frac{\Omega T/2}{\sin(\Omega T/2)}H_r(j\Omega) \tag{7.108}$$

where $H_r(j\Omega)$ is the ideal lowpass reconstruction filter of Eq. (4.128) so that the effect of the discrete-time filter $H(e^{j\Omega T})$ would be undistorted by the zero-order hold. Another approach is to build the compensation into the discrete-time filter by designing a filter $\tilde{H}(e^{j\Omega T})$ such that

$$\tilde{H}(e^{j\Omega T}) = \frac{\Omega T/2}{\sin(\Omega T/2)}H(e^{j\Omega T}). \tag{7.109}$$

A D/A-compensated lowpass filter can be readily designed by the Parks–McClellan algorithm if we simply define the desired response as

$$\tilde{H}_d(e^{j\omega}) = \begin{cases} \dfrac{\omega/2}{\sin(\omega/2)}, & 0 \le \omega \le \omega_p, \\ 0, & \omega_s \le \omega \le \pi. \end{cases} \tag{7.110}$$

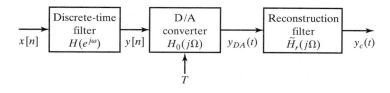

Figure 7.45 Precompensation of a discrete-time filter for the effects of a D/A converter.

Figure 7.46 shows the response functions for such a filter, where the specifications are again $\omega_p = 0.4\pi$, $\omega_s = 0.6\pi$, $\delta_1 = 0.01$, and $\delta_2 = 0.001$. In this case, the specifications are met with $M = 28$ rather than $M = 27$ as in the previous constant-gain case. Thus, for essentially no penalty, we have incorporated compensation for the D/A converter into the discrete-time filter so that the effective passband of the filter will be flat. (To emphasize the sloping nature of the passband, Figure 7.46(c) shows the magnitude response in the passband, rather than the approximation error, as in the frequency response plots for the other FIR examples.)

7.5.3 Bandpass Filter

Section 7.4 focused entirely on the lowpass optimal FIR, for which there are only two approximation bands. However, bandpass and bandstop filters require three approximation bands. To design such filters, it is necessary to generalize the discussion of Section 7.4 to the multiband case. This requires that we explore the implications of the alternation theorem and the properties of the approximating polynomial in the more general context. First, recall that, as stated, the alternation theorem does not assume any limit on the number of disjoint approximation intervals. Therefore, the *minimum* number of alternations for the optimum approximation is still $(L+2)$. However, multiband filters can have more than $(L+3)$ alternations, because there are more band edges. (Problem 7.57 explores this issue.) This means that some of the statements proved in Section 7.4.1 are not true in the multiband case. For example, it is *not* necessary for all the local maxima or minima of $A_e(e^{j\omega})$ to lie inside the approximation intervals. Thus, local extrema can occur in the transition regions, and the approximation need not be equiripple in the approximation regions.

To illustrate this, consider the desired response

$$H_d(e^{j\omega}) = \begin{cases} 0, & 0 \le \omega \le 0.3\pi, \\ 1, & 0.35\pi \le \omega \le 0.6\pi, \\ 0, & 0.7\pi \le \omega \le \pi, \end{cases} \tag{7.111}$$

and the error weighting function

$$W(\omega) = \begin{cases} 1, & 0 \le \omega \le 0.3\pi, \\ 1, & 0.35\pi \le \omega \le 0.6\pi, \\ 0.2, & 0.7\pi \le \omega \le \pi. \end{cases} \tag{7.112}$$

A value of $M + 1 = 75$ was chosen for the length of the impulse response of the filter. Figure 7.47 shows the response functions for the resulting filter. Note that the transition region from the second approximation band to the third is no longer monotonic. However, the use of two local extrema in this unconstrained region does not violate the alternation theorem. Since $M = 74$, the filter is a type I system, and the order of the implicit approximating polynomial is $L = M/2 = 74/2 = 37$. Thus, the alternation theorem requires at least $L+2 = 39$ alternations. It can be readily seen in Figure 7.47(c), which shows the unweighted approximation error, that there are 13 alternations in each band, for a total of 39.

Such approximations as shown in Figure 7.47 are optimal in the sense of the alternation theorem, but they would probably be unacceptable in a filtering application. In general, there is no guarantee that the transition regions of a multiband filter will be

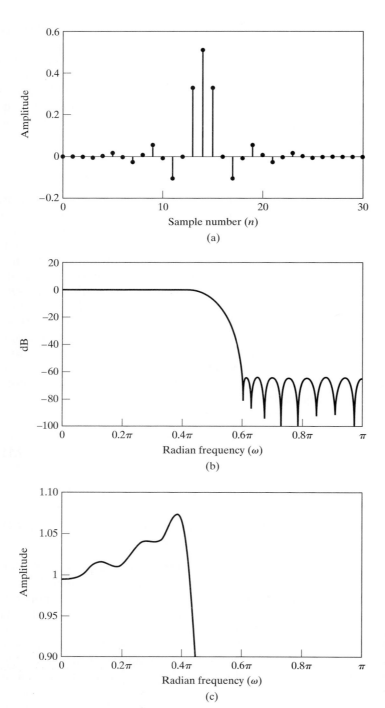

Figure 7.46 Optimum D/A-compensated lowpass filter for $\omega_p = 0.4\pi$, $\omega_s = 0.6\pi$, $K = 10$, and $M = 28$. (a) Impulse response. (b) Log magnitude of the frequency response. (c) Magnitude response in passband.

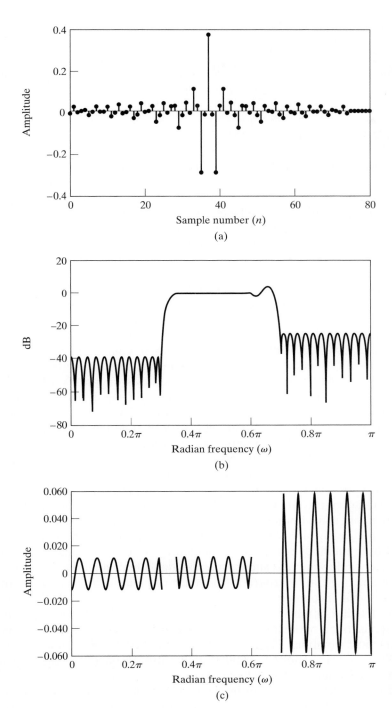

Figure 7.47   Optimum FIR bandpass filter for $M = 74$. (a) Impulse response. (b) Log magnitude of the frequency response. (c) Approximation error (unweighted).

monotonic, because the Parks–McClellan algorithm leaves these regions completely unconstrained. When this kind of response results for a particular choice of the filter parameters, acceptable transition regions can usually be obtained by systematically changing one or more of the band edge frequencies, the impulse-response length, or the error-weighting function and redesigning the filter.

7.6 COMMENTS ON IIR AND FIR DISCRETE-TIME FILTERS

This chapter has been concerned with design methods for linear time-invariant discrete-time systems. We have discussed a wide range of methods of designing both infinite-duration and finite-duration impulse-response filters. Questions that naturally arise are: What type of system is best, IIR or FIR? Why give so many different design methods? and Which method yields the best results? As in any engineering design problem, it is generally not possible to give a precise answer as to what is best. We have discussed a wide variety of methods for both IIR and FIR filters because no single type of filter and no single design method is best for all circumstances.

The choice between an FIR filter and an IIR filter depends on the importance to the design problem of the advantages of each type. IIR filters, for example, have the advantage that a variety of frequency-selective filters can be designed using closed-form design formulas. That is, once the problem has been specified in terms appropriate for a given approximation method (e.g., Butterworth, Chebyshev, or elliptic), then the order of the filter that will meet the specifications can be computed, and the coefficients (or poles and zeros) of the discrete-time filter can be obtained by straightforward substitution into a set of design equations. This kind of simplicity of the design procedure makes it feasible to design IIR filters by manual computation if necessary, and it leads to straightforward noniterative computer programs for IIR filter design. These methods are limited to frequency-selective filters, and they permit only the magnitude response to be specified. If other magnitude shapes are desired, or if it is necessary to approximate a prescribed phase- or group-delay response, an algorithmic procedure will be required.

In contrast, FIR filters can have a precisely (generalized) linear phase. However, closed-form design equations do not exist for FIR filters. Although the window method is straightforward to apply, some iteration may be necessary to meet a prescribed specification. The Parks–McClellan algorithm leads to lower order filters than the window method, does, and both methods can be implemented on a personal computer or a workstation. Also, the window method and most of the algorithmic methods afford the possibility of approximating rather arbitrary frequency-response characteristics with little more difficulty than is encountered in the design of lowpass filters. In addition, the design problem for FIR filters is much more under control than the IIR design problem, because of the existence of an optimality theorem for FIR filters that is meaningful in a wide range of practical situations. Design techniques for FIR filters without linear phase have been given by Chen and Parks (1987), Parks and Burrus (1987), Schüssler and Steffen (1988), and Karam and McClellan (1995).

Questions of economics also arise in implementing a discrete-time filter. Economic concerns are usually measured in terms of hardware complexity, chip area, or computational speed. These factors are more or less directly related to the order of the filter required to meet a given specification. If we put aside phase considerations, it is generally true that a given magnitude-response specification can be met most efficiently with an IIR filter. However, in many cases, the linear phase available with an FIR filter may be well worth the extra cost, and in applications such as decimation or interpolation, FIR filters may be just as efficient as IIR filters. Furthermore, special purpose DSP micro-computers generally have arithmetic capabilities which are designed for accumulating sums of products as required for FIR filters. A detailed consideration of questions of computational efficiency is given in Rabiner, Kaiser, Herrmann, and Dolan (1974).

Thus, a multitude of trade-offs must be considered in designing a discrete-time filter. Clearly, the final choice will most often be made by engineering judgment on such questions as the formulation of the specifications, the method of implementation of the filter, and the computational facilities and software available for carrying out the design.

7.7 SUMMARY

In this chapter, we have considered a variety of design techniques for both infinite-duration and finite-duration impulse-response discrete-time filters. Our emphasis was on the frequency-domain specification of the desired filter characteristics, since this is most common in practice. Our objective was to give a general picture of the wide range of possibilities available for discrete-time filter design, while also giving sufficient detail about some of the techniques so that they may be applied directly, without further reference to the extensive literature on discrete-time filter design. In the FIR case, considerable detail was presented on both the window method and the Parks–McClellan algorithmic method of filter design.

The chapter concluded with some remarks on the choice between the two classes of digital filters. The main point of that discussion was that the choice is not always clear cut and may depend on a multitude of factors that are often difficult to quantify or discuss in general terms. However, it should be clear from this chapter and Chapter 6 that digital filters are characterized by great flexibility in design and implementation. This flexibility makes it possible to implement rather sophisticated signal-processing schemes that in many cases would be difficult, if not impossible, to implement by analog means.

PROBLEMS

Basic Problems with Answers

7.1. Consider a causal continuous-time system with impulse response $h_c(t)$ and system function

$$H_c(s) = \frac{s + a}{(s + a)^2 + b^2}.$$

(a) Use impulse invariance to determine $H_1(z)$ for a discrete-time system such that $h_1[n] = h_c(nT)$.

(b) Use step invariance to determine $H_2(z)$ for a discrete-time system such that $s_2[n] = s_c(nT)$, where

$$s_2[n] = \sum_{k=-\infty}^{n} h_2[k] \quad \text{and} \quad s_c(t) = \int_{-\infty}^{t} h_c(\tau)d\tau.$$

(c) Determine the step response $s_1[n]$ of system 1 and the impulse response $h_2[n]$ of system 2. Is it true that $h_2[n] = h_1[n] = h_c(nT)$? Is it true that $s_1[n] = s_2[n] = s_c(nT)$?

7.2. A discrete-time lowpass filter is to be designed by applying the impulse invariance method to a continuous-time Butterworth filter having magnitude-squared function

$$|H_c(j\Omega)|^2 = \frac{1}{1 + (\Omega/\Omega_c)^{2N}}.$$

The specifications for the discrete-time system are those of Example 7.2, i.e.,

$$0.89125 \le |H(e^{j\omega})| \le 1, \quad 0 \le |\omega| \le 0.2\pi,$$

$$|H(e^{j\omega})| \le 0.17783, \qquad 0.3\pi \le |\omega| \le \pi.$$

Assume, as in that example, that aliasing will not be a problem; i.e., design the continuous-time Butterworth filter to meet passband and stopband specifications as determined by the desired discrete-time filter.

(a) Sketch the tolerance bounds on the magnitude of the frequency response, $|H_c(j\Omega)|$, of the continuous-time Butterworth filter such that after application of the impulse invariance method (i.e., $h[n] = T_d h_c(nT_d)$), the resulting discrete-time filter will satisfy the given design specifications. Do not assume that $T_d = 1$ as in Example 7.2.

(b) Determine the integer order N and the quantity $T_d\Omega_c$ such that the continuous-time Butterworth filter exactly meets the specifications determined in Part (a) at the passband edge.

(c) Note that if $T_d = 1$, your answer in Part (b) should give the values of N and Ω_c obtained in Example 7.2. Use this observation to determine the system function $H_c(s)$ for $T_d \ne 1$ and to argue that the system function $H(z)$ which results from impulse invariance design with $T_d \ne 1$ is the same as the result for $T_d = 1$ given by Eq. (7.19).

7.3. We wish to use impulse invariance or the bilinear transformation to design a discrete-time filter that meets specifications of the following form:

$$1 - \delta_1 \le |H(e^{j\omega})| \le 1 + \delta_1, \quad 0 \le |\omega| \le \omega_p,$$

$$|H(e^{j\omega})| \le \delta_2, \qquad \omega_s \le |\omega| \le \pi. \tag{P7.3-1}$$

For historical reasons, most of the design formulas, tables, or charts for continuous-time filters are normally specified with a peak gain of unity in the passband; i.e.,

$$1 - \hat{\delta}_1 \le |H_c(j\Omega)| \le 1, \quad 0 \le |\Omega| \le \Omega_p,$$

$$|H_c(\Omega)| \le \hat{\delta}_2, \quad \Omega_s \le |\Omega|. \tag{P7.3-2}$$

Useful design charts for continuous-time filters specified in this form were given by Rabiner, Kaiser, Herrmann, and Dolan (1974).

(a) To use such tables and charts to design discrete-time systems with a peak gain of $(1+\delta_1)$, it is necessary to convert the discrete-time specifications into specifications of the form of Eq. (P7.3-2). This can be done by dividing the discrete-time specifications by $(1+\delta_1)$. Use this approach to obtain an expression for $\hat{\delta}_1$ and $\hat{\delta}_2$ in terms of δ_1 and δ_2.

(b) In Example 7.2, we designed a discrete-time filter with a maximum passband gain of unity. This filter can be converted to a filter satisfying a set of specifications such as those in Eq. (P7.3-1) by multiplying by a constant of the form $(1 + \delta_1)$. Find the required value of δ_1 and the corresponding value of δ_2 for this example, and use Eq. (7.19) to determine the coefficients of the system function of the new filter.

(c) Repeat Part (b) for the filter in Example 7.3.

7.4. The system function of a discrete-time system is

$$H(z) = \frac{2}{1 - e^{-0.2}z^{-1}} - \frac{1}{1 - e^{-0.4}z^{-1}}.$$

(a) Assume that this discrete-time filter was designed by the impulse invariance method with $T_d = 2$; i.e., $h[n] = 2h_c(2n)$, where $h_c(t)$ is real. Find the system function $H_c(s)$ of a continuous-time filter that could have been the basis for the design. Is your answer unique? If not, find another system function $H_c(s)$.

(b) Assume that $H(z)$ was obtained by the bilinear transform method with $T_d = 2$. Find the system function $H_c(s)$ that could have been the basis for the design. Is your answer unique? If not, find another $H_c(s)$.

7.5. We wish to use the Kaiser window method to design a discrete-time filter with generalized linear phase that meets specifications of the following form:

$$|H(e^{j\omega})| \leq 0.01, \qquad 0 \leq |\omega| \leq 0.25\pi,$$
$$0.95 \leq |H(e^{j\omega})| \leq 1.05, \qquad 0.35\pi \leq |\omega| \leq 0.6\pi,$$
$$|H(e^{j\omega})| \leq 0.01, \qquad 0.65\pi \leq |\omega| \leq \pi.$$

(a) Determine the minimum length $(M + 1)$ of the impulse response and the value of the Kaiser window parameter β for a filter that meets the preceding specifications.

(b) What is the delay of the filter?

(c) Determine the ideal impulse response $h_d[n]$ to which the Kaiser window should be applied.

7.6. We wish to use the Kaiser window method to design a real-valued FIR filter with generalized linear phase that meets the following specifications:

$$0.9 < H(e^{j\omega}) < 1.1, \qquad 0 \leq |\omega| \leq 0.2\pi,$$
$$-0.06 < H(e^{j\omega}) < 0.06, \qquad 0.3\pi \leq |\omega| \leq 0.475\pi,$$
$$1.9 < H(e^{j\omega}) < 2.1, \qquad 0.525\pi \leq |\omega| \leq \pi.$$

This specification is to be met by applying the Kaiser window to the ideal real-valued impulse response associated with the ideal frequency response $H_d(e^{j\omega})$ given by

$$H_d(e^{j\omega}) = \begin{cases} 1, & 0 \leq |\omega| \leq 0.25\pi, \\ 0, & 0.25\pi \leq |\omega| \leq 0.5\pi, \\ 2, & 0.5\pi \leq |\omega| \leq \pi. \end{cases}$$

(a) What is the maximum value of δ that can be used to meet this specification? What is the corresponding value of β? Clearly explain your reasoning.

(b) What is the maximum value of $\Delta\omega$ that can be used to meet the specification. What is the corresponding value of M? Clearly explain your reasoning.

7.7. We are interested in implementing a continuous-time LTI lowpass filter $H(j\Omega)$ using the system shown in Fig. 4.11 when the discrete-time system has frequency response $H_d(e^{j\omega})$. The sampling time $T = 10^{-4}$ second and the input signal $x_c(t)$ is appropriately bandlimited with $X_c(j\Omega) = 0$ for $|\Omega| \geq 2\pi(5000)$.

Let the specifications on $|H(j\Omega)|$ be

$$0.99 \leq |H_e(j\Omega)| \leq 1.01, \qquad |\Omega| \leq 2\pi(1000),$$
$$|H_e(j\Omega)| \leq 0.01, \qquad |\Omega| \geq 2\pi(1100).$$

Determine the corresponding specifications on the discrete-time frequency response $H_d(e^{j\omega})$.

7.8. We wish to design an optimal (Parks–McClellan) zero-phase Type I FIR lowpass filter with passband frequency $\omega_p = 0.3\pi$ and stopband frequency $\omega_s = 0.6\pi$ with equal error weighting in the passband and stopband. The impulse response of the desired filter has length 11; i.e., $h[n] = 0$ for $n < -5$ or $n > 5$. Figure P7.8-1 shows the frequency response

$H(e^{j\omega})$ for two different filters. For each filter, specify how many alternations the filter has, and state whether it satisfies the alternation theorem as the optimal filter in the minimax sense meeting the preceding specifications.

Figure P7.8-1

7.9. Suppose we design a discrete-time filter using the impulse invariance technique with an ideal continuous-time lowpass filter as a prototype. The prototype filter has a cutoff frequency of $\Omega_c = 2\pi(1000)$ rad/s, and the impulse invariance transformation uses $T = 0.2$ ms. What is the cutoff frequency ω_c for the resulting discrete-time filter?

7.10. We wish to design a discrete-time lowpass filter using the bilinear transformation on a continuous-time ideal lowpass filter. Assume that the continuous-time prototype filter has cutoff frequency $\Omega_c = 2\pi(2000)$ rad/s and we choose the bilinear transformation parameter $T = 0.4$ ms. What is the cutoff frequency ω_c for the resulting discrete-time filter?

7.11. Suppose that we have an ideal discrete-time lowpass filter with cutoff frequency $\omega_c = \pi/4$. In addition, we are told that this filter resulted from applying impulse invariance to a continuous-time prototype lowpass filter using $T = 0.1$ ms. What was the cutoff frequency Ω_c for the prototype continuous-time filter?

7.12. An ideal discrete-time highpass filter with cutoff frequency $\omega_c = \pi/2$ was designed using the bilinear transformation with $T = 1$ ms. What was the cutoff frequency Ω_c for the prototype continuous-time ideal highpass filter?

7.13. An ideal discrete-time lowpass filter with cutoff frequency $\omega_c = 2\pi/5$ was designed using impulse invariance from an ideal continuous-time lowpass filter with cutoff frequency $\Omega_c = 2\pi(4000)$ rad/s. What was the value of T? Is this value unique? If not, find another value of T consistent with the information given.

7.14. The bilinear transformation is used to design an ideal discrete-time lowpass filter with cutoff frequency $\omega_c = 3\pi/5$ from an ideal continuous-time lowpass filter with cutoff frequency $\Omega_c = 2\pi(300)$ rad/s. Give a choice for the parameter T that is consistent with this information. Is this choice unique? If not, give another choice which is consistent with the information.

7.15. We wish to design an FIR lowpass filter satisfying the specifications

$$0.95 < H(e^{j\omega}) < 1.05, \qquad 0 \le |\omega| \le 0.25\pi,$$

$$-0.1 < H(e^{j\omega}) < 0.1, \qquad 0.35\pi \le |\omega| \le \pi,$$

by applying a window $w[n]$ to the impulse response $h_d[n]$ for the ideal discrete-time lowpass filter with cutoff $\omega_c = 0.3\pi$. Which of the filters listed in Section 7.2.1 can be used to meet this specification? For each window that you claim will satisfy this specification, give the minimum length $M + 1$ required for the filter.

7.16. We wish to design an FIR lowpass filter satisfying the specifications

$$0.98 < H(e^{j\omega}) < 1.02, \qquad 0 \le |\omega| \le 0.63\pi,$$

$$-0.15 < H(e^{j\omega}) < 0.15, \qquad 0.65\pi \le |\omega| \le \pi,$$

by applying a Kaiser window to the impulse response $h_d[n]$ for the ideal discrete-time lowpass filter with cutoff $\omega_c = 0.64\pi$. Find the values of β and M required to satisfy this specification.

7.17. Suppose that we wish to design a bandpass filter satisfying the following specification:

$$-0.02 < |H(e^{j\omega})| < 0.02, \qquad 0 \le |\omega| \le 0.2\pi,$$

$$0.95 < |H(e^{j\omega})| < 1.05, \qquad 0.3\pi \le |\omega| \le 0.7\pi,$$

$$-0.001 < |H(e^{j\omega})| < 0.001, \qquad 0.75\pi \le |\omega| \le \pi.$$

The filter will be designed by applying impulse invariance with $T = 5$ ms to a prototype continuous-time filter. State the specifications that should be used to design the prototype continuous-time filter.

7.18. Suppose that we wish to design a highpass filter satisfying the following specification:

$$-0.04 < |H(e^{j\omega})| < 0.04, \qquad 0 \le |\omega| \le 0.2\pi,$$

$$0.995 < |H(e^{j\omega})| < 1.005, \qquad 0.3\pi \le |\omega| \le \pi.$$

The filter will be designed using the bilinear transformation and $T = 2$ ms with a prototype continuous-time filter. State the specifications that should be used to design the prototype continuous-time filter to ensure that the specifications for the discrete-time filter are met.

7.19. We wish to design a discrete-time ideal bandpass filter that has a passband $\pi/4 \le \omega \le \pi/2$ by applying impulse invariance to an ideal continuous-time bandpass filter with passband $2\pi(300) \le \Omega \le 2\pi(600)$. Specify a choice for T that will produce the desired filter. Is your choice of T unique?

7.20. Specify whether the following statement is true or false. Justify your answer.

Statement: If the bilinear transformation is used to transform a continuous-time all-pass system to a discrete-time system, the resulting discrete-time system will also be an all-pass system.

Basic Problems

7.21. Suppose that we are given a continuous-time lowpass filter with frequency response $H_c(j\omega)$ such that

$$1 - \delta_1 \leq |H_c(j\Omega)| \leq 1 + \delta_1, \qquad |\Omega| \leq \Omega_p,$$

$$|H_c(j\Omega)| \leq \delta_2, \qquad |\Omega| \geq \Omega_s.$$

A set of discrete-time lowpass filters can be obtained from $H_c(s)$ by using the bilinear transformation, i.e.,

$$H(z) = H_c(s)\big|_{s=(2/T_d)[(1-z^{-1})/(1+z^{-1})]},$$

with T_d variable.

(a) Assuming that Ω_p is fixed, find the value of T_d such that the corresponding passband cutoff frequency for the discrete-time system is $\omega_p = \pi/2$.

(b) With Ω_p fixed, sketch ω_p as a function of $0 < T_d < \infty$.

(c) With both Ω_p and Ω_s fixed, sketch the transition region $\Delta\omega = (\omega_s - \omega_p)$ as a function of $0 < T_d < \infty$.

7.22. Consider a continuous-time system with system function

$$H_c(s) = \frac{1}{s}.$$

This system is called an *integrator*, since the output $y_c(t)$ is related to the input $x_c(t)$ by

$$y_c(t) = \int_{-\infty}^{t} x_c(\tau)d\tau.$$

Suppose a discrete-time system is obtained by applying the bilinear transformation to $H_c(s)$.

(a) What is the system function $H(z)$ of the resulting discrete-time system? What is the impulse response $h[n]$?

(b) If $x[n]$ is the input and $y[n]$ is the output of the resulting discrete-time system, write the difference equation that is satisfied by the input and output. What problems do you anticipate in implementing the discrete-time system using this difference equation?

(c) Obtain an expression for the frequency response $H(e^{j\omega})$ of the system. Sketch the magnitude and phase of the discrete-time system for $0 \leq |\omega| \leq \pi$. Compare them with the magnitude and phase of the frequency response $H_c(j\Omega)$ of the continuous-time integrator. Under what conditions could the discrete-time "integrator" be considered a good approximation to the continuous-time integrator?

Now consider a continuous-time system with system function

$$G_c(s) = s.$$

This system is called a *differentiator*, since the output is the derivative of the input. Suppose a discrete-time system is obtained by applying the bilinear transformation to $G_c(s)$.

(d) What is the system function $G(z)$ of the resulting discrete-time system? What is the impulse response $g[n]$?

(e) Obtain an expression for the frequency response $G(e^{j\omega})$ of the system. Sketch the magnitude and phase of the discrete-time system for $0 \leq |\omega| \leq \pi$. Compare them with the magnitude and phase of the frequency response $G_c(j\Omega)$ of the continuous-time differentiator. Under what conditions could the discrete-time "differentiator" be considered a good approximation to the continuous-time differentiator?

(f) The continuous-time integrator and differentiator are exact inverses of one another. Is the same true of their discrete-time approximations?

7.23. A continuous-time filter with impulse response $h_c(t)$ and frequency-response magnitude

$$|H_c(j\Omega)| = \begin{cases} |\Omega|, & |\Omega| < 10\pi, \\ 0, & |\Omega| > 10\pi, \end{cases}$$

is to be used as the prototype for the design of a discrete-time filter. The resulting discrete-time system is to be used in the configuration of Figure P7.23-1 to filter the continuous-time signal $x_c(t)$.

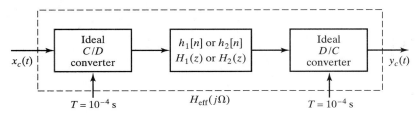

Figure P7.23-1

(a) A discrete-time system with impulse response $h_1[n]$ and system function $H_1(z)$ is obtained from the prototype continuous-time system by impulse invariance with $T_d = 0.01$; i.e., $h_1[n] = 0.01h_c(0.01n)$. Plot the magnitude of the overall effective frequency response $H_{\text{eff}}(j\Omega) = Y_c(j\Omega)/X_c(j\Omega)$ when this discrete-time system is used in Figure P7.23-1.

(b) Alternatively, suppose that a discrete-time system with impulse response $h_2[n]$ and system function $H_2(z)$ is obtained from the prototype continuous-time system by the bilinear transformation with $T_d = 2$; i.e.,

$$H_2(z) = H_c(s)\big|_{s=(1-z^{-1})/(1+z^{-1})}.$$

Plot the magnitude of the overall effective frequency response $H_{\text{eff}}(j\Omega)$ when this discrete-time system is used in Figure P7.23-1.

Advanced Problems

7.24. Assume that $H_c(s)$ has an rth-order pole at $s = s_0$, so that $H_c(s)$ can be expressed as

$$H_c(s) = \sum_{k=1}^{r} \frac{A_k}{(s - s_0)^k} + G_c(s),$$

where $G_c(s)$ has only first-order poles. Assume $H_c(s)$ is causal.

(a) Give a formula for determining the constants A_k from $H_c(s)$.

(b) Obtain an expression for the impulse response $h_c(t)$ in terms of s_0 and $g_c(t)$, the inverse Laplace transform of $G_c(s)$.

7.25. *Impulse invariance* and the *bilinear transformation* are two methods for designing discrete-time filters. Both methods transform a continuous-time system function $H_c(s)$ into a discrete-time system function $H(z)$. Answer the following questions by indicating which method(s) will yield the desired result:

(a) A minimum-phase continuous-time system has all its poles and zeros in the left-half s-plane. If a minimum-phase continuous-time system is transformed into a discrete-time system, which method(s) will result in a minimum-phase discrete-time system?

(b) If the continuous-time system is an all-pass system, its poles will be at locations s_k in the left-half s-plane, and its zeros will be at corresponding locations $-s_k$ in the right-half s-plane. Which design method(s) will result in an all-pass discrete-time system?

(c) Which design method(s) will guarantee that

$$H(e^{j\omega})\big|_{\omega=0} = H_c(j\Omega)\big|_{\Omega=0}?$$

(d) If the continuous-time system is a bandstop filter, which method(s) will result in a discrete-time bandstop filter?

(e) Suppose that $H_1(z)$, $H_2(z)$, and $H(z)$ are transformed versions of $H_{c1}(s)$, $H_{c2}(s)$, and $H_c(s)$, respectively. Which design method(s) will guarantee that $H(z) = H_1(z)H_2(z)$ whenever $H_c(s) = H_{c1}(s)H_{c2}(s)$?

(f) Suppose that $H_1(z)$, $H_2(z)$, and $H(z)$ are transformed versions of $H_{c1}(s)$, $H_{c2}(s)$, and $H_c(s)$, respectively. Which design method(s) will guarantee that $H(z) = H_1(z) + H_2(z)$ whenever $H_c(s) = H_{c1}(s) + H_{c2}(s)$?

(g) Assume that two continuous-time system functions satisfy the condition

$$\frac{H_{c1}(j\Omega)}{H_{c2}(j\Omega)} = \begin{cases} e^{-j\pi/2}, & \Omega > 0, \\ e^{j\pi/2}, & \Omega < 0. \end{cases}$$

If $H_1(z)$ and $H_2(z)$ are transformed versions of $H_{c1}(s)$ and $H_{c2}(s)$, respectively, which design method(s) will result in discrete-time systems such that

$$\frac{H_1(e^{j\omega})}{H_2(e^{j\omega})} = \begin{cases} e^{-j\pi/2}, & 0 < \omega < \pi, \\ e^{j\pi/2}, & -\pi < \omega < 0? \end{cases}$$

(Such systems are called "90-degree phase splitters.")

7.26. A discrete-time filter with system function $H(z)$ is designed by transforming a continuous-time filter with system function $H_c(s)$. It is desired that

$$H(e^{j\omega})\big|_{\omega=0} = H_c(j\Omega)\big|_{\Omega=0}.$$

(a) Could this condition hold for a filter designed by impulse invariance? If so, what condition(s), if any, would $H_c(j\Omega)$ have to satisfy?

(b) Could the condition hold for a filter designed using the bilinear transformation? If so, what conditions, if any, would $H_c(j\Omega)$ have to satisfy?

7.27. Suppose that we are given an ideal lowpass discrete-time filter with frequency response

$$H(e^{j\omega}) = \begin{cases} 1, & |\omega| < \pi/4, \\ 0, & \pi/4 < |\omega| \le \pi. \end{cases}$$

We wish to derive new filters from this prototype by manipulations of the impulse response $h[n]$.

(a) Plot the frequency response $H_1(e^{j\omega})$ for the system whose impulse response is $h_1[n] = h[2n]$.

(b) Plot the frequency response $H_2(e^{j\omega})$ for the system whose impulse response is

$$h_2[n] = \begin{cases} h[n/2], & n = 0, \pm 2, \pm 4, \dots, \\ 0, & \text{otherwise.} \end{cases}$$

(c) Plot the frequency response $H_3(e^{j\omega})$ for the system whose impulse response is $h_3[n] = e^{j\pi n}h[n] = (-1)^n h[n]$.

7.28. Consider a continuous-time lowpass filter $H_c(s)$ with passband and stopband specifications

$$1 - \delta_1 \le |H_c(j\Omega)| \le 1 + \delta_1, \qquad |\Omega| \le \Omega_p,$$
$$|H_c(j\Omega)| \le \delta_2, \qquad \Omega_s \le |\Omega|.$$

This filter is transformed to a lowpass discrete-time filter $H_1(z)$ by the transformation

$$H_1(z) = H_c(s)\big|_{s=(1-z^{-1})/(1+z^{-1})},$$

and the same continuous-time filter is transformed to a highpass discrete-time filter by the transformation

$$H_2(z) = H_c(s)\big|_{s=(1+z^{-1})/(1-z^{-1})}.$$

(a) Determine a relationship between the passband cutoff frequency Ω_p of the continuous-time lowpass filter and the passband cutoff frequency ω_{p1} of the discrete-time lowpass filter.

(b) Determine a relationship between the passband cutoff frequency Ω_p of the continuous-time lowpass filter and the passband cutoff frequency ω_{p2} of the discrete-time highpass filter.

(c) Determine a relationship between the passband cutoff frequency ω_{p1} of the discrete-time lowpass filter and the passband cutoff frequency ω_{p2} of the discrete-time highpass filter.

(d) The network in Figure P7.28-1 depicts an implementation of the discrete-time lowpass filter with system function $H_1(z)$. The coefficients A, B, C, and D are real. How should these coefficients be modified to obtain a network that implements the discrete-time highpass filter with system function $H_2(z)$?

Figure P7.28-1

7.29. A discrete-time system with system function $H(Z)$ and impulse response $h[n]$ has frequency response

$$H(e^{j\theta}) = \begin{cases} A, & |\theta| < \theta_c, \\ 0, & \theta_c < |\theta| \le \pi, \end{cases}$$

where $0 < \theta_c < \pi$. This filter is transformed into a new filter by the transformation $Z = -z^2$; i.e.,

$$H_1(z) = H(Z)\big|_{Z=-z^2} = H(-z^2).$$

(a) Obtain a relationship between the frequency variable θ for the original lowpass system $H(Z)$ and the frequency variable ω for the new system $H_1(z)$.

(b) Sketch and carefully label the frequency response $H_1(e^{j\omega})$ for the new filter.

(c) Obtain a relationship expressing $h_1[n]$ in terms of $h[n]$.

(d) Assume that $H(Z)$ can be realized by the set of difference equations

$$g[n] = x[n] - a_1 g[n-1] - b_1 f[n-2],$$

$$f[n] = a_2 g[n-1] + b_2 f[n-1],$$

$$y[n] = c_1 f[n] - c_2 g[n-1],$$

where $x[n]$ is the input and $y[n]$ is the output of the system. Determine a set of difference equations that will realize the transformed system $H_1(z) = H(-z^2)$.

7.30. Consider designing a discrete-time filter with system function $H(z)$ from a continuous-time filter with rational system function $H_c(s)$ by the transformation

$$H(z) = H_c(s)\big|_{s=\beta[(1-z^{-\alpha})/(1+z^{-\alpha})]},$$

where α is a nonzero integer and β is real.

(a) If $\alpha > 0$, for what values of β does a stable, causal continuous-time filter with rational $H_c(s)$ always lead to a stable, causal discrete-time filter with rational $H(z)$?

(b) If $\alpha < 0$, for what values of β does a stable, causal continuous-time filter with rational $H_c(s)$ always lead to a stable, causal discrete-time filter with rational $H(z)$?

(c) For $\alpha = 2$ and $\beta = 1$, determine to what contour in the z-plane the $j\Omega$-axis of the s-plane maps.

(d) Suppose that the continuous-time filter is a stable lowpass filter with passband frequency response such that

$$1 - \delta_1 \le |H_c(j\Omega)| \le 1 + \delta_1 \qquad \text{for} \quad |\Omega| \le 1.$$

If the discrete-time system $H(z)$ is obtained by the transformation set forth at the beginning of this problem, with $\alpha = 2$ and $\beta = 1$, determine the values of ω in the interval $|\omega| \le \pi$ for which

$$1 - \delta_1 \le |H(e^{j\omega})| \le 1 + \delta_1.$$

7.31. A discrete-time highpass filter can be obtained from a continuous-time lowpass filter by the following transformation:

$$H(z) = H_c(s)\big|_{s=[(1+z^{-1})/(1-z^{-1})]}.$$

(a) Show that this transformation maps the $j\Omega$-axis of the s-plane onto the unit circle of the z-plane.

(b) Show that if $H_c(s)$ is a rational function with all its poles inside the left-half s-plane, then $H(z)$ will be a rational function with all its poles inside the unit circle of the z-plane.

(c) Suppose a desired highpass discrete-time filter has specifications

$$|H(e^{j\omega})| \le 0.01, \quad |\omega| \le \pi/3,$$

$$0.95 \le |H(e^{j\omega})| \le 1.05, \quad \pi/2 \le |\omega| \le \pi.$$

Determine the specifications on the continuous-time lowpass filter so that the desired highpass discrete-time filter results from the transformation specified at the beginning of this problem.

7.32. Let $h_d[n]$ denote the impulse response of an ideal desired system with corresponding frequency response $H_d(e^{j\omega})$, and let $h[n]$ and $H(e^{j\omega})$ denote the impulse response and frequency response, respectively, of an FIR approximation to the ideal system. Assume that $h[n] = 0$ for $n < 0$ and $n > M$. We wish to choose the $(M + 1)$ samples of the impulse response so as to minimize the mean-square error of the frequency response defined as

$$\varepsilon^2 = \frac{1}{2\pi} \int_{-\pi}^{\pi} |H_d(e^{j\omega}) - H(e^{j\omega})|^2 d\omega.$$

(a) Use Parseval's relation to express the error function in terms of the sequences $h_d[n]$ and $h[n]$.

(b) Using the result of Part (a), determine the values of $h[n]$ for $0 \le n \le M$ that minimize ε^2.

(c) The FIR filter determined in Part (b) could have been obtained by a windowing operation. That is, $h[n]$ could have been obtained by multiplying the desired infinite-length sequence $h_d[n]$ by a certain finite-length sequence $w[n]$. Determine the necessary window $w[n]$ such that the optimal impulse response is $h[n] = w[n]h_d[n]$.

7.33. An *ideal discrete-time Hilbert transformer* is a system that introduces -90 degrees ($-\pi/2$ radians) of phase shift for $0 < \omega < \pi$ and $+90$ degrees ($+\pi/2$ radians) of phase shift for $-\pi < \omega < 0$. The magnitude of the frequency response is constant (unity) for $0 < \omega < \pi$ and for $-\pi < \omega < 0$. Such systems are also called *ideal 90-degree phase shifters*.

(a) Give an equation for the ideal frequency response $H_d(e^{j\omega})$ of an ideal discrete-time Hilbert transformer that also includes constant (nonzero) group delay. Plot the phase response of this system for $-\pi < \omega < \pi$.

(b) What type(s) of FIR linear-phase systems (I, II, III, or IV) can be used to approximate the ideal Hilbert transformer in Part (a)?

(c) Suppose that we wish to use the window method to design a linear-phase approximation to the ideal Hilbert transformer. Use $H_d(e^{j\omega})$ given in Part (a) to determine the ideal impulse response $h_d[n]$ if the FIR system is to be such that $h[n] = 0$ for $n < 0$ and $n > M$.

(d) What is the delay of the system if $M = 21$? Sketch the magnitude of the frequency response of the FIR approximation for this case, assuming a rectangular window.

(e) What is the delay of the system if $M = 20$? Sketch the magnitude of the frequency response of the FIR approximation for this case, assuming a rectangular window.

7.34. The commonly used windows presented in Section 7.2.1 can all be expressed in terms of rectangular windows. This fact can be used to obtain expressions for the Fourier transforms of the Bartlett window and the raised-cosine family of windows, which includes the Hanning, Hamming, and Blackman windows.

(a) Show that the $(M+1)$-point Bartlett window, defined by Eq. (7.47b), can be expressed as the convolution of two smaller rectangular windows. Use this fact to show that the Fourier transform of the $(M+1)$-point Bartlett window is

$$W_B(e^{j\omega}) = e^{-j\omega M/2}(2/M)\left(\frac{\sin(\omega M/4)}{\sin(\omega/2)}\right)^2 \quad \text{for } M \text{ even,}$$

or

$$W_B(e^{j\omega}) = e^{-j\omega M/2}(2/M)\left(\frac{\sin[\omega(M+1)/4]}{\sin(\omega/2)}\right)\left(\frac{\sin[\omega(M-1)/4]}{\sin(\omega/2)}\right) \quad \text{for } M \text{ odd.}$$

(b) It can easily be seen that the $(M+1)$-point raised-cosine windows defined by Eqs. (7.47c)–(7.47e) can all be expressed in the form

$$w[n] = [A + B\cos(2\pi n/M) + C\cos(4\pi n/M)]w_R[n],$$

where $w_R[n]$ is an $(M+1)$-point rectangular window. Use this relation to find the Fourier transform of the general raised-cosine window.

(c) Using appropriate choices for A, B, and C and the result determined in Part (b), sketch the magnitude of the Fourier transform of the Hanning window.

7.35. Consider the following ideal frequency response for a multiband filter:

$$H_d(e^{j\omega}) = \begin{cases} e^{-j\omega M/2}, & 0 \le |\omega| < 0.3\pi, \\ 0, & 0.3\pi < |\omega| < 0.6\pi, \\ 0.5e^{-j\omega M/2}, & 0.6\pi < |\omega| \le \pi. \end{cases}$$

The impulse response $h_d[n]$ is multiplied by a Kaiser window with $M = 48$ and $\beta = 3.68$, resulting in a linear-phase FIR system with impulse response $h[n]$.

(a) What is the delay of the filter?

(b) Determine the ideal impulse response $h_d[n]$.

(c) Determine the set of approximation error specifications that is satisfied by the FIR filter; i.e., determine the parameters $\delta_1, \delta_2, \delta_3, B, C, \omega_{p1}, \omega_{s1}, \omega_{s2},$ and ω_{p2} in

$$B - \delta_1 \le |H(e^{j\omega})| \le B + \delta_1, \qquad 0 \le \omega \le \omega_{p1},$$
$$|H(e^{j\omega})| \le \delta_2, \qquad \omega_{s1} \le \omega \le \omega_{s2},$$
$$C - \delta_3 \le |H(e^{j\omega})| \le C + \delta_3, \qquad \omega_{p2} \le \omega \le \pi.$$

7.36. An optimal equiripple FIR linear-phase filter was designed by the Parks–McClellan algorithm. The magnitude of its frequency response is shown in Figure P7.36-1. The maximum approximation error in the passband is $\delta_1 = 0.0531$, and the maximum approximation error in the stopband is $\delta_2 = 0.085$. The passband and stopband cutoff frequencies are $\omega_p = 0.4\pi$ and $\omega_s = 0.58\pi$.

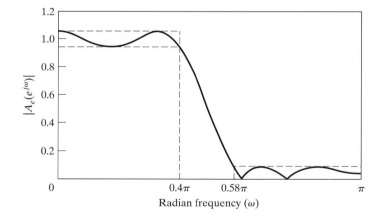

Figure P7.36-1

(a) What type (I, II, III, or IV) of linear-phase system is this? Explain how you can tell.

(b) What was the error-weighting function $W(\omega)$ that was used in the optimization?

(c) Carefully sketch the weighted approximation error; i.e., sketch

$$E(\omega) = W(\omega)[H_d(e^{j\omega}) - A_e(e^{j\omega})].$$

(Note that Figure P7.36-1 shows $|A_e(e^{j\omega})|$.)

(d) What is the length of the impulse response of the system?

(e) If this system is causal, what is the smallest delay that it can have?

(f) Plot the zeros of the system function $H(z)$ as accurately as you can in the z-plane.

7.37. The frequency response of a desired filter $h_d[n]$ is shown in Figure P7.37-1. In this problem, we wish to design an N-point causal linear-phase FIR filter $h[n]$ that minimizes the integral-squared error

$$\epsilon = \frac{1}{2\pi} \int_{-\pi}^{\pi} |A(e^{j\omega}) - H_d(e^{j\omega})|^2 d\omega,$$

Figure P7.37-1

where the frequency response of the filter $h[n]$ is

$$H_d(e^{j\omega}) = A(e^{j\omega})e^{-j\omega(N-1)/2}$$

and N is an odd integer.
(a) Determine $h_d[n]$.
(b) What symmetry should $h_d[n]$ have in the range $0 \le n \le N - 1$? Briefly explain your reasoning.
(c) Determine $h_d[n]$ in the range $0 \le n \le N - 1$.
(d) Determine an expression for the minimum integral-squared error ϵ as a function of $h_d[n]$ and N.

7.38. Consider a type I linear-phase FIR lowpass filter with impulse response $h_{LP}[n]$ of length $(M + 1)$ and frequency response

$$H_{LP}(e^{j\omega}) = A_e(e^{j\omega})e^{-j\omega M/2}.$$

The system has the amplitude function $A_e(e^{j\omega})$ shown in Figure P7.38-1.

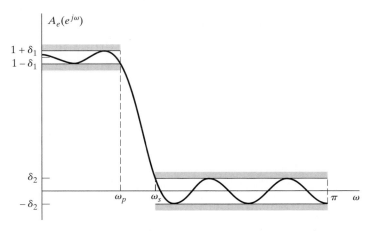

Figure P7.38-1

This amplitude function is the optimal (in the Parks–McClellan sense) approximation to unity in the band $0 \le \omega \le \omega_p$, where $\omega_p = 0.27\pi$, and the optimal approximation to zero in the band $\omega_s \le \omega \le \pi$, where $\omega_s = 0.4\pi$.
(a) What is the value of M?
Suppose now that a highpass filter is derived from this lowpass filter by defining

$$h_{HP}[n] = (-1)^{n+1}h_{LP}[n] = -e^{j\pi n}h_{LP}[n].$$

(b) Show that the resulting frequency response is of the form $H_{HP}(e^{j\omega}) = B_e(e^{j\omega})e^{-j\omega M/2}$.
(c) Sketch $B_e(e^{j\omega})$ for $0 \le \omega \le \pi$.
(d) It is asserted that for the given value of M (as found in Part (a)), the resulting highpass filter is the optimum approximation to zero in the band $0 \le \omega \le 0.6\pi$ and to unity in the band $0.73\pi \le \omega \le \pi$. Is this assertion correct? Justify your answer.

7.39. Design a three-point optimal (in the minimax sense) causal lowpass filter with $\omega_s = \pi/2$, $\omega_p = \pi/3$, and $K = 1$. Specify the impulse response $h[n]$ of the filter you design. *Note:* $\cos(\pi/2) = 0$ and $\cos(\pi/3) = 0.5$.

7.40. Filter C is a stable continuous-time IIR filter with system function $H_e(s)$ and impulse response $h_c(t)$. Filter B is a stable discrete-time filter with system function $H_b(z)$ and impulse response $h_b[n]$. Filter B is related to filter C through the bilinear transformation. Determine whether the following statement is true or false. If it is true, state your reasoning. If it is false, demonstrate with a counterexample.

> **Statement:** Filter B cannot be an FIR filter.

7.41. Suppose that a discrete-time filter is obtained from a prototype continuous-time filter $H_e(s)$ by bilinear transformation. Furthermore, assume that the continuous-time filter has a constant group delay; i.e.,

$$H_e(j\Omega) = A(\Omega)e^{-j\Omega\alpha},$$

where $A(\Omega)$ is purely real. Would the resulting discrete-time filter also have a constant group delay? Explain your reasoning.

Extension Problems

7.42. If a linear time-invariant continuous-time system has a rational system function, then its input and output satisfy an ordinary linear differential equation with constant coefficients. A standard procedure in the simulation of such systems is to use finite-difference approximations to the derivatives in the differential equations. In particular, since, for continuous differentiable functions $y_c(t)$,

$$\frac{dy_c(t)}{dt} = \lim_{T \to 0} \left[\frac{y_c(t) - y_c(t - T)}{T} \right],$$

it seems plausible that if T is "small enough," we should obtain a good approximation if we replace $dy_c(t)/dt$ by $[y_c(t) - y_c(t - T)]/T$.

While this simple approach may be useful for simulating continuous-time systems, it is *not* generally a useful method for designing discrete-time systems for filtering applications. To understand the effect of approximating differential equations by difference equations, it is helpful to consider a specific example. Assume that the system function of a continuous-time system is

$$H_c(s) = \frac{A}{s + c},$$

where A and c are constants.

(a) Show that the input $x_c(t)$ and the output $y_c(t)$ of the system satisfy the differential equation

$$\frac{dy_c(t)}{dt} + cy_c(t) = Ax_c(t).$$

(b) Evaluate the differential equation at $t = nT$, and substitute

$$\left. \frac{dy_c(t)}{dt} \right|_{t=nT} \approx \frac{y_c(nT) - y_c(nT - T)}{T},$$

i.e., replace the first derivative by the *first backward difference.*

(c) Define $x[n] = x_c(nT)$ and $y[n] = y_c(nT)$. With this notation and the result of Part (b), obtain a difference equation relating $x[n]$ and $y[n]$, and determine the system function $H(z) = Y(z)/X(z)$ of the resulting discrete-time system.

(d) Show that, for this example,

$$H(z) = H_c(s)\big|_{s=(1-z^{-1})/T};$$

i.e., show that $H(z)$ can be obtained directly from $H_c(s)$ by the mapping

$$s = \frac{1 - z^{-1}}{T}.$$

(It can be demonstrated that if higher order derivatives are approximated by repeated application of the first backward difference, then the result of Part (d) holds for higher order systems as well.)

(e) For the mapping of Part (d), determine the contour in the z-plane to which the $j\Omega$-axis of the s-plane maps. Also, determine the region of the z-plane that corresponds to the left half of the s-plane. If the continuous-time system with system function $H_c(s)$ is stable, will the discrete-time system obtained by first backward difference approximation also be stable? Will the frequency response of the discrete-time system be a faithful reproduction of the frequency response of the continuous-time system? How will the stability and frequency response be affected by the choice of T?

(f) Assume that the first derivative is approximated by the *first forward difference*; i.e.,

$$\frac{dy_c(t)}{dt}\bigg|_{t=nT} \approx \frac{y_c(nT + T) - y_c(nT)}{T}.$$

Determine the corresponding mapping from the s-plane to the z-plane, and repeat part (e) for this mapping.

7.43. Consider a linear time-invariant continuous-time system with rational system function $H_c(s)$. The input $x_c(t)$ and the output $y_c(t)$ satisfy an ordinary linear differential equation with constant coefficients. One approach to simulating such systems is to use numerical techniques to integrate the differential equation. In this problem, we demonstrate that if the trapezoidal integration formula is used, this approach is equivalent to transforming the continuous-time system function $H_c(s)$ to a discrete-time system function $H(z)$ using the bilinear transformation.

To demonstrate this statement, consider the continuous-time system function

$$H_c(s) = \frac{A}{s + c},$$

where A and c are constants. The corresponding differential equation is

$$\dot{y}_c(t) + cy_c(t) = Ax_c(t),$$

where

$$\dot{y}_c(t) = \frac{dy_c(t)}{dt}.$$

(a) Show that $y_c(nT)$ can be expressed in terms of $\dot{y}_c(t)$ as

$$y_c(nT) = \int_{(nT-T)}^{nT} \dot{y}_c(\tau)d\tau + y_c(nT - T).$$

The definite integral in this equation represents the area beneath the function $\dot{y}_c(t)$ for the interval from $(nT - T)$ to nT. Figure P7.43-1 shows a function $\dot{y}_c(t)$ and a shaded trapezoid-shaped region whose area approximates the area beneath the curve. This approximation to the integral is known as the *trapezoidal approximation*. Clearly, as T approaches zero, the approximation improves. Use the trapezoidal approximation to obtain an expression for $y_c(nT)$ in terms of $y_c(nT - T)$, $\dot{y}_c(nT)$, and $\dot{y}_c(nT - T)$.

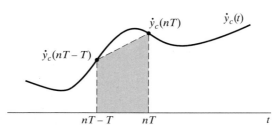

Figure P7.43-1

(b) Use the differential equation to obtain an expression for $\dot{y}_c(nT)$, and substitute this expression into the expression obtained in part (a).

(c) Define $x[n] = x_c(nT)$ and $y[n] = y_c(nT)$. With this notation and the result of part (b), obtain a difference equation relating $x[n]$ and $y[n]$, and determine the system function $H(z) = Y(z)/X(z)$ of the resulting discrete-time system.

(d) Show that, for this example,

$$H(z) = H_c(s)\big|_{s=(2/T)[(1-z^{-1})/(1+z^{-1})]};$$

i.e., show that $H(z)$ can be obtained directly from $H_c(s)$ by the bilinear transformation. (For higher order differential equations, repeated trapezoidal integration applied to the highest order derivative of the output will result in the same conclusion for a general continuous-time system with rational system function.)

7.44. In this problem, we consider a method of filter design that might be called *autocorrelation invariance*. Consider a stable continuous-time system with impulse response $h_c(t)$ and system function $H_c(s)$. The autocorrelation function of the system is defined as

$$\phi_c(\tau) = \int_{-\infty}^{\infty} h_c(t)h_c(t+\tau)d\tau,$$

and for a real impulse response, it is easily shown that the Laplace transform of $\phi_c(\tau)$ is $\Phi_c(s) = H_c(s)H_c(-s)$. Similarly, consider a discrete-time system with impulse response $h[n]$ and system function $H(z)$. The autocorrelation function of a discrete-time system is defined as

$$\phi[m] = \sum_{n=-\infty}^{\infty} h[n]h[n+m],$$

and for a real impulse response, $\Phi(z) = H(z)H(z^{-1})$.

   Autocorrelation invariance implies that a discrete-time filter is defined by equating the autocorrelation function of the discrete-time system to the sampled autocorrelation function of a continuous-time system; i.e.,

$$\phi[m] = T_d\phi_c(mT_d), \qquad -\infty < m < \infty.$$

The following design procedure is proposed for autocorrelation invariance when $H_c(s)$ is a rational function having N first-order poles at s_k, $k = 1, 2, \ldots, N$, and $M < N$ zeros:

1. Obtain a partial fraction expansion of $\Phi_c(s)$ in the form

$$\Phi_c(s) = \sum_{k=1}^{N} \left(\frac{A_k}{s - s_k} + \frac{B_k}{s + s_k} \right).$$

2. Form the z-transform

$$\Phi(z) = \sum_{k=1}^{N} \left(\frac{T_d A_k}{1 - e^{s_k T_d} z^{-1}} + \frac{T_d B_k}{1 - e^{-s_k T_d} z^{-1}} \right).$$

3. Find the poles and zeros of $\Phi(z)$, and form a minimum-phase system function $H(z)$ from the poles and zeros of $\Phi(z)$ that are *inside* the unit circle.

(a) Justify each step in the proposed design procedure; i.e., show that the autocorrelation function of the resulting discrete-time system is a sampled version of the autocorrelation function of the continuous-time system. To verify the procedure, it may be helpful to try it out on the first-order system with impulse response

$$h_c(t) = e^{-\alpha t} u(t)$$

and system function

$$H_c(s) = \frac{1}{s + \alpha}.$$

(b) What is the relationship between $|H(e^{j\omega})|^2$ and $|H_c(j\Omega)|^2$? What types of frequency-response functions would be appropriate for autocorrelation invariance design?

(c) Is the system function obtained in Step 3 unique? If not, describe how to obtain additional autocorrelation-invariant discrete-time systems.

7.45. Let $H_{lp}(Z)$ denote the system function for a discrete-time lowpass filter. The implementations of such a system can be represented by linear signal flow graphs consisting of adders, gains, and unit delay elements as in Figure P7.45-1. We want to implement a lowpass filter for which the cutoff frequency can be varied by changing a single parameter. The proposed strategy is to replace each unit delay element in a flow graph representing $H_{lp}(Z)$ by the network shown in Figure P7.45-2, where α is real and $|\alpha| < 1$.

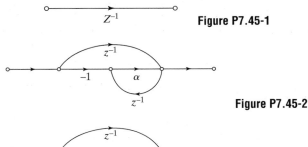

Figure P7.45-1

Figure P7.45-2

Figure P7.45-3

(a) Let $H(z)$ denote the system function for the filter that results when the network of Figure P7.45-2 is substituted for each unit delay branch in the network that implements $H_{lp}(Z)$. Show that $H(z)$ and $H_{lp}(Z)$ are related by a mapping of the Z-plane into the z-plane.

(b) If $H(e^{j\omega})$ and $H_{lp}(e^{j\theta})$ are the frequency responses of the two systems, determine the relationship between the frequency variables ω and θ. Sketch ω as a function of θ for $\alpha = 0, \pm0.5$, and show that $H(e^{j\omega})$ is a lowpass filter. Also, if θ_p is the passband cutoff frequency for the original lowpass filter $H_{lp}(Z)$, obtain an equation for ω_p, the cutoff frequency of the new filter $H(z)$, as a function of α and θ_p.

(c) Assume that the original lowpass filter has system function

$$H_{lp}(Z) = \frac{1}{1 - 0.9Z^{-1}}.$$

Draw the flow graph of an implementation of $H_{lp}(Z)$, and also, draw the flow graph of the implementation of $H(z)$ obtained by replacing the unit delay elements in the first flow graph by the network in Figure P7.45-2. Does the resulting network correspond to a computable difference equation?

(d) If $H_{lp}(Z)$ corresponds to an FIR system implemented in direct form, would the flow graph manipulation lead to a computable difference equation? If the FIR system $H_{lp}(Z)$ was a linear-phase system, would the resulting system $H(z)$ also be a linear-phase system? If the FIR system has an impulse response of length $M+1$ samples what would be the length of the impulse response of the transformed system?

(e) To avoid the difficulties that arose in Part (c), it is suggested that the network of Figure P7.45-2 be cascaded with a unit delay element, as depicted in Figure P7.45-3. Repeat the analysis of Part (a) when the network of Figure P7.45-3 is substituted for each unit delay element. Determine an equation that expresses θ as a function of ω, and show that if $H_{lp}(e^{j\theta})$ is a lowpass filter, then $H(e^{j\omega})$ is not a lowpass filter.

7.46. If we are given a basic filter module (a hardware or computer subroutine), it is sometimes possible to use it repetitively to implement a new filter with sharper frequency-response characteristics. One approach is to cascade the filter with itself two or more times, but it can easily be shown that, while stopband errors are squared (thereby reducing them if they are less than 1), this approach will increase the passband approximation error. Another approach, suggested by Tukey (1977), is shown in the block diagram of Figure P7.46-1. Tukey called this approach "twicing."

Figure P7.46-1

(a) Assume that the basic system has a symmetric finite-duration impulse response; i.e.,

$$h[n] = \begin{cases} h[-n], & -L \le n \le L, \\ 0 & \text{otherwise.} \end{cases}$$

Determine whether the overall impulse response $g[n]$ is (i) an FIR and (ii) symmetric.

(b) Suppose that $H(e^{j\omega})$ satisfies the following approximation error specifications:

$$(1 - \delta_1) \le H(e^{j\omega}) \le (1 + \delta_1), \qquad 0 \le \omega \le \omega_p,$$
$$-\delta_2 \le H(e^{j\omega}) \le \delta_2, \qquad \omega_s \le \omega \le \pi.$$

It can be shown that if the basic system has these specifications, the overall frequency response $G(e^{j\omega})$ satisfies specifications of the form

$$A \le G(e^{j\omega}) \le B, \qquad 0 \le \omega \le \omega_p,$$
$$C \le G(e^{j\omega}) \le D, \qquad \omega_s \le \omega \le \pi.$$

Determine A, B, C, and D in terms of δ_1 and δ_2. If $\delta_1 \ll 1$ and $\delta_2 \ll 1$, what are the approximate maximum passband and stopband approximation errors for $G(e^{j\omega})$?

(c) As determined in Part (b), Tukey's twicing method improves the passband approximation error, but increases the stopband error. Kaiser and Hamming (1977) generalized the twicing method so as to improve *both* the passband and the stopband. They called their approach "sharpening." The simplest sharpening system that improves both passband and stopband is shown in Figure P7.46-2. Assume again that the impulse response of the basic system is as given in Part (a). Repeat Part (b) for the system of Figure P7.46-2.

(d) The basic system was assumed to be noncausal. If the impulse response of the basic system is a causal linear-phase FIR system such that

$$h[n] = \begin{cases} h[M - n], & 0 \leq n \leq M, \\ 0, & \text{otherwise,} \end{cases}$$

how should the systems of Figures P7.46-1 and P7.46-2 be modified? What type(s) (I, II, III, or IV) of causal linear-phase FIR system(s) can be used? What are the lengths of the impulse responses $g[n]$ for the systems in Figures P7.46-1 and P7.46-2 (in terms of L)?

Figure P7.46-2

7.47. Consider the design of a lowpass linear-phase FIR filter by means of the Parks–McClellan algorithm. Use the alternation theorem to argue that the approximation must decrease monotonically in the "don't care" region between the passband and the stopband approximation intervals. *Hint:* Show that all the local maxima and minima of the trigonometric polynomial must be in either the passband or the stopband to satisfy the alternation theorem.

7.48. Figure P7.48-1 shows the frequency response $A_e(e^{j\omega})$ of a discrete-time FIR system for which the impulse response is

$$h_e[n] = \begin{cases} h_e[-n], & -L \leq n \leq L, \\ 0, & \text{otherwise.} \end{cases}$$

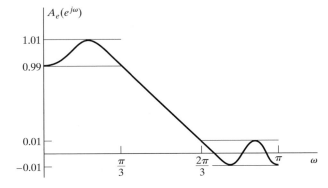

Figure P7.48-1

(a) Show that $A_e(e^{j\omega})$ cannot correspond to an FIR filter generated by the Parks–McClellan algorithm with a passband edge frequency of $\pi/3$, a stopband edge frequency of $2\pi/3$, and an error-weighting function of unity in the passband and stopband. Clearly explain your reasoning. *Hint:* The alternation theorem states that the best approximation is unique.

(b) Based on Figure P7.48-1 and the statement that $A_e(e^{j\omega})$ cannot correspond to an optimal filter, what can be concluded about the value of L?

7.49. Consider the system in Figure P7.49-1.

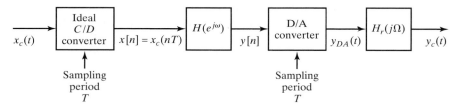

Figure P7.49-1

1. Assume that $X_c(j\Omega) = 0$ for $|\Omega| \geq \pi/T$ and that

$$H_r(j\Omega) = \begin{cases} 1, & |\Omega| < \pi/T, \\ 0, & |\Omega| > \pi/T, \end{cases}$$

denotes an ideal lowpass reconstruction filter.

2. The D/A converter has a built-in zero-order-hold circuit, so that

$$Y_{DA}(t) = \sum_{n=-\infty}^{\infty} y[n]h_0(t - nT),$$

where

$$h_0(t) = \begin{cases} 1, & 0 \leq t < T, \\ 0, & \text{otherwise.} \end{cases}$$

(We neglect quantization in the D/A converter.)

3. The second system in Figure P7.49-1 is a linear-phase FIR discrete-time system with frequency response $H(e^{j\omega})$.

We wish to design the FIR system using the Parks–McClellan algorithm to compensate for the effects of the zero-order-hold system.

(a) The Fourier transform of the output is $Y_c(j\Omega) = H_{\text{eff}}(j\Omega)X_c(j\Omega)$. Determine an expression for $H_{\text{eff}}(j\Omega)$ in terms of $H(e^{j\Omega T})$ and T.

(b) If the linear-phase FIR system is such that $h[n] = 0$ for $n < 0$ and $n > 51$, and $T = 10^{-4}$ s, what is the overall time delay (in ms) between $x_c(t)$ and $y_c(t)$?

(c) Suppose that when $T = 10^{-4}$ s we want the effective frequency response to be equiripple (in both the passband and the stopband) within the following tolerances:

$$0.99 \leq |H_{\text{eff}}(j\Omega)| \leq 1.01, \qquad |\Omega| \leq 2\pi(1000),$$

$$|H_{\text{eff}}(j\Omega)| \leq 0.01, \qquad 2\pi(2000) \leq |\Omega| \leq 2\pi(5000).$$

We want to achieve this by designing an optimum linear-phase filter (using the Parks–McClellan algorithm) that includes compensation for the zero-order hold. Give an equation for the ideal response $H_d(e^{j\omega})$ that should be used. Find and sketch the weighting function $W(\omega)$ that should be used. Sketch a "typical" frequency response $H(e^{j\omega})$ that might result.

(d) How would you modify your results in part (c) to include magnitude compensation for a reconstruction filter $H_r(j\Omega)$ with zero gain above $\Omega = 2\pi(5000)$, but with sloping passband?

7.50. After a discrete-time signal is lowpass filtered, it is often downsampled or decimated, as depicted in Figure P7.50-1. Linear-phase FIR filters are frequently desirable in such applications, but if the lowpass filter in the figure has a narrow transition band, an FIR system will have a long impulse response and thus will require a large number of multiplications and additions per output sample.

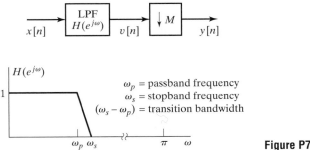

Figure P7.50-1

In this problem, we will study the merits of a multistage implementation of the system in Figure P7.50-1. Such implementations are particularly useful when ω_s is small and the decimation factor M is large. A general multistage implementation is depicted in Figure P7.50-2. The strategy is to use a wider transition band in the lowpass filters of the earlier stages, thereby reducing the length of the required filter impulse responses in those stages. As decimation occurs, the number of signal samples is reduced, and we can progressively decrease the widths of the transition bands of the filters that operate on the decimated signal. In this manner, the overall number of computations required to implement the decimator may be reduced.

Figure P7.50-2

(a) If no aliasing is to occur as a result of the decimation in Figure P7.50-1, what is the maximum allowable decimation factor M in terms of ω_s?

(b) Let $M = 100$, $\omega_s = \pi/100$, and $\omega_p = 0.9\pi/100$ in the system of Figure P7.50-1. If $x[n] = \delta[n]$, sketch $V(e^{j\omega})$ and $Y(e^{j\omega})$.

Now consider a two-stage implementation of the decimator for $M = 100$, as depicted in Figure P7.50-3, where $M_1 = 50$, $M_2 = 2$, $\omega_{p1} = 0.9\pi/100$, $\omega_{p2} = 0.9\pi/2$, and $\omega_{s2} = \pi/2$. We must choose ω_{s1} or, equivalently, the transition band of LPF$_1$, $(\omega_{s1} - \omega_{p1})$, such that the two-stage implementation yields the same equivalent passband and stopband frequencies as the single-stage decimator. (We are not concerned about the detailed shape of the frequency response in the transition band, except that both systems should have a monotonically decreasing response in the transition band.)

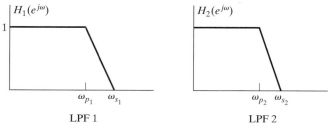

LPF 1   LPF 2   **Figure P7.50-3**

(c) For an arbitrary ω_{s1} and the input $x[n] = \delta[n]$, sketch $V_1(e^{j\omega})$, $W_1(e^{j\omega})$, $V_2(e^{j\omega})$, and $Y(e^{j\omega})$ for the two-stage decimator of Figure P7.50-3.

(d) Find the *largest* value of ω_{s1} such that the two-stage decimator yields the same equivalent passband and stopband cutoff frequencies as the single-stage system in Part (b).

In addition to possessing a nonzero transition bandwidth, the lowpass filters must differ from the ideal by passband and stopband approximation errors of δ_p and δ_s, respectively. Assume that linear-phase equiripple FIR approximations are used. It follows from Eq. (7.104) that, for optimum lowpass filters,

$$N \approx \frac{-10\log_{10}(\delta_p\delta_s) - 13}{2.324\Delta\omega} + 1, \qquad (P7.50\text{-}1)$$

where N is the length of the impulse response and $\Delta\omega = \omega_s - \omega_p$ is the transition band of the lowpass filter. Equation (P7.50-1) provides the basis for comparing the two implementations of the decimator. Equation (7.63) could be used in place of Eq. (P7.50-1) to estimate the impulse-response length if the filters are designed by the Kaiser window method.)

(e) Assume that $\delta_p = 0.01$ and $\delta_s = 0.001$ for the lowpass filter in the single-stage implementation. Compute the length N of the impulse response of the lowpass filter, and determine the number of multiplications required to compute each sample of the output. Take advantage of the symmetry of the impulse response of the linear-phase FIR system. (Note that in this decimation application, only every Mth sample of the output need be computed; i.e., the compressor commutes with the multiplications of the FIR system.)

(f) Using the value of ω_{s1} found in Part (d), compute the impulse response lengths N_1 and N_2 of LPF$_1$ and LPF$_2$, respectively, in the two-stage decimator of Figure P7.50-3. Determine the total number of multiplications required to compute each sample of the output in the two-stage decimator.

(g) If the approximation error specifications $\delta_p = 0.01$ and $\delta_s = 0.001$ are used for both filters in the two-stage decimator, the overall passband ripple may be greater than 0.01, since the passband ripples of the two stages can reinforce each other; e.g., $(1 + \delta_p)(1 + \delta_p) > (1 + \delta_p)$. To compensate for this, the filters in the two-stage implementation can each be designed to have only one-half the passband ripple of the single-stage implementation. Therefore, assume that $\delta_p = 0.005$ and $\delta_s = 0.001$ for each filter in the two-stage decimator. Calculate the impulse response lengths N_1 and N_2 of LPF$_1$ and LPF$_2$, respectively, and determine the total number of multiplications required to compute each sample of the output.

(h) Should we also reduce the specification on the stopband approximation error for the filters in the two-stage decimator?

(i) *Optional.* The combination of $M_1 = 50$ and $M_2 = 2$ may not yield the smallest total number of multiplications per output sample. Other integer choices for M_1 and M_2 are possible such that $M_1 M_2 = 100$. Determine the values of M_1 and M_2 that minimize the number of multiplications per output sample.

7.51. In this problem, we develop a technique for designing discrete-time filters with minimum phase. Such filters have all their poles and zeros inside (or on) the unit circle. (We will allow zeros on the unit circle.) Let us first consider the problem of converting a type I linear-phase FIR equiripple lowpass filter to a minimum-phase system. If $H(e^{j\omega})$ is the frequency response of a type I linear-phase filter, then

1. The corresponding impulse response

$$h[n] = \begin{cases} h[M-n], & 0 \le n \le M, \\ 0, & \text{otherwise}, \end{cases}$$

   is real and M is an even integer.
2. It follows from Part 1 that $H(e^{j\omega}) = A_e(e^{j\omega})e^{-j\omega n_0}$, where $A_e(e^{j\omega})$ is real and $n_0 = M/2$ is an integer.
3. The passband ripple is δ_1; i.e., in the passband, $A_e(e^{j\omega})$ oscillates between $(1 + \delta_1)$ and $(1 - \delta_1)$. (See Figure P7.51-1.)

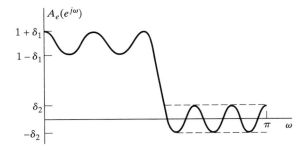

Figure P7.51-1

4. The stopband ripple is δ_2; i.e., in the stopband, $-\delta_2 \le A_e(e^{j\omega}) \le \delta_2$, and $A_e(e^{j\omega})$ oscillates between $-\delta_2$ and $+\delta_2$. (See Figure P7.51-1.)

The following technique was proposed by Herrmann and Schüssler (1970a) for converting this linear-phase system into a minimum-phase system that has a system function $H_{min}(z)$ and unit sample response $h_{min}[n]$ (in this problem, we assume that minimum-phase systems can have zeros *on* the unit circle):

Step 1. Create a new sequence

$$h_1[n] = \begin{cases} h[n], & n \ne n_0, \\ h[n_0] + \delta_2, & n = n_0. \end{cases}$$

Step 2. Recognize that $H_1(z)$ can be expressed in the form

$$H_1(z) = z^{-n_0} H_2(z) H_2(1/z) = z^{-n_0} H_3(z)$$

for some $H_2(z)$, where $H_2(z)$ has all its poles and zeros inside or on the unit circle and $h_2[n]$ is real.

Step 3. Define

$$H_{\min}(z) = \frac{H_2(z)}{a}.$$

The denominator constant where $a = (\sqrt{1 - \delta_1 + \delta_2} + \sqrt{1 + \delta_1 + \delta_2})/2$ normalizes the passband so that the resulting frequency response $H_{\min}(e^{j\omega})$ will oscillate about a value of unity.

(a) Show that if $h_1[n]$ is chosen as in Step 1, then $H_1(e^{j\omega})$ can be written as

$$H_1(e^{j\omega}) = e^{-j\omega n_0} H_3(e^{j\omega}),$$

where $H_3(e^{j\omega})$ is real and nonnegative for all values of ω.

(b) If $H_3(e^{j\omega}) \geq 0$, as was shown in Part (a), show that there exists an $H_2(z)$ such that

$$H_3(z) = H_2(z)H_2(1/z),$$

where $H_2(z)$ is a minimum-phase system function and $h_2[n]$ is real (i.e., justify Step 2).

(c) Demonstrate that the new filter $H_{\min}(e^{j\omega})$ is an equiripple lowpass filter (i.e., that its magnitude characteristic is of the form shown in Figure P7.51-2) by evaluating δ_1' and δ_2'. What is the length of the new impulse response $h_{\min}[n]$?

                                                  Figure P7.51-2

(d) In Parts (a), (b), and (c), we assumed that we started with a type I FIR linear-phase filter. Will this technique work if we remove the linear-phase constraint? Will it work if we use a type II FIR linear-phase system?

7.52. Suppose that we have a program that finds the set of coefficients $a[n], n = 0, 1, \ldots, L$, that minimizes

$$\max_{\omega \in F} \left\{ \left| W(\omega) \left[H_d(e^{j\omega}) - \sum_{n=0}^{L} a[n] \cos \omega n \right] \right| \right\},$$

given L, F, $W(\omega)$, and $H_d(e^{j\omega})$. We have shown that the solution to this optimization problem implies a noncausal FIR zero-phase system with impulse response satisfying $h_e[n] = h_e[-n]$. By delaying $h_e[n]$ by L samples, we obtain a causal type I FIR linear-phase system with frequency response

$$H(e^{j\omega}) = e^{-j\omega M/2} \sum_{n=0}^{L} a[n] \cos \omega n = \sum_{n=0}^{2L} h[n]e^{-j\omega n},$$

where the impulse response is related to the coefficients $a[n]$ by

$$a[n] = \begin{cases} 2h[M/2 - n] & \text{for } 1 \leq n \leq L, \\ h[M/2] & \text{for } n = 0, \end{cases}$$

and $M = 2L$ is the order of the system function polynomial. (The length of the impulse response is $M + 1$.)

The other three types (II, III, and IV) of linear-phase FIR filters can be designed by the available program if we make suitable modifications to the weighting function $W(\omega)$ and the desired frequency response $H_d(e^{j\omega})$. To see how to do this, it is necessary to manipulate the expressions for the frequency response into the standard form assumed by the program.

(a) Assume that we wish to design a causal type II FIR linear-phase system such that $h[n] = h[M - n]$ for $n = 0, 1, \ldots, M$, where M is an odd integer. Show that the frequency response of this type of system can be expressed as

$$H(e^{j\omega}) = e^{-j\omega M/2} \sum_{n=1}^{(M+1)/2} b[n] \cos\omega\left(n - \tfrac{1}{2}\right),$$

and determine the relationship between the coefficients $b[n]$ and $h[n]$.

(b) Show that the summation

$$\sum_{n=1}^{(M+1)/2} b[n] \cos\omega\left(n - \tfrac{1}{2}\right)$$

can be written as

$$\cos(\omega/2) \sum_{n=0}^{(M-1)/2} \tilde{b}[n] \cos\omega n$$

by obtaining an expression for $b[n]$ for $n = 1, 2, \ldots, (M + 1)/2$ in terms of $\tilde{b}[n]$ for $n = 0, 1, \ldots, (M - 1)/2$. *Hint:* Note carefully that $b[n]$ is to be expressed in terms of $\tilde{b}[n]$. Also, use the trigonometric identity $\cos\alpha \cos\beta = \tfrac{1}{2}\cos(\alpha + \beta) + \tfrac{1}{2}\cos(\alpha - \beta)$.

(c) If we wish to use the given program to design type II systems (M odd) for a given F, $W(\omega)$, and $H_d(e^{j\omega})$, show how to obtain $\tilde{L}$, $\tilde{F}$, $\tilde{W}(\omega)$, and $\tilde{H}_d(e^{j\omega})$ in terms of M, F, $W(\omega)$, and $H_d(e^{j\omega})$ such that if we run the program using $\tilde{L}$, $\tilde{F}$, $\tilde{W}(\omega)$, and $\tilde{H}_d(e^{j\omega})$, we may use the resulting set of coefficients to determine the impulse response of the desired type II system.

(d) Parts (a)–(c) can be repeated for types III and IV causal linear-phase FIR systems where $h[n] = -h[M - n]$. For these cases, you must show that, for type III systems (M even), the frequency response can be expressed as

$$H(e^{j\omega}) = e^{-j\omega M/2} \sum_{n=1}^{M/2} c[n] \sin\omega n$$

$$= e^{-j\omega M/2} \sin\omega \sum_{n=0}^{(M-2)/2} \tilde{c}[n] \cos\omega n,$$

and for type IV systems (M odd),

$$H(e^{j\omega}) = e^{-j\omega M/2} \sum_{n=1}^{(M+1)/2} d[n] \sin\omega\left(n - \tfrac{1}{2}\right)$$

$$= e^{-j\omega M/2} \sin(\omega/2) \sum_{n=0}^{(M-1)/2} \tilde{d}[n] \cos\omega n.$$

As in Part (b), it is necessary to express $c[n]$ in terms of $\tilde{c}[n]$ and $d[n]$ in terms of $\tilde{d}[n]$ using the trigonometric identity $\sin\alpha \cos\beta = \tfrac{1}{2}\sin(\alpha + \beta) + \tfrac{1}{2}\sin(\alpha - \beta)$. McClellan and Parks (1973) and Rabiner and Gold (1975) give more details on issues raised in this problem.

7.53. In this problem, we consider a method of obtaining an implementation of a variable-cutoff linear-phase filter. Assume that we are given a zero-phase filter designed by the Parks–McClellan method. The frequency response of this filter can be represented as

$$A_e(e^{j\theta}) = \sum_{k=0}^{L} a_k (\cos \theta)^k,$$

and its system function can therefore be represented as

$$A_e(Z) = \sum_{k=0}^{L} a_k \left(\frac{Z + Z^{-1}}{2} \right)^k,$$

with $e^{j\theta} = Z$. (We use Z for the original system and z for the system to be obtained by transformation of the original system.)

(a) Using the preceding expression for the system function, draw a block diagram or flow graph of an implementation of the system that utilizes multiplications by the coefficients a_k, additions, and elemental systems having system function $(Z + Z^{-1})/2$.

(b) What is the length of the impulse response of the system? The overall system can be made causal by cascading the system with a delay of L samples. Distribute this delay as unit delays so that all parts of the network will be causal.

(c) Suppose that we obtain a new system function from $A_e(Z)$ by the substitution

$$B_e(z) = A_e(Z)\Big|_{(Z+Z^{-1})/2 = \alpha_0 + \alpha_1[(z+z^{-1})/2]}.$$

Using the flow graph obtained in Part (a), draw the flow graph of a system that implements the system function $B_e(z)$. What is the length of the impulse response of this system? Modify the network as in Part (b) to make the overall system and all parts of the network causal.

(d) If $A_e(e^{j\theta})$ is the frequency response of the original filter and $B_e(e^{j\omega})$ is the frequency response of the transformed filter, determine the relationship between θ and ω.

(e) The frequency response of the original optimal filter is shown in Figure P7.53-1. For the case $\alpha_1 = 1 - \alpha_0$ and $0 \le \alpha_0 < 1$, describe how the frequency response $B_e(e^{j\omega})$ changes as α_0 varies. *Hint:* Plot $A_e(e^{j\theta})$ and $B_e(e^{j\omega})$ as functions of $\cos \theta$ and $\cos \omega$. Are the resulting transformed filters also optimal in the sense of having the minimum maximum weighted approximation errors in the transformed passband and stopband?

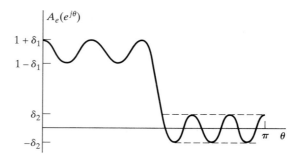

Figure P7.53-1

(f) *Optional.* Repeat Part (e) for the case $\alpha_1 = 1 + \alpha_0$ and $-1 < \alpha_0 \le 0$.

7.54. In this problem, we consider the effect of mapping continuous-time filters to discrete-time filters by replacing derivatives in the differential equation for a continuous-time filter by central differences to obtain a difference equation. The first central difference of a sequence $x[n]$ is defined as

$$\Delta^{(1)}\{x[n]\} = x[n+1] - x[n-1],$$

and the kth central difference is defined recursively as

$$\Delta^{(k)}\{x[n]\} = \Delta^{(1)}\{\Delta^{(k-1)}\{x[n]\}\}.$$

For consistency, the zeroth central difference is defined as

$$\Delta^{(0)}\{x[n]\} = x[n].$$

(a) If $X(z)$ is the z-transform of $x[n]$, determine the z-transform of $\Delta^{(k)}\{x[n]\}$.

The mapping of an LTI continuous-time filter to an LTI discrete-time filter is as follows: Let the continuous-time filter with input $x(t)$ and output $y(t)$ be specified by a differential equation of the form

$$\sum_{k=0}^{N} a_k \frac{d^k y(t)}{dt^k} = \sum_{r=0}^{M} b_r \frac{d^r x(t)}{dt^r}.$$

Then the corresponding discrete-time filter with input $x[n]$ and output $y[n]$ is specified by the difference equation

$$\sum_{k=0}^{N} a_k \Delta^{(k)}\{y[n]\} = \sum_{r=0}^{M} b_r \Delta^{(r)}\{x[n]\}.$$

(b) If $H_c(s)$ is a rational continuous-time system function and $H_d(z)$ is the discrete-time system function obtained by mapping the differential equation to a difference equation as indicated in Part (a), then

$$H_d(z) = H_e(s)\big|_{s=m(z)}.$$

Determine $m(z)$.

(c) Assume that $H_c(s)$ approximates a continuous-time lowpass filter with a cutoff frequency of $\Omega = 1$; i.e.,

$$H(j\Omega) \approx \begin{cases} 1, & |\Omega| < 1, \\ 0, & \text{otherwise.} \end{cases}$$

This filter is mapped to a discrete-time filter using central differences as discussed in Part (a). Sketch the approximate frequency response that you would expect for the discrete-time filter, assuming that it is stable.

7.55. Let $h[n]$ be the optimal type I equiripple lowpass filter shown in Fig. P7.55-1, designed with weighting function $W(e^{j\omega})$ and desired frequency response $H_d(e^{j\omega})$. For simplicity, assume that the filter is zero phase (i.e., noncausal). We will use $h[n]$ to design five different FIR filters as follows:

$$h_1[n] = h[-n],$$

$$h_2[n] = (-1)^n h[n],$$

$$h_3[n] = h[n] * h[n],$$

$$h_4[n] = h[n] - K\delta[n], \quad \text{where } K \text{ is a constant,}$$

$$h_5[n] = \begin{cases} h[n/2] & \text{for } n \text{ even,} \\ 0 & \text{otherwise.} \end{cases}$$

For each filter $h_i[n]$, determine whether $h_i[n]$ is optimal in the minimax sense. That is, determine whether

$$h_i[n] = \min_{h_i[n]} \max_{\omega \in F} \left(W(e^{j\omega}) |H_d(e^{j\omega}) - H_i(e^{j\omega})| \right)$$

for some choices of a piecewise-constant $H_d(e^{j\omega})$ and a piecewise-constant $W(e^{j\omega})$, where F is a union of disjoint closed intervals on $0 \le \omega \le \pi$. If $h_i[n]$ is optimal, determine the corresponding $H_d(e^{j\omega})$ and $W(e^{j\omega})$. If $h_i[n]$ is not optimal, explain why.

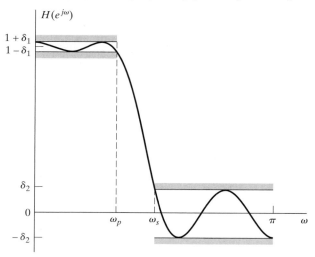

Figure P7.55-1

7.56. Suppose that you have used the Parks–McClellan algorithm to design a causal FIR linear-phase system. The system function of this system is denoted $H(z)$. The length of the impulse response is 25 samples, $h[n] = 0$ for $n < 0$ and for $n > 24$, and $h[0] \ne 0$. For each of the following questions, answer "true", "false", or "insufficient information given":
(a) $h[n + 12] = h[12 - n]$ or $h[n + 12] = -h[12 - n]$ for $-\infty < n < \infty$.
(b) The system has a stable and causal inverse.
(c) We know that $H(-1) = 0$.
(d) The maximum weighted approximation error is the same in all approximation bands.
(e) The system can be implemented by a signal flow graph that has no feedback paths.
(f) The group delay is positive for $0 < \omega < \pi$.

7.57. Consider the design of a type I bandpass linear-phase FIR filter using the Parks–McClellan algorithm. The impulse response length is $M + 1 = 2L + 1$. Recall that for type I systems, the frequency response is of the form $H(e^{j\omega}) = A_e(e^{j\omega})e^{-j\omega M/2}$, and the Parks–McClellan algorithm finds the function $A_e(e^{j\omega})$ that minimizes the maximum value of the error function

$$E(\omega) = W(\omega)[H_d(e^{j\omega}) - A_e(e^{j\omega})], \qquad \omega \in F,$$

where F is a closed subset of the interval $0 \le \omega \le \pi$, $W(\omega)$ is a weighting function, and $H_d(e^{j\omega})$ defines the desired frequency response in the approximation intervals F. The tolerance scheme for a bandpass filter is shown in Figure P7.57-1.
(a) Give the equation for the desired response $H_d(e^{j\omega})$ for the tolerance scheme in Figure P7.57-1.
(b) Give the equation for the weighting function $W(\omega)$ for the tolerance scheme in Figure P7.57-1.
(c) What is the *minimum* number of alternations of the error function for the optimum filter?
(d) What is the *maximum* number of alternations of the error function for the optimum filter?

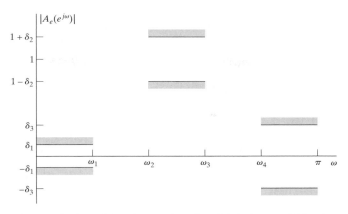

Figure P7.57-1

(e) Sketch a "typical" weighted error function $E(\omega)$ that could be the error function for an optimum bandpass filter if $M = 14$. Assume the *maximum* number of alternations.

(f) Now suppose that M, ω_1, ω_2, ω_3, the weighting function, and the desired function are kept the same, but ω_4 is *increased,* so that the transition band ($\omega_4 - \omega_3$) is increased. Will the optimum filter for these new specifications *necessarily* have a *smaller* value of the maximum approximation error than the optimum filter associated with the original specifications? Clearly show your reasoning.

(g) In the lowpass filter case, all local minima and maxima of $A_e(e^{j\omega})$ must occur in the approximation bands $\omega \in F$; they *cannot* occur in the "don't care" bands. Also, in the lowpass case, the local minima and maxima that occur in the approximation bands must be alternations of the error. Show that this is not necessarily true in the bandpass filter case. Specifically, use the alternation theorem to show (i) that local maxima and minima of $A_e(e^{j\omega})$ are not restricted to the approximation bands and (ii) that local maxima and minima in the approximation bands need not be alternations.

7.58. It is often desirable to transform a prototype discrete-time lowpass filter to another kind of discrete-time frequency-selective filter. In particular, the impulse invariance approach cannot be used to convert continuous-time highpass or bandstop filters to discrete-time highpass or bandstop filters. Consequently, the traditional approach has been to design a prototype lowpass discrete-time filter using either impulse invariance or the bilinear transformation and then to use an algebraic transformation to convert the discrete-time lowpass filter into the desired frequency-selective filter.

To see how this is done, assume that we are given a lowpass system function $H_{\text{lp}}(Z)$ that we wish to transform to a new system function $H(z)$, which has either lowpass, highpass, bandpass, or bandstop characteristics when it is evaluated on the unit circle. Note that we associate the complex variable Z with the prototype lowpass filter and the complex variable z with the transformed filter. Then we define a mapping from the Z-plane to the z-plane of the form

$$Z^{-1} = G(z^{-1}) \qquad (\text{P7.58-1})$$

such that

$$H(z) = H_{\text{lp}}(Z)\Big|_{Z^{-1}=G(z^{-1})}. \qquad (\text{P7.58-2})$$

Instead of expressing Z as a function of z, we have assumed in Eq. (P7.58-1) that Z^{-1} is expressed as a function of z^{-1}. Thus, according to Eq. (P7.58-2), in obtaining $H(z)$ from $H_{\text{lp}}(Z)$, we simply replace Z^{-1} everywhere in $H_{\text{lp}}(Z)$ by the function $G(z^{-1})$. This is a convenient representation because $H_{\text{lp}}(Z)$ is normally expressed as a rational function of Z^{-1}.

If $H_{\mathrm{lp}}(Z)$ is the rational system function of a causal and stable system, we naturally require that the transformed system function $H(z)$ be a rational function of z^{-1} and that the system also be causal and stable. This places the following constraints on the transformation $Z^{-1} = G(z^{-1})$:

1. $G(z^{-1})$ must be a rational function of z^{-1}.
2. The inside of the unit circle of the Z-plane must map to the inside of the unit circle of the z-plane.
3. The unit circle of the Z-plane must map onto the unit circle of the z-plane.

In this problem, you will derive and characterize the algebraic transformations necessary to convert a discrete-time lowpass filter into another lowpass filter with a different cutoff frequency or to a discrete-time highpass filter.

(a) Let θ and ω be the frequency variables (angles) in the Z-plane and z-plane, respectively, i.e., on the respective unit circles $Z = e^{j\theta}$ and $z = e^{j\omega}$. Show that, in order for Condition 3 to hold, $G(z^{-1})$ must be an all-pass system, i.e.,

$$|G(e^{-j\omega})| = 1. \tag{P7.58-3}$$

(b) It is possible to show that the most general form of $G(z^{-1})$ that satisfies all of the preceding three conditions is

$$Z^{-1} = G(z^{-1}) = \pm \prod_{k=1}^{N} \frac{z^{-1} - \alpha_k}{1 - \alpha_k z^{-1}}. \tag{P7.58-4}$$

From our discussion of all-pass systems in Chapter 5, it should be clear that $G(z^{-1})$, as given in Eq. (P7.58-4), satisfies Eq. (P7.58-3), i.e., is an allpass system, and thus meets Condition 3. Eq. (P7.58-4) also clearly meets Condition 1. Demonstrate that Condition 2 is satisfied if and only if $|\alpha_k| < 1$.

(c) A simple first-order $G(z^{-1})$ can be used to map a prototype lowpass filter $H_{\mathrm{lp}}(Z)$ with cutoff θ_p to a new filter $H(z)$ with cutoff ω_p. Demonstrate that

$$G(z^{-1}) = \frac{z^{-1} - \alpha}{1 - \alpha z^{-1}}$$

will produce the desired mapping for some value of α. Solve for α as a function of θ_p and ω_p. Problem 7.45 uses this approach to design lowpass filters with adjustable cutoff frequencies.

(d) Consider the case of a prototype lowpass filter with $\theta_p = \pi/2$. For each of the following choices of α, specify the resulting cutoff frequency ω_p for the transformed filter:
   (i) $\alpha = -0.2679$.
   (ii) $\alpha = 0$.
   (iii) $\alpha = 0.4142$.

(e) It is also possible to find a first-order all-pass system for $G(z^{-1})$ such that the prototype lowpass filter is transformed to a discrete-time highpass filter with cutoff ω_p. Note that such a transformation must map $Z^{-1} = e^{j\theta_p} \to z^{-1} = e^{j\omega_p}$ and also map $Z^{-1} = 1 \to z^{-1} = -1$; i.e., $\theta = 0$ maps to $\omega = \pi$. Find $G(z^{-1})$ for this transformation, and also, find an expression for α in terms of θ_p and ω_p.

(f) Using the same prototype filter and values for α as in Part (d), sketch the frequency responses for the highpass filters resulting from the transformation you specified in Part (e).

Similar, but more complicated, transformations can be used to convert the prototype lowpass filter $H_{\mathrm{lp}}(Z)$ into bandpass and bandstop filters. Constantinides (1970) describes these transformations in more detail.

8

THE DISCRETE FOURIER TRANSFORM

8.0 INTRODUCTION

In Chapters 2 and 3 we discussed the representation of sequences and linear time-invariant systems in terms of the Fourier and z-transforms, respectively. For finite-duration sequences, it is possible to develop an alternative Fourier representation, referred to as the *discrete Fourier transform* (DFT). The DFT is itself a sequence rather than a function of a continuous variable, and it corresponds to samples, equally spaced in frequency, of the Fourier transform of the signal. In addition to its theoretical importance as a Fourier representation of sequences, the DFT plays a central role in the implementation of a variety of digital signal-processing algorithms. This is because efficient algorithms exist for the computation of the DFT. These algorithms will be discussed in detail in Chapter 9. The application of the DFT to spectral analysis will be described in Chapter 10.

Although several points of view can be taken toward the derivation and interpretation of the DFT representation of a finite-duration sequence, we have chosen to base our presentation on the relationship between periodic sequences and finite-length sequences. We will begin by considering the Fourier series representation of periodic sequences. While this representation is important in its own right, we are most often interested in the application of Fourier series results to the representation of finite-length sequences. We accomplish this by constructing a periodic sequence for which each period is identical to the finite-length sequence. As we will see, the Fourier series representation of the periodic sequence corresponds to the DFT of the finite-length sequence. Thus, our approach is to define the Fourier series representation for periodic sequences and to study the properties of such representations. Then we repeat essentially the same derivations, assuming that the sequence to be represented is a finite-length sequence. This

541

approach to the DFT emphasizes the fundamental inherent periodicity of the DFT representation and ensures that this periodicity is not overlooked in applications of the DFT.

8.1 REPRESENTATION OF PERIODIC SEQUENCES: THE DISCRETE FOURIER SERIES

Consider a sequence $\tilde{x}[n]$ that is periodic[1] with period N, so that $\tilde{x}[n] = \tilde{x}[n + rN]$ for any integer values of n and r. As with continuous-time periodic signals, such a sequence can be represented by a Fourier series corresponding to a sum of harmonically related complex exponential sequences, i.e., complex exponentials with frequencies that are integer multiples of the fundamental frequency $(2\pi/N)$ associated with the periodic sequence $\tilde{x}[n]$. These periodic complex exponentials are of the form

$$e_k[n] = e^{j(2\pi/N)kn} = e_k[n + rN], \tag{8.1}$$

where k is an integer, and the Fourier series representation then has the form[2]

$$\tilde{x}[n] = \frac{1}{N} \sum_k \tilde{X}[k] e^{j(2\pi/N)kn}. \tag{8.2}$$

The Fourier series representation of a continuous-time periodic signal generally requires infinitely many harmonically related complex exponentials, whereas the Fourier series for any discrete-time signal with period N requires only N harmonically related complex exponentials. To see this, note that the harmonically related complex exponentials $e_k[n]$ in Eq. (8.1) are identical for values of k separated by N; i.e., $e_0[n] = e_N[n]$, $e_1[n] = e_{N+1}[n]$, and, in general,

$$e_{k+\ell N}[n] = e^{j(2\pi/N)(k+\ell N)n} = e^{j(2\pi/N)kn} e^{j2\pi\ell n} = e^{j(2\pi/N)kn} = e_k[n], \tag{8.3}$$

where ℓ is an integer. Consequently, the set of N periodic complex exponentials $e_0[n]$, $e_1[n], \ldots, e_{N-1}[n]$ defines all the distinct periodic complex exponentials with frequencies that are integer multiples of $(2\pi/N)$. Thus, the Fourier series representation of a periodic sequence $\tilde{x}[n]$ need contain only N of these complex exponentials, and hence, it has the form

$$\tilde{x}[n] = \frac{1}{N} \sum_{k=0}^{N-1} \tilde{X}[k] e^{j(2\pi/N)kn}. \tag{8.4}$$

To obtain the sequence of Fourier series coefficients $\tilde{X}[k]$ from the periodic sequence $\tilde{x}[n]$, we exploit the orthogonality of the set of complex exponential sequences. After multiplying both sides of Eq. (8.4) by $e^{-j(2\pi/N)rn}$ and summing from $n = 0$ to $n = N - 1$, we obtain

$$\sum_{n=0}^{N-1} \tilde{x}[n] e^{-j(2\pi/N)rn} = \sum_{n=0}^{N-1} \frac{1}{N} \sum_{k=0}^{N-1} \tilde{X}[k] e^{j(2\pi/N)(k-r)n}. \tag{8.5}$$

---

[1] Henceforth, we will use the tilde (˜) to denote periodic sequences whenever it is important to clearly distinguish between periodic and aperiodic sequences.

[2] The multiplicative constant $1/N$ is included in Eq. (8.2) for convenience. It could also be absorbed into the definition of $\tilde{X}[k]$.

After interchanging the order of summation on the right-hand side, we see that Eq. (8.5) becomes

$$\sum_{n=0}^{N-1} \tilde{x}[n] e^{-j(2\pi/N)rn} = \sum_{k=0}^{N-1} \tilde{X}[k] \left[\frac{1}{N} \sum_{n=0}^{N-1} e^{j(2\pi/N)(k-r)n} \right]. \tag{8.6}$$

The following identity expresses the orthogonality of the complex exponentials:

$$\frac{1}{N} \sum_{n=0}^{N-1} e^{j(2\pi/N)(k-r)n} = \begin{cases} 1, & k-r = mN, \quad m \text{ an integer,} \\ 0, & \text{otherwise.} \end{cases} \tag{8.7}$$

This identity can easily be proved (see Problem 8.51), and when it is applied to the summation in brackets in Eq. (8.6), the result is

$$\sum_{n=0}^{N-1} \tilde{x}[n] e^{-j(2\pi/N)rn} = \tilde{X}[r]. \tag{8.8}$$

Thus, the Fourier series coefficients $\tilde{X}[k]$ in Eq. (8.4) are obtained from $\tilde{x}[n]$ by the relation

$$\tilde{X}[k] = \sum_{n=0}^{N-1} \tilde{x}[n] e^{-j(2\pi/N)kn}. \tag{8.9}$$

Note that the sequence $\tilde{X}[k]$ is periodic with period N; i.e., $\tilde{X}[0] = \tilde{X}[N]$, $\tilde{X}[1] = \tilde{X}[N+1]$, and, more generally,

$$\tilde{X}[k+N] = \sum_{n=0}^{N-1} \tilde{x}[n] e^{-j(2\pi/N)(k+N)n}$$

$$= \left(\sum_{n=0}^{N-1} \tilde{x}[n] e^{-j(2\pi/N)kn} \right) e^{-j2\pi n} = \tilde{X}[k]$$

for any integer k.

The Fourier series coefficients can be interpreted to be a sequence of finite length, given by Eq. (8.9) for $k = 0, \ldots, (N-1)$, and zero otherwise, or as a periodic sequence defined for all k by Eq. (8.9). Clearly, both of these interpretations are acceptable, since in Eq. (8.4) we use only the values of $\tilde{X}[k]$ for $0 \le k \le (N-1)$. An advantage of interpreting the Fourier series coefficients $\tilde{X}[k]$ as a periodic sequence is that there is then a duality between the time and frequency domains for the Fourier series representation of periodic sequences. Equations (8.9) and (8.4) together are an analysis–synthesis pair and will be referred to as the *discrete Fourier series* (DFS) representation of a periodic sequence. For convenience in notation, these equations are often written in terms of the complex quantity

$$W_N = e^{-j(2\pi/N)}. \tag{8.10}$$

With this notation, the DFS analysis–synthesis pair is expressed as follows:

$$\text{Analysis equation:} \quad \tilde{X}[k] = \sum_{n=0}^{N-1} \tilde{x}[n] W_N^{kn}. \tag{8.11}$$

$$\text{Synthesis equation:} \quad \tilde{x}[n] = \frac{1}{N} \sum_{k=0}^{N-1} \tilde{X}[k] W_N^{-kn}. \tag{8.12}$$

In both of these equations, $\tilde{X}[k]$ and $\tilde{x}[n]$ are periodic sequences. We will sometimes find it convenient to use the notation

$$\tilde{x}[n] \overset{\mathcal{DFS}}{\longleftrightarrow} \tilde{X}[k] \tag{8.13}$$

to signify the relationships of Eqs. (8.11) and (8.12). The following examples illustrate the use of those equations.

Example 8.1 Discrete Fourier Series of a Periodic Impulse Train

We consider the periodic impulse train

$$\tilde{x}[n] = \sum_{r=-\infty}^{\infty} \delta[n - rN] = \begin{cases} 1, & n = rN, \quad r \text{ any integer,} \\ 0, & \text{otherwise,} \end{cases} \tag{8.14}$$

Since $\tilde{x}[n] = \delta[n]$ for $0 \leq n \leq N-1$, the DFS coefficients are found, using Eq. (8.11), to be

$$\tilde{X}[k] = \sum_{n=0}^{N-1} \delta[n] W_N^{kn} = W_N^0 = 1. \tag{8.15}$$

In this case, $\tilde{X}[k]$ is the same for all k. Thus, substituting Eq. (8.15) into Eq. (8.12) leads to the representation

$$\tilde{x}[n] = \sum_{r=-\infty}^{\infty} \delta[n - rN] = \frac{1}{N} \sum_{k=0}^{N-1} W_N^{-kn} = \frac{1}{N} \sum_{k=0}^{N-1} e^{j(2\pi/N)kn}. \tag{8.16}$$

(Note the similarity to the orthogonality relation of Eq. (8.7).)

Example 8.1 produced a useful representation of a periodic impulse train in terms of a sum of complex exponentials, where all the complex exponentials have the same magnitude and phase and add to unity at integer multiples of N and to zero for all other integers. If we look closely at Eqs. (8.11) and (8.12), we see that the two equations are very similar, differing only in a constant multiplier and the sign of the exponents. This duality between the periodic sequence $\tilde{x}[n]$ and its discrete Fourier series coefficients $\tilde{X}[k]$ is illustrated in the following example.

Example 8.2 Duality in the Discrete Fourier Series

Here we let the discrete Fourier series coefficients be the periodic impulse train

$$\tilde{Y}[k] = \sum_{r=-\infty}^{\infty} N\delta[k - rN].$$

Substituting $\tilde{Y}[k]$ into Eq. (8.12) gives

$$\tilde{y}[n] = \frac{1}{N} \sum_{k=0}^{N-1} N\delta[k] W_N^{-kn} = W_N^{-0} = 1.$$

In this case, $\tilde{y}[n] = 1$ for all n. Comparing this result with the results for $\tilde{x}[n]$ and $\tilde{X}[k]$ of Example 8.1, we see that $\tilde{Y}[k] = N\tilde{x}[k]$ and $\tilde{y}[n] = \tilde{X}[n]$. In Section 8.2.3, we will show that this example is a special case of a more general duality property.

If the sequence $\tilde{x}[n]$ is equal to unity over only part of one period, we can also obtain a closed-form expression for the DFS coefficients. This is illustrated by the following example.

Example 8.3   The Discrete Fourier Series of a Periodic Rectangular Pulse Train

For this example, $\tilde{x}[n]$ is the sequence shown in Figure 8.1, whose period is $N = 10$.

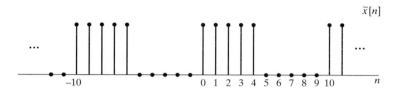

Figure 8.1 Periodic sequence with period $N = 10$ for which the Fourier series representation is to be computed.

From Eq. (8.11),

$$\tilde{X}[k] = \sum_{n=0}^{4} W_{10}^{kn} = \sum_{n=0}^{4} e^{-j(2\pi/10)kn}. \qquad (8.17)$$

This finite sum has the closed form

$$\tilde{X}[k] = \frac{1 - W_{10}^{5k}}{1 - W_{10}^{k}} = e^{-j(4\pi k/10)} \frac{\sin(\pi k/2)}{\sin(\pi k/10)}. \qquad (8.18)$$

The magnitude and phase of the periodic sequence $\tilde{X}[k]$ are shown in Figure 8.2.

(a)

(b)

Figure 8.2 Magnitude and phase of the Fourier series coefficients of the sequence of Figure 8.1.

We have shown that any periodic sequence can be represented as a sum of complex exponential sequences. The key results are summarized in Eqs. (8.11) and (8.12). As we will see, these relationships are the basis for the DFT, which focuses on finite-length sequences. Before discussing the DFT, however, we will consider some of the basic properties of the DFS representation of periodic sequences, and then we will show how we can use the DFS representation to obtain a Fourier transform representation of periodic signals.

8.2 PROPERTIES OF THE DISCRETE FOURIER SERIES

Just as with Fourier series and Fourier and Laplace transforms for continuous-time signals, and with z-transforms for discrete-time aperiodic sequences, certain properties of discrete Fourier series are of fundamental importance to its successful use in signal-processing problems. In this section, we summarize these important properties. It is not surprising that many of the basic properties are analogous to properties of the z-transform and Fourier transform. However, we will be careful to point out where the periodicity of both $\tilde{x}[n]$ and $\tilde{X}[k]$ results in some important distinctions. Furthermore, an exact duality exists between the time and frequency domains in the DFS representation that does not exist in the Fourier transform and z-transform representation of sequences.

8.2.1 Linearity

Consider two periodic sequences $\tilde{x}_1[n]$ and $\tilde{x}_2[n]$, both with period N, such that

$$\tilde{x}_1[n] \overset{\mathcal{DFS}}{\longleftrightarrow} \tilde{X}_1[k] \tag{8.19a}$$

and

$$\tilde{x}_2[n] \overset{\mathcal{DFS}}{\longleftrightarrow} \tilde{X}_2[k] \tag{8.19b}$$

Then

$$a\tilde{x}_1[n] + b\tilde{x}_2[n] \overset{\mathcal{DFS}}{\longleftrightarrow} a\tilde{X}_1[k] + b\tilde{X}_2[k]. \tag{8.20}$$

This linearity property follows immediately from the form of Eqs. (8.11) and (8.12).

8.2.2 Shift of a Sequence

If a periodic sequence $\tilde{x}[n]$ has Fourier coefficients $\tilde{X}[k]$, then $\tilde{x}[n - m]$ is a shifted version of $\tilde{x}[n]$, and

$$\tilde{x}[n - m] \overset{\mathcal{DFS}}{\longleftrightarrow} W_N^{km} \tilde{X}[k]. \tag{8.21}$$

The proof of this property is considered in Problem 8.52. Any shift that is greater than or equal to the period (i.e., $m \geq N$) cannot be distinguished in the time domain from a shorter shift m_1 such that $m = m_1 + m_2 N$, where m_1 and m_2 are integers and $0 \leq m_1 \leq N - 1$. (Another way of stating this is that $m_1 = m$ modulo N or, equivalently, m_1 is the remainder when m is divided by N.) It is easily shown that with this representation of m, $W_N^{km} = W_N^{km_1}$; i.e., as it must be, the ambiguity of the shift in the time domain is also manifest in the frequency-domain representation.

Because the sequence of Fourier series coefficients of a periodic sequence is a periodic sequence, a similar result applies to a shift in the Fourier coefficients by an integer ℓ. Specifically,

$$W_N^{-n\ell}\tilde{x}[n] \overset{\mathcal{DFS}}{\longleftrightarrow} \tilde{X}[k-\ell]. \tag{8.22}$$

Note the difference in the sign of the exponents in Eqs. (8.21) and (8.22).

8.2.3 Duality

Because of the strong similarity between the Fourier analysis and synthesis equations in continuous time, there is a duality between the time domain and frequency domain. However, for the discrete-time Fourier transform of aperiodic signals, no similar duality exists, since aperiodic signals and their Fourier transforms are very different kinds of functions: Aperiodic discrete-time signals are, of course, aperiodic sequences, while their Fourier transforms are always periodic functions of a continuous frequency variable.

From Eqs. (8.11) and (8.12), we see that the DFS analysis and synthesis equations differ only in a factor of $1/N$ and in the sign of the exponent of W_N. Furthermore, a periodic sequence and its DFS coefficients are the same kinds of functions; they are both periodic sequences. Specifically, taking account of the factor $1/N$ and the difference in sign in the exponent between Eqs. (8.11) and (8.12), it follows from Eq. (8.12) that

$$N\tilde{x}[-n] = \sum_{k=0}^{N-1} \tilde{X}[k]W_N^{kn} \tag{8.23}$$

or, interchanging the roles of n and k in Eq. (8.23),

$$N\tilde{x}[-k] = \sum_{n=0}^{N-1} \tilde{X}[n]W_N^{nk}. \tag{8.24}$$

We see that Eq. (8.24) is similar to Eq. (8.11). In other words, the sequence of DFS coefficients of the periodic sequence $\tilde{X}[n]$ is $N\tilde{x}[-k]$, i.e., the original periodic sequence in reverse order and multiplied by N. This duality property is summarized as follows: If

$$\tilde{x}[n] \overset{\mathcal{DFS}}{\longleftrightarrow} \tilde{X}[k], \tag{8.25a}$$

then

$$\tilde{X}[n] \overset{\mathcal{DFS}}{\longleftrightarrow} N\tilde{x}[-k]. \tag{8.25b}$$

8.2.4 Symmetry Properties

As we discussed in Section 2.8, the Fourier transform of an aperiodic sequence has a number of useful symmetry properties. The same basic properties also hold for the DFS representation of a periodic sequence. The derivation of these properties, which is similar in style to the derivations in Chapter 2, is left as an exercise. (See Problem 8.53.) The resulting properties are summarized for reference as properties 9–17 in Table 8.1 in Section 8.2.6.

8.2.5 Periodic Convolution

Let $\tilde{x}_1[n]$ and $\tilde{x}_2[n]$ be two periodic sequences, each with period N and with discrete Fourier series coefficients denoted by $\tilde{X}_1[k]$ and $\tilde{X}_2[k]$, respectively. If we form the product

$$\tilde{X}_3[k] = \tilde{X}_1[k]\tilde{X}_2[k], \tag{8.26}$$

then the periodic sequence $\tilde{x}_3[n]$ with Fourier series coefficients $\tilde{X}_3[k]$ is

$$\tilde{x}_3[n] = \sum_{m=0}^{N-1} \tilde{x}_1[m]\tilde{x}_2[n-m]. \tag{8.27}$$

This result is not surprising, since our previous experience with transforms suggests that multiplication of frequency-domain functions corresponds to convolution of time-domain functions and Eq. (8.27) looks very much like a convolution sum. Equation (8.27) involves the summation of values of the product of $\tilde{x}_1[m]$ with $\tilde{x}_2[n-m]$, which is a time-reversed and time-shifted version of $\tilde{x}_2[m]$, just as in aperiodic discrete convolution. However, the sequences in Eq. (8.27) are all periodic with period N, and the summation is over only one period. A convolution in the form of Eq. (8.27) is referred to as a *periodic convolution*. Just as with aperiodic convolution, periodic convolution is commutative; i.e.,

$$\tilde{x}_3[n] = \sum_{m=0}^{N-1} \tilde{x}_2[m]\tilde{x}_1[n-m]. \tag{8.28}$$

To demonstrate that $\tilde{X}_3[k]$, given by Eq. (8.26), is the sequence of Fourier coefficients corresponding to $\tilde{x}_3[n]$ given by Eq. (8.27), let us first apply the DFS analysis equation (8.11) to Eq. (8.27) to obtain

$$\tilde{X}_3[k] = \sum_{n=0}^{N-1} \left(\sum_{m=0}^{N-1} \tilde{x}_1[m]\tilde{x}_2[n-m] \right) W_N^{kn}, \tag{8.29}$$

which, after we interchange the order of summation, becomes

$$\tilde{X}_3[k] = \sum_{m=0}^{N-1} \tilde{x}_1[m] \left(\sum_{n=0}^{N-1} \tilde{x}_2[n-m]W_N^{kn} \right). \tag{8.30}$$

The inner sum on the index n is the DFS for the shifted sequence $\tilde{x}_2[n-m]$. Therefore, from the shifting property of Section 8.2.2, we obtain

$$\sum_{n=0}^{N-1} \tilde{x}_2[n-m]W_N^{kn} = W_N^{km}\tilde{X}_2[k],$$

which can be substituted into Eq. (8.30) to yield

$$\tilde{X}_3[k] = \sum_{m=0}^{N-1} \tilde{x}_1[m]W_N^{km}\tilde{X}_2[k] = \left(\sum_{m=0}^{N-1} \tilde{x}_1[m]W_N^{km} \right) \tilde{X}_2[k] = \tilde{X}_1[k]\tilde{X}_2[k]. \tag{8.31}$$

In summary,

$$\sum_{m=0}^{N-1} \tilde{x}_1[m]\tilde{x}_2[n-m] \overset{\mathcal{DFS}}{\longleftrightarrow} \tilde{X}_1[k]\tilde{X}_2[k]. \tag{8.32}$$

The periodic convolution of periodic sequences thus corresponds to multiplication of the corresponding periodic sequences of Fourier series coefficients.

Since periodic convolutions are somewhat different from aperiodic convolutions, it is worthwhile to consider the mechanics of evaluating Eq. (8.27). First note that Eq. (8.27) calls for the product of sequences $\tilde{x}_1[m]$ and $\tilde{x}_2[n-m] = \tilde{x}_2[-(m-n)]$ viewed as functions of m with n fixed. This is the same as for an aperiodic convolution, but with the following two major differences:

1. The sum is over the finite interval $0 \leq m \leq N-1$.
2. The values of $\tilde{x}_2[n-m]$ in the interval $0 \leq m \leq N-1$ repeat periodically for m outside of that interval.

These details are illustrated by the following example.

Example 8.4   Periodic Convolution

An illustration of the procedure for forming the periodic convolution of two periodic sequences corresponding to Eq. (8.27) is given in Figure 8.3, where we have illustrated

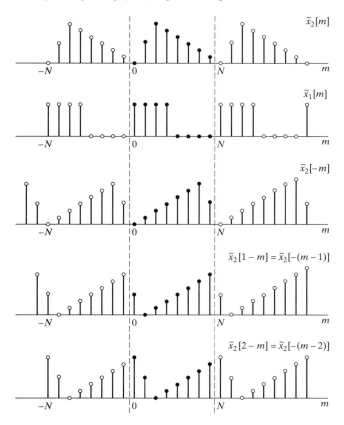

Figure 8.3   Procedure for forming the periodic convolution of two periodic sequences.

the sequences $\tilde{x}_2[m]$, $\tilde{x}_1[m]$, $\tilde{x}_2[-m]$, $\tilde{x}_2[1-m] = \tilde{x}_2[-(m-1)]$, and $\tilde{x}_2[2-m] = \tilde{x}_2[-(m-2)]$. To evaluate $\tilde{x}_3[n]$ in Eq. (8.27) for $n = 2$, for example, we multiply $\tilde{x}_1[m]$ by $\tilde{x}_2[2-m]$ and then sum the product terms $\tilde{x}_1[m]\tilde{x}_2[2-m]$ for $0 \le m \le N-1$, obtaining $\tilde{x}_3[2]$. As n changes, the sequence $\tilde{x}_2[n-m]$ shifts appropriately, and Eq. (8.27) is evaluated for each value of $0 \le n \le N-1$. Note that as the sequence $\tilde{x}_2[n-m]$ shifts to the right or left, values that leave the interval between the dotted lines at one end reappear at the other end because of the periodicity. Because of the periodicity of $\tilde{x}_3[n]$, there is no need to continue to evaluate Eq. (8.27) outside the interval $0 \le n \le N-1$.

The duality theorem (Section 8.2.3) suggests that if the roles of time and frequency are interchanged, we will obtain a result almost identical to the previous result. That is, the periodic sequence

$$\tilde{x}_3[n] = \tilde{x}_1[n]\tilde{x}_2[n], \tag{8.33}$$

where $\tilde{x}_1[n]$ and $\tilde{x}_2[n]$ are periodic sequences, each with period N, has the discrete Fourier series coefficients given by

$$\tilde{X}_3[k] = \frac{1}{N}\sum_{\ell=0}^{N-1}\tilde{X}_1[\ell]\tilde{X}_2[k-\ell], \tag{8.34}$$

corresponding to $1/N$ times the periodic convolution of $\tilde{X}_1[k]$ and $\tilde{X}_2[k]$. This result can also be verified by substituting $\tilde{X}_3[k]$, given by Eq. (8.34), into the Fourier series relation of Eq. (8.12) to obtain $\tilde{x}_3[n]$.

8.2.6 Summary of Properties of the DFS Representation of Periodic Sequences

The properties of the discrete Fourier series representation discussed in this section are summarized in Table 8.1.

TABLE 8.1 SUMMARY OF PROPERTIES OF THE DFS

Periodic Sequence (Period N)	DFS Coefficients (Period N)
1. $\tilde{x}[n]$	$\tilde{X}[k]$ periodic with period N
2. $\tilde{x}_1[n], \tilde{x}_2[n]$	$\tilde{X}_1[k], \tilde{X}_2[k]$ periodic with period N
3. $a\tilde{x}_1[n] + b\tilde{x}_2[n]$	$a\tilde{X}_1[k] + b\tilde{X}_2[k]$
4. $\tilde{X}[n]$	$N\tilde{x}[-k]$
5. $\tilde{x}[n-m]$	$W_N^{km}\tilde{X}[k]$
6. $W_N^{-\ell n}\tilde{x}[n]$	$\tilde{X}[k-\ell]$
7. $\sum_{m=0}^{N-1}\tilde{x}_1[m]\tilde{x}_2[n-m]$ (periodic convolution)	$\tilde{X}_1[k]\tilde{X}_2[k]$
8. $\tilde{x}_1[n]\tilde{x}_2[n]$	$\frac{1}{N}\sum_{\ell=0}^{N-1}\tilde{X}_1[\ell]\tilde{X}_2[k-\ell]$ (periodic convolution)
9. $\tilde{x}^*[n]$	$\tilde{X}^*[-k]$

(*continued*)

TABLE 8.1   (*Continued*)

Periodic Sequence (Period N)	DFS Coefficients (Period N)
10. $\tilde{x}^*[-n]$	$\tilde{X}^*[k]$
11. $\mathcal{R}e\{\tilde{x}[n]\}$	$\tilde{X}_e[k] = \frac{1}{2}(\tilde{X}[k] + \tilde{X}^*[-k])$
12. $j\mathcal{J}m\{\tilde{x}[n]\}$	$\tilde{X}_o[k] = \frac{1}{2}(\tilde{X}[k] - \tilde{X}^*[-k])$
13. $\tilde{x}_e[n] = \frac{1}{2}(\tilde{x}[n] + \tilde{x}^*[-n])$	$\mathcal{R}e\{\tilde{X}[k]\}$
14. $\tilde{x}_o[n] = \frac{1}{2}(\tilde{x}[n] - \tilde{x}^*[-n])$	$j\mathcal{J}m\{\tilde{X}[k]\}$
Properties 15–17 apply only when $x[n]$ is real.	
15. Symmetry properties for $\tilde{x}[n]$ real.	$\begin{cases} \tilde{X}[k] = \tilde{X}^*[-k] \\ \mathcal{R}e\{\tilde{X}[k]\} = \mathcal{R}e\{\tilde{X}[-k]\} \\ \mathcal{J}m\{\tilde{X}[k]\} = -\mathcal{J}m\{\tilde{X}[-k]\} \\ |\tilde{X}[k]| = |\tilde{X}[-k]| \\ \sphericalangle\tilde{X}[k] = -\sphericalangle\tilde{X}[-k] \end{cases}$
16. $\tilde{x}_e[n] = \frac{1}{2}(\tilde{x}[n] + \tilde{x}[-n])$	$\mathcal{R}e\{\tilde{X}[k]\}$
17. $\tilde{x}_0[n] = \frac{1}{2}(\tilde{x}[n] - \tilde{x}[-n])$	$j\mathcal{J}m\{\tilde{X}[k]\}$

8.3 THE FOURIER TRANSFORM OF PERIODIC SIGNALS

As discussed in Section 2.7, uniform convergence of the Fourier transform of a sequence requires that the sequence be absolutely summable, and mean-square convergence requires that the sequence be square summable. Periodic sequences satisfy neither condition, because they do not approach zero as n approaches $\pm\infty$. However, as we discussed briefly in Section 2.7, sequences that can be expressed as a sum of complex exponentials can be considered to have a Fourier transform representation in the form of Eq. (2.152), i.e., as a train of impulses. Similarly, it is often useful to incorporate the discrete Fourier series representation of periodic signals within the framework of the Fourier transform. This can be done by interpreting the Fourier transform of a periodic signal to be an impulse train in the frequency domain with the impulse values proportional to the DFS coefficients for the sequence. Specifically, if $\tilde{x}[n]$ is periodic with period N and the corresponding discrete Fourier series coefficients are $\tilde{X}[k]$, then the Fourier transform of $\tilde{x}[n]$ is defined to be the impulse train

$$\tilde{X}(e^{j\omega}) = \sum_{k=-\infty}^{\infty} \frac{2\pi}{N} \tilde{X}[k]\delta\left(\omega - \frac{2\pi k}{N}\right). \tag{8.35}$$

Note that $\tilde{X}(e^{j\omega})$ has the necessary periodicity with period 2π since $\tilde{X}[k]$ is periodic with period N and the impulses are spaced at integer multiples of $2\pi/N$, where N is an integer. To show that $\tilde{X}(e^{j\omega})$ as defined in Eq. (8.35) is a Fourier transform representation of the periodic sequence $\tilde{x}[n]$, we substitute Eq. (8.35) into the inverse Fourier transform equation (2.133); i.e.,

$$\frac{1}{2\pi}\int_{0-\epsilon}^{2\pi-\epsilon} \tilde{X}(e^{j\omega})e^{j\omega n}d\omega = \frac{1}{2\pi}\int_{0-\epsilon}^{2\pi-\epsilon} \sum_{k=-\infty}^{\infty} \frac{2\pi}{N}\tilde{X}[k]\delta\left(\omega - \frac{2\pi k}{N}\right)e^{j\omega n}d\omega, \tag{8.36}$$

where ϵ satisfies the inequality $0 < \epsilon < (2\pi/N)$. Recall that in evaluating the inverse Fourier transform, we can integrate over *any* interval of length 2π, since the integrand

$\tilde{X}(e^{j\omega})e^{j\omega n}$ is periodic with period 2π. In Eq. (8.36) the integration limits are denoted $0-\epsilon$ and $2\pi-\epsilon$, which means that the integration is from just before $\omega = 0$ to just before $\omega = 2\pi$. These limits are convenient because they include the impulse at $\omega = 0$ and exclude the impulse at $\omega = 2\pi$. Interchanging the order of integration and summation leads to

$$\frac{1}{2\pi}\int_{0-\epsilon}^{2\pi-\epsilon}\tilde{X}(e^{j\omega})e^{j\omega n}d\omega = \frac{1}{N}\sum_{k=-\infty}^{\infty}\tilde{X}[k]\int_{0-\epsilon}^{2\pi-\epsilon}\delta\left(\omega - \frac{2\pi k}{N}\right)e^{j\omega n}d\omega$$

$$= \frac{1}{N}\sum_{k=0}^{N-1}\tilde{X}[k]e^{j(2\pi/N)kn}.$$
(8.37)

The final form of Eq. (8.37) results because only the impulses corresponding to $k = 0, 1, \ldots, (N-1)$ are included in the interval between $\omega = 0 - \epsilon$ and $\omega = 2\pi - \epsilon$.

Comparing Eq. (8.37) and Eq. (8.12), we see that the final right-hand side of Eq. (8.37) is exactly equal to the Fourier series representation for $\tilde{x}[n]$, as specified by Eq. (8.12). Consequently, the inverse Fourier transform of the impulse train in Eq. (8.35) is the periodic signal $\tilde{x}[n]$, as desired.

Although the Fourier transform of a periodic sequence does not converge in the normal sense, the introduction of impulses permits us to include periodic sequences formally within the framework of Fourier transform analysis. This approach was also used in Chapter 2 to obtain a Fourier transform representation of other nonsummable sequences, such as the two-sided constant sequence (Example 2.23) or the complex exponential sequence (Example 2.24). Although the discrete Fourier series representation is adequate for most purposes, the Fourier transform representation of Eq. (8.35) sometimes leads to simpler or more compact expressions and simplified analysis.

Example 8.5   The Fourier Transform of a Periodic Impulse Train

Consider the periodic impulse train

$$\tilde{p}[n] = \sum_{r=-\infty}^{\infty}\delta[n-rN],$$
(8.38)

which is the same as the periodic sequence $\tilde{x}[n]$ considered in Example 8.1. From the results of that example, it follows that

$$\tilde{P}[k] = 1, \qquad \text{for all } k.$$
(8.39)

Therefore, the Fourier transform of $\tilde{p}[n]$ is

$$\tilde{P}(e^{j\omega}) = \sum_{k=-\infty}^{\infty}\frac{2\pi}{N}\delta\left(\omega - \frac{2\pi k}{N}\right).$$
(8.40)

The result of Example 8.5 is the basis for a useful interpretation of the relation between a periodic signal and a finite-length signal. Consider a finite-length signal $x[n]$

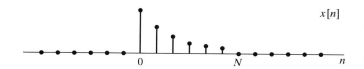

Figure 8.4   Periodic sequence $\tilde{x}[n]$ formed by repeating a finite-length sequence, $x[n]$, periodically. Alternatively, $x[n] = \tilde{x}[n]$ over one period and is zero otherwise.

such that $x[n] = 0$ except in the interval $0 \le n \le N - 1$, and consider the convolution of $x[n]$ with the periodic impulse train $\tilde{p}[n]$ of Example 8.5:

$$\tilde{x}[n] = x[n] * \tilde{p}[n] = x[n] * \sum_{r=-\infty}^{\infty} \delta[n - rN]$$
$$= \sum_{r=-\infty}^{\infty} x[n - rN]. \tag{8.41}$$

Equation (8.41) states that $\tilde{x}[n]$ consists of a set of periodically repeated copies of the finite-length sequence $x[n]$. Figure 8.4 illustrates how a periodic sequence $\tilde{x}[n]$ can be formed from a finite-length sequence $x[n]$ through Eq. (8.41). The Fourier transform of $x[n]$ is $X(e^{j\omega})$, and the Fourier transform of $\tilde{x}[n]$ is

$$\tilde{X}(e^{j\omega}) = X(e^{j\omega})\tilde{P}(e^{j\omega})$$

$$= X(e^{j\omega}) \sum_{k=-\infty}^{\infty} \frac{2\pi}{N} \delta\left(\omega - \frac{2\pi k}{N}\right) \tag{8.42}$$

$$= \sum_{k=-\infty}^{\infty} \frac{2\pi}{N} X(e^{j(2\pi/N)k})\delta\left(\omega - \frac{2\pi k}{N}\right).$$

Comparing Eq. (8.42) with Eq. (8.35), we conclude that

$$\tilde{X}[k] = X(e^{j(2\pi/N)k}) = X(e^{j\omega})\big|_{\omega=(2\pi/N)k}. \tag{8.43}$$

In other words, the periodic sequence $\tilde{X}[k]$ of DFS coefficients in Eq. (8.11) has an interpretation as equally spaced samples of the Fourier transform of the finite-length sequence obtained by extracting one period of $\tilde{x}[n]$; i.e.,

$$x[n] = \begin{cases} \tilde{x}[n], & 0 \le n \le N - 1, \\ 0, & \text{otherwise.} \end{cases} \tag{8.44}$$

This is also consistent with Figure 8.4, where it is clear that $x[n]$ can be obtained from $\tilde{x}[n]$ using Eq. (8.44). We can verify Eq. (8.43) in yet another way. Since $x[n] = \tilde{x}[n]$ for

$0 \le n \le N-1$ and $x[n] = 0$ otherwise,

$$X(e^{j\omega}) = \sum_{n=0}^{N-1} x[n]e^{-j\omega n} = \sum_{n=0}^{N-1} \tilde{x}[n]e^{-j\omega n}. \tag{8.45}$$

Comparing Eq. (8.45) and Eq. (8.11), we see again that

$$\tilde{X}[k] = X(e^{j\omega})|_{\omega=2\pi k/N}. \tag{8.46}$$

This corresponds to sampling the Fourier transform at N equally spaced frequencies between $\omega = 0$ and $\omega = 2\pi$ with a frequency spacing of $2\pi/N$.

Example 8.6   Relationship Between the Fourier Series Coefficients and the Fourier Transform of One Period

We again consider the sequence $\tilde{x}[n]$ of Example 8.3, which is shown in Figure 8.1. One period of $\tilde{x}[n]$ for the sequence in Figure 8.1 is

$$x[n] = \begin{cases} 1, & 0 \le n \le 4, \\ 0, & \text{otherwise.} \end{cases} \tag{8.47}$$

(a)

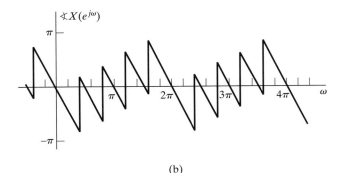

(b)

Figure 8.5   Magnitude and phase of the Fourier transform of one period of the sequence in Figure 8.1.

The Fourier transform of one period of $\tilde{x}[n]$ is given by

$$X(e^{j\omega}) = \sum_{n=0}^{4} e^{-j\omega n} = e^{-j2\omega}\frac{\sin(5\omega/2)}{\sin(\omega/2)}. \tag{8.48}$$

Equation (8.46) can be shown to be satisfied for this example by substituting $\omega = 2\pi k/10$ into Eq. (8.48), giving

$$\tilde{X}[k] = e^{-j(4\pi k/10)}\frac{\sin(\pi k/2)}{\sin(\pi k/10)},$$

which is identical to the result in Eq. (8.18). The magnitude and phase of $X(e^{j\omega})$ are sketched in Figure 8.5. Note that the phase is discontinuous at the frequencies where $X(e^{j\omega}) = 0$. That the sequences in Figures 8.2(a) and (b) correspond to samples of Figures 8.5(a) and (b), respectively, is demonstrated in Figure 8.6, where Figures 8.2 and 8.5 have been superimposed.

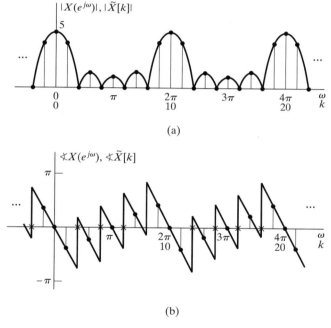

(a)

(b)

Figure 8.6 Overlay of Figures 8.2 and 8.5 illustrating the DFS coefficients of a periodic sequence as samples of the Fourier transform of one period.

8.4 SAMPLING THE FOURIER TRANSFORM

In this section, we discuss with more generality the relationship between an aperiodic sequence with Fourier transform $X(e^{j\omega})$ and the periodic sequence for which the DFS coefficients correspond to samples of $X(e^{j\omega})$ equally spaced in frequency. We will find this relationship to be particularly important when we discuss the discrete Fourier transform and its properties later in the chapter.

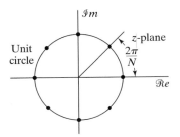

Figure 8.7   Points on the unit circle at which $X(z)$ is sampled to obtain the periodic sequence $\tilde{X}[k]$ ($N = 8$).

Consider an aperiodic sequence $x[n]$ with Fourier transform $X(e^{j\omega})$, and assume that a sequence $\tilde{X}[k]$ is obtained by sampling $X(e^{j\omega})$ at frequencies $\omega_k = 2\pi k/N$; i.e.,

$$\tilde{X}[k] = X(e^{j\omega})|_{\omega=(2\pi/N)k} = X(e^{j(2\pi/N)k}). \tag{8.49}$$

Since the Fourier transform is periodic in ω with period 2π, the resulting sequence is periodic in k with period N. Also, since the Fourier transform is equal to the z-transform evaluated on the unit circle, it follows that $\tilde{X}[k]$ can also be obtained by sampling $X(z)$ at N equally spaced points on the unit circle. Thus,

$$\tilde{X}[k] = X(z)|_{z=e^{j(2\pi/N)k}} = X(e^{j(2\pi/N)k}). \tag{8.50}$$

These sampling points are depicted in Figure 8.7 for $N = 8$. The figure makes it clear that the sequence of samples is periodic, since the N points are equally spaced starting with zero angle. Therefore, the same sequence repeats as k varies outside the range $0 \leq k \leq N - 1$.

Note that the sequence of samples $\tilde{X}[k]$, being periodic with period N, *could* be the sequence of discrete Fourier series coefficients of a sequence $\tilde{x}[n]$. To obtain that sequence, we can simply substitute $\tilde{X}[k]$ obtained by sampling into Eq. (8.12):

$$\tilde{x}[n] = \frac{1}{N} \sum_{k=0}^{N-1} \tilde{X}[k] W_N^{-kn}. \tag{8.51}$$

Since we have made no assumption about $x[n]$ other than that the Fourier transform exists,

$$X(e^{j\omega}) = \sum_{m=-\infty}^{\infty} x[m]e^{-j\omega m}. \tag{8.52}$$

Substituting Eq. (8.52) into Eq. (8.49) and then substituting the resulting expression for $\tilde{X}[k]$ into Eq. (8.51) gives

$$\tilde{x}[n] = \frac{1}{N} \sum_{k=0}^{N-1} \left[\sum_{m=-\infty}^{\infty} x[m]e^{-j(2\pi/N)km} \right] W_N^{-kn}, \tag{8.53}$$

which, after we interchange the order of summation, becomes

$$\tilde{x}[n] = \sum_{m=-\infty}^{\infty} x[m] \left[\frac{1}{N} \sum_{k=0}^{N-1} W_N^{-k(n-m)} \right] = \sum_{m=-\infty}^{\infty} x[m]\tilde{p}[n - m]. \tag{8.54}$$

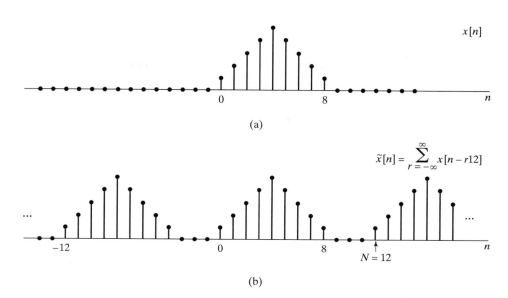

Figure 8.8   (a) Finite-length sequence $x[n]$. (b) Periodic sequence $\tilde{x}[n]$ corresponding to sampling the Fourier transform of $x[n]$ with $N = 12$.

The term in brackets in Eq. (8.54) can be seen from either Eq. (8.7) or Eq. (8.16) to be the Fourier series representation of the periodic impulse train of Examples 8.1 and 8.2. Specifically,

$$\tilde{p}[n - m] = \frac{1}{N} \sum_{k=0}^{N-1} W_N^{-k(n-m)} = \sum_{r=-\infty}^{\infty} \delta[n - m - rN] \qquad (8.55)$$

and therefore,

$$\tilde{x}[n] = x[n] * \sum_{r=-\infty}^{\infty} \delta[n - rN] = \sum_{r=-\infty}^{\infty} x[n - rN], \qquad (8.56)$$

where $*$ denotes aperiodic convolution. That is, $\tilde{x}[n]$ is the periodic sequence that results from the aperiodic convolution of $x[n]$ with a periodic unit-impulse train. Thus, the periodic sequence $\tilde{x}[n]$, corresponding to $\tilde{X}[k]$ obtained by sampling $X(e^{j\omega})$, is formed from $x[n]$ by adding together an infinite number of shifted replicas of $x[n]$. The shifts are all the positive and negative integer multiples of N, the period of the sequence $\tilde{X}[k]$. This is illustrated in Figure 8.8, where the sequence $x[n]$ is of length 9 and the value of N in Eq. (8.56) is $N = 12$. Consequently, the delayed replications of $x[n]$ do not overlap, and one period of the periodic sequence $\tilde{x}[n]$ is recognizable as $x[n]$. This is consistent with the discussion in Section 8.3 and Example 8.6 wherein we showed that the Fourier series coefficients for a periodic sequence are samples of the Fourier transform of one period. In Figure 8.9 the same sequence $x[n]$ is used, but the value of N is now $N = 7$. In this case the replicas of $x[n]$ overlap and one period of $\tilde{x}[n]$ is no longer identical to $x[n]$. In both cases, however, Eq. (8.49) still holds; i.e., in both

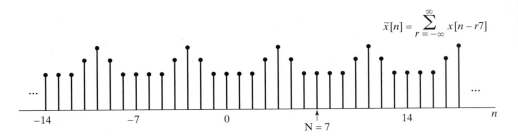

Figure 8.9 Periodic sequence $\tilde{x}[n]$ corresponding to sampling the Fourier transform of $x[n]$ in Figure 8.8(a) with $N = 7$.

cases the DFS coefficients of $\tilde{x}[n]$ are samples of the Fourier transform of $x[n]$ spaced in frequency at integer multiples of $2\pi/N$. This discussion should be reminiscent of our discussion of sampling in Chapter 4. The difference is that here we are sampling in the frequency domain rather than in the time domain. However, the general outlines of the mathematical representations are very similar.

For the example in Figure 8.8, the original sequence $x[n]$ can be recovered from $\tilde{x}[n]$ by extracting one period. Equivalently, the Fourier transform $X(e^{j\omega})$ can be recovered from the samples spaced in frequency by $2\pi/12$. In contrast, in Figure 8.9, $x[n]$ cannot be recovered by extracting one period of $\tilde{x}[n]$, and, equivalently, $X(e^{j\omega})$ cannot be recovered from its samples if the sample spacing is only $2\pi/7$. In effect, for the case illustrated in Figure 8.8, the Fourier transform of $x[n]$ has been sampled at a sufficiently small spacing (in frequency) to be able to recover it from these samples, whereas Figure 8.9 represents a case for which the Fourier transform has been undersampled. The relationship between $x[n]$ and one period of $\tilde{x}[n]$ in the undersampled case can be thought of as a form of aliasing in the time domain, essentially identical to the frequency-domain aliasing (discussed in Chapter 4) that results from undersampling in the time domain. Obviously, time-domain aliasing can be avoided only if $x[n]$ has finite length, just as frequency-domain aliasing can be avoided only for signals that have bandlimited Fourier transforms.

This discussion highlights several important concepts that will play a central role in the remainder of the chapter. We have seen that samples of the Fourier transform of an aperiodic sequence $x[n]$ can be thought of as DFS coefficients of a periodic sequence $\tilde{x}[n]$ obtained through summing periodic replicas of $x[n]$. If $x[n]$ has finite length and we take a sufficient number of equally spaced samples of its Fourier transform (specifically, a number greater than or equal to the length of $x[n]$), then the Fourier transform is recoverable from these samples, and, equivalently, $x[n]$ is recoverable from the corresponding periodic sequence $\tilde{x}[n]$ through the relation

$$x[n] = \begin{cases} \tilde{x}[n], & 0 \le n \le N - 1, \\ 0, & \text{otherwise.} \end{cases} \tag{8.57}$$

A direct relationship between $X(e^{j\omega})$ and its samples $\tilde{X}[k]$, i.e., an interpolation formula for $X(e^{j\omega})$, can be derived (see Problem 8.54). However, the essence of our previous discussion is that to represent or to recover $x[n]$ it is not necessary to know

$X(e^{j\omega})$ at all frequencies if $x[n]$ has finite length. Given a finite-length sequence $x[n]$, we can form a periodic sequence using Eq. (8.56), which in turn can be represented by a discrete Fourier series. Alternatively, given the sequence of Fourier coefficients $\tilde{X}[k]$, we can find $\tilde{x}[n]$ and then use Eq. (8.57) to obtain $x[n]$. When the Fourier series is used in this way to represent finite-length sequences, it is called the discrete Fourier transform (DFT). In developing, discussing, and applying the DFT, it is always important to remember that the representation through samples of the Fourier transform is in effect a representation of the finite-duration sequence by a periodic sequence, one period of which is the finite-duration sequence that we wish to represent.

8.5 FOURIER REPRESENTATION OF FINITE-DURATION SEQUENCES: THE DISCRETE FOURIER TRANSFORM

In this section, we formalize the point of view suggested at the end of the previous section. We begin by considering a finite-length sequence $x[n]$ of length N samples such that $x[n] = 0$ outside the range $0 \leq n \leq N - 1$. In many instances, we will want to assume that a sequence has length N even if its length is $M \leq N$. In such cases, we simply recognize that the last $(N - M)$ samples are zero. To each finite-length sequence of length N, we can always associate a periodic sequence

$$\tilde{x}[n] = \sum_{r=-\infty}^{\infty} x[n - rN]. \qquad (8.58a)$$

The finite-length sequence $x[n]$ can be recovered from $\tilde{x}[n]$ through Eq. (8.57), i.e.,

$$x[n] = \begin{cases} \tilde{x}[n], & 0 \leq n \leq N - 1, \\ 0, & \text{otherwise.} \end{cases} \qquad (8.58b)$$

Recall from Section 8.4 that the DFS coefficients of $\tilde{x}[n]$ are samples (spaced in frequency by $2\pi/N$) of the Fourier transform of $x[n]$. Since $x[n]$ is assumed to have finite length N, there is no overlap between the terms $x[n - rN]$ for different values of r. Thus, Eq. (8.58a) can alternatively be written as

$$\tilde{x}[n] = x[(n \text{ modulo } N)]. \qquad (8.59)$$

For convenience, we will use the notation $((n))_N$ to denote (n modulo N); with this notation, Eq. (8.59) is expressed as

$$\tilde{x}[n] = x[((n))_N]. \qquad (8.60)$$

Note that Eq. (8.60) is equivalent to Eq. (8.58a) *only* when $x[n]$ has length less than or equal to N. The finite-duration sequence $x[n]$ is obtained from $\tilde{x}[n]$ by extracting one period, as in Eq. (8.58b).

One informal and useful way of visualizing Eq. (8.59) is to think of wrapping a plot of the finite-duration sequence $x[n]$ around a cylinder with a circumference equal to the length of the sequence. As we repeatedly traverse the circumference of the cylinder, we see the finite-length sequence periodically repeated. With this interpretation, represen-tation of the finite-length sequence by a periodic sequence corresponds to wrapping the sequence around the cylinder; recovering the finite-length sequence from the periodic sequence using Eq. (8.58b) can be visualized as unwrapping the cylinder and laying it flat so that the sequence is displayed on a linear time axis rather than a circular (modulo N) time axis.

As defined in Section 8.1, the sequence of discrete Fourier series coefficients $\tilde{X}[k]$ of the periodic sequence $\tilde{x}[n]$ is itself a periodic sequence with period N. To maintain a duality between the time and frequency domains, we will choose the Fourier coefficients that we associate with a finite-duration sequence to be a finite-duration sequence cor-responding to one period of $\tilde{X}[k]$. This finite-duration sequence, $X[k]$, will be referred to as the discrete Fourier transform (DFT). Thus, the DFT, $X[k]$, is related to the DFS coefficients, $\tilde{X}[k]$, by

$$X[k] = \begin{cases} \tilde{X}[k], & 0 \leq k \leq N-1, \\ 0, & \text{otherwise,} \end{cases} \tag{8.61}$$

and

$$\tilde{X}[k] = X[(k \text{ modulo } N)] = X[((k))_N]. \tag{8.62}$$

From Section 8.1, $\tilde{X}[k]$ and $\tilde{x}[n]$ are related by

$$\tilde{X}[k] = \sum_{n=0}^{N-1} \tilde{x}[n] W_N^{kn}, \tag{8.63}$$

$$\tilde{x}[n] = \frac{1}{N} \sum_{k=0}^{N-1} \tilde{X}[k] W_N^{-kn}. \tag{8.64}$$

Since the summations in Eqs. (8.63) and (8.64) involve only the interval between zero and $(N-1)$, it follows from Eqs. (8.58b)–(8.64) that

$$X[k] = \begin{cases} \displaystyle\sum_{n=0}^{N-1} x[n] W_N^{kn}, & 0 \leq k \leq N-1, \\ 0, & \text{otherwise,} \end{cases} \tag{8.65}$$

$$x[n] = \begin{cases} \displaystyle\frac{1}{N} \sum_{k=0}^{N-1} X[k] W_N^{-kn}, & 0 \leq n \leq N-1, \\ 0, & \text{otherwise.} \end{cases} \tag{8.66}$$

Generally, the DFT analysis and synthesis equations are written as follows:

$$\text{Analysis equation:} \quad X[k] = \sum_{n=0}^{N-1} x[n] W_N^{kn}, \tag{8.67}$$

$$\text{Synthesis equation:} \quad x[n] = \frac{1}{N} \sum_{k=0}^{N-1} X[k] W_N^{-kn}. \tag{8.68}$$

That is, the fact that $X[k] = 0$ for k outside the interval $0 \leq k \leq N-1$ and that $x[n] = 0$ for n outside the interval $0 \leq n \leq N-1$ is implied, but not always stated explicitly. The relationship between $x[n]$ and $X[k]$ implied by Eqs. (8.67) and (8.68) will sometimes be denoted as

$$x[n] \overset{\mathcal{DFT}}{\longleftrightarrow} X[k]. \tag{8.69}$$

In recasting Eqs. (8.11) and (8.12) in the form of Eqs. (8.67) and (8.68) for finite-duration sequences, we have not eliminated the inherent periodicity. As with the DFS, the DFT $X[k]$ is equal to samples of the periodic Fourier transform $X(e^{j\omega})$, and if Eq. (8.68) is evaluated for values of n outside the interval $0 \leq n \leq N-1$, the result will not be zero, but rather a periodic extension of $x[n]$. The inherent periodicity is always present. Sometimes it causes us difficulty and sometimes we can exploit it, but to totally ignore it is to invite trouble. In defining the DFT representation, we are simply recognizing that we are *interested* in values of $x[n]$ only in the interval $0 \leq n \leq N-1$ because $x[n]$ is really zero outside that interval, and we are *interested* in values of $X[k]$ only in the interval $0 \leq k \leq N-1$ because these are the only values needed in Eq. (8.68).

Example 8.7   The DFT of a Rectangular Pulse

To illustrate the DFT of a finite-duration sequence, consider $x[n]$ shown in Figure 8.10(a). In determining the DFT, we can consider $x[n]$ as a finite-duration sequence with any length greater than or equal to $N = 5$. Considered as a sequence of length $N = 5$, the periodic sequence $\tilde{x}[n]$ whose DFS corresponds to the DFT of $x[n]$ is shown in Figure 8.10(b). Since the sequence in Figure 8.10(b) is constant over the interval $0 \leq n \leq 4$, it follows that

$$\tilde{X}[k] = \sum_{n=0}^{4} e^{-j(2\pi k/5)n} = \frac{1 - e^{-j2\pi k}}{1 - e^{-j(2\pi k/5)}}$$

$$= \begin{cases} 5, & k = 0, \pm 5, \pm 10, \ldots, \\ 0, & \text{otherwise;} \end{cases} \tag{8.70}$$

i.e., the only nonzero DFS coefficients $\tilde{X}[k]$ are at $k = 0$ and integer multiples of $k = 5$ (all of which represent the same complex exponential frequency). The DFS coefficients are shown in Figure 8.10(c). Also shown is the magnitude of the Fourier transform, $|X(e^{j\omega})|$. Clearly, $\tilde{X}[k]$ is a sequence of samples of $X(e^{j\omega})$ at frequencies

(a)

(b)

(c)

(d)

Figure 8.10   Illustration of the DFT. (a) Finite-length sequence $x[n]$. (b) Periodic sequence $\tilde{x}[n]$ formed from $x[n]$ with period $N = 5$. (c) Fourier series coefficients $\tilde{X}[k]$ for $\tilde{x}[n]$. To emphasize that the Fourier series coefficients are samples of the Fourier transform, $|X(e^{j\omega})|$ is also shown. (d) DFT of $x[n]$.

$\omega_k = 2\pi k/5$. According to Eq. (8.61), the five-point DFT of $x[n]$ corresponds to the finite-length sequence obtained by extracting one period of $\tilde{X}[k]$. Consequently, the five-point DFT of $x[n]$ is shown in Figure 8.10(d).

If, instead, we consider $x[n]$ to be of length $N = 10$, then the underlying periodic sequence is that shown in Figure 8.11(b), which is the periodic sequence considered

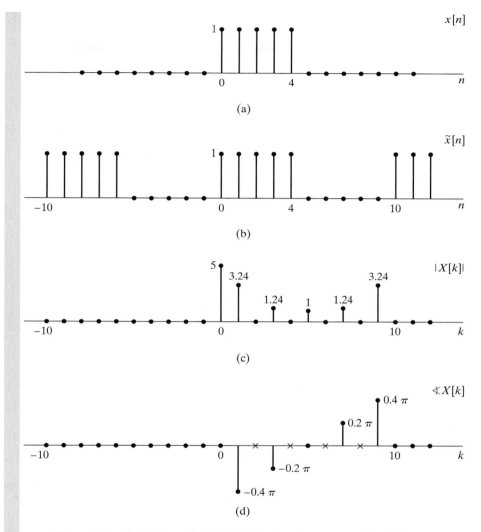

Figure 8.11   Illustration of the DFT. (a) Finite-length sequence $x[n]$. (b) Periodic sequence $\tilde{x}[n]$ formed from $x[n]$ with period $N = 10$. (c) DFT magnitude. (d) DFT phase. (x's indicate indeterminate values.)

in Example 8.3. Therefore, $\tilde{X}[k]$ is as shown in Figures 8.2 and 8.6, and the 10-point DFT $X[k]$ shown in Figures 8.11(c) and 8.11(d) is one period of $\tilde{X}[k]$.

The distinction between the finite-duration sequence $x[n]$ and the periodic sequence $\tilde{x}[n]$ related through Eqs. (8.57) and (8.60) may seem minor, since, by using these equations, it is straightforward to construct one from the other. However, the distinction becomes important in considering properties of the DFT and in considering the effect on $x[n]$ of modifications to $X[k]$. This will become evident in the next section, where we discuss the properties of the DFT representation.

8.6 PROPERTIES OF THE DISCRETE FOURIER TRANSFORM

In this section, we consider a number of properties of the DFT for finite-duration sequences. Our discussion parallels the discussion of Section 8.2 for periodic sequences. However, particular attention is paid to the interaction of the finite-length assumption and the implicit periodicity of the DFT representation of finite-length sequences.

8.6.1 Linearity

If two finite-duration sequences $x_1[n]$ and $x_2[n]$ are linearly combined, i.e., if

$$x_3[n] = ax_1[n] + bx_2[n], \tag{8.71}$$

then the DFT of $x_3[n]$ is

$$X_3[k] = aX_1[k] + bX_2[k]. \tag{8.72}$$

Clearly, if $x_1[n]$ has length N_1 and $x_2[n]$ has length N_2, then the maximum length of $x_3[n]$ will be $N_3 = \max[N_1, N_2]$. Thus, in order for Eq. (8.72) to be meaningful, both DFTs must be computed with the same length $N \geq N_3$. If, for example, $N_1 < N_2$, then $X_1[k]$ is the DFT of the sequence $x_1[n]$ augmented by $(N_2 - N_1)$ zeros. That is, the N_2-point DFT of $x_1[n]$ is

$$X_1[k] = \sum_{n=0}^{N_1-1} x_1[n]W_{N_2}^{kn}, \qquad 0 \leq k \leq N_2 - 1, \tag{8.73}$$

and the N_2-point DFT of $x_2[n]$ is

$$X_2[k] = \sum_{n=0}^{N_2-1} x_2[n]W_{N_2}^{kn}, \qquad 0 \leq k \leq N_2 - 1. \tag{8.74}$$

In summary, if

$$x_1[n] \overset{\mathcal{DFS}}{\longleftrightarrow} X_1[k] \tag{8.75a}$$

and

$$x_2[n] \overset{\mathcal{DFS}}{\longleftrightarrow} X_2[k], \tag{8.75b}$$

then

$$ax_1[n] + bx_2[n] \overset{\mathcal{DFS}}{\longleftrightarrow} aX_1[k] + bX_2[k], \tag{8.76}$$

where the lengths of the sequences and their discrete Fourier transforms are all equal to the maximum of the lengths of $x_1[n]$ and $x_2[n]$. Of course, DFTs of greater length can be computed by augmenting *both* sequences with zero-valued samples.

8.6.2 Circular Shift of a Sequence

According to Section 2.9.2 and property 2 in Table 2.2, if $X(e^{j\omega})$ is the Fourier transform of $x[n]$, then $e^{-j\omega m}X(e^{j\omega})$ is the Fourier transform of the time-shifted sequence $x[n-m]$. In other words, a shift in the time domain by m points (with positive m corresponding to a time delay and negative m to a time advance) corresponds in the frequency domain to multiplication of the Fourier transform by the linear phase factor $e^{-j\omega m}$. In Section 8.2.2, we discussed the corresponding property for the DFS coefficients of a periodic sequence; specifically, if a periodic sequence $\tilde{x}[n]$ has Fourier series coefficients $\tilde{X}[k]$, then the shifted sequence $\tilde{x}[n-m]$ has Fourier series coefficients $e^{-j(2\pi k/N)m}\tilde{X}[k]$. Now we will consider the operation in the time domain that corresponds to multiplying the DFT coefficients of a finite-length sequence $x[n]$ by the linear phase factor $e^{-j(2\pi k/N)m}$. Specifically, let $x_1[n]$ denote the finite-length sequence for which the DFT is $e^{-j(2\pi k/N)m}X[k]$; i.e., if

$$x[n] \overset{\mathcal{DFJ}}{\longleftrightarrow} X[k], \tag{8.77}$$

then we are interested in $x_1[n]$ such that

$$x_1[n] \overset{\mathcal{DFJ}}{\longleftrightarrow} X_1[k] = e^{-j(2\pi k/N)m}X[k]. \tag{8.78}$$

Since the N-point DFT represents a finite-duration sequence of length N, both $x[n]$ and $x_1[n]$ must be zero outside the interval $0 \le n \le N-1$, and consequently, $x_1[n]$ cannot result from a simple time shift of $x[n]$. The correct result follows directly from the result of Section 8.2.2 and the interpretation of the DFT as the Fourier series coefficients of the periodic sequence $x_1[((n))_N]$. In particular, from Eqs. (8.59) and (8.62) it follows that

$$\tilde{x}[n] = x[((n))_N] \overset{\mathcal{DFS}}{\longleftrightarrow} \tilde{X}[k] = X[((k))_N], \tag{8.79}$$

and similarly, we can define a periodic sequence $\tilde{x}_1[n]$ such that

$$\tilde{x}_1[n] = x_1[((n))_N] \overset{\mathcal{DFS}}{\longleftrightarrow} \tilde{X}_1[k] = X_1[((k))_N], \tag{8.80}$$

where, by assumption,

$$X_1[k] = e^{-j(2\pi k/N)m}X[k]. \tag{8.81}$$

Therefore, the discrete Fourier series coefficients of $\tilde{x}_1[n]$ are

$$\tilde{X}_1[k] = e^{-j[2\pi((k))_N/N]m}X[((k))_N]. \tag{8.82}$$

Note that

$$e^{-j[2\pi((k))_N/N]m} = e^{-j(2\pi k/N)m}. \tag{8.83}$$

That is, since $e^{-j(2\pi k/N)m}$ is periodic with period N in both k and m, we can drop the notation $((k))_N$. Hence, Eq. (8.82) becomes

$$\tilde{X}_1[k] = e^{-j(2\pi k/N)m}\tilde{X}[k], \tag{8.84}$$

so that it follows from Section 8.2.2 that

$$\tilde{x}_1[n] = \tilde{x}[n-m] = x[((n-m))_N]. \tag{8.85}$$

Thus, the finite-length sequence $x_1[n]$ whose DFT is given by Eq. (8.81) is

$$x_1[n] = \begin{cases} \tilde{x}_1[n] = x[((n-m))_N], & 0 \le n \le N-1, \\ 0, & \text{otherwise.} \end{cases} \tag{8.86}$$

Equation (8.86) tells us how to construct $x_1[n]$.

Example 8.8   Circular Shift of a Sequence

The circular shift procedure is illustrated in Figure 8.12 for $m = -2$; i.e., we want to determine $x_1[n] = x[((n + 2))_N]$ for $N = 6$, which we have shown will have DFT $X_1[k] = W_6^{-2k} X[k]$. Specifically, from $x[n]$, we construct the periodic sequence $\tilde{x}[n] = x[((n))_6]$, as indicated in Figure 8.12(b). According to Eq. (8.85), we then shift $\tilde{x}[n]$ by 2 to the left, obtaining $\tilde{x}_1[n] = \tilde{x}[n + 2]$ as in Figure 8.12(c). Finally, using Eq. (8.86), we extract one period of $\tilde{x}_1[n]$ to obtain $x_1[n]$, as indicated in Figure 8.12(d).

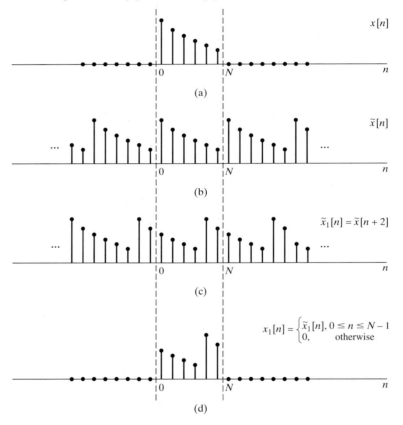

Figure 8.12   Circular shift of a finite-length sequence; i.e., the effect in the time domain of multiplying the DFT of the sequence by a linear phase factor.

A comparison of Figures 8.12(a) and (d) indicates clearly that $x_1[n]$ does *not* correspond to a linear shift of $x[n]$, and in fact, both sequences are confined to the interval between 0 and $(N - 1)$. By reference to Figure 8.12, we see that $x_1[n]$ can be formed by shifting $x[n]$ so that as a sequence value leaves the interval 0 to $(N - 1)$ at one end, it enters at the other end. Another interesting point is that, for the example shown in Figure 8.12(a), if we form $x_2[n] = x[((n - 4))_6]$ by shifting the sequence by 4 to the right modulo 6, we obtain the same sequence as $x_1[n]$. In terms of the DFT, this results because $W_6^{4k} = W_6^{-2k}$ or, more generally, $W_N^{mk} = W_N^{-(N-m)k}$, which implies that an N-point circular shift in one direction by m is the same as a circular shift in the opposite direction by $N - m$.

In Section 8.4, we suggested the interpretation of forming the periodic sequence $\tilde{x}[n]$ from the finite-length sequence $x[n]$ by displaying $x[n]$ around the circumference of a cylinder with a circumference of exactly N points. As we repeatedly traverse the circumference of the cylinder, the sequence that we see is the periodic sequence $\tilde{x}[n]$. A linear shift of this sequence corresponds, then, to a *rotation* of the cylinder. In the context of finite-length sequences and the DFT, such a shift is called a *circular* shift or a *rotation* of the sequence in the interval $0 \leq n \leq N - 1$.

In summary, the circular shift property of the DFT is

$$x[((n - m))_N], \quad 0 \leq n \leq N - 1 \overset{\mathcal{DFT}}{\longleftrightarrow} e^{-j(2\pi k/N)m} X[k]. \tag{8.87}$$

8.6.3 Duality

Since the DFT is so closely associated with the DFS, we would expect the DFT to exhibit a duality property similar to that of the DFS discussed in Section 8.2.3. In fact, from an examination of Eqs. (8.67) and (8.68), we see that the analysis and synthesis equations differ only in the factor $1/N$ and the sign of the exponent of the powers of W_N.

The DFT duality property can be derived by exploiting the relationship between the DFT and the DFS as in our derivation of the circular shift property. Toward this end, consider $x[n]$ and its DFT $X[k]$, and construct the periodic sequences

$$\tilde{x}[n] = x[((n))_N], \tag{8.88a}$$

$$\tilde{X}[k] = X[((k))_N], \tag{8.88b}$$

so that

$$\tilde{x}[n] \overset{\mathcal{DFS}}{\longleftrightarrow} \tilde{X}[k]. \tag{8.89}$$

From the duality property given in Eqs. (8.25),

$$\tilde{X}[n] \overset{\mathcal{DFS}}{\longleftrightarrow} N\tilde{x}[-k]. \tag{8.90}$$

If we define the periodic sequence $\tilde{x}_1[n] = \tilde{X}[n]$, one period of which is the finite-length sequence $x_1[n] = X[n]$, then the DFS coefficients of $\tilde{x}_1[n]$ are $\tilde{X}_1[k] = N\tilde{x}[-k]$. Therefore, the DFT of $x_1[n]$ is

$$X_1[k] = \begin{cases} N\tilde{x}[-k], & 0 \leq k \leq N - 1, \\ 0, & \text{otherwise}, \end{cases} \tag{8.91}$$

or, equivalently,

$$X_1[k] = \begin{cases} Nx[((-k))_N], & 0 \leq k \leq N - 1, \\ 0, & \text{otherwise}. \end{cases} \tag{8.92}$$

Consequently, the duality property for the DFT can be expressed as follows: If

$$x[n] \overset{\mathcal{DFT}}{\longleftrightarrow} X[k], \tag{8.93a}$$

then

$$X[n] \overset{\mathcal{DFT}}{\longleftrightarrow} Nx[((-k))_N], \quad 0 \leq k \leq N - 1. \tag{8.93b}$$

The sequence $Nx[((-k))_N]$ is $Nx[k]$ index reversed, modulo N. As in the case of shifting modulo N, the process of index reversing modulo N is usually best visualized in terms of the underlying periodic sequences.

Example 8.9   The Duality Relationship for the DFT

To illustrate the duality relationship in Eqs. (8.93), let us consider the sequence $x[n]$ of Example 8.7. Figure 8.13(a) shows the finite-length sequence $x[n]$, and Figures 8.13(b) and 8.13(c) are the real and imaginary parts, respectively, of the corresponding 10-point DFT $X[k]$. By simply relabeling the horizontal axis, we obtain the complex sequence $x_1[n] = X[n]$, as shown in Figures 8.13(d) and 8.13(e). According to the duality relation in Eqs. (8.93), the 10-point DFT of the (complex-valued) sequence $X[n]$ is the sequence shown in Figure 8.13(f).

8.6.4 Symmetry Properties

Since the DFT of $x[n]$ is identical to the DFS coefficients of the periodic sequence $\tilde{x}[n] = x[((n))_N]$, symmetry properties associated with the DFT can be inferred from the symmetry properties of the DFS summarized in Table 8.1 in Section 8.2.6. Specifically, using Eqs. (8.88) together with properties 9 and 10 in Table 8.1, we have

$$x^*[n] \overset{\mathcal{DFS}}{\longleftrightarrow} X^*[((-k))_N], \qquad 0 \le n \le N-1, \tag{8.94}$$

and

$$x^*[((-n))_N] \overset{\mathcal{DFS}}{\longleftrightarrow} X^*[k], \qquad 0 \le n \le N-1. \tag{8.95}$$

Properties 11–14 in Table 8.1 refer to the decomposition of a periodic sequence into the sum of a conjugate-symmetric and a conjugate-antisymmetric sequence. This suggests the decomposition of the finite-duration sequence $x[n]$ into the two finite-duration sequences of duration N corresponding to one period of the conjugate-symmetric and one period of the conjugate-antisymmetric components of $\tilde{x}[n]$. We will denote these components of $x[n]$ as $x_{\text{ep}}[n]$ and $x_{\text{op}}[n]$. Thus, with

$$\tilde{x}[n] = x[((n))_N] \tag{8.96}$$

and the conjugate-symmetric part being

$$\tilde{x}_e[n] = \tfrac{1}{2}\{\tilde{x}[n] + \tilde{x}^*[-n]\}, \tag{8.97}$$

and the conjugate-antisymmetric part being

$$\tilde{x}_o[n] = \tfrac{1}{2}\{\tilde{x}[n] - \tilde{x}^*[-n]\}, \tag{8.98}$$

we define $x_{\text{ep}}[n]$ and $x_{\text{op}}[n]$ as

$$x_{\text{ep}}[n] = \tilde{x}_e[n], \qquad 0 \le n \le N-1, \tag{8.99}$$

$$x_{\text{op}}[n] = \tilde{x}_o[n], \qquad 0 \le n \le N-1, \tag{8.100}$$

or, equivalently,

$$x_{\text{ep}}[n] = \tfrac{1}{2}\{x[((n))_N] + x^*[((-n))_N]\}, \qquad 0 \le n \le N-1, \tag{8.101a}$$

$$x_{\text{op}}[n] = \tfrac{1}{2}\{x[((n))_N] - x^*[((-n))_N]\}, \qquad 0 \le n \le N-1, \tag{8.101b}$$

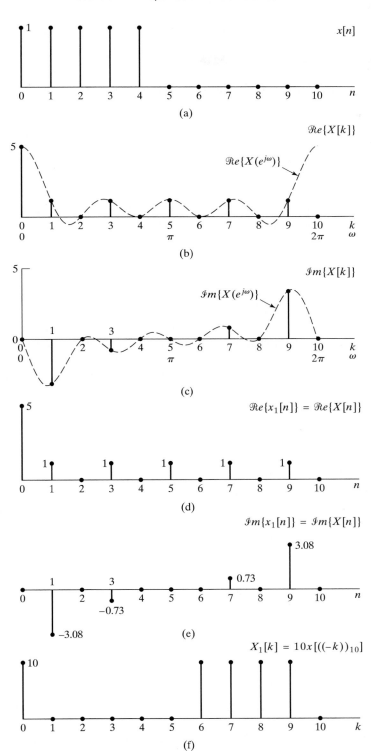

Figure 8.13 Illustration of duality.
(a) Real finite-length sequence $x[n]$.
(b) and (c) Real and imaginary
parts of corresponding DFT $X[k]$.
(d) and (e) The real and imaginary parts
of the dual sequence $x_1[n] = X[n]$.
(f) The DFT of $x_1[n]$.

with both $x_{\mathrm{ep}}[n]$ and $x_{\mathrm{op}}[n]$ being finite-length sequences, i.e., both zero outside the interval $0 \le n \le N-1$. Since $((-n))_N = (N-n)$ and $((n))_N = n$ for $0 \le n \le N-1$, we can also express Eqs. (8.101) as

$$x_{\mathrm{ep}}[n] = \tfrac{1}{2}\{x[n] + x^*[N-n]\}, \qquad 1 \le n \le N-1, \tag{8.102a}$$

$$x_{\mathrm{ep}}[0] = \mathcal{R}e\{x[0]\}, \tag{8.102b}$$

$$x_{\mathrm{op}}[n] = \tfrac{1}{2}\{x[n] - x^*[N-n]\}, \qquad 1 \le n \le N-1, \tag{8.102c}$$

$$x_{\mathrm{op}}[0] = j\,\mathcal{J}m\{x[0]\}. \tag{8.102d}$$

This form of the equations is convenient, since it avoids the modulo N computation of indices.

Clearly, $x_{\mathrm{ep}}[n]$ and $x_{\mathrm{op}}[n]$ are not equivalent to $x_e[n]$ and $x_o[n]$ as defined by Eq. (2.154). However, it can be shown (see Problem 8.56) that

$$x_{\mathrm{ep}}[n] = \{x_e[n] + x_e[n-N]\}, \qquad 0 \le n \le N-1, \tag{8.103}$$

and

$$x_{\mathrm{op}}[n] = \{x_o[n] + x_o[n-N]\}, \qquad 0 \le n \le N-1. \tag{8.104}$$

In other words, $x_{\mathrm{ep}}[n]$ and $x_{\mathrm{op}}[n]$ can be generated by aliasing $x_e[n]$ and $x_o[n]$ into the interval $0 \le n \le N-1$. The sequences $x_{\mathrm{ep}}[n]$ and $x_{\mathrm{op}}[n]$ will be referred to as *the periodic conjugate-symmetric* and *periodic conjugate-antisymmetric* components, respectively, of $x[n]$. When $x_{\mathrm{ep}}[n]$ and $x_{\mathrm{op}}[n]$ are real, they will be referred to as the *periodic even* and *periodic odd* components, respectively. Note that the sequences $x_{\mathrm{ep}}[n]$ and $x_{\mathrm{op}}[n]$ are *not* periodic sequences; they are, however, finite-length sequences that are equal to one period of the periodic sequences $\tilde{x}_e[n]$ and $\tilde{x}_o[n]$, respectively.

Equations (8.101) and (8.102) define $x_{\mathrm{ep}}[n]$ and $x_{\mathrm{op}}[n]$ in terms of $x[n]$. The inverse relation, expressing $x[n]$ in terms of $x_{\mathrm{ep}}[n]$ and $x_{\mathrm{op}}[n]$, can be obtained by using Eqs. (8.97) and (8.98) to express $\tilde{x}[n]$ as

$$\tilde{x}[n] = \tilde{x}_e[n] + \tilde{x}_o[n]. \tag{8.105}$$

Thus,

$$x[n] = \tilde{x}[n] = \tilde{x}_e[n] + \tilde{x}_o[n], \qquad 0 \le n \le N-1. \tag{8.106}$$

Combining Eqs. (8.106) with Eqs. (8.99) and (8.100), we obtain

$$x[n] = x_{\mathrm{ep}}[n] + x_{\mathrm{op}}[n]. \tag{8.107}$$

Alternatively, Eqs. (8.102), when added, also lead to Eq. (8.107). The symmetry properties of the DFT associated with properties 11–14 in Table 8.1 now follow in a straightforward way:

$$\mathcal{R}e\{x[n]\} \overset{\mathcal{DFJ}}{\longleftrightarrow} X_{\mathrm{ep}}[k], \tag{8.108}$$

$$j\,\mathcal{J}m\{x[n]\} \overset{\mathcal{DFJ}}{\longleftrightarrow} X_{\mathrm{op}}[k], \tag{8.109}$$

$$x_{\mathrm{ep}}[n] \overset{\mathcal{DFJ}}{\longleftrightarrow} \mathcal{R}e\{X[k]\}, \tag{8.110}$$

$$x_{\mathrm{op}}[n] \overset{\mathcal{DFJ}}{\longleftrightarrow} j\,\mathcal{J}m\{X[k]\}. \tag{8.111}$$

8.6.5 Circular Convolution

In Section 8.2.5, we showed that multiplication of the DFS coefficients of two periodic sequences corresponds to a periodic convolution of the sequences. Here we consider two *finite-duration* sequences $x_1[n]$ and $x_2[n]$, both of length N, with DFTs $X_1[k]$ and $X_2[k]$, respectively, and we wish to determine the sequence $x_3[n]$ for which the DFT is $X_3[k] = X_1[k]X_2[k]$. To determine $x_3[n]$, we can apply the results of Section 8.2.5. Specifically, $x_3[n]$ corresponds to one period of $\tilde{x}_3[n]$, which is given by Eq. (8.27). Thus,

$$x_3[n] = \sum_{m=0}^{N-1} \tilde{x}_1[m]\tilde{x}_2[n-m], \qquad 0 \leq n \leq N-1, \tag{8.112}$$

or, equivalently,

$$x_3[n] = \sum_{m=0}^{N-1} x_1[((m))_N]x_2[((n-m))_N], \qquad 0 \leq n \leq N-1. \tag{8.113}$$

Since $((m))_N = m$ for $0 \leq m \leq N-1$, Eq. (8.113) can be written

$$x_3[n] = \sum_{m=0}^{N-1} x_1[m]x_2[((n-m))_N], \qquad 0 \leq n \leq N-1. \tag{8.114}$$

Equations (8.112) and (8.114) differ from a linear convolution of $x_1[n]$ and $x_2[n]$ as defined by Eq. (2.52) in some important respects. In linear convolution, the computation of the sequence value $x_3[n]$ involves multiplying one sequence by a time-reversed and linearly shifted version of the other and then summing the values of the product $x_1[m]x_2[n-m]$ over all m. To obtain successive values of the sequence formed by the convolution operation, the two sequences are successively shifted relative to each other. In contrast, for the convolution defined by Eq. (8.114), the second sequence is circularly time reversed and circularly shifted with respect to the first. For this reason, the operation of combining two finite-length sequences according to Eq. (8.114) is called *circular convolution*. More specifically, we refer to Eq. (8.114) as an *N-point circular convolution*, explicitly identifying the fact that both sequences have length N (or less) and that the sequences are shifted modulo N. Sometimes the operation of forming a sequence $x_3[n]$ for $0 \leq n \leq N-1$ using Eq. (8.114) will be denoted

$$x_3[n] = x_1[n] \,\text{Ⓝ}\, x_2[n]. \tag{8.115}$$

Since the DFT of $x_3[n]$ is $X_3[k] = X_1[k]X_2[k]$ and since $X_1[k]X_2[k] = X_2[k]X_1[k]$, it follows with no further analysis that

$$x_3[n] = x_2[n] \,\text{Ⓝ}\, x_1[n], \tag{8.116}$$

or, more specifically,

$$x_3[n] = \sum_{m=0}^{N-1} x_2[m]x_1[((n-m))_N]. \tag{8.117}$$

That is, circular convolution, like linear convolution, is a commutative operation.

Since circular convolution is really just periodic convolution, Example 8.4 and Figure 8.3 are also illustrative of circular convolution. However, if we utilize the notion of circular shifting, it is not necessary to construct the underlying periodic sequences as in Figure 8.3. This is illustrated in the following examples.

Example 8.10   Circular Convolution with a Delayed Impulse Sequence

An example of circular convolution is provided by the result of Section 8.6.2. Let $x_2[n]$ be a finite-duration sequence of length N and

$$x_1[n] = \delta[n - n_0], \tag{8.118}$$

where $0 < n_0 < N$. Clearly, $x_1[n]$ can be considered as the finite-duration sequence

$$x_1[n] = \begin{cases} 0, & 0 \leq n < n_0, \\ 1, & n = n_0, \\ 0, & n_0 < n \leq N - 1. \end{cases} \tag{8.119}$$

as depicted in Figure 8.14 for $n_0 = 1$.

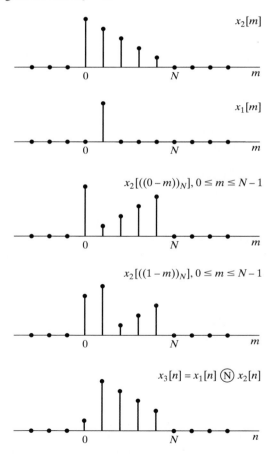

Figure 8.14   Circular convolution of a finite-length sequence $x_2[n]$ with a single delayed impulse, $x_1[n] = \delta[n - 1]$.

The DFT of $x_1[n]$ is

$$X_1[k] = W_N^{kn_0}. \tag{8.120}$$

If we form the product

$$X_3[k] = W_N^{kn_0} X_2[k], \tag{8.121}$$

we see from Section 8.6.2 that the finite-duration sequence corresponding to $X_3[k]$ is the sequence $x_2[n]$ rotated to the right by n_0 samples in the interval $0 \leq n \leq N - 1$. That is, the circular convolution of a sequence $x_2[n]$ with a single delayed unit impulse results in a rotation of $x_2[n]$ in the interval $0 \leq n \leq N - 1$. This example is illustrated in Figure 8.14 for $N = 5$ and $n_0 = 1$. Here we show the sequences $x_2[m]$ and $x_1[m]$ and then $x_2[((0 - m))_N]$ and $x_2[((1 - m))_N]$. It is clear from these two cases that the result of circular convolution of $x_2[n]$ with a single shifted unit impulse will be to circularly shift $x_2[n]$. The last sequence shown is $x_3[n]$, the result of the circular convolution of $x_1[n]$ and $x_2[n]$.

Example 8.11   Circular Convolution of Two Rectangular Pulses

As another example of circular convolution, let

$$x_1[n] = x_2[n] = \begin{cases} 1, & 0 \leq n \leq L - 1, \\ 0, & \text{otherwise,} \end{cases} \tag{8.122}$$

where, in Figure 8.15, $L = 6$. If we let N denote the DFT length, then, for $N = L$, the N-point DFTs are

$$X_1[k] = X_2[k] = \sum_{n=0}^{N-1} W_N^{kn} = \begin{cases} N, & k = 0, \\ 0, & \text{otherwise.} \end{cases} \tag{8.123}$$

If we explicitly multiply $X_1[k]$ and $X_2[k]$, we obtain

$$X_3[k] = X_1[k]X_2[k] = \begin{cases} N^2, & k = 0, \\ 0, & \text{otherwise,} \end{cases} \tag{8.124}$$

from which it follows that

$$x_3[n] = N, \qquad 0 \leq n \leq N - 1. \tag{8.125}$$

This result is depicted in Figure 8.15. Clearly, as the sequence $x_2[((n - m))_N]$ is rotated with respect to $x_1[m]$, the sum of products $x_1[m]x_2[((n - m))_N]$ will always be equal to N.

It is, of course, possible to consider $x_1[n]$ and $x_2[n]$ as $2L$-point sequences by augmenting them with L zeros. If we then perform a $2L$-point circular convolution of the augmented sequences, we obtain the sequence in Figure 8.16, which can be seen to be identical to the linear convolution of the finite-duration sequences $x_1[n]$ and $x_2[n]$. This important observation will be discussed in much more detail in Section 8.7.

Note that for $N = 2L$, as in Figure 8.16,

$$X_1[k] = X_2[k] = \frac{1 - W_N^{Lk}}{1 - W_N^k},$$

so the DFT of the triangular-shaped sequence $x_3[n]$ in Figure 8.16(e) is

$$X_3[k] = \left(\frac{1 - W_N^{Lk}}{1 - W_N^k} \right)^2,$$

with $N = 2L$.

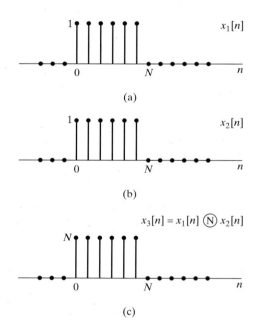

(a)

(b)

$$x_3[n] = x_1[n] \ \text{Ⓝ} \ x_2[n]$$

(c)

Figure 8.15 N-point circular convolution of two constant sequences of length N.

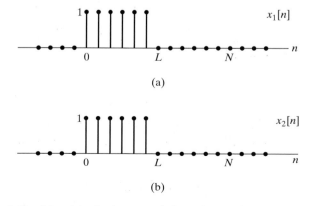

(a)

(b)

Figure 8.16 $2L$-point circular convolution of two constant sequences of length L.

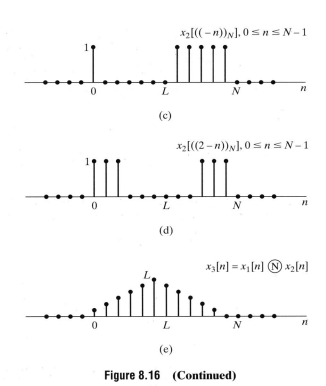

Figure 8.16   (Continued)

The circular convolution property is represented as

$$x_1[n] \; \textcircled{N} \; x_2[n] \; \overset{\mathcal{DFT}}{\longleftrightarrow} \; X_1[k]X_2[k]. \tag{8.126}$$

In view of the duality of the DFT relations, it is not surprising that the DFT of a product of two N-point sequences is the circular convolution of their respective discrete Fourier transforms. Specifically, if $x_3[n] = x_1[n]x_2[n]$, then

$$X_3[k] = \frac{1}{N} \sum_{\ell=0}^{N-1} X_1[\ell]X_2[((k - \ell))_N] \tag{8.127}$$

or

$$x_1[n]x_2[n] \; \overset{\mathcal{DFT}}{\longleftrightarrow} \; \frac{1}{N} X_1[k] \; \textcircled{N} \; X_2[k]. \tag{8.128}$$

8.6.6 Summary of Properties of the Discrete Fourier Transform

The properties of the discrete Fourier transform that we discussed in Section 8.6 are summarized in Table 8.2. Note that for all of the properties, the expressions given specify $x[n]$ for $0 \le n \le N - 1$ and $X[k]$ for $0 \le k \le N - 1$. Both $x[n]$ and $X[k]$ are equal to zero outside those ranges.

TABLE 8.2

Finite-Length Sequence (Length N)	N-point DFT (Length N)
1. $x[n]$	$X[k]$
2. $x_1[n], x_2[n]$	$X_1[k], X_2[k]$
3. $ax_1[n] + bx_2[n]$	$a X_1[k] + b X_2[k]$
4. $X[n]$	$Nx[((-k))_N]$
5. $x[((n - m))_N]$	$W_N^{km} X[k]$
6. $W_N^{-\ell n} x[n]$	$X[((k - \ell))_N]$
7. $\displaystyle\sum_{m=0}^{N-1} x_1(m)x_2[((n-m))_N]$	$X_1[k]X_2[k]$
8. $x_1[n]x_2[n]$	$\dfrac{1}{N}\displaystyle\sum_{\ell=0}^{N-1} X_1(\ell)X_2[((k-\ell))_N]$
9. $x^*[n]$	$X^*[((-k))_N]$
10. $x^*[((-n))_N]$	$X^*[k]$
11. $\mathcal{R}e\{x[n]\}$	$X_{\mathrm{ep}}[k] = \frac{1}{2}\{X[((k))_N] + X^*[((-k))_N]\}$
12. $j\mathcal{J}m\{x[n]\}$	$X_{\mathrm{op}}[k] = \frac{1}{2}\{X[((k))_N] - X^*[((-k))_N]\}$
13. $x_{\mathrm{ep}}[n] = \frac{1}{2}\{x[n] + x^*[((-n))_N]\}$	$\mathcal{R}e\{X[k]\}$
14. $x_{\mathrm{op}}[n] = \frac{1}{2}\{x[n] - x^*[((-n))_N]\}$	$j\mathcal{J}m\{X[k]\}$

Properties 15–17 apply only when $x[n]$ is real.

15. Symmetry properties	$\begin{cases} X[k] = X^*[((-k))_N] \\ \mathcal{R}e\{X[k]\} = \mathcal{R}e\{X[((-k))_N]\} \\ \mathcal{J}m\{X[k]\} = -\mathcal{J}m\{X[((-k))_N]\} \\ |X[k]| = |X[((-k))_N]| \\ \sphericalangle\{X[k]\} = -\sphericalangle\{X[((-k))_N]\} \end{cases}$
16. $x_{\mathrm{ep}}[n] = \frac{1}{2}\{x[n] + x[((-n))_N]\}$	$\mathcal{R}e\{X[k]\}$
17. $x_{\mathrm{op}}[n] = \frac{1}{2}\{x[n] - x[((-n))_N]\}$	$j\mathcal{J}m\{X[k]\}$

8.7 LINEAR CONVOLUTION USING THE DISCRETE FOURIER TRANSFORM

We will show in Chapter 9 that efficient algorithms are available for computing the discrete Fourier transform of a finite-duration sequence. These are known collectively as *fast Fourier transform* (FFT) algorithms. Because these algorithms are available, it is computationally efficient to implement a convolution of two sequences by the following procedure:

(a) Compute the N-point discrete Fourier transforms $X_1[k]$ and $X_2[k]$ of the two sequences $x_1[n]$ and $x_2[n]$, respectively.

(b) Compute the product $X_3[k] = X_1[k]X_2[k]$ for $0 \leq k \leq N - 1$.

(c) Compute the sequence $x_3[n] = x_1[n] \; \text{Ⓝ} \; x_2[n]$ as the inverse DFT of $X_3[k]$.

In most applications, we are interested in implementing a linear convolution of two sequences; i.e., we wish to implement a linear time-invariant system. This is certainly true, for example, in filtering a sequence such as a speech waveform or a radar signal or in computing the autocorrelation function of such signals. As we saw in Section 8.6.5, the multiplication of discrete Fourier transforms corresponds to a circular convolution of the sequences. To obtain a linear convolution, we must ensure that circular convolution has the effect of linear convolution. The discussion at the end of Example 8.11 hints at how this might be done. We now present a more detailed analysis.

8.7.1 Linear Convolution of Two Finite-Length Sequences

Consider a sequence $x_1[n]$ whose length is L points and a sequence $x_2[n]$ whose length is P points, and suppose that we wish to combine these two sequences by linear convolution to obtain a third sequence

$$x_3[n] = \sum_{m=-\infty}^{\infty} x_1[m]x_2[n-m]. \tag{8.129}$$

Figure 8.17(a) shows a typical sequence $x_1[m]$ and Figure 8.17(b) shows a typical sequence $x_2[n-m]$ for several values of n. Clearly, the product $x_1[m]x_2[n-m]$ is zero for all m whenever $n < 0$ and $n > L + P - 2$; i.e., $x_3[n] \neq 0$ for $0 \leq n \leq L + P - 2$. Therefore, $(L + P - 1)$ is the maximum length of the sequence $x_3[n]$ resulting from the linear convolution of a sequence of length L with a sequence of length P.

8.7.2 Circular Convolution as Linear Convolution with Aliasing

As Examples 8.10 and 8.11 show, whether a circular convolution corresponding to the product of two N-point DFTs is the same as the linear convolution of the corresponding finite-length sequences depends on the length of the DFT in relation to the length of the finite-length sequences. An extremely useful interpretation of the relationship between circular convolution and linear convolution is in terms of time aliasing. Since this interpretation is so important and useful in understanding circular convolution, we will develop it in several ways.

In Section 8.4 we observed that if the Fourier transform $X(e^{j\omega})$ of a sequence $x[n]$ is sampled at frequencies $\omega_k = 2\pi k/N$, then the resulting sequence corresponds to the DFS coefficients of the periodic sequence

$$\tilde{x}[n] = \sum_{r=-\infty}^{\infty} x[n-rN]. \tag{8.130}$$

From our discussion of the DFT, it follows that the finite-length sequence

$$X[k] = \begin{cases} X(e^{j(2\pi k/N)}), & 0 \leq k \leq N - 1, \\ 0, & \text{otherwise}, \end{cases} \tag{8.131}$$

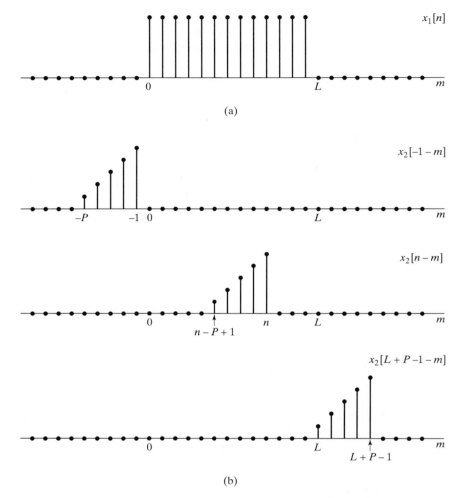

Figure 8.17 Example of linear convolution of two finite-length sequences showing that the result is such that $x_3[n] = 0$ for $n \leq -1$ and for $n \geq L + P - 1$. (a) Finite-length sequence $x_1[n]$. (b) $x_2[n - m]$ for several values of n.

is the DFT of one period of $\tilde{x}[n]$, as given by Eq. (8.130); i.e.,

$$x_p[n] = \begin{cases} \tilde{x}[n], & 0 \leq n \leq N - 1, \\ 0, & \text{otherwise.} \end{cases} \qquad (8.132)$$

Obviously, if $x[n]$ has length less than or equal to N, no time aliasing occurs and $x_p[n] = x[n]$. However, if the length of $x[n]$ is greater than N, $x_p[n]$ may not be equal to $x[n]$ for some or all values of n. We will henceforth use the subscript p to denote that a sequence is one period of a periodic sequence resulting from an inverse DFT of a sampled Fourier transform. The subscript can be dropped if it is clear that time aliasing is avoided.

The sequence $x_3[n]$ in Eq. (8.129) has Fourier transform

$$X_3(e^{j\omega}) = X_1(e^{j\omega})X_2(e^{j\omega}).\tag{8.133}$$

If we define a DFT

$$X_3[k] = X_3(e^{j(2\pi k/N)}), \qquad 0 \le k \le N-1,\tag{8.134}$$

then it is clear from Eqs. (8.133) and (8.134) that, also

$$X_3[k] = X_1(e^{j(2\pi k/N)})X_2(e^{j(2\pi k/N)}), \qquad 0 \le k \le N-1,\tag{8.135}$$

and therefore,

$$X_3[k] = X_1[k]X_2[k].\tag{8.136}$$

That is, the sequence resulting as the inverse DFT of $X_3[k]$ is

$$x_{3p}[n] = \begin{cases} \displaystyle\sum_{r=-\infty}^{\infty} x_3[n - rN], & 0 \le n \le N-1, \\ 0, & \text{otherwise,} \end{cases}\tag{8.137}$$

and from Eq. (8.136), it follows that

$$x_{3p}[n] = x_1[n] \,\textcircled{N}\, x_2[n].\tag{8.138}$$

Thus, the circular convolution of two finite-length sequences is equivalent to linear convolution of the two sequences, followed by time aliasing according to Eq. (8.137).

Note that if N is greater than or equal to either L or P, $X_1[k]$ and $X_2[k]$ represent $x_1[n]$ and $x_2[n]$ exactly, but $x_{3p}[n] = x_3[n]$ for all n only if N is greater than or equal to the length of the sequence $x_3[n]$. As we showed in Section 8.7.1, if $x_1[n]$ has length L and $x_2[n]$ has length P, then $x_3[n]$ has maximum length $(L+P-1)$. Therefore, the circular convolution corresponding to $X_1[k]X_2[k]$ is identical to the linear convolution corresponding to $X_1(e^{j\omega})X_2(e^{j\omega})$ if N, the length of the DFTs, satisfies $N \ge L + P - 1$.

Example 8.12   Circular Convolution as Linear Convolution with Aliasing

The results of Example 8.15 are easily understood in light of the interpretation just discussed. Note that $x_1[n]$ and $x_2[n]$ are identical constant sequences of length $L = P = 6$, as shown in Figure 8.18(a). The linear convolution of $x_1[n]$ and $x_2[n]$ is of length $L + P - 1 = 11$ and has the triangular shape shown in Figure 8.18(b). In Figures 8.18(c) and (d) are shown two of the shifted versions $x_3[n - rN]$ in Eq. (8.137), $x_3[n - N]$ and $x_3[n + N]$ for $N = 6$. The N-point circular convolution

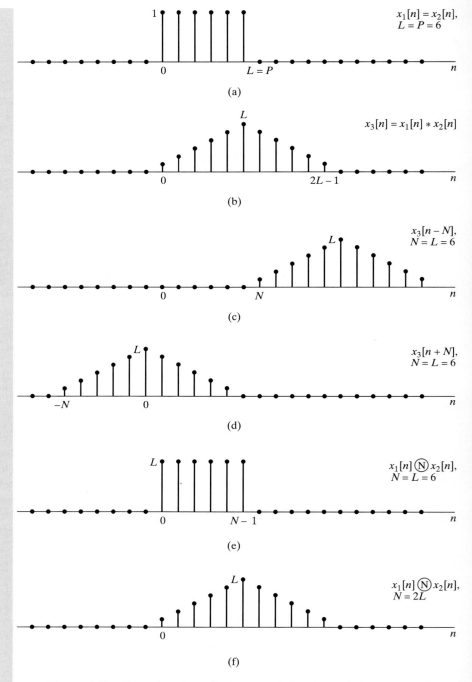

Figure 8.18 Illustration that circular convolution is equivalent to linear convolution followed by aliasing. (a) The sequences $x_1[n]$ and $x_2[n]$ to be convolved. (b) The linear convolution of $x_1[n]$ and $x_2[n]$. (c) $x_3[n-N]$ for $N = 6$. (d) $x_3[n+N]$ for $N = 6$. (e) $x_1[n] \; ⑥ \; x_2[n]$, which is equal to the sum of (b), (c), and (d) in the interval $0 \leq n \leq 5$. (f) $x_1[n] \; ⑫ \; x_2[n]$.

of $x_1[n]$ and $x_2[n]$ can be formed by using Eq. (8.137). This is shown in Figure 8.18(e) for $N = L = 6$ and in Figure 8.18(f) for $N = 2L = 12$. Note that for $N = L = 6$, only $x_3[n]$ and $x_3[n + N]$ contribute to the result. For $N = 2L = 12$, only $x_3[n]$ contributes to the result. Since the length of the linear convolution is $(2L - 1)$, the result of the circular convolution for $N = 2L$ is identical to the result of linear convolution for all $0 \le n \le N - 1$. In fact, this would be true for $N = 2L - 1 = 11$ as well.

As Example 8.12 points out, time aliasing in the circular convolution of two finite-length sequences can be avoided if $N \ge L + P - 1$. Also, it is clear that if $N = L = P$, all of the sequence values of the circular convolution may be different from those of the linear convolution. However, if $P < L$, some of the sequence values in an L-point circular convolution will be equal to the corresponding sequence values of the linear convolution. The time-aliasing interpretation is useful for showing this.

Consider two finite-duration sequences $x_1[n]$ and $x_2[n]$, with $x_1[n]$ of length L and $x_2[n]$ of length P, where $P < L$, as indicated in Figures 8.19(a) and 8.19(b), respectively. Let us first consider the L-point circular convolution of $x_1[n]$ and $x_2[n]$ and inquire as to which sequence values in the circular convolution are identical to values that would be obtained from a linear convolution and which are not. The linear convolution of $x_1[n]$ with $x_2[n]$ will be a finite-length sequence of length $(L + P - 1)$, as indicated in Figure 8.19(c). To determine the L-point circular convolution, we use Eqs. (8.137) and

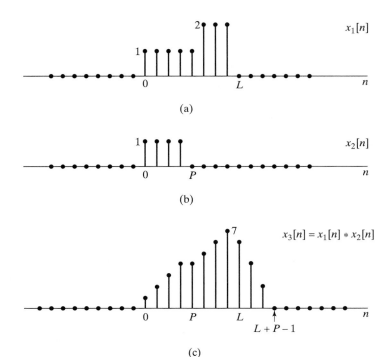

(a)

(b)

(c)

Figure 8.19 An example of linear convolution of two finite-length sequences.

(8.138) so that

$$
x_{3p}[n] = \begin{cases} x_1[n] \, ⓁＬ \, x_2[n] = \displaystyle\sum_{r=-\infty}^{\infty} x_3[n - rL], & 0 \le n \le L - 1, \\[2ex] 0, & \text{otherwise.} \end{cases} \tag{8.139}
$$

Figure 8.20(a) shows the term in Eq. (8.139) for $r = 0$, and Figures 8.20(b) and 8.20(c) show the terms for $r = -1$ and $r = +1$, respectively. From Figure 8.20, it should be clear that in the interval $0 \le n \le L - 1$, $x_{3p}[n]$ is influenced only by $x_3[n]$ and $x_3[n + L]$.

In general, whenever $P < L$, only the term $x_3[n + L]$ will alias into the interval $0 \le n \le L - 1$. More specifically, when these terms are summed, the last $(P - 1)$ points of $x_3[n + L]$, which extend from $n = 0$ to $n = P - 2$, will be added to the first $(P - 1)$ points of $x_3[n]$, and the last $(P - 1)$ points of $x_3[n]$, extending from $n = L$ to $n = L + P - 2$, will be discarded. Then $x_{3p}[n]$ is formed by extracting the portion for $0 \le n \le L - 1$. Since the last $(P - 1)$ points of $x_3[n + L]$ and the last $(P - 1)$ points of $x_3[n]$ are identical, we can alternatively view the process of forming the circular convolution $x_{3p}[n]$ through linear convolution plus aliasing as taking the $(P - 1)$ values of $x_3[n]$ from $n = L$ to $n = L + P - 2$ and adding them to the first $(P - 1)$ values of $x_3[n]$. This process is illustrated in Figure 8.21 for the case $P = 4$ and $L = 8$. Figure 8.21(a) shows the linear convolution $x_3[n]$, with the points for $n \ge L$ denoted by open symbols. Note that only $(P - 1)$ points for $n \ge L$ are nonzero. Figure 8.21(b) shows the formation of $x_{3p}[n]$ by "wrapping $x_3[n]$ around on itself." The first $(P - 1)$ points are corrupted by the time aliasing, and the remaining points from $n = P - 1$ to $n = L - 1$ (i.e., the last $L - P + 1$ points) are *not* corrupted; that is, they are identical to what would be obtained with a linear convolution.

From this discussion, it should be clear that if the circular convolution is of sufficient length relative to the lengths of the sequences $x_1[n]$ and $x_2[n]$, then aliasing with nonzero values can be avoided, in which case the circular convolution and linear convolution will be identical. Specifically, if, for the case just considered, $x_3[n]$ is replicated with period $N \ge L + P - 1$, then no nonzero overlap will occur. Figures 8.21(c) and 8.21(d) illustrate this case, again for $P = 4$ and $L = 8$, with $N = 11$.

8.7.3 Implementing Linear Time-Invariant Systems Using the DFT

The previous discussion focused on ways of obtaining a linear convolution from a circular convolution. Since linear time-invariant systems can be implemented by convolution, this implies that circular convolution (implemented by the procedure suggested at the beginning of Section 8.7) can be used to implement these systems. To see how this can be done, let us first consider an L-point input sequence $x[n]$ and a P-point impulse response $h[n]$. The linear convolution of these two sequences, which will be denoted by $y[n]$, has finite duration with length $(L + P - 1)$. Consequently, as discussed in Section 8.7.2, for the circular convolution and linear convolution to be identical, the circular convolution must have a length of at least $(L + P - 1)$ points. The circular convolution can be achieved by

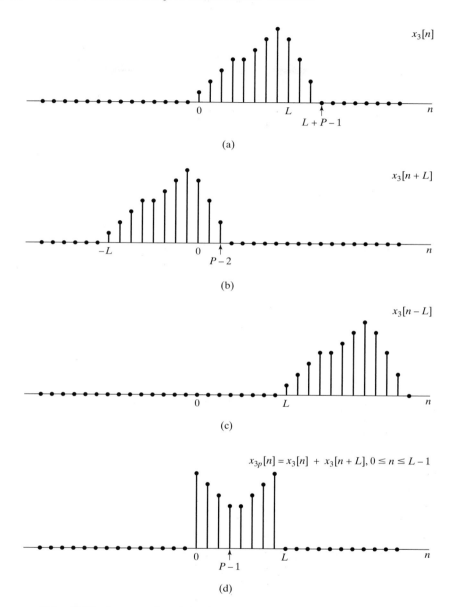

Figure 8.20   Interpretation of circular convolution as linear convolution followed by aliasing for the circular convolution of the two sequences $x_1[n]$ and $x_2[n]$ in Figure 8.19.

multiplying the DFTs of $x[n]$ and $h[n]$. Since we want the product to represent the DFT of the linear convolution of $x[n]$ and $h[n]$, which has length $(L+P-1)$, the DFTs that we compute must also be of at least that length, i.e., both $x[n]$ and $h[n]$ must be augmented with sequence values of zero amplitude. This process is often referred to as *zero-padding*.

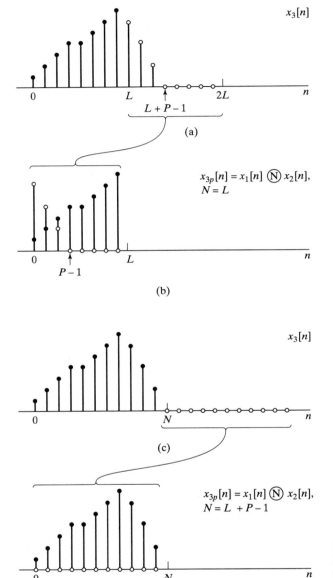

Figure 8.21   Illustration of how the result of a circular convolution "wraps around." (a) and (b) $N = L$, so the aliased "tail" overlaps the first $(P - 1)$ points. (c) and (d) $N = (L + P - 1)$, so no overlap occurs.

This procedure permits the computation of the linear convolution of two finite-length sequences using the discrete Fourier transform; i.e., the output of an FIR system whose input also has finite length can be computed with the DFT. In many applications, such as filtering a speech waveform, the input signal is of indefinite duration. While, theoretically, we might be able to store the entire waveform and then implement the procedure just discussed using a DFT for a large number of points, such a DFT is

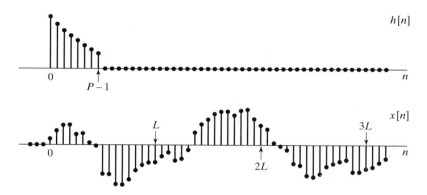

Figure 8.22   Finite-length impulse response $h[n]$ and indefinite-length signal $x[n]$ to be filtered.

generally impractical to compute. Another consideration is that for this method of filtering, no filtered samples can be computed until all the input samples have been collected. Generally, we would like to avoid such a large delay in processing. The solution to both of these problems is to use *block convolution,* in which the signal to be filtered is segmented into sections of length L. Each section can then be convolved with the finite-length impulse response and the filtered sections fitted together in an appropriate way. The linear filtering of each block can then be implemented using the DFT.

To illustrate the procedure and to develop the procedure for fitting the filtered sections together, consider the impulse response $h[n]$ of length P and the signal $x[n]$ depicted in Figure 8.22. Henceforth, we will assume that $x[n] = 0$ for $n < 0$ and that the length of $x[n]$ is much greater than P. The sequence $x[n]$ can be represented as a sum of shifted finite-length segments of length L; i.e.,

$$x[n] = \sum_{r=0}^{\infty} x_r[n - rL], \tag{8.140}$$

where

$$x_r[n] = \begin{cases} x[n + rL], & 0 \le n \le L - 1, \\ 0, & \text{otherwise.} \end{cases} \tag{8.141}$$

Figure 8.23(a) illustrates this segmentation for $x[n]$ in Figure 8.22. Note that within each segment, the first sample is at $n = 0$; however, the zeroth sample of $x_r[n]$ is the rLth sample of the sequence $x[n]$. This is shown in Figure 8.23(a) by plotting the segments in their shifted positions.

Because convolution is a linear time-invariant operation, it follows from Eq. (8.140) that

$$y[n] = x[n] * h[n] = \sum_{r=0}^{\infty} y_r[n - rL], \tag{8.142}$$

where

$$y_r[n] = x_r[n] * h[n]. \tag{8.143}$$

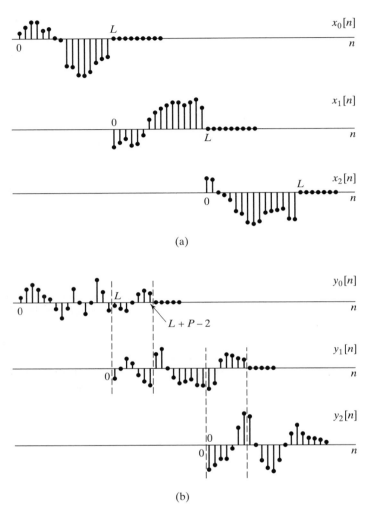

Figure 8.23   (a) Decomposition of $x[n]$ in Figure 8.22 into nonoverlapping sections of length L. (b) Result of convolving each section with $h[n]$.

Since the sequences $x_r[n]$ have only L nonzero points and $h[n]$ is of length P, each of the terms $y_r[n] = x_r[n] * h[n]$ has length $(L + P - 1)$. Thus, the linear convolution $x_r[n] * h[n]$ can be obtained by the procedure described earlier using N-point DFTs, where $N \geq L + P - 1$. Since the beginning of each input section is separated from its neighbors by L points and each filtered section has length $(L + P - 1)$, the nonzero points in the filtered sections will overlap by $(P - 1)$ points, and these overlap samples must be added in carrying out the sum required by Eq. (8.142). This is illustrated in Figure 8.23(b), which shows the filtered sections, $y_r[n] = x_r[n] * h[n]$. Just as the input waveform is reconstructed by adding the delayed waveforms in Figure 8.23(a), the filtered result $x[n] * h[n]$ is constructed by adding the delayed filtered sections depicted in Figure 8.23(b). This procedure for constructing the filtered output from filtered sections is often referred to as the *overlap–add method,* because the filtered sections are overlapped and added to construct the output. The overlapping occurs because the linear convolution of each section with the impulse response is, in general, longer than the length of the section. The overlap–add method of block convolution is *not* tied to the DFT and circular convolution. Clearly, all that is required is that the smaller convolutions be computed and the results combined appropriately.

An alternative block convolution procedure, commonly called the *overlap–save method,* corresponds to implementing an L-point circular convolution of a P-point impulse response $h[n]$ with an L-point segment $x_r[n]$ and identifying the part of the circular convolution that corresponds to a linear convolution. The resulting output segments are then "patched together" to form the output. Specifically, we showed that if an L-point sequence is circularly convolved with a P-point sequence ($P < L$), then the first $(P-1)$ points of the result are incorrect, while the remaining points are identical to those that would be obtained had we implemented a linear convolution. Therefore, we can divide $x[n]$ into sections of length L so that each input section overlaps the preceding section by $(P-1)$ points. That is, we define the sections as

$$x_r[n] = x[n + r(L - P + 1) - P + 1], \qquad 0 \le n \le L - 1, \qquad (8.144)$$

where, as before, we have defined the time origin for each section to be at the beginning of that section rather than at the origin of $x[n]$. This method of sectioning is depicted in Figure 8.24(a). The circular convolution of each section with $h[n]$ is denoted $y_{rp}[n]$, the extra subscript p indicating that $y_{rp}[n]$ is the result of a circular convolution in which time aliasing has occurred. These sequences are depicted in Figure 8.24(b). The portion of each output section in the region $0 \le n \le P - 2$ is the part that must be discarded. The remaining samples from successive sections are then abutted to construct the final filtered output. That is,

$$y[n] = \sum_{r=0}^{\infty} y_r[n - r(L - P + 1) + P - 1], \qquad (8.145)$$

where

$$y_r[n] = \begin{cases} y_{rp}[n], & P - 1 \le n \le L - 1, \\ 0, & \text{otherwise.} \end{cases} \qquad (8.146)$$

This procedure is called the overlap–save method because the input segments overlap, so that each succeeding input section consists of $(L - P + 1)$ new points and $(P - 1)$ points saved from the previous section.

The utility of the overlap–add and the overlap–save methods of block convolution may not be immediately apparent. In Chapter 9, we consider highly efficient algorithms for computing the DFT. These algorithms, collectively called the fast Fourier transform (FFT), are so efficient that, for FIR impulse responses of even modest length (on the order of 25 or 30), it may be more efficient to carry out block convolution using the DFT than to implement the linear convolution directly. The length P at which the DFT method becomes more efficient is, of course, dependent on the hardware and software available to implement the computations. (See Stockham, 1966, and Helms, 1967.)

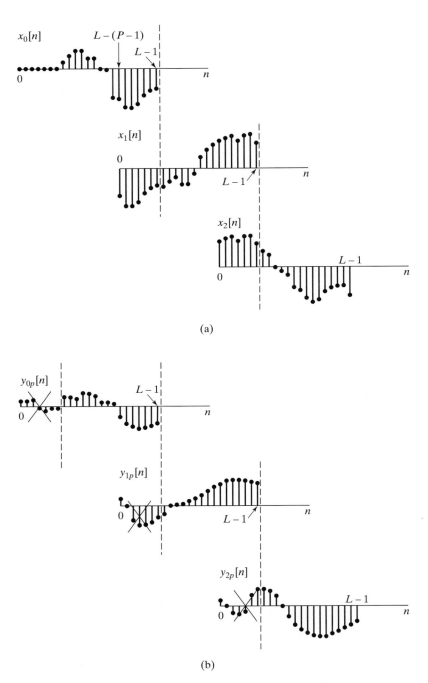

Figure 8.24   (a) Decomposition of $x[n]$ in Figure 8.22 into overlapping sections of length L. (b) Result of convolving each section with $h[n]$. The portions of each filtered section to be discarded in forming the linear convolution are indicated.

8.8 THE DISCRETE COSINE TRANSFORM (DCT)

The DFT is perhaps the most common example of a general class of finite-length transform representations of the form

$$A[k] = \sum_{n=0}^{N-1} x[n]\phi_k^*[n], \tag{8.147}$$

$$x[n] = \frac{1}{N} \sum_{k=0}^{N-1} A[k]\phi_k[n], \tag{8.148}$$

where the sequences $\phi_k[n]$, referred to as the *basis sequences,* are orthogonal to one another; i.e.,

$$\frac{1}{N} \sum_{n=0}^{N-1} \phi_k[n]\phi_m^*[n] = \begin{cases} 1, & m = k, \\ 0, & m \neq k. \end{cases} \tag{8.149}$$

In the case of the DFT, the basis sequences are the complex periodic sequences $e^{j2\pi kn/N}$, and the sequence $A[k]$ is, in general, complex even if the sequence $x[n]$ is real. It is natural to inquire as to whether there exist sets of real-valued basis sequences that would yield a real-valued transform sequence $A[k]$ when $x[n]$ is real. This has led to the definition of a number of other orthogonal transform representations, such as Haar transforms, Hadamard transforms (Elliott and Rao, 1982), and Hartley transforms (Bracewell, 1983, 1984, 1986) (The definition and properties of the Hartley transform are explored in Problem 8.65.) Another orthogonal transform for real sequences is the DCT, or discrete cosine transform. (See Ahmed et al., 1974 and Rao and Yip, 1990.) The DCT is closely related to the DFT and has become especially useful and important in a number of signal-processing applications, particularly speech and image compression. In this section, we conclude our discussion of the DFT by introducing the DCT and showing its relationship to the DFT.

8.8.1 Definitions of the DCT

The DCT is a transform in the form of Eqs. (8.147) and (8.148) with basis sequences $\phi_k[n]$ that are cosines. Since cosines are both periodic and have even symmetry, the extension of $x[n]$ outside the range $0 \leq n \leq (N-1)$ in the synthesis equation (8.148) will be both periodic and symmetric. In other words, just as the DFT involves an implicit assumption of periodicity, the DCT involves implicit assumptions of both periodicity and *even symmetry*.

In the development of the DFT, we represented finite-length sequences by first forming periodic sequences from which the finite-length sequence can be uniquely recovered and then utilizing an expansion in terms of periodic complex exponentials. In a similar style, the DCT corresponds to forming a periodic, symmetric sequence from a finite-length sequence in such a way that the original finite-length sequence can be uniquely recovered. Because there are many ways to do this, there are many definitions of the DCT. In Figure 8.25, we show 17 samples for each of four examples of symmetric periodic extensions of a four-point sequence. The original finite-length sequence is shown in each subfigure as the samples with solid dots. These sequences are all periodic (with period 16 or less) and also have even symmetry. In each case, the finite length

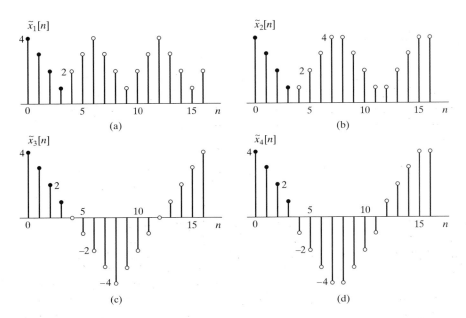

Figure 8.25 Four ways to extend a four-point sequence $x[n]$ both periodically and symmetrically. The finite-length sequence $x[n]$ is plotted with solid dots. (a) Type-1 periodic extension for DCT-1. (b) Type-2 periodic extension for DCT-2. (c) Type-3 periodic extension for DCT-3. (d) Type-4 periodic extension for DCT-4.

sequence is easily extracted as the first four points of one period. For convenience, we denote the periodic sequences obtained by replicating with period 16 each of the four subsequences in Figure 8.25(a), (b), (c) and (d) as $\tilde{x}_1[n]$, $\tilde{x}_2[n]$, $\tilde{x}_3[n]$, and $\tilde{x}_4[n]$, respectively. We note that $\tilde{x}_1[n]$ has period $(2N-2) = 6$ and has even symmetry about both $n = 0$ and $n = (N-1) = 3$. The sequence $\tilde{x}_2[n]$ has period $2N = 8$ and has even symmetry about the "half sample" points $n = -\frac{1}{2}$ and $\frac{7}{2}$. The sequence $\tilde{x}_3[n]$ has period $4N = 16$ and has even symmetry about $n = 0$ and $n = 8$. The sequence $\tilde{x}_4[n]$ also has period $4N = 16$ and even symmetry about the "half sample" points $n = -\frac{1}{2}$ and $n = (2N - \frac{1}{2}) = \frac{15}{2}$.

The four different cases shown in Figure 8.25 illustrate the periodicity that is implicit in the four common forms of the DCT, which are referred to as DCT-1, DCT-2, DCT-3, and DCT-4 respectively. It can be shown (see Martucci, 1994) that there are four more ways to create an even periodic sequence from $x[n]$. This implies four other possible DCT representations. Furthermore, it is also possible to create eight odd-symmetric periodic real sequences from $x[n]$, leading to eight different versions of the *discrete sine transform* (or DST), where the basis sequences in the orthonormal representation are sine functions. These transforms make up a family of 16 orthonormal transforms for real sequences. Of these, the DCT-1 and DCT-2 representations are the most used, and we shall therefore focus on them in the remainder of our discussion.

8.8.2 Definition of the DCT-1 and DCT-2

All of the periodic extensions leading to different forms of the DCT can be thought of as a sum of shifted copies of the N-point sequences $\pm x[n]$ and $\pm x[-n]$. The differences between the extensions for the DCT-1 and DCT-2 depend on whether the endpoints overlap with shifted versions of themselves and, if so, which of the endpoints overlap. For the DCT-1, $x[n]$ is first modified at the endpoints and then extended to have period $2N - 2$. The resulting periodic sequence is

$$\tilde{x}_1[n] = x_\alpha[((n))_{2N-2}] + x_\alpha[((-n))_{2N-2}], \tag{8.150}$$

where $x_\alpha[n]$ is the modified sequence $x_\alpha[n] = \alpha[n]x[n]$, with

$$\alpha[n] = \begin{cases} \frac{1}{2}, & n = 0 \text{ and } N - 1, \\ 1, & 1 \le n \le N - 2. \end{cases} \tag{8.151}$$

The weighting of the endpoints compensates for the doubling that occurs when the two terms in Equation (8.150) overlap at $n = 0$, $n = (N - 1)$, and at the corresponding points spaced from these by integer multiples of $(2N-2)$. With this weighting, it is easily verified that $x[n] = \tilde{x}_1[n]$ for $n = 0, 1, \ldots, N - 1$. The resulting periodic sequence $\tilde{x}_1[n]$ has even periodic symmetry about the points $n = 0$ and $n = N - 1, 2N - 2$, etc., which we refer to as *Type-1* periodic symmetry. The DCT-1 is defined by the transform pair

$$X^{c1}[k] = 2 \sum_{n=0}^{N-1} \alpha[n]x[n] \cos\left(\frac{\pi kn}{N-1}\right), \qquad 0 \le k \le N - 1, \tag{8.152}$$

$$x[n] = \frac{1}{N-1} \sum_{k=0}^{N-1} \alpha[k] X^{c1}[k] \cos\left(\frac{\pi kn}{N-1}\right), \qquad 0 \le n \le N - 1, \tag{8.153}$$

where $\alpha[n]$ is defined in Eq. (8.151).

For the DCT-2, $x[n]$ is extended to have period $2N$, and the periodic sequence is given by

$$\tilde{x}_2[n] = x[((n))_{2N}] + x[((-n-1))_{2N}], \tag{8.154}$$

Because the endpoints do not overlap, no modification of them is required to ensure that $x[n] = \tilde{x}_2[n]$ for $n = 0, 1, \ldots, N-1$. In this case, which we call *Type-2* periodic symmetry, the periodic sequence $\tilde{x}_2[n]$ has even periodic symmetry about the "half sample" points $-1/2, N - 1/2, 2N - 1/2$, etc. The DCT-2 is defined by the transform pair

$$X^{c2}[k] = 2 \sum_{n=0}^{N-1} x[n] \cos\left(\frac{\pi k(2n + 1)}{2N}\right), \qquad 0 \le k \le N - 1, \tag{8.155}$$

$$x[n] = \frac{1}{N} \sum_{k=0}^{N-1} \beta[k] X^{c2}[k] \cos\left(\frac{\pi k(2n + 1)}{2N}\right), \qquad 0 \le n \le N - 1, \tag{8.156}$$

where the inverse DCT-2 involves the weighting function

$$\beta[k] = \begin{cases} \frac{1}{2}, & k = 0 \\ 1, & 1 \le k \le N - 1. \end{cases} \tag{8.157}$$

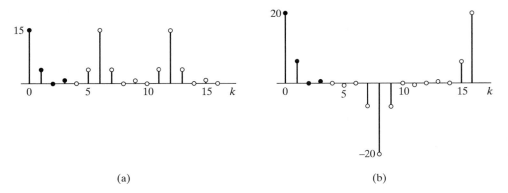

Figure 8.26   DCT-1 and DCT-2 for the four-point sequence used in Figure 8.25.
(a) DCT-1. (b) DCT-2.

In many treatments, the DCT definitions include normalization factors that make the transforms *unitary*.[3] For example, the DCT-2 form is often defined as

$$\tilde{X}^{c2}[k] = \sqrt{\frac{2}{N}} \tilde{\beta}[k] \sum_{n=0}^{N-1} x[n] \cos\left(\frac{\pi k(2n+1)}{2N}\right), \qquad 0 \le k \le N-1, \qquad (8.158)$$

$$x[n] = \sqrt{\frac{2}{N}} \sum_{k=0}^{N-1} \tilde{\beta}[k] \tilde{X}^{c2}[k] \cos\left(\frac{\pi k(2n+1)}{2N}\right), \qquad 0 \le n \le N-1, \quad (8.159)$$

where

$$\tilde{\beta}[k] = \begin{cases} \frac{1}{\sqrt{2}}, & k = 0, \\ 1, & k = 1, 2, \ldots, N-1. \end{cases} \qquad (8.160)$$

Comparing these equations with Equations (8.155) and (8.156), we see that the multiplicative factors 2, $1/N$, and $\beta[k]$ have been redistributed between the direct and inverse transforms. (A similar normalization can be applied to define a normalized version of the DCT-1.) While this normalization creates a unitary transform representation, the definitions in Eqs. (8.152) and (8.153) and Eqs. (8.155) and (8.156) are simpler to relate to the DFT as we have defined it in this chapter. Therefore, in the following discussions, we use our definitions rather than the normalized definitions that are found, for example, in Rao and Yip (1990) and many other texts.

Although we normally evaluate the DCT only for $0 \le k \le N-1$, nothing prevents our evaluating the DCT equations outside that interval, as illustrated in Figure 8.26, where the DCT values for $0 \le k \le N-1$ are shown as solid dots. These figures illustrate that the DCTs also are even periodic sequences. However, the symmetry of the transform sequence is not always the same as the symmetry of the implicit periodic input sequence: While $\tilde{x}_1[n]$ and the extension of $X^{c1}[k]$ both have Type-1 symmetry, we

---

[3]The DCT would be a unitary transform if it is orthonormal and also has the property that

$$\sum_{n=0}^{N-1} (x[n])^2 = \sum_{k=0}^{N-1} (X^{c2}[k])^2.$$

see from a comparison of Figures 8.25(c) and 8.26(b) that $X^{c2}[k]$ has the same symmetry as $\tilde{x}_3[n]$ rather than $\tilde{x}_2[n]$.

Since the DCTs are orthogonal transform representations, they have properties similar in form to those of the DFT. These properties are elaborated on in some detail in Ahmed et al (1974) and Rao and Yip (1990).

8.8.3 Relationship between the DFT and the DCT-1

As we would expect, there is a close relationship between the DFT and the various classes of the DCT of a finite-length sequence. To develop this relationship, we note that, since, for the DCT-1, $\tilde{x}_1[n]$ is constructed from $x_1[n]$ through Eqs. (8.150) and (8.151), one period of the periodic sequence $\tilde{x}_1[n]$ defines the finite-length sequence

$$x_1[n] = x_\alpha[((n))_{2N-2}] + x_\alpha[((-n))_{2N-2}] = \tilde{x}_1[n], \qquad n = 0, 1, \ldots, 2N-3, \qquad (8.161)$$

where $x_\alpha[n] = \alpha[n]x[n]$ is the N-point real sequence with endpoints divided by 2. From Equation (8.161), it follows that the $(2N-2)$-point DFT of the $(2N-2)$-point sequence $x_1[n]$ is

$$X_1[k] = X_\alpha[k] + X_\alpha^*[k] = 2\mathcal{R}e\{X_\alpha[k]\}, \qquad k = 0, 1, \ldots, 2N-3, \qquad (8.162)$$

where $X_\alpha[k]$ is the $(2N-2)$-point DFT of the N-point sequence $\alpha[n]x[n]$; i.e., $\alpha[n]x[n]$ is padded with $(N-1)$ zero samples. Using the definition of the $(2N-2)$-point DFT of the padded sequence, we obtain

$$X_1[k] = 2\mathcal{R}e\{X_\alpha[k]\} = 2\sum_{n=0}^{N-1} \alpha[n]x[n] \cos\left(\frac{2\pi kn}{2N-2}\right) = X^{c1}[k]. \qquad (8.163)$$

Therefore, the DCT-1 of an N-point sequence is identical to the $(2N-2)$-point DFT of the symmetrically extended sequence $x_1[n]$, and it is also identical to twice the real part of the first N points of the $(2N-2)$-point DFT of the weighted sequence $x_\alpha[n]$.

Since, as discussed in Chapter 9, fast computational algorithms exist for the DFT, they can be used to compute the DFTs $X_\alpha[k]$ or $X_1[k]$ in Equation (8.163), thus providing a convenient and readily available fast computation of the DCT-1. Since the definition of the DCT-1 involves only real-valued coefficients, there are also efficient algorithms for computing the DCT-1 of real sequences more directly, without requiring the use of complex multiplications and additions. (See Ahmed et al., 1974 and Chen et al., 1977.)

The inverse DCT-1 can also be computed using the inverse DFT. It is only necessary to use Equation (8.163) to construct $X_1[k]$ from $X^{c1}[k]$ and then compute the inverse $(2N-2)$-point DFT. Specifically,

$$X_1[k] = \begin{cases} X^{c1}[k], & k = 0, \ldots, N-1, \\ X^{c1}[2N-2-k], & k = N, \ldots, 2N-3, \end{cases} \qquad (8.164)$$

and, using the definition of the $(2N-2)$-point inverse DFT, we can compute the symmetrically extended sequence

$$x_1[n] = \frac{1}{2N-2} \sum_{k=0}^{2N-3} X_1[k]e^{j2\pi kn/(2N-2)}, \qquad n = 0, 1, \ldots, 2N-3, \qquad (8.165)$$

from which we can obtain $x[n]$ by extracting the first N points, i.e., $x[n] = x_1[n]$ for $n = 0, 1, \ldots, N - 1$. By substitution of Eq. (8.164) into Eq. (8.165), it also follows that the inverse DCT-1 relation can be expressed in terms of $X^{c1}[k]$ and cosine functions, as in Equation (8.153). (See Problem 8.68.)

8.8.4 Relationship between the DFT and the DCT-2

It is also possible to express the DCT-2 of a finite-length sequence $x[n]$ in terms of the DFT. To develop this relationship, observe that one period of the periodic sequence $\tilde{x}_2[n]$ defines the $2N$-point sequence

$$x_2[n] = x[((n))_{2N}] + x[((-n-1))_{2N}] = \tilde{x}_2[n], \qquad n = 0, 1, \ldots, 2N - 1, \qquad (8.166)$$

where $x[n]$ is the original N-point real sequence. From Equation (8.166), it follows that the $2N$-point DFT of the $2N$-point sequence $x_2[n]$ is

$$X_2[k] = X[k] + X^*[k]e^{j2\pi k/(2N)}, \qquad k = 0, 1, \ldots, 2N - 1, \qquad (8.167)$$

where $X[k]$ is the $2N$-point DFT of the N-point sequence $x[n]$; i.e., in this case, $x[n]$ is padded with N zero samples. From Equation (8.167), we obtain

$$\begin{aligned} X_2[k] &= X[k] + X^*[k]e^{j2\pi k/(2N)} \\ &= e^{j\pi k/(2N)}\left(X[k]e^{-j\pi k/(2N)} + X^*[k]e^{j\pi k/(2N)} \right) \\ &= e^{j\pi k/(2N)}2\mathcal{Re}\left\{ X[k]e^{-j\pi k/(2N)} \right\}. \end{aligned} \qquad (8.168)$$

From the definition of the $2N$-point DFT of the padded sequence, it follows that

$$\mathcal{Re}\left\{ X[k]e^{-j\pi k/(2N)} \right\} = \sum_{n=0}^{N-1} x[n]\cos\left(\frac{\pi k(2n+1)}{2N} \right). \qquad (8.169)$$

Therefore, using Equations (8.155), (8.167), and (8.169), we can express $X^{c2}[k]$ in terms of $X[k]$, the $2N$-point DFT of the N-point sequence $x[n]$, as

$$X^{c2}[k] = 2\mathcal{Re}\left\{ X[k]e^{-j\pi k/(2N)} \right\}, \qquad k = 0, 1, \ldots, N - 1, \qquad (8.170)$$

or in terms of the $2N$-point DFT of the $2N$-point symmetrically extended sequence $x_2[n]$ defined by Equation (8.166) as

$$X^{c2}[k] = e^{-j\pi k/(2N)}X_2[k], \qquad k = 0, 1, \ldots, N - 1, \qquad (8.171)$$

and equivalently,

$$X_2[k] = e^{j\pi k/(2N)}X^{c2}[k], \qquad k = 0, 1, \ldots, N - 1. \qquad (8.172)$$

As in the case of the DCT-1, fast algorithms can be used to compute the DFTs in Equations (8.170) and (8.171). Makhoul (1980) discusses other ways that the DFT can be used to compute the DCT-2. (See also Problem 8.69.) In addition, special fast algorithms for the computation of the DCT-2 have been developed (Rao and Yip, 1990).

The inverse DCT-2 can also be computed using the inverse DFT. The procedure utilizes Eq. (8.172) together with a symmetry property of the DCT-2. Specifically, it is easily verified by direct substitution into Eq. (8.155) that

$$X^{c2}[2N - k] = -X^{c2}[k], \qquad k = 0, 1, \ldots, 2N - 1, \tag{8.173}$$

from which it follows that

$$X_2[k] = \begin{cases} X^{c2}[0], & k = 0, \\ e^{j\pi k/(2N)} X^{c2}[k], & k = 1, \ldots, N - 1, \\ 0, & k = N, \\ -e^{j\pi k/(2N)} X^{c2}[2N - k], & k = N+1, N+2, \ldots, 2N - 1. \end{cases} \tag{8.174}$$

Using the definition of the inverse DFT, we can compute the symmetrically extended sequence

$$x_2[n] = \frac{1}{2N} \sum_{k=0}^{2N-1} X_2[k] e^{j2\pi kn/(2N)}, \qquad n = 0, 1, \ldots, 2N - 1, \tag{8.175}$$

from which we can obtain $x[n] = x_2[n]$ for $n = 0, 1, \ldots, N-1$. By substituting Eq. (8.174) into Eq. (8.175), we can easily show that the inverse DCT-2 relation is that given by Eq. (8.156). (See Problem 8.70.)

8.8.5 Energy Compaction Property of the DCT-2

The DCT-2 is used in many data compression applications in preference to the DFT because of a property that is frequently referred to as "energy compaction." Specifically, the DCT-2 of a finite-length sequence often has its coefficients more highly concentrated at low indices than the DFT does. The importance of this flows from Parseval's theorem, which, for the DCT-1, is

$$\sum_{n=0}^{N-1} \alpha[n]|x[n]|^2 = \frac{1}{2N - 2} \sum_{k=0}^{N-1} \alpha[k]|X^{c1}[k]|^2, \tag{8.176}$$

and, for the DCT-2, is

$$\sum_{n=0}^{N-1} |x[n]|^2 = \frac{1}{N} \sum_{k=0}^{N-1} \beta[k]|X^{c2}[k]|^2, \tag{8.177}$$

where $\beta[k]$ is defined in Eq. (8.157). The DCT can be said to be concentrated in the low indices of the DCT if the remaining DCT coefficients can be set to zero without a significant impact on the energy of the signal. We illustrate the energy compaction property in the following example.

Example 8.13   Energy Compaction in the DCT-2

Consider a test input of the form

$$x[n] = a^n \cos(\omega_0 n + \phi), \qquad n = 0, 1, \ldots, N-1. \tag{8.178}$$

Such a signal is illustrated in Figure 8.27 for $a = .9$, $\omega_0 = 0.1\pi$, $\phi = 0$, and $N = 32$.

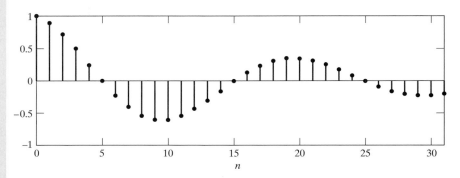

Figure 8.27   Test signal for comparing DFT and DCT.

The real and imaginary parts of the 32-point DFT of the 32-point sequence in Figure 8.27 are shown in Figures 8.28(a) and (b), respectively, and the DCT-2 of the sequence is shown in Figure 8.28(c). In the case of the DFT, the real and imaginary parts are shown for $k = 0, 1, \ldots, 16$. Since the signal is real, $X[0]$ and $X[16]$ are real. The remaining values are complex and conjugate symmetric. Thus, the 32 real numbers shown in Figures 8.28(a) and (b) completely specify the 32-point DFT. In the case of the DCT-2, we show all 32 of the real DCT-2 values. Clearly, the DCT-2 values are highly concentrated at low indices, so Parseval's theorem suggests that the energy of the sequence is more concentrated in the DCT-2 representation than in the DFT representation.

This energy concentration property can be quantified by truncating the two representations and comparing the mean-squared approximation error for the two representations when both use the same number of real coefficient values. To do this, we define

$$x_m^{\text{dft}}[n] = \frac{1}{N} \sum_{k=0}^{N-1} T_m[k] X[k] e^{j2\pi kn/N}, \qquad n = 0, 1, \ldots, N-1, \tag{8.179}$$

where, in this case, $X[k]$ is the N-point DFT of $x[n]$ and

$$T_m[k] = \begin{cases} 1, & 0 \le k \le (N-1-m)/2, \\ 0, & (N+1-m)/2 \le k \le (N-1+m)/2, \\ 1, & (N+1+m)/2 \le k \le N-1. \end{cases}$$

If $m = 1$, the term $X[N/2]$ is removed. If $m = 3$, then the terms $X[N/2]$ and $X[N/2-1]$ and its corresponding complex conjugate $X[N/2+1]$ are removed, and so forth; i.e., $x_m^{\text{dft}}[n]$ for $m = 1, 3, 5, \ldots, N-1$ is the sequence that is synthesized by symmetrically omitting m DFT coefficients.[4] With the exception of the DFT value, $X[N/2]$, which

---

[4]For simplicity, we assume that N is an even integer.

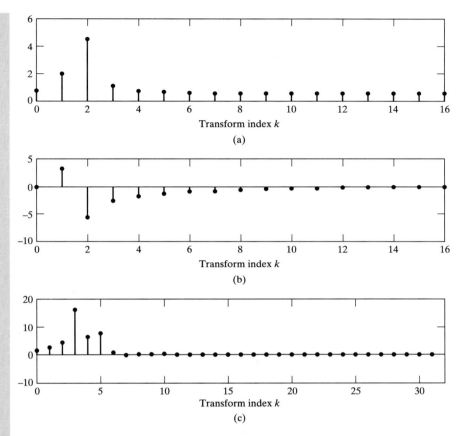

Figure 8.28   (a) Real part of N-point DFT; (b) Imaginary part of N-point DFT; (c) N-point DCT-2 of the test signal plotted in Figure 8.27.

is real, each omitted complex DFT value and its corresponding complex conjugate actually corresponds to omitting two real numbers. For example, $m = 5$ would correspond to setting the coefficients $X[14]$, $X[15]$, $X[16]$, $X[17]$, and $X[18]$ to zero in synthesizing $x_5^{\text{dft}}[n]$ from the 32-point DFT shown in Figures 8.28(a) and (b).

Likewise, we can truncate the DCT-2 representation, obtaining

$$x_m^{\text{dct}}[n] = \frac{1}{N} \sum_{k=0}^{N-1-m} \beta[k] X^{c2}[k] \cos\left(\frac{\pi k(2n+1)}{2N}\right), \qquad 0 \le n \le N-1. \qquad (8.180)$$

In this case, if $m = 5$, we omit the DCT-2 coefficients $X^{c2}[27], \ldots, X^{c2}[31]$ in the synthesis of $x_m^{\text{dct}}[n]$ from the DCT-2 shown in Figure 8.28(c). Since these coefficients are very small, $x_5^{\text{dct}}[n]$ should differ only slightly from $x[n]$.

To show how the approximation errors depend on m for the DFT and the DCT-2, we define

$$E^{\text{dft}}[m] = \frac{1}{N} \sum_{n=0}^{N-1} |x[n] - x_m^{\text{dft}}[n]|^2$$

Figure 8.29   Comparison of truncation errors for DFT and DCT-2.

and

$$E^{\text{dct}}[m] = \frac{1}{N} \sum_{n=0}^{N-1} |x[n] - x_m^{\text{dct}}[n]|^2$$

to be the mean-squared approximation errors for the truncated DFT and DCT, respectively. These errors are plotted in Figure 8.29, with $E^{\text{dft}}[m]$ indicated with $\circ$ and $E^{\text{dct}}[m]$ shown with $\bullet$. For the special cases $m = 0$ (no truncation) and $m = N-1$ (only the DC value is retained), the DFT truncation function is $T_0[k] = 1$ for $0 \leq k \leq N-1$ and $T_{N-1}[k] = 0$ for $1 \leq k \leq N-1$ and $T_{N-1}[0] = 1$. In these cases, both representations give the same error. For values $1 \leq m \leq 30$, the DFT error grows steadily as m increases, while the DCT error remains very small up to about $m = 25$, implying that the 32 numbers of the sequence $x[n]$ can be represented with slight error by only seven DCT-2 coefficients.

The signal in Example 8.13 is a low frequency exponentially decaying signal with zero phase. We have chosen this example very carefully to emphasize the energy compaction property. Not every choice of $x[n]$ will give such dramatic results. Highpass signals and even some signals of the form of Equation (8.178) with different parameters do not show this dramatic difference. Nevertheless, in many cases of interest in data compression, the DCT-2 provides a distinct advantage over the DFT. It can be shown (Rao and Yip, 1990) that the DCT-2 is nearly optimum in the sense of minimum mean-squared truncation error for sequences with exponential correlation functions.

8.8.6 Applications of the DCT

The major application of the DCT-2 is in signal compression, where it is a key part of many standardized algorithms. (See Jayant and Noll, 1984 and Rao and Hwang, 1996.) In this application, the blocks of the signal are represented by their cosine transforms. The popularity of the DCT in signal compression is mainly due to its energy concentration property, which we demonstrated by a simple example in the previous section.

The DCT representations, being orthogonal transforms like the DFT, have many properties similar to those of the DFT that make them very flexible for manipulating the signals that they represent. One of the most important properties of the DFT is that periodic convolution of two finite-length sequences corresponds to multiplication of their corresponding DFTs. We have seen in Section 8.7 that it is possible to exploit this property to compute linear convolutions by doing only DFT computations. In the case of the DCT, the corresponding result is that multiplication of DCTs corresponds to periodic convolution of the underlying symmetrically extended sequences. However, there are additional complications. For example, the periodic convolution of two Type-2 symmetric periodic sequences is not a Type-2 sequence, but rather, a Type-1 sequence. Alternatively, periodic convolution of a Type-1 sequence with a Type-2 sequence of the same implied period is a Type-2 sequence. Thus, a mixture of DCTs is required to effect periodic symmetric convolution by inverse transformation of the product of DCTs. There are many more ways to do this because we have many different DCT definitions from which to choose. Each different combination would correspond to periodic convolution of a pair of symmetrically extended finite sequences. Martucci (1994) provides a complete discussion of the use of DCT and DST transforms in implementing symmetric periodic convolution.

Multiplication of DCTs corresponds to a special type of periodic convolution that has some features that may be useful in some applications. As we have seen for the DFT, periodic convolution is characterized by end effects, or "wrap around" effects. Indeed, even linear convolution of two finite-length sequences has end effects as the impulse response engages and disengages from the input. The end effects of periodic symmetric convolution are different from ordinary convolution and from periodic convolution as implemented by multiplying DFTs. The symmetric extension creates symmetry at the endpoints. The "smooth" boundaries that this implies often mitigate the end effects encountered in convolving finite-length sequences. One area in which symmetric convolution is particularly useful is image filtering, where objectionable edge effects are perceived as blocking artifacts. In such representations, the DCT may be superior to the DFT or even ordinary linear convolution. In doing periodic symmetric convolution by multiplication of DCTs, we can force the same result as ordinary convolution by extending the sequences with a sufficient number of zero samples placed at both the beginning and the end of each sequence.

8.9 SUMMARY

In this chapter, we have discussed discrete Fourier representations of finite-length sequences. Most of our discussion focused on the discrete Fourier transform (DFT), which is based on the discrete Fourier series representation of periodic sequences. By defining a periodic sequence for which each period is identical to the finite-length sequence, the DFT becomes identical to one period of the discrete Fourier series coefficients. Because of the importance of this underlying periodicity, we first examined the properties of discrete Fourier series representations and then interpreted those properties in terms of finite-length sequences. An important result is that the DFT values are equal to samples of the z-transform at equally spaced points on the unit circle. This leads to the notion of time aliasing in the interpretation of DFT properties, a concept we utilized extensively in the study of circular convolution and its relation to linear convolution. We then used the results of this study to show how the DFT could be employed to implement the linear convolution of a finite-length impulse response with an indefinitely long input signal.

The chapter concluded with an introduction to the discrete cosine transform. It was shown that the DCT and DFT are closely related and that they share an implicit assumption of periodicity. The energy compaction property, which is the main reason for the popularity of the DCT in data compression, was demonstrated with an example.

PROBLEMS

Basic Problems with Answers

8.1. Suppose $x_c(t)$ is a periodic continuous-time signal with period 1 ms and for which the Fourier series is

$$x_c(t) = \sum_{k=-9}^{9} a_k e^{j(2\pi kt/10^{-3})}.$$

The Fourier series coefficients a_k are zero for $|k| > 9$. $x_c(t)$ is sampled with a sample spacing $T = \frac{1}{6} \times 10^{-3}$ s to form $x[n]$. That is,

$$x[n] = x_c\left(\frac{n10^{-3}}{6}\right).$$

(a) Is $x[n]$ periodic and, if so, with what period?
(b) Is the sampling rate above the Nyquist rate? That is, is T sufficiently small to avoid aliasing?
(c) Find the discrete Fourier series coefficients of $x[n]$ in terms of a_k.

8.2. Suppose $\tilde{x}[n]$ is a periodic sequence with period N. Then $\tilde{x}[n]$ is also periodic with period $3N$. Let $\tilde{X}[k]$ denote the DFS coefficients of $\tilde{x}[n]$ considered as a periodic sequence with period N, and let $\tilde{X}_3[k]$ denote the DFS coefficients of $\tilde{x}[n]$ considered as a periodic sequence with period $3N$.
(a) Express $\tilde{X}_3[k]$ in terms of $\tilde{X}[k]$.
(b) By explicitly calculating $\tilde{X}[k]$ and $\tilde{X}_3[k]$, verify your result in Part (a) when $\tilde{x}[n]$ is as given in Figure P8.2-1.

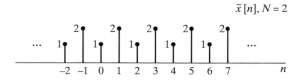

$\tilde{x}[n], N = 2$

-2 -1 0 1 2 3 4 5 6 7 n **Figure P8.2-1**

8.3. Figure P8.3-1 shows three periodic sequences $\tilde{x}_1[n]$ through $\tilde{x}_3[n]$. These sequences can be expressed in a Fourier series as

$$\tilde{x}[n] = \frac{1}{N} \sum_{k=0}^{N-1} \tilde{X}[k] e^{j(2\pi/N)kn}.$$

(a) For which sequences can the time origin be chosen such that all the $\tilde{X}[k]$ are real?
(b) For which sequences can the time origin be chosen such that all the $\tilde{X}[k]$ (except for k an integer multiple of N) are imaginary?
(c) For which sequences does $\tilde{X}[k] = 0$ for $k = \pm 2, \pm 4, \pm 6$?

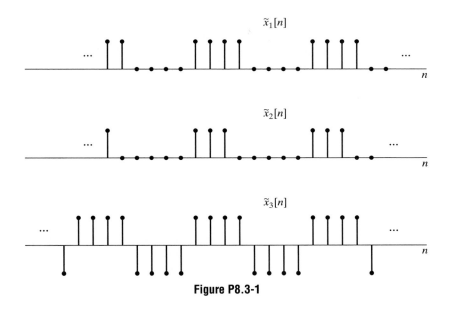

Figure P8.3-1

8.4. Consider the sequence $x[n]$ given by $x[n] = \alpha^n u[n]$. A periodic sequence $\tilde{x}[n]$ is constructed from $x[n]$ in the following way:

$$\tilde{x}[n] = \sum_{r=-\infty}^{\infty} x[n + rN].$$

(a) Determine the Fourier transform $X(e^{j\omega})$ of $x[n]$.
(b) Determine the discrete Fourier series $\tilde{X}[k]$ of $\tilde{x}[n]$.
(c) How is $\tilde{X}[k]$ related to $X(e^{j\omega})$?

8.5. Compute the DFT of each of the following finite-length sequences considered to be of length N (where N is even):
(a) $x[n] = \delta[n]$,
(b) $x[n] = \delta[n - n_0]$, $0 \le n_0 \le N - 1$,
(c) $x[n] = \begin{cases} 1, & n \text{ even}, \quad 0 \le n \le N - 1, \\ 0, & n \text{ odd}, \quad 0 \le n \le N - 1, \end{cases}$
(d) $x[n] = \begin{cases} 1, & 0 \le n \le N/2 - 1, \\ 0, & N/2 \le n \le N - 1, \end{cases}$
(e) $x[n] = \begin{cases} a^n, & 0 \le n \le N - 1, \\ 0, & \text{otherwise}. \end{cases}$

8.6. Consider the complex sequence

$$x[n] = \begin{cases} e^{j\omega_0 n}, & 0 \le n \le N-1, \\ 0, & \text{otherwise.} \end{cases}$$

(a) Find the Fourier transform $X(e^{j\omega})$ of $x[n]$.
(b) Find the N-point DFT $X[k]$ of the finite-length sequence $x[n]$.
(c) Find the DFT of $x[n]$ for the case $\omega_0 = 2\pi k_0/N$, where k_0 is an integer.

8.7. Consider the finite-length sequence $x[n]$ in Figure P8.7-1. Let $X(z)$ be the z-transform of $x[n]$. If we sample $X(z)$ at $z = e^{j(2\pi/4)k}$, $k = 0, 1, 2, 3$, we obtain

$$X_1[k] = X(z)\big|_{z=e^{j(2\pi/4)k}}, \qquad k = 0, 1, 2, 3.$$

Sketch the sequence $x_1[n]$ obtained as the inverse DFT of $X_1[k]$.

Figure P8.7-1

8.8. Let $X(e^{j\omega})$ denote the Fourier transform of the sequence $x[n] = \left(\frac{1}{2}\right)^n u[n]$. Let $y[n]$ denote a finite-duration sequence of length 10; i.e., $y[n] = 0, n < 0$, and $y[n] = 0, n \ge 10$. The 10-point DFT of $y[n]$, denoted by $Y[k]$, corresponds to 10 equally spaced samples of $X(e^{j\omega})$; i.e., $Y[k] = X(e^{j2\pi k/10})$. Determine $y[n]$.

8.9. Consider a 20-point finite-duration sequence $x[n]$ such that $x[n] = 0$ outside $0 \le n \le 19$, and let $X(e^{j\omega})$ represent the Fourier transform of $x[n]$.
(a) If it is desired to evaluate $X(e^{j\omega})$ at $\omega = 4\pi/5$ by computing one M-point DFT, determine the smallest possible M, and develop a method to obtain $X(e^{j\omega})$ at $\omega = 4\pi/5$ using the smallest M.
(b) If it is desired to evaluate $X(e^{j\omega})$ at $\omega = 10\pi/27$ by computing one L-point DFT, determine the smallest possible L, and develop a method to obtain $X(e^{j10\pi/27})$ using the smallest L.

8.10. The two eight-point sequences $x_1[n]$ and $x_2[n]$ shown in Figure P8.10-1 have DFTs $X_1[k]$ and $X_2[k]$, respectively. Determine the relationship between $X_1[k]$ and $X_2[k]$.

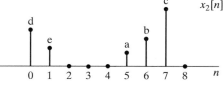

Figure P8.10-1

8.11. Figure P8.11-1 shows two finite-length sequences $x_1[n]$ and $x_2[n]$. Sketch their six-point circular convolution.

Figure P8.11-1

8.12. Suppose we have two four-point sequences $x[n]$ and $h[n]$ as follows:

$$x[n] = \cos\left(\frac{\pi n}{2}\right), \quad n = 0, 1, 2, 3,$$
$$h[n] = 2^n, \qquad\quad n = 0, 1, 2, 3.$$

(a) Calculate the four-point DFT $X[k]$.
(b) Calculate the four-point DFT $H[k]$.
(c) Calculate $y[n] = x[n]\,\textcircled{4}\,h[n]$ by doing the circular convolution directly.
(d) Calculate $y[n]$ of Part (c) by multiplying the DFTs of $x[n]$ and $h[n]$ and performing an inverse DFT.

8.13. Consider the finite-length sequence $x[n]$ in Figure P8.13-1. The five-point DFT of $x[n]$ is denoted by $X[k]$. Plot the sequence $y[n]$ whose DFT is

$$Y[k] = W_5^{-2k} X[k].$$

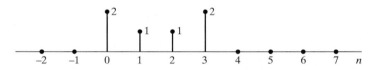

Figure P8.13-1

8.14. Two finite-length signals, $x_1[n]$ and $x_2[n]$, are sketched in Figure P8.14-1. Assume that $x_1[n]$ and $x_2[n]$ are zero outside of the region shown in the figure. Let $x_3[n]$ be the eight-point circular convolution of $x_1[n]$ with $x_2[n]$; i.e., $x_3[n] = x_1[n]\,\textcircled{8}\,x_2[n]$. Determine $x_3[2]$.

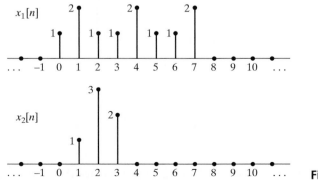

Figure P8.14-1

8.15. Figure P8.15-1 shows two sequences $x_1[n]$ and $x_2[n]$. The value of $x_2[n]$ at time $n = 3$ is not known, but is shown as a variable a. Figure P8.15-2 shows $y[n]$, the four-point circular convolution of $x_1[n]$ and $x_2[n]$. Based on the graph of $y[n]$, can you specify a uniquely? If so, what is a? If not, give two possible values of a that would yield the sequence $y[n]$ as shown.

Figure P8.15-1

Figure P8.15-2

8.16. Figure P8.16-1 shows a six-point discrete-time sequence $x[n]$. Assume that $x[n] = 0$ outside the interval shown. The value of $x[4]$ is not known and is represented as b. Note that the sample shown for b in the figure is not necessarily to scale. Let $X(e^{j\omega})$ be the DTFT of $x[n]$ and $X_1[k]$ be samples of $X(e^{j\omega})$ every $\pi/2$; i.e.,

$$X_1[k] = X(e^{j\omega})|_{\omega=(\pi/2)k}, \qquad 0 \le k \le 3.$$

The four-point sequence $x_1[n]$ that results from taking the four-point inverse DFT of $X_1[k]$ is shown in Figure P8.16-2. Based on this figure, can you determine b uniquely? If so, give the value for b.

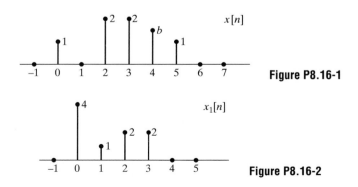

Figure P8.16-1

Figure P8.16-2

8.17. Figure P8.17-1 shows two finite-length sequences $x_1[n]$ and $x_2[n]$. What is the smallest N such that the N-point circular convolution of $x_1[n]$ and $x_2[n]$ are equal to the linear convolution of these sequences, i.e., such that $x_1[n] \, \textcircled{N} \, x_2[n] = x_1[n] * x_2[n]$?

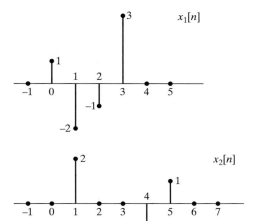

Figure P8.17-1

8.18. Figure P8.18-1 shows a sequence $x[n]$ for which the value of $x[3]$ is an unknown constant c. The sample with amplitude c is not necessarily drawn to scale. Let

$$X_1[k] = X[k]e^{j2\pi 3k/5},$$

where $X[k]$ is the five-point DFT of $x[n]$. The sequence $x_1[n]$ plotted in Figure P8.18-2 is the inverse DFT of $X_1[k]$. What is the value of c?

Figure P8.18-1

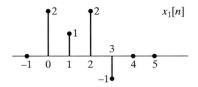

Figure P8.18-2

8.19. Two finite-length sequences $x[n]$ and $x_1[n]$ are shown in Figure P8.19-1. The DFTs of these sequences, $X[k]$ and $X_1[k]$, respectively, are related by the equation

$$X_1[k] = X[k]e^{-j(2\pi km/6)},$$

where m is an unknown constant. Can you determine a value of m consistent with Figure P8.19-1? Is your choice of m unique? If so, justify your answer. If not, find another choice of m consistent with the information given.

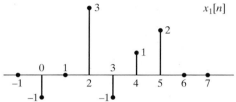

Figure P8.19-1

8.20. Two finite-length sequences $x[n]$ and $x_1[n]$ are shown in Figure P8.20-1. The N-point DFTs of these sequences, $X[k]$ and $X_1[k]$, respectively, are related by the equation

$$X_1[k] = X[k]e^{j2\pi k2/N},$$

where N is an unknown constant. Can you determine a value of N consistent with Figure P8.20-1? Is your choice for N unique? If so, justify your answer. If not, find another choice of N consistent with the information given.

Figure P8.20-1

Basic Problems

8.21. (a) Figure 8.21-1 shows two periodic sequences, $\tilde{x}_1[n]$ and $\tilde{x}_2[n]$, with period $N = 7$. Find a sequence $\tilde{y}_1[n]$ whose DFS is equal to the product of the DFS of $\tilde{x}_1[n]$ and the DFS of $\tilde{x}_2[n]$, i.e.,

$$\tilde{Y}_1[k] = \tilde{X}_1[k]\tilde{X}_2[k].$$

Figure P8.21-1

(b) Figure 8.21-2 shows a periodic sequence $\tilde{x}_3[n]$ with period $N = 7$. Find a sequence $\tilde{y}_2[n]$ whose DFS is equal to the product of the DFS of $\tilde{x}_1[n]$ and the DFS of $\tilde{x}_3[n]$, i.e.,

$$\tilde{Y}_2[k] = \tilde{X}_1[k]\tilde{X}_3[k].$$

Figure P8.21-2

8.22. $x[n]$ denotes a finite-length sequence of length N. Show that

$$x[((-n))_N] = x[((N-n))_N].$$

8.23. Consider a finite-duration sequence $x[n]$ of length P such that $x[n] = 0$ for $n < 0$ and $n \geq P$. We want to compute samples of the Fourier transform at the N equally spaced frequencies

$$\omega_k = \frac{2\pi k}{N}, \qquad k = 0, 1, \ldots, N-1.$$

Determine and justify procedures for computing the N samples of the Fourier transform using only one N-point DFT for the following two cases:
(a) $N > P$.
(b) $N < P$.

8.24. Consider a real finite-length sequence $x[n]$ with Fourier transform $X(e^{j\omega})$ and DFT $X[k]$. If

$$\mathcal{I}m\{X[k]\} = 0, \qquad k = 0, 1, \ldots, N-1,$$

can we conclude that

$$\mathcal{I}m\{X(e^{j\omega})\} = 0, \qquad -\pi \leq \omega \leq \pi?$$

State your reasoning if your answer is yes. Give a counterexample if your answer is no.

8.25. Consider the finite-length sequence $x[n]$ in Figure P8.25-1. The four-point DFT of $x[n]$ is denoted $X[k]$. Plot the sequence $y[n]$ whose DFT is

$$Y[k] = W_4^{3k} X[k].$$

Figure P8.25-1

8.26. Consider the real finite-length sequence $x[n]$ shown in Figure P8.26-1.

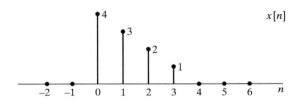

Figure P8.26-1

(a) Sketch the finite-length sequence $y[n]$ whose six-point DFT is

$$Y[k] = W_6^{4k} X[k],$$

where $X[k]$ is the six-point DFT of $x[n]$.

(b) Sketch the finite-length sequence $w[n]$ whose six-point DFT is

$$W[k] = \mathcal{Re}\{X[k]\}.$$

(c) Sketch the finite-length sequence $q[n]$ whose three-point DFT is

$$Q[k] = X[2k], \qquad k = 0, 1, 2.$$

8.27. Figure P8.27-1 shows two sequences,

$$x_1[n] = \begin{cases} 1, & 0 \le n \le 99, \\ 0, & \text{otherwise,} \end{cases}$$

and

$$x_2[n] = \begin{cases} 1, & 0 \le n \le 9, \\ 0, & \text{otherwise.} \end{cases}$$

(a)

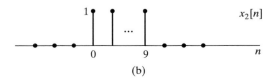

(b)

Figure P8.27-1

(a) Determine and sketch the linear convolution $x_1[n] * x_2[n]$.

(b) Determine and sketch the 100-point circular convolution $x_1[n] \circledⓝ x_2[n]$.

(c) Determine and sketch the 110-point circular convolution $x_1[n] \circledⓝ x_2[n]$.

8.28. Figure P8.28-1 shows a finite-length sequence $x[n]$. Sketch the sequences

$$x_1[n] = x[((n-2))_4], \qquad 0 \le n \le 3,$$

and

$$x_2[n] = x[((-n))_4], \qquad 0 \le n \le 3.$$

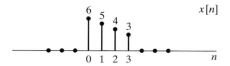

Figure P8.28-1

8.29. Figure P8.29-1 shows two finite-length sequences. Sketch their N-point circular convolution for $N = 6$ and for $N = 10$.

Figure P8.29-1

Advanced Problems

8.30. Consider a finite-length sequence $x[n]$ of length N; i.e.,

$$x[n] = 0 \qquad \text{outside} \quad 0 \le n \le N-1.$$

$X(e^{j\omega})$ denotes the Fourier transform of $x[n]$. $\tilde{X}[k]$ denotes the sequence of 64 equally spaced samples of $X(e^{j\omega})$, i.e.,

$$\tilde{X}[k] = X(e^{j\omega})|_{\omega=2\pi k/64}.$$

It is known that in the range $0 \le k \le 63$, $\tilde{X}[32] = 1$ and all the other values of $\tilde{X}[k]$ are zero.

(a) If the sequence length is $N = 64$, determine one sequence $x[n]$ consistent with the given information. Indicate whether the answer is unique. If it is, clearly explain why. If it is not, give a second *distinct* choice.

(b) If the sequence length is $N = 192 = 3 \times 64$, determine one sequence $x[n]$ consistent with the constraint that in the range $0 \le k \le 63$, $\tilde{X}[32] = 1$ and all other values in that range are zero. Indicate whether the answer is unique. If it is, clearly explain why. If it is not, give a second *distinct* choice.

8.31. The DFT of a finite-duration sequence corresponds to samples of its z-transform on the unit circle. For example, the DFT of a 10-point sequence $x[n]$ corresponds to samples of $X(z)$ at the 10 equally spaced points indicated in Figure P8.31-1. We wish to find the equally spaced samples of $X(z)$ on the contour shown in Figure P8.31-2; i.e., we wish to obtain

$$X(z)\big|_{z=0.5e^{j[(2\pi k/10)+(\pi/10)]}}.$$

Show how to modify $x[n]$ to obtain a sequence $x_1[n]$ such that the DFT of $x_1[n]$ corresponds to the desired samples of $X(z)$.

Figure P8.31-1

Figure P8.31-2

8.32. A finite-duration sequence $x[n]$ of length 8 has the eight-point DFT $X[k]$ shown in Figure P8.32-1. A new sequence of length 16 is defined by

$$y[n] = \begin{cases} x[n/2], & n \text{ even,} \\ 0, & n \text{ odd.} \end{cases}$$

From the list in Figure P8.32-2, choose the sketch corresponding to $Y[k]$, the sixteen-point DFT of $y[n]$.

Figure P8.32-1

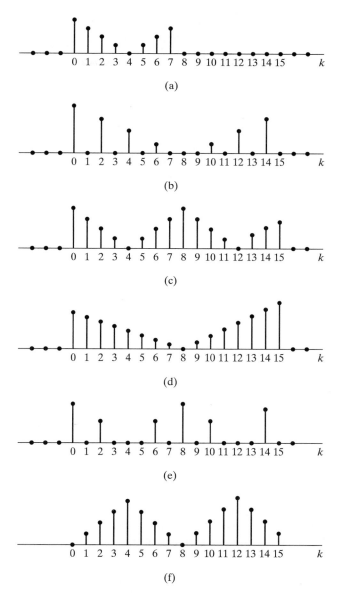

(a)

(b)

(c)

(d)

(e)

(f)   **Figure P8.32-2**

8.33. Consider a finite-length sequence $x[n]$ of length N as indicated in Figure P8.33-1. (The solid line is used to suggest the envelope of the sequence values between 0 and $N - 1$.) Two finite-length sequences $x_1[n]$ and $x_2[n]$ of length $2N$ are constructed from $x[n]$ as indicated in Figure P8.33-2, with $x_1[n]$ and $x_2[n]$ given as follows:

$$x_1[n] = \begin{cases} x[n], & 0 \le n \le N - 1, \\ 0, & \text{otherwise,} \end{cases}$$

$$x_2[n] = \begin{cases} x[n], & 0 \le n \le N - 1, \\ -x[n - N], & N \le n \le 2N - 1, \\ 0, & \text{otherwise.} \end{cases}$$

The N-point DFT of $x[n]$ is denoted by $X[k]$, and the $2N$-point DFTs of $x_1[n]$ and $x_2[n]$ are denoted by $X_1[k]$ and $X_2[k]$, respectively.

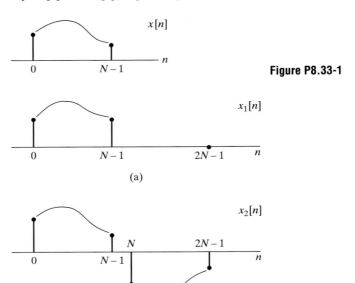

Figure P8.33-1

(a)

Figure P8.33-2

(b)

(a) Specify whether $X_2[k]$ can be obtained if $X[k]$ is given. Clearly indicate your reasoning.
(b) Determine the simplest possible relationship whereby one can obtain $X[k]$ from $X_1[k]$.

8.34. The even part of a real sequence $x[n]$ is defined by

$$x_e[n] = \frac{x[n] + x[-n]}{2}.$$

Suppose that $x[n]$ is a real finite-length sequence defined such that $x[n] = 0$ for $n < 0$ and $n \geq N$. Let $X[k]$ denote the N-point DFT of $x[n]$.
(a) Is $\mathcal{R}e\{X[k]\}$ the DFT of $x_e[n]$?
(b) What is the inverse DFT of $\mathcal{R}e\{X[k]\}$ in terms of $x[n]$?

8.35. Determine a sequence $x[n]$ that satisfies all of the following three conditions:

Condition 1: The Fourier transform of $x[n]$ has the form

$$X(e^{j\omega}) = 1 + A_1 \cos\omega + A_2 \cos 2\omega,$$

where A_1 and A_2 are some unknown constants.

Condition 2: The sequence $x[n]*\delta[n-3]$ evaluated at $n = 2$ is 5.

Condition 3: For the three-point sequence $w[n]$ shown in Figure P8.35-1, the result of the eight-point circular convolution of $w[n]$ and $x[n-3]$ is 11 when $n = 2$; i.e.,

$$\left.\sum_{m=0}^{7} w[m]x[((n-3-m))_8]\right|_{n=2} = 11.$$

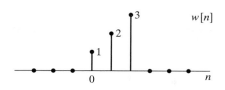

Figure P8.35-1

8.36. Consider the finite-length sequence

$$x[n] = 2\delta[n] + \delta[n-1] + \delta[n-3].$$

We perform the following operation on this sequence:
(i) We compute the five-point DFT $X[k]$.
(ii) We compute a five-point inverse DFT of $Y[k] = X[k]^2$ to obtain a sequence $y[n]$.
  (a) Determine the sequence $y[n]$ for $n = 0, 1, 2, 3, 4$.
  (b) If N-point DFTs are used in the two-step procedure, how should we choose N so that $y[n] = x[n]*x[n]$ for $0 \le n \le N - 1$?

8.37. Consider a finite-duration sequence $x[n]$ that is zero for $n < 0$ and $n \ge N$, where N is even. The z-transform of $x[n]$ is denoted by $X(z)$. Table P8.37-1 lists seven sequences obtained from $x[n]$. Table P8.37-2 lists nine sequences obtained from $X(z)$. For each sequence in Table P8.37-1, find its DFT in Table P8.37-2. The size of the transform considered must be greater than or equal to the length of the sequence $g_k[n]$. *For purposes of illustration only,* assume that $x[n]$ can be represented by the envelope shown in Figure P8.37-1.

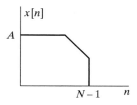

Figure P8.37-1

TABLE P8.37-1

$g_1[n] = x[N-1-n]$

$g_2[n] = (-1)^n x[n]$

$g_3[n] = \begin{cases} x[n], & 0 \le n \le N-1, \\ x[n-N], & N \le n \le 2N-1, \\ 0, & \text{otherwise} \end{cases}$

(*Continued*)

TABLE P8.37-1   *(Continued)*

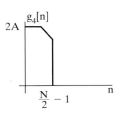

$$g_4[n] = \begin{cases} x[n] + x[n + N/2], & 0 \le n \le N/2 - 1, \\ 0, & \text{otherwise} \end{cases}$$

$$g_5[n] = \begin{cases} x[n], & 0 \le n \le N - 1, \\ 0, & N \le n \le 2N - 1, \\ 0, & \text{otherwise} \end{cases}$$

$$g_6[n] = \begin{cases} x[n/2], & n \text{ even}, \\ 0, & n \text{ odd} \end{cases}$$

$$g_7[n] = x[2n]$$

TABLE P8.37-2

$$H_1[k] = X(e^{j2\pi k/N})$$

$$H_2[k] = X(e^{j2\pi k/2N})$$

$$H_3[k] = \begin{cases} 2X(e^{j2\pi k/2N}), & k \text{ even}, \\ 0, & k \text{ odd} \end{cases}$$

$$H_4[k] = X(e^{j2\pi k/(2N-1)})$$

$$H_5[k] = 0.5\{X(e^{j2\pi k/N}) + X(e^{j2\pi(k+N/2)/N})\}$$

$$H_6[k] = X(e^{j4\pi k/N})$$

$$H_7[k] = e^{j2\pi k/N} X(e^{-j2\pi k/N})$$

$$H_8[k] = X(e^{j(2\pi/N)(k+N/2)})$$

$$H_9[k] = X(e^{-j2\pi k/N})$$

8.38. $x[n]$ is a real-valued finite-length sequence of length 10 and is nonzero in the interval from 0 to 9, i.e.,

$$x[n] = 0, \qquad n < 0, n \ge 10,$$

$$x[n] \ne 0, \qquad 0 \le n \le 9.$$

$X(e^{j\omega})$ denotes the Fourier transform of $x[n]$, and $X[k]$ denotes the 10-point DFT of $x[n]$.

Determine a choice for $x[n]$ so that $X[k]$ is real valued for all k and

$$X(e^{j\omega}) = A(\omega)e^{j\alpha\omega}, \qquad |\omega| < \pi,$$

where $A(\omega)$ is real and α is a nonzero real constant.

8.39. Two finite-length sequences $x_1[n]$ and $x_2[n]$, which are zero outside the interval $0 \le n \le 99$, are circularly convolved to form a new sequence $y[n]$; i.e.,

$$y[n] = x_1[n]\,\textcircled{\scriptsize 100}\,x_2[n] = \sum_{k=0}^{99} x_1[k]x_2[((n-k))_{100}], \qquad 0 \le n \le 99.$$

If, in fact, $x_1[n]$ is nonzero only for $10 \le n \le 39$, determine the set of values of n for which $y[n]$ is guaranteed to be identical to the *linear* convolution of $x_1[n]$ and $x_2[n]$.

8.40. Consider two finite-length sequences $x[n]$ and $h[n]$ for which $x[n] = 0$ outside the interval $0 \le n \le 49$ and $h[n] = 0$ outside the interval $0 \le n \le 9$.
   (a) What is the maximum possible number of nonzero values in the *linear* convolution of $x[n]$ and $h[n]$?
   (b) The 50-point *circular* convolution of $x[n]$ and $h[n]$ is

$$x[n]\,\textcircled{\scriptsize 50}\,h[n] = 10, \qquad 0 \le n \le 49.$$

   The first 5 points of the *linear* convolution of $x[n]$ and $h[n]$ are

$$x[n] * h[n] = 5, \qquad 0 \le n \le 4.$$

   Determine as many points as possible of the linear convolution of $x[n] * h[n]$.

8.41. Consider two finite-duration sequences $x[n]$ and $y[n]$. $x[n]$ is zero for $n < 0$, $n \ge 40$, and $9 < n < 30$, and $y[n]$ is zero for $n < 10$ and $n > 19$, as indicated in Figure P8.41-1.

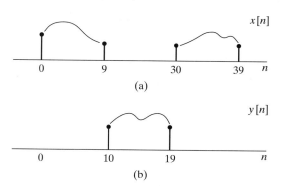

Figure P8.41-1

Let $w[n]$ denote the linear convolution of $x[n]$ and $y[n]$. Let $g[n]$ denote the 40-point circular convolution of $x[n]$ and $y[n]$:

$$w[n] = x[n] * y[n] = \sum_{k=-\infty}^{\infty} x[k]y[n-k],$$

$$g[n] = x[n]\,\textcircled{\scriptsize 40}\,y[n] = \sum_{k=0}^{39} x[k]y[((n-k))_{40}].$$

   (a) Determine the values of n for which $w[n]$ can be nonzero.
   (b) Determine the values of n for which $w[n]$ can be obtained from $g[n]$. Explicitly specify at what index values n in $g[n]$ these values of $w[n]$ appear.

8.42. Two finite-duration sequences $h_1[n]$ and $h_2[n]$ of length 8 are sketched in Figure P8.42-1. The two sequences are related by a circular shift, i.e., $h_1[n] = h_2[((n-m))_8]$.

(a)

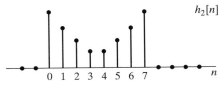

(b)   **Figure P8.42-1**

(a) Specify whether the magnitudes of the eight-point DFTs are equal.
(b) We wish to implement a lowpass FIR filter and must use either $h_1[n]$ or $h_2[n]$ as the impulse response. Which one of the following statements is correct?
   (i) $h_1[n]$ is a better lowpass filter than $h_2[n]$.
   (ii) $h_2[n]$ is a better lowpass filter than $h_1[n]$.
   (iii) The two sequences are both about equally good (or bad) as lowpass filters.

8.43. We want to implement the linear convolution of a 10,000-point sequence with an FIR impulse response that is 100 points long. The convolution is to be implemented by using DFTs and inverse DFTs of length 256.
   (a) If the overlap–add method is used, what is the minimum number of 256-point DFTs and the minimum number of 256-point inverse DFTs needed to implement the convolution for the entire 10,000-point sequence? Justify your answer.
   (b) If the overlap–save method is used, what is the minimum number of 256-point DFTs and the minimum number of 256-point inverse DFTs needed to implement the convolution for the entire 10,000-point sequence? Justify your answer.
   (c) We will see in Chapter 9 that when N is a power of 2, an N-point DFT or inverse DFT requires $(N/2)\log_2 N$ complex multiplications and $N\log_2 N$ complex additions. For the same filter and impulse response length considered in Parts (a) and (b), compare the number of arithmetic operations (multiplications and additions) required in the overlap–add method, the overlap–save method, and direct convolution.

8.44. Let $x_1[n]$ be a sequence obtained by expanding the sequence $x[n] = \left(\frac{1}{4}\right)^n u[n]$ by a factor of 4; i.e.,

$$x_1[n] = \begin{cases} x[n/4], & k = 0, \pm 4, \pm 8, \ldots, \\ 0, & \text{otherwise.} \end{cases}$$

Find and sketch a six-point sequence $q[n]$ whose six-point DFT $Q[k]$ satisfies the two constraints

$$Q[0] = X_1(1),$$

$$Q[3] = X_1(-1),$$

where $X_1(z)$ represents the z-transform of $x_1[n]$.

8.45. Let $x_2[n]$ be a real-valued five-point sequence whose seven-point DFT is denoted by $X_2[k]$. If $\mathrm{Real}\{X_2[k]\}$ is the seven-point DFT of $g[n]$, show that $g[0] = x_2[0]$, and determine the relationship between $g[1]$ and $x_2[1]$. Justify your answer.

8.46. Shown in Figure P8.46-1 are three finite-length sequences of length 5. $X_i(e^{j\omega})$ denotes the DTFT of $x_i[n]$, and $X_i[k]$ denotes the five-point DFT of $x_i[n]$. For each of the following properties, indicate which sequences satisfy the property and which do not. Clearly justify your answers for each sequence and each property.

  (i) $X_i[k]$ is real for all k.
  (ii) $X_i(e^{j\omega}) = A_i(\omega)e^{j\alpha_i\omega}$, where $A_i(\omega)$ is real and α_i is constant.
  (iii) $X_i[k] = B_i[k]e^{j\gamma_i k}$ where $B_i[k]$ is real and γ_i is a constant.

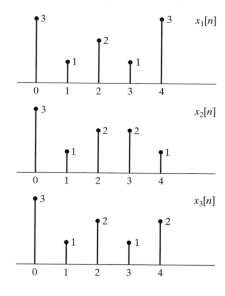

Figure P8.46-1

8.47. Let

$$x[n] = 0, \quad n < 0, n > 7,$$

be a real eight-point sequence, and let $X[k]$ be its eight-point DFT.
  (a) Evaluate

$$\left(\frac{1}{8}\sum_{k=0}^{7} X[k]e^{j(2\pi/8)kn}\right)\bigg|_{n=9}$$

  in terms of $x[n]$.
  (b) Let

$$v[n] = 0, \quad n < 0, n > 7,$$

  be an eight-point sequence, and let $V[k]$ be its eight-point DFT.
  If $V[k] = X(z)$ at $z = 2\exp(j(2\pi k + \pi)/8)$ for $k = 0, \ldots, 7$, where $X(z)$ is the z-transform of $x[n]$, express $v[n]$ in terms of $x[n]$.
  (c) Let

$$w[n] = 0, \quad n < 0, n > 3$$

  be a four-point sequence, and let $W[k]$ be its four-point DFT.
  If $W[k] = X[k] + X[k+4]$, express $w[n]$ in terms of $x[n]$.

(d) Let

$$y[n] = 0, \quad n < 0, n > 7,$$

be an eight-point sequence, and let $Y[k]$ be its eight-point DFT. If

$$Y[k] = \begin{cases} 2X[k], & k = 0, 2, 4, 6, \\ 0, & k = 1, 3, 5, 7, \end{cases}$$

express $y[n]$ in terms of $x[n]$.

8.48. **(a)** Suppose

$$x[n] = 0, \qquad n < 0, \ n > (N-1),$$

is an N-point sequence having at least one nonzero sample. Is it possible for such a sequence to have a DTFT

$$X(e^{j2\pi k/M}) = 0, \qquad k = 0, 1, \ldots, M-1,$$

where M is an integer greater than or equal to N? If your answer is yes, construct an example. If your answer is no, explain your reasoning.

(b) Suppose $M < N$. Repeat Part (a).

8.49. Suppose $x_1[n]$ is an infinite-length, stable (i.e., absolutely summable) sequence with z-transform given by

$$X_1(z) = \frac{1}{1 - \frac{1}{3}z^{-1}}.$$

Suppose $x_2[n]$ is a finite-length sequence of length N, and the N-point DFT of $x_2[n]$ is

$$X_2[k] = X_1(z)\big|_{z=e^{j2\pi k/N}}, \qquad k = 0, 1, \ldots, N-1.$$

Determine $x_2[n]$.

8.50. Read each part of this problem carefully to note the differences among parts.

(a) Consider the signal

$$x[n] = \begin{cases} 1 + \cos(\pi n/4) - 0.5\cos(3\pi n/4), & 0 \le n \le 7, \\ 0, & \text{otherwise,} \end{cases}$$

which can be represented by the IDFT equation as

$$x[n] = \begin{cases} \frac{1}{8}\sum_{k=0}^{7} X_8[k]e^{j(2\pi k/8)n}, & 0 \le n \le 7, \\ 0, & \text{otherwise,} \end{cases}$$

where $X_8[k]$ is the eight-point DFT of $x[n]$. Plot $X_8[k]$ for $0 \le k \le 7$.

(b) Determine $V_{16}[k]$, the sixteen-point DFT of the sixteen-point sequence

$$v[n] = \begin{cases} 1 + \cos(\pi n/4) - 0.5\cos(3\pi n/4), & 0 \le n \le 15, \\ 0, & \text{otherwise.} \end{cases}$$

Plot $V_{16}[k]$ for $0 \le k \le 15$.

(c) Finally, consider $|X_{16}[k]|$, the magnitude of the sixteen-point DFT of the eight-point sequence

$$x[n] = \begin{cases} 1 + \cos(\pi n/4) - 0.5\cos(3\pi n/4), & 0 \le n \le 7, \\ 0, & \text{otherwise.} \end{cases}$$

Sketch $|X_{16}[k]|$ for $0 \le k \le 15$ *without explicitly evaluating the DFT expression*. You will not be able to find all values of $|X_{16}[k]|$ by inspection as in Parts (a) and (b), but you should be able to find some of the values exactly. Plot all the values you know exactly with a solid circle, and plot estimates of the other values with an open circle.

Extension Problems

8.51. In deriving the DFS analysis equation (8.11), we used the identity of Eq. (8.7). To verify this identity, we will consider the two conditions $k - r = mN$ and $k - r \neq mN$ separately.

(a) For $k - r = mN$, show that $e^{j(2\pi/N)(k-r)n} = 1$ and, from this, that

$$\frac{1}{N} \sum_{n=0}^{N-1} e^{j(2\pi/N)(k-r)n} = 1 \qquad \text{for } k - r = mN.$$

Since k and r are both integers in Eq. (8.7), we can make the substitution $k - r = \ell$ and consider the summation

$$\frac{1}{N} \sum_{n=0}^{N-1} e^{j(2\pi/N)\ell n} = \frac{1}{N} \sum_{n=0}^{N-1} [e^{j(2\pi/N)\ell}]^n.$$

Because this is the sum of a finite number of terms in a geometric series, it can be expressed in closed form as

$$\frac{1}{N} \sum_{n=0}^{N-1} [e^{j(2\pi/N)\ell}]^n = \frac{1}{N} \frac{1 - e^{j(2\pi/N)\ell N}}{1 - e^{j(2\pi/N)\ell}}.$$

(b) For what values of ℓ is the right-hand side of this equation indeterminate? That is, are the numerator and denominator both zero?

(c) From the result in Part (b), show that if $k - r \neq mN$, then

$$\frac{1}{N} \sum_{n=0}^{N-1} e^{j(2\pi/N)(k-r)n} = 0.$$

8.52. In Section 8.2, we stated the property that if

$$\tilde{x}_1[n] = \tilde{x}[n - m],$$

then

$$\tilde{X}_1[k] = W_N^{km} \tilde{X}[k],$$

where $\tilde{X}[k]$ and $\tilde{X}_1[k]$ are the DFS coefficients of $\tilde{x}[n]$ and $\tilde{x}_1[n]$, respectively. In this problem, we consider the proof of that property.

(a) Using Eq. (8.11) together with an appropriate substitution of variables, show that $\tilde{X}_1[k]$ can be expressed as

$$\tilde{X}_1[k] = W_N^{km} \sum_{r=-m}^{N-1-m} \tilde{x}[r] W_N^{kr}. \qquad \text{(P8.52-1)}$$

(b) The summation in Eq. (P8.52-1) can be rewritten as

$$\sum_{r=-m}^{N-1-m} \tilde{x}[r] W_N^{kr} = \sum_{r=-m}^{-1} \tilde{x}[r] W_N^{kr} + \sum_{r=0}^{N-1-m} \tilde{x}[r] W_N^{kr}. \qquad \text{(P8.52-2)}$$

Using the fact that $\tilde{x}[r]$ and W_N^{kr} are both periodic, show that

$$\sum_{r=-m}^{-1} \tilde{x}[r] W_N^{kr} = \sum_{r=N-m}^{N-1} \tilde{x}[r] W_N^{kr}. \qquad \text{(P8.52-3)}$$

(c) From your results in Parts (a) and (b), show that

$$\tilde{X}_1[k] = W_N^{km} \sum_{r=0}^{N-1} \tilde{x}[r] W_N^{kr} = W_N^{km} \tilde{X}[k].$$

8.53. (a) Table 8.1 lists a number of symmetry properties of the discrete Fourier series for periodic sequences, several of which we repeat here. Prove that each of these properties is true. In carrying out your proofs, you may use the definition of the discrete Fourier series and any previous property in the list. (For example, in proving property 3, you may use properties 1 and 2.)

Sequence	Discrete Fourier Series
1. $\tilde{x}^*[n]$	$\tilde{X}^*[-k]$
2. $\tilde{x}^*[-n]$	$\tilde{X}^*[k]$
3. $\mathcal{R}e\{\tilde{x}[n]\}$	$\tilde{X}_e[k]$
4. $j\mathcal{J}m\{\tilde{x}[n]\}$	$\tilde{X}_o[k]$

(b) From the properties proved in Part (a), show that for a real periodic sequence $\tilde{x}[n]$, the following symmetry properties of the discrete Fourier series hold:

1. $\mathcal{R}e\{\tilde{X}[k]\} = \mathcal{R}e\{\tilde{X}[-k]\}$
2. $\mathcal{J}m\{\tilde{X}[k]\} = -\mathcal{J}m\{\tilde{X}[-k]\}$
3. $|\tilde{X}[k]| = |\tilde{X}[-k]|$
4. $\sphericalangle\tilde{X}[k] = -\sphericalangle\tilde{X}[-k]$

8.54. We stated in Section 8.4 that a direct relationship between $X(e^{j\omega})$ and $\tilde{X}[k]$ can be derived, where $\tilde{X}[k]$ is the DFS coefficients of a periodic sequence and $X(e^{j\omega})$ is the Fourier transform of one period. Since $\tilde{X}[k]$ corresponds to samples of $X(e^{j\omega})$, the relationship then corresponds to an interpolation formula.

One approach to obtaining the desired relationship is to rely on the discussion of Section 8.4, the relationship of Eq. (8.54), and the modulation property of Section 2.9.7. The procedure is as follows:

1. With $\tilde{X}[k]$ denoting the DFS coefficients of $\tilde{x}[n]$, express the Fourier transform $\tilde{X}(e^{j\omega})$ of $\tilde{x}[n]$ as an impulse train.
2. From Eq. (8.57), $x[n]$ can be expressed as $x[n] = \tilde{x}[n]w[n]$, where $w[n]$ is an appropriate finite-length window.
3. Since $x[n] = \tilde{x}[n]w[n]$, from Section 2.9.7, $X(e^{j\omega})$ can be expressed as the (periodic) convolution of $\tilde{X}(e^{j\omega})$ and $W(e^{j\omega})$.

By carrying out the details in this procedure, show that $X(e^{j\omega})$ can be expressed as

$$X(e^{j\omega}) = \frac{1}{N} \sum_k \tilde{X}[k] \frac{\sin[(\omega N - 2\pi k)/2]}{\sin\{[\omega - (2\pi k/N)]/2\}} e^{-j[(N-1)/2](\omega - 2\pi k/N)}.$$

Specify explicitly the limits on the summation.

8.55. Let $X[k]$ denote the N-point DFT of the N-point sequence $x[n]$.
(a) Show that if

$$x[n] = -x[N - 1 - n],$$

then $X[0] = 0$. Consider separately the cases of N even and N odd.
(b) Show that if N is even and if

$$x[n] = x[N - 1 - n],$$

then $X[N/2] = 0$.

8.56. In Section 2.8, the conjugate-symmetric and conjugate-antisymmetric components of a sequence $x[n]$ were defined, respectively, as

$$x_e[n] = \frac{1}{2}(x[n] + x^*[-n]),$$

$$x_o[n] = \frac{1}{2}(x[n] - x^*[-n]).$$

In Section 8.6.4, we found it convenient to respectively define the periodic conjugate-symmetric and periodic conjugate-antisymmetric components of a sequence of finite duration N as

$$x_{ep}[n] = \frac{1}{2}\{x[((n))_N] + x^*[((-n))_N]\}, \qquad 0 \le n \le N-1,$$

$$x_{op}[n] = \frac{1}{2}\{x[((n))_N] - x^*[((-n))_N]\}, \qquad 0 \le n \le N-1.$$

(a) Show that $x_{ep}[n]$ can be related to $x_e[n]$ and that $x_{op}[n]$ can be related to $x_o[n]$ by the relations

$$x_{ep}[n] = (x_e[n] + x_e[n - N]), \qquad 0 \le n \le N-1,$$

$$x_{op}[n] = (x_o[n] + x_o[n - N]), \qquad 0 \le n \le N-1.$$

(b) $x[n]$ is considered to be a sequence of length N, and in general, $x_e[n]$ cannot be recovered from $x_{ep}[n]$, and $x_o[n]$ cannot be recovered from $x_{op}[n]$. Show that with $x[n]$ considered as a sequence of length N, but with $x[n] = 0$, $n > N/2$, $x_e[n]$ can be obtained from $x_{ep}[n]$, and $x_o[n]$ can be obtained from $x_{op}[n]$.

8.57. Show from Eqs. (8.65) and (8.66) that with $x[n]$ as an N-point sequence and $X[k]$ as its N-point DFT,

$$\sum_{n=0}^{N-1} |x[n]|^2 = \frac{1}{N} \sum_{k=0}^{N-1} |X[k]|^2.$$

This equation is commonly referred to as *Parseval's relation* for the DFT.

8.58. $x[n]$ is a real-valued, nonnegative, finite-length sequence of length N; i.e., $x[n]$ is real and nonnegative for $0 \le n \le N-1$ and is zero otherwise. The N-point DFT of $x[n]$ is $X[k]$, and the Fourier transform of $x[n]$ is $X(e^{j\omega})$.

Determine whether each of the following statements is true or false. For each statement, if you indicate that it is true, clearly show your reasoning. If you state that it is false, construct a counterexample.

(a) If $X(e^{j\omega})$ is expressible in the form

$$X(e^{j\omega}) = B(\omega)e^{j\alpha\omega},$$

where $B(\omega)$ is real and α is a real constant, then $X[k]$ can be expressed in the form

$$X[k] = A[k]e^{j\gamma k},$$

where $A[k]$ is real and γ is a real constant.

(b) If $X[k]$ is expressible in the form

$$X[k] = A[k]e^{j\gamma k},$$

where $A[k]$ is real and γ is a real constant, then $X(e^{j\omega})$ can be expressed in the form

$$X(e^{j\omega}) = B(\omega)e^{j\alpha\omega},$$

where $B(\omega)$ is real and α is a real constant.

8.59. $x[n]$ and $y[n]$ are two real-valued, positive, finite-length sequences of length 256; i.e.,

$$x[n] > 0, \qquad 0 \le n \le 255,$$
$$y[n] > 0, \qquad 0 \le n \le 255,$$
$$x[n] = y[n] = 0, \qquad \text{otherwise.}$$

$r[n]$ denotes the *linear* convolution of $x[n]$ and $y[n]$. $R(e^{j\omega})$ denotes the Fourier transform of $r[n]$. $R_s[k]$ denotes 128 equally spaced samples of $R(e^{j\omega})$; i.e.,

$$R_s[k] = R(e^{j\omega})|_{\omega=2\pi k/128}, \qquad k = 0, 1, \ldots, 127.$$

Given $x[n]$ and $y[n]$, we want to obtain $R_s[k]$ as efficiently as possible. The *only* modules available are those shown in Figure P8.59-1. The costs associated with each module are as follows:

Modules I and II are free.
Module III costs 10 units.
Module IV costs 50 units.
Module V costs 100 units.

Module I

(a)

Module II

(b)

Module III

(c)

Module IV

(d)

Module V

(e)

Figure P8.59-1

By appropriately connecting one or several of each module, construct a system for which the inputs are $x[n]$ and $y[n]$ and the output is $R_s[k]$. The important considerations are (a) whether the system works and (b) how efficient it is. The lower the *total* cost, the more efficient the system is.

8.60. $y[n]$ is the output of a stable LTI system with system function $H(z) = 1/(z - bz^{-1})$, where b is a known constant. We would like to recover the input signal $x[n]$ by operating on $y[n]$.

The following procedure is proposed for recovering part of $x[n]$ from the data $y[n]$:

1. Using $y[n], 0 \leq n \leq N - 1$, calculate $Y[k]$, the N-point DFT of $y[n]$.
2. Form

$$V[k] = (W_N^{-k} - bW_N^k)Y[k].$$

3. Calculate the inverse DFT of $V[k]$ to obtain $v[n]$.
   For which values of the index n in the range $n = 0, 1, \ldots, N-1$ are we guaranteed that

$$x[n] = v[n]?$$

8.61. A modified discrete Fourier transform (MDFT) was proposed (Vernet, 1971) that computes samples of the z-transform on the unit circle offset from those computed by the DFT. In particular, with $X_M[k]$ denoting the MDFT of $x(n)$,

$$X_M[k] = X(z)\Big|_{z=e^{j[2\pi k/N + \pi/N]}}, \qquad k = 0, 1, 2, \ldots, N - 1.$$

Assume that N is even.

(a) The N-point MDFT of a sequence $x[n]$ corresponds to the N-point DFT of a sequence $x_M[n]$, which is easily constructed from $x[n]$. Determine $x_M[n]$ in terms of $x[n]$.

(b) If $x[n]$ is real, not all the points in the DFT are independent, since the DFT is conjugate symmetric; i.e., $X[k] = X^*[((-k))_N]$ for $0 \leq k \leq N - 1$. Similarly, if $x[n]$ is real, not all the points in the MDFT are independent. Determine, for $x[n]$ real, the relationship between points in $X_M[k]$.

(c) (i) Let $R[k] = X_M[2k]$; that is, $R[k]$ contains the even-numbered points in $X_M[k]$. From your answer in part (b), show that $X_M[k]$ can be recovered from $R[k]$.

   (ii) $R[k]$ can be considered to be the $N/2$-point MDFT of an $N/2$-point sequence $r[n]$. Determine a simple expression relating $r[n]$ directly to $x[n]$.

   According to Parts (b) and (c), the N-point MDFT of a real sequence $x[n]$ can be computed by forming $r[n]$ from $x[n]$ and then computing the $N/2$-point MDFT of $r[n]$. The next two parts are directed at showing that the MDFT can be used to implement a linear convolution.

(d) Consider three sequences $x_1[n]$, $x_2[n]$, and $x_3[n]$, all of length N. Let $X_{1M}[k]$, $X_{2M}[k]$, and $X_{3M}[k]$, respectively, denote the MDFTs of the three sequences. If

$$X_{3M}[k] = X_{1M}[k]X_{2M}[k],$$

express $x_3[n]$ in terms of $x_1[n]$ and $x_2[n]$. Your expression must be of the form of a single summation over a "combination" of $x_1[n]$ and $x_2[n]$ in the same style as (but not identical to) a circular convolution.

(e) It is convenient to refer to the result in Part (d) as a modified circular convolution. If the sequences $x_1[n]$ and $x_2[n]$ are both zero for $n \geq N/2$, show that the modified circular convolution of $x_1[n]$ and $x_2[n]$ is identical to the linear convolution of $x_1[n]$ and $x_2[n]$.

8.62. In some applications in coding theory, it is necessary to compute a 63-point circular convolution of two 63-point sequences $x[n]$ and $h[n]$. Suppose that the only computational devices available are multipliers, adders, and processors that compute N-point DFTs, with N restricted to be a power of 2.

(a) It is possible to compute the 63-point circular convolution of $x[n]$ and $h[n]$ using a number of 64-point DFTs, inverse DFTs, and the overlap–add method. How many DFTs are needed? *Hint:* Consider each of the 63-point sequences as the sum of a 32-point sequence and 31-point sequence.

(b) Specify an algorithm that computes the 63-point circular convolution of $x[n]$ and $h[n]$ using two 128-point DFTs and one 128-point inverse DFT.

(c) We could also compute the 63-point circular convolution of $x[n]$ and $h[n]$ by computing their linear convolution in the time domain and then aliasing the results. In terms of multiplications, which of these three methods is most efficient? Which is least efficient? (Assume that one complex multiplication requires four real multiplications and that $x[n]$ and $h[n]$ are real.)

8.63. We want to filter a very long string of data with an FIR filter whose impulse response is 50 samples long. We wish to implement this filter with a DFT using the overlap–save technique. The procedure is as follows:

1. The input sections must be overlapped by V samples.
2. From the output of each section, we must extract M samples such that when these samples from each section are abutted, the resulting sequence is the desired filtered output.

Assume that the input segments are 100 samples long and that the size of the DFT is $128 (= 2^7)$ points. Assume further that the output sequence from the circular convolution is indexed from point 0 to point 127.

(a) Determine V.
(b) Determine M.
(c) Determine the index of the beginning and the end of the M points extracted; i.e., determine which of the 128 points from the circular convolution are extracted to be abutted with the result from the previous section.

8.64. A problem that often arises in practice is one in which a distorted signal $y[n]$ is the output that results when a desired signal $x[n]$ has been filtered by an LTI system. We wish to recover the original signal $x[n]$ by processing $y[n]$. In theory, $x[n]$ can be recovered from $y[n]$ by passing $y[n]$ through an inverse filter having a system function equal to the reciprocal of the system function of the distorting filter.

Suppose that the distortion is caused by an FIR filter with impulse response

$$h[n] = \delta[n] - \frac{1}{2}\delta[n - n_0],$$

where n_0 is a positive integer, i.e., the distortion of $x[n]$ takes the form of an echo at delay n_0.

(a) Determine the z-transform $H(z)$ and the N-point DFT $H[k]$ of the impulse response $h[n]$. Assume that $N = 4n_0$.
(b) Let $H_i(z)$ denote the system function of the inverse filter, and let $h_i[n]$ be the corresponding impulse response. Determine $h_i[n]$. Is this an FIR or an IIR filter? What is the duration of $h_i[n]$?
(c) Suppose that we use an FIR filter of length N in an attempt to implement the inverse filter, and let the N-point DFT of the FIR filter be

$$G[k] = 1/H[k], \qquad k = 0, 1, \ldots, N - 1.$$

What is the impulse response $g[n]$ of the FIR filter?
(d) It might appear that the FIR filter with DFT $G[k] = 1/H[k]$ implements the inverse filter perfectly. After all, one might argue that the FIR distorting filter has an N-point DFT $H[k]$ and the FIR filter in cascade has an N-point DFT $G[k] = 1/H[k]$, and since $G[k]H[k] = 1$ for all k, we have implemented an all-pass, nondistorting filter. Briefly explain the fallacy in this argument.
(e) Perform the convolution of $g[n]$ with $h[n]$, and thus determine how well the FIR filter with N-point DFT $G[k] = 1/H[k]$ implements the inverse filter.

8.65. A sequence $x[n]$ of length N has a discrete Hartley transform (DHT) defined as

$$X_H[k] = \sum_{n=0}^{N-1} x[n]H_N[nk], \qquad k = 0, 1, \ldots, N - 1, \qquad \text{(P8.65-1)}$$

where

$$H_N[a] = C_N[a] + S_N[a],$$

with

$$C_N[a] = \cos(2\pi a/N), \qquad S_N[a] = \sin(2\pi a/N).$$

Originally proposed by R.V.L. Hartley in 1942 for the continuous-time case, the Hartley transform has properties that make it useful and attractive in the discrete-time case as well (Bracewell, 1983, 1984). Specifically, from Eq. (P8.65-1), it is apparent that the DHT of a real sequence is also a real sequence. In addition, the DHT has a convolution property, and fast algorithms exist for its computation.

In complete analogy with the DFT, the DHT has an implicit periodicity that must be acknowledged in its use. That is, if we consider $x[n]$ to be a finite-length sequence such that $x[n] = 0$ for $n < 0$ and $n > N - 1$, then we can form a periodic sequence

$$\tilde{x}[n] = \sum_{r=-\infty}^{\infty} x[n + rN]$$

such that $x[n]$ is simply one period of $\tilde{x}[n]$. The periodic sequence $\tilde{x}[n]$ can be represented by a discrete Hartley series (DHS), which in turn can be interpreted as the DHT by focusing attention on only one period of the periodic sequence.

(a) The DHS analysis equation is defined by

$$\tilde{X}_H[k] = \sum_{n=0}^{N-1} \tilde{x}[n] H_N[nk]. \tag{P8.65-2}$$

Show that the DHS coefficients form a sequence that is also periodic with period N; i.e.,

$$\tilde{X}_H[k] = \tilde{X}_H[k + N] \qquad \text{for all } k.$$

(b) It can also be shown that the sequences $H_N[nk]$ are orthogonal; i.e.,

$$\sum_{k=0}^{N-1} H_N[nk] H_N[mk] = \begin{cases} N, & ((n))_N = ((m))_N, \\ 0, & \text{otherwise.} \end{cases}$$

Using this property and the DHS analysis formula of Eq. (P8.65-2), show that the DHS synthesis formula is

$$\tilde{x}[n] = \frac{1}{N} \sum_{k=0}^{N-1} \tilde{X}_H[k] H_N[nk]. \tag{P8.65-3}$$

Note that the DHT is simply one period of the DHS coefficients, and likewise, the DHT synthesis (inverse) equation is identical to the DHS synthesis equation (P8.65-3), except that we simply extract one period of $\tilde{x}[n]$; i.e., the DHT synthesis expression is

$$x[n] = \frac{1}{N} \sum_{k=0}^{N-1} X_H[k] H_N[nk], \qquad n = 0, 1, \ldots, N - 1. \tag{P8.65-4}$$

With Eqs. (P8.65-1) and (P8.65-4) as definitions of the analysis and synthesis relations, respectively, for the DHT, we may now proceed to derive the useful properties of this representation of a finite-length discrete-time signal.

(c) Verify that $H_N[a] = H_N[a + N]$, and verify the following useful property of $H_N[a]$:

$$H_N[a + b] = H_N[a] C_N[b] + H_N[-a] S_N[b]$$
$$= H_N[b] C_N[a] + H_N[-b] S_N[a].$$

(d) Consider a circularly shifted sequence

$$x_1[n] = \begin{cases} \tilde{x}[n - n_0] = x[((n - n_0))_N], & n = 0, 1, \ldots, N - 1, \\ 0, & \text{otherwise.} \end{cases} \qquad \text{(P8.65-5)}$$

In other words, $x_1[n]$ is the sequence that is obtained by extracting one period from the shifted periodic sequence $\tilde{x}[n - n_0]$. Using the identity verified in Part (c), show that the DHS coefficients for the shifted periodic sequence are

$$\tilde{x}[n - n_0] \overset{\mathcal{DHS}}{\longleftrightarrow} \tilde{X}_H[k]C_N[n_0k] + \tilde{X}_H[-k]S_N[n_0k]. \qquad \text{(P8.65-6)}$$

From this, we conclude that the DHT of the finite-length circularly shifted sequence $x[((n - n_0))_N]$ is

$$x[((n - n_0))_N] \overset{\mathcal{DHT}}{\longleftrightarrow} X_H[k]C_N[n_0k] + X_H[((-k))_N]S_N[n_0k]. \qquad \text{(P8.65-7)}$$

(e) Suppose that $x_3[n]$ is the N-point circular convolution of two N-point sequences $x_1[n]$ and $x_2[n]$; i.e.,

$$x_3[n] = x_1[n] \; \text{\textcircled{N}} \; x_2[n]$$

$$= \sum_{m=0}^{N-1} x_1[m]x_2[((n - m))_N], \qquad n = 0, 1, \ldots, N - 1. \qquad \text{(P8.65-8)}$$

By applying the DHT to both sides of Eq. (P8.65-8) and using Eq. (P8.65-7), show that

$$\begin{aligned} X_{H3}[k] &= \tfrac{1}{2}X_{H1}[k](X_{H2}[k] + X_{H2}[((-k))_N]) \\ &\quad + \tfrac{1}{2}X_{H1}[((-k))_N](X_{H2}[k] - X_{H2}[((-k))_N]) \end{aligned} \qquad \text{(P8.65-9)}$$

for $k = 0, 1, \ldots, N - 1$. This is the desired convolution property.

Note that a linear convolution can be computed using the DHT in the same way that the DFT can be used to compute a linear convolution. While computing $X_{H3}[k]$ from $X_{H1}[k]$ and $X_{H2}[k]$ requires the same amount of computation as computing $X_3[k]$ from $X_1[k]$ and $X_2[k]$, the computation of the DHT requires only half the number of real multiplications required to compute the DFT.

(f) Suppose that we wish to compute the DHT of an N-point sequence $x[n]$ and we have available the means to compute the N-point DFT. Describe a technique for obtaining $X_H[k]$ from $X[k]$ for $k = 0, 1, \ldots, N - 1$.

(g) Suppose that we wish to compute the DFT of an N-point sequence $x[n]$ and we have available the means to compute the N-point DHT. Describe a technique for obtaining $X[k]$ from $X_H[k]$ for $k = 0, 1, \ldots, N - 1$.

8.66. Let $x[n]$ be an N-point sequence such that $x[n] = 0$ for $n < 0$ and for $n > N - 1$. Let $\hat{x}[n]$ be the $2N$-point sequence obtained by repeating $x[n]$; i.e.,

$$\hat{x}[n] = \begin{cases} x[n], & 0 \leq n \leq N - 1, \\ x[n - N], & N \leq n \leq 2N - 1, \\ 0, & \text{otherwise.} \end{cases}$$

Consider the implementation of a discrete-time filter shown in Figure P8.66b-1. This system has an impulse response $h[n]$ that is $2N$ points long; i.e., $h[n] = 0$ for $n < 0$ and for $n > 2N - 1$.

(a) In Figure P8.66b-1, what is $\hat{X}[k]$, the $2N$-point DFT of $\hat{x}[n]$, in terms of $X[k]$, the N-point DFT of $x[n]$?

(b) The system shown in Figure P8.66b-1 can be implemented using only N-point DFTs as indicated in Figure P8.66-2 for appropriate choices for System A and System B. Specify

System A and System B so that $\hat{y}[n]$ in Figure P8.66b-1 and $y[n]$ in Figure P8.66-2 are equal for $0 \leq n \leq 2N - 1$. Note that $h[n]$ and $y[n]$ in Figure P8.66-2 are $2N$-point sequences and $w[n]$ and $g[n]$ are N-point sequences.

Figure P8.66-1

Figure P8.66-2

8.67. In this problem, you will examine the use of the DFT to implement the filtering necessary for the discrete-time interpolation, or upsampling, of a signal. Assume that the discrete-time signal $x[n]$ was obtained by sampling a continuous-time signal $x_c(t)$ with a sampling period T. Moreover, the continuous-time signal is appropriately bandlimited; i.e., $X_c(j\Omega) = 0$ for $|\Omega| \geq 2\pi / T$. For this problem, we will assume that $x[n]$ has length N; i.e., $x[n] = 0$ for $n < 0$ or $n > N - 1$, where N is even. It is not strictly possible to have a signal that is both perfectly bandlimited and of finite duration, but this is often assumed in practical systems processing finite-length signals which have very little energy outside the band $|\Omega| \leq 2\pi / T$.

We wish to implement a 1:4 interpolation, i.e., increase the sampling rate by a factor of 4. As seen in Figure 4.24, we can perform this sampling rate conversion using a sampling rate expander followed by an appropriate lowpass filter. In this chapter, we have seen that the lowpass filter could be implemented using the DFT if the filter is an FIR impulse response. For this problem, assume that this filter has an impulse response $h[n]$ that is $N + 1$ points long. Figure P8.67-1 depicts such a system, where $H[k]$ is the $4N$-point DFT of the impulse response of the lowpass filter. Note that both $v[n]$ and $y[n]$ are $4N$-point sequences.

Figure P8.67-1

(a) Specify the DFT $H[k]$ such that the system in Figure P8.67-1 implements the desired upsampling system. Think carefully about the phase of the values of $H[k]$.

(b) It is also possible to upsample $x[n]$ using the system in Figure P8.67-2. Specify System A in the middle box so that the $4N$-point signal $y_2[n]$ in this figure is the same as $y[n]$ in Figure P8.67-2. Note that System A may consist of more than one operation.

(c) Is there a reason that the implementation in Figure P8.67-2 might be preferable to Figure P8.67-1?

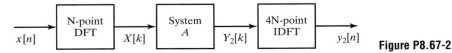

Figure P8.67-2

8.68. Derive Eq. (8.153) using Eqs. (8.164) and (8.165).

8.69. Consider the following procedure

(a) Form the sequence $v[n] = x_2[2n]$ where $x_2[n]$ is given by Eq. (8.166). This yields

$$v[n] = x[2n] \qquad n = 0, 1, \ldots, N/2 - 1$$
$$v[N - 1 - n] = x[2n + 1], \quad n = 0, 1, \ldots, N/2 - 1.$$

(b) Compute $V[k]$, the N-point DFT of $v[n]$. Demonstrate that the following is true:

$$X^{c2}[k] = 2\mathcal{R}e\{e^{-j2\pi k/(4N)}V[k]\}, \qquad k = 0, 1, \ldots, N - 1,$$

$$= 2\sum_{n=0}^{N-1} v[n]\cos\left[\frac{\pi k(4n + 1)}{2N}\right], \qquad k = 0, 1, \ldots, N - 1,$$

$$= 2\sum_{n=0}^{N-1} x[n]\cos\left[\frac{\pi k(2n + 1)}{2N}\right], \qquad k = 0, 1, \ldots, N - 1.$$

Note that this algorithm uses N-point rather than $2N$-point DFTs as required in Eq. (8.167). In addition, since $v[n]$ is a real sequence, we can exploit even and odd symmetries to do the computation of $V[k]$ in one $N/4$-point complex DFT.

8.70. Derive Eq. (8.156) using Eqs. (8.174) and (8.175).

8.71. (a) Use Parseval's theorem for the DFT to derive a relationship between $\sum |X^{c1}[k]|^2$ and $\sum |x[n]|^2$.

(b) Use Parseval's theorem for the DFT to derive a relationship between $\sum |X^{c2}[k]|^2$ and $\sum |x[n]|^2$.

9

COMPUTATION OF THE DISCRETE FOURIER TRANSFORM

9.0 INTRODUCTION

The discrete Fourier transform (DFT) plays an important role in the analysis, design, and implementation of discrete-time signal-processing algorithms and systems. The basic properties of the Fourier transform and discrete Fourier transform, discussed in Chapters 2 and 8, respectively, make it particularly convenient to analyze and design systems in the Fourier domain. Equally important is the fact that efficient algorithms exist for explicitly computing the DFT. As a result, the DFT is an important component in many practical applications of discrete-time systems.

As discussed in Chapter 8, the DFT is identical to samples of the Fourier transform at equally spaced frequencies. Consequently, computation of the N-point DFT corresponds to the computation of N samples of the Fourier transform at N equally spaced frequencies $\omega_k = 2\pi k/N$, i.e., at N points on the unit circle in the z-plane. In this chapter, we discuss several methods for computing values of the DFT. The major focus of the chapter is a particularly efficient class of algorithms for the digital computation of the N-point DFT. Collectively, these efficient algorithms are called *fast Fourier transform* (FFT) algorithms, and we discuss them in Sections 9.3 and 9.4. To achieve the highest efficiency, the FFT algorithms must compute all N values of the DFT. When we require values of the DFT over only a portion of the frequency range $0 \leq \omega < 2\pi$, other algorithms may be more efficient and flexible, even though they are less efficient than the FFT algorithms for computation of all the values of the DFT. Examples of such algorithms are the Goertzel algorithm, discussed in Section 9.2, and the chirp transform algorithm, discussed in Section 9.6.2.

There are many ways to measure the complexity and efficiency of an implementation or algorithm, and a final assessment depends on both the available technology and the intended application. We will use the number of arithmetic multiplications and additions as a measure of computational complexity. This measure is simple to apply, and the number of multiplications and additions is directly related to the computational speed when algorithms are implemented on general-purpose digital computers or special-purpose microprocessors. However, other measures are sometimes more appropriate. For example, in custom VLSI implementations, the area of the chip and power requirements are important considerations and may not be directly related to the number of arithmetic operations.

In terms of multiplications and additions, the class of FFT algorithms can be orders of magnitude more efficient than competing algorithms. The efficiency of FFT algorithms is so high, in fact, that in many cases the most efficient procedure for implementing a convolution is to compute the transform of the sequences to be convolved, multiply their transforms, and then compute the inverse transform of the product of transforms. The details of this technique were discussed in Section 8.7. In seeming contradiction to this, there is a set of algorithms (mentioned briefly in Section 9.6) for evaluation of the DFT (or a more general set of samples of the Fourier transform) that derive their efficiency from a reformulation of the Fourier transform in terms of a convolution and thereby implement the Fourier transform computation by using efficient procedures for evaluating the associated convolution. This suggests the possibility of implementing a convolution by multiplication of DFTs, where the DFTs have been implemented by first expressing them as convolutions and then taking advantage of efficient procedures for implementing the associated convolutions. While this seems on the surface to be a basic contradiction, we will see in Section 9.6 that in certain cases it is an entirely reasonable approach and is not at all a contradiction.

9.1 EFFICIENT COMPUTATION OF THE DISCRETE FOURIER TRANSFORM

As defined in Chapter 8, the DFT of a finite-length sequence of length N is

$$X[k] = \sum_{n=0}^{N-1} x[n] W_N^{kn}, \qquad k = 0, 1, \ldots, N-1, \tag{9.1}$$

where $W_N = e^{-j(2\pi/N)}$. The inverse discrete Fourier transform is given by

$$x[n] = \frac{1}{N} \sum_{k=0}^{N-1} X[k] W_N^{-kn}, \qquad n = 0, 1, \ldots, N-1. \tag{9.2}$$

In Eqs. (9.1) and (9.2), both $x[n]$ and $X[k]$ may be complex. Since the expressions on the right-hand sides of those equations differ only in the sign of the exponent of W_N and in the scale factor $1/N$, a discussion of computational procedures for Eq. (9.1) applies with straightforward modifications to Eq. (9.2). (See Problem 9.1.)

To create a frame of reference for our discussion of computation of the DFT, let us first consider direct evaluation of the DFT expression in Eq. (9.1). Since $x[n]$ may be complex, N complex multiplications and $(N - 1)$ complex additions are required to compute each value of the DFT if we use Eq. (9.1) directly as a formula for computation. To compute all N values therefore requires a total of N^2 complex multiplications and $N(N-1)$ complex additions. Expressing Eq. (9.1) in terms of operations on real numbers, we obtain

$$X[k] = \sum_{n=0}^{N-1} \left[(\mathcal{R}e\{x[n]\}\mathcal{R}e\{W_N^{kn}\} - \mathcal{I}m\{x[n]\}\mathcal{I}m\{W_N^{kn}\}) \right.$$

$$\left. + j(\mathcal{R}e\{x[n]\}\mathcal{I}m\{W_N^{kn}\} + \mathcal{I}m\{x[n]\}\mathcal{R}e\{W_N^{kn}\}) \right], \qquad (9.3)$$

$$k = 0, 1, \ldots, N-1,$$

which shows that each complex multiplication requires four real multiplications and two real additions, and each complex addition requires two real additions. Therefore, for each value of k, the direct computation of $X[k]$ requires $4N$ real multiplications and $(4N-2)$ real additions.[1] Since $X[k]$ must be computed for N different values of k, the direct computation of the discrete Fourier transform of a sequence $x[n]$ requires $4N^2$ real multiplications and $N(4N-2)$ real additions. Besides the multiplications and additions called for by Eq. (9.3), the digital computation of the DFT on a general-purpose digital computer or with special-purpose hardware also requires provision for storing and accessing the N complex input sequence values $x[n]$ and values of the complex coefficients W_N^{kn}. Since the amount of computation, and thus the computation time, is approximately proportional to N^2, it is evident that the number of arithmetic operations required to compute the DFT by the direct method becomes very large for large values of N. For this reason, we are interested in computational procedures that reduce the number of multiplications and additions.

Most approaches to improving the efficiency of the computation of the DFT exploit the symmetry and periodicity properties of W_N^{kn}; specifically,

1. $W_N^{k[N-n]} = W_N^{-kn} = (W_N^{kn})^*$ (complex conjugate symmetry);
2. $W_N^{kn} = W_N^{k(n+N)} = W_N^{(k+N)n}$ (periodicity in n and k).

As an illustration, using the first property, i.e., the symmetry of the implicit cosine and sine functions, we can group terms in the summation in Eq. (9.3) for n and $(N-n)$. For example,

$$\mathcal{R}e\{x[n]\}\mathcal{R}e\{W_N^{kn}\} + \mathcal{R}e\{x[N-n]\}\mathcal{R}e\{W_N^{k[N-n]}\}$$

$$= (\mathcal{R}e\{x[n]\} + \mathcal{R}e\{x[N-n]\})\mathcal{R}e\{W_N^{kn}\}$$

and

$$-\mathcal{I}m\{x[n]\}\mathcal{I}m\{W_N^{kn}\} - \mathcal{I}m\{x[N-n]\}\mathcal{I}m\{W_N^{k[N-n]}\}$$

$$= -(\mathcal{I}m\{x[n]\} - \mathcal{I}m\{x[N-n]\})\mathcal{I}m\{W_N^{kn}\}.$$

Similar groupings can be used for the other terms in Eq. (9.3). In this way, the number of multiplications can be reduced by approximately a factor of 2. We can also take advantage of the fact that for certain values of the product kn, the sine and cosine functions

---

[1] Throughout the discussion, the figure for the number of computations is only approximate. Multiplication by W_N^0, for example, does not require a multiplication. Nevertheless, the estimate of computational complexity obtained by including such multiplications is sufficiently accurate to permit comparisons between different classes of algorithms.

take on the value 1 or 0, thereby eliminating the need for multiplications. However, reductions of this type still leave us with an amount of computation that is proportional to N^2. Fortunately, the second property, the periodicity of the complex sequence W_N^{kn}, can be exploited in achieving significantly greater reductions of the computation.

Computational algorithms that exploit both the symmetry and the periodicity of the sequence W_N^{kn} were known long before the era of high-speed digital computation. At that time, any scheme that reduced manual computation by even a factor of 2 was welcomed. Heideman et al. (1984) have traced the origins of the FFT back to Gauss, as early as 1805. Runge (1905) and later Danielson and Lanczos (1942) described algorithms for which computation was roughly proportional to $N \log N$ rather than N^2. However, the distinction was not of great importance for the small values of N that were feasible for hand computation. The possibility of greatly reduced computation was generally overlooked until about 1965, when Cooley and Tukey (1965) published an algorithm for the computation of the discrete Fourier transform that is applicable when N is a composite number, i.e., the product of two or more integers. The publication of their paper touched off a flurry of activity in the application of the discrete Fourier transform to signal processing and resulted in the discovery of a number of highly efficient computational algorithms. Collectively, the entire set of such algorithms has come to be known as the *fast Fourier transform,* or the FFT.[2]

FFT algorithms are based on the fundamental principle of decomposing the computation of the discrete Fourier transform of a sequence of length N into successively smaller discrete Fourier transforms. The manner in which this principle is implemented leads to a variety of different algorithms, all with comparable improvements in computational speed. In this chapter, we are concerned with two basic classes of FFT algorithms. The first, called decimation in time, derives its name from the fact that in the process of arranging the computation into smaller transformations, the sequence $x[n]$ (generally thought of as a time sequence) is decomposed into successively smaller subsequences. In the second general class of algorithms, the sequence of discrete Fourier transform coefficients $X[k]$ is decomposed into smaller subsequences—hence its name, decimation in frequency.

In the sections that follow, we consider a number of algorithms for computing the discrete Fourier transform. The algorithms vary in efficiency, but all of them require fewer multiplications and additions than does direct evaluation of Eq. (9.3). We begin in the next section with a discussion of Goertzel's algorithm (Goertzel, 1958), which requires computation proportional to N^2, but with a smaller constant of proportionality than that of the direct method. One of the principal advantages of Goertzel's algorithm is that it is not restricted to computation of the DFT, but is in fact equally valid for the computation of any desired set of samples of the Fourier transform of a sequence.

In Sections 9.3 and 9.4 we present a detailed discussion of FFT algorithms for which computation is proportional to $N \log N$. This class of algorithms is considerably more efficient in terms of arithmetic operations than the Goertzel algorithm, but is specifically oriented toward computation of *all* the values of the DFT. We do not attempt to be exhaustive in our coverage of that class of algorithms, but we illustrate the general principles common to all algorithms of this type by considering in detail only a few of the more commonly used schemes.

---

[2]See Cooley et al. (1967) for a historical summary of results related to the FFT.

In Section 9.5, we consider some of the practical issues that arise in implementing the FFT. In Section 9.6, we discuss algorithms that rely on formulating the computation of the DFT in terms of a convolution. In Section 9.7, we consider effects of arithmetic roundoff.

9.2 THE GOERTZEL ALGORITHM

The Goertzel algorithm (Goertzel, 1958) is an example of how the periodicity of the sequence W_N^{kn} can be used to reduce computation. To derive the algorithm, we begin by noting that

$$W_N^{-kN} = e^{j(2\pi/N)Nk} = e^{j2\pi k} = 1. \tag{9.4}$$

This is, of course, a direct result of the periodicity of W_N^{-kn}. Because of Eq. (9.4), we may multiply the right side of Eq. (9.1) by W_N^{-kN} without affecting the equation. Thus,

$$X[k] = W_N^{-kN} \sum_{r=0}^{N-1} x[r]W_N^{kr} = \sum_{r=0}^{N-1} x[r]W_N^{-k(N-r)}. \tag{9.5}$$

To suggest the final result, let us define the sequence

$$y_k[n] = \sum_{r=-\infty}^{\infty} x[r]W_N^{-k(n-r)}u[n-r]. \tag{9.6}$$

From Eqs. (9.5) and (9.6) and the fact that $x[n] = 0$ for $n < 0$ and $n \geq N$, it follows that

$$X[k] = y_k[n]\Big|_{n=N}. \tag{9.7}$$

Equation (9.6) can be interpreted as a discrete convolution of the finite-duration sequence $x[n], 0 \leq n \leq N-1$, with the sequence $W_N^{-kn}u[n]$. Consequently, $y_k[n]$ can be viewed as the response of a system with impulse response $W_N^{-kn}u[n]$ to a finite-length input $x[n]$. In particular, $X[k]$ is the value of the output when $n = N$.

A system with impulse response $W_N^{-kn}u[n]$ is depicted in Figure 9.1, where initial rest conditions are assumed. Since the general input $x[n]$ and the coefficient W_N^{-k} are both complex, the computation of each new value of $y_k[n]$ using the system of Figure 9.1 requires 4 real multiplications and 4 real additions. All the intervening values $y_k[1], y_k[2], \ldots, y_k[N-1]$ must be computed in order to compute $y_k[N] = X[k]$, so the use of the system in Figure 9.1 as a computational algorithm requires $4N$ real multiplications and $4N$ real additions to compute $X[k]$ for a particular value of k. Thus, this procedure is slightly less efficient than the direct method. However, it avoids the computation or storage of the coefficients W_N^{kn}, since these quantities are implicitly computed by the recursion implied by Figure 9.1.

Figure 9.1   Flow graph of first-order complex recursive computation of $X[k]$.

It is possible to retain this simplification while reducing the number of multiplications by a factor of 2. To see how this may be done, note that the system function of the system of Figure 9.1 is

$$H_k(z) = \frac{1}{1 - W_N^{-k}z^{-1}}. \tag{9.8}$$

Multiplying both the numerator and the denominator of $H_k[z]$ by the factor $(1 - W_N^k z^{-1})$, we obtain

$$\begin{aligned} H_k(z) &= \frac{1 - W_N^k z^{-1}}{(1 - W_N^{-k}z^{-1})(1 - W_N^k z^{-1})} \\[2mm] &= \frac{1 - W_N^k z^{-1}}{1 - 2\cos(2\pi k/N)z^{-1} + z^{-2}}. \end{aligned} \tag{9.9}$$

The signal flow graph of Figure 9.2 corresponds to the system of Eq. (9.9).

If the input is complex, only two real multiplications per sample are required to implement the poles of this system, since the coefficients are real and the factor -1 need not be counted as a multiplication. As in the case of the first-order system, for a complex input, four real additions per sample are required to implement the poles (if the input is complex). Since we only need to bring the system to a state from which $y_k[N]$ can be computed, the complex multiplication by $-W_N^k$ required to implement the zero of the system function need not be performed at every iteration of the difference equation, but only after the Nth iteration. Thus, the total computation is $2N$ real multiplications and $4N$ real additions for the poles, plus 4 real multiplications and 4 real additions for the zero. The total computation is therefore $2(N+2)$ real multiplications and $4(N+1)$ real additions, about half the number of real multiplications required with the direct method. In this more efficient scheme, we still have the advantage that $\cos(2\pi k/N)$ and W_N^k are the only coefficients that must be computed and stored. The coefficients W_N^{kn} are again computed implicitly in the iteration of the recursion formula implied by Figure 9.2.

As an additional advantage of the use of this network, let us consider the computation of the z-transform of $x[n]$ at conjugate locations on the unit circle, that is, the computation of $X[k]$ and $X[N-k]$. It is straightforward to verify that the network in the form of Figure 9.2 required to compute $X[N-k]$ has exactly the same poles as that in Figure 9.2, but the coefficient for the zero is the complex conjugate of that in Figure 9.2. (See Problem 9.21.) Since the zero is implemented only on the final iteration, the $2N$ multiplications and $4N$ additions required for the poles can be used for the computation of two DFT values. Thus, for the computation of all N values of the discrete Fourier

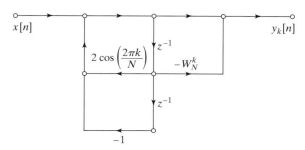

Figure 9.2 Flow graph of second-order recursive computation of $X[k]$ (Goertzel algorithm).

transform using the Goertzel algorithm, the number of real multiplications required is approximately N^2 and the number of real additions is approximately $2N^2$. While this is more efficient than the direct computation of the discrete Fourier transform, the amount of computation is still proportional to N^2.

In either the direct method or the Goertzel method we do not need to evaluate $X[k]$ at all N values of k. Indeed, we can evaluate $X[k]$ for any M values of k, with each DFT value being computed by a recursive system of the form of Figure 9.2 with appropriate coefficients. In this case, the total computation is proportional to NM. The Goertzel method and the direct method are attractive when M is small; however, as indicated previously, algorithms are available for which the computation is proportional to $N \log_2 N$ when N is a power of 2. Therefore, when M is less than $\log_2 N$, either the Goertzel algorithm or the direct method may in fact be the most efficient method, but when all N values of $X[k]$ are required, the decimation-in-time algorithms, to be considered next, are roughly $(N/\log_2 N)$ times more efficient than either the direct method or the Goertzel method.

9.3 DECIMATION-IN-TIME FFT ALGORITHMS

In computing the DFT, dramatic efficiency results from decomposing the computation into successively smaller DFT computations. In this process, we exploit both the symmetry and the periodicity of the complex exponential $W_N^{kn} = e^{-j(2\pi/N)kn}$. Algorithms in which the decomposition is based on decomposing the sequence $x[n]$ into successively smaller subsequences are called *decimation-in-time algorithms*.

The principle of the decimation-in-time algorithm is most conveniently illustrated by considering the special case of N an integer power of 2, i.e., $N = 2^\nu$. Since N is an even integer, we can consider computing $X[k]$ by separating $x[n]$ into two $(N/2)$-point[3] sequences consisting of the even-numbered points in $x[n]$ and the odd-numbered points in $x[n]$. With $X[k]$ given by

$$X[k] = \sum_{n=0}^{N-1} x[n] W_N^{nk}, \qquad k = 0, 1, \ldots, N-1, \tag{9.10}$$

and separating $x[n]$ into its even- and odd-numbered points, we obtain

$$X[k] = \sum_{n \text{ even}} x[n] W_N^{nk} + \sum_{n \text{ odd}} x[n] W_N^{nk}, \tag{9.11}$$

or, with the substitution of variables $n = 2r$ for n even and $n = 2r + 1$ for n odd,

$$X[k] = \sum_{r=0}^{(N/2)-1} x[2r] W_N^{2rk} + \sum_{r=0}^{(N/2)-1} x[2r+1] W_N^{(2r+1)k}$$

$$= \sum_{r=0}^{(N/2)-1} x[2r] (W_N^2)^{rk} + W_N^k \sum_{r=0}^{(N/2)-1} x[2r+1] (W_N^2)^{rk}. \tag{9.12}$$

---

[3]When discussing FFT algorithms, we use the words *sample* and *point* interchangeably to mean *sequence value*. Also, we refer to a sequence of length N as an N-point sequence, and the DFT of a sequence of length N will be called an N-point DFT.

But $W_N^2 = W_{N/2}$, since

$$W_N^2 = e^{-2j(2\pi/N)} = e^{-j2\pi/(N/2)} = W_{N/2}. \tag{9.13}$$

Consequently, Eq. (9.12) can be rewritten as

$$X[k] = \sum_{r=0}^{(N/2)-1} x[2r]W_{N/2}^{rk} + W_N^k \sum_{r=0}^{(N/2)-1} x[2r+1]W_{N/2}^{rk} \tag{9.14}$$

$$= G[k] + W_N^k H[k], \qquad k = 0, 1, \ldots, N-1.$$

Each of the sums in Eq. (9.14) is recognized as an $(N/2)$-point DFT, the first sum being the $(N/2)$-point DFT of the even-numbered points of the original sequence and the second being the $(N/2)$-point DFT of the odd-numbered points of the original sequence. Although the index k ranges over N values, $k = 0, 1, \ldots, N-1$, each of the sums must be computed only for k between 0 and $(N/2) - 1$, since $G[k]$ and $H[k]$ are each periodic in k with period $N/2$. After the two DFTs are computed, they are combined according to Eq. (9.14) to yield the N-point DFT $X[k]$. Figure 9.3 depicts this computation for $N = 8$. In this figure, we have used the signal flow graph conventions that were introduced in Chapter 6 for representing difference equations. That is, branches entering a node are summed to produce the node variable. When no coefficient is indicated, the branch transmittance is assumed to be unity. For other branches, the transmittance of a branch is an integer power of W_N.

In Figure 9.3, two 4-point DFTs are computed, with $G[k]$ designating the 4-point DFT of the even-numbered points and $H[k]$ designating the 4-point DFT of the odd-numbered points. $X[0]$ is then obtained by multiplying $H[0]$ by W_N^0 and adding the product to $G[0]$. $X[1]$ is obtained by multiplying $H[1]$ by W_N^1 and adding that result to $G[1]$. Equation (9.14) states that, to compute $X[4]$, we should multiply $H[4]$ by W_N^4 and add the result of $G[4]$. However, since $G[k]$ and $H[k]$ are both periodic in k with

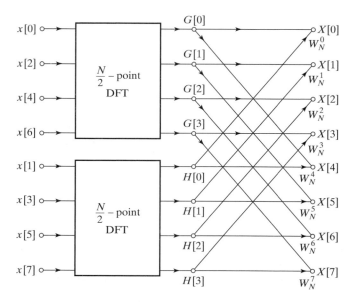

Figure 9.3 Flow graph of the decimation-in-time decomposition of an N-point DFT computation into two $(N/2)$-point DFT computations $(N = 8)$.

period 4, $H[4] = H[0]$ and $G[4] = G[0]$. Thus, $X[4]$ is obtained by multiplying $H[0]$ by W_N^4 and adding the result to $G[0]$. As shown in Figure 9.3, the values $X[5]$, $X[6]$, and $X[7]$ are obtained similarly.

With the computation restructured according to Eq. (9.14), we can compare the number of multiplications and additions required with those required for a direct computation of the DFT. Previously we saw that, for direct computation without exploiting symmetry, N^2 complex multiplications and additions were required.[4] By comparison, Eq. (9.14) requires the computation of two $(N/2)$-point DFTs, which in turn requires $2(N/2)^2$ complex multiplications and approximately $2(N/2)^2$ complex additions if we do the $(N/2)$-point DFTs by the direct method. Then the two $(N/2)$-point DFTs must be combined, requiring N complex multiplications, corresponding to multiplying the second sum by W_N^k, and N complex additions, corresponding to adding the product obtained to the first sum. Consequently, the computation of Eq. (9.14) for all values of k requires at most $N + 2(N/2)^2$ or $N + N^2/2$ complex multiplications and complex additions. It is easy to verify that for $N > 2$, the total $N + N^2/2$ will be less than N^2.

Equation (9.14) corresponds to breaking the original N-point computation into two $(N/2)$-point DFT computations. If $N/2$ is even, as it is when N is equal to a power of 2, then we can consider computing each of the $(N/2)$-point DFTs in Eq. (9.14) by breaking each of the sums in that equation into two $(N/4)$-point DFTs, which would then be combined to yield the $(N/2)$-point DFTs. Thus, $G[k]$ in Eq. (9.14) would be represented as

$$G[k] = \sum_{r=0}^{(N/2)-1} g[r]W_{N/2}^{rk} = \sum_{\ell=0}^{(N/4)-1} g[2\ell]W_{N/2}^{2\ell k} + \sum_{\ell=0}^{(N/4)-1} g[2\ell+1]W_{N/2}^{(2\ell+1)k}, \qquad (9.15)$$

or

$$G[k] = \sum_{\ell=0}^{(N/4)-1} g[2\ell]W_{N/4}^{\ell k} + W_{N/2}^{k} \sum_{\ell=0}^{(N/4)-1} g[2\ell+1]W_{N/4}^{\ell k}. \qquad (9.16)$$

Similarly, $H[k]$ would be represented as

$$H[k] = \sum_{\ell=0}^{(N/4)-1} h[2\ell]W_{N/4}^{\ell k} + W_{N/2}^{k} \sum_{\ell=0}^{(N/4)-1} h[2\ell+1]W_{N/4}^{\ell k}. \qquad (9.17)$$

Consequently, the $(N/2)$-point DFT $G[k]$ can be obtained by combining the $(N/4)$-point DFTs of the sequences $g[2\ell]$ and $g[2\ell + 1]$. Similarly, the $(N/2)$-point DFT $H[k]$ can be obtained by combining the $(N/4)$-point DFTs of the sequences $h[2\ell]$ and $h[2\ell + 1]$. Thus, if the 4-point DFTs in Figure 9.3 are computed according to Eqs. (9.16) and (9.17), then that computation would be carried out as indicated in Figure 9.4. Inserting the computation of Figure 9.4 into the flow graph of Figure 9.3, we obtain the complete flow graph of Figure 9.5, where we have expressed the coefficients in terms of powers of W_N rather than powers of $W_{N/2}$, using the fact that $W_{N/2} = W_N^2$.

For the 8-point DFT that we have been using as an illustration, the computation has been reduced to a computation of 2-point DFTs. For example, the 2-point DFT of the sequence consisting of $x[0]$ and $x[4]$ is depicted in Figure 9.6. With the computation

---

[4]For simplicity, we shall assume that N is large, so that $(N - 1)$ can be approximated by N.

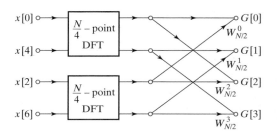

Figure 9.4   Flow graph of the decimation-in-time decomposition of an $(N/2)$-point DFT computation into two $(N/4)$-point DFT computations $(N = 8)$.

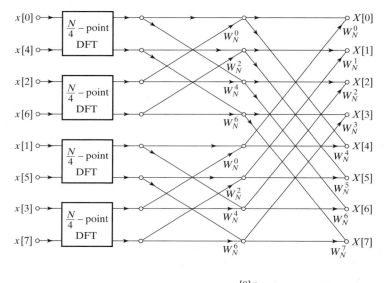

Figure 9.5   Result of substituting the structure of Figure 9.4 into Figure 9.3.

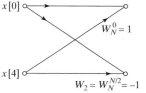

Figure 9.6   Flow graph of a 2-point DFT.

of Figure 9.6 inserted in the flow graph of Figure 9.5, we obtain the complete flow graph for computation of the 8-point DFT, as shown in Figure 9.7.

For the more general case, but with N still a power of 2, we would proceed by decomposing the $(N/4)$-point transforms in Eqs. (9.16) and (9.17) into $(N/8)$-point transforms and continue until we were left with only 2-point transforms. This requires $v = \log_2 N$ stages of computation. Previously, we found that in the original decomposition of an N-point transform into two $(N/2)$-point transforms, the number of complex multiplications and additions required was $N + 2(N/2)^2$. When the $(N/2)$-point transforms are decomposed into $(N/4)$-point transforms, the factor of $(N/2)^2$ is replaced by $N/2 + 2(N/4)^2$, so the overall computation then requires $N + N + 4(N/4)^2$ complex multiplications and additions. If $N = 2^v$, this can be done at most $v = \log_2 N$ times, so that after carrying out this decomposition as many times as possible, the number of complex multiplications and additions is equal to $Nv = N\log_2 N$.

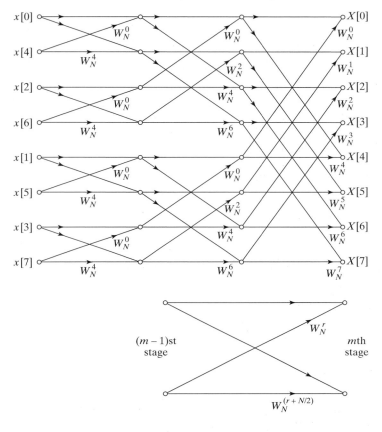

Figure 9.7   Flow graph of complete decimation-in-time decomposition of an 8-point DFT computation.

Figure 9.8   Flow graph of basic butterfly computation in Figure 9.7.

The flow graph of Figure 9.7 displays the operations explicitly. By counting branches with transmittances of the form W_N^r, we note that each stage has N complex multiplications and N complex additions. Since there are $\log_2 N$ stages, we have a total of $N \log_2 N$ complex multiplications and additions. This is the substantial computational savings that we have previously indicated was possible. For example, if $N = 2^{10} = 1024$, then $N^2 = 2^{20} = 1,048,576$, and $N \log_2 N = 10,240$, a reduction of more than two orders of magnitude!

The computation in the flow graph of Figure 9.7 can be reduced further by exploiting the symmetry and periodicity of the coefficients W_N^r. We first note that, in proceeding from one stage to the next in Figure 9.7, the basic computation is in the form of Figure 9.8, i.e., it involves obtaining a pair of values in one stage from a pair of values in the preceding stage, where the coefficients are always powers of W_N and the exponents are separated by $N/2$. Because of the shape of the flow graph, this elementary computation is called a *butterfly*. Since

$$W_N^{N/2} = e^{-j(2\pi/N)N/2} = e^{-j\pi} = -1, \tag{9.18}$$

the factor $W_N^{r+N/2}$ can be written as

$$W_N^{r+N/2} = W_N^{N/2} W_N^r = -W_N^r. \tag{9.19}$$

With this observation, the butterfly computation of Figure 9.8 can be simplified to the form shown in Figure 9.9, which requires only one complex multiplication instead of two. Using the basic flow graph of Figure 9.9 as a replacement for butterflies of the form of Figure 9.8, we obtain from Figure 9.7 the flow graph of Figure 9.10. In particular, the number of complex multiplications has been reduced by a factor of 2 over the number in Figure 9.7.

Figure 9.9   Flow graph of simplified butterfly computation requiring only one complex multiplication.

9.3.1 In-Place Computations

The flow graph of Figure 9.10 describes an algorithm for the computation of the discrete Fourier transform. The essential features of the flow graph are the branches connecting the nodes and the transmittance of each of these branches. No matter how the nodes in the flow graph are rearranged, it will always represent the same computation, provided that the connections between the nodes and the transmittances of the connections are maintained. The particular form for the flow graph in Figure 9.10 arose out of deriving the algorithm by separating the original sequence into the even-numbered and odd-numbered points and then continuing to create smaller and smaller subsequences in the same way. An interesting by-product of this derivation is that this flow graph, in addition to describing an efficient procedure for computing the discrete Fourier transform, also suggests a useful way of storing the original data and storing the results of the computation in intermediate arrays.

To see this, it is useful to note that according to Figure 9.10, each stage of the computation takes a set of N complex numbers and transforms them into another set of

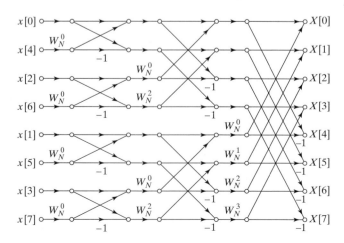

Figure 9.10   Flow graph of 8-point DFT using the butterfly computation of Figure 9.9.

N complex numbers through basic butterfly computations of the form of Figure 9.9. This process is repeated $v = \log_2 N$ times, resulting in the computation of the desired discrete Fourier transform. When implementing the computations depicted in Figure 9.10, we can imagine the use of two arrays of (complex) storage registers, one for the array being computed and one for the data being used in the computation. For example, in computing the first array in Figure 9.10, one set of storage registers would contain the input data and the second set would contain the computed results for the first stage. While the validity of Figure 9.10 is not tied to the order in which the input data are stored, we can order the set of complex numbers in the same order that they appear in the figure (from top to bottom). We denote the sequence of complex numbers resulting from the *m*th stage of computation as $X_m[\ell]$, where $\ell = 0, 1, \ldots, N-1$, and $m = 1, 2, \ldots, v$. Furthermore, for convenience, we define the set of input samples as $X_0[\ell]$. We can think of $X_{m-1}[\ell]$ as the input array and $X_m[\ell]$ as the output array for the *m*th stage of the computations. Thus, for the case of $N = 8$, as in Figure 9.10,

$$
\begin{aligned}
X_0[0] &= x[0], \\
X_0[1] &= x[4], \\
X_0[2] &= x[2], \\
X_0[3] &= x[6], \\
X_0[4] &= x[1], \\
X_0[5] &= x[5], \\
X_0[6] &= x[3], \\
X_0[7] &= x[7].
\end{aligned}
\tag{9.20}
$$

Using this notation, we can label the input and output of the butterfly computation in Figure 9.9 as indicated in Figure 9.11, with the associated equations

$$
X_m[p] = X_{m-1}[p] + W_N^r X_{m-1}[q], \tag{9.21a}
$$

$$
X_m[q] = X_{m-1}[p] - W_N^r X_{m-1}[q]. \tag{9.21b}
$$

In Eqs. (9.21), *p*, *q*, and *r* vary from stage to stage in a manner that is readily inferred from Figure 9.10 and from Eqs. (9.11), (9.14), (9.16), etc. It is clear from Figures 9.10 and 9.11 that only the complex numbers in locations *p* and *q* of the $(m-1)$st array are required to compute the elements *p* and *q* of the *m*th array. Thus, only one complex array of *N* storage registers is physically necessary to implement the complete computation if $X_m[p]$ and $X_m[q]$ are stored in the same storage registers as $X_{m-1}[p]$ and $X_{m-1}[q]$, respectively. This kind of computation is commonly referred to as an *in-place*

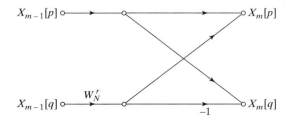

Figure 9.11   Flow graph of Eqs. (9.21).

computation. The fact that the flow graph of Figure 9.10 (or Figure 9.7) represents an in-place computation is tied to the fact that we have associated nodes in the flow graph that are on the same horizontal line with the same storage location and the fact that the computation between two arrays consists of a butterfly computation in which the input nodes and the output nodes are horizontally adjacent.

In order that the computation may be done in place as just discussed, the input sequence must be stored (or at least accessed) in a nonsequential order, as shown in the flow graph of Figure 9.10. In fact, the order in which the input data are stored and accessed is referred to as *bit-reversed* order. To see what is meant by this terminology, we note that for the 8-point flow graph that we have been discussing, three binary digits are required to index through the data. Writing the indices in Eqs. (9.20) in binary form, we obtain the following set of equations:

$$
\begin{aligned}
X_0[000] &= x[000],\\
X_0[001] &= x[100],\\
X_0[010] &= x[010],\\
X_0[011] &= x[110],\\
X_0[100] &= x[001],\\
X_0[101] &= x[101],\\
X_0[110] &= x[011],\\
X_0[111] &= x[111].
\end{aligned}
\tag{9.22}
$$

If (n_2, n_1, n_0) is the binary representation of the index of the sequence $x[n]$, then the sequence value $x[n_2, n_1, n_0]$ is stored in the array position $X_0[n_0, n_1, n_2]$. That is, in determining the position of $x[n_2, n_1, n_0]$ in the input array, we must reverse the order of the bits of the index n.

Let us first consider the process depicted in Figure 9.12 for sorting a data sequence in normal order by successive examination of the bits representing the data index. If the most significant bit of the data index is zero, $x[n]$ belongs in the top half of the

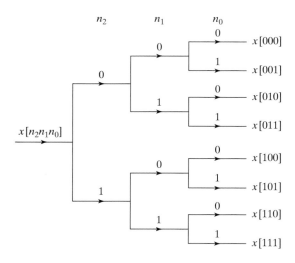

Figure 9.12   Tree diagram depicting normal-order sorting.

sorted array; otherwise it belongs in the bottom half. Next, the top half and bottom half subsequences can be sorted by examining the second most significant bit, and so on.

To see why bit-reversed order is necessary for in-place computation, recall the process that resulted in Figure 9.7 and Figure 9.10. The sequence $x[n]$ was first divided into the even-numbered samples, with the even-numbered samples occurring in the top half of Figure 9.3 and the odd-numbered samples occurring in the bottom half. Such a separation of the data can be carried out by examining the least significant bit $[n_0]$ in the index n. If the least significant bit is 0, the sequence value corresponds to an even-numbered sample and therefore will appear in the top half of the array $X_0[\ell]$, and if the least significant bit is 1, the sequence value corresponds to an odd-numbered sample and consequently will appear in the bottom half of the array. Next, the even- and odd-indexed subsequences are sorted into their even- and odd-indexed parts, and this can be done by examining the second least significant bit in the index. Considering first the even-indexed subsequence, if the second least significant bit is 0, the sequence value is an even-numbered term in the subsequence, and if the second least significant bit is 1, then the sequence value has an odd-numbered index in this subsequence. The same process is carried out for the subsequence formed from the original odd-indexed sequence values. This process is repeated until N subsequences of length 1 are obtained. This sorting into even- and odd-indexed subsequences is depicted by the tree diagram of Figure 9.13.

The tree diagrams of Figures 9.12 and 9.13 are identical, except that for normal sorting, we examine the bits representing the index from left to right, whereas for the sorting leading naturally to Figure 9.7 or 9.10, we examine the bits in reverse order, right to left, resulting in bit-reversed sorting. Thus, the necessity for bit-reversed ordering of the sequence $x[n]$ results from the manner in which the DFT computation is decomposed into successively smaller DFT computations in arriving at Figures 9.7 and 9.10.

9.3.2 Alternative Forms

Although it is reasonable to store the results of each stage of the computation in the order in which the nodes appear in Figure 9.10, it is certainly not necessary to do so. No matter how the nodes of Figure 9.10 are rearranged, the result will always be

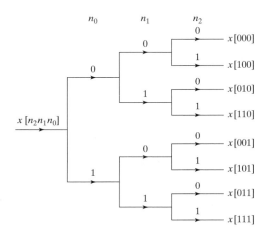

Figure 9.13   Tree diagram depicting bit-reversed sorting.

a valid computation of the discrete Fourier transform of $x[n]$, as long as the branch transmittances are unchanged. Only the order in which data are accessed and stored will change. If we associate the nodes with indexing of an array of complex storage locations, it is clear from our previous discussion that a flow graph corresponding to an in-place computation results only if the rearrangement of nodes is such that the input and output nodes for each butterfly computation are horizontally adjacent. Otherwise two complex storage arrays will be required. Figure 9.10, is, of course, such an arrangement. Another is depicted in Figure 9.14. In this case, the input sequence is in normal order and the sequence of DFT values is in bit-reversed order. Figure 9.14 can be obtained from Figure 9.10 as follows: All the nodes that are horizontally adjacent to $x[4]$ in Figure 9.10 are interchanged with all the nodes horizontally adjacent to $x[1]$. Similarly, all the nodes that are horizontally adjacent to $x[6]$ in Figure 9.10 are interchanged with those that are horizontally adjacent to $x[3]$. The nodes horizontally adjacent to $x[0]$, $x[2]$, $x[5]$, and $x[7]$ are not disturbed. The resulting flow graph in Figure 9.14 corresponds to the form of the decimation-in-time algorithm originally given by Cooley and Tukey (1965).

The only difference between Figures 9.10 and 9.14 is in the ordering of the nodes. The branch transmittances (powers of W_N) remain the same. There are, of course, a large variety of possible orderings. However, most do not make much sense from a computational viewpoint. As one example, suppose that the nodes are ordered such that the input and output both appear in normal order. A flow graph of this type is shown in Figure 9.15. In this case, however, the computation cannot be carried out in place because the butterfly structure does not continue past the first stage. Thus, two complex arrays of length N would be required to perform the computation depicted in Figure 9.15.

In realizing the computations depicted by Figures 9.10, 9.14, and 9.15, it is clearly necessary to access elements of intermediate arrays in nonsequential order. Thus, for greater computational speed, the complex numbers must be stored in random-access memory. For example, in the computation of the first array in Figure 9.10 from the input array, the inputs to each butterfly computation are adjacent node variables and are thought of as being stored in adjacent storage locations. In the computation of the second intermediate array from the first, the inputs to a butterfly are separated by two storage locations; and in the computation of the third array from the second, the inputs to a butterfly computation are separated by four storage locations. If $N > 8$, the separation between butterfly inputs is 8 for the fourth stage, 16 for the fifth stage, and so on. The separation in the last (vth) stage is $N/2$.

In Figure 9.14 the situation is similar in that, to compute the first array from the input data we use data separated by 4, to compute the second array from the first array we use input data separated by 2, and then finally, to compute the last array we use adjacent data. Although it is straightforward to imagine simple algorithms for modifying index registers to access the data in the flow graph of either Figure 9.10 or Figure 9.14, the data are not accessed sequentially, so random-access memory would be very desirable. In the flow graph of Figure 9.15, the data are accessed nonsequentially, the computation is not in place, and a scheme for indexing the data is considerably more complicated than in either of the two previous cases. Consequently, this structure has no apparent advantages.

Some forms have advantages even if in-place computation is not possible. A rearrangement of the flow graph in Figure 9.10 that is particularly useful when random-

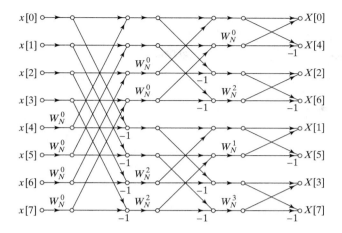

Figure 9.14   Rearrangement of Figure 9.10 with input in normal order and output in bit-reversed order.

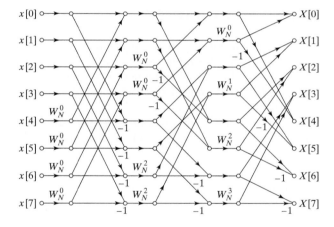

Figure 9.15   Rearrangement of Figure 9.10 with both input and output in normal order.

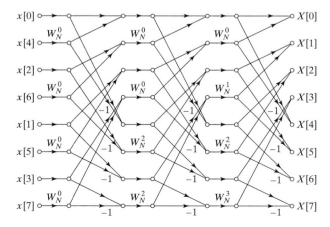

Figure 9.16   Rearrangement of Figure 9.10 having the same geometry for each stage, thereby permitting sequential data accessing and storage.

access memory is not available is shown in Figure 9.16. This flow graph represents the decimation-in-time algorithm originally given by Singleton (1969). (See DSP Committee, 1979, for a program using serial memory.) Note first that in this flow graph the input is in bit-reversed order and the output is in normal order. The important feature of the flow graph is that the geometry is identical for each stage; only the branch transmittances change from stage to stage. This makes it possible to access data sequentially. Suppose, for example that we have four separate disk files, and suppose that the first half of the input sequence (in bit-reversed order) is stored in one file and the second half is stored in a second file. Then the sequence can be accessed sequentially in files 1 and 2 and the results written sequentially on files 3 and 4, with the first half of the new array being written to file 3 and the second half to file 4. Then at the next stage of computation, files 3 and 4 are the input, and the output is written to files 1 and 2. This is repeated for each of the v stages. Such an algorithm could be useful in computing the DFT of extremely long sequences.

9.4 DECIMATION-IN-FREQUENCY FFT ALGORITHMS

The decimation-in-time FFT algorithms are all based on structuring the DFT computation by forming smaller and smaller subsequences of the input sequence $x[n]$. Alternatively, we can consider dividing the output sequence $X[k]$ into smaller and smaller subsequences in the same manner. FFT algorithms based on this procedure are commonly called *decimation-in-frequency* algorithms.

To develop these FFT algorithms, let us again restrict the discussion to N a power of 2 and consider computing separately the even-numbered frequency samples and the odd-numbered frequency samples. Since

$$X[k] = \sum_{n=0}^{N-1} x[n] W_N^{nk}, \qquad k = 0, 1, \ldots, N-1, \tag{9.23}$$

the even-numbered frequency samples are

$$X[2r] = \sum_{n=0}^{N-1} x[n] W_N^{n(2r)}, \qquad r = 0, 1, \ldots, (N/2) - 1, \tag{9.24}$$

which can be expressed as

$$X[2r] = \sum_{n=0}^{(N/2)-1} x[n] W_N^{2nr} + \sum_{n=N/2}^{N-1} x[n] W_N^{2nr}. \tag{9.25}$$

With a substitution of variables in the second summation in Eq. (9.25), we obtain

$$X[2r] = \sum_{n=0}^{(N/2)-1} x[n] W_N^{2nr} + \sum_{n=0}^{(N/2)-1} x[n + (N/2)] W_N^{2r[n+(N/2)]}. \tag{9.26}$$

Finally, because of the periodicity of W_N^{2rn},

$$W_N^{2r[n+(N/2)]} = W_N^{2rn} W_N^{rN} = W_N^{2rn}, \tag{9.27}$$

and since $W_N^2 = W_{N/2}$, Eq. (9.26) can be expressed as

$$X[2r] = \sum_{n=0}^{(N/2)-1} (x[n] + x[n + (N/2)]) W_{N/2}^{rn}, \qquad r = 0, 1, \ldots, (N/2) - 1. \tag{9.28}$$

Equation (9.28) is the $(N/2)$-point DFT of the $(N/2)$-point sequence obtained by adding the first half and the last half of the input sequence. Adding the two halves of the input sequence represents time aliasing, consistent with the fact that in computing only the even-numbered frequency samples, we are undersampling the Fourier transform of $x[n]$.

We can now consider obtaining the odd-numbered frequency points, given by

$$X[2r + 1] = \sum_{n=0}^{N-1} x[n] W_N^{n(2r+1)}, \qquad r = 0, 1, \ldots, (N/2) - 1. \tag{9.29}$$

As before, we can rearrange Eq. (9.29) as

$$X[2r + 1] = \sum_{n=0}^{(N/2)-1} x[n] W_N^{n(2r+1)} + \sum_{n=N/2}^{N-1} x[n] W_N^{n(2r+1)}. \tag{9.30}$$

An alternative form for the second summation in Eq. (9.30) is

$$\sum_{n=N/2}^{N-1} x[n] W_N^{n(2r+1)} = \sum_{n=0}^{(N/2)-1} x[n + (N/2)] W_N^{[n+(N/2)](2r+1)}$$

$$= W_N^{(N/2)(2r+1)} \sum_{n=0}^{(N/2)-1} x[n + (N/2)] W_N^{n(2r+1)}$$

$$= - \sum_{n=0}^{(N/2)-1} x[n + (N/2)] W_N^{n(2r+1)}, \tag{9.31}$$

where we have used the fact that $W_N^{(N/2)2r} = 1$ and $W_N^{(N/2)} = -1$. Substituting Eq. (9.31) into Eq. (9.30) and combining the two summations, we obtain

$$X[2r + 1] = \sum_{n=0}^{(N/2)-1} (x[n] - x[n + (N/2)]) W_N^{n(2r+1)}, \tag{9.32}$$

or, since $W_N^2 = W_{N/2}$,

$$X[2r+1] = \sum_{n=0}^{(N/2)-1} (x[n] - x[n + (N/2)]) W_N^n W_{N/2}^{nr},$$

$$r = 0, 1, \ldots, (N/2) - 1.$$

(9.33)

Equation (9.33) is the $(N/2)$-point DFT of the sequence obtained by subtracting the second half of the input sequence from the first half and multiplying the resulting sequence by W_N^n. Thus, on the basis of Eqs. (9.28) and (9.33), with $g[n] = x[n] + x[n + N/2]$ and $h[n] = x[n] - x[n + N/2]$, the DFT can be computed by first forming the sequences $g[n]$ and $h[n]$, then computing $h[n]W_N^n$, and finally computing the $(N/2)$-point DFTs of these two sequences to obtain the even-numbered output points and the odd-numbered output points, respectively. The procedure suggested by Eqs. (9.28) and (9.33) is illustrated for the case of an 8-point DFT in Figure 9.17.

Proceeding in a manner similar to that followed in deriving the decimation-in-time algorithm, we note that since N is a power of 2, $N/2$ is even; consequently, the $(N/2)$-point DFTs can be computed by computing the even-numbered and odd numbered output points for those DFTs separately. As in the case of the procedure leading to Eqs. (9.28) and (9.33), this is accomplished by combining the first half and the last half of the input points for each of the $(N/2)$-point DFTs and then computing $(N/4)$-point DFTs. The flow graph resulting from taking this step for the 8-point example is shown in Figure 9.18. For the 8-point example, the computation has now been reduced to the computation of 2-point DFTs, which are implemented by adding and subtracting the input points, as discussed previously. Thus, the 2-point DFTs in Figure 9.18 can be replaced by the computation shown in Figure 9.19, so the computation of the 8-point DFT can be accomplished by the algorithm depicted in Figure 9.20.

By counting the arithmetic operations in Figure 9.20 and generalizing to $N = 2^v$, we see that the computation of Figure 9.20 requires $(N/2) \log_2 N$ complex multiplications and $N \log_2 N$ complex additions. Thus, the total number of computations is the same for the decimation-in-frequency and the decimation-in-time algorithms.

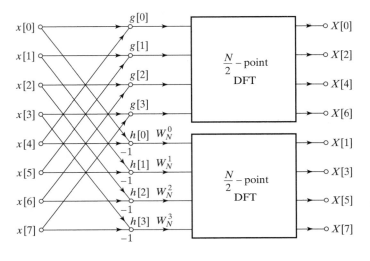

Figure 9.17 Flow graph of decimation-in-frequency decomposition of an N-point DFT computation into two $(N/2)$-point DFT computations $(N = 8)$.

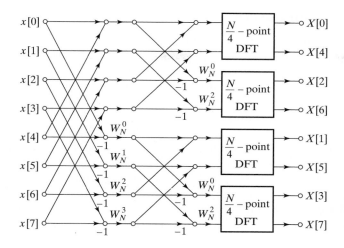

Figure 9.18   Flow graph of decimation-in-frequency decomposition of an 8-point DFT into four 2-point DFT computations.

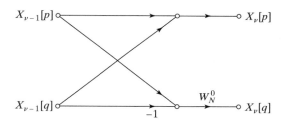

Figure 9.19   Flow graph of a typical 2-point DFT as required in the last stage of decimation-in-frequency decomposition.

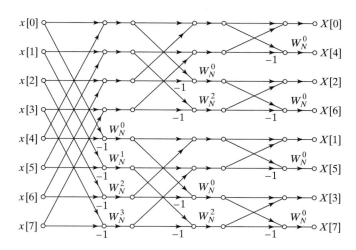

Figure 9.20   Flow graph of complete decimation-in-frequency decomposition of an 8-point DFT computation.

9.4.1 In-Place Computation

The flow graph in Figure 9.20 depicts one FFT algorithm based on decimation in frequency. We can observe a number of similarities and also a number of differences in comparing this graph with the flow graphs derived on the basis of decimation in time. As with decimation in time, of course, the flow graph of Figure 9.20 corresponds to a computation of the discrete Fourier transform, regardless of how the graph is drawn, as long as the same nodes are connected to each other with the proper branch transmittances. In other words, the flow graph of Figure 9.20 is not based on any assumption about the order in which the input sequence values are stored. However, as was done with the decimation-in-time algorithms, we can interpret successive vertical nodes in the flow graph of Figure 9.20 as corresponding to successive storage registers in a digital memory. In this case, the flow graph in Figure 9.20 begins with the input sequence in normal order and provides the output DFT in bit-reversed order. The basic computation again has the form of a butterfly computation, although the butterfly is different from that arising in the decimation-in-time algorithms. However, because of the butterfly nature of the computation, the flow graph of Figure 9.20 can be interpreted as an in-place computation of the discrete Fourier transform.

9.4.2 Alternative Forms

A variety of alternative forms for the decimation-in-frequency algorithm can be obtained by transposing the decimation-in-time forms developed in Section 9.3.2. If we denote the sequence of complex numbers resulting from the mth stage of the computation as $X_m[\ell]$, where $\ell = 0, 1, \ldots, N-1$, and $m = 1, 2, \ldots, \nu$, then the basic butterfly computation shown in Figure 9.21 has the form

$$X_m[p] = X_{m-1}[p] + X_{m-1}[q], \tag{9.34a}$$

$$X_m[q] = (X_{m-1}[p] - X_{m-1}[q])W_N^r. \tag{9.34b}$$

By comparing Figures 9.11 and 9.21 or Eqs. (9.21) and (9.34), we see that the butterfly computations are different for the two classes of FFT algorithms. However, we also note a resemblance between the basic butterfly flow graphs of Figures 9.11 and 9.21 and between the FFT flow graphs of Figures 9.10 and 9.20. Specifically, Figure 9.20 can be obtained from Figure 9.10 and Figure 9.21 from Figure 9.11 by reversing the direction of signal flow and interchanging the input and output. That is, in the terminology of Chapter 6, Figure 9.20 is the transpose of the flow graph in Figure 9.10, and Figure 9.21 is the transpose of Figure 9.11. The application of the transposition theorem is not straightforward in this case, because the FFT flow graphs have more than one input

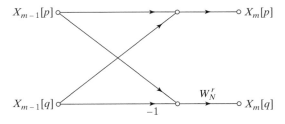

Figure 9.21 Flow graph of a typical butterfly computation required in Figure 9.20.

and output node. Nevertheless, it is true that the input-output characteristics of the flow graphs in Figures 9.10 and 9.20 are the same. This can be shown by noting that the butterfly equations in Eqs. (9.34) can be solved backward, starting with the output array. (Problem 9.23 outlines a proof of this result.) More generally, it is true that for each decimation-in-time FFT algorithm, there exists a decimation-in-frequency FFT algorithm that corresponds to interchanging the input and output and reversing the direction of all the arrows in the flow graph.

This result implies that all the flow graphs of Section 9.3 have counterparts in the class of decimation-in-frequency algorithms. This, of course, also corresponds to the fact that, as before, it is possible to rearrange the nodes of a flow graph without altering the final result.

Applying the transposition procedure to Figure 9.14 leads to Figure 9.22. In this flow graph, the output is in normal order and the input is in bit-reversed order. Alternatively, the transpose of the flow graph of Figure 9.15 is the flow graph of Figure 9.23,

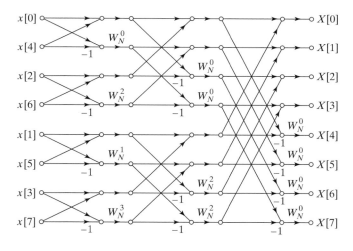

Figure 9.22   Flow graph of a decimation-in-frequency DFT algorithm obtained from Figure 9.20. Input in bit-reversed order and output in normal order. (Transpose of Figure 9.14.)

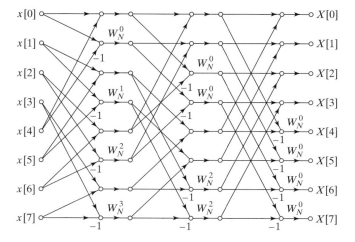

Figure 9.23   Rearrangement of Figure 9.20 with both input and output in normal order. (Transpose of Figure 9.15.)

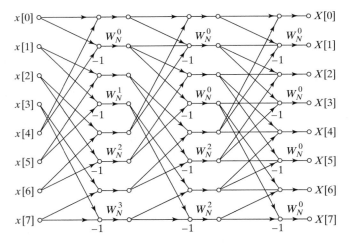

Figure 9.24   Rearrangement of Figure 9.20 having the same geometry for each stage, thereby permitting sequential data accessing and storage. (Transpose of Figure 9.16.)

where both the input and the output are in normal order. As in the case of Figure 9.15, this flow graph does not correspond to an in-place computation.

The transpose of Figure 9.16 is shown in Figure 9.24. Each stage of Figure 9.24 has the same geometry, a property that is desirable for computing large transforms using sequential data storage, as discussed before.

9.5 PRACTICAL CONSIDERATIONS

In Sections 9.3 and 9.4, we discussed the basic principles of efficient computation of the DFT when N is an integer power of 2. In these discussions, we favored the use of signal flow graph representations rather than explicitly writing out in detail the equations that such flow graphs represent. Of necessity, we have shown flow graphs for specific values of N. However, by considering a flow graph such as that in Figure 9.10, for a specific value of N, it is possible to see how to structure a general computational algorithm that would apply to any $N = 2^v$.

Although it is true that the flow graphs of the previous sections capture the essence of the FFT algorithms that they depict, a variety of details must be considered in the implementation of a given algorithm. In this section, we briefly suggest some of these. Specifically, in Section 9.5.1 we discuss issues associated with accessing and storing data in the intermediate arrays of the FFT. In Section 9.5.2 we discuss issues associated with computing or accessing the branch coefficients in the flow graph. Our emphasis is on algorithms for N a power of 2, but much of the discussion applies to the general case as well. For purposes of illustration, we focus primarily on the decimation-in-time algorithm of Figure 9.10. In Section 9.5.3 we briefly comment on algorithms for more general values of N.

9.5.1 Indexing

In the algorithm depicted in Figure 9.10, the input must be in bit-reversed order so that the computation can be performed in place. The resulting DFT is then in normal order. Generally, sequences do not originate in bit-reversed order, so the first step in the implementation of Figure 9.10 is to sort the input sequence into bit-reversed order. As can be seen from that figure and Eqs. (9.20) and (9.22), bit-reversed sorting can be done in place, since samples are only pairwise interchanged; i.e., a sample at a given index is interchanged with the sample in the location specified by the bit-reversed index. This is conveniently done in place by using two counters, one in normal order and the other in bit-reversed order. The data in the two positions specified by the two counters are simply interchanged. Once the input is in bit-reversed order, we can proceed with the first stage of computation. In this case, the inputs to the butterflies are adjacent elements of the array $X_0[\cdot]$. In the second stage, the inputs to the butterflies are separated by 2. In the mth stage, the butterfly inputs are separated by 2^{m-1}. The coefficients are powers of $W_N^{(N/2^m)}$ in the mth stage and are required in normal order if computation of butterflies begins at the top of the flow graph of Figure 9.10. The preceding statements define the manner in which data must be accessed at a given stage, which, of course, depends on the flow graph that is implemented. For example, in the mth stage of Figure 9.14, the butterfly spacing is 2^{v-m}, and in this case the coefficients are required in bit-reversed order. The input is in normal order; however, the output is in bit-reversed order, so it generally would be necessary to sort the output into normal order by using a normal-order counter and a bit-reversed counter, as discussed previously.

In general, if we consider all the flow graphs in Sections 9.3 and 9.4, we see that each algorithm has its own characteristic indexing problems. The choice of a particular algorithm depends on a number of factors. The algorithms utilizing an in-place computation have the advantage of making efficient use of memory. Two disadvantages, however, are that the kind of memory required is random-access rather than sequential memory and that either the input sequence or the output DFT sequence is in bit-reversed order. Furthermore, depending on whether a decimation-in-time or a decimation-in-frequency algorithm is chosen and whether the inputs or the outputs are in bit-reversed order, the coefficients are required to be accessed in either normal order or bit-reversed order. If non-random-access sequential memory is used, some fast Fourier transform algorithms utilize sequential memory, as we have shown, but either the inputs or the outputs must be in bit-reversed order. While the flow graph for the algorithm can be arranged so that the inputs, the outputs, and the coefficients are in normal order, the indexing structure required to implement these algorithms is complicated, and twice as much random access memory is required. Consequently, the use of such algorithms does not appear to be advantageous.

The in-place FFT algorithms of Figures 9.10, 9.14, 9.20, and 9.22 are among the most commonly used. If a sequence is to be transformed only once, then bit-reversed sorting must be implemented on either the input or the output. However, in some situations a sequence is transformed, the result is modified in some way, and then the inverse DFT is computed. For example, in implementing FIR digital filters by block convolution using the discrete Fourier transform, the DFT of a section of the input sequence is multiplied by the DFT of the filter impulse response, and the result is inverse transformed to obtain a segment of the output of the filter. Similarly, in computing an autocorrelation function or cross-correlation function using the discrete Fourier transform, a sequence will be transformed, the DFTs will be multiplied, and then the resulting product will be inverse transformed. When two transforms are cascaded in this way, it is possible, by appropriate choice of the FFT algorithms, to avoid the need for bit reversal. For example, in implementing an FIR digital filter using the DFT, we can choose an algorithm for the direct transform that utilizes the data in normal order and provides a DFT in bit-reversed order. Either the flow graph corresponding to Figure 9.14, based on decimation in time, or that of Figure 9.20, based on decimation in frequency, could be used in this way. The difference between these two forms is that the decimation-in-time form requires the coefficients in bit-reversed order, whereas the decimation-in-frequency form requires the coefficients in normal order.

In using either of these algorithms, the transform occurs in bit-reversed order. Consequently, in using the DFT for block convolution, if the DFT of the impulse response of the filter has been stored in bit-reversed order, the DFTs can simply be multiplied point by point in the order stored. For the inverse DFT, we can choose a form of the algorithm that requires bit-reversed data at the input and provides normally ordered results. Here, either the flow graph of Figure 9.10, based on decimation in time, or that of Figure 9.22, based on decimation in frequency, can be used. Figure 9.10, however, utilizes coefficients in normal order, whereas Figure 9.22 requires the coefficients in bit-reversed order. If the decimation-in-time form of the algorithm is chosen for the direct transform, then the decimation-in-frequency form of the algorithm should be chosen for the inverse transform, requiring coefficients in bit-reversed order. Likewise, the decimation-in-frequency algorithm for the direct transform should be paired with the decimation-in-time algorithm for the inverse transform, which would then utilize normally ordered coefficients.

9.5.2 Coefficients

We have observed that the coefficients W_N^r may be required in either bit-reversed order or in normal order. In either case we must store a table sufficient to look up all required values, or we must compute the values as needed. The first alternative has the advantage of speed, but of course requires extra storage. We observe from the flow graphs that we require W_N^r for $r = 0, 1, \ldots, (N/2) - 1$. Thus, we require $N/2$ complex storage registers for a complete table of values of W_N^r.[5] In the case of algorithms in which the coefficients are required in bit-reversed order, we can simply store the table in bit-reversed order.

The computation of the coefficients as they are needed saves storage, but is less efficient than storing a lookup table. If the coefficients are to be computed, it is generally most efficient to use a recursion formula. At any given stage, the required coefficients are all powers of a complex number of the form W_N^q, where q depends on the algorithm and the stage. Thus, if the coefficients are required in normal order, we can use the recursion formula

$$W_N^{q\ell} = W_N^q \cdot W_N^{q(\ell-1)} \tag{9.35}$$

to obtain the ℓth coefficient from the $(\ell - 1)$st coefficient. Clearly, algorithms that require coefficients in bit-reversed order are not well suited to this approach. It should be noted that Eq. (9.35) is essentially the coupled-form oscillator of Problem 6.21. When using finite-precision arithmetic, errors can build up in the iteration of this difference

---

[5]This number can be reduced using symmetry at the cost of greater complexity in accessing desired values.

equation. Therefore, it is generally necessary to reset the value at prescribed points (e.g., $W_N^{N/4} = -j$) so that errors do not become unacceptable.

9.5.3 Algorithms for More General Values of *N*

Although the special case of N a power of 2 leads to algorithms that have a simple structure, this is not the only restriction on N that can lead to a reduction in computation in the DFT. Indeed, in many cases it is desirable to evaluate the DFT efficiently for other values of N, and the same principles that were applied in the power of 2 decimation-in-time and decimation-in-frequency algorithms can be employed when N is a composite integer, i.e., the product of two or more integer factors. For example, if $N = RQ$, it is possible to express an N-point DFT as either the sum of R Q-point DFTs or as the sum of Q R-point DFTs and thereby obtain reductions in the number of computations. If N has many factors, the process can be repeated for each of the factors. Algorithms for general composite N involve more complicated indexing than the power of 2 case. In some cases these algorithms are straightforward generalizations of the algorithms discussed in Sections 9.3 and 9.4, in which case they are called "Cooley–Tukey" type algorithms. If the factors of N are relatively prime, the number of multiplications can be reduced at the cost of the more complicated indexing of a "prime factor" algorithm. The details of the more general Cooley–Tukey and prime factor algorithms are discussed in Burrus and Parks (1985), Burrus (1988), and Blahut (1985).

9.6 IMPLEMENTATION OF THE DFT USING CONVOLUTION

Because of the dramatic efficiency of the FFT, convolution is often implemented by explicitly computing the inverse DFT of the product of the DFTs of each sequence to be convolved, where an FFT algorithm is used to compute both the forward and the inverse DFTs. In contrast, and even in apparent (but, of course, not actual) contradiction, it is sometimes preferable to compute the DFT by first reformulating it as a convolution. We have already seen an example of this in the Goertzel algorithm. A number of other, more sophisticated, procedures are based on this approach.

9.6.1 Overview of the Winograd Fourier Transform Algorithm

One procedure proposed and developed by S. Winograd (1978), often referred to as the Winograd Fourier transform algorithm (WFTA), achieves its efficiency by expressing the DFT in terms of polynomial multiplication or, equivalently, convolution. The WFTA uses an indexing scheme corresponding to the decomposition of the DFT into a multiplicity of short-length DFTs where the lengths are relatively prime. Then the short DFTs are converted into periodic convolutions. A scheme for converting a DFT into a convolution when the number of input samples is prime was proposed by Rader (1968), but its application awaited the development of efficient methods for computing periodic convolutions. Winograd combined all of the foregoing procedures together with highly efficient algorithms for computing cyclic convolutions into a new approach to computing the DFT. The techniques for deriving efficient algorithms for computing short convolutions are based on relatively advanced number-theoretic concepts, such as the Chinese remainder theorem for polynomials, and consequently, we do not explore the details here. However, excellent discussions of the details of the WFTA are available in McClellan and Rader (1979), Blahut (1985), and Burrus (1988).

With the WFTA approach, the number of multiplications required for an N-point DFT is proportional to N rather than $N \log N$. Although this approach leads to algorithms that are optimal in terms of minimizing multiplications, the number of additions is significantly increased in comparison with the FFT. Therefore, the WFTA is most advantageous when multiplication is significantly slower than addition, as is often the case with fixed-point digital arithmetic. However, in processors where multiplication and accumulation are tied together, the Cooley–Tukey or prime factor algorithms are generally preferable. Additional difficulties with the WFTA are that indexing is more complicated, in-place computation is not possible, and there are major structural differences in algorithms for different values of N.

Thus, although the WFTA is extremely important as a benchmark for determining how efficient the DFT computation can be (in terms of number of multiplications), other factors often dominate in determining the speed and efficiency of a hardware or software implementation of the DFT computation.

9.6.2 The Chirp Transform Algorithm

Another algorithm based on expressing the DFT as a convolution is referred to as the chirp transform algorithm (CTA). This algorithm is not optimal in minimizing any measure of computational complexity, but it has been useful in a variety of applications, particularly when implemented in technologies that are well suited to doing convolution with a fixed, prespecified impulse response. The CTA is also more flexible than the FFT, since it can be used to compute *any* set of equally spaced samples of the Fourier transform on the unit circle.

To derive the CTA, we let $x[n]$ denote an N-point sequence and $X(e^{j\omega})$ its Fourier transform. We consider the evaluation of M samples of $X(e^{j\omega})$ that are equally spaced in angle on the unit circle, as indicated in Figure 9.25, i.e., at frequencies

$$\omega_k = \omega_0 + k\Delta\omega, \qquad k = 0, 1, \ldots, M-1, \tag{9.36}$$

where the starting frequency ω_0 and the frequency increment $\Delta\omega$ can be chosen arbitrarily. (For the specific case of the DFT, $\omega_0 = 0$, $M = N$, and $\Delta\omega = 2\pi/N$.) The Fourier transform corresponding to this more general set of frequency samples is given by

$$X(e^{j\omega_k}) = \sum_{n=0}^{N-1} x[n]e^{-j\omega_k n}, \qquad k = 0, 1, \ldots, M-1, \tag{9.37}$$

or, with W defined as

$$W = e^{-j\Delta\omega} \tag{9.38}$$

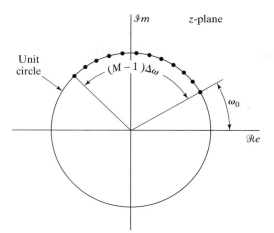

Figure 9.25 Frequency samples for chirp transform algorithm.

and using Eq. (9.36),

$$X(e^{j\omega_k}) = \sum_{n=0}^{N-1} x[n]e^{-j\omega_0 n}W^{nk}. \tag{9.39}$$

To express $X(e^{j\omega_k})$ as a convolution, we use the identity

$$nk = \tfrac{1}{2}[n^2 + k^2 - (k-n)^2] \tag{9.40}$$

to express Eq. (9.39) as

$$X(e^{j\omega_k}) = \sum_{n=0}^{N-1} x[n]e^{-j\omega_0 n}W^{n^2/2}W^{k^2/2}W^{-(k-n)^2/2}. \tag{9.41}$$

Letting

$$g[n] = x[n]e^{-j\omega_0 n}W^{n^2/2}, \tag{9.42}$$

we can then write

$$X(e^{j\omega_k}) = W^{k^2/2}\left(\sum_{n=0}^{N-1} g[n]W^{-(k-n)^2/2}\right), \qquad k = 0, 1, \ldots, M-1. \tag{9.43}$$

In preparation for interpreting Eq. (9.43) as the output of a linear time-invariant system, we obtain more familiar notation by replacing k by n and n by k in Eq. (9.43):

$$X(e^{j\omega_n}) = W^{n^2/2}\left(\sum_{k=0}^{N-1} g[k]W^{-(n-k)^2/2}\right), \qquad n = 0, 1, \ldots, M-1. \tag{9.44}$$

In the form of Eq. (9.44), $X(e^{j\omega_n})$ corresponds to the convolution of the sequence $g[n]$ with the sequence $W^{-n^2/2}$, followed by multiplication by the sequence $W^{n^2/2}$. The output sequence, indexed on the independent variable n, is the sequence of frequency samples $X(e^{j\omega_n})$. With this interpretation, the computation of Eq. (9.44) is as depicted in Figure 9.26. The sequence $W^{-n^2/2}$ can be thought of as a complex exponential sequence with linearly increasing frequency $n\Delta w$. In radar systems, such signals are called chirp signals—hence the name *chirp transform*. A system similar to Figure 9.26 is commonly used in radar and sonar signal processing for pulse compression (Skolnik, 1986).

For the evaluation of the Fourier transform samples specified in Eq. (9.44), we need only compute the output of the system in Figure 9.26 over a finite interval. In Figure 9.27, we depict illustrations of the sequences $g[n]$, $W^{-n^2/2}$, and $g[n] * W^{-n^2/2}$. Since $g[n]$ is of finite duration, only a finite portion of the sequence $W^{-n^2/2}$ is used in obtaining $g[n] * W^{-n^2/2}$ over the interval $n = 0, 1, \ldots, M - 1$, specifically, that portion from $n = -(N - 1)$ to $n = M - 1$. Let us define

$$h[n] = \begin{cases} W^{-n^2/2}, & -(N-1) \le n \le M - 1, \\ 0, & \text{otherwise,} \end{cases} \qquad (9.45)$$

as illustrated in Figure 9.28. It is easily verified by considering the graphical represen-

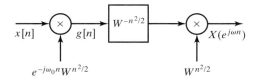

Figure 9.26 Block diagram of chirp transform algorithm.

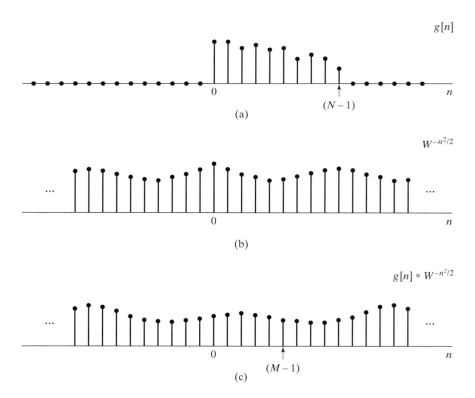

Figure 9.27 An illustration of the sequences used in the chirp transform algorithm. Note that the actual sequences involved are complex valued. (a) $g[n] = x[n]e^{-j\omega_0 n} W^{n^2/2}$. (b) $W^{-n^2/2}$. (c) $g[n] * W^{-n^2/2}$.

tation of the process of convolution that

$$g[n] * W^{-n^2/2} = g[n] * h[n], \qquad n = 0, 1, \ldots, M - 1. \tag{9.46}$$

Consequently, the infinite-duration impulse response $W^{-n^2/2}$ in the system of Figure 9.26 can be replaced by the finite-duration impulse response of Figure 9.28. The system is now as indicated in Figure 9.29, where $h[n]$ is specified by Eq. (9.45) and the frequency samples are given by

$$X(e^{j\omega_n}) = y[n], \qquad n = 0, 1, \ldots, M - 1. \tag{9.47}$$

Evaluation of frequency samples using the procedure indicated in Figure 9.29 has a number of potential advantages. In general, we do not require $N = M$ as in the FFT algorithms, and neither N nor M need be composite numbers. In fact, they may be prime numbers if desired. Furthermore, the parameter ω_0 is arbitrary. This increased flexibility over the FFT does not preclude efficient computation, since the convolution in Figure 9.29 can be implemented efficiently using an FFT algorithm with the technique of Section 8.7 to compute the convolution. As discussed in that section, the FFT size must be greater than or equal to $(M + N - 1)$ in order that the circular convolution will be equal to $g[n]*h[n]$ for $0 \le n \le M-1$. The FFT size is otherwise arbitrary and can, for example, be chosen to be a power of 2. It is interesting to note that the FFT algorithms used to compute the convolution implied by the CTA could be of the Winograd type. These algorithms themselves use convolution to implement the DFT computation.

In the system of Figure 9.29 $h[n]$ is noncausal, and for certain real-time implementations it must be modified to obtain a causal system. Since $h[n]$ is of finite duration, this modification is easily accomplished by delaying $h[n]$ by $(N - 1)$ to obtain a causal impulse response:

$$h_1[n] = \begin{cases} W^{-(n-N+1)^2/2}, & n = 0, 1, \ldots, M + N - 2, \\ 0, & \text{otherwise.} \end{cases} \tag{9.48}$$

Since both the chirp demodulation factor at the output and the output signal are also

$h[n]$

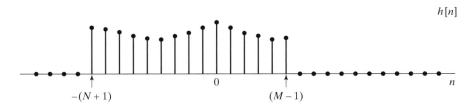

Figure 9.28 An illustration of the region of support for the FIR chirp filter. Note that the actual values of $h[n]$ as given by Eq. (9.45) are complex.

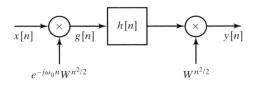

Figure 9.29 Block diagram of chirp transform system for finite-length impulse response.

delayed by $(N-1)$ samples, the Fourier transform values are

$$X(e^{j\omega_n}) = y_1[n+N-1], \qquad n = 0, 1, \ldots, M-1. \tag{9.49}$$

Modifying the system of Figure 9.29 to obtain a causal system results in the system of Figure 9.30. An advantage of this system stems from the fact that it involves the convolution of the input signal (modulated with a chirp) with a fixed, causal impulse response. Certain technologies, such as charge-coupled devices (CCD) and surface acoustic wave (SAW) devices, are particularly useful for implementing convolution with a fixed, pre-specified impulse response. These devices can be used to implement FIR filters, with the filter impulse response being specified at the time of fabrication by a geometric pattern of electrodes. A similar approach was followed by Hewes et al. (1979) in implementing the CTA with CCDs.

Further simplification of the CTA results when the frequency samples to be computed correspond to the DFT, i.e., when $\omega_0 = 0$ and $W = e^{-j2\pi/N}$, so that $\omega_n = 2\pi n/N$. In this case, it is convenient to modify the system of Figure 9.30. Specifically, with $\omega_0 = 0$ and $W = e^{-j2\pi/N} = W_N$, consider applying an additional unit of delay to the impulse response in Figure 9.30. With N even, $W_N^N = e^{j2\pi} = 1$, so

$$W_N^{-(n-N)^2/2} = W_N^{-n^2/2}. \tag{9.50}$$

Therefore, the system now is as shown in Figure 9.31, where

$$h_2[n] = \begin{cases} W_N^{-n^2/2}, & n = 1, 2, \ldots, M+N-1, \\ 0, & \text{otherwise}. \end{cases} \tag{9.51}$$

In this case, the chirp signal modulating $x[n]$ and the chirp signal modulating the output of the FIR filter are identical, and

$$X(e^{j2\pi n/N}) = y_2[n+N], \qquad n = 0, 1, \ldots, M-1. \tag{9.52}$$

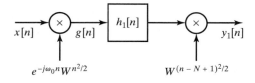

Figure 9.30   Block diagram of chirp transform system for causal finite-length impulse response.

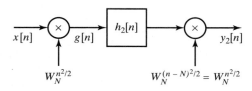

Figure 9.31   Block diagram of chirp transform system for obtaining DFT samples.

Example 9.1   Chirp Transform Parameters

Suppose we have a finite-length sequence $x[n]$ that is nonzero only on the interval $n = 0, \ldots, 25$, and we wish to compute 16 samples of the DTFT $X(e^{j\omega})$ at the frequencies $\omega_k = 2\pi/27 + 2\pi k/1024$ for $k = 0, \ldots, 15$. We can compute the desired frequency samples through convolution with a causal impulse response using the system in Figure 9.30 with an appropriate choice of parameters. We set $M = 16$, the number of samples desired, and $N = 26$, the length of the sequence. The frequency of the initial sample, ω_0, is $2\pi/27$, while the interval between adjacent frequency samples, $\Delta\omega$, is $2\pi/1024$. With these choices for the parameters, we know from Eq. (9.38) that $W = e^{-j\Delta\omega}$, and so the causal impulse response we desire is from Eq. (9.48)

$$h_1[n] = \begin{cases} [e^{-j2\pi/1024}]^{-(n-25)^2/2}, & n = 0, \ldots, 40, \\ 0, & \text{otherwise.} \end{cases}$$

For this causal impulse response, the output $y_1[n]$ will be the desired frequency samples beginning at $y_1[25]$, i.e.,

$$y_1[n + 25] = X(e^{j\omega_n})|_{\omega_n=2\pi/27+2\pi n/1024}, \qquad n = 0, \ldots, 15.$$

An algorithm similar to the CTA was first proposed by Bluestein (1970), who showed that a recursive realization of Figure 9.30 can be obtained for the case $\Delta\omega = 2\pi/N$, N a perfect square. (See Problem 9.41.) Rabiner et al. (1969) generalized this algorithm to obtain samples of the z-transform equally spaced in angle on a spiral contour in the z-plane. This more general form of the CTA was called the chirp z-transform (CZT) algorithm. The algorithm that we have called the CTA is a special case of the CZT algorithm.

9.7 EFFECTS OF FINITE REGISTER LENGTH

Since the fast Fourier transform algorithm is widely used for digital filtering and spectrum analysis, it is important to understand the effects of finite register length in the computation. As in the case of digital filters, a precise analysis of the effects is difficult. However, a simplified analysis is often sufficient for the purpose of choosing the required register length. The analysis that we will present is similar in style to that carried out in Section 6.8. Specifically, we analyze arithmetic round-off by means of a linear-noise model obtained by inserting an additive noise source at each point in the computation algorithm where round-off occurs. Furthermore, we will make a number of assumptions to simplify the analysis. The results that we obtain lead to several simplified, but useful, estimates of the effect of arithmetic round-off. Although the analysis is for rounding, it is generally easy to modify the results for truncation.

We have seen several different algorithmic structures for the FFT. However, the effects of round-off noise are very similar among the different classes of algorithms. Therefore, even though we consider only the radix-2 decimation-in-time algorithm, our results are representative of other forms as well.

The flow graph depicting a decimation-in-time algorithm for $N = 8$ was shown in Figure 9.10 and is reproduced in Figure 9.32. Some key aspects of this diagram are common to all standard radix-2 algorithms. The DFT is computed in $v = \log_2 N$ stages. At each stage a new array of N numbers is formed from the previous array by linear

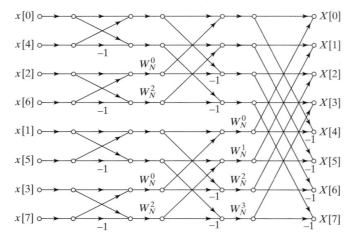

Figure 9.32 Flow graph for decimation-in-time FFT algorithm.

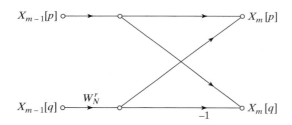

Figure 9.33 Butterfly computation for decimation in time.

combinations of the elements, taken two at a time. The νth array contains the desired DFT. For radix-2 decimation-in-time algorithms, the basic 2-point DFT computation is of the form

$$X_m[p] = X_{m-1}[p] + W_N^r X_{m-1}[q], \tag{9.53a}$$

$$X_m[q] = X_{m-1}[p] - W_N^r X_{m-1}[q]. \tag{9.53b}$$

Here the subscripts m and $(m-1)$ refer to the mth array and the $(m-1)$st array, respectively, and p and q denote the location of the numbers in each array. (Note that $m = 0$ refers to the input array and $m = \nu$ refers to the output array.) A flow graph representing the butterfly computation is shown in Figure 9.33.

At each stage, $N/2$ separate butterfly computations are carried out to produce the next array. The integer r varies with p, q, and m in a manner that depends on the specific form of the FFT algorithm used. However, our analysis is not tied to the specific way in which r varies. Also, the specific relationship among p, q, and m, which determines how we index through the mth array, is not important for the analysis. The details of the analysis for decimation in time and decimation in frequency differ somewhat due to the different butterfly forms, but the basic results do not change significantly. In our analysis we assume a butterfly of the form of Eqs. (9.53a) and (9.53b), corresponding to decimation in time.

We model the round-off noise by associating an additive noise generator with each fixed-point multiplication. With this model, the butterfly of Figure 9.33 is replaced by

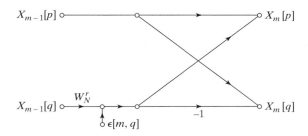

Figure 9.34   Linear-noise model for fixed-point round-off noise in a decimation-in-time butterfly computation.

that of Figure 9.34 for analyzing the round-off noise effects. The notation $\varepsilon[m, q]$ represents the complex-valued error introduced in computing the mth array from the $(m-1)$st array; specifically, it indicates the error resulting from quantization of multiplication of the qth element of the $(m-1)$st array by a complex coefficient.

Since we assume that, in general, the input to the FFT is a complex sequence, each of the multiplications is complex and thus consists of four real multiplications. We assume that the errors due to each real multiplication have the following properties:

1. The errors are uniformly distributed random variables over the range $-(1/2) \cdot 2^{-B}$ to $(1/2) \cdot 2^{-B}$, where, as defined in Section 6.6.1, numbers are represented as $(B+1)$-bit signed fractions. Therefore, each error source has variance $2^{-2B}/12$.

2. The errors are uncorrelated with one another.

3. All the errors are uncorrelated with the input and, consequently, also with the output.

Since each of the four noise sequences is uncorrelated zero-mean white noise and all have the same variance,

$$\mathcal{E}\{|\varepsilon[m, q]|^2\} = 4 \cdot \frac{2^{-2B}}{12} = \tfrac{1}{3} \cdot 2^{-2B} = \sigma_B^2. \tag{9.54}$$

To determine the mean-square value of the output noise at any output node, we must account for the contribution from each of the noise sources that propagate to that node. We can make the following observations from the flow graph of Figure 9.32:

1. The transmission function from any node in the flow graph to any other node to which it is connected is multiplication by a complex constant of unity magnitude (because each branch transmittance is either unity or an integer power of W_N).

2. Each output node connects to seven butterflies in the flow graph. (In general, each output node would connect to $(N-1)$ butterflies.) For example, Figure 9.35(a) shows the flow graph with all the butterflies removed that do not connect to $X[0]$, and Figure 9.35(b) shows the flow graph with all the butterflies removed that do not connect to $X[2]$.

These observations can be generalized to the case of N an arbitrary power of 2.

As a consequence of the first observation, the mean-square value of the magnitude of the component of the output noise due to each elemental noise source is the same and equal to σ_B^2. The total output noise at each output node is equal to the sum of the noise

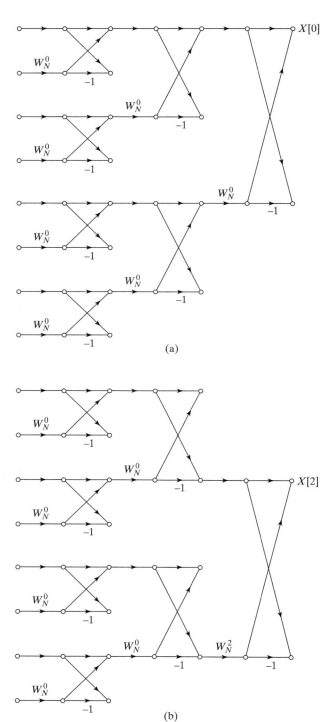

Figure 9.35 (a) Butterflies that affect $X[0]$; (b) butterflies that affect $X[2]$.

propagated to that node. Since we assume that all the noise sources are uncorrelated, the mean-square value of the magnitude of the output noise is equal to σ_B^2 times the number of noise sources that propagate to that node. At most one complex noise source is introduced at each butterfly; consequently, from observation 2, at most $(N-1)$ noise sources propagate to each output node. In fact, not all the butterflies generate round-off noise, since some (for example, all those in the first and second stages for $N = 8$) involve only multiplication by unity. However, if for simplicity we assume that round-off occurs for each butterfly, we can consider the result as an upper bound on the output noise. With this assumption, then, the mean square value of the output noise in the kth DFT value, $F[k]$, is given by

$$\mathcal{E}\{|F[k]|^2\} = (N-1)\sigma_B^2, \tag{9.55}$$

which, for large N, we approximate as

$$\mathcal{E}\{|F[k]|^2\} \cong N\sigma_B^2. \tag{9.56}$$

According to this result, the mean-square value of the output noise is proportional to N, the number of points transformed. The effect of doubling N, or adding another stage in the FFT, is to double the mean-square value of the output noise. In Problem 9.45, we consider the modification of this result when we do not insert noise sources for those butterflies that involve only multiplication by unity or j. Note that for FFT algorithms, a double-length accumulator does not help us reduce round-off noise, since the outputs of the butterfly computation must be stored in $(B+1)$-bit registers at the output of each stage.

   In implementing an FFT algorithm with fixed-point arithmetic, we must ensure against overflow. From Eqs. (9.53a) and (9.53b), it follows that

$$\max(|X_{m-1}[p]|, |X_{m-1}[q]|) \le \max(|X_m[p]|, |X_m[q]|) \tag{9.57}$$

and also

$$\max(|X_m[p]|, |X_m[q]|) \le 2\max(|X_{m-1}[p]|, |X_{m-1}[q]|). \tag{9.58}$$

(See Problem 9.44.) Equation (9.57) implies that the maximum magnitude is non decreasing from stage to stage. If the magnitude of the output of the FFT is less than unity, then the magnitude of the points in each array must be less than unity, i.e., there will be no overflow in any of the arrays.[6]

   To express this constraint as a bound on the input sequence, we note that the condition

$$|x[n]| < \frac{1}{N}, \qquad 0 \le n \le N-1, \tag{9.59}$$

is both necessary and sufficient to guarantee that

$$|X[k]| < 1, \qquad 0 \le k \le N-1. \tag{9.60}$$

This follows from the definition of the DFT, since

$$|X[k]| = |\sum_{n=0}^{N-1} x[n]W_N^{kn}| \le \sum_{n=0}^{N-1} |x[n]| \quad k = 0, 1, \ldots N-1. \tag{9.61}$$

---

[6] Actually, one should discuss overflow in terms of the real and imaginary parts of the data rather than the magnitude. However, $|x| < 1$ implies that $|\mathcal{R}e\{x\}| < 1$ and $|\mathcal{J}m\{x\}| < 1$, and only a slight increase in allowable signal level is achieved by scaling on the basis of real and imaginary parts.

Thus, Eq. (9.59) is sufficient to guarantee that there will be no overflow for all stages of the algorithm.

To obtain an explicit expression for the noise-to-signal ratio at the output of the FFT algorithm, consider an input in which successive sequence values are uncorrelated, i.e., a white-noise input signal. Also, assume that the real and imaginary parts of the input sequence are uncorrelated and that each has an amplitude density that is uniform between $-1/(\sqrt{2}N)$ and $+1/(\sqrt{2}N)$. (Note that this signal satisfies Eq. 9.59.) Then the average squared magnitude of the complex input sequence is

$$\mathcal{E}\{|x[n]|^2\} = \frac{1}{3N^2} = \sigma_x^2. \tag{9.62}$$

The DFT of the input sequence is

$$X[k] = \sum_{n=0}^{N-1} x[n]W^{kn}, \tag{9.63}$$

from which it can be shown that, under the foregoing assumptions on the input,

$$\mathcal{E}\{|X[k]|^2\} = \sum_{n=0}^{N-1} \mathcal{E}\{|x[n]|^2\}|W^{kn}|^2$$
$$= N\sigma_x^2 = \frac{1}{3N}. \tag{9.64}$$

Combining Eqs. (9.56) and (9.64), we obtain

$$\frac{\mathcal{E}\{|F[k]|^2\}}{\mathcal{E}\{|X[k]|^2\}} = 3N^2\sigma_B^2 = N^2 2^{-2B}. \tag{9.65}$$

According to Eq. (9.65), the noise-to-signal ratio increases as N^2, or 1 bit per stage. That is, if N is doubled, corresponding to adding one additional stage to the FFT, then to maintain the same noise-to-signal ratio, 1 bit must be added to the register length. The assumption of a white-noise input signal is, in fact, not critical here. For a variety of other inputs, the noise-to-signal ratio is still proportional to N^2, with only the constant of proportionality changing.

Equation (9.58) suggests an alternative scaling procedure. Since the maximum magnitude increases by no more than a factor of 2 from stage to stage, we can prevent overflow by requiring that $|x[n]| < 1$ and incorporating an attenuation of $\frac{1}{2}$ at the input to each stage. In this case, the output will consist of the DFT scaled by $1/N$. Although the mean-square output signal will be $1/N$ times what it would be if no scaling were introduced, the input amplitude can be N times larger without causing overflow. For the white-noise input signal, this means that we can assume that the real and imaginary parts are uniformly distributed from $-1/\sqrt{2}$ to $1/\sqrt{2}$, so that $|x[n]| < 1$. Thus, with the ν divisions by 2, the maximum expected value of the magnitude squared of the DFT that can be attained (for the white input signal) is the same as that given in Eq. (9.64). However, the output noise level will be much less than in Eq. (9.56), since the noise introduced at early stages of the FFT will be attenuated by the scaling that takes place in the later arrays. Specifically, with scaling by $1/2$ introduced at the input to each butterfly, we modify the butterfly of Figure 9.34 to that of Figure 9.36, where,

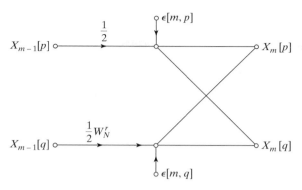

Figure 9.36 Butterfly showing scaling multipliers and associated fixed-point round-off noise.

in particular, two noise sources are now associated with each butterfly. As before, we assume that the real and imaginary parts of these noise sources are uncorrelated and are also uncorrelated with the other noise sources and that the real and imaginary parts are uniformly distributed between $\pm(1/2) \cdot 2^{-B}$. Thus, as before,

$$\mathcal{E}\{|\varepsilon[m, q]|^2\} = \sigma_B^2 = \tfrac{1}{3} \cdot 2^{-2B} = \mathcal{E}\{|\varepsilon[m, p]|^2\}. \tag{9.66}$$

Because the noise sources are all uncorrelated, the mean-squared magnitude of the noise at each output node is again the sum of the mean-squared contributions of each noise source in the flow graph. However, unlike the previous case, the attenuation that each noise source experiences through the flow graph depends on the array at which it originates. A noise source originating at the mth array will propagate to the output with multiplication by a complex constant with magnitude $(1/2)^{\nu-m-1}$. By examination of Figure 9.32, we see that for the case $N = 8$, each output node connects to:

  1 butterfly originating at the $(\nu - 1)$st array,
  2 butterflies originating at the $(\nu - 2)$nd array,
  4 butterflies originating at the $(\nu - 3)$rd array, etc.

For the general case with $N = 2^\nu$, each output node connects to $2^{\nu-m-1}$ butterflies and therefore to $2^{\nu-m}$ noise sources that originate at the mth array. Thus, at each output node, the mean-square magnitude of the noise is

$$\mathcal{E}\{|F[k]|^2\} = \sigma_B^2 \sum_{m=0}^{\nu-1} 2^{\nu-m} \cdot (0.5)^{2\nu-2m-2}$$

$$= \sigma_B^2 \sum_{m=0}^{\nu-1} (0.5)^{\nu-m-2}$$

$$= \sigma_B^2 \cdot 2 \sum_{k=0}^{\nu-1} 0.5^k \tag{9.67}$$

$$= 2\sigma_B^2 \frac{1 - 0.5^\nu}{1 - 0.5} = 4\sigma_B^2(1 - 0.5^\nu).$$

For large N, we assume that 0.5^ν (i.e., $1/N$) is negligible compared with unity, so

$$\mathcal{E}\{|F[k]|^2\} \cong 4\sigma_B^2 = \tfrac{4}{3} \cdot 2^{-2B}, \tag{9.68}$$

which is much less than the noise variance resulting when all the scaling is carried out on the input data.

Now we can combine Eq. (9.68) with Eq. (9.64) to obtain the output noise-to-signal ratio for the case of step-by-step scaling and white input. We obtain

$$\frac{\mathcal{E}\{|F[k]|^2\}}{\mathcal{E}\{|X[k]|^2\}} = 12N\sigma_B^2 = 4N \cdot 2^{-2B}, \tag{9.69}$$

a result proportional to N rather than to N^2. An interpretation of Eq. (9.69) is that the output noise-to-signal ratio increases as N, corresponding to half a bit per stage, a result first obtained by Welch (1969). It is important to note again that the assumption of a white-noise signal is not essential in the analysis. The basic result of an increase of half a bit per stage holds for a broad class of signals, with only the constant multiplier in Eq. (9.69) being dependent on the signal.

We should also note that the dominant factor that causes the increase of the noise-to-signal ratio with N is the decrease in signal level (required by the overflow constraint) as we pass from stage to stage. According to Eq. (9.68), very little noise (only a bit or two) is present in the final array. Most of the noise has been shifted out of the binary word by the scalings.

We have assumed straight fixed-point computation in the preceding discussion; i.e., only preset attenuations were allowed, and we were not permitted to rescale on the basis of an overflow test. Clearly, if the hardware or programming facility is such that straight fixed-point computation must be used, we should, if possible, incorporate attenuators of $1/2$ at each array rather than use a large attenuation of the input array.

A third approach to avoiding overflow is the use of *block floating point*. In this procedure the original array is normalized to the far left of the computer word, with the restriction that $|x[n]| < 1$; the computation proceeds in a fixed-point manner, except that after every addition there is an overflow test. If overflow is detected, the entire array is divided by 2 and the computation continues. The number of necessary divisions by 2 are counted to determine a scale factor for the entire final array. The output noise-to-signal ratio depends strongly on how many overflows occur and at what stages of the computation they occur. The positions and timing of overflows are determined by the signal being transformed; thus, to analyze the noise-to-signal ratio in a block floating-point implementation of the FFT, we would need to know the input signal.

The preceding analysis shows that scaling to avoid overflow is the dominant factor in determining the noise-to-signal ratio of fixed-point implementations of FFT algorithms. Therefore, floating-point arithmetic should improve the performance of these algorithms. The effect of floating-point round-off on the FFT was analyzed both theoretically and experimentally by Gentleman and Sande (1966), Weinstein (1969), and Kaneko and Liu (1970). These investigations show that, since scaling is no longer necessary, the decrease of noise-to-signal ratio with increasing N is much less dramatic than for fixed-point arithmetic.

For example, Weinstein (1969a) showed theoretically that the noise-to-signal ratio is proportional to v for $N = 2^v$, rather than proportional to N as in the fixed-point case. Therefore, quadrupling v (raising N to the fourth power) increases the noise-to-signal ratio by only 1 bit.

9.8 SUMMARY

In this chapter we have considered techniques for computation of the discrete Fourier transform, and we have seen how the periodicity and symmetry of the complex factor $e^{-j(2\pi/N)kn}$ can be exploited to increase the efficiency of DFT computations.

   We considered the Goertzel algorithm and the direct evaluation of the DFT expression because of the importance of these techniques when not all N of the DFT values are required. However, our major emphasis was on fast Fourier transform (FFT) algorithms. We described the decimation-in-time and decimation-in-frequency classes of FFT algorithms in some detail and some of the implementation considerations, such as indexing and coefficient quantization. Much of the detailed discussion concerned algorithms that require N to be a power of 2, since these algorithms are easy to understand, simple to program, and most often used.

   The use of convolution as the basis for computing the DFT was briefly discussed. We presented a brief overview of the Winograd Fourier transform algorithm, and in somewhat more detail we discussed an algorithm called the chirp transform algorithm.

   The final section of the chapter discussed effects of finite word length in DFT computations. We used linear-noise models to show that the noise-to-signal ratio of a DFT computation varies differently with the length of the sequence, depending on how scaling is done. We also commented briefly on the use of floating-point representations.

PROBLEMS

Basic Problems with Answers

9.1. Suppose that a computer program is available for computing the DFT

$$X[k] = \sum_{n=0}^{N-1} x[n]e^{-j(2\pi/N)kn}, \qquad k = 0, 1, \ldots, N-1;$$

i.e., the input to the program is the sequence $x[n]$ and the output is the DFT $X[k]$. Show how the input and/or output sequences may be rearranged such that the program can also be used to compute the inverse DFT

$$x[n] = \frac{1}{N}\sum_{k=0}^{N-1} X[k]e^{j(2\pi/N)kn}, \qquad n = 0, 1, \ldots, N-1;$$

i.e., the input to the program should be $X[k]$ or a sequence simply related to $X[k]$, and the output should be either $x[n]$ or a sequence simply related to $x[n]$. There are several possible approaches.

9.2. Figure P9.2-1 shows the graph representation of a decimation-in-time FFT algorithm for $N = 8$. The heavy line shows a path from sample $x[7]$ to DFT sample $X[2]$.

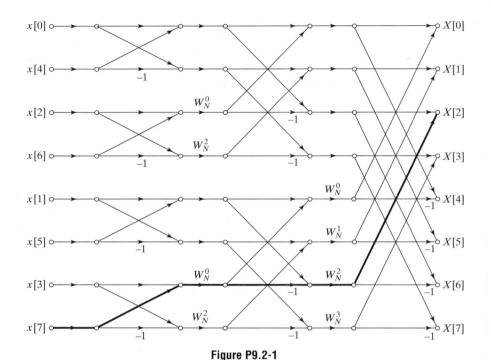

Figure P9.2-1

(a) What is the "gain" along the path that is emphasized in Figure P9.2-1?

(b) How many other paths in the flow graph begin at $x[7]$ and end at $X[2]$? Is this true in general? That is, how many paths are there between each input sample and each output sample?

(c) Now consider the DFT sample $X[2]$. By tracing paths in the flow graph of Figure P9.2-1, show that each input sample contributes the proper amount to the output DFT sample; i.e., verify that

$$X[2] = \sum_{n=0}^{N-1} x[n] e^{-j(2\pi/N)2n}.$$

9.3. Figure P9.3-1 shows the flow graph for an 8-point decimation-in-time FFT algorithm. Let $x[n]$ be the sequence whose DFT is $X[k]$. In the flow graph, $A[\cdot]$, $B[\cdot]$, $C[\cdot]$, and $D[\cdot]$ represent separate arrays that are indexed consecutively in the same order as the indicated nodes.

(a) Specify how the elements of the sequence $x[n]$ should be placed in the array $A[r]$, $r = 0, 1, \ldots, 7$. Also, specify how the elements of the DFT sequence should be extracted from the array $D[r]$, $r = 0, 1, \ldots, 7$.

(b) Without determining the values in the intermediate arrays, $B[\cdot]$ and $C[\cdot]$, determine and sketch the array sequence $D[r]$, $r = 0, 1, \ldots, 7$, if the input sequence is $x[n] = (-W_N)^n$, $n = 0, 1, \ldots, 7$.

(c) Determine and sketch the sequence $C[r]$, $r = 0, 1, \ldots, 7$, if the output Fourier transform is $X[k] = 1$, $k = 0, 1, \ldots, 7$.

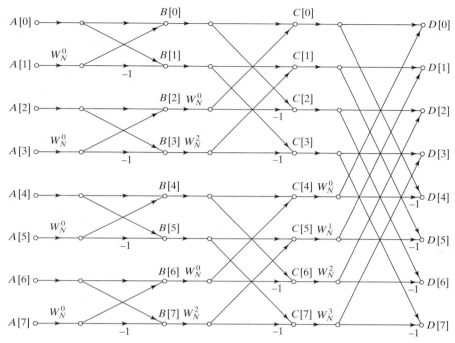

Figure P9.3-1

9.4. In implementing an FFT algorithm, it is sometimes useful to generate the powers of W_N with a recursive difference equation, or oscillator. In this problem we consider a radix-2 decimation-in-time algorithm for $N = 2^v$. Figure 9.10 depicts this type of algorithm for $N = 8$. To generate the coefficients efficiently, the frequency of the oscillator would change from stage to stage.

      Assume that the arrays are numbered 0 through $v = \log_2 N$, so the array holding the initial input sequence is the zeroth array and the DFT is in the vth array. In computing the butterflies in a given stage, all butterflies requiring the same coefficients W_N^r are evaluated before obtaining new coefficients. In indexing through the array, we assume that the data in the array are stored in consecutive complex registers numbered 0 through $(N-1)$. All the following questions are concerned with the computation of the mth array from the $(m-1)$st array, where $1 \le m \le v$. Answers should be expressed in terms of m.

    (a) How many butterflies must be computed in the mth stage? How many different coefficients are required in the mth stage?

    (b) Write a difference equation whose impulse response $h[n]$ contains the coefficients W_N^r required by the butterflies in the mth stage.

    (c) The difference equation from Part (b) should have the form of an oscillator, i.e., $h[n]$ should be periodic for $n \ge 0$. What is the period of $h[n]$? Based on this, write an expression for the frequency of this oscillator as a function of m.

9.5. Computing the DFT generally requires complex multiplications. Consider the product $X + jY = (A + jB)(C + jD) = (AC - BD) + j(BC + AD)$. In this form, a complex multiplication requires four real multiplications and two real additions. Verify that a complex multiplication can be performed with three real multiplications and five additions using

the algorithm

$$X = (A - B)D + (C - D)A,$$

$$Y = (A - B)D + (C + D)B.$$

9.6. Consider the butterfly in Figure P9.6-1. This butterfly was extracted from a signal flow graph implementing an FFT algorithm. Choose the most accurate statement from the following list:

1. The butterfly was extracted from a decimation-in-time FFT algorithm.
2. The butterfly was extracted from a decimation-in-frequency FFT algorithm.
3. It is not possible to say from the figure which kind of FFT algorithm the butterfly came from.

Figure P9.6-1

9.7. Consider the system shown in Figure P9.7-1. If the input to the system, $x[n]$, is a 32-point sequence in the interval $0 \le n \le 31$, the output $y[n]$ at $n = 32$ is equal to $X(e^{j\omega})$ evaluated at a specific frequency ω_k. What is ω_k for the coefficients shown in Figure P9.7-1?

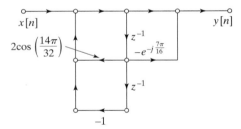

Figure P9.7-1

9.8. A finite-length signal $x[n]$ is nonzero in the interval $0 \le n \le 19$. This signal is the input to the system shown in Figure P9.8-1, where

$$h[n] = \begin{cases} e^{j(2\pi/21)(n-19)^2/2}, & n = 0, 1, \ldots, 28, \\ 0, & \text{otherwise.} \end{cases}$$

$$W = e^{-j(2\pi/21)}$$

The output of the system, $y[n]$, for the interval $n = 19, \ldots, 28$ can be expressed in terms of the DTFT $X(e^{j\omega})$ for appropriate values of ω. Write an expression for $y[n]$ in this interval in terms of $X(e^{j\omega})$.

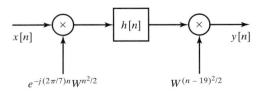

Figure P9.8-1

9.9. The butterfly flow graph in Figure 9.9 can be used to compute the DFT of a sequence of length $N = 2^\nu$ "in-place," i.e., using a single array of complex-valued registers. Assume this array of registers $A[\ell]$ is indexed on $0 \le l \le N - 1$. The input sequence is initially stored in $A[\ell]$ in bit-reversed order. The array is then processed by ν stages of butterflies. Each butterfly takes two array elements $A[\ell_0]$ and $A[\ell_1]$ as inputs, then stores its outputs into those same array locations. The values of ℓ_0 and ℓ_1 depend on the stage number and the location of the butterfly in the signal flow graph. The stages of the computation are indexed by $m = 1, \ldots, \nu$.

(a) What is $|\ell_1 - \ell_0|$ as a function of the stage number m?

(b) Many stages contain butterflies with the same "twiddle" factor W_N^r. For these stages, how far apart are the values of ℓ_0 for the butterflies with the same W_N^r?

9.10. Consider the system shown in Figure P9.10-1, with

$$h[n] = \begin{cases} e^{j(2\pi/10)(n-11)^2/2}, & n = 0, 1, \ldots, 15, \\ 0, & \text{otherwise.} \end{cases}$$

It is desired that the output of the system, $y[n + 11] = X(e^{j\omega_n})$, where $\omega_n = (2\pi/19) + n(2\pi/10)$ for $n = 0, \ldots, 4$. Give the correct value for the sequence $r[n]$ in Figure P9.10-1 such that the output $y[n]$ provides the desired samples of the discrete-time Fourier transform.

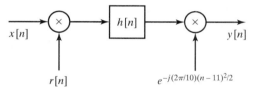

$$x[n] \qquad r[n] \qquad \qquad e^{-j(2\pi/10)(n-11)^2/2} \qquad y[n]$$

Figure P9.10-1

9.11. Assume that you wish to sort a sequence $x[n]$ of length $N = 16$ into bit-reversed order for input to an FFT algorithm. Give the new sample order for the bit-reversed sequence.

9.12. For the following statement, assume that the sequence $x[n]$ has length $N = 2^\nu$ and that $X[k]$ is the N-point DFT of $x[n]$. Indicate whether the statement is true or false, and justify your answer.

> **Statement:** It is impossible to construct a signal flow graph to compute $X[k]$ from $x[n]$ such that both $x[n]$ and $X[k]$ are in normal sequential (not bit-reversed) order.

9.13. The butterfly in Figure P9.13-1 was taken from a decimation-in-frequency FFT with $N = 16$, where the input sequence was arranged in normal order. Note that a 16-point FFT will have four stages, indexed $m = 1, \ldots, 4$. Which of the four stages have butterflies of this form? Justify your answer.

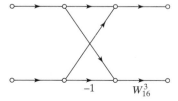

$$-1 \qquad W_{16}^3$$

Figure P9.13-1

9.14. The butterfly in Figure P9.14-1 was taken from a decimation-in-time FFT with $N = 16$. Assume that the four stages of the signal flow graph are indexed by $m = 1, \ldots, 4$. What are the possible values of r for each of the four stages?

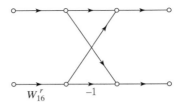

Figure P9.14-1

9.15. Suppose you have two programs for computing the DFT of a sequence $x[n]$ that has $N = 2^\nu$ nonzero samples. Program A computes the DFT by directly implementing the definition of the DFT sum from Eq. (8.67) and takes N^2 seconds to run. Program B implements the decimation-in-time FFT algorithm and takes $10N\log_2 N$ seconds to run. What is the shortest sequence N such that Program B runs faster than Program A?

9.16. The butterfly in Figure P9.16-1 was taken from a decimation-in-time FFT with $N = 16$. Assume that the four stages of the signal flow graph are indexed by $m = 1, \ldots, 4$. Which of the four stages have butterflies of this form?

Figure P9.16-1

9.17. Suppose you are told that an $N = 32$ FFT algorithm has a "twiddle" factor of W_{32}^2 for one of the butterflies in its fifth (last) stage. Is the FFT a decimation-in-time or decimation-in-frequency algorithm?

9.18. Suppose you have a signal $x[n]$ with 1021 nonzero samples whose discrete-time Fourier transform you wish to estimate by computing the DFT. You find that it takes your computer 100 seconds to compute the 1021-point DFT of $x[n]$. You then add three zero-valued samples at the end of the sequence to form a 1024-point sequence $x_1[n]$. The same program on your computer requires only 1 second to compute $X_1[k]$. Reflecting, you realize that by using $x_1[n]$, you are able to compute more samples of $X(e^{j\omega})$ in a much shorter time by adding some zeros to the end of $x[n]$ and pretending that the sequence is longer. How do you explain this apparent paradox?

9.19. Consider the signal flow graph in Figure P9.19-1. Suppose that the input to the system $x[n]$ is an 8-point sequence. Choose the values of a and b such that $y[8] = X(e^{j6\pi/8})$.

Figure P9.19-1

9.20. Suppose that you time-reverse and delay a real-valued 32-point sequence $x[n]$ to obtain $x_1[n] = x[32 - n]$. If $x_1[n]$ is used as the input for the system in Figure P9.7-1, find an expression for $y[32]$ in terms of $X(e^{j\omega})$, the discrete-time Fourier transform of the original sequence $x[n]$.

Basic Problems

9.21. In Section 9.2, we used the fact that $W_N^{-kN} = 1$ to derive a recurrence algorithm for computing a specific DFT value $X[k]$ for a finite-length sequence $x[n], n = 0, 1, \ldots, N-1$.

(a) Using the fact that $W_N^{kN} = W_N^{Nn} = 1$, show that $X[N-k]$ can be obtained as the output after N iterations of the difference equation depicted in Figure P9.21-1. That is, show that

$$X[N - k] = y_k[N].$$

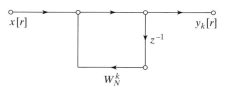

$x[r]$ $y_k[r]$

z^{-1}

W_N^k **Figure P9.21-1**

(b) Show that $X[N - k]$ is also equal to the output after N iterations of the difference equation depicted in Figure P9.21-2. Note that the system of Figure P9.21-2 has the same poles as the system in Figure 9.2, but the coefficient required to implement the complex zero in Figure P9.21-2 is the complex conjugate of the corresponding coefficient in Figure 9.2; i.e., $W_N^{-k} = (W_N^k)^*$.

$x[r]$ $y_k[r]$

$2\cos\left(\dfrac{2\pi k}{N}\right)$ z^{-1} $-W_N^k$

z^{-1}

-1 **Figure P9.21-2**

9.22. Construct a flow graph for a 16-point radix-2 decimation-in-time FFT algorithm. Label all multipliers in terms of powers of W_{16}, and also label any branch transmittances that are equal to -1. Label the input and output nodes with the appropriate values of the input and DFT sequences, respectively. Determine the number of real multiplications and the number of real additions required to implement the flow graph.

Advanced Problems

9.23. In Section 9.4.2, it was asserted that the transpose of the flow graph of an FFT algorithm is also the flow graph of an FFT algorithm. The purpose of this problem is to develop that result for radix-2 FFT algorithms.

(a) The basic butterfly for the decimation-in-frequency radix-2 FFT algorithm is depicted in Figure P9.23-1. This flow graph represents the equations

$$X_m[p] = X_{m-1}[p] + X_{m-1}[q],$$

$$X_m[q] = (X_{m-1}[p] - X_{m-1}[q])W_N^r.$$

Starting with these equations, show that $X_{m-1}[p]$ and $X_{m-1}[q]$ can be computed from $X_m[p]$ and $X_m[q]$, respectively, using the butterfly shown in Figure P9.23-2.

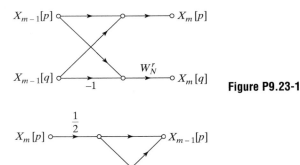

Figure P9.23-1

Figure P9.23-2

(b) In the decimation-in-frequency algorithm of Figure 9.20, $X_v[r], r = 0, 1, \ldots, N-1$ is the DFT $X[k]$ arranged in bit-reversed order, and $X_0[r] = x[r], r = 0, 1, \ldots, N-1$; i.e., the zeroth array is the input sequence arranged in normal order. If each butterfly in Figure 9.20 is replaced by the appropriate butterfly of the form of Figure P9.23-2, the result would be a flow graph for computing the sequence $x[n]$ (in normal order) from the DFT $X[k]$ (in bit-reversed order). Draw the resulting flow graph for $N = 8$.

(c) The flow graph obtained in Part (b) represents an *inverse* DFT algorithm, i.e., an algorithm for computing

$$x[n] = \frac{1}{N} \sum_{n=0}^{N-1} X[k]W_N^{-kn}, \qquad n = 0, 1, \ldots, N-1.$$

Modify the flow graph obtained in Part (b) so that it computes the DFT

$$X[k] = \sum_{n=0}^{N-1} x[n]W_N^{kn}, \qquad k = 0, 1, \ldots, N-1,$$

rather than the inverse DFT.

(d) Observe that the result in Part (c) is the transpose of the decimation-in-frequency algorithm of Figure 9.20 and that it is identical to the decimation-in-time algorithm depicted in Figure 9.10. Does it follow that, to each decimation-in-time algorithm (e.g., Figures 9.14–9.16), there corresponds a decimation-in-frequency algorithm that is the transpose of the decimation-in-time algorithm and vice versa? Explain.

9.24. We have seen that an FFT algorithm can be viewed as an interconnection of computational elements called butterflies. For example, the butterfly for a radix-2 decimation-in-frequency FFT algorithm is shown in Figure P9.24-1. The butterfly takes two complex numbers as input and produces two complex numbers as output. Its implementation requires a complex multiplication by W_N^r, where r is an integer that depends on the location of the butterfly in the flow graph of the algorithm. Since the complex multiplier is of the form $W_N^r = e^{j\theta}$, the CORDIC rotator algorithm (see Problem 9.39) can be used to implement the complex multiplication efficiently. Unfortunately, while the CORDIC rotator algorithm accomplishes the desired change of angle, it also introduces a fixed magnification that is independent of the angle θ. Thus, if the CORDIC rotator algorithm were used to implement the multiplications by W_N^r, the butterfly of Figure P9.24-1 would be replaced by the butterfly of Figure P9.24-2, where G represents the fixed magnification factor of the CORDIC rotator. (We assume no error in approximating the angle of rotation.) If each butterfly in the flow graph of the decimation-in-frequency FFT algorithm is replaced by the butterfly of Figure P9.24-2, we obtain a modified FFT algorithm for which the flow graph would be as shown in Figure P9.24-3 for $N = 8$. The output of this modified algorithm would not be the desired DFT.

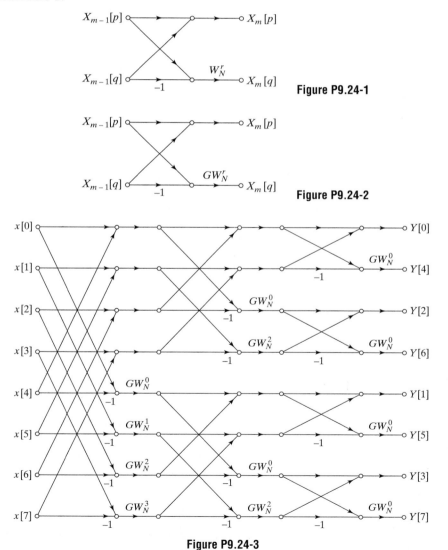

Figure P9.24-1

Figure P9.24-2

Figure P9.24-3

(a) Show that the output for the modified FFT algorithm is $Y[k] = W[k]X[k]$, where $X[k]$ is the correct DFT of the input sequence $x[n]$ and $W[k]$ is a function of G, N, and k.

(b) The sequence $W[k]$ can be described by a particularly simple rule. Find this rule and indicate its dependence on G, N, and k.

(c) Suppose that we wish to preprocess the input sequence $x[n]$ to compensate for the effect of the modified FFT algorithm. Determine a procedure for obtaining a sequence $\hat{x}[n]$ from $x[n]$ such that if $\hat{x}[n]$ is the input to the modified FFT algorithm, then the output will be $X[k]$, the correct DFT of the original sequence $x[n]$.

9.25. This problem deals with the efficient computation of samples of the z-transform of a finite-length sequence. Using the chirp transform algorithm, develop a procedure for computing values of $X(z)$ at 25 points spaced uniformly on an arc of a circle of radius 0.5, beginning at an angle of $-\pi/6$ and ending at an angle of $2\pi/3$. The length of the sequence is 100 samples.

9.26. The N-point DFT of the N-point sequence $x[n] = e^{-j(\pi/N)n^2}$, for N even, is

$$X[k] = \sqrt{N}e^{-j\pi/4}e^{j(\pi/N)k^2}.$$

Determine the $2N$-point DFT of the $2N$-point sequence $y[n] = e^{-j(\pi/N)n^2}$, assuming that N is even.

9.27. We are given a finite-length sequence $x[n]$ of length 627 (i.e., $x[n] = 0$ for $n < 0$ and $n > 626$), and we have available an FFT program that will compute the DFT of a sequence of any length $N = 2^v$.

For the given sequence, we want to compute samples of the discrete-time Fourier transform at frequencies

$$\omega_k = \frac{2\pi}{627} + \frac{2\pi k}{256}, \qquad k = 0, 1, \ldots, 255.$$

Specify how to obtain a new sequence $y[n]$ from $x[n]$ such that the desired frequency samples can be obtained by applying the available FFT program to $y[n]$ with v *as small as possible*.

9.28. A finite-length signal of length $L = 500$ ($x[n] = 0$ for $n < 0$ and $n > L - 1$) is obtained by sampling a continuous-time signal with sampling rate 10,000 samples per second. We wish to compute samples of the z-transform of $x[n]$ at the N equally spaced points $z_k = (0.8)e^{j2\pi k/N}$, for $0 \le k \le N - 1$, with an effective frequency spacing of 50 Hz or less.

(a) Determine the minimum value for N if $N = 2^v$.

(b) Determine a sequence $y[n]$ of length N, where N is as determined in Part (a), such that its DFT $Y[k]$ is equal to the desired samples of the z-transform of $x[n]$.

9.29. Suppose that a finite-length sequence $x[n]$ has the N-point DFT $X[k]$, and suppose that the sequence satisfies the symmetry condition

$$x[n] = -x[((n + N/2))_N], \qquad 0 \le n \le N - 1,$$

where N is even and $x[n]$ is complex.

(a) Show that $X[k] = 0$ for $k = 0, 2, \ldots, N - 2$.

(b) Show how to compute the odd-indexed DFT values $X[k], k = 1, 3, \ldots, N - 1$ using only one $N/2$-point DFT plus a small amount of extra computation.

9.30. Consider an N-point sequence $x[n]$ with DFT $X[k], k = 0, 1, \ldots, N - 1$. The following algorithm computes the even-indexed DFT values $X[k], k = 0, 2, \ldots, N - 2$, for N even, using only a single $N/2$-point DFT:

1. Form the sequence $y[n]$ by time aliasing, i.e.,

$$y[n] = \begin{cases} x[n] + x[n + N/2], & 0 \le n \le N/2 - 1, \\ 0, & \text{otherwise.} \end{cases}$$

2. Compute $Y[r], r = 0, 1, \ldots, (N/2) - 1$, the $N/2$-point DFT of $y[n]$.
3. Then the even-indexed values of $X[k]$ are $X[k] = Y[k/2]$, for $k = 0, 2, \ldots, N - 2$.

(a) Show that the preceding algorithm produces the desired results.

(b) Now suppose that we form a finite-length sequence $y[n]$ from a sequence $x[n]$ by

$$y[n] = \begin{cases} \displaystyle\sum_{r=-\infty}^{\infty} x[n + rM], & 0 \leq n \leq M - 1, \\ 0, & \text{otherwise.} \end{cases}$$

Determine the relationship between the M-point DFT $Y[k]$ and $X(e^{j\omega})$, the Fourier transform of $x[n]$. Show that the result of Part (a) is a special case of the result of Part (b).

(c) Develop an algorithm similar to the one in Part (a) to compute the odd-indexed DFT values $X[k], k = 1, 3, \ldots, N - 1$, for N even, using only a single $N/2$-point DFT.

9.31. Suppose that an FFT program is available that computes the DFT of a complex sequence. If we wish to compute the DFT of a real sequence, we may simply specify the imaginary part to be zero and use the program directly. However, the symmetry of the DFT of a real sequence can be used to reduce the amount of computation.

(a) Let $x[n]$ be a real-valued sequence of length N, and let $X[k]$ be its DFT with real and imaginary parts denoted $X_R[k]$ and $X_I[k]$, respectively; i.e.,

$$X[k] = X_R[k] + j X_I[k].$$

Show that if $x[n]$ is real, then $X_R[k] = X_R[N - k]$ and $X_I[k] = -X_I[N - k]$ for $k = 1, \ldots, N - 1$.

(b) Now consider two real-valued sequences $x_1[n]$ and $x_2[n]$ with DFTs $X_1[k]$ and $X_2[k]$, respectively. Let $g[n]$ be the complex sequence $g[n] = x_1[n] + j x_2[n]$, with corresponding DFT $G[k] = G_R[k] + j G_I[k]$. Also, let $G_{OR}[k], G_{ER}[k], G_{OI}[k]$ and $G_{EI}[k]$ denote, respectively, the odd part of the real part, the even part of the real part, the odd part of the imaginary part, and the even part of the imaginary part of $G[k]$, as defined through Eqs. (8.102). Specifically, for $1 \leq k \leq N - 1$,

$$G_{OR}[k] = \tfrac{1}{2}\{G_R[k] - G_R[N - k]\},$$

$$G_{ER}[k] = \tfrac{1}{2}\{G_R[k] + G_R[N - k]\},$$

$$G_{OI}[k] = \tfrac{1}{2}\{G_I[k] - G_I[N - k]\},$$

$$G_{EI}[k] = \tfrac{1}{2}\{G_I[k] + G_I[N - k]\},$$

and $G_{OR}[0] = G_{OI}[0] = 0$, $G_{ER}[0] = G_R[0]$, $G_{EI}[0] = G_I[0]$. Determine expressions for $X_1[k]$ and $X_2[k]$ in terms of $G_{OR}[k], G_{ER}[k], G_{OI}[k]$, and $G_{EI}[k]$.

(c) Assume that $N = 2^\nu$ and that a radix-2 FFT program is available to compute the DFT. Determine the number of real multiplications and the number of real additions required to compute both $X_1[k]$ and $X_2[k]$ by (i) using the program twice (with the imaginary part of the input set to zero) to compute the two complex N-point DFTs $X_1[k]$ and $X_2[k]$ separately and (ii) using the scheme suggested in Part (b), which requires only one N-point DFT to be computed.

(d) Assume that we have only one real N-point sequence $x[n]$, where N is a power of 2. Let $x_1[n]$ and $x_2[n]$ be the two real $N/2$-point sequences $x_1[n] = x[2n]$ and $x_2[n] = x[2n+1]$, where $n = 0, 1, \ldots, (N/2) - 1$. Determine $X[k]$ in terms of the $(N/2)$-point DFTs $X_1[k]$ and $X_2[k]$.

(e) Using the results of Parts (b), (c), and (d), describe a procedure for computing the DFT of the real N-point sequence $x[n]$ utilizing only one $N/2$-point FFT computation.

Determine the numbers of real multiplications and real additions required by this procedure, and compare these numbers with the numbers required if the $X[k]$ is computed using one N-point FFT computation with the imaginary part set to zero.

9.32. Let $x[n]$ and $h[n]$ be two real finite-length sequences such that

$$x[n] = 0 \quad \text{for } n \text{ outside the interval } 0 \leq n \leq L - 1,$$

$$h[n] = 0 \quad \text{for } n \text{ outside the interval } 0 \leq n \leq P - 1.$$

We wish to compute the sequence $y[n] = x[n] * h[n]$, where $*$ denotes ordinary convolution.
   (a) What is the length of the sequence $y[n]$?
   (b) For direct evaluation of the convolution sum, how many real multiplications are required to compute all of the nonzero samples of $y[n]$? The following identity may be useful:

$$\sum_{k=1}^{N} k = \frac{N(N+1)}{2}.$$

   (c) State a procedure for using the DFT to compute all of the nonzero samples of $y[n]$. Determine the minimum size of the DFTs and inverse DFTs in terms of L and P.
   (d) Assume that $L = P = N/2$, where $N = 2^v$ is the size of the DFT. Determine a formula for the number of real multiplications required to compute all the nonzero values of $y[n]$ using the method of Part (c) if the DFTs are computed using a radix-2 FFT algorithm. Use this formula to determine the minimum value of N for which the FFT method requires fewer real multiplications than the direct evaluation of the convolution sum.

9.33. In Section 8.7.3, we showed that linear time-invariant filtering can be implemented by sectioning the input signal into finite-length segments and using the DFT to implement circular convolutions on these segments. The two methods discussed were called the overlap–add and the overlap–save methods. If the DFTs are computed using an FFT algorithm, these sectioning methods can require fewer complex multiplications per output sample than the direct evaluation of the convolution sum.
   (a) Assume that the complex input sequence $x[n]$ is of infinite duration and that the complex impulse response $h[n]$ is of length P samples, so that $h[n] \neq 0$ only for $0 \leq n \leq P-1$. Also, assume that the output is computed using the overlap–save method, with the DFTs of length $L = 2^v$, and suppose that these DFTs are computed using a radix-2 FFT algorithm. Determine an expression for the number of complex multiplications required per output sample as a function of v and P.
   (b) Suppose that the length of the impulse response is $P = 500$. By evaluating the formula obtained in Part (a), plot the number of multiplications per output sample as a function of v for the values of $v \leq 20$ such that the overlap–save method applies. For what value of v is the number of multiplications minimal? Compare the number of complex multiplications per output sample for the overlap–save method using the FFT with the number of complex multiplications per output sample required for direct evaluation of the convolution sum.
   (c) Show that for large FFT lengths, the number of complex multiplications per output sample is approximately v. Thus, beyond a certain FFT length, the overlap–save method is less efficient than the direct method. If $P = 500$, for what value of v will the direct method be more efficient?
   (d) Assume that the FFT length is twice the length of the impulse response (i.e., $L = 2P$), and assume that $L = 2^v$. Using the formula obtained in Part (a), determine the smallest value of P such that the overlap–save method using the FFT requires fewer complex multiplications than the direct convolution method.

9.34. $x[n]$ is a 1024-point sequence that is nonzero only for $0 \leq n \leq 1023$. Let $X[k]$ be the 1024-point DFT of $x[n]$. Given $X[k]$, we want to compute $x[n]$ in the ranges $0 \leq n \leq 3$ and $1020 \leq n \leq 1023$ using the system in Figure P9.34-1. Note that the input to the system is the sequence of DFT coefficients. By selecting $m_1[k]$, $m_2[k]$, and $h[k]$, show how the system can be used to compute the desired samples of $x[n]$. Note that the samples $y[n]$ for $0 \leq n \leq 7$ must contain the desired samples of $x[n]$.

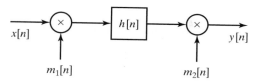

Figure P9.34-1

9.35. Consider a class of DFT-based algorithms for implementing a causal FIR filter with impulse response $h[n]$ that is zero outside the interval $0 \leq n \leq 63$. The input signal (for the FIR filter) $x[n]$ is segmented into an infinite number of possibly overlapping 128-point blocks $x_i[n]$, for i an integer and $-\infty \leq i \leq \infty$, such that

$$x_i[n] = \begin{cases} x[n], & iL \leq n \leq iL + 127, \\ 0, & \text{otherwise}, \end{cases}$$

where L is a positive integer.

Specify a method for computing

$$y_i[n] = x_i[n] * h[n]$$

for any i. Your answer should be in the form of a block diagram utilizing only the types of modules shown in Figures P9.35-1 and P9.35-2. A module may be used more than once or not at all.

The four modules in Figure P9.35-2 either use radix-2 FFTs to compute $X[k]$, the N-point DFT of $x[n]$, or use radix-2 inverse FFTs to compute $x[n]$ from $X[k]$.

Your specification must include the lengths of the FFTs and IFFTs used. For each "shift by n_0" module, you should also specify a value for n_0, the amount by which the input sequence is to be shifted.

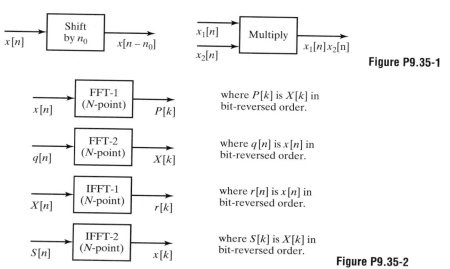

Figure P9.35-1

where $P[k]$ is $X[k]$ in bit-reversed order.

where $q[n]$ is $x[n]$ in bit-reversed order.

where $r[n]$ is $x[n]$ in bit-reversed order.

where $S[k]$ is $X[k]$ in bit-reversed order.

Figure P9.35-2

Extension Problems

9.36. In many applications (such as evaluating frequency responses and interpolation), it is of interest to compute the DFT of a short sequence that is "zero-padded." In such cases, a specialized "pruned" FFT algorithm can be used to increase the efficiency of computation (Markel, 1971). In this problem, we will consider pruning of the radix-2 decimation-in-frequency algorithm when the length of the input sequence is $M \leq 2^\mu$ and the length of the DFT is $N = 2^\nu$, where $\mu < \nu$.

    (a) Draw the complete flow graph of a decimation-in-frequency radix-2 FFT algorithm for $N = 16$. Label all branches appropriately.

    (b) Assume that the input sequence is of length $M = 2$; i.e., $x[n] \neq 0$ only for $N = 0$ and $N = 1$. Draw a new flow graph for $N = 16$ that shows how the nonzero input samples propagate to the output DFT; i.e., eliminate or prune all branches in the flow graph of Part (a) that represent operations on zero-inputs.

    (c) In Part (b), all of the butterflies in the first three stages of computation should have been effectively replaced by a half-butterfly of the form shown in Figure P9.36-1, and in the last stage, all the butterflies should have been of the regular form. For the general case where the length of the input sequence is $M \leq 2^\mu$ and the length of the DFT is $N = 2^\nu$, where $\mu < \nu$, determine the number of stages in which the pruned butterflies can be used. Also, determine the number of complex multiplications required to compute the N-point DFT of an M-point sequence using the pruned FFT algorithm. Express your answers in terms of ν and μ.

Figure P9.36-1

9.37. In Section 9.3, we showed that if N is divisible by 2, an N-point DFT may be expressed as

$$X[k] = G[k] + W_N^k[k]H[k], \qquad 0 \leq k \leq N - 1, \tag{P9.37-1}$$

where $G[k]$ is the $N/2$-point DFT of the sequence of even-indexed samples,

$$g[n] = x[2n], \qquad 0 \leq n \leq (N/2) - 1,$$

and $H[k]$ is the $N/2$-point DFT of the odd-indexed samples,

$$h[n] = x[2n + 1], \qquad 0 \leq n \leq (N/2) - 1.$$

When $N = 2^\nu$, repeated application of this decomposition leads to the decimation-in-time FFT algorithm depicted for $N = 8$ in Figure 9.10. As we have seen, such algorithms require complex multiplications by the "twiddle" factors W_N^k. Rader and Brenner (1976) derived a new algorithm in which the multipliers are purely imaginary, thus requiring only two real multiplications and no real additions. In this algorithm, Eq. (P9.37-1) is replaced by the equations

$$X[0] = G[0] + F[0], \tag{P9.37-2a}$$

$$X[N/2] = G[0] - F[0], \tag{P9.37-2b}$$

$$X[k] = G[k] - \frac{1}{2}j\,\frac{F[k]}{\sin(2\pi k/N)}, \qquad k \neq 0, N/2. \tag{P9.37-2c}$$

Here, $F[k]$ is the $N/2$-point DFT of the sequence

$$f[n] = x[2n+1] - x[2n-1] + Q,$$

where

$$Q = \frac{2}{N} \sum_{n=0}^{(N/2)-1} x[2n+1]$$

is a quantity that need be computed only once.

(a) Show that $F[0] = H[0]$ and therefore that Eqs. (P9.37-2a) and (P9.37-2b) give the same result as Eq. (P9.37-1) for $k = 0, N/2$.

(b) Show that

$$F[k] = H[k]W_N^k(W_N^{-k} - W_N^k)$$

for $k = 1, 2, \ldots, (N/2) - 1$. Use this result to obtain Eq. (P9.37-2c). Why must we compute $X[0]$ and $X[N/2]$ using separate equations?

(c) When $N = 2^\nu$, we can apply Eqs. (P9.37-2a)–(P9.37-2c) repeatedly to obtain a complete decimation-in-time FFT algorithm. Determine formulas for the number of real multiplications and for the number of real additions as a function of N. In counting operations due to Eq. (P9.37-2c), take advantage of any symmetries and periodicities, but do not exclude "trivial" multiplications by $\pm j/2$.

(d) Rader and Brenner (1976) state that FFT algorithms based on Eqs. (P9.37-2a)–(P9.37-2c) have "poor noise properties." Explain why this might be true.

9.38. A modified FFT algorithm called the *split-radix* FFT, or SRFFT, was proposed by Duhamel and Hollman (1984) and Duhamel (1986). The flow graph for the split-radix algorithm is similar to the radix-2 flow graph, but it requires fewer real multiplications. In this problem, we illustrate the principles of the SRFFT for computing the DFT $X[k]$ of a sequence $x[n]$ of length N.

(a) Show that the even-indexed terms of $X[k]$ can be expressed as the $N/2$-point DFT

$$X[2k] = \sum_{n=0}^{(N/2)-1} (x[n] + x[n + N/2])W_N^{2kn}$$

for $k = 0, 1, \ldots, (N/2) - 1$.

(b) Show that the odd-indexed terms of the DFT $X[k]$ can be expressed as the $N/4$-point DFTs

$$X[4k+1]$$

$$= \sum_{n=0}^{(N/4)-1} \{(x[n] - x[n + N/2]) - j(x[n + N/4] - x[n + 3N/4])\}W_N^n W_N^{4kn}$$

for $k = 0, 1, \ldots, (N/4) - 1$, and

$$X[4k+3]$$

$$= \sum_{n=0}^{(N/4)-1} \{(x[n] - x[n + N/2]) + j(x[n + N/4] - x[n + 3N/4])\}W_N^{3n} W_N^{4kn}$$

for $k = 0, 1, \ldots, (N/4) - 1$.

(c) The flow graph in Figure P9.38-1 represents the preceding decomposition of the DFT for a 16-point transform. Redraw this flow graph, labeling each branch with the appropriate multiplier coefficient.

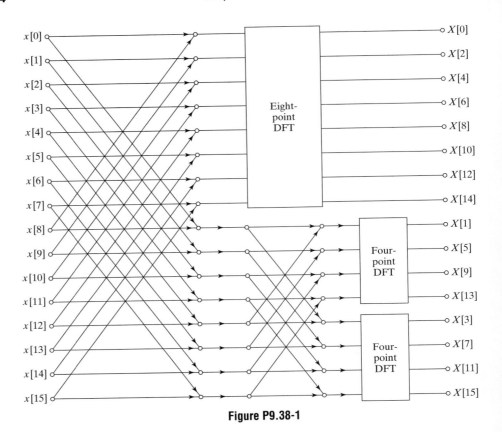

Figure P9.38-1

(d) Determine the number of real multiplications required to implement the 16-point transform when the SRFFT principle is applied to compute the other DFTs in Figure P9.38-1. Compare this number with the number of real multiplications required to implement a 16-point radix-2 decimation-in-frequency algorithm. In both cases, assume that multiplications by W_N^0 are not done.

9.39. In computing the DFT, it is necessary to multiply a complex number by another complex number whose magnitude is unity, i.e., $(X + jY)e^{j\theta}$. Clearly, such a complex multiplication changes only the angle of the complex number, leaving the magnitude unchanged. For this reason, multiplications by a complex number $e^{j\theta}$ are sometimes called *rotations*. In DFT or FFT algorithms, many different angles θ may be needed. However, it may be undesirable to store a table of all required values of $\sin\theta$ and $\cos\theta$, and computing these functions by a power series requires many multiplications and additions. With the CORDIC algorithm given by Volder (1959), the product $(X + jY)e^{j\theta}$ can be evaluated efficiently by a combination of additions, binary shifts, and table lookups from a small table.

(a) Define $\theta_i = \arctan(2^{-i})$. Show that any angle $0 < \theta < \pi/2$ can be represented as

$$\theta = \sum_{i=0}^{M-1} \alpha_i \theta_i + \epsilon = \hat{\theta} + \epsilon,$$

where $\alpha_i = \pm 1$ and the error ϵ is bounded by

$$|\epsilon| \leq \arctan(2^{-M}).$$

(b) The angles θ_i may be computed in advance and stored in a small table of length M. State an algorithm for obtaining the sequence $\{\alpha_i\}$ for $i = 0, 1, \ldots, M - 1$, such that $\alpha_i = \pm 1$. Use your algorithm to determine the sequence $\{\alpha_i\}$ for representing the angle $\theta = 100\pi/512$ when $M = 11$.

(c) Using the result of Part (a), show that the recursion

$$X_0 = X,$$
$$Y_0 = Y,$$
$$X_i = X_{i-1} - \alpha_{i-1}Y_{i-1}2^{-i+1}, \qquad i = 1, 2, \ldots, M,$$
$$Y_i = Y_{i-1} + \alpha_{i-1}X_{i-1}2^{-i+1}, \qquad i = 1, 2, \ldots, M,$$

will produce the complex number

$$(X_M + jY_M) = (X + jY)G_Me^{j\hat\theta},$$

where $\hat\theta = \sum_{i=0}^{M-1} \alpha_i\theta_i$ and G_M is real, is positive, and does not depend on θ. That is, the original complex number is rotated in the complex plane by an angle $\hat\theta$ and magnified by the constant G_M.

(d) Determine the magnification constant G_M as a function of M.

9.40. In Section 9.4, we developed the decimation-in-frequency FFT algorithm for radix 2, i.e., $N = 2^v$. It is possible to formulate a similar algorithm for the general case of $N = m^v$, where m is an integer. Such an algorithm is known as a radix-m FFT algorithm. In this problem, we will examine the radix-3 decimation-in-frequency FFT for the case when $N = 9$, i.e., the input sequence $x[n] = 0$ for $n < 0$ and $n > 8$.

(a) Formulate a method of computing the DFT samples $X[3k]$ for $k = 0, 1, 2$. Consider defining $X_1[k] = X(e^{j\omega_k})|_{\omega_k=2\pi k/3}$. How can you define a time sequence $x_1[n]$ in terms of $x[n]$ such that the 3-point DFT of $x_1[n]$ is $X_1[k] = X[3k]$?

(b) Now define a sequence $x_2[n]$ in terms of $x[n]$ such that the 3-point DFT of $x_2[n]$ is $X_2[k] = X[3k + 1]$ for $k = 0, 1, 2$. Similarly, define $x_3[n]$ such that its 3-point DFT $X_3[k] = X[3k + 2]$ for $k = 0, 1, 2$. Note that we have now defined the 9-point DFT as three 3-point DFTs from appropriately constructed 3-point sequences.

(c) Draw the signal flow graph for the $N = 3$ DFT, i.e., the radix-3 butterfly.

(d) Using the results for Parts (a) and (b), sketch the signal flow graph for the system that constructs the sequences $x_1[n]$, $x_2[n]$, and $x_3[n]$, and then use 3-point DFT boxes on these sequences to produce $X[k]$ for $k = 0, \ldots, 8$. Note that in the interest of clarity, you should not draw the signal flow graph for the $N = 3$ DFTs, but simply use boxes labeled "$N = 3$ DFT." The interior of these boxes is the system you drew for Part (c).

(e) Appropriate factoring of the powers of W_9 in the system you drew in Part (d) allows these systems to be drawn as $N = 3$ DFTs, followed by "twiddle" factors analogous to those in the radix-2 algorithm. Redraw the system in Part (d) such that it consists entirely of $N = 3$ DFTs with "twiddle" factors. This is the complete formulation of the radix-3 decimation-in-frequency FFT for $N = 9$.

(f) How many complex multiplications are required to compute a 9-point DFT using a direct implementation of the DFT equation? Contrast this with the number of complex multiplications required by the system you drew in Part (e). In general, how many complex multiplications are required for the radix-3 FFT of a sequence of length $N = 3^v$?

9.41. Bluestein (1970) showed that if $N = M^2$, then the chirp transform algorithm has a recursive implementation.

(a) Show that the DFT can be expressed as the convolution

$$X[k] = h^*[k] \sum_{n=0}^{N-1} (x[n]h^*[n])h[k-n],$$

where * denotes complex conjugation and

$$h[n] = e^{j(\pi/N)n^2}, \qquad -\infty < n < \infty.$$

(b) Show that the desired values of $X[k]$ (i.e., for $k = 0, 1, \ldots, N-1$) can also be obtained by evaluating the convolution of Part (a) for $k = N, N+1, \ldots, 2N-1$.

(c) Use the result of Part (b) to show that $X[k]$ is also equal to the output of the system shown in Figure P9.41-1 for $k = N, N+1, \ldots, 2N-1$, where $\hat{h}[k]$ is the finite-duration sequence

$$\hat{h}[k] = \begin{cases} e^{j(\pi/N)k^2}, & 0 \le k \le 2N-1, \\ 0, & \text{otherwise.} \end{cases}$$

(d) Using the fact that $N = M^2$, show that the system function corresponding to the impulse response $\hat{h}[k]$ is

$$\hat{H}(z) = \sum_{k=0}^{2N-1} e^{j(\pi/N)k^2} z^{-k}$$

$$= \sum_{r=0}^{M-1} e^{j(\pi/N)r^2} z^{-r} \frac{1 - z^{-2M^2}}{1 + e^{j(2\pi/M)r} z^{-M}}.$$

Hint: Express k as $k = r + \ell M$.

(e) The expression for $\hat{H}(z)$ obtained in Part (d) suggests a recursive realization of the FIR system. Draw the flow graph of such an implementation.

(f) Use the result of Part (e) to determine the total numbers of complex multiplications and additions required to compute all of the N desired values of $X[k]$. Compare those numbers with the numbers required for direct computation of $X[k]$.

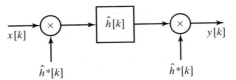

$x[k]$ $\hat{h}^*[k]$ $\hat{h}[k]$ $\hat{h}^*[k]$ $y[k]$ **Figure P9.41-1**

9.42. In the Goertzel algorithm for computation of the discrete Fourier transform, $X[k]$ is computed as

$$X[k] = y_k[N],$$

where $y_k[n]$ is the output of the network shown in Figure P9.42-1. Consider the implementation of the Goertzel algorithm using fixed-point arithmetic with rounding. Assume that

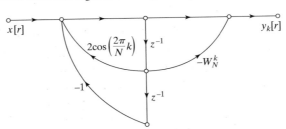

Figure P9.42-1

the register length is B bits plus the sign, and assume that the products are rounded before additions. Also, assume that round-off noise sources are independent.

(a) Assuming that $x[n]$ is real, draw a flow graph of the linear-noise model for the finite-precision computation of the real and imaginary parts of $X[k]$. Assume that multiplication by ± 1 produces no round-off noise.

(b) Compute the variance of the round-off noise in both the real part and the imaginary part of $X[k]$.

9.43. Consider direct computation of the DFT using fixed-point arithmetic with rounding. Assume that the register length is B bits plus the sign (i.e., a total of $B + 1$ bits) and that the round-off noise introduced by any real multiplication is independent of that produced by any other real multiplication. Assuming that $x[n]$ is real, determine the variance of the round-off noise in both the real part and the imaginary part of each DFT value $X[k]$.

9.44. In implementing a decimation-in-time FFT algorithm, the basic butterfly computation is

$$X_m[p] = X_{m-1}[p] + W_N^r X_{m-1}[q],$$

$$X_m[q] = X_{m-1}[p] - W_N^r X_{m-1}[q].$$

In using fixed-point arithmetic to implement the computations, it is commonly assumed that all numbers are scaled to be less than unity. Therefore, to avoid overflow, it is necessary to ensure that the real numbers that result from the butterfly computations do not exceed unity.

(a) Show that if we require

$$|X_{m-1}[p]| < \tfrac{1}{2} \quad \text{and} \quad |X_{m-1}[q]| < \tfrac{1}{2},$$

then overflow cannot occur in the butterfly computation; i.e.,

$$|\mathcal{R}e\{X_m[p]\}| < 1, \quad |\mathcal{J}m\{X_m[p]\}| < 1,$$

and

$$|\mathcal{R}e\{X_m[q]\}| < 1, \quad |\mathcal{J}m\{X_m[q]\}| < 1.$$

(b) In practice, it is easier and most convenient to require

$$|\mathcal{R}e\{X_{m-1}[p]\}| < \tfrac{1}{2}, \quad |\mathcal{J}m\{X_{m-1}[p]\}| < \tfrac{1}{2},$$

and

$$|\mathcal{R}e\{X_{m-1}[q]\}| < \tfrac{1}{2}, \quad |\mathcal{J}m\{X_{m-1}[q]\}| < \tfrac{1}{2}.$$

Are these conditions sufficient to guarantee that overflow cannot occur in the decimation-in-time butterfly computation? Explain.

9.45. In deriving formulas for the noise-to-signal ratio for the fixed-point radix-2 decimation-in-time FFT algorithm, we assumed that each output node was connected to $(N - 1)$ butterfly computations, each of which contributed an amount $\sigma_B^2 = \tfrac{1}{3} \cdot 2^{-2B}$ to the output noise variance. However, when $W_N^r = \pm 1$ or $\pm j$, the multiplications can in fact be done without error. Thus, if the results derived in Section 9.7 are modified to account for this fact, we obtain a less pessimistic estimate of quantization noise effects.

(a) For the decimation-in-time algorithm discussed in Section 9.7, determine, for each stage, the number of butterflies that involve multiplication by either ± 1 or $\pm j$.

(b) Use the result of Part (a) to find improved estimates of the output noise variance, Eq. (9.55), and noise-to-signal ratio, Eq. (9.65), for odd values of k. Discuss how these estimates are different for even values of k. Do not attempt to find a closed form expression of these quantities for even values of k.

(c) Repeat Parts (a) and (b) for the case where the output of each stage is attenuated by a factor of $\tfrac{1}{2}$; i.e., derive modified expressions corresponding to Eq. (9.68) for the output noise variance and Eq. (9.69) for the output noise-to-signal ratio, assuming that multiplications by ± 1 and $\pm j$ do not introduce error.

9.46. In Section 9.7 we considered a noise analysis of the decimation-in-time FFT algorithm of Figure 9.10. Carry out a similar analysis for the decimation-in-frequency algorithm of Figure 9.20, obtaining equations for the output noise variance and noise-to-signal ratio for scaling at the input and also for scaling by $\frac{1}{2}$ at each stage of computation.

9.47. In this problem, we consider a procedure for computing the DFT of four real symmetric or antisymmetric N-point sequences using only one N-point DFT computation. Since we are considering only finite-length sequences, by *symmetric* and *antisymmetric*, we explicitly mean *periodic symmetric* and *periodic antisymmetric*, as defined in Section 8.6.4. Let $x_1[n]$, $x_2[n]$, $x_3[n]$, and $x_4[n]$ denote the four real sequences of length N, and let $X_1[k]$, $X_2[k]$, $X_3[k]$, and $X_4[k]$ denote the corresponding DFTs. We assume first that $x_1[n]$ and $x_2[n]$ are symmetric and $x_3[n]$ and $x_4[n]$ are antisymmetric; i.e.,

$$x_1[n] = x_1[N-n], \qquad x_2[n] = x_2[N-n],$$

$$x_3[n] = -x_3[N-n], \qquad x_4[n] = -x_4[N-n],$$

for $n = 1, 2, \ldots, N-1$ and $x_3[0] = x_4[0] = 0$.

(a) Define $y_1[n] = x_1[n] + x_3[n]$ and let $Y_1[k]$ denote the DFT of $y_1[n]$. Determine how $X_1[k]$ and $X_2[k]$ can be recovered from $Y_1[k]$.

(b) $y_1[n]$ as defined in Part (a) is a real sequence with symmetric part $x_1[n]$ and antisymmetric part $x_3[n]$. Similarly, we define the real sequence $y_2[n] = x_2[n] + x_4[n]$, and we let $y_3[n]$ be the complex sequence

$$y_3[n] = y_1[n] + jy_2[n].$$

First determine how $Y_1[k]$ and $Y_2[k]$ can be determined from $Y_3[k]$, and then, using the results of Part (a), show how to obtain $X_1[k]$, $X_2[k]$, $X_3[k]$, and $X_4[k]$ from $Y_3[k]$.

The result of Part (b) shows that we can compute the DFT of four real sequences simultaneously with only one N-point DFT computation if two sequences are symmetric and the other two are antisymmetric. Now consider the case when all four are symmetric; i.e.,

$$x_i[n] = x_i[N-n], \qquad i = 1, 2, 3, 4,$$

for $n = 0, 1, \ldots, N-1$.

(c) Consider a real symmetric sequence $x_3[n]$. Show that the sequence

$$u_3[n] = x_3[((n+1))_N] - x_3[((n-1))_N]$$

is an antisymmetric sequence; i.e., $u_3[n] = -u_3[N-n]$ for $n = 1, 2, \ldots, N-1$ and $u_3[0] = 0$.

(d) Let $U_3[k]$ denote the N-point DFT of $u_3[n]$. Determine an expression for $U_3[k]$ in terms of $X_3[k]$.

(e) By using the procedure of Part (c), we can form the real sequence $y_1[n] = x_1[n] + u_3[n]$, where $x_1[n]$ is the symmetric part and $u_3[n]$ is the antisymmetric part of $y_1[n]$. Determine how $X_1[k]$ and $X_3[k]$ can be recovered from $Y_1[k]$.

(f) Now let $y_3[n] = y_1[n] + jy_2[n]$, where

$$y_1[n] = x_1[n] + u_3[n], \qquad y_2[n] = x_2[n] + u_4[n],$$

with

$$u_3[n] = x_3[((n+1))_N] - x_3[((n-1))_N],$$

$$u_4[n] = x_4[((n+1))_N] - x_4[((n-1))_N],$$

for $n = 0, 1, \ldots, N - 1$. Determine how to obtain $X_1[k]$, $X_2[k]$, $X_3[k]$, and $X_4[k]$ from $Y_3[k]$. (Note that $X_3[0]$ and $X_4[0]$ cannot be recovered from $Y_3[k]$, and if N is even, $X_3[N/2]$ and $X_4[N/2]$ also cannot be recovered from $Y_3[k]$.)

9.48. The input and output of a linear time-invariant system satisfy a difference equation of the form

$$y[n] = \sum_{k=1}^{N} a_k y[n-k] + \sum_{k=0}^{M} b_k x[n-k].$$

Assume that an FFT program is available for computing the DFT of any finite-length sequence of length $N = 2^\nu$. Describe a procedure that utilizes the available FFT program to compute

$$H(e^{j(2\pi/512)k}) \qquad \text{for } k = 0, 1, \ldots, 511,$$

where $H(z)$ is the system function of the system.

9.49. Suppose that we wish to multiply two very large numbers (possibly thousands of bits long) on a 16-bit computer. In this problem, we will investigate a technique for doing this using FFTs.

(a) Let $p(x)$ and $q(x)$ be the two polynomials

$$p(x) = \sum_{i=0}^{L-1} a_i x^i, \qquad q(x) = \sum_{i=0}^{M-1} b_i x^i.$$

Show that the coefficients of the polynomial $r(x) = p(x)q(x)$ can be computed using circular convolution.

(b) Show how to compute the coefficients of $r(x)$ using a radix-2 FFT program. For what orders of magnitude of $(L + M)$ is this procedure more efficient than direct computation? Assume that $L + M = 2^\nu$ for some integer ν.

(c) Now suppose that we wish to compute the product of two very long positive binary integers u and v. Show that their product can be computed using polynomial multiplication, and describe an algorithm for computing the product using an FFT algorithm. If u is an 8000-bit number and v is a 1000-bit number, approximately how many real multiplications and additions are required to compute the product $u \cdot v$ using this method?

(d) Give a qualitative discussion of the effect of finite-precision arithmetic in implementing the algorithm of Part (c).

9.50. The discrete Hartley transform (DHT) of a sequence $x[n]$ of length N is defined as

$$X_H[k] = \sum_{n=0}^{N-1} x[n] H_N[nk], \qquad k = 0, 1, \ldots, N - 1,$$

where

$$H_N[a] = C_N[a] + S_N[a],$$

with

$$C_N[a] = \cos(2\pi a/N), \qquad S_N[a] = \sin(2\pi a/N).$$

Problem 8.65 explores the properties of the discrete Hartley transform in detail, particularly its circular convolution property.

(a) Verify that $H_N[a] = H_N[a + N]$, and verify the following useful property of $H_N[a]$:

$$H_N[a + b] = H_N[a]C_N[b] + H_N[-a]S_N[b]$$
$$= H_N[b]C_N[a] + H_N[-b]S_N[a].$$

(b) By decomposing $x[n]$ into its even-numbered points and odd-numbered points, and by using the identity derived in Part (a), derive a fast DHT algorithm based on the decimation-in-time principle.

9.51. In this problem, we will write the FFT as a sequence of matrix operations. Consider the 8-point decimation-in-time FFT algorithm shown in Figure P9.51-1. Let **a** and **f** denote the input and output vectors, respectively. Assume that the input is in bit-reversed order and that the output is in normal order (compare with Figure 9.10). Let **b**, **c**, **d**, and **e** denote the intermediate vectors shown on the flow graph.

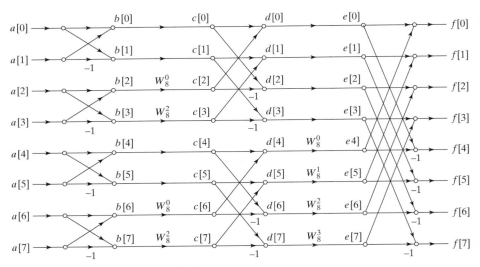

Figure P9.51-1

(a) Determine the matrices $\mathbf{F}_1$, $\mathbf{T}_1$, $\mathbf{F}_2$, $\mathbf{T}_2$, and $\mathbf{F}_3$ such that

$$\mathbf{b} = \mathbf{F}_1 \mathbf{a},$$

$$\mathbf{c} = \mathbf{T}_1 \mathbf{b},$$

$$\mathbf{d} = \mathbf{F}_2 \mathbf{c},$$

$$\mathbf{e} = \mathbf{T}_2 \mathbf{d},$$

$$\mathbf{f} = \mathbf{F}_3 \mathbf{e}.$$

(b) The overall FFT, taking input **a** and yielding output **f** can be described in matrix notation as $\mathbf{f} = \mathbf{Q}\mathbf{a}$, where

$$\mathbf{Q} = \mathbf{F}_3 \mathbf{T}_2 \mathbf{F}_2 \mathbf{T}_1 \mathbf{F}_1.$$

Let $\mathbf{Q}^H$ be the complex (Hermitian) transpose of the matrix **Q**. Draw the flow graph for the sequence of operations described by $\mathbf{Q}^H$. What does this structure compute?

(c) Determine $(1/N)\mathbf{Q}^H\mathbf{Q}$.

9.52. In many applications, there is a need to convolve long sequences $x[n]$ and $h[n]$ whose samples are integers. Since the sequences have integer coefficients, the result of the convolution $y[n] = x[n] * h[n]$ will naturally also have integer coefficients as well.

A major drawback of computing the convolution of integer sequences with FFTs is that floating-point arithmetic chips are more expensive than integer arithmetic chips. Also, rounding noise introduced during a floating-point computation may corrupt the result. In this problem, we consider a class of FFT algorithms known as *number-theoretic transforms* (NTTs), which overcome these drawbacks.

(a) Let $x[n]$ and $h[n]$ be N-point sequences and denote their DFTs by $X[k]$ and $H[k]$, respectively. Derive the circular convolution property of the DFT. Specifically, show that $Y[k] = X[k]H[k]$, where $y[n]$ is the N-point circular convolution of $x[n]$ and $h[n]$. Show that the circular convolution property holds as long as W_N in the DFT satisfies

$$\sum_{n=0}^{N-1} W_N^{nk} = \begin{cases} N, & k = 0, \\ 0, & k \neq 0. \end{cases} \qquad \text{(P9.52-1)}$$

The key to defining NTTs is to find an integer-valued W_N that satisfies Eq. (P9.52-1). This will enforce the orthogonality of the basis vectors required for the DFT to function properly. Unfortunately, no integer-valued W_N exists that has this property for standard integer arithmetic.

In order to overcome this problem, NTTs use integer arithmetic defined modulo some integer P. Throughout the current problem, we will assume that $P = 17$. That is, addition and multiplication are defined as standard integer addition and multiplication, followed by modulo $P = 17$ reduction. For example, $((23+18))_{17} = 7$, $((10+7))_{17} = 0$, $((23 \times 18))_{17} = 6$, and $((10 \times 7))_{17} = 2$. (Just compute the sum or product in the normal way, and then take the remainder modulo 17.)

(b) Let $P = 17$, $N = 4$, and $W_N = 4$. Verify that

$$\left(\left(\sum_{n=0}^{N-1} W_N^{nk} \right) \right)_P = \begin{cases} N, & k = 0, \\ 0, & k \neq 0. \end{cases}$$

(c) Let $x[n]$ and $h[n]$ be the sequences

$$x[n] = \delta[n] + 2\delta[n-1] + 3\delta[n-2],$$

$$h[n] = 3\delta[n] + \delta[n-1].$$

Compute the 4-point NTT $X[k]$ of $x[n]$ as follows:

$$X[k] = \left(\left(\sum_{n=0}^{N-1} x[n] W_N^{nk} \right) \right)_P.$$

Compute $H[k]$ in a similar fashion. Also, compute $Y[k] = H[k]X[k]$. Assume the values of P, N, and W_N given in Part (a). *Be sure to use modulo 17 arithmetic for each operation throughout the computation, not just for the final result!*

(d) The inverse NTT of $Y[k]$ is defined by the equation

$$y[n] = \left(\left((N)^{-1} \sum_{k=0}^{N-1} Y[k] W_N^{-nk} \right) \right)_P. \qquad \text{(P9.52-2)}$$

In order to compute this quantity properly, we must determine the *integers* $(1/N)^{-1}$ and W_N^{-1} such that

$$\left(\left((N)^{-1} N \right) \right)_P = 1,$$

$$\left(\left(W_N W_N^{-1} \right) \right)_P = 1.$$

Use the values of P, N, and W_N given in Part (a), and determine the aforesaid integers.

(e) Compute the inverse NTT shown in Eq. (P9.52-2) using the values of $(N)^{-1}$ and W_N^{-1} determined in Part (d). Check your result by manually computing the convolution $y[n] = x[n] * h[n]$.

9.53. Sections 9.3 and 9.4 focus on the fast Fourier transform for sequences where N is a power of 2. However, it is also possible to find efficient algorithms to compute the DFT when the length N has more than one prime factor, i.e., cannot be expressed as $N = m^v$ for some integer m. In this problem, you will examine the case where $N = 6$. The techniques described extend easily to other composite numbers. Burrus and Parks (1985) discuss such algorithms in more detail.

(a) The key to decomposing the FFT for $N = 6$ is to use the concept of an *index map*, proposed by Cooley and Tukey (1965) in their original paper on the FFT. Specifically, for the case of $N = 6$, we will represent the indices n and k as

$$n = 3n_1 + n_2 \quad \text{for } n_1 = 0, 1; n_2 = 0, 1, 2; \qquad \text{(P9.53-1)}$$

$$k = k_1 + 2k_2 \quad \text{for } k_1 = 0, 1; k_2 = 0, 1, 2; \qquad \text{(P9.53-2)}$$

Verify that using each possible value of n_1 and n_2 produces each value of $n = 0, \ldots, 5$ once and only once. Demonstrate that the same holds for k with each choice of k_1 and k_2.

(b) Substitute Eqs. (P9.53-1) and (P9.53-2) into the definition of the DFT to get a new expression for the DFT in terms of n_1, n_2, k_1, and k_2. The resulting equation should have a double summation over n_1 and n_2 instead of a single summation over n.

(c) Examine the W_6 terms in your equation carefully. You can rewrite some of these as equivalent expressions in W_2 and W_3.

(d) Based on Part (c), group the terms in your DFT such that the n_2 summation is outside and the n_1 summation is inside. You should be able to write this expression so that it can be interpreted as three DFTs with $N = 2$, followed by some "twiddle" factors (powers of W_6), followed by two $N = 3$ DFTs.

(e) Draw the signal flow graph implementing your expression from Part (d). How many complex multiplications does this require? How does this compare with the number of complex multiplications required by a direct implementations of the DFT equation for $N = 6$.

(f) Find an alternative indexing similar to Eqs. (P9.53-1) and (P9.53-2) that results in a signal flow graph that is two $N = 3$ DFTs followed by three $N = 2$ DFTs.

10

FOURIER ANALYSIS OF SIGNALS USING THE DISCRETE FOURIER TRANSFORM

10.0 INTRODUCTION

In Chapter 8, we developed the discrete Fourier transform (DFT) as a Fourier representation of finite-length signals. Because the DFT can be explicitly computed by efficient algorithms discussed in Chapter 9, it plays a central role in a wide variety of signal-processing applications, including filtering and spectral analysis. In this chapter, we take an introductory look at Fourier analysis of signals using the DFT.

In applications and algorithms based on explicit evaluation of the Fourier transform, it is ideally the discrete-time Fourier transform that is desired, while it is the DFT that can actually be computed. For finite-length signals, the DFT provides frequency-domain samples of the discrete-time Fourier transform, and the implications of this sampling must be clearly understood and accounted for. For example, as considered in Section 8.7, in linear filtering or convolution implemented by multiplying DFTs rather than discrete-time Fourier transforms, a circular convolution is implemented and special care must be taken to ensure that the results will be equivalent to a linear convolution. In addition, in many filtering and spectral analysis applications, the signals do not inherently have finite length. As we will discuss, this inconsistency between the finite-length requirement of the DFT and the reality of indefinitely long signals can be accommodated exactly or approximately through the concepts of *windowing, block processing,* and the *time-dependent Fourier transform.*

10.1 FOURIER ANALYSIS OF SIGNALS USING THE DFT

One of the major applications of the DFT is in analyzing the frequency content of continuous-time signals. For example, as we describe in Section 10.5.1, in speech analysis and processing, frequency analysis of speech signals is particularly useful in identifying and modeling the resonances of the vocal cavity. Another example, introduced in Section 10.5.2, is Doppler radar systems, in which the velocity of a target is represented by the frequency shift between the transmitted and received signals.

The basic steps in applying the DFT to continuous-time signals are indicated in Figure 10.1. The antialiasing filter is incorporated to eliminate or minimize the effect of aliasing when the continuous-time signal is converted to a sequence. The need for multiplication of $x[n]$ by $w[n]$, i.e., windowing, is a consequence of the finite-length requirement of the DFT. In many cases of practical interest, $s_c(t)$ and, consequently, $x[n]$ are very long or even indefinitely long (such as with speech or music). Therefore, a finite-duration window $w[n]$ is applied to $x[n]$ prior to computation of the DFT. Figure 10.2 illustrates the Fourier transforms of the signals in Figure 10.1. Figure 10.2(a) shows a continuous-time spectrum that tapers off at high frequencies but is not band-limited. It also indicates the presence of some narrowband signal energy, represented by the narrow peaks. The frequency response of an antialiasing filter is illustrated in Figure 10.2(b). As indicated in Figure 10.2(c), the resulting continuous-time Fourier transform $X_c(j\Omega)$ contains little useful information about $S_c(j\Omega)$ for frequencies above the cutoff frequency of the filter. Since $H_{aa}(j\Omega)$ cannot be ideal, the Fourier components of the input in the passband and the transition band also will be modified by the frequency response of the filter.

The conversion of $x_c(t)$ to the sequence of samples $x[n]$ is represented in the frequency domain by periodic replication and frequency normalization, i.e.,

$$X(e^{j\omega}) = \frac{1}{T} \sum_{r=-\infty}^{\infty} X_c\left(j\frac{\omega}{T} + j\frac{2\pi r}{T}\right), \tag{10.1}$$

as illustrated in Figure 10.2(d). Since, in a practical implementation, the antialiasing filter cannot have infinite attenuation in the stopband, some nonzero overlap of the terms in Eq. (10.1), i.e., aliasing, can be expected; however, this source of error can be made negligibly small either with a high-quality continuous-time filter or through the use of initial oversampling followed by more effective lowpass filtering and decimation, as discussed in Section 4.8.1. If $x[n]$ is a digital signal, so that A/D conversion is incorporated in the second system in Figure 10.1, then quantization error is also introduced.

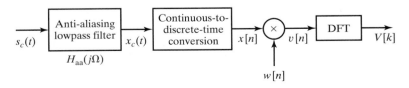

Figure 10.1 Processing steps in the discrete-time Fourier analysis of a continuous-time signal.

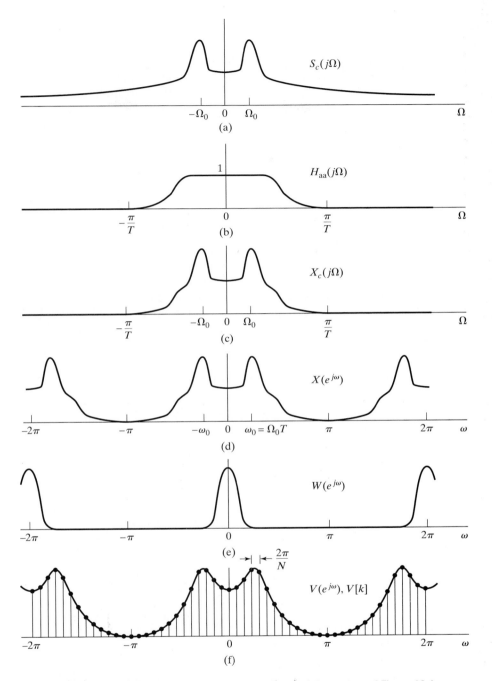

Figure 10.2 Illustration of the Fourier transforms of the system of Figure 10.1. (a) Fourier transform of continuous-time input signal. (b) Frequency response of antialiasing filter. (c) Fourier transform of output of antialiasing filter. (d) Fourier transform of sampled signal. (e) Fourier transform of window sequence. (f) Fourier transform of windowed signal segment and frequency samples obtained using DFT samples.

As we have seen in Section 4.8.2, this error can be modeled as a noise sequence added to $x[n]$. The noise can be made negligible through the use of fine-grained quantization.

As indicated, the sequence $x[n]$ is typically multiplied by a finite-duration window $w[n]$, since the input to the DFT must be of finite duration. This produces the finite-length sequence $v[n] = w[n]x[n]$. The effect in the frequency domain is a periodic convolution, i.e.,

$$V(e^{j\omega}) = \frac{1}{2\pi} \int_{-\pi}^{\pi} X(e^{j\theta})W(e^{j(\omega-\theta)})d\theta. \tag{10.2}$$

Figure 10.2(e) illustrates the Fourier transform of a typical window sequence. Note that the main lobe is assumed to be concentrated around $\omega = 0$. If $w[n]$ is constant over the range of n for which it is nonzero, it is referred to as a *rectangular window*. However, as we will see, there are good reasons to taper the window at its edges. The properties of windows such as the Bartlett, Hamming, Hanning, Blackman, and Kaiser windows are discussed in Chapter 7 and in Section 10.2. At this point, it is sufficient to observe that convolution of $W(e^{j\omega})$ with $X(e^{j\omega})$ will tend to smooth sharp peaks and discontinuities in $X(e^{j\omega})$. This is depicted by the continuous curve plotted in Figure 10.2(f).

The final operation in Figure 10.1 is determining the DFT. The DFT of the windowed sequence $v[n] = w[n]x[n]$ is

$$V[k] = \sum_{n=0}^{N-1} v[n]e^{-j(2\pi/N)kn}, \qquad k = 0, 1, \ldots, N-1, \tag{10.3}$$

where we assume that the window length L is less than or equal to the DFT length N. $V[k]$, the DFT of the finite-length sequence $v[n]$, corresponds to equally spaced samples of the Fourier transform of $v[n]$; i.e.,

$$V[k] = V(e^{j\omega})\big|_{\omega=2\pi k/N}. \tag{10.4}$$

Figure 10.2(f) also shows $V[k]$ as the samples of $V(e^{j\omega})$. Since the spacing between DFT frequencies is $2\pi/N$, and the relationship between the normalized discrete-time frequency variable and the continuous-time frequency variable is $\omega = \Omega T$, the DFT frequencies correspond to the continuous-time frequencies

$$\Omega_k = \frac{2\pi k}{NT}. \tag{10.5}$$

Many commercial real-time spectrum analyzers are based on the principles embodied in Figures 10.1 and 10.2. It should be clear from the preceding discussion, however, that numerous factors affect the interpretation of the continuous-time Fourier transform of the input in terms of the DFT of a windowed segment of the sampled signal. To accommodate these factors, care must be taken in filtering and sampling the input signal. Furthermore, to interpret the results correctly, the effects of the time-domain windowing and of the frequency-domain sampling inherent in the DFT must be clearly understood. For the remainder of the discussion, we will assume that the issues of antialiasing filtering and continuous-to-discrete-time conversion have been satisfactorily handled and are negligible. In the next section we concentrate specifically on the effects of windowing and of the frequency-domain sampling imposed by the

DFT. We choose sinusoidal signals as the specific class of examples to discuss, but most of the issues raised apply more generally.

Example 10.1   Fourier Analysis Using the DFT

Consider a bandlimited continous-time signal $x_c(t)$ such that $X_c(j\Omega) = 0$ for $|\Omega| \geq 2\pi(2500)$. We wish to use the system of Figure 10.1 to estimate the continous-time spectrum $X_c(j\Omega)$. Assume that the antialiasing filter $H_{aa}(j\Omega)$ is ideal and the sampling rate for the C/D converter is $1/T = 5000$ samples/sec. If we want the DFT samples $V[k]$ to be equivalent to samples of $X_c(j\Omega)$ that are at most $2\pi(10)$ rad/sec or 10 Hz apart, what is the minimum value that we should use for the DFT size N?

From (Eq. 10.5), we see that adjacent samples in the DFT correspond to continuous-time frequencies separated by $2\pi/(NT)$. Therefore, we require that

$$\frac{2\pi}{NT} \leq 20\pi,$$

which implies that

$$N \geq 500$$

satisfies the condition. If we wish to use a radix-2 FFT algorithm to compute the DFT in Figure 10.1, we would choose $N = 512$ for an equivalent continuous-time frequency spacing of $\Delta\Omega = 2\pi(5000/512) = 2\pi(9.77)$ rad/sec.

Example 10.2   Relationship Between DFT Values

Consider the problem posed in Example 10.1, where $1/T = 5000$, $N = 512$, and $x_c(t)$ is real-valued and is sufficiently bandlimited to avoid aliasing with the given sampling rate. If it is determined that $V[11] = 2000(1 + j)$, what can be said about other values of $V[k]$ or about $X_c(j\Omega)$?

Referring to the symmetry properties of the DFT given in Table 8.2, $V[k] = V^*[((-k))_N]$, $k = 0, 1, \ldots, N - 1$ and consequently $V[N - k] = V^*[k]$, so it follows in this case that

$$V[512 - 11] = V[501] = V^*[11] = 2000(1 - j).$$

We also know that the DFT sample $k = 11$ corresponds to the continuous-time frequency $\Omega_{11} = 2\pi(11)(5000)/512 = 2\pi(107.4)$, and similarly, $k = 501$ corresponds to the frequency $-2\pi(11)(5000)/512 = -2\pi(107.4)$. Although windowing will smooth the spectrum, we can say that

$$X_c(j\Omega_{11}) = X_c(j2\pi(107.4)) \approx T \cdot V[11] = 0.4(1 + j).$$

Note that the factor T is required to compensate for the factor $1/T$ introduced by sampling, as in Eq. (10.1). We can again exploit symmetry to conclude that

$$X_c(-j\Omega_{11}) = X_c(-j2\pi(107.4)) \approx T \cdot V^*[11] = 0.4(1 - j).$$

10.2 DFT ANALYSIS OF SINUSOIDAL SIGNALS

The discrete-time Fourier transform of a sinusoidal signal $A\cos(\omega_0 n + \phi)$ is a pair of impulses at $+\omega_0$ and $-\omega_0$ (repeating periodically with period 2π). In analyzing sinusoidal signals using the DFT, windowing and spectral sampling have an important effect. As we will see in Section 10.2.1, windowing smears or broadens the impulses in the theoretical Fourier representation, and thus, the exact frequency is less sharply defined. Windowing also reduces the ability to resolve sinusoidal signals that are closely spaced in frequency. The spectral sampling inherent in the DFT has the effect of potentially giving a misleading or inaccurate picture of the true spectrum of the sinusoidal signal. This effect is discussed in Section 10.2.2.

10.2.1 The Effect of Windowing

Let us consider a continuous-time signal consisting of the sum of two sinusoidal components; i.e.,

$$s_c(t) = A_0 \cos(\Omega_0 t + \theta_0) + A_1 \cos(\Omega_1 t + \theta_1), \qquad -\infty < t < \infty. \quad (10.6)$$

Assuming ideal sampling with no aliasing and no quantization error, we obtain the discrete-time signal

$$x[n] = A_0 \cos(\omega_0 n + \theta_0) + A_1 \cos(\omega_1 n + \theta_1), \qquad -\infty < n < \infty, \quad (10.7)$$

where $\omega_0 = \Omega_0 T$ and $\omega_1 = \Omega_1 T$. The windowed sequence $v[n]$ in Figure 10.1 is th

$$v[n] = A_0 w[n] \cos(\omega_0 n + \theta_0) + A_1 w[n] \cos(\omega_1 n + \theta_1). \quad (10.8)$$

To obtain the discrete-time Fourier transform of $v[n]$, we can expand Eq. (10.8) in terms of complex exponentials and utilize the frequency-shifting property of Eq. (2.163) in Section 2.9.2. Specifically, we rewrite $v[n]$ as

$$v[n] = \frac{A_0}{2} w[n] e^{j\theta_0} e^{j\omega_0 n} + \frac{A_0}{2} w[n] e^{-j\theta_0} e^{-j\omega_0 n}$$
$$+ \frac{A_1}{2} w[n] e^{j\theta_1} e^{j\omega_1 n} + \frac{A_1}{2} w[n] e^{-j\theta_1} e^{-j\omega_1 n}, \quad (10.9)$$

from which, with Eq. (2.163), it follows that the Fourier transform of the windowed sequence is

$$V(e^{j\omega}) = \frac{A_0}{2} e^{j\theta_0} W(e^{j(\omega - \omega_0)}) + \frac{A_0}{2} e^{-j\theta_0} W(e^{j(\omega + \omega_0)})$$
$$+ \frac{A_1}{2} e^{j\theta_1} W(e^{j(\omega - \omega_1)}) + \frac{A_1}{2} e^{-j\theta_1} W(e^{j(\omega + \omega_1)}). \quad (10.10)$$

According to Eq. (10.10), the Fourier transform of the windowed signal consists of the Fourier transform of the window, replicated at the frequencies $\pm\omega_0$ and $\pm\omega_1$ and scaled by the complex amplitudes of the individual complex exponentials that make up the signal.

Example 10.3   Effect of Windowing on Fourier Analysis of Sinusoidal Signals

We consider the system of Figure 10.1 and, in particular, $W(e^{j\omega})$ and $V(e^{j\omega})$ for the specific case of a sampling rate $1/T = 10$ kHz and a rectangular window $w[n]$ of length 64. The signal amplitude and phase parameters are $A_0 = 1$ and $A_1 = 0.75$ and

$\theta_0 = \theta_1 = 0$. To illustrate the essential features, we specifically display only the magnitudes of the Fourier transforms.

In Figure 10.3(a) we show $|W(e^{j\omega})|$ and in Figures 10.3(b), (c), (d), and (e) we show $|V(e^{j\omega})|$ for several choices of Ω_0 and Ω_1 in Eq. (10.6) or, equivalently, ω_0 and ω_1 in Eq. (10.7). In Figure 10.3(b), $\Omega_0 = (2\pi/6) \times 10^4$ and $\Omega_1 = (2\pi/3) \times 10^4$, or, equivalently, $\omega_0 = 2\pi/6$ and $\omega_1 = 2\pi/3$. In Figure 10.3(c)–(e), the frequencies become progressively closer. For the parameters in Figure 10.3(b), the frequency and amplitude of the individual components are evident. Specifically, Eq. (10.10) suggests that, with no overlap between the replicas of $W(e^{j\omega})$ at ω_0 and ω_1, there will be a peak of height $32A_0$ at ω_0 and $32A_1$ at ω_1, since $W(e^{j\omega})$ has a peak height of 64. In Figure 10.3(b), the two peaks are at approximately $\omega_0 = 2\pi/6$ and $\omega_1 = 2\pi/3$, with peak amplitudes

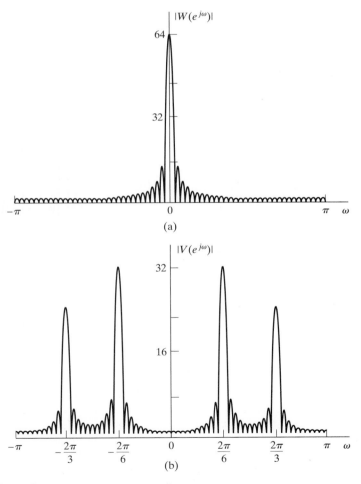

Figure 10.3 Illustration of Fourier analysis of windowed cosines with a rectangular window. (a) Fourier transform of window. (b)–(e) Fourier transform of windowed cosines as $\Omega_1 - \Omega_0$ becomes progressively smaller. (b) $\Omega_0 = (2\pi/6) \times 10^4$, $\Omega_1 = (2\pi/3) \times 10^4$.

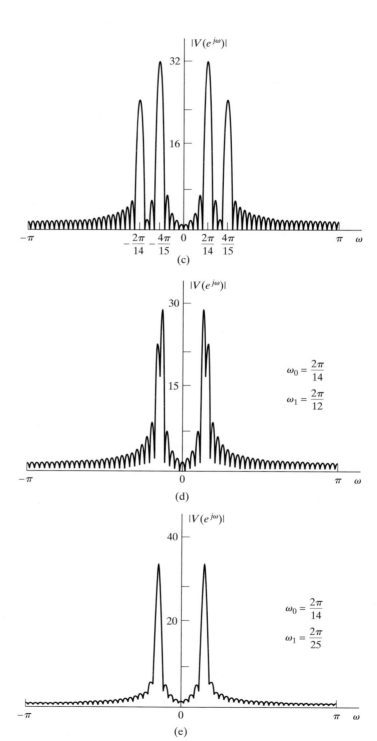

Figure 10.3 (*continued*) (c) $\Omega_0 = (2\pi/14) \times 10^4$, $\Omega_1 = (4\pi/15) \times 10^4$. (d) $\Omega_0 = (2\pi/14) \times 10^4$, $\Omega_1 = (2\pi/12) \times 10^4$. (e) $\Omega_0 = (2\pi/14) \times 10^4$, $\Omega_1 = (4\pi/25) \times 10^4$.

in the correct ratio. In Figure 10.3(c), there is more overlap between the window replicas at ω_0 and ω_1, and while two distinct peaks are present, the amplitude of the spectrum at $\omega = \omega_0$ is affected by the amplitude of the sinusoidal signal at frequency ω_1 and vice versa. This interaction is called *leakage:* The component at one frequency leaks into the vicinity of another component due to the spectral smearing introduced by the window. Figure 10.3(d) shows the case where the leakage is even greater. Notice how side lobes adding out of phase can *reduce* the heights of the peaks. In Figure 10.3(e), the overlap between the spectral windows at ω_0 and ω_1 is so significant that the two peaks visible in (b)–(d) have merged into one. In other words, with this window, the two frequencies corresponding to Figure 10.3(e) will not be *resolved* in the spectrum.

Reduced resolution and leakage are the two primary effects on the spectrum as a result of applying a window to the signal. The resolution is influenced primarily by the width of the main lobe of $W(e^{j\omega})$, while the degree of leakage depends on the relative amplitude of the main lobe and the side lobes of $W(e^{j\omega})$. In Chapter 7, in a filter design context, we showed that the width of the main lobe and the relative side-lobe amplitude depend primarily on the window length L and the shape (amount of tapering) of the window. The rectangular window, which has Fourier transform

$$W_r(e^{j\omega}) = \sum_{n=0}^{L-1} e^{-j\omega n} = e^{-j\omega(L-1)/2} \frac{\sin(\omega L/2)}{\sin(\omega/2)}, \tag{10.11}$$

has the narrowest main lobe for a given length, but it has the largest side lobes of all the commonly used windows. As defined in Chapter 7, the Kaiser window is

$$w_K[n] = \begin{cases} \dfrac{I_0[\beta(1 - [(n-\alpha)/\alpha]^2)^{1/2}]}{I_0(\beta)}, & 0 \leq n \leq L-1, \\ 0, & \text{otherwise,} \end{cases} \tag{10.12}$$

where $\alpha = (L-1)/2$ and $I_0(\cdot)$ is the zeroth-order modified Bessel function of the first kind. (Note that the notation of Eq. (10.12) differs slightly from that of Eq. (7.59) in that L denotes the length of the window in Eq. (10.12) while the length of the filter design window in Eq. (7.59) is denoted $M+1$.) We have already seen in the context of the filter design problem that this window has two parameters, β and L, which can be used to trade between main-lobe width and relative side-lobe amplitude. (Recall that the Kaiser window reduces to the rectangular window when $\beta = 0$.) The main-lobe width Δ_{ml} is defined as the symmetric distance between the central zero-crossings. The relative side-lobe level A_{sl} is defined as the ratio in dB of the amplitude of the main lobe to the amplitude of the largest side lobe. Figure 10.4, which is a duplicate of Figure 7.24, shows Fourier transforms of Kaiser windows for different lengths and different values of β. In designing a Kaiser window for spectrum analysis, we want to specify a desired value of A_{sl} and determine the required value of β. Figure 10.4(c) shows that the relative side-lobe amplitude is essentially independent of the window length and thus depends only on β. This was confirmed by Kaiser and Schafer (1980), who obtained the following least squares approximation to β as a function of A_{sl}:

$$\beta = \begin{cases} 0, & A_{sl} < 13.26, \\ 0.76609(A_{sl} - 13.26)^{0.4} + 0.09834(A_{sl} - 13.26), & 13.26 < A_{sl} < 60, \\ 0.12438(A_{sl} + 6.3), & 60 < A_{sl} < 120. \end{cases} \tag{10.13}$$

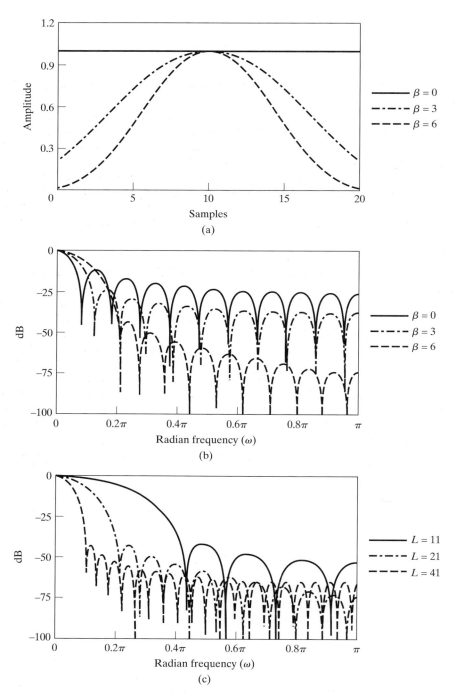

Figure 10.4 (a) Kaiser windows for $\beta = 0, 3$, and 6 and $L = 21$. (b) Fourier transform corresponding to windows in (a). (c) Fourier transforms of Kaiser windows with $\beta = 6$ and $L = 11, 21$, and 41.

Using values of β from Eq. (10.13) gives windows with actual values of A_{sl} that differ by less than 0.36% from the desired value across the range $13.26 < A_{sl} < 120$. (Note that the value 13.26 is the relative side-lobe amplitude of the rectangular window, to which the Kaiser window reduces for $\beta = 0$.)

Figure 10.4(c) also shows that the main-lobe width is inversely proportional to the length of the window. The trade-off between main-lobe width, relative side-lobe amplitude, and window length is displayed by the approximate relationship

$$L \simeq \frac{24\pi(A_{sl} + 12)}{155\Delta_{ml}} + 1, \tag{10.14}$$

which was also given by Kaiser and Schafer (1980).

Equations (10.12), (10.13), and (10.14) are the necessary equations for determining a Kaiser window with desired values of main-lobe width and relative side-lobe amplitude. To design a window for prescribed values of A_{sl} and Δ_{ml} requires simply the computation of β from Eq. (10.13), the computation of L from Eq. (10.14), and the computation of the window using Eq. (10.12). Many of the remaining examples of this chapter use the Kaiser window. Other spectral analysis windows are considered by Harris (1978).

10.2.2 The Effect of Spectral Sampling

As mentioned previously, the DFT of the windowed sequence $v[n]$ provides samples of $V(e^{j\omega})$ at the N equally spaced discrete-time frequencies $\omega_k = 2\pi k/N$, $k = 0, 1, \ldots,$ $N - 1$. These are equivalent to the continuous-time frequencies $\Omega_k = (2\pi k)/(NT)$, for $k = 0, 1, \ldots, N/2$ (assuming that N is even). The indices $k = N/2 + 1, \ldots, N - 1$ correspond to the negative continuous-time frequencies $-2\pi(N - k)/(NT)$. Spectral sampling, as imposed by the DFT, can sometimes produce misleading results. This effect is best illustrated by example.

Example 10.4   Illustration of the Effect of Spectral Sampling

Let us consider the same parameters as in Figure 10.3(c) in Example 10.3, i.e., $A_0 = 1$, $A_1 = 0.75$, $\omega_0 = 2\pi/14$, $\omega_1 = 4\pi/15$, and $\theta_1 = \theta_2 = 0$ in Eq. (10.8). $w[n]$ is a rectangular window of length 64. Then

$$v[n] = \begin{cases} \cos\left(\dfrac{2\pi}{14}n\right) + 0.75\cos\left(\dfrac{4\pi}{15}n\right), & 0 \le n \le 63, \\ 0, & \text{otherwise.} \end{cases} \tag{10.15}$$

Figure 10.5(a) shows the windowed sequence $v[n]$. Figures 10.5(b), (c), (d), and (e) show the corresponding real part, imaginary part, magnitude, and phase, respectively, of the DFT of length $N = 64$. Figure 10.5(f) shows $|V(e^{j\omega})|$, corresponding to Figure 10.3(c), but plotted from $\omega = 0$ to $\omega = 2\pi$ for comparison with the DFT in Figure 10.5(d).

In Figures 10.5(b)–(e), the horizontal (frequency) axis is labeled in terms of the DFT index or frequency sample number k. The value $k = 32$ corresponds to $\omega = \pi$

(a)

(b)

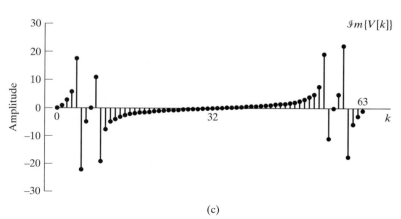

(c)

Figure 10.5   Cosine sequence and discrete Fourier transform with a rectangular window. (a) Windowed signal. (b) Real part of DFT. (c) Imaginary part of DFT.

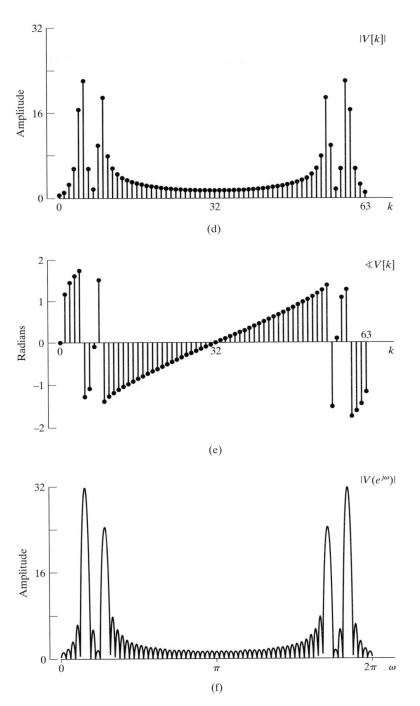

Figure 10.5   (*continued*) (d) Magnitude of DFT. (e) Phase of DFT. (f) Magnitude of discrete-time Fourier transform.

or, equivalently, $\Omega = \pi/T$. As is the usual convention in displaying the DFT of a time sequence, we display the DFT values in the range from $k = 0$ to $k = N - 1$, corresponding to displaying samples of the discrete-time Fourier transform in the frequency range 0 to 2π. Because of the inherent periodicity of the discrete-time Fourier transform, the first half of this range corresponds to the positive continuous-time frequencies, i.e., Ω between zero and π/T, and the second half of the range to the negative frequencies, i.e., Ω between $-\pi/T$ and zero. Note the even periodic symmetry of the real part and the magnitude and the odd periodic symmetry of the imaginary part and the phase.

The magnitude of the DFT in Figure 10.5(d) corresponds to samples of the magnitude of the spectrum displayed in Figure 10.5(f) and shows the expected concentration around $\omega = 2\pi/7.5$ and $\omega = 2\pi/14$, the frequencies of the two sinusoidal components of the input. Specifically, the frequency $\omega_1 = 4\pi/15 = 2\pi(8.533\ldots)/64$ lies between the DFT samples corresponding to $k = 8$ and $k = 9$. Likewise, the frequency $\omega_0 = 2\pi/14 = 2\pi(4.5714\ldots)/64$ lies between the DFT samples corresponding to $k = 4$ and $k = 5$. Note that the frequency locations of the peaks in Figure 10.5(f) are between spectral samples obtained from the DFT. In general, the locations of peaks in the DFT values do not necessarily coincide with the exact frequency locations of the peaks in the Fourier transform, since the true spectral peaks can lie between spectral samples. Correspondingly, as evidenced in comparing Figures 10.5(f) and 10.5(d), the relative amplitudes of peaks in the DFT will not necessarily reflect the relative amplitudes of the true spectral peaks.

Example 10.5   Spectral Sampling with Frequencies Matching DFT Frequencies

Consider the sequence

$$
v[n] = \begin{cases} \cos\left(\dfrac{2\pi}{16}n\right) + 0.75\cos\left(\dfrac{2\pi}{8}n\right), & 0 \le n \le 63, \\ 0, & \text{otherwise,} \end{cases} \tag{10.16}
$$

as shown in Figure 10.6(a). Again, a rectangular window is used with $N = L = 64$. This is very similar to the previous example, except that in this case, the frequencies of the cosines coincide exactly with two of the DFT frequencies. Specifically, the frequency $\omega_1 = 2\pi/8 = 2\pi 8/64$ corresponds exactly to the DFT sample $k = 8$ and the frequency $\omega_0 = 2\pi/16 = 2\pi 4/64$ to the DFT sample $k = 4$.

The magnitude of the 64-point DFT of $v[n]$ for this example is shown in Figure 10.6(b) and corresponds to samples of $|V(e^{j\omega})|$ (which is plotted in Figure 10.6(c)) at a frequency spacing of $2\pi/64$. Although the signal parameters for this example and Example 10.4 are very similar, the appearance of the DFT is strikingly different. In particular, for this example, the DFT has two strong spectral lines at the frequencies of the two sinusoidal components in the signal and no frequency content at the other DFT values. In fact, this clean appearance of the DFT in Figure 10.6(b) is largely an illusion resulting from the sampling of the spectrum. Comparing Figures 10.6(b) and (c), we can see that the reason for the clean appearance of Figure 10.6(b) is that for this choice of parameters, the Fourier transform is exactly zero at the frequencies that are sampled by the DFT except those corresponding to $k = 4, 8, 56$, and 60. Although the

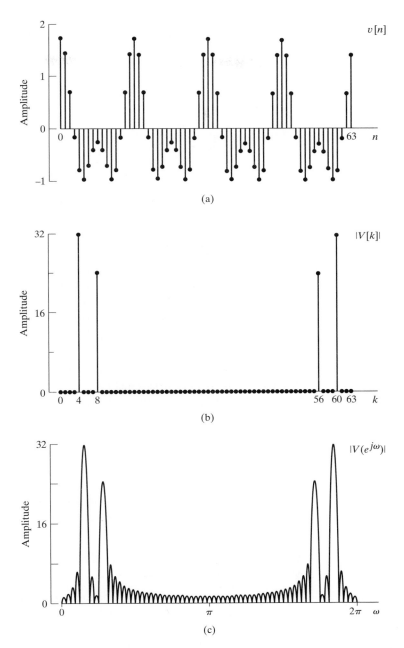

Figure 10.6   Discrete Fourier analysis of the sum of two sinusoids for a case in which the Fourier transform is zero at all DFT frequencies except those corresponding to the frequencies of the two sinusoidal components. (a) Windowed signal. (b) Magnitude of DFT. (c) Magnitude of discrete-time Fourier transform ($|V(e^{j\omega})|$).

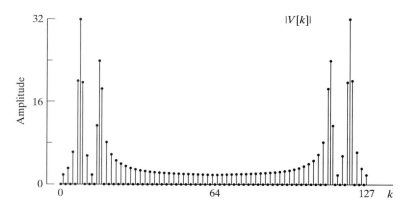

Figure 10.7   DFT of the signal as in Figure 10.6(a), but with twice the number of frequency samples used in Figure 10.6(b).

signal of Figure 10.6(a) has significant content at almost all frequencies, as evidenced by Figure 10.6(c), we do not see this in the DFT because of the sampling of the spectrum. Another way of understanding this is to note that, since the signal frequencies are all multiples of $2\pi/64$, the signal is periodic with period $N = 64$. Thus, the 64-point rectangular window selects exactly one period, which is what is needed to compute the DFS of the periodic signal. (In fact we see from Figure 10.6(a) that $v[n]$ is also periodic with period 8.) This is an example of how the inherent assumption of periodicity gives a correct answer to a different problem. We are interested in the finite-length case and the results are quite misleading.

    To illustrate this point further, we can extend $v[n]$ in Eq. (10.16) by zero-padding to obtain a 128-point sequence. The corresponding 128-point DFT is shown in Figure 10.7. With this finer sampling of the spectrum, the presence of significant content at other frequencies becomes apparent. In this case, the windowed signal is *not* naturally periodic with period 128.

In Figures 10.5, 10.6, and 10.7, the windows were rectangular. In the next set of examples, we illustrate the effect of different choices for the window.

Example 10.6   DFT Analysis of Sinusoidal Signals Using a Kaiser Window

Let us return to the frequency, amplitude, and phase parameters of Example 10.3, but now with a Kaiser window applied, so that

$$v[n] = w_K[n]\cos\left(\frac{2\pi}{14}n\right) + 0.75w_K[n]\cos\left(\frac{4\pi}{15}n\right), \qquad (10.17)$$

where $w_K[n]$ is the Kaiser window as given by Eq. (10.12). We will select the Kaiser window parameter β to be equal to 5.48, which, according to Eq. (10.13), results in a window for which the relative side-lobe amplitude is $A_{sl} = -40$ dB. Figure 10.8(a) shows the windowed sequence $v[n]$ for a window length of $L = 64$, and Figure 10.8(b) shows the magnitude of the corresponding DFT. From Eq. (10.15), we see that the difference between the two frequencies is $\omega_1 - \omega_0 = 2\pi/7.5 - 2\pi/14 = 0.389$. From Eq. (10.14), it follows that the width of the main lobe of the Fourier transform of the Kaiser window with $L = 64$ and $\beta = 5.48$ is $\Delta_{ml} = 0.401$. Thus, the main lobes of the two replicas of $W_K(e^{j\omega})$ centered at ω_0 and ω_1 will just slightly overlap in the frequency interval between the two frequencies. This is evident in Figure 10.8(b), where we see that the two frequency components are clearly resolved.

Figure 10.8(c) shows the same signal, multiplied by a Kaiser window with $L = 32$ and $\beta = 5.48$. Since the window is half as long, we expect the width of the main lobe of the Fourier transform of the window to double, and Figure 10.8(d) confirms this.

(a)

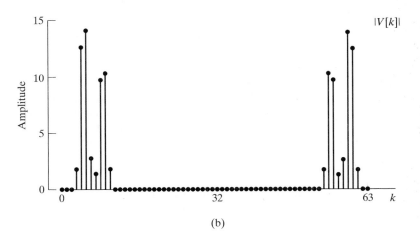

(b)

Figure 10.8 Discrete Fourier analysis with Kaiser window. (a) Windowed sequence for $L = 64$. (b) Magnitude of DFT for $L = 64$.

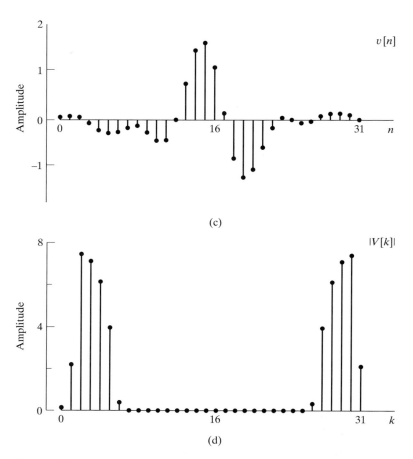

Figure 10.8 (*continued*) (c) Windowed sequence for $L = 32$. (d) Magnitude of DFT for $L = 32$.

Specifically, Eqs. (10.13) and (10.14) confirm that for $L = 32$ and $\beta = 5.48$, the main-lobe width is $\Delta_{ml} = 0.815$. Now the main lobes of the two copies of the Fourier transform of the window overlap throughout the region between the two cosine frequencies, and we do not see two distinct peaks.

In all the previous examples except in Figure 10.7, the DFT length N was equal to the window length L. In Figure 10.7, zero-padding was applied to the windowed sequence before computing the DFT to obtain the Fourier transform on a more finely divided set of frequencies. However, we must realize that this zero-padding will not improve the resolution, which depends on the length and shape of the window. This is illustrated by the next example.

Example 10.7   DFT Analysis with 32-point Kaiser Window and Zero-Padding

In this example we repeat Example 10.6 using the Kaiser window with $L = 32$ and $\beta = 5.48$, and with the DFT length varying. Figure 10.9(a) shows the DFT magnitude for $N = L = 32$ as in Figure 10.8(d), and Figures 10.9(b), (c), and (d) show the DFT magnitude again with $L = 32$, but with DFT lengths $N = 64$, $N = 128$, and $N = 1024$, respectively. As with Example 10.5, this zero-padding of the 32-point sequence results in finer spectral sampling of the discrete-time Fourier transform. The underlying envelope of each DFT magnitude in Figure 10.9 is the same. Consequently, increasing the DFT size by zero-padding does not change the ability to resolve the two sinusoidal frequency components, but it does change the spacing of the frequency samples.

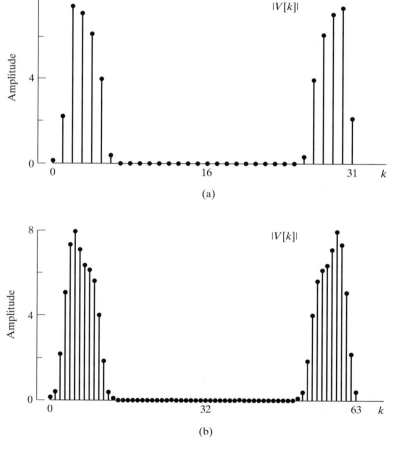

(a)

(b)

Figure 10.9   Illustration of effect of DFT length for Kaiser window of length $L = 32$. (a) Magnitude of DFT for $N = 32$. (b) Magnitude of DFT for $N = 64$.

Figure 10.9 (*continued*) (c) Magnitude of DFT for $N = 128$. (d) Magnitude of DFT for $N = 1024$. (DFT values are linearly interpolated to obtain a smooth curve.)

For a complete representation of a sequence of length L, the L-point DFT is sufficient, since the original sequence can be recovered exactly from it. However, as we saw in the preceding examples, simple examination of the L-point DFT can result in misleading interpretations. For this reason, it is common to apply zero-padding so that the spectrum is sufficiently oversampled and important features are therefore readily apparent. With a high degree of time-domain zero-padding or frequency-domain oversampling, simple interpolation (e.g., linear interpolation) between the DFT values provides a reasonably accurate picture of the Fourier spectrum, which can then be used, for example, to estimate the locations and amplitudes of spectral peaks. This is illustrated in the following example.

Example 10.8   Oversampling and Linear Interpolation for Frequency Estimation

Figure 10.10 shows how a 1024-point DFT can be used to obtain a finely spaced evaluation of the Fourier transform of a windowed signal and how increasing the window width improves the ability to resolve closely spaced sinusoidal components. The signal of Example 10.6 was windowed with Kaiser windows of lengths $L = 32, 42,$

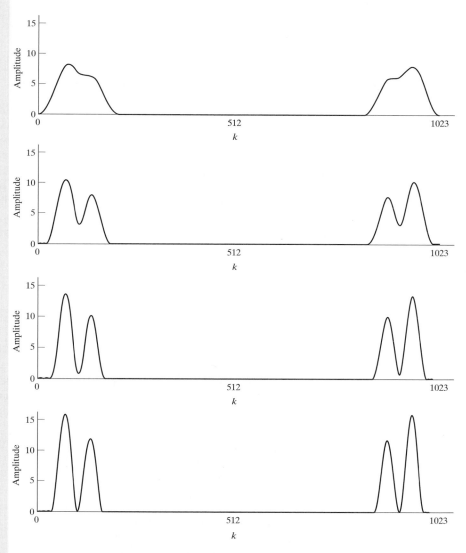

Figure 10.10   Illustration of the computation of the DFT for $N \gg L$ with linear interpolation to create a smooth curve (a) $N = 1024,\ L = 32$ (b) $N = 1024,\ L = 42$ (c) $N = 1024,\ L = 54$ (d) $N = 1024,\ L = 64$.

54, and 64 with $\beta = 5.48$. First note that in all cases, the 1024-point DFT gives a smooth result when the points are connected by straight lines. In Figure 10.10(a), where $L = 32$, the two sinusoidal components are not resolved, and, of course, increasing the DFT length will only result in a smoother curve. As the window length increases, however, we see steady improvement in our ability to distinguish the two frequencies and the approximate amplitudes of each sinusoidal component. Note that the 1024-point DFT in Figure 10.10(d) would be much more effective for precisely locating the peak of the windowed Fourier transform than the coarsely sampled DFT in Figure 10.8(b), which is also computed with a 64-point Kaiser window. Note also that the amplitudes of the two peaks in Figure 10.10 are very close to being in the correct ratio of 0.75 to 1.

10.3 THE TIME-DEPENDENT FOURIER TRANSFORM

The previous section illustrated the use of the DFT for obtaining a frequency-domain representation of a signal composed of sinusoidal components. In that discussion, we assumed that the frequencies of the cosines did not change with time so that no matter how long the window, the signal properties would be the same from the beginning to the end of the window. Often, in practical applications of sinusoidal signal models, the signal properties (amplitudes, frequencies, and phases) will change with time. For example, nonstationary signal models of this type are required to describe radar, sonar, speech, and data communication signals. A single DFT estimate is not sufficient to describe such signals, and as a result, we are led to the concept of the *time-dependent Fourier transform,* also referred to as the short-time Fourier transform.[1]

The time-dependent Fourier transform of a signal $x[n]$ is defined as

$$X[n, \lambda) = \sum_{m=-\infty}^{\infty} x[n + m]w[m]e^{-j\lambda m}, \tag{10.18}$$

where $w[n]$ is a window sequence. In the time-dependent Fourier representation, the one-dimensional sequence $x[n]$, a function of a single discrete variable, is converted into a two-dimensional function of the time variable n, which is discrete, and the frequency variable λ, which is continuous.[2] Note that the time-dependent Fourier transform is periodic in λ with period 2π, and therefore, we need consider only values of λ for $0 \leq \lambda < 2\pi$ or any other interval of length 2π.

Equation (10.18) can be interpreted as the Fourier transform of the shifted signal $x[n+m]$, as viewed through the window $w[m]$. The window has a stationary origin, and as n changes, the signal slides past the window so that, at each value of n, a different portion of the signal is viewed.

---

[1]Further discussion of the time-dependent Fourier transform can be found in a variety of references, including Allen and Rabiner (1977), Rabiner and Schafer (1978), Crochiere and Rabiner (1983), and Nawab and Quatieri (1988).

[2]We denote the frequency variable of the time-dependent Fourier transform by λ to maintain a distinction from the frequency variable of the conventional discrete-time Fourier transform, which will be denoted ω. We use the mixed bracket–parenthesis notation $X[n, \lambda)$ as a reminder that n is a discrete variable and λ a continuous variable.

Example 10.9   Time-Dependent Fourier Transform of a Linear Chirp Signal

The relationship of the window to the shifted signal is illustrated in Figure 10.11 for the signal

$$x[n] = \cos(\omega_0 n^2), \qquad \omega_0 = 2\pi \times 7.5 \times 10^{-6}, \tag{10.19}$$

corresponding to a linear frequency modulation (i.e., the "instantaneous frequency" is $2\omega_0 n$). As we saw in Chapter 9 in the context of the chirp transform algorithm, a signal of this type is often referred to as a linear chirp. Typically, $w[m]$ in Eq. (10.18) has finite length around $m = 0$, so that $X[n, \lambda)$ displays the frequency characteristics of the signal around time n. For example, in Figure 10.12 we show a display of the magnitude of the

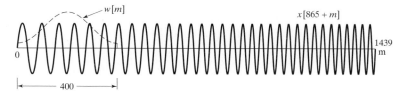

Figure 10.11   Two segments of the linear chirp signal $x[n] = \cos(\omega_0 n^2)$ with the window superimposed. $X[n, \lambda)$ at $n = 225$ is the discrete-time Fourier transform of the top trace multiplied by the window. $X[865, \lambda)$ is the discrete-time Fourier transform of the bottom trace multiplied by the window.

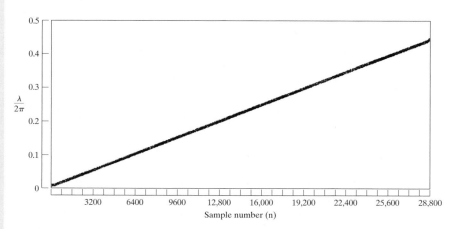

Figure 10.12   The magnitude of the time-dependent Fourier transform of $x[n] = \cos(\omega_0 n^2)$ using a Hamming window of length 400.

time-dependent Fourier transform of the signal of Eq. (10.19) and Figure 10.11 with $w[m]$ a Hamming window of length 400. In this display, referred to as a *spectrogram,* the vertical dimension $\lambda/2\pi$ is proportional to frequency and the horizontal dimension (n) is proportional to time. The magnitude of the time-dependent Fourier transform is represented by the darkness of the markings. In Figure 10.12, the linear progression of the frequency with time is clear.

Since $X[n, \lambda)$ is the discrete-time Fourier transform of $x[n + m]w[m]$, the time-dependent Fourier transform is invertible if the window has at least one nonzero sample. Specifically, from the Fourier transform synthesis equation (2.133),

$$x[n + m]w[m] = \frac{1}{2\pi} \int_0^{2\pi} X[n, \lambda)e^{j\lambda m}d\lambda, \qquad -\infty < m < \infty, \qquad (10.20)$$

from which it follows that

$$x[n] = \frac{1}{2\pi w[0]} \int_0^{2\pi} X[n, \lambda)d\lambda \qquad (10.21)$$

if $w[0] \neq 0$.[3] Not just the single sample $x[n]$, but all of the samples that are multiplied by nonzero samples of the window, can be recovered in a similar manner using Eq. (10.20).

A rearrangement of the sum in Eq. (10.18) leads to another useful interpretation of the time-dependent Fourier transform. If we make the substitution $m' = n + m$ in Eq. (10.18), then $X[n, \lambda)$ can be written as

$$X[n, \lambda) = \sum_{m'=-\infty}^{\infty} x[m']w[-(n - m')]e^{j\lambda(n-m')}. \qquad (10.22)$$

Equation (10.22) can be interpreted as the convolution

$$X[n, \lambda) = x[n] * h_\lambda[n], \qquad (10.23a)$$

where

$$h_\lambda[n] = w[-n]e^{j\lambda n}. \qquad (10.23b)$$

From Eq. (10.23a), we see that the time-dependent Fourier transform as a function of n with λ fixed can be interpreted as the output of a linear time-invariant filter with impulse response $h_\lambda[n]$ or, equivalently, with frequency response

$$H_\lambda(e^{j\omega}) = W(e^{j(\lambda-\omega)}). \qquad (10.24)$$

In general, a window that is nonzero for positive time will be called a *noncausal window,* since the computation of $X[n, \lambda)$ using Eq. (10.18) requires samples that *follow* sample n in the sequence. Equivalently, in the linear-filtering interpretation, the impulse response $h_\lambda[n] = w[-n]e^{j\lambda n}$ is noncausal.

In the definition of Eq. (10.18), the time origin of the window is held fixed and the signal is shifted past the interval of support of the window. This effectively redefines the time origin for Fourier analysis to be at sample n of the signal. Another possibility is to shift the window as n changes, keeping the time origin for Fourier analysis fixed at

---

[3]Since $X[n, \lambda)$ is periodic in λ with period 2π, the integration in Eqs. (10.20) and (10.21) can be over any interval of length 2π.

the original time origin of the signal. This leads to a definition for the time-dependent Fourier transform of the form

$$\check{X}[n, \lambda] = \sum_{m=-\infty}^{\infty} x[m]w[m-n]e^{-j\lambda m}. \tag{10.25}$$

The relationship between the definitions of Eqs. (10.18) and (10.25) is easily shown to be

$$\check{X}[n, \lambda] = e^{-j\lambda n} X[n, \lambda] \tag{10.26}$$

The definition of Eq. (10.18) is particularly convenient when we consider using the DFT to obtain samples in λ of the time-dependent Fourier transform, since, if $w[m]$ is of finite length in the range $0 \leq m \leq (L-1)$, then so is $x[n+m]w[m]$. On the other hand, the definition of Eq. (10.25) has some advantages for the interpretation of Fourier analysis in terms of filter banks. Since our primary interest is in applications of the DFT, we will base our discussion on Eq. (10.18).

10.3.1 The Effect of the Window

The primary purpose of the window in the time-dependent Fourier transform is to limit the extent of the sequence to be transformed so that the spectral characteristics are reasonably stationary over the duration of the window. The more rapidly the signal characteristics change, the shorter the window should be. We saw in Section 10.2 that as the window becomes shorter, frequency resolution decreases. The same effect is true, of course, for $X[n, \lambda]$. On the other hand, as the window length decreases, the ability to resolve changes with time increases. Consequently, the choice of window length becomes a trade-off between frequency resolution and time resolution.

The effect of the window on the properties of the time-dependent Fourier transform can be seen by assuming that the signal $x[n]$ has a conventional discrete-time Fourier transform $X(e^{j\omega})$. First let us assume that the window is unity for all m; i.e., assume that there is no window at all. Then, from Eq. (10.18),

$$X[n, \lambda] = X(e^{j\lambda})e^{j\lambda n}. \tag{10.27}$$

Of course, a typical window for spectrum analysis tapers to zero so as to select only a portion of the signal for analysis. As discussed in Section 10.2, the length and shape of the window are chosen so that the Fourier transform of the window is narrow in frequency compared with variations with frequency of the Fourier transform of the signal. The Fourier transform of a typical window is illustrated in Figure 10.13(a).

If we consider the time-dependent Fourier transform for fixed n, then it follows from the properties of Fourier transforms that

$$X[n, \lambda] = \frac{1}{2\pi} \int_0^{2\pi} e^{j\theta n} X(e^{j\theta})W(e^{j(\lambda-\theta)})d\theta; \tag{10.28}$$

i.e., the Fourier transform of the shifted signal is convolved with the Fourier transform of the window. This is similar to Eq. (10.2), except that in Eq. (10.2) we assumed that the signal was not successively shifted relative to the window. Here we compute a Fourier transform for each value of n. In Section 10.2 we saw that the ability to resolve two narrowband signal components depends on the width of the main lobe of the Fourier

(a)

(b)

Figure 10.13 (a) Fourier transform of window in time-dependent Fourier analysis. (b) Equivalent bandpass filter for time-dependent Fourier analysis.

transform of the window, while the degree of leakage of one component into the vicinity of the other depends on the relative side-lobe amplitude. The case of no window at all corresponds to $w[n] = 1$ for all n. In this case $W(e^{j\omega}) = 2\pi\delta(\omega)$ for $-\pi \leq \omega \leq \pi$, which gives precise frequency resolution, but no time resolution.

In the linear-filtering interpretation of Eqs. (10.23a), (10.23b), and (10.24), $W(e^{j\omega})$ typically has the lowpass characteristics depicted in Figure 10.13(a), and consequently, $H_\lambda(e^{j\omega})$ is a bandpass filter whose passband is centered at $\omega = \lambda$, as depicted in Figure 10.13(b). Clearly, the width of the passband of this filter is approximately equal to the width of the main lobe of the Fourier transform of the window. The degree of rejection of adjacent frequency components depends on the relative side-lobe amplitude.

The preceding discussion suggests that if we are using the time-dependent Fourier transform to obtain a time-dependent estimate of the frequency spectrum of a signal, it is desirable to taper the window to lower the side lobes and to use as long a window as feasible to improve the frequency resolution. We will consider some examples in Section 10.5. However, before doing so, we discuss the use of the DFT in explicitly evaluating the time-dependent Fourier transform.

10.3.2 Sampling in Time and Frequency

Explicit computation of $X[n, \lambda)$ can be done only on a finite set of values of λ, corresponding to sampling the time-dependent Fourier transform in the frequency variable domain. Just as finite-length signals can be exactly represented through samples of the discrete-time Fourier transform, signals of indeterminate length can be represented through samples of the time-dependent Fourier transform if the window in Eq. (10.18) has finite length. As an example, suppose that the window has length L with samples beginning at $m = 0$; i.e.,

$$w[m] = 0 \qquad \text{outside the interval } 0 \le m \le L - 1. \qquad (10.29)$$

If we sample $X[n, \lambda)$ at N equally spaced frequencies $\lambda_k = 2\pi k/N$, with $N \ge L$, then we can still recover the original sequence from the sampled time-dependent Fourier transform. Specifically, if we define $X[n, k]$ to be

$$X[n, k] = X[n, 2\pi k/N) = \sum_{m=0}^{L-1} x[n + m]w[m]e^{-j(2\pi/N)km}, \qquad 0 \le k \le N - 1, \quad (10.30)$$

then $X[n, k]$ is the DFT of the windowed sequence $x[n + m]w[m]$. Using the inverse DFT, we obtain

$$x[n + m]w[m] = \frac{1}{N} \sum_{k=0}^{N-1} X[n, k]e^{j(2\pi/N)km}, \qquad 0 \le m \le L - 1. \qquad (10.31)$$

Since we assume that the window $w[m] \ne 0$ for $0 \le m \le L - 1$, the sequence values can be recovered in the interval from n through $(n + L - 1)$ using the equation

$$x[n + m] = \frac{1}{Nw[m]} \sum_{k=0}^{N-1} X[n, k]e^{j(2\pi/N)km}, \qquad 0 \le m \le L - 1, \qquad (10.32)$$

where it is assumed that $w[m] \ne 0$ for $0 \le m \le L - 1$. The important point is that the window has finite length and that we take at least as many samples in the λ dimension as there are nonzero samples in the window; i.e., $N \ge L$. While Eq. (10.29) corresponds to a noncausal window, we could have used a causal window with $w[m] \ne 0$ for $-(L - I) \le m \le 0$ or a symmetric window such that $w[m] = w[-m]$ for $|m| \le (L - 1)/2$, with L an odd integer. The use of a noncausal window in Eq. (10.30) is simply more convenient for our analysis, since it leads very naturally to the interpretation of the sampled time-dependent Fourier transform as the DFT of the windowed sequence beginning with sample n.

Since Eq. (10.30) corresponds to sampling Eq. (10.18) in λ, it also corresponds to sampling Eqs. (10.22), (10.23a), and (10.23b) in λ. Specifically, Eq. (10.30) can be rewritten as

$$X[n, k] = x[n] * h_k[n], \qquad 0 \le k \le N - 1, \qquad (10.33a)$$

where

$$h_k[n] = w[-n]e^{j(2\pi/N)kn}. \qquad (10.33b)$$

Equations (10.33a) and (10.33b) can be viewed as a bank of N filters, as depicted in Figure 10.14, with the kth filter having frequency response

$$H_k(e^{j\omega}) = W(e^{j[(2\pi k/N) - \omega]}). \qquad (10.34)$$

Our discussion suggests that $x[n]$ for $-\infty < n < \infty$ can be reconstructed if $X[n, \lambda)$ or $X[n, k]$ is sampled in the time dimension as well. Specifically, using Eq. (10.32), we can reconstruct the signal in the interval $n_0 \le n \le n_0 + L - 1$ from $X[n_0, k]$, and we can reconstruct the signal in the interval $n_0 + L \le n \le n_0 + 2L - 1$ from $X[n_0 + L, k]$, etc. Thus, $x[n]$ can be reconstructed exactly from the time-dependent Fourier transform sampled in both the frequency and the time dimension. In general, for the region of

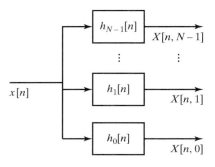

Figure 10.14 Filter bank representation of the time-dependent Fourier transform.

support of the window as specified in Eq. (10.29), we define this sampled time-dependent Fourier transform as

$$X[rR, k] = X[rR, 2\pi k/N) = \sum_{m=0}^{L-1} x[rR + m]w[m]e^{-j(2\pi/N)km}, \quad (10.35)$$

where r and k are integers such that $-\infty < r < \infty$ and $0 \le k \le N - 1$. To further simplify our notation, we define

$$X_r[k] = X[rR, k] = X[rR, \lambda_k), \qquad -\infty < r < \infty, \quad 0 \le k \le N - 1, \quad (10.36)$$

where $\lambda_k = 2\pi k/N$. This notation denotes explicitly that the sampled time-dependent Fourier transform is simply a sequence of N-point DFTs of the windowed signal segments

$$x_r[m] = x[rR + m]w[m], \qquad -\infty < r < \infty, \quad 0 \le m \le L - 1, \quad (10.37)$$

with the window position moving in jumps of R samples in time. Figure 10.15 shows lines in the $[n, \lambda]$-plane corresponding to the region of support of $X[n, \lambda]$ and the grid of sampling points in the $[n, \lambda]$-plane for the case $N = 10$ and $R = 3$. As we have shown, it is possible to uniquely reconstruct the original signal from such a two-dimensional discrete representation.

Equation (10.35) involves the following integer parameters: the window length L; the number of samples in the frequency dimension, or the DFT length N; and the sampling interval in the time dimension, R. However, not all choices of these parameters will permit exact reconstruction of the signal. The choice $L \le N$ guarantees that we *can* reconstruct the windowed segments $x_r[m]$ from the block transforms $X_r[k]$. If $R < L$, the segments overlap, but if $R > L$, some of the samples of the signal are not used and therefore cannot be reconstructed from $X_r[k]$. Thus, in general, the three sampling parameters should satisfy the relation $N \ge L \ge R$. Notice that each block of R samples of the signal is represented by N complex numbers in the sampled time-dependent Fourier representation; or, if the signal is real, only N real numbers are required, due to the symmetry of the DFT. As previously shown, the signal can be reconstructed exactly from the sampled time-dependent Fourier transform for the special case $N = L = R$. In this case, N samples of a real signal are represented by N real numbers, and this is the minimum that we could expect to achieve for an arbitrarily chosen signal.

Another way to see that the time-dependent Fourier transform can be sampled in the time dimension is to recall that for fixed λ, or equivalently, for fixed k, the time-dependent Fourier transform is a one-dimensional sequence that is the output of a

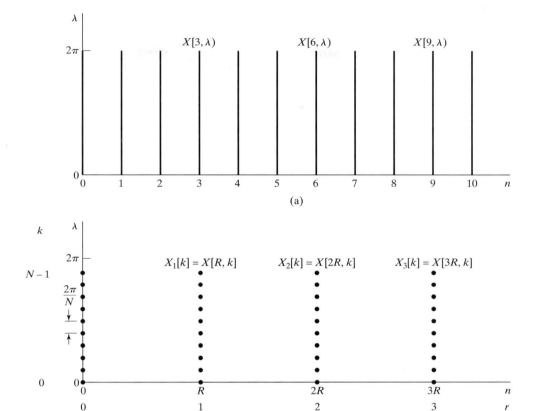

Figure 10.15   (a) Region of support for $X[n, \lambda]$. (b) Grid of sampling points in the $[n, \lambda]$-plane for the sampled time-dependent Fourier transform with $N = 10$ and $R = 3$.

bandpass filter with frequency response as in Eq. (10.24). Thus, we expect that the sampling rate of the sequences representing each of the frequencies could be reduced by a factor $2\pi/\Delta_{ml}$, where Δ_{ml} is the width of the main lobe of the Fourier transform of the window. Figure 10.16(a) shows the set of bandpass filters corresponding to a rectangular window with $L = N = 16$. Note that the passbands of the filters overlap significantly, and their frequency selectivity is not ideal by any standard. In fact the sidelobes of any one of the bandpass filters overlap completely with the passbands on either side. This suggests that we might encounter a problem with aliasing in the time dimension, since the Fourier transform of any other finite-length window will not be an ideal filter response either. For example, Figure 10.16(b) shows the case for a Kaiser window of the same length, i.e., $L = N = 16$. The sidelobes are much smaller, but the mainlobe is much wider, so the filters overlap considerably. Although it is not obvious by looking at Figure 10.16, the previous argument based on block processing ideas shows conclusively that we can nevertheless reconstruct the original signal exactly from the time- and frequency-sampled time-dependent Fourier transform. A more detailed

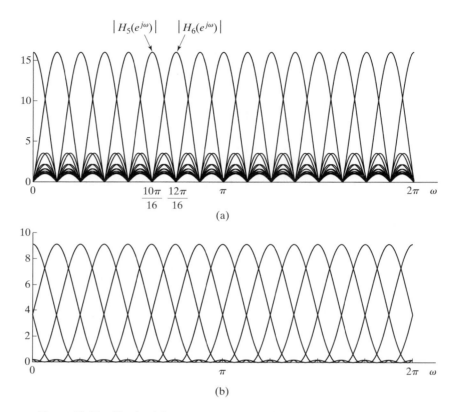

Figure 10.16 Filterbank frequency response. (a) Rectangular window. (b) Kaiser window.

analysis of the linear-filtering point of view shows that the aliasing distortion due to the nonideal frequency responses can be canceled in the reconstruction process. This point is considered in Problem 10.40 and is discussed in detail in Rabiner and Schafer (1978) and Crochiere and Rabiner (1983).

10.4 BLOCK CONVOLUTION USING THE TIME-DEPENDENT FOURIER TRANSFORM

One of the uses of the time-dependent Fourier transform is as a basis for processing a discrete-time signal by performing the modifications in the frequency domain. This is done by computing a time-dependent Fourier transform representation, modifying that representation, and then reconstructing a discrete-time signal. This approach is widely used in digital speech coding, where, instead of quantizing the samples of the speech signal, the sampled time-dependent Fourier transform is quantized and coded for either transmission or storage. A discussion of applications of this type would take us too far afield; however, these kinds of block-processing techniques for discrete-time signals were also introduced in Chapter 8 when we discussed the use of the DFT for implementing the convolution of a finite-length impulse response with an input signal of indefinite length. This method of implementation of linear time-invariant systems has a useful interpretation in terms of the definitions and concepts of Section 10.3.

Assume that $x[n] = 0$ for $n < 0$, and suppose that we compute the time-dependent Fourier transform for $R = L$ and a rectangular window. In other words, the sampled time-dependent Fourier transform $X_r[k]$ consists of a set of N-point DFTs of segments of the input sequence

$$x_r[m] = x[rL + m], \qquad 0 \le m \le L - 1. \tag{10.38}$$

Since each sample of the signal $x[n]$ is included and the blocks do not overlap, it follows that

$$x[n] = \sum_{r=0}^{\infty} x_r[n - rL]. \tag{10.39}$$

Now suppose that we define a new time-dependent Fourier transform

$$Y_r[k] = H[k]X_r[k], \qquad 0 \le k \le N - 1, \tag{10.40}$$

where $H[k]$ is the N-point DFT of a finite-length unit sample sequence $h[n]$ such that $h[n] = 0$ for $n < 0$ and for $n > P - 1$. If we compute the inverse DFT of $Y_r[k]$, we obtain

$$y_r[m] = \frac{1}{N} \sum_{k=0}^{N-1} Y_r[k] e^{j(2\pi/N)km} = \sum_{\ell=0}^{N-1} x_r[\ell] h[((m - \ell))_N]. \tag{10.41}$$

That is, $y_r[m]$ is the N-point circular convolution of $h[m]$ and $x_r[m]$. Since $h[m]$ has length P samples and $x_r[m]$ has length L samples, it follows from the discussion of Section 8.7 that if $N \ge L + P - 1$, then $y_r[m]$ will be identical to the linear convolution of $h[m]$ with $x_r[m]$ in the interval $0 \le m \le L + P - 2$, and it will be zero otherwise. Thus, it follows that if we construct an output signal

$$y[n] = \sum_{r=0}^{\infty} y_r[n - rL], \tag{10.42}$$

then $y[n]$ is the output of a linear time-invariant system with impulse response $h[n]$. The procedure just described corresponds exactly to the *overlap–add* method of block convolution. The overlap–save method discussed in Section 8.7 can also be applied within the framework of the time-dependent Fourier transform.

10.5 FOURIER ANALYSIS OF NONSTATIONARY SIGNALS

In Section 10.4, we considered a simple example of how the time-dependent Fourier transform can be used to implement linear filtering. In such applications, we are not so much interested in spectral resolution as in whether it is possible to reconstruct a modified signal from the modified time-dependent Fourier transform. On the other hand, the concept of the time-dependent Fourier transform is perhaps most widely used as a framework for a variety of techniques for obtaining spectral estimates for nonstationary discrete-time signals, and in these applications spectral resolution, time variation, and other issues are the most important.

A nonstationary signal is a signal whose properties vary with time, for example, a sum of sinusoidal components with time-varying amplitudes, frequencies, or phases. As we will illustrate in Section 10.5.1 for speech signals and in Section 10.5.2 for Doppler radar signals, the time-dependent Fourier transform often provides a useful description of how the signal properties change with time.

When we apply time-dependent Fourier analysis to a sampled signal, the entire discussion of Section 10.1 holds for each DFT that is computed. In other words, for each segment $x_r[n]$ of the signal, the sampled time-dependent Fourier transform $X_r[k]$ would be related to the Fourier transform of the original continuous-time signal by the processes described in Section 10.1. Furthermore, if we were to apply the time-dependent Fourier transform to sinusoidal signals with constant (i.e., non-time-varying) parameters, the discussion of Section 10.2 should also apply to each of the DFTs that we compute. When the signal frequencies do not change with time, it is tempting to assume that the time-dependent Fourier transform would vary only in the frequency dimension in the manner described in Section 10.2, but this would be true only in very special cases. For example, the time-dependent Fourier transform will be constant in the time dimension if the signal is periodic with period N_p and $L = \ell_0 N_p$ and $R = r_0 N_p$, where ℓ_0 and r_0 are integers; i.e., the window includes exactly ℓ_0 periods, and the window is moved by exactly r_0 periods between computations of the DFT. In general, even if the signal is exactly periodic, the varying phase relationships that would result as different segments of the waveform are shifted into the analysis window would cause the time-dependent Fourier transform to vary in the time dimension. However, for stationary signals, if we use a window that tapers to zero at its ends, the magnitude $|X_r[k]|$ will vary only slightly from segment to segment, with most of the variation of the complex time-dependent Fourier transform occurring in the phase.

10.5.1 Time-Dependent Fourier Analysis of Speech Signals

Speech is produced by excitation of an acoustic tube, the *vocal tract,* which is terminated on one end by the lips and on the other end by the glottis. There are three basic classes of speech sounds:

- *Voiced sounds* are produced by exciting the vocal tract with quasi-periodic pulses of airflow caused by the opening and closing of the glottis.
- *Fricative sounds* are produced by forming a constriction somewhere in the vocal tract and forcing air through the constriction so that turbulence is created, thereby producing a noiselike excitation.
- *Plosive sounds* are produced by completely closing off the vocal tract, building up pressure behind the closure, and then abruptly releasing the pressure.

Detailed discussions of models for the speech signal and applications of the time-dependent Fourier transform are found in texts such as Flanagan (1972), Rabiner and Schafer (1978), O'Shaughnessy (1987), Parsons (1986), and Deller et al. (1993).

With a constant vocal tract shape, speech can be modeled as the response of a linear time-invariant system (the vocal tract) to a quasi-periodic pulse train for voiced sounds or wideband noise for unvoiced sounds. The vocal tract is an acoustic transmission system characterized by its natural frequencies, called *formants,* which correspond to resonances in its frequency response. In normal speech, the vocal tract changes shape relatively slowly with time as the tongue and lips perform the gestures of speech, and thus it can be modeled as a slowly time-varying filter that imposes its frequency-response properties on the spectrum of the excitation. A typical speech waveform is shown in Figure 10.17.

From this brief description of the process of speech production and from Figure 10.17, we see that speech is clearly a nonstationary signal. However, as illustrated in the figure, the characteristics of the signal can be assumed to remain essentially constant over time intervals on the order of 30 or 40 ms. The frequency content of the speech signal may range up to 15 kHz or higher, but speech is highly intelligible even when bandlimited to frequencies below about 3 kHz. Commercial telephone systems, for example, typically limit the highest transmitted frequency to about 3 kHz. A standard sampling rate for digital telephone communication systems is 8000 samples/s. At this sampling rate, a 40-ms time interval is spanned by 320 samples.

Figure 10.17 shows that the waveform consists of a sequence of quasi-periodic *voiced* segments interspersed with noiselike *unvoiced* segments. This figure suggests that if the window length L is not too long, the properties of the signal will not change appreciably from the beginning of the segment to the end. Thus, the DFT of a windowed speech segment should display the frequency-domain properties of the signal at the time corresponding to the window location. For example, if the window length is long enough so that the harmonics are resolved, the DFT of a windowed segment of voiced speech should show a series of peaks at integer multiples of the fundamental frequency of the signal in that interval. This would normally require that the window span several periods of the waveform. If the window is too short, then the harmonics will not be resolved, but the general spectral shape will still be evident. This is typical of the trade-off between frequency resolution and time resolution that is required in the analysis of nonstationary signals. If the window is too long, the signal properties may change too much across the window; if the window is too short, resolution of narrowband components will be sacrificed. This trade-off is illustrated in the following example.

Example 10.10   Spectrogram Display of the Time-Dependent Fourier Transform of Speech

Figure 10.18(a) shows a spectrogram display of the time-dependent Fourier transform of the sentence in Figure 10.17. The time waveform is also shown on the same time scale, below the spectrogram. More specifically, Figure 10.18(a) is a *wideband spectrogram*. A wideband spectrogram representation results from a window that is relatively short in time and is characterized by poor resolution in the frequency dimension and good resolution in the time dimension. The frequency axis is labeled in terms of continuous-time frequency. Since the sampling rate of the signal was 16,000 samples/s, it follows that the frequency $\lambda = \pi$ corresponds to 8 kHz. The specific window used in Figure 10.18(a) was a Hamming window of duration 6.7 ms, corresponding to $L = 108$. The value of R was 16, representing 1-ms time increments. The broad, dark bars that move horizontally across the spectrogram correspond to the resonance frequencies of the vocal tract, which, as we see, change with time. The vertically striated appearance of the spectrogram is due to the quasi-periodic nature of voiced portions of the waveform, as is evident by comparing the variations in the waveform display and the spectrogram. Since the length of the

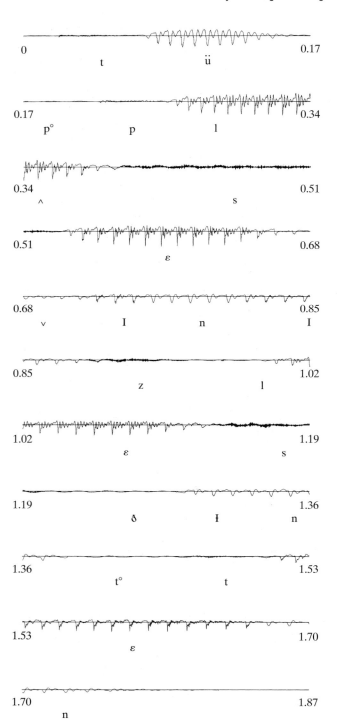

Figure 10.17 Waveform of the speech utterance "Two plus seven is less than ten." Each line is 0.17 s in duration. The time-aligned phonemic transcript is indicated below the waveform.

analysis window is on the order of the length of a period of the waveform, as the window slides along in time it alternately covers high-energy segments of the waveform and then lower energy segments in between, thereby producing the vertical striations in the plot during voiced intervals.

In a *narrowband* time-dependent Fourier analysis, a longer window is used to provide higher frequency resolution, with a corresponding decrease in time resolution. Such a narrowband analysis of speech is illustrated by the display in Figure 10.18(b). In this case, the window was a Hamming window of duration 45 ms. This corresponds to $L = 720$. The value of R was again 16.

Figure 10.18   (a) Wideband spectrogram of waveform of Figure 10.17. (b) Narrowband spectrogram.

This example only hints at the many reasons that the time-dependent Fourier transform is so important in speech analysis and processing. Indeed, the concept is used directly and indirectly as the basis for acoustic–phonetic analysis and for many fundamental speech-processing applications, such as digital coding, noise and reverberation removal, speech recognition, speaker verification, and speaker identification. For present purposes, our discussion simply serves as an introductory illustration.

10.5.2 Time-Dependent Fourier Analysis of Radar Signals

Another application area in which the time-dependent Fourier transform plays an important role is radar signal analysis. The following are elements of a typical radar system:

- Antennas for transmitting and receiving (often the same).
- A transmitter that generates an appropriate signal at microwave frequencies. In our discussion, we will assume that the signal consists of sinusoidal pulses. While this is often the case, other signals may be used, depending on the specific radar objectives and design.
- A receiver that amplifies and detects echoes of the transmitted pulses that have been reflected from objects illuminated by the antenna.

In such a radar system, the transmitted sinusoidal signal propagates at the speed of light, reflects off the object, and returns at the speed of light to the antenna, thereby undergoing a time delay of the round-trip travel time from the antenna to the object. If we assume that the transmitted signal is a sinusoidal pulse of the form $\cos(\Omega_0 t)$ and the distance from the antenna to the object is $\rho(t)$, then the received signal is a pulse of the form

$$s(t) = \cos[\Omega_0(t - 2\rho(t)/c)], \tag{10.43}$$

where c is the velocity of light. If the object is not moving relative to the antenna, then $\rho(t) = \rho_0$, where ρ_0 is the *range*. Since the time delay between the transmitted and received pulses is $2\rho_0/c$, a measurement of the time delay may be used to estimate the range. If, however, $\rho(t)$ is not constant, the received signal is an angle-modulated sinusoid and the phase difference contains information about both the range and the relative motion of the object with respect to the antenna. Specifically, let us represent the time-varying range in a Taylor's series expansion as

$$\rho(t) = \rho_0 + \dot{\rho}_0 t + \frac{1}{2!}\ddot{\rho}_0 t^2 + \cdots, \tag{10.44}$$

where ρ_0 is the nominal range, $\dot{\rho}_0$ is the velocity, $\ddot{\rho}_0$ is the acceleration, and so on. Assuming that the object moves with constant velocity (i.e., $\ddot{\rho}_0 = 0$), and substituting Eq. (10.44) into Eq. (10.43), we obtain

$$s(t) = \cos[(\Omega_0 - 2\Omega_0\dot{\rho}_0/c)t - 2\Omega_0\rho_0/c]. \tag{10.45}$$

In this case, the frequency of the received signal differs from the frequency of the transmitted signal by the *Doppler frequency,* defined as

$$\Omega_d = -2\Omega_0 \dot{\rho}_0/c. \tag{10.46}$$

Thus, the time delay can still be used to estimate the range, and we can determine the speed of the object relative to the antenna if we can determine the Doppler frequency.

In a practical setting, the received signal is generally very weak, and thus a noise term should be added to Eq. (10.45). We will neglect the effects of noise in the simple analysis of this section. Also, in most radar systems, the signal of Eq. (10.45) would be frequency shifted to a lower nominal frequency in the detection process. However, the Doppler shift will still satisfy Eq. (10.46), even if $s(t)$ is demodulated to a lower center frequency.

To apply time-dependent Fourier analysis to such signals, we first bandlimit the signal to a frequency band that includes the expected Doppler frequency shifts and then sample the resulting signal with an appropriate sampling period T, thereby obtaining a discrete-time signal of the form

$$x[n] = \cos[(\omega_0 - 2\omega_0 \dot{\rho}_0/c)n - 2\omega_0 \rho_0/c], \tag{10.47}$$

where $\omega_0 = \Omega_0 T$. In many cases, the object's motion would be more complicated than we have assumed, requiring the incorporation of higher order terms from Eq. (10.44) and thereby producing a more complicated angle modulation in the received signal. Another way to represent this more complicated variation of the frequency of the echoes is to use the time-dependent Fourier transform with a window that is short enough so that the assumption of constant Doppler-shifted frequency is valid across the entire window interval, but not so short as to sacrifice adequate resolution when two or more moving objects create Doppler-shifted return signals that are superimposed at the receiver.

Example 10.11

An example of time-dependent Fourier analysis of Doppler radar signals is shown in Figure 10.19. (See Schaefer et al., 1979.) The radar data had been preprocessed to remove low-velocity Doppler shifts, leaving the variations displayed in the figure. The window for the time-dependent Fourier transform was a Kaiser window with $N = L = 64$ and $\beta = 4$. In the figure, $|X_r[k]|$ is plotted with time as the vertical dimension (increasing upward) and frequency as the horizontal dimension.[4] In this case, the successive DFTs are plotted close together. A hidden-line elimination algorithm is used to create a two-dimensional view of the time-dependent Fourier transform. To the left of the centerline is a strong peak that moves in a smooth path through the time-frequency plane. This corresponds to a moving object whose velocity is varying in a regular manner. The other broad peaks in the time-dependent Fourier transform are due to noise and spurious returns called *clutter* in radar terminology. An example of motion that might create such a variation of the Doppler frequency is a rocket moving at constant velocity, but rotating about its longitudinal axis. A peak moving through the

---

[4]The plot shows the negative frequencies on the left of the line through the center of the plot and positive frequencies on the right. This can be achieved by computing the DFT of $(-1)^n x_r[n]$ and noting that the computation effectively shifts the origin of the DFT index to $k = N/2$. Alternatively, the DFT of $x_r[n]$ can be computed and then reindexed.

time-dependent Fourier transform might correspond to reflections from a fin on the rocket that is alternately moving toward and then away from the antenna because of the spinning of the rocket. Figure 10.19(b) shows an estimate of the Doppler frequency as a function of time. This estimate was obtained simply by locating the highest peak in each DFT.

Figure 10.19   Illustration of time-dependent Fourier analysis of Doppler radar signal. (a) Sequence of Fourier transforms of Doppler radar signal. (b) Doppler frequency estimated by picking the largest peak in the time-dependent Fourier transform.

10.6 FOURIER ANALYSIS OF STATIONARY RANDOM SIGNALS: THE PERIODOGRAM

In the previous sections, we discussed and illustrated Fourier analysis for sinusoidal signals with stationary (non-time-varying) parameters and for nonstationary signals such as speech and radar. In cases where the signal can be modeled by a sum of sinusoids or a linear system excited by a periodic pulse train, the Fourier transforms of finite-length segments of the signal have a convenient and natural interpretation in terms of Fourier transforms, windowing, and linear system theory. However, more noiselike signals, such as the example of unvoiced speech in Section 10.5.1, are best modeled as random signals.

As we discussed in Section 2.10 and as shown in Appendix A, random processes are often used to model signals when the process that generates the signal is too complex for a reasonable deterministic model. Typically, when the input to a linear time-invariant system is modeled as a stationary random process, many of the essential characteristics of the input and output are adequately represented by averages, such as the mean value (dc level), variance (average power), autocorrelation function, or power density spectrum. Consequently, it is of particular interest to estimate these for a given signal. As discussed in Appendix A, a typical estimate of the mean value of a stationary random process from a finite-length segment of data is the *sample mean,* defined as

$$\hat{m}_x = \frac{1}{L} \sum_{n=0}^{L-1} x[n]. \tag{10.48}$$

Similarly, a typical estimate of the variance is the *sample variance,* defined as

$$\hat{\sigma}_x^2 = \frac{1}{L} \sum_{n=0}^{L-1} (x[n] - \hat{m}_x)^2. \tag{10.49}$$

The sample mean and the sample variance, which are themselves random variables, are *unbiased* and *asymptotically unbiased* estimators, respectively; i.e., the expected value of $\hat{m}_x$ is the true mean m_x and the expected value of $\hat{\sigma}_x^2$ approaches the true variance σ_x^2 as L approaches ∞. Furthermore, they are both *consistent* estimators; i.e., they improve with increasing L, since their variances approach zero as L approaches ∞.

In the remainder of this chapter, we study the estimation of the power spectrum[5] of a random signal using the DFT. As we will see, there are two basic approaches to estimating the power spectrum. One approach, which we develop in this section, is referred to as *periodogram analysis* and is based on direct Fourier transformation of finite-length segments of the signal. The second approach, developed in Section 10.7, is to first estimate the autocovariance sequence and then compute the Fourier transform of this estimate. In either case, we are typically interested in obtaining unbiased consistent estimators. Unfortunately, the analysis of such estimators is very difficult, and generally, only approximate analyses can be accomplished. Even approximate analyses are beyond the scope of this text, and we refer to the results of such analyses only in a qualitative way. Detailed discussions are given in Blackman and Tukey (1958), Hannan (1960), Jenkins and Watts (1968), Koopmans (1995), Kay and Marple (1981), Marple (1987), and Kay (1988).

10.6.1 The Periodogram

Let us consider the problem of estimating the power density spectrum $P_{ss}(\Omega)$ of a continuous-time signal $s_c(t)$. An intuitive approach to the estimation of the power spectrum is suggested by Figure 10.1 and the associated discussion in Section 10.1; based on that approach, we now assume that the input signal $s_c(t)$ is a stationary random signal. The antialiasing lowpass filter creates a new stationary random signal whose power spectrum is bandlimited, so that the signal can be sampled without aliasing. Then $x[n]$ is a stationary discrete-time random signal whose power density spectrum $P_{xx}(\omega)$ is proportional to $P_{ss}(\Omega)$ over the bandwidth of the antialiasing filter; i.e.,

$$P_{xx}(\omega) = \frac{1}{T} P_{ss}\left(\frac{\omega}{T}\right), \qquad |\omega| < \pi, \tag{10.50}$$

where we have assumed that the cutoff frequency of the antialiasing filter is π/T and that T is the sampling period. (See Problem 10.33 for a further consideration of

---

[5]The term *power spectrum* is commonly used interchangeably with the more precise term *power density spectrum.*

sampling of random signals.) Consequently, a reasonable estimate of $P_{xx}(\omega)$ will provide a reasonable estimate of $P_{ss}(\Omega)$. The window $w[n]$ in Figure 10.1 selects a finite-length segment (L samples) of $x[n]$, which we denote $v[n]$, the Fourier transform of which is

$$V(e^{j\omega}) = \sum_{n=0}^{L-1} w[n]x[n]e^{-j\omega n}. \tag{10.51}$$

Consider as an estimate of the power spectrum the quantity

$$I(\omega) = \frac{1}{LU}|V(e^{j\omega})|^2, \tag{10.52}$$

where the constant U anticipates a need for normalization to remove bias in the spectral estimate. When the window $w[n]$ is the rectangular window sequence, this estimator for the power spectrum is called the *periodogram*. If the window is not rectangular, $I(\omega)$ is called the *modified periodogram*. Clearly, the periodogram has some of the basic properties of the power spectrum. It is nonnegative, and for real signals, it is a real and even function of frequency. Furthermore, it can be shown (Problem 10.26) that

$$I(\omega) = \frac{1}{LU} \sum_{m=-(L-1)}^{L-1} c_{vv}[m]e^{-j\omega m}, \tag{10.53}$$

where

$$c_{vv}[m] = \sum_{n=0}^{L-1} x[n]w[n]x[n+m]w[n+m]. \tag{10.54}$$

We note that the sequence $c_{vv}[m]$ is the aperiodic correlation sequence for the finite-length sequence $v[n] = w[n]x[n]$. Consequently, the periodogram is in fact the Fourier transform of the aperiodic correlation of the windowed data sequence.

Explicit computation of the periodogram can be carried out only at discrete frequencies. From Eqs. (10.51) and (10.52), we see that if the discrete-time Fourier transform of $w[n]x[n]$ is replaced by its DFT, we will obtain samples at the DFT frequencies $\omega_k = 2\pi k/N$ for $k = 0, 1, \ldots, N-1$. Specifically, samples of the periodogram are given by

$$I(\omega_k) = \frac{1}{LU}|V[k]|^2, \tag{10.55}$$

where $V[k]$ is the N-point DFT of $w[n]x[n]$. If we want to choose N to be greater than the window length L, appropriate zero-padding would be applied to the sequence $w[n]x[n]$.

If a random signal has a nonzero mean, its power spectrum has an impulse at zero frequency. If the mean is relatively large, this component will dominate the spectrum estimate, causing low-amplitude, low-frequency components to be obscured by leakage. Therefore, in practice the mean is often estimated using Eq. (10.48), and the resulting estimate is subtracted from the random signal before computing the power spectrum estimate. Although the sample mean is only an approximate estimate of the zero-frequency component, subtracting it from the signal often leads to better estimates at neighboring frequencies.

10.6.2 Properties of the Periodogram

The nature of the periodogram estimate of the power spectrum can be determined by recognizing that, for each value of ω, $I(\omega)$ is a random variable. By computing the mean and variance of $I(\omega)$, we can determine whether the estimate is biased and whether it is consistent.

From Eq. (10.53), the expected value of $I(\omega)$ is

$$\mathcal{E}\{I(\omega)\} = \frac{1}{LU} \sum_{m=-(L-1)}^{L-1} \mathcal{E}\{c_{vv}[m]\}e^{-j\omega m}. \tag{10.56}$$

The expected value of $c_{vv}[m]$ can be expressed as

$$\begin{aligned}
\mathcal{E}\{c_{vv}[m]\} &= \sum_{n=0}^{L-1} \mathcal{E}\{x[n]w[n]x[n+m]w[n+m]\} \\
&= \sum_{n=0}^{L-1} w[n]w[n+m]\mathcal{E}\{x[n]x[n+m]\}.
\end{aligned} \tag{10.57}$$

Since we are assuming that $x[n]$ is stationary,

$$\mathcal{E}\{x[n]x[n+m]\} = \phi_{xx}[m], \tag{10.58}$$

and Eq. (10.57) can then be rewritten as

$$\mathcal{E}\{c_{vv}[m]\} = c_{ww}[m]\phi_{xx}[m], \tag{10.59}$$

where $c_{ww}[m]$ is the aperiodic autocorrelation of the window, i.e.,

$$c_{ww}[m] = \sum_{n=0}^{L-1} w[n]w[n+m]. \tag{10.60}$$

From Eq. (10.56), Eq. (10.59), and the modulation–windowing property of Fourier transforms (Section 2.9.7), it follows that

$$\mathcal{E}\{I(\omega)\} = \frac{1}{2\pi LU} \int_{-\pi}^{\pi} P_{xx}(\theta)C_{ww}(e^{j(\omega-\theta)})d\theta, \tag{10.61}$$

where $C_{ww}(e^{j\omega})$ is the Fourier transform of the aperiodic autocorrelation of the window, i.e.,

$$C_{ww}(e^{j\omega}) = |W(e^{j\omega})|^2. \tag{10.62}$$

According to Eq. (10.61), the (modified) periodogram is a biased estimate of the power spectrum, since $\mathcal{E}\{I(\omega)\}$ is not equal to $P_{xx}(\omega)$. Indeed, we see that the bias arises as a result of convolution of the true power spectrum with the Fourier transform of the aperiodic autocorrelation of the data window. If we increase the window length, we expect that $W(e^{j\omega})$ should become more concentrated around $\omega = 0$, and thus $C_{ww}(e^{j\omega})$ should look increasingly like a periodic impulse train. If the scale factor $1/(LU)$ is correctly chosen, then $\mathcal{E}\{I(\omega)\}$ should approach $P_{xx}(\omega)$ as $W(e^{j\omega})$ approaches a periodic impulse train. The scale can be adjusted by choosing the normalizing constant U so that

$$\frac{1}{2\pi LU} \int_{-\pi}^{\pi} |W(e^{j\omega})|^2 d\omega = \frac{1}{LU} \sum_{n=0}^{L-1} (w[n])^2 = 1, \tag{10.63}$$

or

$$U = \frac{1}{L} \sum_{n=0}^{L-1} (w[n])^2. \tag{10.64}$$

For the rectangular window, we would then choose $U = 1$, while other data windows would require a value of $0 < U < 1$ if $w[n]$ is normalized to a maximum value of 1. Alternatively, the normalization can be absorbed into the amplitude of $w[n]$. Therefore, if properly normalized, the (modified) periodogram is asymptotically unbiased; i.e., the bias approaches zero as the window length increases.

To examine whether the periodogram is a consistent estimate or becomes a consistent estimate as the window length increases, it is necessary to consider the behavior of the variance of the periodogram. An expression for the variance of the periodogram is very difficult to obtain even in the simplest cases. However, it has been shown (see Jenkins and Watts, 1968) that over a wide range of conditions, as the window length increases,

$$\text{var}[I(\omega)] \simeq P_{xx}^2(\omega). \tag{10.65}$$

That is, the variance of the periodogram estimate is approximately the same size as the square of the power spectrum that we are estimating. Therefore, since the variance does not asymptotically approach zero with increasing window length, the periodogram is not a consistent estimate.

The properties of the periodogram estimate of the power spectrum just discussed are illustrated in Figure 10.20, which shows periodogram estimates of white noise using rectangular windows of lengths $L = 16, 64, 256$, and 1024. The sequence $x[n]$ was obtained from a pseudorandom-number generator whose output was scaled so that $|x[n]| \leq \sqrt{3}$. A good random-number generator produces a uniform distribution of amplitudes, and the sample-to-sample correlation is small. Thus, the power spectrum of the output of the random-number generator could be modeled in this case by $P_{xx}(\omega) = \sigma_x^2 = 1$ for all ω. For each of the four rectangular windows, the periodogram was computed with normalizing constant $U = 1$ and at frequencies $\omega_k = 2\pi k/N$ for

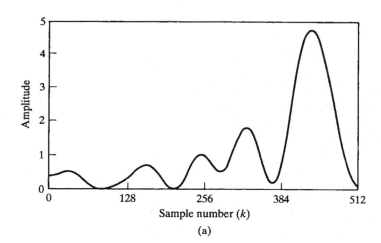

(a)

Figure 10.20 Periodograms of pseudorandom white-noise sequence. (a) Window length $L = 16$ and DFT length $N = 1024$.

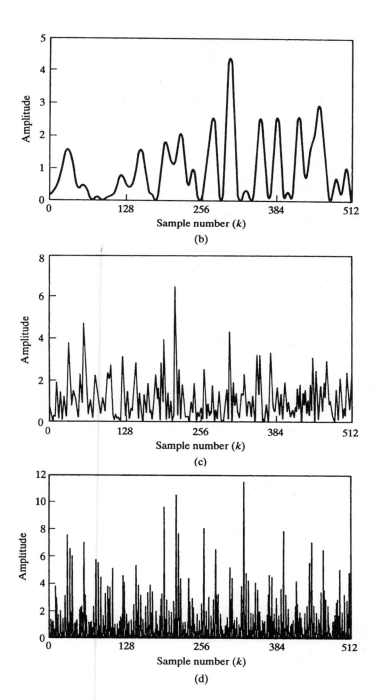

Figure 10.20 (*continued*) (b) $L = 64$ and $N = 1024$. (c) $L = 256$ and $N = 1024$. (d) $L = 1024$ and $N = 1024$.

$N = 1024$ using the DFT. That is,

$$I[k] = I(\omega_k) = \frac{1}{L}|V[k]|^2 = \frac{1}{L}\left|\sum_{n=0}^{L-1} w[n]x[n]e^{-j(2\pi/N)kn}\right|^2. \qquad (10.66)$$

In Figure 10.20, the DFT values are connected by straight lines for purposes of display. Recall that $I(\omega)$ is real and an even function of ω so we only need to plot $I[k]$ for $0 \leq k \leq N/2$ corresponding to $0 \leq \omega \leq \pi$. We note that the spectral estimate fluctuates more rapidly as the window length L increases. This behavior can be understood by recalling that, although we view the periodogram method as a direct computation of the spectral estimate, we have seen that the underlying correlation estimate of Eq. (10.54) is, in effect, Fourier transformed to obtain the periodogram. Figure 10.21 illustrates a windowed sequence, $x[n]w[n]$, and a shifted version, $x[n+m]w[n+m]$, as required in Eq. (10.54). From this figure, we see that $L - m - 1$ signal values are involved in computing a particular correlation lag value $c_{vv}[m]$. Thus, when m is close to L, only a few values of $x[n]$ are involved in the computation, and we expect that the estimate of the correlation sequence will be considerably more inaccurate for these values of m and consequently will also show considerable variation between adjacent values of m. On the other hand, when m is small, many more samples are involved, and the variability of $c_{vv}[m]$ with m should not be as great. The variability at large values of m manifests itself in the Fourier transform as fluctuations at all frequencies, and thus, for large L, the periodogram estimate tends to vary rapidly with frequency. Indeed, it can be shown (see Jenkins and Watts, 1968) that if $N = L$, the periodogram estimates at the DFT frequencies $2\pi k/N$ become uncorrelated. Since, as N increases, the DFT frequencies get closer together, this behavior is inconsistent with our goal of obtaining a good estimate of the power spectrum. We would prefer to obtain a smooth spectrum

(a)

(b)

Figure 10.21 Illustration of sequences involved in Eq. (10.54). (a) A finite-length sequence. (b) Shifted sequence for $m > 0$.

estimate without random variations resulting from the estimation process. This can be accomplished by averaging multiple independent periodogram estimates to reduce the fluctuations.

10.6.3 Periodogram Averaging

The averaging of periodograms in spectrum estimation was first studied extensively by Bartlett (1953); later, after fast algorithms for computing the DFT were developed, Welch (1970) combined these computational algorithms with the use of a data window $w[n]$ to develop the method of averaging modified periodograms. In periodogram averaging, a data sequence $x[n], 0 \le n \le Q-1$, is divided into segments of length-L samples, with a window of length L applied to each; i.e., we form the segments

$$x_r[n] = x[rR+n]w[n], \qquad 0 \le n \le L-1. \tag{10.67}$$

If $R < L$ the segments overlap, and for $R = L$ the segments are contiguous. Note that Q denotes the length of the available data. The total number of segments depends on the values of, and relationship among, R, L, and Q. Specifically, there will be K full-length segments, where K is the largest integer for which $(K-1)R + (L-1) \le Q-1$. The periodogram of the rth segment is

$$I_r(\omega) = \frac{1}{LU}|X_r(e^{j\omega})|^2, \tag{10.68}$$

where $X_r(e^{j\omega})$ is the discrete-time Fourier transform of $x_r[n]$. Each $I_r(\omega)$ has the properties of a periodogram, as described previously. Periodogram averaging consists of averaging together the K periodogram estimates $I_r(\omega)$; i.e., we form the time-averaged periodogram defined as

$$\bar{I}(\omega) = \frac{1}{K}\sum_{r=0}^{K-1} I_r(\omega). \tag{10.69}$$

To examine the bias and variance of $\bar{I}(\omega)$, let us take $L = R$, so that the segments do not overlap, and assume that $\phi_{xx}[m]$ is small for $m > L$; i.e., signal samples more than L apart are approximately uncorrelated. Then it is reasonable to assume that the periodograms $I_r(\omega)$ will be identically distributed independent random variables. Under this assumption, the expected value of $\bar{I}(\omega)$ is

$$\mathcal{E}\{\bar{I}(\omega)\} = \frac{1}{K}\sum_{r=0}^{K-1} \mathcal{E}\{I_r(\omega)\}, \tag{10.70}$$

or, since we assume that the periodograms are independent and identically distributed,

$$\mathcal{E}\{\bar{I}(\omega)\} = \mathcal{E}\{I_r(\omega)\} \qquad \text{for any } r. \tag{10.71}$$

From Eq. (10.61), it follows that

$$\mathcal{E}\{\bar{I}(\omega)\} = \mathcal{E}\{I_r(\omega)\} = \frac{1}{2\pi LU}\int_{-\pi}^{\pi} P_{xx}(\theta)C_{ww}(e^{j(\omega-\theta)})d\theta, \tag{10.72}$$

where L is the window length. When the window $w[n]$ is the rectangular window, the method of averaging periodograms is called *Bartlett's procedure,* and in this case it can be shown that

$$c_{ww}[m] = \begin{cases} L - |m|, & |m| \leq (L-1), \\ 0 & \text{otherwise,} \end{cases} \tag{10.73}$$

and, therefore,

$$C_{ww}(e^{j\omega}) = \left(\frac{\sin(\omega L/2)}{\sin(\omega/2)} \right)^2. \tag{10.74}$$

That is, the expected value of the average periodogram spectrum estimate is the convolution of the true power spectrum with the Fourier transform of the triangular sequence $c_{ww}[n]$ that results as the autocorrelation of the rectangular window. Thus, the average periodogram is also a biased estimate of the power spectrum.

To examine the variance, we use the fact that, in general, the variance of the average of K independent identically distributed random variables is $1/K$ times the variance of each individual random variable. (See Papoulis, 1991.) Therefore, the variance of the average periodogram is

$$\text{var}[\bar{I}(\omega)] = \frac{1}{K} \text{var}[I_r(\omega)], \tag{10.75}$$

or, with Eq. (10.65), it follows that

$$\text{var}[\bar{I}(\omega)] \simeq \frac{1}{K} P_{xx}^2(\omega). \tag{10.76}$$

Consequently, the variance of $\bar{I}(\omega)$ is inversely proportional to the number of periodograms averaged, and as K increases, the variance approaches zero.

From Eq. (10.74), we see that as L, the length of the segment $x_r[n]$, increases, the main lobe of $C_{ww}(e^{j\omega})$ decreases in width, and consequently, from Eq. (10.72), $\mathcal{E}\{\bar{I}(\omega)\}$ more closely approximates $P_{xx}(\omega)$. However, for fixed total data length Q, the total number of segments (assuming that $L = R$) is Q/L; therefore, as L increases, K decreases. Correspondingly, from Eq. (10.76), the variance of $\bar{I}(\omega)$ will increase. Thus, as is typical in statistical estimation problems, for a fixed data length there is a trade-off between bias and variance. However, as the data length Q increases, both L and K can be allowed to increase, so that as Q approaches ∞, the bias and variance of $\bar{I}(\omega)$ can approach zero. Consequently, periodogram averaging provides an asymptotically unbiased, consistent estimate of $P_{xx}(\omega)$.

The preceding discussion assumed that nonoverlapping rectangular windows were used in computing the time-dependent periodograms. Welch (1970) showed that if a different window shape is used, the variance of the average periodogram still behaves as in Eq. (10.76). Welch also considered the case of overlapping windows and showed that if the overlap is one-half the window length, the variance is further reduced by almost a factor of 2, due to the doubling of the number of sections. Greater overlap does not continue to reduce the variance, because the segments become less and less independent as the overlap increases.

10.6.4 Computation of Average Periodograms Using the DFT

As with the periodogram, the average periodogram can be explicitly evaluated only at a discrete set of frequencies. Because of the availability of the fast Fourier transform algorithms for computing the DFT, a particularly convenient and widely used choice is the set of frequencies $\omega_k = 2\pi k/N$ for an appropriate choice of N. From Eq. (10.69), we see that if the DFT of $x_r[n]$ is substituted for the Fourier transform of $x_r[n]$ in Eq. (10.68), we obtain samples of $\bar{I}(\omega)$ at the DFT frequencies $\omega_k = 2\pi k/N$ for $k = 0, 1, \ldots, N-1$. Specifically, with $X_r[k]$ denoting the DFT of $x_r[n]$,

$$I_r[k] = I_r(\omega_k) = \frac{1}{LU}|X_r[k]|^2, \tag{10.77a}$$

$$\bar{I}[k] = \bar{I}(\omega_k) = \frac{1}{K}\sum_{r=0}^{K-1} I_r[k]. \tag{10.77b}$$

We denote $I_r(2\pi k/N)$ as the sequence $I_r[k]$ and $\bar{I}(2\pi k/N)$ as the sequence $\bar{I}[k]$. According to Eqs. (10.77a) and (10.77b), the average periodogram estimate of the power spectrum is computed at N equally spaced frequencies by averaging the DFTs of the windowed data segments with the normalizing factor LU. This method of power spectrum estimation provides a very convenient framework within which to trade between resolution and variance of the spectral estimate. It is particularly simple and efficient to implement using the fast Fourier transform algorithms discussed in Chapter 9. An important advantage of the method over those to be discussed in Section 10.7 is that the spectrum estimate is always nonnegative.

10.6.5 An Example of Periodogram Analysis

Power spectrum analysis is a valuable tool for modeling signals, and it also can be used to detect signals, particularly when it comes to finding hidden periodicities in sampled signals. As an example of this type of application of the average periodogram method, consider the sequence

$$x[n] = A\cos(\omega_0 n + \theta) + e[n], \tag{10.78}$$

where θ is a random variable uniformly distributed between 0 and 2π and $e[n]$ is a zero-mean white-noise sequence that has a constant power spectrum; i.e., $P_{ee}(\omega) = \sigma_e^2$ for all ω. In signal models of this form, the cosine is generally the desired component and $e[n]$ is an undesired noise component. Often, in practical signal detection problems we are interested in the case for which the power in the cosine signal is small compared with the noise power. It can be shown (see Problem 10.34) that over one period in frequency, the power spectrum for this signal is

$$P_{xx}(\omega) = \frac{A^2\pi}{2}[\delta(\omega - \omega_0) + \delta(\omega + \omega_0)] + \sigma_e^2 \qquad \text{for } |\omega| \leq \pi. \tag{10.79}$$

From Eqs. (10.72) and (10.79), it follows that the expected value of the average periodogram is

$$\mathcal{E}\{\bar{I}(\omega)\} = \frac{A^2}{4LU}[C_{ww}(e^{j(\omega-\omega_0)}) + C_{ww}(e^{j(\omega+\omega_0)})] + \sigma_e^2. \tag{10.80}$$

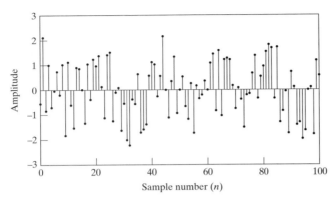

Figure 10.22   Cosine sequence with white noise, as in Eq. (10.78).

Figures 10.22 and 10.23 show the use of the averaging method for a signal of the form of Eq. (10.78), with $A = 0.5$, $\omega_0 = 2\pi/21$, and random phase $0 \le \theta < 2\pi$. The noise was uniformly distributed in amplitude such that $-\sqrt{3} < e[n] \le \sqrt{3}$. Therefore, it is easily shown that $\sigma_e^2 = 1$. The mean of the noise component is zero. Figure 10.22 shows 101 samples of the sequence $x[n]$. Since the noise component $e[n]$ has maximum amplitude $\sqrt{3}$, the cosine component in the sequence $x[n]$ is not visually apparent.

Figure 10.23 shows average periodogram estimates of the power spectrum for rectangular windows with amplitude 1, so that $U = 1$, and of lengths $L = 1024, 256, 64$, and 16, with the total record length $Q = 1024$ in all cases. Except for Figure 10.23(a), the windows overlap by one-half the window length. Figure 10.23(a) is the periodogram of the entire record, and Figures 10.23(b), (c), and (d) show the average periodogram for $K = 7, 31$, and 127 segments, respectively. In all cases, the average periodogram was evaluated using 1024-point DFTs at frequencies $\omega_k = 2\pi k/1024$. (For window lengths $L < 1024$, we must pad the windowed sequence with zero-samples before computing the DFT.) Therefore, the frequency $\omega_0 = 2\pi/21$ lies between DFT frequencies $\omega_{48} = 2\pi 48/1024$ and $\omega_{49} = 2\pi 49/1024$.

In using such estimates of the power spectrum to detect the presence and/or the frequency of the cosine component, we might search for the highest peaks in the spectral estimate and compare their size with that of the surrounding spectral values. From Eqs. (10.74) and (10.80), the expected value of the average periodogram at the frequency ω_0 is

$$\mathcal{E}\{\bar{I}(\omega_0)\} = \frac{A^2 L}{4} + \sigma_e^2. \tag{10.81}$$

Thus, if the peak due to the cosine component is to stand out against the variability of the average periodogram, then in this special case we must choose L so that $A^2 L/4 \gg \sigma_e^2$. This is illustrated by Figure 10.23(a), where L is as large as it can be for the record length Q. We see that $L = 1024$ gives a very narrow main lobe of the Fourier transform of the autocorrelation of the rectangular window, so it would be possible to resolve very closely spaced sinusoidal signals. Note that for the parameters of this example ($A = 0.5$, $\sigma_e^2 = 1$) and with $L = 1024$, the peak amplitude in the periodogram at frequency $2\pi/21$ is close, but not equal, to the expected value of 65. We also observe additional peaks in the periodogram with amplitudes greater than 10. Clearly, if the cosine amplitude A had been smaller by only a factor of 2, it is possible that its peak would have been confused with the inherent variability of the periodogram.

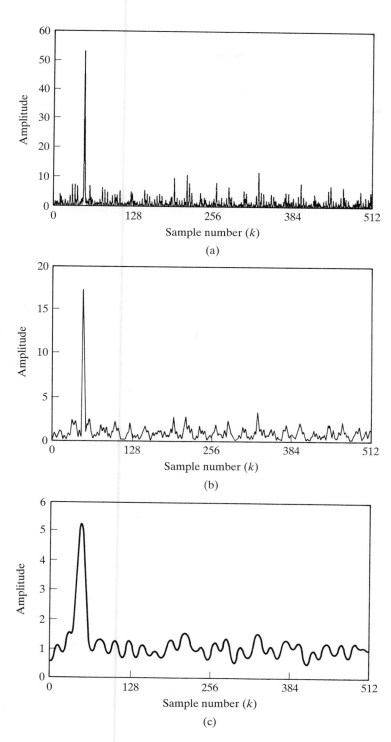

Figure 10.23 Example of average periodogram for signal of length $Q = 1024$. (a) Periodogram for window length $L = Q = 1024$ (only one segment). (b) $K = 7$ and $L = 256$ (overlap by $L/2$). (c) $K = 31$ and $L = 64$.

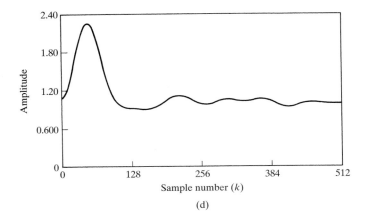

Figure 10.23 (*continued*)
(d) $K = 127$ and $L = 16$.

We have seen that the only sure way to reduce the variance of the spectrum estimate is to increase the record length of the signal. This is not always possible, and even if it is possible, longer records require more processing. We can reduce the variability of the estimate while keeping the record length constant if we use shorter windows and average over more sections. The cost of doing this is illustrated by parts (b), (c), and (d) of Figure 10.23. Note that as more sections are used, the variance of the spectral estimate decreases, but in accordance with Eq. (10.81), so does the amplitude of the peak due to the cosine. Thus, we again face a trade-off. That the shorter windows reduce variability is clear, especially if we compare the high-frequency regions away from the peak in parts (a), (b) and (c) of Figure 10.23. Recall that the idealized power spectrum of the model for the pseudorandom-noise generator is a constant ($\sigma_e^2 = 1$) for all frequencies. In Figure 10.23(a) there are peaks as high as about 10 when the true spectrum is 1. In Figure 10.23(b) the variation away from 1 is less than about 3, and in Figure 10.23(c) the variation around 1 is less than .5. However, shorter windows also reduce the peak amplitude of any narrowband component, and they also degrade our ability to resolve closely spaced sinusoids. This reduction in peak amplitude is also clear from Figure 10.23. Again, if we were to reduce A by a factor of 2 in Figure 10.23(b), the peak height would be approximately 4, which is not much different from many of the other peaks in the high-frequency region. In Figure 10.23(c) a reduction of A by a factor of 2 would make the peak approximately 1.25, which would be indistinguishable from the other ripples in the estimate. In Figure 10.23(d) the window is very short, and thus the fluctuations of the spectrum estimate are greatly reduced, but the spectral peak due to the cosine is very broad and barely above the noise even for $A = .5$. If the length were any smaller, spectral leakage from the negative-frequency component would cause there to be no distinct peak in the low-frequency region.

This example confirms that the average periodogram provides a straightforward method of trading off between spectral resolution and reduction of the variance of the spectral estimate. Although the theme of the example was the detection of a sinusoid in noise, the average periodogram could also be used in signal modeling. The spectral estimates of Figure 10.23 clearly suggest a signal model of the form of Eq. (10.78), and most of the parameters of the model could be estimated from the average periodogram power spectrum estimate.

10.7 SPECTRUM ANALYSIS OF RANDOM SIGNALS USING ESTIMATES OF THE AUTOCORRELATION SEQUENCE

In the previous section, we considered the periodogram as a direct estimate of the power spectrum of a random signal. The periodogram or the average periodogram is a direct estimate in the sense that it is obtained directly by Fourier transformation of the samples of the random signal. Another approach, based on the fact that the power density spectrum is the Fourier transform of the autocorrelation function, is to first estimate the autocorrelation function $\phi_{xx}[m]$ and then compute the Fourier transform of this estimate. In this section, we explore some of the important facets of this approach and show how the DFT can be used to implement it.

Let us assume, as before, that we are given a finite record of a random signal $x[n]$. This sequence is denoted

$$v[n] = \begin{cases} x[n] & \text{for } 0 \leq n \leq Q-1, \\ 0 & \text{otherwise.} \end{cases} \tag{10.82}$$

Consider an estimate of the autocorrelation sequence as

$$\hat{\phi}_{xx}[m] = \frac{1}{Q} c_{vv}[m], \tag{10.83}$$

where, since $c_{vv}[-m] = c_{vv}[m]$,

$$c_{vv}[m] = \sum_{n=0}^{Q-1} v[n]v[n+m] = \begin{cases} \displaystyle\sum_{n=0}^{Q-|m|-1} x[n]x[n+|m|], & |m| \leq Q-1, \\ 0 & \text{otherwise,} \end{cases} \tag{10.84}$$

corresponding to the aperiodic correlation of a rectangularly windowed segment of $x[n]$.

To determine the properties of this estimate of the autocorrelation sequence, we consider the mean and variance of the random variable $\hat{\phi}_{xx}[m]$. From Eqs. (10.83) and (10.84), it follows that

$$\mathcal{E}\{\hat{\phi}_{xx}[m]\} = \frac{1}{Q} \sum_{n=0}^{Q-|m|-1} \mathcal{E}\{x[n]x[n+|m|]\} = \frac{1}{Q} \sum_{n=0}^{Q-|m|-1} \phi_{xx}[m], \tag{10.85}$$

and since $\phi_{xx}[m]$ does not depend on n for a stationary random process,

$$\mathcal{E}\{\hat{\phi}_{xx}[m]\} = \begin{cases} \left(\dfrac{Q-|m|}{Q}\right)\phi_{xx}[m], & |m| \leq Q-1, \\ 0 & \text{otherwise.} \end{cases} \tag{10.86}$$

From Eq. (10.86), we see that $\hat{\phi}_{xx}[m]$ is a biased estimate of $\phi_{xx}[m]$, since $\mathcal{E}\{\hat{\phi}_{xx}[m]\}$ is not equal to $\phi_{xx}[m]$, but the bias is small if $|m| \ll Q$. We see also that an unbiased estimator of the autocorrelation sequence for $|m| \leq Q-1$ is

$$\check{\phi}_{xx}[m] = \left(\frac{1}{Q-|m|}\right) c_{vv}[m]; \tag{10.87}$$

i.e., the estimator is unbiased if we divide by the number of nonzero terms in the sum of lagged products rather than by the total number of samples in the data record.

The variance of the autocorrelation function estimates is difficult to compute, even with simplifying assumptions. However, approximate formulas for the variance of both $\hat{\phi}_{xx}[m]$ and $\check{\phi}_{xx}[m]$ can be found in Jenkins and Watts (1968). For our purposes here, it is sufficient to observe from Eq. (10.84) that as $|m|$ approaches Q, fewer and fewer samples of $x[n]$ are involved in the computation of the autocorrelation estimate, and therefore, the variance of the autocorrelation estimate can be expected to increase with increasing $|m|$. In the case of the periodogram, this increased variance affects the spectrum estimate at all frequencies because all the autocorrelation lag values are implicitly involved in the computation of the periodogram. However, by explicitly computing the autocorrelation estimate, we are free to choose which correlation lag values to include in estimating the power spectrum. Thus, we define the power spectrum estimate

$$S(\omega) = \sum_{m=-(M-1)}^{M-1} \hat{\phi}_{xx}[m]w_c[m]e^{-j\omega m}, \qquad (10.88)$$

where $w_c[m]$ is a symmetric window of length $(2M - 1)$ applied to the estimated autocorrelation function. We require that the product of the autocorrelation sequence and the window be an even sequence when $x[n]$ is real, so that the power spectrum estimate will be a real even function of ω. Therefore, the correlation window must be an even sequence. By limiting the length of the correlation window so that $M \ll Q$, we include only autocorrelation estimates for which the variance is low.

The mechanism by which windowing the autocorrelation sequence reduces the variance of the power spectrum estimate is best understood in the frequency domain. From Eqs. (10.53), (10.54), and (10.84), it follows that, with $w[n] = 1$ for $0 \le n \le (Q - 1)$, i.e., a rectangular window, the periodogram is the Fourier transform of the autocorrelation estimate $\hat{\phi}_{xx}[m]$; i.e.,

$$\hat{\phi}_{xx}[m] = \frac{1}{Q}c_{vv}[m] \overset{\mathcal{F}}{\longleftrightarrow} \frac{1}{Q}|V(e^{j\omega})|^2 = I(\omega). \qquad (10.89)$$

Therefore, from Eq. (10.88), the spectrum estimate obtained by windowing of $\hat{\phi}_{xx}[m]$ is the convolution

$$S(\omega) = \frac{1}{2\pi} \int_{-\pi}^{\pi} I(\theta)W_c(e^{j(\omega-\theta)})d\theta. \qquad (10.90)$$

From Eq. (10.90), we see that the effect of applying the window $w_c[m]$ to the autocorrelation estimate is to convolve the periodogram with the Fourier transform of the autocorrelation window. This will smooth the rapid fluctuations of the periodogram spectrum estimate. The shorter the correlation window, the smoother will be the spectrum estimate, and vice versa.

The power spectrum $P_{xx}(\omega)$ is a nonnegative function of frequency, and the periodogram and the average periodogram automatically have this property by definition. In contrast, from Eq. (10.90), it is evident that nonnegativity is not guaranteed for $S(\omega)$, unless we impose the further condition that

$$W_c(e^{j\omega}) \ge 0 \qquad \text{for } -\pi < \omega \le \pi. \qquad (10.91)$$

This condition is satisfied by the Fourier transform of the triangular (Bartlett) window, but it is not satisfied by the rectangular, Hanning, Hamming, or Kaiser windows. Therefore, although these latter windows have smaller side lobes than the triangular window, spectral leakage may cause negative spectral estimates in low-level regions of the spectrum.

The expected value of the smoothed periodogram is

$$\mathcal{E}\{S(\omega)\} = \sum_{m=-(M-1)}^{M-1} \mathcal{E}\{\hat{\phi}_{xx}[m]\}w_c[m]e^{-j\omega m}$$

$$= \sum_{m=-(M-1)}^{M-1} \phi_{xx}[m] \left(\frac{Q-|m|}{Q} \right) w_c[m]e^{-j\omega m}. \qquad (10.92)$$

If $Q \gg M$, the term $(Q-|m|)/Q$ in Eq. (10.92) can be neglected,[6] so we obtain

$$\mathcal{E}\{S(\omega)\} \cong \sum_{m=-(M-1)}^{M-1} \phi_{xx}[m]w_c[m]e^{-j\omega m} = \frac{1}{2\pi}\int_{-\pi}^{\pi} P_{xx}(\theta)W_c[e^{j(\omega-\theta)}]d\theta. \qquad (10.93)$$

Thus, the windowed autocorrelation estimate leads to a biased estimate of the power spectrum. Just as with the average periodogram, it is possible to trade spectral resolution for reduced variance of the spectrum estimate. If the length of the data record is fixed, we can have lower variance if we are willing to accept poorer resolution of closely spaced narrowband spectral components, or we can have better resolution if we can accept higher variance. If we are free to observe the signal for a longer time (i.e., increase the length Q of the data record), then both the resolution and the variance can be improved. The spectrum estimate $S(\omega)$ is asymptotically unbiased if the correlation window is normalized so that

$$\frac{1}{2\pi}\int_{-\pi}^{\pi} W_c(e^{j\omega})d\omega = 1 = w_c[0]. \qquad (10.94)$$

With this normalization, as we increase Q together with the length of the correlation window, the Fourier transform of the correlation window approaches a periodic impulse train and the convolution of Eq. (10.93) duplicates $P_{xx}(\omega)$.

The variance of $S(\omega)$ has been shown (see Jenkins and Watts, 1968) to be of the form

$$\text{var}[S(\omega)] \simeq \left(\frac{1}{Q} \sum_{m=-(M-1)}^{M-1} w_c^2[m] \right) P_{xx}^2(\omega). \qquad (10.95)$$

Comparing Eq. (10.95) with the corresponding result in Eq. (10.65) for the periodogram leads to the conclusion that, to reduce the variance of the spectrum estimate, we should choose M and the window shape, possibly subject to the condition of Eq. (10.91), so that the factor

$$\left(\frac{1}{Q} \sum_{m=-(M-1)}^{M-1} w_c^2[m] \right) \qquad (10.96)$$

is as small as possible. Problem 10.29 deals with the computation of this variance reduction factor for several commonly used windows.

---

[6]More precisely, we could define an effective correlation window $w_e[m] = w_c[m](Q-|m|)/Q$.

Estimation of the power spectrum based on the Fourier transform of an estimate of the autocorrelation function is a clear alternative to the method of averaging periodograms. It is not necessarily better in any general sense; it simply has different features and its implementation would be different. In some situations it may be desirable to compute estimates of both the autocorrelation sequence and the power spectrum, in which case it would be natural to use the method of this section. Problem 10.37 explores the issue of determining an autocorrelation estimate from the average periodogram.

10.7.1 Computing Correlation and Power Spectrum Estimates Using the DFT

The autocorrelation estimate

$$\hat{\phi}_{xx}[m] = \frac{1}{Q} \sum_{n=0}^{Q-|m|-1} x[n]x[n+|m|] \tag{10.97}$$

is required for $|m| \leq M - 1$ in the method of power spectrum estimation that we are considering. Since $\hat{\phi}_{xx}[-m] = \hat{\phi}_{xx}[m]$, it is necessary to compute Eq. (10.97) only for nonnegative values of m, i.e., for $0 \leq m \leq M - 1$. The DFT and its associated fast computational algorithms can be used to advantage in the computation of $\hat{\phi}_{xx}[m]$ if we observe that $\hat{\phi}_{xx}[m]$ is the aperiodic discrete convolution of the finite-length sequence $x[n]$ with $x[-n]$. If we compute $X[k]$, the N-point DFT of $x[n]$, and multiply by $X^*[k]$, we obtain $|X[k]|^2$, which corresponds to the circular convolution of the finite-length sequence $x[n]$ with $x[((-n))_N]$, i.e., a *circular autocorrelation*. As our discussion in Section 8.7 suggests, and as developed in Problem 10.27, it should be possible to augment the sequence $x[n]$ with zero-valued samples and force the circular autocorrelation to be equal to the desired aperiodic autocorrelation over the interval $0 \leq m \leq M - 1$.

To see how to choose N for the DFT, consider Figure 10.24. Figure 10.24(a) shows the two sequences $x[n]$ and $x[n + m]$ as functions of n for a particular positive value of m. Figure 10.24(b) shows the sequences $x[n]$ and $x[((n + m))_N]$ that are involved in the circular autocorrelation corresponding to $|X[k]|^2$. Clearly, the circular autocorrelation will be equal to $Q\hat{\phi}_{xx}[m]$ for $0 \leq m \leq M - 1$ if $x[((n + m))_N]$ does not wrap around and overlap $x[n]$ when $0 \leq m \leq M - 1$. From Figure 10.24(b), it follows that this will be the case whenever $N - (M - 1) \geq Q$ or $N \geq Q + M - 1$.

In summary, we can compute $\hat{\phi}_{xx}[m]$ for $0 \leq m \leq M-1$ by the following procedure:

1. Form an N-point sequence by augmenting $x[n]$ with $(M - 1)$ zero-samples.

2. Compute the N-point DFT,

$$X[k] = \sum_{n=0}^{N-1} x[n]e^{-j(2\pi/N)kn} \qquad \text{for } k = 0, 1, \ldots, N - 1.$$

3. Compute

$$|X[k]|^2 = X[k]X^*[k] \qquad \text{for } k = 0, 1, \ldots, N - 1.$$

4. Compute the inverse DFT of $|X[k]|^2$ to obtain

$$\tilde{c}_{vv}[m] = \frac{1}{N} \sum_{k=0}^{N-1} |X[k]|^2 e^{j(2\pi/N)km} \qquad \text{for } m = 0, 1, \ldots, N - 1.$$

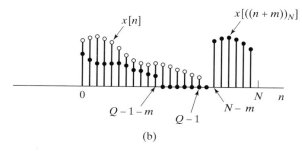

Figure 10.24 Computation of the circular autocorrelation. (a) $x[n]$ and $x[n+m]$ for a finite-length sequence of length Q. (b) $x[n]$ and $x[((n+m))_N]$ as in circular correlation.

5. Divide the resulting sequence by Q to obtain the autocorrelation estimate

$$\hat{\phi}_{xx}[m] = \frac{1}{Q}\tilde{c}_{vv}[m] \qquad \text{for } m = 0, 1, \ldots, M-1.$$

This is the desired set of autocorrelation values, which can be extended symmetrically for negative values of m.

If M is small, it may be more efficient simply to evaluate Eq. (10.97) directly. In this case, the amount of computation is proportional to $Q \cdot M$. In contrast, if the DFTs in this procedure are computed using one of the FFT algorithms discussed in Chapter 9 with $N \geq Q + M - 1$, the amount of computation will be approximately proportional to $N \log_2 N$ for N a power of 2. Consequently, for sufficiently large values of M, use of the FFT is more efficient than direct evaluation of Eq. (10.97). The exact break-even value of M will depend on the particular implementation of the DFT computations; however, as shown by Stockham (1966), this value would probably be less than $M = 100$.

We should remember that in order to reduce the variance of the estimate of the autocorrelation sequence or the power spectrum estimated from it, we must use large values of the record length Q. In such cases, it may be inconvenient or impossible to efficiently compute the ($N = Q + M - 1$)-point DFTs called for by the procedure just outlined. However, M is generally much less than Q. In such cases, it is possible to section the sequence $x[n]$ in a manner similar to the procedures that were discussed in Section 8.7.3 for convolution of a finite-length impulse response with an indefinitely long input sequence. Rader (1970) presented a particularly efficient and flexible procedure that uses many of the properties of the DFT of real sequences to reduce the amount of computation required. The development of this technique is the basis for Problem 10.38.

Once the autocorrelation estimate has been computed, samples of the power spectrum estimate $S(\omega)$ can be computed at frequencies $\omega_k = 2\pi k/N$ by forming the finite-length sequence

$$s[m] = \begin{cases} \hat{\phi}_{xx}[m]w_c[m], & 0 \le m \le M-1, \\ 0, & M \le m \le N-M, \\ \hat{\phi}_{xx}[N-m]w_c[N-m], & N-M+1 \le m \le N-1, \end{cases} \qquad (10.98)$$

where $w_c[m]$ is the symmetric correlation window. Then the DFT of $s[m]$ is

$$S[k] = S(\omega)|_{\omega=2\pi k/N}, \qquad k = 0, 1, \ldots, N-1, \qquad (10.99)$$

where $S(\omega)$ is the Fourier transform of the windowed autocorrelation sequence as defined by Eq. (10.88). Note that N can be chosen as large as is convenient and practical, thereby providing samples of $S(\omega)$ at closely spaced frequencies. However, the frequency resolution is always determined by the length and shape of the window $w_c[m]$.

10.7.2 An Example of Power Spectrum Estimation Based on Estimation of the Autocorrelation Sequence

In Chapter 4, we assumed that the error introduced by quantization is a white-noise random process. We can use the techniques of this section to illustrate the validity of this assumption by estimating the autocorrelation sequence and power spectrum of quantization noise.

Consider the experiment depicted in Figure 10.25. A lowpass-filtered speech signal $x_c(t)$ was sampled at a 10-KHz rate, yielding a sequence of samples $x[n]$. (Although the samples were quantized to 12 bits by the A/D converter, for purposes of this experiment we assume that the samples are unquantized.) The samples were rescaled so that $|x[n]| \le 16{,}000$. These samples were first quantized with an 8-bit linear quantizer, and the corresponding error sequence $e[n] = Q[x[n]] - x[n]$ was computed as shown in Figure 10.25. Figure 10.26(a) shows 400 consecutive samples of the speech signal, and Figure 10.26(b) shows the corresponding error sequence. (The samples are connected with straight lines for convenience in plotting.) Visual inspection and comparison of these two plots tends to strengthen our belief in the previously assumed model; however, the flatness of the quantization noise spectrum can be verified only by estimating the power spectrum of the quantization noise $e[n]$.

Figure 10.27 shows estimates of the normalized autocorrelation and power spectrum of the quantization noise for a record length of $Q = 2000$ samples. The mean of the noise sequence $e[n]$ was estimated and subtracted before the autocorrelation sequence estimate was calculated for $M = 512$. The resulting autocorrelation estimate was divided by $\hat{\phi}_{ee}[0]$ to obtain the normalized estimate $\hat{\rho}_{ee}[m] = \hat{\phi}_{ee}[m]/\hat{\phi}_{ee}[0]$, which is plotted in Figures 10.27(a) and (c). Note that the normalized autocorrelation is 1.0 at $m = 0$ and much smaller elsewhere. Indeed, $-0.0548 \le \hat{\rho}_{xx}[m] \le 0.0579$ for $1 \le m \le 512$. This

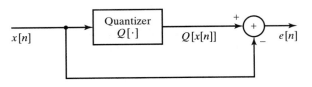

Figure 10.25   Procedure for obtaining quantization noise sequence.

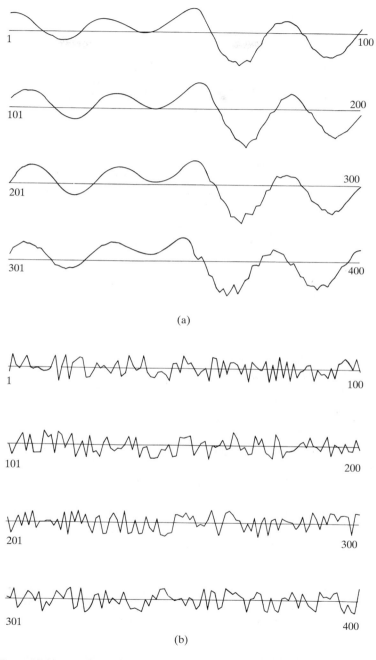

(a)

(b)

Figure 10.26   (a) Speech waveform and (b) the corresponding quantization error for 8-bit quantization (magnified 66 times with respect to part a). Each line corresponds to 100 consecutive samples connected by straight lines for convenience in plotting.

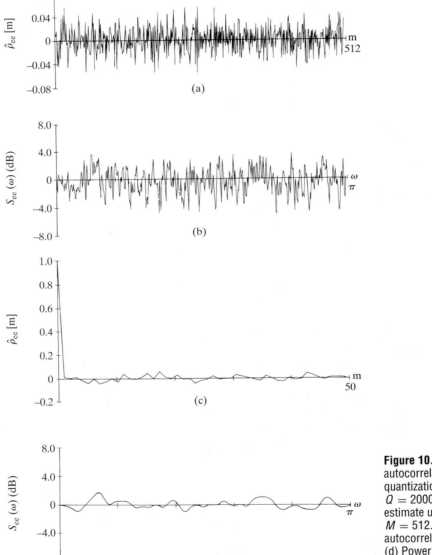

Figure 10.27   (a) Normalized autocorrelation estimate for 8-bit quantization noise; record length $Q = 2000$. (b) Power spectrum estimate using Bartlett window with $M = 512$. (c) Normalized autocorrelation estimate, $0 \leq m \leq 50$. (d) Power spectrum estimate using Bartlett window with $M = 50$.

seems to support our assumption that the error sequence is uncorrelated from sample to sample.

The power spectrum of the noise was estimated by windowing the normalized autocorrelation with a Bartlett window, as discussed in Section 10.7.1, and with $M = 512$. The result, depicted in Figure 10.27(b), plotted in dB, shows rather erratic fluctuations about 0 dB (the value of the normalized power spectrum of white noise). A smoother estimate is shown in Figure 10.27(d). In this case, a Bartlett window with $M = 50$ was used. The resulting smoothing, corresponding to a loss of resolution, is quite evident in comparing Figure 10.27(b) with Figure 10.27(d). We see from Figure 10.27(d) that the spectrum estimate is between -1.097 dB and $+1.631$ dB for all frequencies. Thus, we are again encouraged to believe that the white-noise model is appropriate for this case of quantization.

Although we have computed quantitative estimates of the autocorrelation and the power spectrum, our interpretation of these measurements has been only qualitative. It is reasonable now to ask how small the autocorrelation would be if $e[n]$ were really a white-noise process? To give quantitative answers to such questions, confidence intervals for our estimates could be computed and statistical decision theory applied. (See Jenkins and Watts, 1968, for some tests for white noise.) In many cases, however, this additional statistical treatment is not necessary. In a practical setting, we are often comfortable and content simply with the observation that the normalized autocorrelation is very small everywhere, except at $m = 0$.

One of the most important insights of this chapter is that the estimate of the autocorrelation and power spectrum of a stationary random process should improve if the record length is increased. This is illustrated by Figure 10.28, which corresponds to Figure 10.27, except that Q is increased to 14,000 samples. Recall that the variance of the autocorrelation estimate is proportional to $1/Q$. Thus, increasing Q from 2000 to 14,000 should bring about a sevenfold reduction in the variance of the estimate. A comparison of Figures 10.27(a) and 10.28(a) seems to verify this result. For $Q = 2000$, the estimate falls between the limits -0.0548 and $+0.0579$, while for $Q = 14,000$, the limits are -0.0254 and $+0.0231$. Comparing the range of variation for $Q = 2000$ with the range for $Q = 14,000$ indicates that the reduction is consistent with the sevenfold reduction

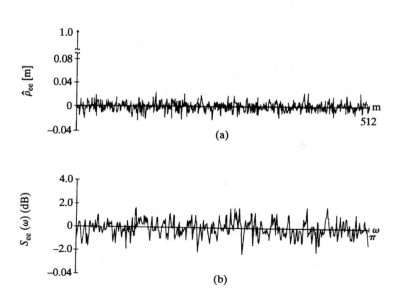

Figure 10.28   (a) Normalized autocorrelation estimate for 8-bit quantization noise; record length $Q = 14,000$. (b) Power spectrum estimate using Bartlett window with $M = 512$.

Figure 10.28 (*continued*)
(c) Normalized autocorrelation estimate, $0 \le m \le 50$. (d) Power spectrum estimate using Bartlett window with $M = 50$.

Figure 10.29 Quantization error waveform for 3-bit quantization. (Same scale as original signal, which is shown in Figure 10.26(a).)

in variance that we expected.[7] We note from Eq. (10.96) that a similar reduction in variance of the spectrum estimate is also expected. This is again evident in comparing Figures 10.27(b) and (d) with Figures 10.28(b) and (d), respectively. (Be sure to note that the scales are different between the two sets of plots.)

In Chapter 4 we argued that the white-noise model was reasonable, as long as the quantization step size was small. When the number of bits is small, this condition does not hold. To see the effect on the quantization noise spectrum, the previous experiment was repeated using only 8 quantization levels, or 3 bits. Figure 10.29 shows the quantization error for 3-bit quantization of the speech waveform segment shown in Figure 10.26(a). Note that portions of the error waveform tend to look very much like the original speech waveform. We would expect this to be reflected in the estimate of the power spectrum.

---

[7]Recall that a reduction in variance by a factor of 7 corresponds to a reduction in amplitude by a factor of $\sqrt{7} \approx 2.65$.

Figure 10.30 shows the autocorrelation and power spectrum estimates of the error sequence for 3-bit quantization for a record length of 14,000 samples. In this case, the autocorrelation shown in Figures 10.30 (a) and 10.30 (c) is much less like the ideal autocorrelation for white noise.

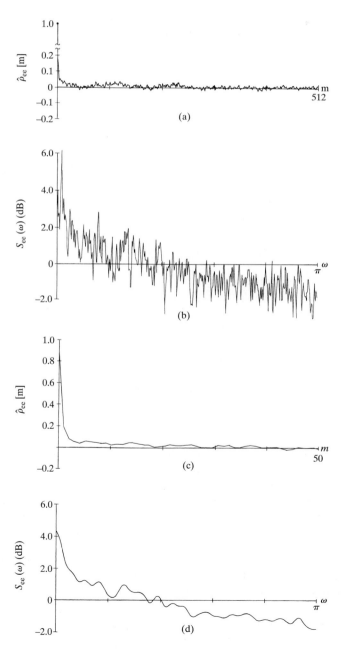

Figure 10.30   (a) Normalized autocorrelation estimate for 3-bit quantization noise; record length $Q = 14{,}000$. (b) Power spectrum estimate using Bartlett window with $M = 512$. (c) Normalized autocorrelation estimate, $0 \leq m \leq 50$. (d) Power spectrum estimate using Bartlett window with $M = 50$.

Figures 10.30 (b) and (d) show the power spectrum estimates for Bartlett windows with $M = 512$ and $M = 50$, respectively. Clearly, the spectrum is not flat. (In fact, it tends to have the general shape of the speech spectrum.) Thus, the white-noise model for quantization noise can be viewed only as a rather crude approximation in this case.

This example illustrates how autocorrelation and power spectrum estimates are often used to bolster our confidence in theoretical models. Specifically, we have demonstrated the validity of some of our basic assumptions in Chapter 4, and we have given an indication of how these assumptions break down for very crude quantization. This is only a rather simple, but useful, example of how the techniques of the current chapter are often applied in practice.

10.8 SUMMARY

One of the important applications of signal processing is the spectral analysis of signals. Because of the computational efficiency of the FFT, many of the techniques for spectral analysis of continuous-time or discrete-time signals utilize the DFT either directly or indirectly. In this chapter, we explored and illustrated some of these techniques.

Many of the issues associated with spectral analysis are best understood in the context of the analysis of sinusoidal signals. Since the use of the DFT requires finite-length signals, windowing must be applied in advance of the analysis. For sinusoidal signals, the width of the spectral peak observed in the DFT is dependent on the window length, with an increasing window length resulting in the sharpening of the peak. Consequently, the ability to resolve closely spaced sinusoids in the spectral estimate decreases as the window becomes shorter. A second, independent effect inherent in spectral analysis using the DFT is the associated spectral sampling. Specifically, since the spectrum can be computed only at a set of sample frequencies, the observed spectrum can be misleading if we are not careful in its interpretation. For example, important features in the spectrum may not be directly evident in the sampled spectrum. To avoid this, the spectral sample spacing can be reduced by increasing the DFT size in one of two ways. One method is to increase the DFT size while keeping the window length fixed (requiring zero-padding of the windowed sequence). This does not increase resolution. The second method is to increase both the window length and the DFT size. In this case spectral sample spacing is decreased and the ability to resolve closely spaced sinusoidal components is increased.

While increased window length and resolution are typically beneficial in the spectral analysis of stationary data, for time-varying data it is generally preferable to keep the window length sufficiently short so that over the window duration, the signal characteristics are approximately stationary. This leads to the concept of the time-dependent Fourier transform, which, in effect, is a sequence of Fourier transforms obtained as the signal slides past a finite-duration window. A common and useful interpretation of the time-dependent Fourier transform is as a bank of filters, with the frequency response of each filter corresponding to the transform of the window, frequency shifted to one of the DFT frequencies. The time-dependent Fourier transform has important applications both as an intermediate step in filtering signals and for analyzing and interpreting time-varying signals such as speech and radar signals. Spectral analysis of nonstationary signals typically involves a trade-off between time and frequency resolution. Specifically, our ability to track spectral characteristics in time increases as the length of the analysis window decreases. However, a shorter analysis window results in decreased frequency resolution.

The DFT also plays an important role in the analysis of stationary random signals. An intuitive approach to estimating the power spectrum of random signals is to compute the squared magnitude of the DFT of a segment of the signal. The resulting estimate, called the periodogram, is asymptotically unbiased. The variance of the periodogram estimate, however, does not decrease to zero as the length of the segment increases, and consequently, the periodogram is not a good estimate. However, by dividing the available signal sequence into shorter segments and averaging the associated periodograms, we can obtain a well-behaved estimate. An alternative approach is to first estimate the autocorrelation function. This can be done either directly or with the DFT. If a window is then applied to the autocorrelation estimates followed by the DFT, the result, referred to as the smoothed periodogram, is a good spectral estimate.

PROBLEMS

Basic Problems with Answers

10.1. A real continuous-time signal $x_c(t)$ is bandlimited to frequencies below 5 kHz; i.e., $X_c(j\Omega) = 0$ for $|\Omega| \geq 2\pi(5000)$. The signal $x_c(t)$ is sampled with a sampling rate of 10,000 samples per second (10 kHz) to produce a sequence $x[n] = x_c(nT)$ with $T = 10^{-4}$. Let $X[k]$ be the 1000-point DFT of $x[n]$.
   (a) To what continuous-time frequency does the index $k = 150$ in $X[k]$ correspond?
   (b) To what continuous-time frequency does the index $k = 800$ in $X[k]$ correspond?

10.2. A continuous-time signal $x_c(t)$ is bandlimited to 5 kHz; i.e., $X_c(j\Omega) = 0$ for $|\Omega| \geq 2\pi(5000)$. $x_c(t)$ is sampled with period T, producing the sequence $x[n] = x_c(nT)$. To examine the spectral properties of the signal, we compute the N-point DFT of a segment of N samples of $x[n]$ using a computer program that requires $N = 2^v$, where v is an integer. Determine the *minimum* value for N and the range of sampling rates

$$F_{\min} < \frac{1}{T} < F_{\max}$$

such that aliasing is avoided and the effective spacing between DFT values is *less* than 5 Hz; i.e., the equivalent continuous-time frequencies at which the Fourier transform is evaluated are separated by less than 5 Hz.

10.3. A speech signal is sampled with a sampling rate of 16,000 samples/s (16 kHz). A window of 20-ms duration is used in time-dependent Fourier analysis of the signal, as described in Section 10.3, with the window being advanced by 40 samples between computations of the DFT. Assume that the length of each DFT is $N = 2^v$.
   (a) How many samples are there in each segment of speech selected by the window?
   (b) What is the "frame rate" of the time-dependent Fourier analysis; i.e., how many DFT computations are done per second of input signal?

(c) What is the minimum size N of the DFT such that the original input signal can be reconstructed from the time-dependent Fourier transform?

(d) What is the spacing (in Hz) between the DFT samples for the minimum N from Part (c)?

10.4. A real-valued continuous-time segment of a signal $x_c(t)$ is sampled at a rate of 20,000 samples/sec, yielding a 1000-point finite-length discrete-time sequence $x[n]$ that is nonzero in the interval $0 \leq n \leq 999$. It is known that $x_c(t)$ is also bandlimited such that $X_c(j\Omega) = 0$ for $|\Omega| \geq 2\pi(10,000)$; i.e., assume that the sampling operation does not introduce any distortion due to aliasing.

$X[k]$ denotes the 1000-point DFT of $x[n]$. $X[800]$ is known to have the value $X[800] = 1 + j$.

(a) From the information given, can you determine $X[k]$ at any other values of k? If so, state which value(s) of k and what the corresponding value of $X[k]$ is. If not, explain why not.

(b) From the information given, state the value(s) of Ω for which $X_c(j\Omega)$ is known and the corresponding value(s) of $X_c(j\Omega)$.

10.5. A continuous-time signal $x_c(t) = \cos(\Omega_0 t)$ is sampled with period T to produce the sequence $x[n] = x_c(nT)$. An N-point rectangular window is applied to $x[n]$ for $0, 1, \ldots, N-1$, and $X[k]$, for $k = 0, 1, \ldots, N-1$, is the N-point DFT of the resulting sequence.

(a) Assuming that Ω_0, N, and k_0 are fixed, how should T be chosen so that $X[k_0]$ and $X[N - k_0]$ are nonzero and $X[k] = 0$ for all other values of k?

(b) Is your answer unique? If not, give another value of T that satisfies the conditions of Part (a).

10.6. Let $x_c(t)$ be a real-valued, bandlimited signal whose Fourier transform $X_c(j\Omega)$ is zero for $|\Omega| \geq 2\pi(5000)$. The sequence $x[n]$ is obtained by sampling $x_c(t)$ at 10 kHz. Assume that the sequence $x[n]$ is zero for $n < 0$ and $n > 999$.

Let $X[k]$ denote the 1000-point DFT of $x[n]$. It is known that $X[900] = 1$ and $X[420] = 5$. Determine $X_c(j\Omega)$ for as many values of Ω as you can in the region $|\Omega| < 2\pi(5000)$.

10.7. Consider estimating the spectrum of a discrete-time signal $x[n]$ using the DFT with a Hamming window for $w[n]$. A conservative rule of thumb for the frequency resolution of windowed DFT analysis is that the frequency resolution is equal to the width of the main lobe of $W(e^{j\omega})$. You wish to be able to resolve sinusoidal signals that are separated by as little as $\pi/100$ in ω. In addition, your window length L is constrained to be a power of 2. What is the minimum length $L = 2^\nu$ that will meet your resolution requirement?

10.8. Let $x[n]$ be a discrete-time signal whose spectrum you wish to estimate using a windowed DFT. You are required to obtain a frequency resolution of at least $\pi/25$ and are also required to use a window length $N = 256$. A safe estimate of the frequency resolution of a spectral estimate is the main-lobe width of the window used. Which of the windows in Table 7.1 will satisfy the criteria given for frequency resolution?

10.9. The following are three different signals $x_i[n]$ that are the sum of two sinusoids:

$$x_1[n] = \cos(\pi n/4) + \cos(17\pi n/64),$$

$$x_2[n] = \cos(\pi n/4) + 0.8\cos(21\pi n/64),$$

$$x_3[n] = \cos(\pi n/4) + 0.001\cos(21\pi n/64).$$

We wish to estimate the spectrum of each of these signals using a 64-point DFT with a

64-point rectangular window $w[n]$. Indicate which of the signals' 64-point DFTs you would expect to have two distinct spectral peaks after windowing.

10.10. Let $x[n]$ be a discrete-time signal obtained by sampling a continuous-time signal $x_c(t)$ with some sampling period T so that $x[n] = x_c(nT)$. Assume $x_c(t)$ is bandlimited to 100 Hz, i.e, $X_c(j\Omega) = 0$ for $|\Omega| \geq 2\pi(100)$. We wish to estimate the continuous-time spectrum $X_c(j\Omega)$ by computing a 1024-point DFT of $x[n]$, $X[k]$. What is the smallest value of T such that the equivalent frequency spacing between consecutive DFT samples $X[k]$ corresponds to 1 Hz or less in continuous-time frequency?

10.11. Let $x[n]$ be a 5000-point sequence obtained by sampling a continuous-time signal $x_c(t)$ at $T = 50\ \mu s$. Suppose $X[k]$ is the 8192-point DFT of $x[n]$. What is the equivalent frequency spacing in continuous time of adjacent DFT samples?

10.12. Assume that $x[n]$ is a 1000-point sequence obtained by sampling a continuous-time signal $x_c(t)$ at 8 kHz and that $X_c(j\Omega)$ is sufficiently bandlimited to avoid aliasing. What is the minimum DFT length N such that adjacent samples of $X[k]$ correspond to a frequency spacing of 5 Hz or less in the original continuous-time signal?

10.13. Let $X_r[k]$ be the time-dependent Fourier transform (TDFT) defined in Eq. (10.36). For this problem, consider the TDFT when both the DFT length $N = 36$ and the sampling interval $R = 36$. Let the window $w[n]$ be a rectangular window. Compute the TDFT $X_r[k]$ for $-\infty < r < \infty$ and $0 \leq k \leq N - 1$ for the signal

$$x[n] = \begin{cases} \cos(\pi n/6), & 0 \leq n \leq 35, \\ \cos(\pi n/2), & 36 \leq n \leq 71, \\ 0, & \text{otherwise.} \end{cases}$$

10.14. Figure P10.14-1 shows the magnitude $|V[k]|$ of the 128-point DFT $V[k]$ for a signal $v[n]$. The signal $v[n]$ was obtained by multiplying $x[n]$ by a 128-point rectangular window $w[n]$; i.e., $v[n] = x[n]w[n]$. Note that Figure P10.14-1 shows $|V[k]|$ only for the interval $0 \leq k \leq 64$. Which of the following signals could be $x[n]$? That is, which are consistent with

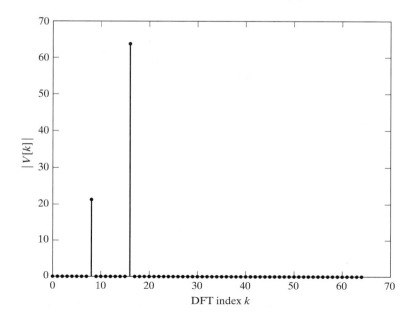

Figure P10.14-1

the information shown in the figure?

$$x_1[n] = \cos(\pi n/4) + \cos(0.26\pi n),$$

$$x_2[n] = \cos(\pi n/4) + (1/3)\sin(\pi n/8),$$

$$x_3[n] = \cos(\pi n/4) + (1/3)\cos(\pi n/8),$$

$$x_4[n] = \cos(\pi n/8) + (1/3)\cos(\pi n/16),$$

$$x_5[n] = (1/3)\cos(\pi n/4) + \cos(\pi n/8),$$

$$x_6[n] = \cos(\pi n/4) + (1/3)\cos(\pi n/8 + \pi/3).$$

10.15. Figure P??-1 shows the spectrogram of a chirp signal of the form

$$x[n] = \sin\left(\omega_0 n + \frac{1}{2}\lambda n^2\right).$$

Note that the spectrogram is a representation of the magnitude of $X[n, k]$, as defined in Eq. (??), where the dark regions indicate large values of $|X[n, k]|$. Based on the figure, estimate ω_0 and λ.

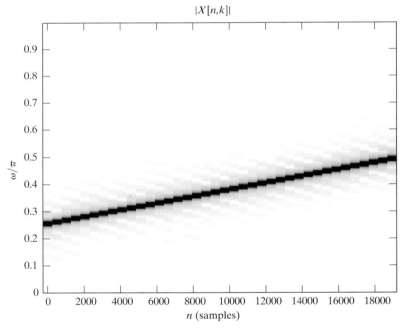

Figure P10.15-1

10.16. A continuous-time signal is sampled at a sampling rate of 10 kHz, and the DFT of 1024 samples is computed. Determine the continuous-time frequency spacing between spectral samples. Justify your answer.

10.17. A signal $x[n]$ is analyzed using the time-dependent Fourier transform $X_r[k]$, as defined in Eq. (10.36). Initially, the analysis is performed with an $N = 128$ DFT using an $L = 128$-point Hamming window $w[n]$. The time-domain sampling of adjacent blocks is $R = 128$; i.e., the windowed segments are offset by 128 samples in time. The frequency resolution obtained with this analysis is not sufficient, and it is desired to improve the resolution.

Several methods of modifying the analysis are suggested to accomplish this goal. Which of the following methods will improve the frequency resolution of the time-dependent Fourier transform $X_r[k]$?

METHOD 1:   Increase N to 256 while maintaining L and R at the same values.

METHOD 2:   Increase both N and L to 256, while maintaining R the same.

METHOD 3:   Decrease R to 64 while maintaining the same N and L.

METHOD 4:   Decrease L to 64 while maintaining the same N and R.

METHOD 5:   Maintain N, R and L the same, but change $w[n]$ to be a rectangular window.

10.18. Let $x[n]$ be a signal with a single sinusoidal component. The signal $x[n]$ is windowed with an L-point Hamming window $w[n]$ to obtain $v_1[n]$ before computing $V_1(e^{j\omega})$. The signal is then windowed with an L-point rectangular window to obtain $v_2[n]$, which is used to compute $V_2(e^{j\omega})$. Will the peaks in $V_2(e^{j\omega})$ and $V_1(e^{j\omega})$ have the same height? If so, justify your answer. If not, which should have a larger peak.

10.19. Assume that you wish to estimate the spectrum of $x[n]$ by applying a Kaiser window to the signal before computing the DTFT. You require that the side lobe of the window be 30 dB below the main lobe and that the frequency resolution be $\pi/40$. The width of the main lobe of the window is a safe estimate of the frequency resolution. Estimate the minimum window length L that will meet these requirements.

10.20. It is desired to estimate the spectrum of $x[n]$ by applying a 512-point Kaiser window to the signal before computing $X(e^{j\omega})$.

(a) The requirements for the frequency resolution of the system specify that the largest allowable main lobe for the Kaiser window is $\pi/100$. What is the best side lobe attenuation expected under these constraints?

(b) Suppose that you know that $x[n]$ contains two sinusoidal components at least $\pi/50$ apart, and that the amplitude of the stronger component is 1. Based on your answer to Part (a), give a threshold on the smallest value of the weaker component you would expect to see over the side lobe of the stronger sinusoid.

Basic Problems

10.21. Let $x[n] = \cos(2\pi n/5)$ and $v[n]$ be the sequence obtained by applying a 32-point rectangular window to $x[n]$ before computing $V(e^{j\omega})$. Sketch $|V(e^{j\omega})|$ for $-\pi \le \omega \le \pi$, labeling the frequencies of all peaks and the first nulls on either side of the peak. In addition, label the amplitudes of the peaks and the strongest side lobe of each peak.

10.22. Sketch the spectrogram obtained by using a 256-point rectangular window and 256-point DFTs with no overlap ($R = 256$) on the signal

$$x[n] = \cos\left[\frac{\pi n}{4} + 1000\sin\left(\frac{\pi n}{8000}\right)\right]$$

for the interval $0 \le n \le 16{,}000$.

Advanced Problems

10.23. Consider a real time-limited continuous-time signal $x_c(t)$ whose duration is 100 ms. Assume that this signal has a bandlimited Fourier transform such that $X_c(j\Omega) = 0$ for $|\Omega| \ge 2\pi(10{,}000)$ rad/s; i.e., assume that aliasing is negligible. We want to compute samples of $X_c(j\Omega)$ with 5-Hz spacing over the interval $0 \le \Omega \le 2\pi(10{,}000)$. This can be done with a

4000-point DFT. Specifically, we want to obtain a 4000-point sequence $x[n]$ for which the 4000-point DFT is related to $X_c(j\Omega)$ by

$$X[k] = \alpha X_c(j2\pi \cdot 5 \cdot k), \qquad k = 0, 1, \ldots, 1999,$$

where α is a known scale factor. Three methods are proposed to obtain a 4000-point sequence whose DFT gives the desired samples of $X_c(j\Omega)$.

METHOD 1: $x_c(t)$ is sampled with a sampling period $T = 25$ μs; i.e., we compute $X_1[k]$, the DFT of the sequence

$$x_1[n] = \begin{cases} x_c(nT), & n = 0, 1, \ldots, 3999, \\ 0, & \text{otherwise}. \end{cases}$$

Since $x_c(t)$ is time limited to 100 ms, $x_1[n]$ is a finite-length sequence of length 4000 (100 ms/25 μs).

METHOD 2: $x_c(t)$ is sampled with a sampling period of $T = 50$ μs. Since $x_c(t)$ is time limited to 100 ms, the resulting sequence will have only 2000 (100 ms/50 μs) nonzero samples; i.e.,

$$x_2[n] = \begin{cases} x_c(nT), & n = 0, 1, \ldots, 1999, \\ 0, & \text{otherwise}. \end{cases}$$

In other words, the sequence is padded with zero-samples to create a 4000-point sequence for which the 4000-point DFT $X_2[k]$ is computed.

METHOD 3: $x_c(t)$ is sampled with a sampling period of $T = 50$ μs, as in Method 2. The resulting 2000-point sequence is used to form the sequence $x_3[n]$ as follows:

$$x_3[n] = \begin{cases} x_c(nT), & 0 \le n \le 1999, \\ x_c((n - 2000)T), & 2000 \le n \le 3999, \\ 0, & \text{otherwise}. \end{cases}$$

The 4000-point DFT $X_3[k]$ of this sequence is computed.

For each of the three methods, determine how each 4000-point DFT is related to $X_c(j\Omega)$. Indicate this relationship in a sketch for a "typical" Fourier transform $X_c(j\Omega)$. State explicitly which method(s) provide the desired samples of $X_c(j\Omega)$.

10.24. A continuous-time finite-duration signal $x_c(t)$ is sampled at a rate of 20,000 samples/s, yielding a 1000-point finite-length sequence $x[n]$ that is nonzero in the interval $0 \le n \le 999$. Assume for this problem that the continuous-time signal is also bandlimited such that $X_c(j\Omega) = 0$ for $|\Omega| \ge 2\pi(10,000)$; i.e., assume that negligible aliasing distortion occurs in sampling. Assume also that a device or program is available for computing 1000-point DFTs and inverse DFTs.

(a) If $X[k]$ denotes the 1000-point DFT of the sequence $x[n]$, how is $X[k]$ related to $X_c(j\Omega)$? What is the effective continuous-time frequency spacing between DFT samples?

The following procedure is proposed for obtaining an expanded view of the Fourier transform $X_c(j\Omega)$ in the interval $|\Omega| \le 2\pi(5000)$, starting with the 1000-point DFT $X[k]$.

Step 1. Form the new 1000-point DFT

$$W[k] = \begin{cases} X[k], & 0 \le k \le 250, \\ 0, & 251 \le k \le 749, \\ X[k], & 750 \le k \le 999. \end{cases}$$

Step 2. Compute the inverse 1000-point DFT of $W[k]$, obtaining $w[n]$ for $n = 0, 1, \ldots, 999$.

Step 3. Decimate the sequence $w[n]$ by a factor of 2 and augment the result with 500 consecutive zero samples, obtaining the sequence

$$y[n] = \begin{cases} w[2n], & 0 \le n \le 499, \\ 0, & 500 \le n \le 999. \end{cases}$$

Step 4. Compute the 1000-point DFT of $y[n]$, obtaining $Y[k]$.

(b) The designer of this procedure asserts that

$$Y[k] = \alpha X_c(j2\pi \cdot 10 \cdot k), \qquad k = 0, 1, \ldots, 500,$$

where α is a constant of proportionality. Is this assertion correct? If not, explain why not.

10.25. Suppose that $y[n]$ is the output of a linear time-invariant FIR system with input $x[n]$; i.e.,

$$y[n] = \sum_{k=0}^{M} h[k]x[n - k].$$

(a) Obtain a relationship between the time-dependent Fourier transform $Y[n, \lambda)$ of the output of the linear system and the time-dependent Fourier transform $X[n, \lambda)$ of the input.

(b) Show that if the window is long compared to M, then

$$\check{Y}[n, \lambda) \simeq H(e^{j\lambda})\check{X}[n, \lambda),$$

where $H(e^{j\omega})$ is the frequency response of the linear system.

10.26. The periodogram $I(\omega)$ of a discrete-time random signal $x[n]$ was defined in Eq. (10.52) as

$$I(\omega) = \frac{1}{LU}|V(e^{j\omega})|^2,$$

where $V(e^{j\omega})$ is the discrete-time Fourier transform of the finite-length sequence $v[n] = w[n]x[n]$, with $w[n]$ a finite-length window sequence of length L, and U is a normalizing constant.

Show that the periodogram is also equal to $1/LU$ times the Fourier transform of the aperiodic autocorrelation sequence of $v[n]$; i.e.,

$$I(\omega) = \frac{1}{LU} \sum_{m=-(L-1)}^{L-1} c_{vv}[m]e^{-j\omega m},$$

where

$$c_{vv}[m] = \sum_{n=0}^{L-1} v[n]v[n + m].$$

10.27. Consider a finite-length sequence $x[n]$ such that $x[n] = 0$ for $n < 0$ and $n \ge L$. Let $X[k]$ be the N-point DFT of the sequence $x[n]$, where $N > L$. Define $c_{xx}[m]$ to be the aperiodic autocorrelation function of $x[n]$; i.e.,

$$c_{xx}[m] = \sum_{n=-\infty}^{\infty} x[n]x[n + m].$$

Define

$$\tilde{c}_{xx}[m] = \frac{1}{N} \sum_{m=0}^{N-1} |X[k]|^2 e^{j(2\pi/N)km}, \qquad m = 0, 1, \ldots, N - 1.$$

(a) Determine the minimum value of N that can be used for the DFT if we require that
$$c_{xx}[m] = \tilde{c}_{xx}[m], \qquad 0 \leq m \leq L-1.$$

(b) Determine the minimum value of N that can be used for the DFT if we require that
$$c_{xx}[m] = \tilde{c}_{xx}[m], \qquad 0 \leq m \leq M-1,$$

where $M < L$.

10.28. The symmetric Bartlett window, which arises in many aspects of power spectrum estimation, it is defined as

$$w_B[m] = \begin{cases} 1 - |m|/M, & |m| \leq M-1, \\ 0, & \text{otherwise.} \end{cases} \qquad \text{(P10.28-1)}$$

The Bartlett window is particularly attractive for obtaining estimates of the power spectrum by windowing an estimated autocorrelation function, as discussed in Section 10.7. This is because its Fourier transform is nonnegative, which guarantees that the smoothed spectrum estimate will be nonnegative at all frequencies.

(a) Show that the Bartlett window as defined in Eq. (P10.28-1) is $(1/M)$ times the aperiodic autocorrelation of the sequence $(u[n] - u[n-M])$.

(b) From the result of Part (a), show that the Fourier transform of the Bartlett window is

$$W_B(e^{j\omega}) = \frac{1}{M} \left[\frac{\sin(\omega M/2)}{\sin(\omega/2)} \right]^2, \qquad \text{(P10.28-2)}$$

which is clearly nonnegative.

(c) Describe a procedure for generating other finite-length window sequences that have nonnegative Fourier transforms.

10.29. In Section 10.7, we showed that a smoothed estimate of the power spectrum can be obtained by windowing an estimate of the autocorrelation sequence. It was stated (see Eq. (10.95)) that the variance of the smoothed spectrum estimate is

$$\text{var}[S(\omega)] \simeq F P_{xx}^2(\omega),$$

where F, the *variance ratio* or *variance reduction factor*, is

$$F = \frac{1}{Q} \sum_{m=-(M-1)}^{M-1} (w_c[m])^2 = \frac{1}{2\pi Q} \int_{-\pi}^{\pi} |W_c(e^{j\omega})|^2 d\omega.$$

As discussed in Section 10.7, Q is the length of the sequence $x[n]$ and $(2M-1)$ is the length of the symmetric window $w_c[m]$ that is applied to the autocorrelation estimate. Thus, if Q is fixed, the variance of the smoothed spectrum estimate can be reduced by adjusting the shape and duration of the window applied to the correlation function.

In this problem we will show that F decreases as the window length decreases, but we also know from the previous discussion of windows in Chapter 7 that the width of the main lobe of $W_c(e^{j\omega})$ increases with decreasing window length, so that the ability to resolve two adjacent frequency components is reduced as the window width decreases. Thus, there is a trade-off between variance reduction and resolution. We will study this trade-off for the following commonly used windows:

Rectangular

$$w_R[m] = \begin{cases} 1, & |m| \leq M-1, \\ 0, & \text{otherwise.} \end{cases}$$

Bartlett (triangular)

$$w_B[m] = \begin{cases} 1 - |m|/M, & |m| \le M-1, \\ 0, & \text{otherwise.} \end{cases}$$

Hanning/Hamming

$$w_H[m] = \begin{cases} \alpha + \beta \cos[\pi m/(M-1)], & |m| \le M-1, \\ 0, & \text{otherwise.} \end{cases}$$

($\alpha = \beta = 0.5$ for the Hanning window, and $\alpha = 0.54$ and $\beta = 0.46$ for the Hamming window.)

(a) Find the Fourier transform of each of the foregoing windows; i.e., compute $W_R(e^{j\omega})$, $W_B(e^{j\omega})$, and $W_R(e^{j\omega})$. Sketch each of these Fourier transforms as functions of ω.

(b) For each of the windows, show that the entries in the following table are approximately true when $M \gg 1$:

Window Name	Approximate Main-lobe Width	Approximate Variance Ratio (F)
Rectangular	$2\pi/M$	$2M/Q$
Bartlett	$4\pi/M$	$2M/(3Q)$
Hanning/Hamming	$3\pi/M$	$2M(\alpha^2 + \beta^2/2)/Q$

10.30. Consider a signal

$$x[n] = \left[\sin\left(\frac{\pi n}{2}\right) \right]^2 u[n]$$

whose time-dependent discrete Fourier transform is computed using the analysis window

$$w[n] = \begin{cases} 1, & 0 \le n \le 13, \\ 0, & \text{otherwise.} \end{cases}$$

Let $X[n,k] = X[n, 2\pi k/7)$ for $0 \le k \le 6$, where $X[n, \lambda)$ is defined as in Section 10.3.

(a) Determine $X[0, k]$ for $0 \le k \le 6$.

(b) Evaluate $\sum_{k=0}^{6} X[n, k]$ for $0 \le n < \infty$.

10.31. (a) Consider the system of Figure P10.31-1 with input $x(t) = e^{j(3\pi/8)10^4 t}$, sampling period $T = 10^{-4}$, and

$$w[n] = \begin{cases} 1, & 0 \le n \le N-1, \\ 0, & \text{otherwise.} \end{cases}$$

What is the smallest nonzero value of N such that $X_w[k]$ is nonzero at exactly one value of k?

(b) Suppose now that $N = 32$, the input signal is $x(t) = e^{j\Omega_0 t}$, and the sampling period T is chosen such that no aliasing occurs during the sampling process. Figures P10.31-2 and P10.31-3 show the magnitude of the sequence $X_w[k]$ for $k = 0, \ldots, 31$ for the

Figure P10.31-1

Figure P10.31-2

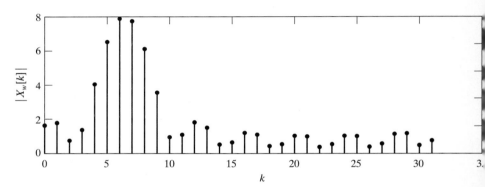

Figure P10.31-3

following two different choices of $w[n]$:

$$w_1[n] = \begin{cases} 1, & 0 \leq n \leq 31, \\ 0, & \text{otherwise,} \end{cases}$$

$$w_2[n] = \begin{cases} 1, & 0 \leq n \leq 7, \\ 0, & \text{otherwise.} \end{cases}$$

Indicate which figure corresponds to which choice of $w[n]$. State your reasoning clearly.

(c) For the input signal and system parameters of Part (b), we would like to estimate the value of Ω_0 from Figure P10.31-2 when the sampling period is $T = 10^{-4}$. Assuming that the sequence

$$w[n] = \begin{cases} 1, & 0 \leq n \leq 31, \\ 0, & \text{otherwise,} \end{cases}$$

and that the sampling period is sufficient to ensure that no aliasing occurs during sampling, estimate the value of Ω_0. Is your estimate exact? If it is not, what is the maximum possible error of your frequency estimate?

(d) Suppose you were provided with the exact values of the 32-point DFT $X_w[k]$ for the window choices $w_1[n]$ and $w_2[n]$. Briefly describe a procedure to obtain a precise estimate of Ω_0.

Extension Problems

10.32. Show that the time-dependent Fourier transform, as defined by Eq. (10.18), has the following properties:

(a) *Linearity:*

$$\text{If}\quad x[n] = ax_1[n] + bx_2[n], \quad \text{then}\quad X[n, \lambda] = a X_1[n, \lambda] + bX_2[n, \lambda].$$

(b) *Shifting:* If $y[n] = x[n - n_0]$, then $Y[n, \lambda] = X[n - n_0, \lambda]$.

(c) *Modulation:* If $y[n] = e^{j\omega_0 n}x[n]$, then $Y[n, \lambda] = e^{j\omega_0 n} X[n, \lambda - \omega_0]$.

(d) *Conjugate Symmetry:* If $x[n]$ is real, then $X[n, \lambda] = X^*[n, -\lambda]$.

10.33. Suppose that $x_c(t)$ is a continuous-time stationary random signal with autocorrelation function

$$\phi_c(\tau) = \mathcal{E}\{x_c(t)x_c(t + \tau)\}$$

and power density spectrum

$$P_c(\Omega) = \int_{-\infty}^{\infty} \phi_c(\tau)e^{-j\Omega\tau}\, d\tau.$$

Consider a discrete-time stationary random signal $x[n]$ that is obtained by sampling $x_c(t)$ with sampling period T; i.e., $x[n] = x_c(nT)$.

(a) Show that $\phi[m]$, the autocorrelation sequence for $x[n]$, is

$$\phi[m] = \phi_c(mT).$$

(b) What is the relationship between the power density spectrum $P_c(\Omega)$ for the continuous-time random signal and the power density spectrum $P(\omega)$ for the discrete-time random signal?

(c) What condition is necessary such that

$$P(\omega) = \frac{1}{T}P_c\left(\frac{\omega}{T}\right), \qquad |\omega| < \pi?$$

10.34. In Section 10.6.5, we considered the estimation of the power spectrum of a sinusoid plus white noise. In this problem, we will determine the true power spectrum of such a signal. Suppose that

$$x[n] = A\cos(\omega_0 n + \theta) + e[n],$$

where θ is a random variable that is uniformly distributed on the interval from 0 to 2π and $e[n]$ is a sequence of zero-mean random variables that are uncorrelated with each other and also uncorrelated with θ. In other words, the cosine component has a randomly selected phase, and $e[n]$ represents white noise.

(a) Show that for the preceding assumptions, the autocorrelation function for $x[n]$ is

$$\phi_{xx}[m] = \mathcal{E}\{x[n]x[m + n]\} = \frac{A^2}{2}\cos(\omega_0 m) + \sigma_e^2\delta[m],$$

where $\sigma_e^2 = \mathcal{E}\{(e[n])^2\}$.

(b) From the result of Part (a), show that over one period in frequency, the power spectrum of $x[n]$ is

$$P_{xx}(\omega) = \frac{A^2\pi}{2}[\delta(\omega - \omega_0) + \delta(\omega + \omega_0)] + \sigma_e^2, \qquad |\omega| \le \pi.$$

10.35. Consider a discrete-time signal $x[n]$ of length N samples that was obtained by sampling a stationary, white, zero-mean continuous-time signal. It follows that

$$\mathcal{E}\{x[n]x[m]\} = \sigma_x^2\delta[n - m],$$

$$\mathcal{E}\{x[n]\} = 0.$$

Suppose that we compute the DFT of the finite-length sequence $x[n]$, thereby obtaining $X[k]$ for $k = 0, 1, \ldots, N - 1$.

(a) Determine the approximate variance of $|X[k]|^2$ using Eqs. (10.65) and (10.66).

(b) Determine the cross-correlation between values of the DFT; i.e., determine $\mathcal{E}\{X[k]X^*[r]\}$ as a function of k and r.

10.36. A bandlimited continuous-time signal has a bandlimited power spectrum that is zero for $|\Omega| \geq 2\pi(10^4)$ rad/s. The signal is sampled at a rate of 20,000 samples/s over a time interval of 10 s. The power spectrum of the signal is estimated by the method of averaging periodograms as described in Section 10.6.3.

(a) What is the length Q (number of samples) of the data record?

(b) If a radix-2 FFT program is used to compute the periodograms, what is the minimum length N if we wish to obtain estimates of the power spectrum at equally spaced frequencies no more than 10 Hz apart?

(c) If the segment length L is equal to the FFT length N in Part (b), how many segments K are available if the segments do not overlap?

(d) Suppose that we wish to reduce the variance of the spectrum estimates by a factor of 10 while maintaining the frequency spacing of Part (b). Give two methods of doing this. Do these two methods give the same results? If not, explain how they differ.

10.37. Suppose that an estimate of the power spectrum of a signal is obtained by the method of averaging periodograms, as discussed in Section 10.6.3. That is, the spectrum estimate is

$$\bar{I}(\omega) = \frac{1}{K}\sum_{r=0}^{K-1} I_r(\omega),$$

where the K periodograms $I_r(\omega)$ are computed from L-point segments of the signal using Eqs. (10.67) and (10.68). We define an estimate of the autocorrelation function as the inverse Fourier transform of $\bar{I}(\omega)$; i.e.,

$$\bar{\phi}[m] = \frac{1}{2\pi}\int_{-\pi}^{\pi} \bar{I}(\omega)e^{j\omega m}d\omega.$$

(a) Show that

$$\mathcal{E}\{\bar{\phi}[m]\} = \frac{1}{LU}c_{ww}[m]\phi_{xx}[m],$$

where L is the length of the segments, U is a normalizing factor given by Eq. (10.64), and $c_{ww}[m]$, given by Eq. (10.60), is the aperiodic autocorrelation function of the window that is applied to the signal segments.

(b) In the application of periodogram averaging, we normally use an FFT algorithm to compute $\bar{I}(\omega)$ at N equally spaced frequencies; i.e.,

$$\bar{I}[k] = \bar{I}(2\pi k/N), \qquad k = 0, 1, \ldots, N - 1,$$

where $N \geq L$. Suppose that we compute an estimate of the autocorrelation function by computing the inverse DFT of $\bar{I}[k]$, as in

$$\bar{\phi}_p[m] = \frac{1}{N} \sum_{k=0}^{N-1} \bar{I}[k]e^{j(2\pi/N)km}, \qquad m = 0, 1, \ldots, N-1.$$

Obtain an expression for $\mathcal{E}\{\bar{\phi}_p[m]\}$.

(c) How should N be chosen so that

$$\mathcal{E}\{\bar{\phi}_p[m]\} = \mathcal{E}\{\bar{\phi}[m]\}, \qquad m = 0, 1, \ldots, L-1?$$

10.38. Consider the computation of the autocorrelation estimate

$$\hat{\phi}_{xx}[m] = \frac{1}{Q} \sum_{n=0}^{Q-|m|-1} x[n]x[n+|m|], \tag{P10.38-1}$$

where $x[n]$ is a real sequence. Since $\hat{\phi}_{xx}[-m] = \hat{\phi}_{xx}[m]$, it is necessary only to evaluate Eq. (P10.38-1) for $0 \le m \le M-1$ to obtain $\hat{\phi}_{xx}[m]$ for $-(M-1) \le m \le M-1$, as is required to estimate the power density spectrum using Eq. (10.88).

(a) When $Q \gg M$, it may not be feasible to compute $\hat{\phi}_{xx}[m]$ using a single FFT computation. In such cases, it is convenient to express $\hat{\phi}_{xx}[m]$ as a sum of correlation estimates based on shorter sequences. Show that if $Q = KM$,

$$\hat{\phi}_{xx}[m] = \frac{1}{Q} \sum_{i=0}^{K-1} c_i[m],$$

where

$$c_i[m] = \sum_{n=0}^{M-1} x[n+iM]x[n+iM+m],$$

for $0 \le m \le M-1$.

(b) Show that the correlations $c_i[m]$ can be obtained by computing the N-point *circular* correlations

$$\tilde{c}_i[m] = \sum_{n=0}^{N-1} x_i[n]y_i[((n+m))_N],$$

where the sequences

$$x_i[n] = \begin{cases} x[n+iM], & 0 \le n \le M-1, \\ 0, & M \le n \le N-1, \end{cases}$$

and

$$y_i[n] = x[n+iM], \qquad 0 \le n \le N-1. \tag{P10.38-2}$$

What is the *minimum* value of N (in terms of M) such that $c_i[m] = \tilde{c}_i[m]$ for $0 \le m \le M-1$?

(c) State a procedure for computing $\hat{\phi}_{xx}[m]$ for $0 \le m \le M-1$ that involves the computation of $2K$ N-point DFTs of real sequences and *one* N-point inverse DFT. How many complex multiplications are required to compute $\hat{\phi}_{xx}[m]$ for $0 \le m \le M-1$ if a radix-2 FFT is used?

(d) What modifications to the procedure developed in Part (c) would be necessary to compute the cross-correlation estimate

$$\hat{\phi}_{xy}[m] = \frac{1}{Q} \sum_{n=0}^{Q-|m|-1} x[n]y[n+m], \qquad -(M-1) \le m \le M-1,$$

where $x[n]$ and $y[n]$ are real sequences known for $0 \le n \le Q-1$?

(e) Rader (1970) showed that, for computing the autocorrelation estimate $\hat{\phi}_{xx}[m]$ for $0 \le m \le M-1$, significant savings of computation can be achieved if $N = 2M$. Show that the N-point DFT of a segment $y_i[n]$ as defined in Eq. (P10.38-2) can be expressed as

$$Y_i[k] = X_i[k] + (-1)^k X_{i+1}[k], \qquad k = 0, 1, \ldots, N-1.$$

State a procedure for computing $\hat{\phi}_{xx}[m]$ for $0 \le m \le M-1$ that involves the computation of K N-point DFTs and one N-point inverse DFT. Determine the total number of complex multiplications in this case if a radix-2 FFT is used.

10.39. In Section 10.3 we defined the time-dependent Fourier transform of the signal $x[m]$ so that, for fixed n, it is equivalent to the regular discrete-time Fourier transform of the sequence $x[n+m]w[m]$, where $w[m]$ is a window sequence. It is also useful to define a time-dependent autocorrelation function for the sequence $x[n]$ such that, for fixed n, its regular Fourier transform is the magnitude squared of the time-dependent Fourier transform. Specifically, the time-dependent autocorrelation function is defined as

$$c[n, m] = \frac{1}{2\pi} \int_{-\pi}^{\pi} |X[n, \lambda)|^2 e^{j\lambda m} d\lambda,$$

where $X[n, \lambda)$ is defined by Eq. (10.18).

(a) Show that if $x[n]$ is real

$$c[n, m] = \sum_{r=-\infty}^{\infty} x[n+r]w[r]x[m+n+r]w[m+r];$$

i.e., for fixed n, $c[n, m]$ is the aperiodic autocorrelation of the sequence $x[n+r]w[r]$, $-\infty < r < \infty$.

(b) Show that the time-dependent autocorrelation function is an even function of m for n fixed, and use this fact to obtain the equivalent expression

$$c[n, m] = \sum_{r=-\infty}^{\infty} x[r]x[r-m]h_m[n-r],$$

where

$$h_m[r] = w[-r]w[-(m+r)]. \tag{P10.39-1}$$

(c) What condition must the window $w[r]$ satisfy so that Eq. (P10.39-1) can be used to compute $c[n, m]$ for fixed m and $-\infty < n < \infty$ by causal operations?

(d) Suppose that

$$w[-r] = \begin{cases} a^r, & r \ge 0, \\ 0, & r < 0. \end{cases} \tag{P10.39-2}$$

Find the impulse response $h_m[r]$ for computing the mth autocorrelation lag value, and find the corresponding system function $H_m(z)$. From the system function, draw the block diagram of a causal system for computing the mth autocorrelation lag value $c[n, m]$ for $-\infty < n < \infty$ for the window of Eq. (P10.39-2).

(e) Repeat Part (d) for

$$w[-r] = \begin{cases} ra^r, & r \geq 0, \\ 0, & r < 0. \end{cases}$$

10.40. Time-dependent Fourier analysis is sometimes implemented as a bank of filters, and even when FFT methods are used, the filter bank interpretation may provide useful insight. This problem examines that interpretation, the basis of which is the fact that when λ is fixed, the time-dependent Fourier transform $X[n, \lambda]$, defined by Eq. (10.18), is simply a sequence that can be viewed as the result of a combination of filtering and modulation operations.

(a) Show that $X[n, \lambda]$ is the output of the system of Figure P10.40-1 if the impulse response of the linear time-invariant system is $h_0[n] = w[-n]$. Show also that if λ is fixed, the overall system in Figure P10.40-1 behaves as a linear time-invariant system, and determine the impulse response and frequency response of the equivalent LTI system.

Figure P10.40-1

(b) Assuming λ fixed in Figure P10.40-1, show that, for typical window sequences and for fixed λ, the sequence $s[n] = \check{X}[n, \lambda]$ has a lowpass discrete-time Fourier transform. Show also that, for typical window sequences, the frequency response of the overall system in Figure P10.40-1 is a bandpass filter centered at $\omega = \lambda$.

(c) Figure P10.40-2 shows a bank of N bandpass filter channels, where each channel is implemented as in Figure P10.40-1. The center frequencies of the channels are $\lambda_k = 2\pi k/N$, and $h_0[n] = w[-n]$ is the impulse response of a lowpass filter. Show that the individual outputs $y_k[n]$ are samples (in the λ-dimension) of the time-dependent Fourier transform. Show also that the overall output is $y[n] = w[0]x[n]$; i.e., show that the system of Figure P10.40-2 reconstructs the input exactly (within a scale factor) from the sampled time-dependent Fourier transform.

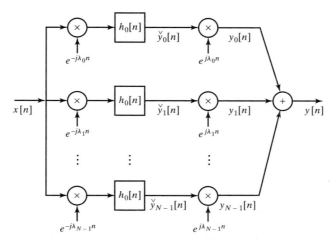

Figure P10.40-2

The system of Figure P10.40-2 converts the single input sequence $x[n]$ into N sequences, thereby increasing the total number of samples per second by the factor N. As shown in Part (b), for typical window sequences, the channel signals $\check{y}_k[n]$ have lowpass Fourier transforms. Thus, it should be possible to reduce the sampling rate of these signals, as shown in Figure P10.40-3. In particular, if the sampling rate is reduced by a factor $R = N$, the total number of samples per second is the same as for $x[n]$. In this case, the filter bank is said to be *critically sampled*. (See Crochiere and Rabiner, 1983.) Reconstruction of the original signal from the decimated channel signals requires interpolation as shown. Clearly, it is of interest to determine how well the original input $x[n]$ can be reconstructed by the system.

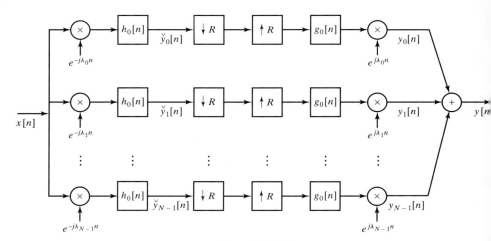

Figure P10.40-3

(d) For the system of Figure P10.40-3, show that the regular discrete-time Fourier transform of the output is given by the relation

$$Y(e^{j\omega}) = \frac{1}{R} \sum_{\ell=0}^{R-1} \sum_{k=0}^{N-1} G_0(e^{j(\omega-\lambda_k)}) H_0(e^{j(\omega-\lambda_k-2\pi\ell/R)}) X(e^{j(\omega-2\pi\ell/R)}),$$

where $\lambda_k = 2\pi k/N$. This expression clearly shows the aliasing resulting from the decimation of the channel signals $\check{y}[n]$. From the expression for $Y(e^{j\omega})$, determine a relation or set of relations that must be satisfied jointly by $H_0(e^{j\omega})$ and $G_0(e^{j\omega})$ such that the aliasing cancels and $y[n] = x[n]$.

(e) Assume that $R = N$ and the frequency response of the lowpass filter is an ideal lowpass filter with frequency response

$$H_0(e^{j\omega}) = \begin{cases} 1, & |\omega| < \pi/N, \\ 0, & \pi/N < |\omega| \le \pi. \end{cases}$$

For this frequency response $H_0(e^{j\omega})$, determine whether it is possible to find a frequency response of the interpolation filter $G_0(e^{j\omega})$ such that the condition derived in Part (d) is satisfied. If so, determine $G_0(e^{j\omega})$.

(f) *Optional:* Explore the possibility of exact reconstruction when the frequency response of the lowpass filter $H_0(e^{j\omega})$ (the Fourier transform of $w[-n]$) is nonideal and nonzero in the interval $|\omega| < 2\pi/N$.

(g) Show that the output of the system of Figure P10.40-3 is

$$y[n] = N \sum_{r=-\infty}^{\infty} x[n - rN] \sum_{\ell=-\infty}^{\infty} g_0[n - \ell R] h_0[\ell R + rN - n].$$

From this expression, determine a relation or set of relations that must be satisfied jointly by $h_0[n]$ and $g_0[n]$ such that $y[n] = x[n]$.

(h) Assume that $R = N$ and the impulse response of the lowpass filter is

$$h_0[n] = \begin{cases} 1, & -(N-1) \le n \le 0, \\ 0, & \text{otherwise.} \end{cases}$$

For this impulse response $h_0[n]$, determine whether it is possible to find an impulse response of the interpolation filter $g_0[n]$ such that the condition derived in Part (g) is satisfied. If so, determine $g_0[n]$.

(i) *Optional:* Explore the possibility of exact reconstruction when the impulse response of the lowpass filter $h_0[n] = w[-n]$ is a tapered window with length greater than N.

10.41. Consider a stable linear time-invariant system with a real input $x[n]$, a real impulse response $h[n]$, and output $y[n]$. Assume that the input $x[n]$ is white noise with zero mean and variance σ_x^2. The system function is

$$H(z) = \frac{\displaystyle\sum_{k=0}^{M} b_k z^{-k}}{1 - \displaystyle\sum_{k=1}^{N} a_k z^{-k}},$$

where we assume the a_k's and b_k's are real for this problem. The input and output satisfy the following difference equation with constant coefficients:

$$y[n] = \sum_{k=1}^{N} a_k y[n-k] + \sum_{k=0}^{M} b_k x[n-k].$$

If all the a_k's are zero, $y[n]$ is called a *moving-average* (MA) linear random process. If all the b_k's are zero, except for b_0, then $y[n]$ is called an *autoregressive* (AR) linear random process. If both N and M are nonzero, then $y[n]$ is an *autoregressive moving-average* (ARMA) linear random process.

(a) Express the autocorrelation of $y[n]$ in terms of the impulse response $h[n]$ of the linear system.

(b) Use the result of Part (a) to express the power density spectrum of $y[n]$ in terms of the frequency response of the system.

(c) Show that the autocorrelation sequence $\phi_{yy}[m]$ of an MA process is nonzero only in the interval $|m| \le M$.

(d) Find a general expression for the autocorrelation sequence for an AR process.

(e) Show that if $b_0 = 1$, the autocorrelation function of an AR process satisfies the difference equation

$$\phi_{yy}[0] = \sum_{k=1}^{N} a_k \phi_{yy}[k] + \sigma_x^2,$$

$$\phi_{yy}[m] = \sum_{k=1}^{N} a_k \phi_{yy}[m-k], \qquad m \ge 1.$$

(f) Use the result of Part (e) and the symmetry of $\phi_{yy}[m]$ to show that

$$\sum_{k=1}^{N} a_k \phi_{yy}[|m-k|] = \phi_{yy}[m], \qquad m = 1, 2, \ldots, N.$$

It can be shown that, given $\phi_{yy}[m]$ for $m = 0, 1, \ldots, N$, we can always solve uniquely for the values of the a_k's and σ_x^2 for the random-process model. These values may be used in the result in Part (b) to obtain an expression for the power density spectrum of $y[n]$. This approach is the basis for a number of parametric spectrum estimation techniques. (For further discussion of these methods, see Gardner, 1988; Kay, 1988; and Marple, 1987.)

10.42. This problem illustrates the basis for an FFT-based procedure for interpolating the samples (obtained at a rate satisfying the Nyquist theorem) of a periodic continuous-time signal. Let

$$x_c(t) = \frac{1}{16} \sum_{k=-4}^{4} \left(\frac{1}{2}\right)^{|k|} e^{jkt}$$

be a periodic signal that is processed by the system in Figure P10.42-1.
(a) Sketch the 16-point sequence $G[k]$.
(b) Specify how you would change $G[k]$ into a 32-point sequence $Q[k]$ so that the 32-point inverse DFT of $Q[k]$ is a sequence

$$q[n] = \alpha x_c \left(\frac{n2\pi}{32}\right), \qquad 0 \le n \le 31,$$

for some nonzero constant α. You need not specify the value of α.

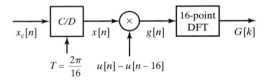

Figure P10.42-1

10.43. In many real applications, practical constraints do not allow long time sequences to be processed. However, significant information can be gained from a windowed section of the sequence. In this problem, you will look at computing the Fourier transform of an infinite-duration signal $x[n]$, given only a block of 256 samples in the range $0 \le n \le 255$. You decide to use a 256-point DFT to estimate the transform by defining the signal

$$\hat{x}[n] = \begin{cases} x[n], & 0 \le n \le 255, \\ 0, & \text{otherwise}, \end{cases}$$

and computing the 256-point DFT of $\hat{x}[n]$.
(a) Suppose the signal $x[n]$ came from sampling a continuous-time signal $x_c(t)$ with sampling frequency $f_s = 20$ kHz; i.e.,

$$x[n] = x_c(nT_s),$$

$$1/T_s = 20 \text{ kHz}.$$

Assume that $x_c(t)$ is bandlimited to 10 kHz. If the DFT of $\hat{x}[n]$ is written $\hat{X}[k]$, $k = 0, 1, \ldots, 255$, what are the continuous-time frequencies corresponding to the DFT indices $k = 32$ and $k = 231$? Be sure to express your answers in Hertz.

(b) Express the DTFT of $\hat{x}[n]$ in terms of the DTFT of $x[n]$ and the DTFT of a 256-point rectangular window $w_R[n]$. Use the notation $X(e^{j\omega})$ and $W_R(e^{j\omega})$ to represent the DTFTs of $x[n]$ and $w_R[n]$, respectively.

(c) Suppose you try an averaging technique to estimate the transform for $k = 32$:

$$X_{avg}[32] = \alpha \hat{X}[31] + \hat{X}[32] + \alpha \hat{X}[33].$$

Averaging in this manner is equivalent to multiplying the signal $\hat{x}[n]$ by a new window $w_{avg}[n]$ before computing the DFT. Show that $W_{avg}(e^{j\omega})$ must satisfy

$$W_{avg}(e^{j\omega}) = \begin{cases} 1, & \omega = 0, \\ \alpha, & \omega = \pm 2\pi/L, \\ 0, & \omega = 2\pi k/L, \quad \text{for } k = 2, 3, \ldots, L-2, \end{cases}$$

where $L = 256$.

(d) Show that the DTFT of this new window can be written in terms of $W_R(e^{j\omega})$ and two shifted versions of $W_R(e^{j\omega})$.

(e) Derive a simple formula for $w_{avg}[n]$, and sketch the window for $\alpha = -0.5$ and $0 \le n \le 255$.

10.44. It is often of interest to zoom in on a region of a DFT of a signal to examine it in more detail. In this problem, you will explore two algorithms for implementing this process of obtaining additional samples of $X(e^{j\omega})$ in a frequency region of interest.

Suppose $X_N[k]$ is the N-point DFT of a finite-length signal $x[n]$. Recall that $X_N[k]$ consists of samples of $X(e^{j\omega})$ every $2\pi/N$ in ω. Given $X_N[k]$, we would like to compute N samples of $X(e^{j\omega})$ between $\omega = \omega_c - \Delta\omega$ and $\omega = \omega_c + \Delta\omega$ with spacing $2\Delta\omega/N$, where

$$\omega_c = \frac{2\pi k_c}{N}$$

and

$$\Delta\omega = \frac{2\pi k_\Delta}{N}.$$

This is equivalent to zooming in on $X(e^{j\omega})$ in the region $\omega_c - \Delta\omega < \omega < \omega_c + \Delta\omega$. One system used to implement the zoom is shown in Figure P10.44-1. Assume that $x_z[n]$ is zero-padded as necessary before the N-point DFT and $h[n]$ is an ideal lowpass filter with a cutoff frequency $\Delta\omega$.

Figure P10.44-1

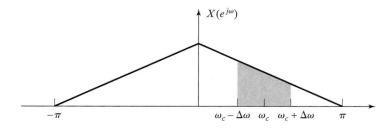

Figure P10.44-2

(a) In terms of k_Δ and the transform length N, what is the largest (possibly noninteger) value of M that can be used if aliasing is to be avoided in the downsampler.

(b) Consider $x[n]$ with the Fourier transform shown in Figure P10.44-2. Using the maximum value of M from Part (a), sketch the Fourier transforms of the intermediate signals $x_\ell[n]$ and $x_z[n]$ when $\omega_c = \pi/2$ and $\Delta\omega = \pi/6$. Demonstrate that the system provides the desired frequency samples.

Another procedure for obtaining the desired samples can be developed by viewing the finite-length sequence $X_N[k]$ indexed on k as a discrete-time data sequence to be processed as shown in Figure P10.44-3. The impulse response of the first system is

$$p[n] = \sum_{r=-\infty}^{\infty} \delta[n + rN],$$

and the filter $h[n]$ has the frequency response

$$H(e^{j\omega}) = \begin{cases} 1, & |\omega| \leq \pi/M, \\ 0, & \text{otherwise.} \end{cases}$$

The zoomed output signal is defined as

$$X_z[n] = \tilde{X}_{NM}[k_c - k_\Delta + n], \quad 0 \leq n \leq N - 1,$$

for appropriate values of k_c and k_Δ.

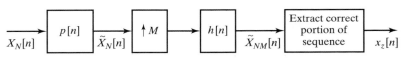

Figure P10.44-3

Assume that k_Δ is chosen so that M is an integer in the following parts.

(c) Suppose that the ideal lowpass filter $h[n]$ is approximated by a causal Type I linear-phase filter of length 513 (nonzero for $0 \leq n \leq 512$). Indicate which samples of $\tilde{X}_{NM}[n]$ provide the desired frequency samples.

(d) Using sketches of a typical spectrum for $X_N[k]$ and $X(e^{j\omega})$, demonstrate that the system in Figure P10.44-3 produces the desired samples.

11

Dᴵˢᶜᴿᴱᵀᴱ HILBERT
TRANSFORMS

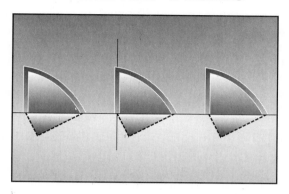

11.0 INTRODUCTION

In general, the specification of the Fourier transform of a sequence requires complete knowledge of both the real and imaginary parts of the magnitude and phase for all frequencies in the range $-\pi < \omega \leq \pi$. However, we have seen that under certain conditions there are constraints on the Fourier transform. For example, in Section 2.8 we saw that if $x[n]$ is real, then its Fourier transform is conjugate symmetric, i.e., $X(e^{j\omega}) = X^*(e^{-j\omega})$. From this, it follows that for real sequences, specification of $X(e^{j\omega})$ for $0 \leq \omega \leq \pi$ also specifies it for $-\pi \leq \omega \leq 0$. Similarly, we saw in Section 5.4 that under the constraint of minimum phase, the Fourier transform magnitude and phase are not independent; i.e., specification of magnitude determines the phase and specification of phase determines the magnitude to within a scale factor. In Section 8.5 we saw that for sequences of finite length N, specification of $X(e^{j\omega})$ at N equally spaced frequencies determines $X(e^{j\omega})$ at all frequencies.

In this chapter, we will see that the constraint of *causality* of a sequence implies unique relationships between the real and imaginary parts of the Fourier transform. Relationships of this type between the real and imaginary parts of complex functions arise in many fields besides signal processing, and they are commonly known as *Hilbert transform relationships*. In addition to developing these relationships for the Fourier transform of causal sequences, we will develop related results for the DFT and for sequences with one-sided Fourier transforms. Also, in Section 11.3 we will indicate how the relationship between magnitude and phase for minimum-phase sequences can be interpreted in terms of the Hilbert transform.

Although we will take an intuitive approach in this chapter, it is important to be aware that the Hilbert transform relationships follow formally from the properties of analytic functions. (See Problem 11.21.) Specifically, the complex functions that arise in the mathematical representation of discrete-time signals and systems are generally very well behaved functions. With few exceptions, the z-transforms that have concerned us have had well-defined regions in which the power series is absolutely convergent. Since a power series represents an analytic function within its region of convergence, it follows that z-transforms are analytic functions inside their regions of convergence. By the definition of an analytic function, this means that the z-transform has a well-defined derivative at every point inside the region of convergence. Furthermore, analyticity implies that the z-transform and all its derivatives are continuous functions within the region of convergence.

The properties of analytic functions imply some rather powerful constraints on the behavior of the z-transform within its region of convergence. Since the Fourier transform is the z-transform evaluated on the unit circle, these constraints also restrict the behavior of the Fourier transform. One such constraint is that the real and imaginary parts satisfy the Cauchy–Riemann conditions, which relate the partial derivatives of the real and imaginary parts of an analytic function. (See, for example, Churchill and Brown, 1990.) Another constraint is the Cauchy integral theorem, through which the value of a complex function is specified everywhere inside a region of analyticity in terms of the values of the function on the boundary of the region. On the basis of these relations for analytic functions, it is possible, under certain conditions, to derive explicit integral relationships between the real and imaginary parts of a z-transform on a closed contour within the region of convergence. In the mathematics literature, these relations are often referred to as *Poisson's formulas*. In the context of system theory, they are known as the *Hilbert transform relations*.

Rather than following the mathematical approach just discussed, we will develop the Hilbert transform relations by exploiting the fact that, on the unit circle, the real and imaginary parts of the z-transform of a causal sequence are the transforms of the even and odd components, respectively, of the sequence (properties 5 and 6, Table 2.1). As we will show, a causal sequence is completely specified by its even part, implying that the z-transform of the original sequence is completely specified by its real part on the unit circle. In addition to applying this argument to specifying the z-transform of a particular causal sequence in terms of its real part on the unit circle, we can also apply it, under certain conditions, to specify the z-transform of a sequence in terms of its magnitude on the unit circle.

The notion of an analytic signal is an important concept in continuous-time signal processing. An analytic signal is a complex time function (which is analytic) having a Fourier transform that vanishes for negative frequencies. A complex sequence cannot be considered in any formal sense to be analytic, since it is a function of an integer variable. However, in a style similar to that described in the previous paragraph, it is possible to relate the real and imaginary parts of a complex sequence whose spectrum is zero on the unit circle for $-\pi < \omega < 0$. A similar approach can also be taken in relating the real and imaginary parts of the discrete Fourier transform for a periodic or, equivalently, a finite-length sequence. In this case, the "causality" condition is that the periodic sequence be zero in the second half of each period.

Thus, in this chapter, a notion of causality will be applied to relate the even and odd components of a function or, equivalently, the real and imaginary parts of its transform. We will apply this approach in four situations. First, we relate the real and imaginary parts of the Fourier transform $X(e^{j\omega})$ of a sequence $x[n]$ that is zero for $n < 0$. In the second situation, we obtain a relationship between the real and imaginary parts of the DFT for periodic sequences or, equivalently, for a finite-length sequence considered to be of length N, but with the last $(N/2) - 1$ points restricted to zero. In the third case, we relate the real and imaginary parts of the *logarithm* of the Fourier transform under the condition that the inverse transform of the logarithm of the transform is zero for $n < 0$. Relating the real and imaginary parts of the logarithm of the Fourier transform corresponds to relating the log magnitude and phase of $X(e^{j\omega})$. Finally, we relate the real and imaginary parts of a complex sequence whose Fourier transform, considered as a periodic function of ω, is zero in the second half of each period.

11.1 REAL- AND IMAGINARY-PART SUFFICIENCY OF THE FOURIER TRANSFORM FOR CAUSAL SEQUENCES

Any sequence can be expressed as the sum of an even sequence and an odd sequence. Specifically, with $x_e[n]$ and $x_o[n]$ denoting the even and odd parts, respectively, of $x[n]$,[1] we have

$$x[n] = x_e[n] + x_o[n], \tag{11.1}$$

where

$$x_e[n] = \frac{x[n] + x[-n]}{2} \tag{11.2}$$

and

$$x_o[n] = \frac{x[n] - x[-n]}{2}. \tag{11.3}$$

Equations (11.1)–(11.3) apply to an arbitrary sequence, whether or not it is causal and whether or not it is real. However, if $x[n]$ is causal, i.e., if $x[n] = 0, n < 0$, then it is possible to recover $x[n]$ from $x_e[n]$ or to recover $x[n]$ for $n \neq 0$ from $x_o[n]$. Consider, for example, the causal sequence $x[n]$ and its even and odd components, as shown in Figure 11.1. Because $x[n]$ is causal, $x[n] = 0$ for $n < 0$ and $x[-n] = 0$ for $n > 0$. Therefore, the nonzero portions of $x[n]$ and $x[-n]$ do not overlap except at $n = 0$. For this reason, it follows from Eqs. (11.2) and (11.3) that

$$x[n] = 2x_e[n]u[n] - x_e[0]\delta[n] \tag{11.4}$$

and

$$x[n] = 2x_o[n]u[n] + x[0]\delta[n]. \tag{11.5}$$

The validity of these relationships is easily seen in Figure 11.1. Note that $x[n]$ is completely determined by $x_e[n]$. On the other hand, $x_o[0] = 0$, so we can recover $x[n]$ from $x_o[n]$ only for $n \neq 0$.

---

[1]If $x[n]$ is real, then $x_e[n]$ and $x_o[n]$ in Eqs. (11.2) and (11.3) are the even and odd parts, respectively, of $x[n]$ as considered in Chapter 2. If $x[n]$ is complex, for the purposes of this discussion we still define $x_e[n]$ and $x_o[n]$ as in Eqs. (11.2) and (11.3), which do not correspond to the conjugate-symmetric and conjugate-antisymmetric parts of a complex sequence as considered in Chapter 2.

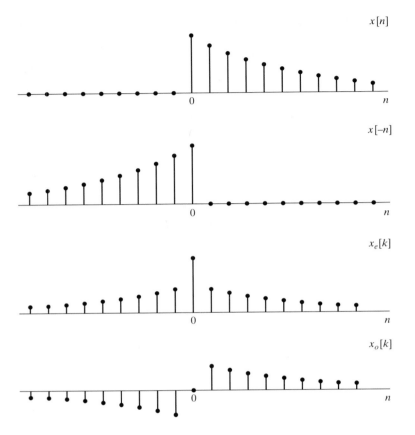

Figure 11.1 Even and odd parts of a real causal sequence.

Now, if $x[n]$ is also stable, i.e., absolutely summable, then its Fourier transform exists. We denote the Fourier transform of $x[n]$ as

$$X(e^{j\omega}) = X_R(e^{j\omega}) + j X_I(e^{j\omega}), \tag{11.6}$$

where $X_R(e^{j\omega})$ is the real part and $X_I(e^{j\omega})$ is the imaginary part of $X(e^{j\omega})$. Recall that if $x[n]$ is a *real* sequence, then $X_R(e^{j\omega})$ is the Fourier transform of $x_e[n]$ and $j X_I(e^{j\omega})$ is the Fourier transform of $x_o[n]$. Therefore, for a *causal, stable, real* sequence, $X_R(e^{j\omega})$ completely determines $X(e^{j\omega})$, since, if we are given $X_R(e^{j\omega})$, we can find $X(e^{j\omega})$ by the following process:

1. Find $x_e[n]$ as the inverse Fourier transform of $X_R(e^{j\omega})$.

2. Find $x[n]$ using Eq. (11.4).

3. Find $X(e^{j\omega})$ as the Fourier transform of $x[n]$.

This also implies, of course, that $X_I(e^{j\omega})$ can be determined from $X_R(e^{j\omega})$. In Example 11.1, we illustrate how this procedure can be applied to obtain $X(e^{j\omega})$ and $X_I(e^{j\omega})$ from $X_R(e^{j\omega})$.

Example 11.1   Finite-Length Sequence

Consider a real, causal sequence $x[n]$ for which $X_R(e^{j\omega})$, the real part of the DTFT, is

$$X_R(e^{j\omega}) = 1 + \cos 2\omega. \tag{11.7}$$

We would like to determine the original sequence $x[n]$, its Fourier transform $X(e^{j\omega})$, and the imaginary part of the Fourier transform, $X_I(e^{j\omega})$. As a first step, we rewrite Eq. (11.7) expressing the cosine as a sum of complex exponentials:

$$X_R(e^{j\omega}) = 1 + \frac{1}{2}e^{-j2\omega} + \frac{1}{2}e^{j2\omega}. \tag{11.8}$$

We know that $X_R(e^{j\omega})$ is the Fourier transform of $x_e[n]$, the even part of $x[n]$ as defined in Eq. (11.2). Comparing Eq. (11.8) with the definition of the Fourier transform, Eq. (2.134), we can match terms to obtain

$$x_e[n] = \delta[n] + \frac{1}{2}\delta[n-2] + \frac{1}{2}\delta[n+2].$$

Now that we have obtained the even part, we can use the relation in Eq. (11.4) to obtain

$$x[n] = \delta[n] + \delta[n-2]. \tag{11.9}$$

From $x[n]$, we get

$$X(e^{j\omega}) = 1 + e^{-j2\omega}$$
$$= 1 + \cos 2\omega - j \sin 2\omega. \tag{11.10}$$

From Eq. (11.10), we can both confirm that $X_R(e^{j\omega})$ is as specified in Eq. (11.7) and also obtain

$$X_I(e^{j\omega}) = -\sin 2\omega. \tag{11.11}$$

As an alternative path to obtaining $X_I(e^{j\omega})$, we can first use Eq. (11.3) to get $x_o[n]$ from $x[n]$. Substituting Eq. (11.9) into Eq. (11.3) then yields

$$x_o[n] = \frac{1}{2}\delta[n-2] - \frac{1}{2}\delta[n+2].$$

The Fourier transform of $x_o[n]$ is $j X_I(e^{j\omega})$, so we find

$$j X_I(e^{j\omega}) = \frac{1}{2}e^{-j2\omega} - \frac{1}{2}e^{j2\omega} = -j \sin 2\omega,$$

so that

$$X_I(e^{j\omega}) = -\sin 2\omega,$$

which is consistent with Eq. (11.11).

Example 11.2   Exponential Sequence

Let

$$X_R(e^{j\omega}) = \frac{1 - \alpha \cos \omega}{1 - 2\alpha \cos \omega + \alpha^2}, \qquad |\alpha| < 1, \tag{11.12}$$

or equivalently,

$$X_R(e^{j\omega}) = \frac{1 - (\alpha/2)(e^{j\omega} + e^{-j\omega})}{1 - \alpha(e^{j\omega} + e^{-j\omega}) + \alpha^2}, \qquad |\alpha| < 1, \tag{11.13}$$

with α real. We first determine $x_e[n]$ and then $x[n]$ using Eq. (11.4).

To obtain $x_e[n]$, the inverse Fourier transform of $X_R(e^{j\omega})$, it is convenient to first obtain $X_R(z)$, the z-transform of $x_e[n]$. This follows directly from Eq. (11.13), given that

$$X_R(e^{j\omega}) = X_R(z)\big|_{z=e^{j\omega}}.$$

Consequently, by replacing $e^{j\omega}$ by z in Eq. (11.13), we obtain

$$X_R(z) = \frac{1 - (\alpha/2)(z + z^{-1})}{1 - \alpha(z + z^{-1}) + \alpha^2} \tag{11.14}$$

$$= \frac{1 - \frac{\alpha}{2}(z + z^{-1})}{(1 - \alpha z^{-1})(1 - \alpha z)}. \tag{11.15}$$

Since we began with the Fourier transform $X_R(e^{j\omega})$ and obtained $X_R(z)$ by extending $X_R(e^{j\omega})$ into the z-plane, the ROC of $X_R(z)$ must, of course, include the unit circle and is then bounded on the inside by the pole at $z = \alpha$ and on the outside by the pole at $z = 1/\alpha$.

From Eq. (11.15), we now want to obtain $x_e[n]$, the inverse z-transform of $X_R(z)$. We do this by expanding Eq. (11.15) in partial fractions, yielding

$$X_R(z) = \frac{1}{2}\left[\frac{1}{1 - \alpha z^{-1}} + \frac{1}{1 - \alpha z}\right], \tag{11.16}$$

with the ROC specified to include the unit circle. The inverse z-transform of Eq. (11.16) can then be applied separately to each term to obtain

$$x_e[n] = \frac{1}{2}\alpha^n u[n] + \frac{1}{2}\alpha^{-n} u[-n]. \tag{11.17}$$

Consequently, from Eq. (11.4),

$$x[n] = \alpha^n u[n] + \alpha^{-n} u[-n]u[n] - \delta[n]$$

$$= \alpha^n u[n].$$

$X(e^{j\omega})$ is then given by

$$X(e^{j\omega}) = \frac{1}{1 - \alpha e^{-j\omega}}, \tag{11.18}$$

and $X(z)$ is given by

$$X(z) = \frac{1}{1 - \alpha z^{-1}} \qquad |z| > |\alpha|. \tag{11.19}$$

The constructive procedure illustrated in Example 11.1 can be interpreted analytically to obtain a general relationship that expresses $X_I(e^{j\omega})$ directly in terms of $X_R(e^{j\omega})$. From Eq. (11.4), the complex convolution theorem, and the fact that $x_e[0] = x[0]$, it follows that

$$X(e^{j\omega}) = \frac{1}{\pi}\int_{-\pi}^{\pi} X_R(e^{j\theta})U(e^{j(\omega-\theta)})d\theta - x[0], \tag{11.20}$$

where $U(e^{j\omega})$ is the Fourier transform of the unit step sequence. As stated in Section 2.7, although the unit step is neither absolutely summable nor square summable, it can be represented by the Fourier transform

$$U(e^{j\omega}) = \sum_{K=-\infty}^{\infty} \pi\delta(\omega - 2\pi k) + \frac{1}{1 - e^{-j\omega}}, \tag{11.21}$$

or, since the term $1/(1 - e^{-j\omega})$ can be rewritten as

$$\frac{1}{1 - e^{-j\omega}} = \frac{1}{2} - \frac{j}{2}\cot\left(\frac{\omega}{2}\right), \tag{11.22}$$

Eq. (11.21) becomes

$$U(e^{j\omega}) = \sum_{k=-\infty}^{\infty} \pi\delta(\omega - 2\pi k) + \frac{1}{2} - \frac{j}{2}\cot\left(\frac{\omega}{2}\right). \tag{11.23}$$

Using Eq. (11.23), we can express Eq. (11.20) as

$$X(e^{j\omega}) = X_R(e^{j\omega}) + jX_I(e^{j\omega})$$

$$= X_R(e^{j\omega}) + \frac{1}{2\pi}\int_{-\pi}^{\pi} X_R(e^{j\theta})d\theta \tag{11.24}$$

$$- \frac{j}{2\pi}\int_{-\pi}^{\pi} X_R(e^{j\theta})\cot\left(\frac{\omega - \theta}{2}\right)d\theta - x[0].$$

Equating real and imaginary parts in Eq. (11.24) and noting that

$$x[0] = \frac{1}{2\pi}\int_{-\pi}^{\pi} X_R(e^{j\theta})d\theta, \tag{11.25}$$

we obtain the relationship

$$X_I(e^{j\omega}) = -\frac{1}{2\pi}\int_{-\pi}^{\pi} X_R(e^{j\theta})\cot\left(\frac{\omega - \theta}{2}\right)d\theta. \tag{11.26}$$

A similar procedure can be followed to obtain $x[n]$ and $X(e^{j\omega})$ from $X_I(e^{j\omega})$ and $x[0]$ using Eq. (11.5). This process results in the following equation for $X_R(e^{j\omega})$ in terms of $X_I(e^{j\omega})$:

$$X_R(e^{j\omega}) = x[0] + \frac{1}{2\pi}\int_{-\pi}^{\pi} X_I(e^{j\theta})\cot\left(\frac{\omega - \theta}{2}\right)d\theta. \tag{11.27}$$

Equations (11.26) and (11.27), which are called *discrete Hilbert transform relationships,* hold for the real and imaginary parts of the Fourier transform of a causal, stable, real sequence. They are improper integrals, since the integrand is singular at $\omega - \theta = 0$. Such integrals must be evaluated carefully to obtain a consistent finite result. This can be done formally by interpreting the integrals as *Cauchy principal values.* That is, Eq. (11.26) becomes

$$X_I(e^{j\omega}) = -\frac{1}{2\pi}\mathcal{P}\int_{-\pi}^{\pi} X_R(e^{j\theta})\cot\left(\frac{\omega - \theta}{2}\right)d\theta, \tag{11.28}$$

and Eq. (11.27) becomes

$$X_R(e^{j\omega}) = x[0] + \frac{1}{2\pi}\mathcal{P}\int_{-\pi}^{\pi} X_I(e^{j\theta})\cot\left(\frac{\omega - \theta}{2}\right)d\theta, \tag{11.29}$$

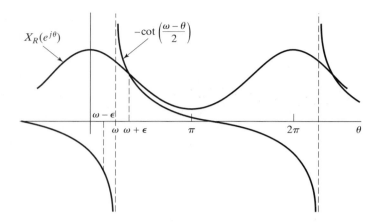

Figure 11.2   Interpretation of the Hilbert transform as a periodic convolution.

where $\mathcal{P}$ denotes the Cauchy principal value of the integral that follows. The meaning of the Cauchy principal value in Eq. (11.28), for example, is

$$
X_I(e^{j\omega}) = -\frac{1}{2\pi} \lim_{\varepsilon \to 0} \left[\int_{\omega+\varepsilon}^{\pi} X_R(e^{j\theta}) \cot\left(\frac{\omega - \theta}{2}\right) d\theta \right.
$$

$$
\left. + \int_{-\pi}^{\omega-\varepsilon} X_R(e^{j\theta}) \cot\left(\frac{\omega - \theta}{2}\right) d\theta \right].
$$

(11.30)

Equation (11.30) shows that $X_I(e^{j\omega})$ is obtained by the periodic convolution of $-\cot(\omega/2)$ with $X_R(e^{j\omega})$, with special care being taken in the vicinity of the singularity at $\theta = \omega$. In a similar manner, Eq. (11.29) involves the periodic convolution of $\cot(\omega/2)$ with $X_I(e^{j\omega})$.

The two functions involved in the convolution integral of Eq. (11.28) (or, equivalently, Eq. (11.30) are illustrated in Figure 11.2. The limit in Eq. (11.30) exists because the function $\cot[(\omega - \theta)/2]$ is antisymmetric at the singular point $\theta = \omega$ and the limit is taken symmetrically about the singularity.

11.2 SUFFICIENCY THEOREMS FOR FINITE-LENGTH SEQUENCES

In Section 11.1, we showed that causality or one-sidedness of a real sequence implies some strong constraints on the Fourier transform of the sequence. The results of the previous section apply, of course, to finite-length causal sequences, but since the finite-length property is more restrictive, it is perhaps reasonable to expect the Fourier transform of a finite-length sequence to be more constrained. We will see that this is indeed the case.

One way to take advantage of the finite-length property is to recall that finite-length sequences can be represented by the discrete Fourier transform. Since the DFT involves sums rather than integrals, the problems associated with improper integrals disappear.

Since the DFT is, in reality, a representation of a periodic sequence, any results we obtain must be based on corresponding results for periodic sequences. Indeed, it is important to keep the inherent periodicity of the DFT firmly in mind in deriving the desired Hilbert transform relation for finite-length sequences. Therefore, we will consider the periodic case first and then discuss the application to the finite-length case.

Consider a periodic sequence $\tilde{x}[n]$ with period N that is related to a finite-length sequence $x[n]$ of length N by

$$\tilde{x}[n] = x[((n))_N]. \tag{11.31}$$

As in Section 11.1, $\tilde{x}[n]$ can be represented as the sum of an even and odd periodic sequence,

$$\tilde{x}[n] = \tilde{x}_e[n] + \tilde{x}_o[n], \qquad n = 0, 1, \ldots, (N-1), \tag{11.32}$$

where

$$\tilde{x}_e[n] = \frac{\tilde{x}[n] + \tilde{x}[-n]}{2}, \qquad n = 0, 1, \ldots, (N-1), \tag{11.33a}$$

and

$$\tilde{x}_o[n] = \frac{\tilde{x}[n] - \tilde{x}[-n]}{2}, \qquad n = 0, 1, \ldots, (N-1). \tag{11.33b}$$

A periodic sequence cannot, of course, be causal in the sense used in Section 11.1. We can, however, define a "periodically causal" periodic sequence to be a periodic sequence for which $\tilde{x}[n] = 0$ for $N/2 < n < N$. That is, $\tilde{x}[n]$ is identically zero over the last half of the period. We assume henceforth that N is even; the case of N odd is considered in Problem 11.23. Note that because of the periodicity of $\tilde{x}[n]$, it is also true that $\tilde{x}[n] = 0$ for $-N/2 < n < 0$. For finite-length sequences, this restriction means that although the sequence is considered to be of length N, the last $(N/2) - 1$ points are in fact zero. In Figure 11.3, we show an example of a periodically causal sequence and its even and odd parts with $N = 8$. Because $\tilde{x}[n]$ is zero in the second half of each period, $\tilde{x}[-n]$ is zero in the first half of each period, and, consequently, except for $n = 0$ and $n = N/2$, there is no overlap between the nonzero portions of $\tilde{x}[n]$ and $\tilde{x}[-n]$. Therefore, for periodically causal periodic sequences,

$$\tilde{x}[n] = \begin{cases} 2\tilde{x}_e[n], & n = 1, 2, \ldots, (N/2) - 1, \\ \tilde{x}_e[n], & n = 0, N/2, \\ 0, & n = (N/2) + 1, \ldots, N - 1, \end{cases} \tag{11.34}$$

and

$$\tilde{x}[n] = \begin{cases} 2\tilde{x}_o[n], & n = 1, 2, \ldots, (N/2) - 1, \\ 0, & n = (N/2) + 1, \ldots, N - 1. \end{cases} \tag{11.35}$$

If we define the periodic sequence

$$\tilde{u}_N[n] = \begin{cases} 1, & n = 0, N/2, \\ 2, & n = 1, 2, \ldots, (N/2) - 1, \\ 0, & n = (N/2) + 1, \ldots, N - 1, \end{cases} \tag{11.36}$$

then it follows that, for N even, we can express $\tilde{x}[n]$ as

$$\tilde{x}[n] = \tilde{x}_e[n]\tilde{u}_N[n] \tag{11.37}$$

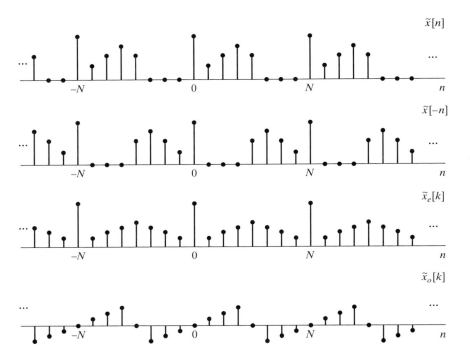

Figure 11.3 Even and odd parts of a periodically causal, real, periodic sequence of period $N = 8$.

and

$$\tilde{x}[n] = \tilde{x}_o[n]\tilde{u}_N[n] + x[0]\tilde{\delta}[n] + x[N/2]\tilde{\delta}[n - (N/2)], \qquad (11.38)$$

where $\tilde{\delta}[n]$ is a periodic unit-impulse sequence with period N. Thus, the sequence $\tilde{x}[n]$ can be completely recovered from $\tilde{x}_e[n]$. On the other hand, $\tilde{x}_o[n]$ will always be zero at $n = 0$ and $n = N/2$, and consequently, $\tilde{x}[n]$ can be recovered from $\tilde{x}_o[n]$ only for $n \neq 0$ and $n \neq N/2$.

If $\tilde{x}[n]$ is a real periodic sequence of period N with discrete Fourier series $\tilde{X}[k]$, then $\tilde{X}_R[k]$, the real part of $\tilde{X}[k]$, is the DFS of $\tilde{x}_e[n]$ and $j\tilde{X}_I[k]$ is the DFS of $\tilde{x}_o[n]$. Hence, Eqs. (11.37) and (11.38) imply that, for a periodic sequence of period N, which is periodically causal in the sense defined earlier, $\tilde{X}[k]$ can be recovered from its real part or (almost) from its imaginary part. Equivalently, $\tilde{X}_I[k]$ can be obtained from $\tilde{X}_R[k]$, and $\tilde{X}_R[k]$ can (almost) be obtained from $\tilde{X}_I[k]$.

Specifically, suppose that we are given $\tilde{X}_R[k]$. Then we can obtain $\tilde{X}[k]$ and $\tilde{X}_I[k]$ by the following procedure:

1. Compute $\tilde{x}_e[n]$ using the DFS synthesis equation

$$\tilde{x}_e[n] = \frac{1}{N} \sum_{k=0}^{N-1} \tilde{X}_R[k] e^{j(2\pi/N)kn}. \qquad (11.39)$$

2. Compute $\tilde{x}[n]$ using Eq. (11.37).

3. Compute $\tilde{X}[k]$ using the DFS analysis equation

$$\tilde{X}[k] = \sum_{k=0}^{N-1} \tilde{x}[n]e^{-j(2\pi/N)kn} = \tilde{X}_R[k] + j\tilde{X}_I[k]. \tag{11.40}$$

In contrast to the general causal case discussed in Section 11.1, the procedure just outlined can be implemented on a computer, since Eqs. (11.39) and (11.40) can be evaluated accurately and efficiently using an FFT algorithm.

To obtain an explicit relation between $\tilde{X}_R[k]$, and $\tilde{X}_I[k]$, we can carry out the procedure analytically. From Eq. (11.37) and Eq. (8.34), it follows that

$$\tilde{X}[k] = \tilde{X}_R[k] + j\tilde{X}_I[k]$$

$$= \frac{1}{N}\sum_{m=0}^{N-1} \tilde{X}_R[m]\tilde{U}_N[k-m]; \tag{11.41}$$

i.e., $\tilde{X}[k]$ is the periodic convolution of $\tilde{X}_R[k]$, the DFS of $\tilde{x}_e[n]$, with $\tilde{U}_N[k]$ the DFS of $\tilde{u}_N[n]$. The DFS of $\tilde{u}_N[n]$ can be shown to be (see Problem 11.22)

$$\tilde{U}_N[k] = \begin{cases} N, & k=0, \\ -j2\cot(\pi k/N), & k \text{ odd}, \\ 0, & k \text{ even}. \end{cases} \tag{11.42}$$

If we define

$$\tilde{V}_N[k] = \begin{cases} -j2\cot(\pi k/N), & k \text{ odd}, \\ 0, & k \text{ even}, \end{cases} \tag{11.43}$$

then Eq. (11.41) can be expressed as

$$\tilde{X}[k] = \tilde{X}_R[k] + \frac{1}{N}\sum_{m=0}^{N-1} \tilde{X}_R[m]\tilde{V}_N[k-m]. \tag{11.44}$$

Therefore,

$$j\tilde{X}_I[k] = \frac{1}{N}\sum_{m=0}^{N-1} \tilde{X}_R[m]\tilde{V}_N[k-m], \tag{11.45}$$

which is the desired relation between the real and imaginary parts of the DFS of a periodically causal periodic sequence. Similarly, beginning with Eq. (11.38) we can show that

$$\tilde{X}_R[k] = \frac{1}{N}\sum_{m=0}^{N-1} j\tilde{X}_I[m]\tilde{V}_N[k-m] + \tilde{x}[0] + (-1)^k\tilde{x}[N/2]. \tag{11.46}$$

Equations (11.45) and (11.46) relate the real and imaginary parts of the DFS representation of the periodic sequence $\tilde{x}[n]$. If $\tilde{x}[n]$ is thought of as the periodic repetition of a finite-length sequence $x[n]$ as in Eq. (11.31), then

$$x[n] = \begin{cases} \tilde{x}[n], & 0 \le n \le N-1, \\ 0, & \text{otherwise.} \end{cases} \tag{11.47}$$

If $x[n]$ has the "periodic causality" property with respect to a period N (i.e., $x[n] = 0$ for $n < 0$ and for $n > N/2$), then all of the preceding discussion applies to the discrete Fourier transform of $x[n]$. In other words, we can remove the tildes from Eqs. (11.45) and (11.46), thereby obtaining the DFT relations

$$j X_I[k] = \begin{cases} \dfrac{1}{N} \displaystyle\sum_{m=0}^{N-1} X_R[m] V_N[k-m], & 0 \le k \le N-1, \\ 0, & \text{otherwise,} \end{cases} \tag{11.48}$$

and

$$X_R[k] = \begin{cases} \dfrac{1}{N} \displaystyle\sum_{m=0}^{N-1} j X_I[m] V_N[k-m] + x[0] + (-1)^k x[N/2], & 0 \le k \le N-1, \\ 0, & \text{otherwise.} \end{cases} \tag{11.49}$$

Note that the sequence $V_N[k-m]$ given by Eq. (11.43) is periodic with period N, so we do not need to worry about computing $((k-m))_N$ in Eqs. (11.48) and (11.49), which are the desired relations between the real and imaginary parts of the N-point DFT of a real sequence whose actual length is less than or equal to $(N/2)+1$ (for N even). These equations are circular convolutions, and, for example, Eq. (11.48) can be evaluated efficiently by the following procedure:

1. Compute the inverse DFT of $X_R[k]$ to obtain the sequence

$$x_{ep}[n] = \frac{x[n] + x[((-n))_N]}{2}, \qquad 0 \le n \le N-1. \tag{11.50}$$

2. Compute the periodic odd part of $x[n]$ by

$$x_{op}[n] = \begin{cases} x_{ep}[n], & 0 < n < N/2, \\ -x_{ep}[n], & N/2 < n \le N-1, \\ 0, & \text{otherwise.} \end{cases} \tag{11.51}$$

3. Compute the DFT of $x_{op}[n]$ to obtain $j X_I[k]$.

Note that if, instead of computing the odd part of $x[n]$ in step 2, we compute

$$x[n] = \begin{cases} x_{ep}[0], & n = 0, \\ 2x_{ep}[n], & 0 < n < N/2, \\ x_{ep}[N/2], & n = N/2, \\ 0, & \text{otherwise,} \end{cases} \tag{11.52}$$

then the DFT of the resulting sequence would be $X[k]$, the complete DFT of $x[n]$.

Example 11.3   Periodic Sequence

Consider a sequence that is periodically causal with period $N = 4$ and that has

$$X_R[k] = \begin{cases} 2, & k = 0, \\ 3, & k = 1, \\ 4, & k = 2, \\ 3, & k = 3. \end{cases}$$

We can find the imaginary part of the DFT in one of two ways. The first way is to use Eq. (11.48). For $N = 4$,

$$V_4[k] = \begin{cases} 2j, & k = -1 + 4m, \\ -2j, & k = 1 + 4m, \\ 0, & \text{otherwise}, \end{cases}$$

where m is an integer. Implementing the convolution in Eq. (11.48) yields

$$j X_I[k] = \frac{1}{4} \sum_{m=0}^{3} X_R[k] V_4[k - m]$$

$$= \begin{cases} j, & k = 1, \\ -j, & k = 3, \\ 0, & \text{otherwise}. \end{cases}$$

Alternatively, we can follow the three-step procedure that includes Eqs. (11.50) and (11.51). Computing the inverse DFT $X_R[k]$ yields

$$x_e[n] = \frac{1}{4} \sum_{h=0}^{3} X_R[k] W_4^{-kn}$$

$$= \frac{1}{4} [2 + 3(j)^n + 4(-1)^n + 3(-j)^n]$$

$$= \begin{cases} 3, & n = 0, \\ -\frac{1}{2}, & n = 1, 3, \\ 0, & n = 2. \end{cases}$$

Note that although this sequence is not itself even symmetric, a periodic replication of $x_e[n]$ is even symmetric. Thus, the DFT $X_R[k]$ of $x_e[n]$ is purely real. Equation (11.51) allows us to find the periodically odd part $x_{op}[n]$; specifically,

$$x_{op}[n] = \begin{cases} -\frac{1}{2}, & n = 1, \\ \frac{1}{2}, & n = 3. \end{cases}$$

Finally, we obtain $j X_I[k]$ from the DFT of $x_{op}[n]$:

$$j X_I[k] = \sum_{n=0}^{3} x_{op}[n] W_4^{nk}$$

$$= -\frac{1}{2} W_4^{k} + \frac{1}{2} W_4^{3k}$$

$$= \begin{cases} j, & k = 1, \\ -j, & k = 3, \\ 0, & \text{otherwise}, \end{cases}$$

which is, of course, the same as was obtained from Eq. (11.48).

11.3 RELATIONSHIPS BETWEEN MAGNITUDE AND PHASE

So far, we have focused on the relationships between the real and imaginary parts of the Fourier transform of a sequence. Often, we are interested in relationships between the magnitude and phase of the Fourier transform. In this section, we consider the conditions under which these functions might be uniquely related. While it might appear on the surface that a relationship between real and imaginary parts implies a relationship between magnitude and phase, that is not the case. This is clearly demonstrated by Example 5.11 in Section 5.4. The two system functions $H_1(z)$ and $H_2(z)$ in that example were assumed to correspond to causal, stable systems. Therefore, the real and imaginary parts of $H_1(e^{j\omega})$ are related through the Hilbert transform relations of Eqs. (11.28) and (11.29), as are the real and imaginary parts of $H_2(e^{j\omega})$. However, $\sphericalangle H_1(e^{j\omega})$ could not be obtained from $|H_1(e^{j\omega})|$, since $H_1(e^{j\omega})$ and $H_2(e^{j\omega})$ have the same magnitude but a different phase.

The Hilbert transform relationship between the real and imaginary parts of the Fourier transform of a sequence $x[n]$ was based on the causality of $x[n]$. We can obtain a Hilbert transform relationship between magnitude and phase by imposing causality on a sequence $\hat{x}[n]$ derived from $x[n]$ for which the Fourier transform $\hat{X}(e^{j\omega})$ is the logarithm of the Fourier transform of $x[n]$. Specifically, we define $\hat{x}[n]$ so that

$$x[n] \xleftrightarrow{\mathcal{F}} X(e^{j\omega}) = |X(e^{j\omega})|e^{j\arg[X(e^{j\omega})]}, \tag{11.53a}$$

$$\hat{x}[n] \xleftrightarrow{\mathcal{F}} \hat{X}(e^{j\omega}), \tag{11.53b}$$

where

$$\hat{X}(e^{j\omega}) = \log[X(e^{j\omega})] = \log|X(e^{j\omega})| + j\arg[X(e^{j\omega})] \tag{11.54}$$

and, as defined in Section 5.1, $\arg[X(e^{j\omega})]$ denotes the continuous phase of $X(e^{j\omega})$. The sequence $\hat{x}[n]$ is commonly referred to as the *complex cepstrum* of $x[n]$. (See Oppenheim et al., 1968 and Oppenheim and Schafer, 1975.)

If we now require that $\hat{x}[n]$ be causal, then the real and imaginary parts of $\hat{X}(e^{j\omega})$, corresponding to $\log|X(e^{j\omega})|$ and $\arg[X(e^{j\omega})]$, respectively, will be related through Eqs. (11.28) and (11.29); i.e.,

$$\arg[X(e^{j\omega})] = -\frac{1}{2\pi}\mathcal{P}\int_{-\pi}^{\pi}\log|X(e^{j\theta})|\cot\left(\frac{\omega-\theta}{2}\right)d\theta \tag{11.55}$$

and

$$\log|X(e^{j\omega})| = \hat{x}[0] + \frac{1}{2\pi}\mathcal{P}\int_{-\pi}^{\pi}\arg[X(e^{j\theta})]\cot\left(\frac{\omega-\theta}{2}\right)d\theta, \tag{11.56a}$$

where, in Eq. (11.56a), $\hat{x}[0]$ is

$$\hat{x}[0] = \frac{1}{2\pi}\int_{-\pi}^{\pi}\log|X(e^{j\omega})|d\omega. \tag{11.56b}$$

Although it is not at all obvious at this point, in Problem 11.32 we develop the fact that the minimum-phase condition defined in Section 5.6, namely, that $X(z)$ have all its poles and zeros inside the unit circle, guarantees causality of the complex cepstrum. Thus, the minimum-phase condition in Section 5.6 and the condition of causality of the complex cepstrum turn out to be the same constraint developed from different perspectives. Note that when $\hat{x}[n]$ is causal, $\arg[X(e^{j\omega})]$ is completely determined through Eq. (11.55) by $\log|X(e^{j\omega})|$; however, the complete determination of $\log|X(e^{j\omega})|$ by Eq. (11.56) requires both the phase $\arg[X(e^{j\omega})]$ and the quantity $\hat{x}[0]$. If $\hat{x}[0]$ is not known, then $\log|X(e^{j\omega})|$ is determined only to within an additive constant, or equivalently, $|X(e^{j\omega})|$ is determined only to within a multiplicative (gain) constant.

Minimum phase and causality of the complex cepstrum are not the only constraints that provide a unique relationship between the magnitude and phase of the Fourier transform. As one example of another type of constraint, it has been shown (Hayes et al., 1980) that if a sequence is of finite length and if its z-transform has no zeros in conjugate reciprocal pairs, then, to within a scale factor, the sequence (and consequently, also the magnitude of the Fourier transform) is uniquely determined by the phase of the Fourier transform.

11.4 HILBERT TRANSFORM RELATIONS
FOR COMPLEX SEQUENCES

Thus far, we have considered Hilbert transform relations for the Fourier transform of causal sequences and the discrete Fourier transform of periodic sequences that are "periodically causal" in the sense that they are zero in the second half of each period. In this section, we consider *complex sequences* for which the real and imaginary components can be related through a discrete convolution similar to the Hilbert transform relations derived in the previous sections. These relations are particularly useful in representing bandpass signals as complex signals in a manner completely analogous to the analytic signals of continuous-time signal theory.

As mentioned previously, it is possible to base the derivation of the Hilbert transform relations on a notion of causality or one-sidedness. Since we are interested in relating the real and imaginary parts of a complex sequence, one-sidedness will be applied to the Fourier transform of the sequence. We cannot, of course, require that the Fourier transform be zero for $\omega < 0$, since it must be periodic. Instead, we consider sequences for which the Fourier transform is zero in the second half of each period; i.e., the z-transform is zero on the bottom half $(-\pi \leq \omega < 0)$ of the unit circle. Thus, with $x[n]$ denoting the sequence and $X(e^{j\omega})$ its Fourier transform, we require that

$$X(e^{j\omega}) = 0, \qquad -\pi \leq \omega < 0. \tag{11.57}$$

(We could just as well assume that $X(e^{j\omega})$ is zero for $0 < \omega \leq \pi$.) The sequence $x[n]$ corresponding to $X(e^{j\omega})$ must be complex, since, if $x[n]$ were real, $X(e^{j\omega})$ would be conjugate symmetric, i.e., $X(e^{j\omega}) = X^*(e^{-j\omega})$. Therefore, we express $x[n]$ as

$$x[n] = x_r[n] + jx_i[n], \tag{11.58}$$

where $x_r[n]$ and $x_i[n]$ are real sequences. In continuous-time signal theory, the comparable signal is an analytic function and thus is called an *analytic signal*. Although analyticity has no formal meaning for sequences, we will nevertheless apply the same terminology to complex sequences whose Fourier transforms are one sided.

If $X_r(e^{j\omega})$ and $X_i(e^{j\omega})$ denote the Fourier transforms of the real sequences $x_r[n]$ and $x_i[n]$, respectively, then

$$X(e^{j\omega}) = X_r(e^{j\omega}) + jX_i(e^{j\omega}), \tag{11.59a}$$

and it follows that

$$X_r(e^{j\omega}) = \frac{1}{2}[X(e^{j\omega}) + X^*(e^{-j\omega})] \tag{11.59b}$$

and

$$j X_i(e^{j\omega}) = \frac{1}{2}[X(e^{j\omega}) - X^*(e^{-j\omega})]. \tag{11.59c}$$

Note that Eq. (11.59c) gives an expression for $j X_i(e^{j\omega})$, which is the Fourier transform of the imaginary signal $j x_i[n]$. Note also that $X_r(e^{j\omega})$ and $X_i(e^{j\omega})$ are both complex-valued functions in general, and the complex transforms $X_r(e^{j\omega})$ and $j X_i(e^{j\omega})$ play a role similar to that played in the previous sections by the even and odd parts, respectively, of causal sequences. However, $X_r(e^{j\omega})$ is conjugate symmetric, i.e., $X_r(e^{j\omega}) = X_r^*(e^{-j\omega})$. Similarly, $j X_i(e^{j\omega})$ is conjugate antisymmetric, i.e., $j X_i(e^{j\omega}) = -j X_i^*(e^{-j\omega})$.

Figure 11.4 depicts an example of a complex one-sided Fourier transform of a complex sequence $x[n] = x_r[n] + j x_i[n]$ and the corresponding two-sided transforms of the real sequences $x_r[n]$ and $x_i[n]$. This figure shows pictorially the cancellation implied by Eqs. (11.59).

If $X(e^{j\omega})$ is zero for $-\pi \le \omega < 0$, then there is no overlap between the nonzero portions of $X(e^{j\omega})$ and $X^*(e^{-j\omega})$. Thus, $X(e^{j\omega})$ can be recovered from either $X_r(e^{j\omega})$ or $X_i(e^{j\omega})$. Since $X(e^{j\omega})$ is assumed to be zero at $\omega = \pm\pi$, $X(e^{j\omega})$ is totally recoverable from $j X_i(e^{j\omega})$. This is in contrast to the situation in Section 11.2, in which the causal sequence could be recovered from its odd part except at the endpoints.

In particular,

$$X(e^{j\omega}) = \begin{cases} 2X_r(e^{j\omega}), & 0 \le \omega < \pi, \\ 0, & -\pi \le \omega < 0, \end{cases} \tag{11.60}$$

and

$$X(e^{j\omega}) = \begin{cases} 2j X_i(e^{j\omega}), & 0 \le \omega < \pi, \\ 0, & -\pi \le \omega < 0. \end{cases} \tag{11.61}$$

Alternatively, we can relate $X_r(e^{j\omega})$ and $X_i(e^{j\omega})$ directly by

$$X_i(e^{j\omega}) = \begin{cases} -j X_r(e^{j\omega}), & 0 < \omega < \pi, \\ j X_r(e^{j\omega}), & -\pi \le \omega < 0, \end{cases} \tag{11.62}$$

or

$$X_i(e^{j\omega}) = H(e^{j\omega}) X_r(e^{j\omega}), \tag{11.63a}$$

where

$$H(e^{j\omega}) = \begin{cases} -j, & 0 < \omega < \pi, \\ j, & -\pi < \omega < 0. \end{cases} \tag{11.63b}$$

Equations (11.63) are illustrated by comparing Figures 11.4(c) and 11.4(d). Now $X_i(e^{j\omega})$ is the Fourier transform of $x_i[n]$, the imaginary part of $x[n]$, and $X_r(e^{j\omega})$ is the Fourier transform of $x_r[n]$, the real part of $x[n]$. Thus, according to Eqs. (11.63), $x_i[n]$ can be obtained by processing $x_r[n]$ with a linear time-invariant discrete-time system with frequency response $H(e^{j\omega})$, as given by Eq. (11.63b). This frequency response has unity magnitude, a phase angle of $-\pi/2$ for $0 < \omega < \pi$, and a phase angle of $+\pi/2$ for $-\pi < \omega < 0$. Such a system is called an ideal 90-degree phase shifter. Alternatively,

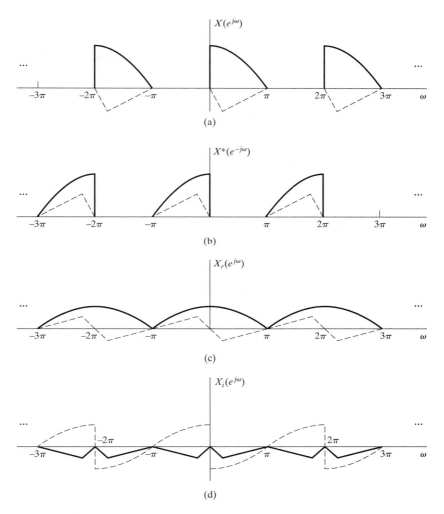

Figure 11.4   Illustration of decomposition of a one-sided Fourier transform. (Solid curves are real parts and dashed curves are imaginary parts.)

when it is clear that we are considering an operation on a sequence, the 90-degree phase shifter is also called a *Hilbert transformer*. From Eqs. (11.63), it follows that

$$X_r(e^{j\omega}) = \frac{1}{H(e^{j\omega})} X_i(e^{j\omega}) = -H(e^{j\omega}) X_i(e^{j\omega}). \tag{11.64}$$

Thus, $-x_r[n]$ can also be obtained from $x_i[n]$ with a 90-degree phase shifter.

The impulse response $h[n]$ of a 90-degree phase shifter, corresponding to the frequency response $H(e^{j\omega})$ given in Eq. (11.63b), is

$$h[n] = \frac{1}{2\pi} \int_{-\pi}^{0} je^{j\omega n}d\omega - \frac{1}{2\pi} \int_{0}^{\pi} je^{j\omega n}d\omega,$$

or

$$h[n] = \begin{cases} \dfrac{2}{\pi} \dfrac{\sin^2(\pi n/2)}{n}, & n \neq 0, \\ 0, & n = 0. \end{cases} \tag{11.65}$$

The impulse response is plotted in Figure 11.5. Using Eqs. (11.63) and (11.64), we obtain the expressions

$$x_i[n] = \sum_{m=-\infty}^{\infty} h[n-m]x_r[m] \tag{11.66a}$$

and

$$x_r[n] = -\sum_{m=-\infty}^{\infty} h[n-m]x_i[m]. \tag{11.66b}$$

Equations (11.66) are the desired Hilbert transform relations between the real and imaginary parts of a discrete-time analytic signal.

Figure 11.6 shows how a discrete-time Hilbert transformer system can be used to form a complex analytic signal, which is simply a pair of real signals.

Figure 11.5   Impulse response of an ideal Hilbert transformer or 90-degree phase shifter.

Figure 11.6   Block diagram representation of the creation of a complex sequence whose Fourier transform is one sided.

11.4.1 Design of Hilbert Transformers

The impulse response of the Hilbert transformer, as given in Eq. (11.65), is not absolutely summable. Consequently,

$$H(e^{j\omega}) = \sum_{n=-\infty}^{\infty} h[n]e^{-j\omega n} \tag{11.67}$$

converges to Eq. (11.63) only in the mean-square sense. Thus, the ideal Hilbert transformer or 90-degree phase shifter takes its place alongside the ideal lowpass filter and ideal bandlimited differentiator as a valuable theoretical concept that corresponds to a noncausal system and for which the system function exists only in a restricted sense.

Approximations to the ideal Hilbert transformer can, of course, be obtained. FIR approximations with constant group delay can be designed using either the window method or the equiripple approximation method. In such approximations the 90-degree phase shift is realized exactly, with an additional linear phase component required for a causal FIR system. The properties of these approximations are illustrated by examples of Hilbert transformers designed with Kaiser windows.

Example 11.4   Kaiser Window Design of Hilbert Transformers

The Kaiser window approximation for an FIR discrete Hilbert transformer of order M (length $M + 1$) would be of the form

$$h[n] = \begin{cases} \dfrac{I_o\{\beta(1 - [(n - n_d)/n_d]^2)^{1/2}\}}{I_o(\beta)} \left\{ \dfrac{2}{\pi} \dfrac{\sin^2[\pi(n - n_d)/2]}{n - n_d} \right\}, & 0 \le n \le M, \\ 0, & \text{otherwise,} \end{cases}$$

(11.68)

Sample number (n)

(a)

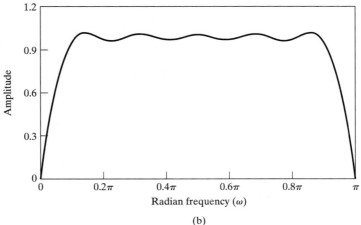

Radian frequency (ω)

(b)

Figure 11.7   (a) Impulse response and (b) magnitude response of an FIR Hilbert transformer designed using the Kaiser window. ($M = 18$ and $\beta = 2.629$.)

where $n_d = M/2$. If M is even, the system is a type III FIR generalized linear-phase system, as discussed in Section 5.7.3.

Figure 11.7(a) shows the impulse response, and Figure 11.7(b) shows the magnitude of the frequency response, for $M = 18$ and $\beta = 2.629$. Because $h[n]$ satisfies the symmetry condition $h[n] = -h[M - n]$ for $0 \leq n \leq M$, the phase is exactly 90° plus a linear component corresponding to a delay of $n_d = 18/2 = 9$ samples; i.e.,

$$\sphericalangle H(e^{j\omega}) = \frac{-\pi}{2} - 9\omega, \qquad 0 < \omega < \pi. \tag{11.69}$$

From Figure 11.7(b), we see that, as required for a type III system, the frequency response is zero at $z = 1$ and $z = -1$ ($\omega = 0$ and $\omega = \pi$). Thus, the magnitude response cannot approximate unity very well except in some middle band $\omega_L < |\omega| < \omega_H$.

If M is an odd integer, we obtain a type IV system, as shown in Figure 11.8, where $M = 17$ and $\beta = 2.44$. For type IV systems, the frequency response is forced

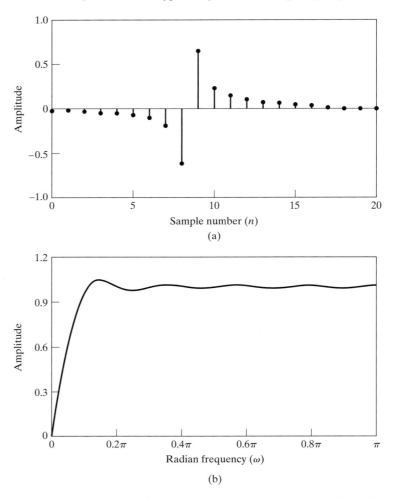

(a)

(b)

Figure 11.8   (a) Impulse response and (b) magnitude response of an FIR Hilbert transformer designed using the Kaiser window. ($M = 17$ and $\beta = 2.44$.)

to be zero only at $z = 1$ ($\omega = 0$). Therefore, a better approximation to a constant-magnitude response is obtained for frequencies around $\omega = \pi$. The phase response is exactly 90° at all frequencies plus a linear phase component corresponding to $n_d = 17/2 = 8.5$ samples delay; i.e.,

$$\sphericalangle H(e^{j\omega}) = \frac{-\pi}{2} - 8.5\omega. \tag{11.70}$$

From a comparison of Figures 11.7(a) and 11.8(a), we see that type III FIR Hilbert transformers have a significant computational advantage over type IV systems when it is not necessary to approximate constant magnitude at $\omega = \pi$. This is because, for type III systems, the even-indexed samples of the impulse response are all exactly zero. Thus, taking advantage of the antisymmetry in both cases, the system with $M = 17$ would require eight multiplications to compute each output sample, while the system with $M = 18$ would require only five multiplications per output sample.

Type III and IV FIR linear-phase Hilbert transformer approximations with equiripple magnitude approximation and exactly 90° phase can be designed using the Parks–McClellan algorithm as described in Sections 7.4 and 7.5, with the expected improvements in magnitude approximation error over window-designed filters of the same length (Rabiner and Schafer, 1974).

The exactness of the phase of type III and IV FIR systems is a compelling motivation for their use in approximating Hilbert transformers. IIR systems must have some phase response error as well as magnitude response error in approximating a Hilbert transformer. The most successful approach to designing IIR Hilbert transformers is to design a "phase splitter," which consists of two all-pass systems whose phase responses differ by approximately 90° over some portion of the band $0 < |\omega| < \pi$. Such systems can be designed by using the bilinear transformation to transform a continuous-time phase-splitting system to a discrete-time system. (For an example of such a system, see Gold et al. (1970).)

Figure 11.9 depicts a 90-degree phase-splitting system. If $x_r[n]$ denotes a real input signal and $x_i[n]$ its Hilbert transform, then the complex sequence $x[n] = x_r[n] + jx_i[n]$ has a Fourier transform that is identically zero for $-\pi \leq \omega < 0$; i.e., $X(z)$ is zero on the bottom half of the unit circle of the z-plane. In the system of Figure 11.6, a Hilbert transformer was used to form the signal $x_i[n]$ from $x_r[n]$. In Figure 11.9, we process $x_r[n]$ through two systems: $H_1(e^{j\omega})$ and $H_2(e^{j\omega})$. Now, if $H_1(e^{j\omega})$ and $H_2(e^{j\omega})$ are all-pass systems whose phase responses differ by 90°, then the complex signal $y[n] = y_r[n] + jy_i[n]$ has a Fourier transform that also vanishes for $-\pi \leq \omega < 0$. Furthermore, $|Y(e^{j\omega})| = |X(e^{j\omega})|$, since the phase-splitting systems are all-pass systems. The phases of $Y(e^{j\omega})$ and $X(e^{j\omega})$ will differ by the phase component common to $H_1(e^{j\omega})$ and $H_2(e^{j\omega})$.

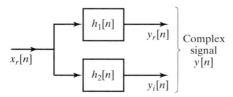

Figure 11.9 Block diagram representation of the all-pass phase splitter method for the creation of a complex sequence whose Fourier transform is one sided.

11.4.2 Representation of Bandpass Signals

Many of the applications of analytic signals concern narrowband communication. In such applications, it is sometimes convenient to represent a bandpass signal in terms of a lowpass signal. To see how this may be done, consider the complex lowpass signal

$$x[n] = x_r[n] + jx_i[n],$$

where $x_i[n]$ is the Hilbert transform of $x_r[n]$ and

$$X(e^{j\omega}) = 0, \qquad -\pi \leq \omega < 0.$$

The Fourier transforms $X_r(e^{j\omega})$ and $jX_i(e^{j\omega})$ are depicted in Figures 11.10(a) and 11.10(b), respectively, and the resulting transform $X(e^{j\omega}) = X_r(e^{j\omega}) + jX_i(e^{j\omega})$ is shown in Figure 11.10(c). (Solid curves are real parts and dashed curves are imaginary parts.) Now, consider the sequence

$$s[n] = x[n]e^{j\omega_c n} = s_r[n] + js_i[n], \tag{11.71}$$

where $s_r[n]$ and $s_i[n]$ are real sequences. The corresponding Fourier transform is

$$S(e^{j\omega}) = X(e^{j(\omega - \omega_c)}), \tag{11.72}$$

which is depicted in Figure 11.10(d). Applying Eqs. (11.59) to $S(e^{j\omega})$ leads to the equations

$$S_r(e^{j\omega}) = \tfrac{1}{2}[S(e^{j\omega}) + S^*(e^{-j\omega})], \tag{11.73a}$$

$$jS_i(e^{j\omega}) = \tfrac{1}{2}[S(e^{j\omega}) - S^*(e^{-j\omega})]. \tag{11.73b}$$

For the example of Figure 11.10, $S_r(e^{j\omega})$ and $jS_i(e^{j\omega})$ are illustrated in Figures 11.10(e) and 11.10(f), respectively. It is straightforward to show that if $X_r(e^{j\omega}) = 0$ for $\Delta\omega < |\omega| \leq \pi$, and if $\omega_c + \Delta\omega < \pi$, then $S(e^{j\omega})$ will be a one-sided bandpass signal such that $S(e^{j\omega}) = 0$ *except* in the interval $\omega_c < \omega < \omega_c + \Delta\omega$. As the example of Figure 11.10 illustrates, and as can be shown using Eqs. (11.58) and (11.59), $S_i(e^{j\omega}) = H(e^{j\omega})S_r(e^{j\omega})$, or, in other words, $s_i[n]$ is the Hilbert transform of $s_r[n]$.

An alternative representation of a complex signal is in terms of magnitude and phase; i.e., $x[n]$ can be expressed as

$$x[n] = A[n]e^{j\phi[n]}, \tag{11.74a}$$

where

$$A[n] = (x_r^2[n] + x_i^2[n])^{1/2} \tag{11.74b}$$

and

$$\phi[n] = \arctan\left(\frac{x_i[n]}{x_r[n]}\right). \tag{11.74c}$$

Therefore, from Eqs. (11.71) and (11.74), we can express $s[n]$ as

$$s[n] = (x_r[n] + jx_i[n])e^{j\omega_c n} \tag{11.75a}$$

$$= A[n]e^{j(\omega_c n + \phi[n])}, \tag{11.75b}$$

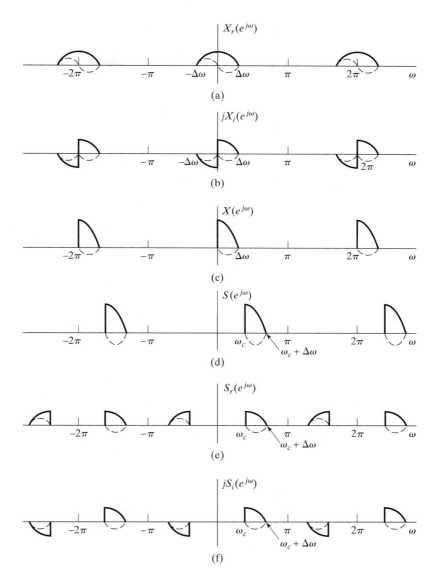

Figure 11.10   Fourier transforms for representation of bandpass signals. (Solid curves are real parts and dashed curves are imaginary parts.) (Note that in parts b and f the functions $jX_i(e^{j\omega})$ and $jS_i(e^{j\omega})$ are plotted, where $X_i(e^{j\omega})$ and $S_i(e^{j\omega})$ are the Fourier transforms of the Hilbert transforms of $x_i[n]$ and $s_i[n]$, respectively.)

from which we obtain the expressions

$$s_r[n] = x_r[n]\cos\omega_c n - x_i[n]\sin\omega_c n, \tag{11.76a}$$

or

$$s_r[n] = A[n]\cos(\omega_c n + \phi[n]), \tag{11.76b}$$

and

$$s_i[n] = x_r[n] \sin \omega_c n + x_i[n] \cos \omega_c n, \qquad (11.77\text{a})$$

or

$$s_i[n] = A[n] \sin(\omega_c n + \phi[n]). \qquad (11.77\text{b})$$

Equations (11.76a) and (11.77a) are depicted in Figures 11.11(a) and 11.11(b), respectively. These diagrams illustrate how a complex bandpass (single-sideband) signal can be formed from a real lowpass signal.

Taken together, Eqs. (11.76) and (11.77) are the desired time-domain representations of a general complex bandpass signal $s[n]$ in terms of the real and imaginary parts of a complex lowpass signal $x[n]$. Generally, this complex representation is a convenient mechanism for representing a real bandpass signal. For example, Eq. (11.76a) provides a time-domain representation of the real bandpass signal in terms of an "in-phase" component $x_r[n]$ and a "quadrature" (90-degree phase-shifted) component $x_i[n]$. Indeed, as illustrated in Figure 11.10(e), Eq. (11.76a) permits the representation of real bandpass signals (or filter impulse responses) whose Fourier transforms are not conjugate symmetric about the center of the passband (as would be the case for signals of the form $x_r[n] \cos \omega_c n$).

It is clear from the form of Eqs. (11.76) and (11.77) and from Figure 11.11 that a general bandpass signal has the form of a sinusoid that is both amplitude and phase modulated. The sequence $A[n]$ is called the envelope and $\phi[n]$ the phase. This narrowband signal representation can be used to represent a variety of amplitude and phase modulation systems. The example of Figure 11.10 is an illustration of single-sideband modulation. If we consider the real signal $s_r[n]$ as resulting from single-sideband modulation with the lowpass real signal $x_r[n]$ as the input, then Figure 11.11(a) represents a

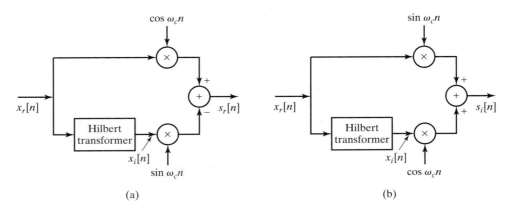

(a)   (b)

Figure 11.11   Block diagram representation of Eqs. (11.76a) and (11.77a) for obtaining a single-sideband signal.

scheme for implementing the single-sideband modulation system. Single-sideband modulation systems are useful in frequency-division multiplexing, since they can represent a real bandpass signal with minimum bandwidth.

11.4.3 Bandpass Sampling

Another important use of analytic signals is in the sampling of bandpass signals. In Chapter 4 we saw that, in general, if a continuous-time signal has a bandlimited Fourier transform such that $S_c(j\Omega) = 0$ for $|\Omega| \geq \Omega_N$, then the signal is exactly represented by its samples if the sampling rate satisfies the inequality $2\pi/T \geq 2\Omega_N$. The key to the proof of this result is to avoid overlapping the replicas of $S_c(j\Omega)$ that form the discrete-time Fourier transform of the sequence of samples. A bandpass continuous-time signal has a Fourier transform such that $S_c(j\Omega) = 0$ for $0 \leq |\Omega| \leq \Omega_c$ and for $|\Omega| \geq \Omega_c + \Delta\Omega$. Thus, its bandwidth, or region of support, is really only $2\Delta\Omega$ rather than $2(\Omega_c + \Delta\Omega)$, and with a proper sampling strategy, the region $-\Omega_c \leq \Omega \leq \Omega_c$ can be filled with images of the nonzero part of $S_c(j\Omega)$ without overlapping. This is greatly facilitated by using a complex representation of the bandpass signal.

As an illustration, consider the system of Figure 11.12 and the signal shown in Figure 11.13(a). The highest frequency of the input signal is $\Omega_c + \Delta\Omega$. If this signal is sampled at exactly the Nyquist rate, $2\pi/T = 2(\Omega_c + \Delta\Omega)$, then the resulting sequence of samples, $s_r[n] = s_c(nT)$, has the Fourier transform $S_r(e^{j\omega})$ plotted in Figure 11.13(b). Using a discrete-time Hilbert transformer, we can form the complex sequence $s[n] = s_r[n] + js_i[n]$ whose Fourier transform is $S(e^{j\omega})$ in Figure 11.13(c). The width of the nonzero region of $S(e^{j\omega})$ is $\Delta\omega = (\Delta\Omega)T$. Defining M as the largest integer less than or equal to $2\pi/\Delta\omega$, we see that M copies of $S(e^{j\omega})$ would fit into the interval $-\pi < \omega < \pi$. (In the example of Figure 11.13(c), $2\pi/\Delta\omega = 5$.) Thus, the sampling rate of $s[n]$ can be reduced by decimation as shown in Figure 11.12, yielding the reduced-rate complex sequence $s_d[n] = s_{rd}[n] + js_{id}[n] = s[Mn]$ whose Fourier transform is

$$S_d(e^{j\omega}) = \frac{1}{M} \sum_{k=0}^{M-1} S(e^{j[(\omega-2\pi k)/M]}). \tag{11.78}$$

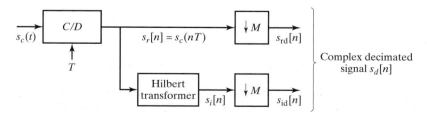

Figure 11.12   System for reduced-rate sampling of a real bandpass signal by decimation of the equivalent complex bandpass signal.

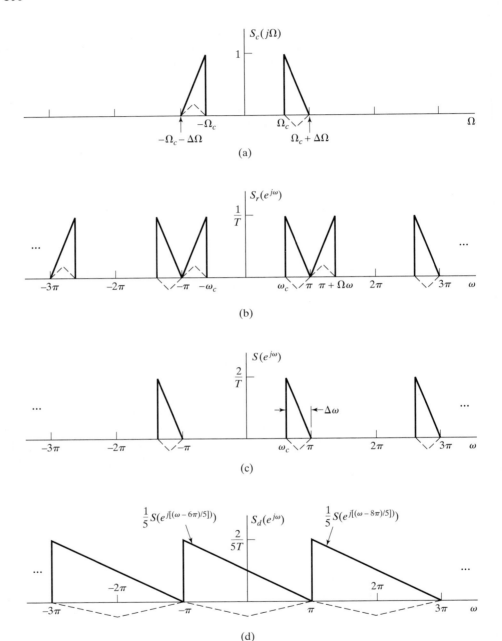

Figure 11.13   Example of reduced-rate sampling of a bandpass signal using the system of Figure 11.12. (a) Fourier transform of continuous-time bandpass signal. (b) Fourier transform of sampled signal. (c) Fourier transform of complex bandpass discrete-time signal derived from the signal of part (a). (d) Fourier transform of decimated complex bandpass of part (c). (Solid curves are real parts and dashed curves are imaginary parts.)

Figure 11.13(d) shows $S_d(e^{j\omega})$ with $M = 5$ in Eq. (11.78). $S(e^{j\omega})$ and two of the frequency-scaled and translated copies of $S(e^{j\omega})$ are indicated explicitly in Figure 11.13(d). It is clear that aliasing has been avoided and that all the information necessary to reconstruct the original sampled real bandpass signal now resides in the discrete-time frequency interval $-\pi < \omega \le \pi$. A complex filter applied to $s_d[n]$ can transform this information in useful ways, such as by further bandlimiting, amplitude or phase compensation, etc., or the complex signal can be coded for transmission or digital storage. This processing takes place at the low sampling rate, and this is, of course, the motivation for reducing the sampling rate.

The original real bandpass signal $s_r[n]$ can be reconstructed ideally by the following procedure:

1. Expand the complex sequence by a factor M; i.e., obtain

$$s_e[n] = \begin{cases} s_{rd}[n/M] + js_{id}[n/M], & n = 0, \pm M, \pm 2M, \ldots, \\ 0, & \text{otherwise.} \end{cases} \qquad (11.79)$$

2. Filter the signal $s_e[n]$ using an ideal *complex* bandpass filter with impulse response $h_i[n]$ and frequency response

$$H_i(e^{j\omega}) = \begin{cases} 0, & -\pi < \omega < \omega_c, \\ M, & \omega_c < \omega < \omega_c + \Delta\omega, \\ 0, & \omega_c + \Delta\omega < \omega < \pi. \end{cases} \qquad (11.80)$$

(In our example, $\omega_c + \Delta\omega = \pi$.)

3. Obtain $s_r[n] = \mathcal{Re}\{s_e[n] * h_i[n]\}$.

A useful exercise is to plot the Fourier transform $S_e(e^{j\omega})$ for the example of Figure 11.13 and verify that the filter of Eq. (11.80) does indeed recover $s[n]$.

Another useful exercise is to consider a complex continuous-time signal with a one-sided Fourier transform equal to $S_c(j\Omega)$ for $\Omega \ge 0$. It can be shown that such a signal can be sampled with sampling rate $2\pi/T = \Delta\Omega$, directly yielding the complex sequence $s_d[n]$.

11.5 SUMMARY

In this chapter, we have discussed a variety of relations between the real and imaginary parts of Fourier transforms and the real and imaginary parts of complex sequences. These relationships are collectively referred to as *Hilbert transform relationships.* Our approach to deriving all the Hilbert transform relations was to apply a basic causality principle that allows a sequence or function to be recovered from its even part. We showed that, for a causal sequence, the real and imaginary parts of the Fourier transform are related through a convolution-type integral. Also, for the special case when the complex cepstrum of a sequence is causal or, equivalently, both the poles and zeros of its z-transform lie inside the unit circle (the minimum-phase condition), the logarithm of the magnitude and the phase of the Fourier transform are a Hilbert transform pair of each other.

Hilbert transform relations were derived for periodic sequences that satisfy a modified causality constraint and for complex sequences whose Fourier transforms vanish on the bottom half of the unit circle. Applications of complex analytic signals to the representation and efficient sampling of handpass signals were also discussed.

PROBLEMS

Basic Problems with Answers

11.1. Consider a sequence $x[n]$ with discrete-time Fourier transform $X(e^{j\omega})$. The sequence $x[n]$ is real valued and causal, and

$$\mathcal{R}e\{X(e^{j\omega})\} = 2 - 2a\cos\omega.$$

Determine $\mathcal{I}m\{X(e^{j\omega})\}$.

11.2. Consider a sequence $x[n]$ and its discrete-time Fourier transform $X(e^{j\omega})$. The following is known:

$$x[n] \text{ is real and causal,}$$
$$\mathcal{R}e\{X(e^{j\omega})\} = \tfrac{5}{4} - \cos\omega.$$

Determine a sequence $x[n]$ consistent with the given information.

11.3. Consider a sequence $x[n]$ and its discrete-time Fourier transform $X(e^{j\omega})$. The following is known:

$$x[n] \text{ is real,}$$
$$x[0] = 0,$$
$$x[1] > 0,$$
$$|X(e^{j\omega})|^2 = \tfrac{5}{4} - \cos\omega.$$

Determine two distinct sequences $x_1[n]$ and $x_2[n]$ consistent with the given information.

11.4. Consider a complex sequence $x[n] = x_r[n] + jx_i[n]$, where $x_r[n]$ and $x_i[n]$ are the real part and imaginary part, respectively. The z-transform $X(z)$ of the sequence $x[n]$ is zero on the bottom half of the unit circle; i.e., $X(e^{j\omega}) = 0$ for $\pi \le \omega < 2\pi$. The real part of $x[n]$ is

$$x_r[n] = \begin{cases} 1/2, & n = 0, \\ -1/4, & n = \pm 2, \\ 0, & \text{otherwise.} \end{cases}$$

Determine the real and imaginary parts of $X(e^{j\omega})$.

11.5. Find the Hilbert transforms $x_i[n] = \mathcal{H}\{x_r[n]\}$ of the following sequences:
(a) $x_r[n] = \cos\omega_0 n$
(b) $x_r[n] = \sin\omega_0 n$
(c) $x_r[n] = \dfrac{\sin(\omega_c n)}{\pi n}$

11.6. The imaginary part of $X(e^{j\omega})$ for a casual, real sequence $x[n]$ is

$$X_I(e^{j\omega}) = 2\sin\omega - 3\sin 4\omega.$$

Additionally, it is known that $X(e^{j\omega})|_{\omega=0} = 6$. Find $x[n]$.

11.7. (a) $x[n]$ is a real, causal sequence with the imaginary part of its discrete-time Fourier transform $X(e^{j\omega})$ given by

$$\mathcal{I}m\{X(e^{j\omega})\} = \sin\omega + 2\sin 2\omega.$$

Determine a choice for $x[n]$.

(b) Is your answer to Part (a) unique? If so, explain why. If not, determine a second, distinct choice for $x[n]$ satisfying the relationship given in Part (a).

11.8. Consider a real, causal sequence $x[n]$ with discrete-time Fourier transform $X(e^{j\omega}) = X_R(e^{j\omega}) + j X_I(e^{j\omega})$. The imaginary part of the discrete-time Fourier transform is

$$X_I(e^{j\omega}) = 3\sin(2\omega).$$

Which of the real parts $X_{Rm}(e^{j\omega})$ listed below are consistent with this information:

$$X_{R1}(e^{j\omega}) = \frac{3}{2}\cos(2\omega),$$

$$X_{R2}(e^{j\omega}) = -3\cos(2\omega) - 1,$$

$$X_{R3}(e^{j\omega}) = -3\cos(2\omega),$$

$$X_{R4}(e^{j\omega}) = 2\cos(3\omega),$$

$$X_{R5}(e^{j\omega}) = \frac{3}{2}\cos(2\omega) + 1.$$

11.9. The following information is known about a real, causal sequence $x[n]$ and its discrete-time Fourier transform $X(e^{j\omega})$:

$$\mathcal{I}m\{X(e^{j\omega})\} = 3\sin(\omega) + \sin(3\omega),$$

$$X(e^{j\omega})|_{\omega=\pi} = 3.$$

Determine a sequence $x[n]$ consistent with this information. Is the sequence unique?

11.10. Consider $h[n]$, the real-valued impulse response of a stable, causal LTI system with frequency response $H(e^{j\omega})$. The following is known:
  (i) The system has a stable, causal inverse.
  (ii) $\left| H(e^{j\omega}) \right|^2 = \dfrac{\frac{5}{4} - \cos\omega}{5 + 4\cos\omega}.$

Determine $h[n]$ in as much detail as possible.

11.11. Let $x[n] = x_r[n] + j x_i[n]$ be a complex-valued sequence such that $X(e^{j\omega}) = 0$ for $-\pi \le \omega < 0$. The imaginary part is

$$x_i[n] = \begin{cases} 4, & n = 3, \\ -4, & n = -3. \end{cases}$$

Specify the real and imaginary parts of $X(e^{j\omega})$.

11.12. $h[n]$ is a causal, real-valued sequence with $h[0]$ nonzero and positive. The magnitude squared of the frequency response of $h[n]$ is given by

$$\left| H(e^{j\omega}) \right|^2 = \frac{10}{9} - \frac{2}{3}\cos(\omega).$$

(a) Determine a choice for $h[n]$.
(b) Is your answer to Part (b) unique? If so, explain why. If not, determine a second, distinct choice for $h[n]$ satisfying the given conditions.

11.13. Let $x[n]$ denote a causal, complex-valued sequence with Fourier transform

$$X(e^{j\omega}) = X_R(e^{j\omega}) + j X_I(e^{j\omega}).$$

If $X_R(e^{j\omega}) = 1 + \cos(\omega) + \sin(\omega) - \sin(2\omega)$, determine $X_I(e^{j\omega})$.

11.14. Consider a real, anticausal sequence $x[n]$ with discrete-time Fourier transform $X(e^{j\omega})$. The real part of $X(e^{j\omega})$ is

$$X_R(e^{j\omega}) = \sum_{k=0}^{\infty} (1/2)^k \cos(k\omega).$$

Find $X_I(e^{j\omega})$, the imaginary part of $X(e^{j\omega})$. (Remember that a sequence is said to be anticausal if $x[n] = 0$ for $n > 0$.)

11.15. $x[n]$ is a real, causal sequence with discrete-time Fourier transform $X(e^{j\omega})$. The imaginary part of $X(e^{j\omega})$ is

$$\mathcal{I}m\{X(e^{j\omega})\} = \sin\omega,$$

and it is also known that

$$\sum_{n=-\infty}^{\infty} x[n] = 3.$$

Determine $x[n]$.

11.16. Consider a real, causal sequence $x[n]$ with discrete-time Fourier transform $X(e^{j\omega})$, where the following two facts are given about $X(e^{j\omega})$:

$$X_R(e^{j\omega}) = 2 - 4\cos(3\omega),$$

$$X(e^{j\omega})|_{\omega=\pi} = 7.$$

Are these facts consistent? That is, can a sequence $x[n]$ satisfy both? If so, give one choice for $x[n]$. If not, explain why not.

11.17. Consider a real, causal, finite-length signal $x[n]$ with length $N = 2$ and with a 2-point discrete-Fourier transform $X[k] = X_R[k] + j X_I[k]$ for $k = 0, 1$. If $X_R[k] = 2\delta[k] - 4\delta[k-1]$, is it possible to determine $x[n]$ uniquely? If so, give $x[n]$. If not, give several choices for $x[n]$ satisfying the stated condition on $X_R[k]$.

11.18. Let $x[n]$ be a real-valued, causal, finite-length sequence with length $N = 3$. Find two choices for $x[n]$ such that the real part of the discrete Fourier transform $X_R[k]$ matches that shown in Figure P11.18-1. Note that only one of your sequences is "periodically causal" according to the definition in Section 11.2, where $x[n] = 0$ for $N/2 < n \leq N - 1$.

Figure P11.18-1

11.19. Let $x[n]$ be a real, causal, finite-length sequence with length $N = 4$ that is also periodically causal. The real part of the 4-point discrete Fourier transform $X_R[k]$ for this sequence is shown in Figure P11.19-1. Determine the imaginary part of the DFT $j X_I[k]$.

Figure P11.19-1

11.20. Consider a sequence $x[n]$ that is real, causal, and of finite length with $N = 6$. The imaginary part of the 6-point discrete Fourier transform of this sequence is

$$j X_I[k] = \begin{cases} -j2/\sqrt{3}, & k = 2, \\ j2/\sqrt{3}, & k = 4, \\ 0, & \text{otherwise.} \end{cases}$$

Additionally, it is known that

$$\frac{1}{6}\sum_{k=0}^{5}X[k]=1.$$

Which of the sequences shown in Figure P11.20-1 are consistent with the information given?

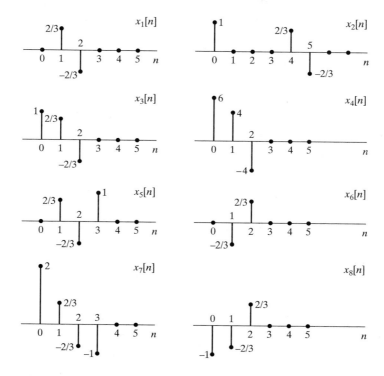

Figure P11.20-1

Basic Problems

11.21. Let $x[n]$ be a real causal sequence for which $|x[n]| < \infty$. The z-transform of $x[n]$ is

$$X(z)=\sum_{n=0}^{\infty}x[n]z^{-n},$$

which is a Taylor's series in the variable z^{-1} and therefore converges to an analytic function everywhere outside some circular disc centered at $z = 0$. (The region of convergence includes the point $z = \infty$, and, in fact, $X(\infty) = x[0]$.) The statement that $X(z)$ is analytic (in its region of convergence) implies strong constraints on the function $X(z)$. (See Churchill and Brown, 1990.) Specifically, its real and imaginary parts each satisfy Laplace's equation, and the real and imaginary parts are related by the Cauchy–Riemann equations. We will use these properties to determine $X(z)$ from its real part when $x[n]$ is a real, finite-valued, causal sequence.

Let the z-transform of such a sequence be

$$X(z)=X_R(z)+jX_I(z).$$

where $X_R(z)$ and $X_I(z)$ are real-valued functions of z. Suppose that $X_R(z)$ is

$$X_R(\rho e^{j\omega}) = \frac{\rho + \alpha \cos \omega}{\rho}, \qquad \alpha \text{ real},$$

for $z = \rho e^{j\omega}$. Then find $X(z)$ (as an explicit function of z), assuming that $X(z)$ is analytic everywhere except at $z = 0$. Do this using both of the following methods.

(a) *Method 1, Frequency Domain.* Use the fact that the real and imaginary parts of $X(z)$ must satisfy the Cauchy–Riemann equations everywhere that $X(z)$ is analytic. The Cauchy–Riemann equations are the following:

1. In Cartesian coordinates,

$$\frac{\partial U}{\partial x} = \frac{\partial V}{\partial y}, \qquad \frac{\partial V}{\partial x} = -\frac{\partial U}{\partial y},$$

where $z = x + jy$ and $X(x + jy) = U(x, y) + jV(x, y)$.

2. In polar coordinates,

$$\frac{\partial U}{\partial \rho} = \frac{1}{\rho}\frac{\partial V}{\partial \omega}, \qquad \frac{\partial V}{\partial \rho} = -\frac{1}{\rho}\frac{\partial U}{\partial \omega},$$

where $z = \rho e^{j\omega}$ and $X(\rho e^{j\omega}) = U(\rho, \omega) + jV(\rho, \omega)$.

Since we know that $U = X_R$, we can integrate these equations to find $V = X_I$ and hence X. (Be careful to treat the constant of integration properly.)

(b) *Method 2, Time Domain.* The sequence $x[n]$ can be represented as $x[n] = x_e[n] + x_o[n]$, where $x_e[n]$ is real and even with Fourier transform $X_R(e^{j\omega})$ and the sequence $x_o[n]$ is real and odd with Fourier transform $jX_I(e^{j\omega})$. Find $x_e[n]$ and, using causality, find $x_o[n]$ and hence $x[n]$ and $X(z)$.

11.22. Show that the sequence of discrete Fourier series coefficients for the sequence

$$\tilde{u}_N[n] = \begin{cases} 1, & n = 0, \quad N/2, \\ 2, & n = 1, 2, \ldots, N/2 - 1, \\ 0, & n = N/2 + 1, \ldots, N - 1, \end{cases}$$

is

$$\tilde{U}_N[k] = \begin{cases} N, & k = 0, \\ -j2\cot(\pi k/N), & k \text{ odd}, \\ 0, & k \text{ even}, k \neq 0. \end{cases}$$

Hint: Find the z-transform of the sequence

$$u_N[n] = 2u[n] - 2u[n - N/2] - \delta[n] + \delta[n - N/2],$$

and sample it to obtain $\tilde{U}[k]$.

Advanced Problems

11.23. Consider a real-valued finite-duration sequence $x[n]$ of length M. Specifically, $x[n] = 0$ for $n < 0$ and $n > M - 1$. Let $X[k]$ denote the N-point DFT of $x[n]$ with $N \geq M$ and N odd. The real part of $X[k]$ is denoted $X_R[k]$.

(a) Determine, in terms of M, the smallest value of N that will permit $X[k]$ to be uniquely determined from $X_R[k]$.

(b) With N satisfying the condition determined in Part (a), $X[k]$ can be expressed as the circular convolution of $X_R[k]$ with a sequence $U_N[k]$. Determine $U_N[k]$.

11.24. Consider a complex sequence $h[n] = h_r[n] + jh_i[n]$, where $h_r[n]$ and $h_i[n]$ are both real sequences, and let $H(e^{j\omega}) = H_R(e^{j\omega}) + jH_I(e^{j\omega})$ denote the Fourier transform of $h[n]$, where $H_R(e^{j\omega})$ and $H_I(e^{j\omega})$ are the real and imaginary parts, respectively, of $H(e^{j\omega})$.

Let $H_{ER}(e^{j\omega})$ and $H_{OR}(e^{j\omega})$ denote the even and odd parts, respectively, of $H_R(e^{j\omega})$, and let $H_{EI}(e^{j\omega})$, and $H_{OI}(e^{j\omega})$ denote the even and odd parts, respectively, of $H_I(e^{j\omega})$. Furthermore, let $H_A(e^{j\omega})$ and $H_B(e^{j\omega})$ denote the real and imaginary parts of the Fourier transform of $h_r[n]$, and let $H_C(e^{j\omega})$ and $H_D(e^{j\omega})$ denote the real and imaginary parts of the Fourier transform of $h_i[n]$. Express $H_A(e^{j\omega})$, $H_B(e^{j\omega})$, $H_C(e^{j\omega})$, and $H_D(e^{j\omega})$ in terms of $H_{ER}(e^{j\omega})$, $H_{OR}(e^{j\omega})$, $H_{EI}(e^{j\omega})$, and $H_{OI}(e^{j\omega})$.

11.25. The ideal Hilbert transformer (90-degree phase shifter) has frequency response (over one period)

$$H(e^{j\omega}) = \begin{cases} -j, & \omega > 0, \\ j, & \omega < 0. \end{cases}$$

Figure P11.25-1 shows $H(e^{j\omega})$, and Figure P11.25-2 shows the frequency response of an ideal lowpass filter $H_{lp}(e^{j\omega})$ with cutoff frequency $\omega_c = \pi/2$. These frequency responses are clearly similar, each having discontinuities separated by π.

Figure P11.25-1

Figure P11.25-2

(a) Obtain a relationship that expresses $H(e^{j\omega})$ in terms of $H_{lp}(e^{j\omega})$. Solve this equation for $H_{lp}(e^{j\omega})$ in terms of $H(e^{j\omega})$.

(b) Use the relationships in part (a) to obtain expressions for $h[n]$ in terms of $h_{lp}[n]$ and for $h_{lp}[n]$ in terms of $h[n]$.

The relationships obtained in Parts (a) and (b) were based on definitions of ideal systems with zero phase. However, similar relationships hold for nonideal systems with generalized linear phase.

(c) Use the results of Part (b) to obtain a relationship between the impulse response of a causal FIR approximation to the Hilbert transformer and the impulse response of a causal FIR approximation to the lowpass filter, both of which are designed by (1) incorporating an appropriate linear phase, (2) determining the corresponding ideal impulse response, and (3) multiplying by the same window of length $(M+1)$ samples, i.e., by the window method discussed in Chapter 7. (If necessary, consider the cases of M even and M odd separately.)

(d) For the Hilbert transformer approximations of Example 11.4, sketch the magnitude of the frequency responses of the corresponding lowpass filters.

11.26. In Section 11.4.3, we discussed an efficient scheme for sampling a bandpass continuous-time signal with Fourier transform such that

$$S_c(j\Omega) = 0 \qquad \text{for} \quad |\Omega| \le \Omega_c \quad \text{and} \quad |\Omega| \ge \Omega_c + \Delta\Omega.$$

In that discussion, it was assumed that the signal was initially sampled with sampling frequency $2\pi/T = 2(\Omega_c + \Delta\Omega)$, i.e., at the lowest possible frequency that avoids aliasing. The bandpass sampling scheme is depicted in Figure 11.12. After we form a complex bandpass discrete-time signal $s[n]$ with one-sided Fourier transform $S(e^{j\omega})$, the complex signal is decimated by a factor M, which is assumed to be the largest integer less than or equal to $2\pi/(\Delta\Omega T)$.

(a) By carrying through an example such as the one depicted in Figure 11.13, show that if the quantity $2\pi/(\Delta\Omega T)$ is not an integer for the initial sampling rate chosen, then the resulting decimated signal $s_d[n]$ will have regions of nonzero length where its Fourier transform $S_d(e^{j\omega})$ is identically zero.

(b) How should the initial sampling frequency $2\pi/T$ be chosen so that a decimation factor M can be found such that the decimated sequence $s_d[n]$ in the system of Figure 11.12 will have a Fourier transform $S_d(e^{j\omega})$ that is not aliased, yet has no regions where it is zero over an interval of nonzero length?

11.27. Consider an LTI system with frequency response,

$$H(e^{j\omega}) = \begin{cases} 1, & 0 \le \omega \le \pi, \\ 0, & -\pi < \omega < 0. \end{cases}$$

The input $x[n]$ to the system is restricted to be real valued and to have a Fourier transform (i.e., $x[n]$ is absolutely summable). Determine whether or not it is possible to always uniquely recover the system input from the system output. If it is possible, describe how. If it is not possible, explain why not.

Extension Problems

11.28. Derive an integral expression for $H(z)$ *inside* the unit circle in terms of $\mathcal{Re}\{H(e^{j\omega})\}$ when $h[n]$ is a real, stable sequence such that $h[n] = 0$ for $n > 0$.

11.29. Let $\mathcal{H}\{\cdot\}$ denote the (ideal) operation of Hilbert transformation; that is,

$$\mathcal{H}\{x[n]\} = \sum_{k=-\infty}^{\infty} x[k]h[n-k],$$

where $h[n]$ is

$$h[n] = \begin{cases} \dfrac{2\sin^2(\pi n/2)}{\pi n}, & n \ne 0, \\ 0, & n = 0. \end{cases}$$

Prove the following properties of the ideal Hilbert transform operator.

(a) $\mathcal{H}\{\mathcal{H}\{x[n]\}\} = -x[n]$

(b) $\displaystyle\sum_{n=-\infty}^{\infty} x[n]\mathcal{H}\{x[n]\} = 0$ [*Hint:* Use Parseval's theorem.]

(c) $\mathcal{H}\{x[n]*y[n]\} = \mathcal{H}\{x[n]\}*y[n] = x[n]*\mathcal{H}\{y[n]\}$, where $x[n]$ and $y[n]$ are any sequences.

11.30. An ideal Hilbert transformer with impulse response

$$h[n] = \begin{cases} \dfrac{2\sin^2(\pi n/2)}{\pi n}, & n \neq 0, \\ 0, & n = 0, \end{cases}$$

has input $x_r[n]$ and output $x_i[n] = x_r[n] * h[n]$, where $x_r[n]$ is a discrete-time random signal.

(a) Find an expression for the autocorrelation sequence $\phi_{x_i x_i}[m]$ in terms of $h[n]$ and $\phi_{x_r x_r}[m]$.

(b) Find an expression for the cross-correlation sequence $\phi_{x_r x_i}[m]$. Show that in this case $\phi_{x_r x_i}[m]$ is an odd function of m.

(c) Find an expression for the autocorrelation function of the complex analytic signal $x[n] = x_r[n] + jx_i[n]$.

(d) Determine the power spectrum $P_{xx}(\omega)$ for the complex signal in Part (c).

11.31. In Section 11.4.3, we discussed an efficient scheme for sampling a bandpass continuous-time signal with Fourier transform such that

$$S_c(j\Omega) = 0 \qquad \text{for} \quad |\Omega| \leq \Omega_c \quad \text{and} \quad |\Omega| \geq \Omega_c + \Delta\Omega.$$

The bandpass sampling scheme is depicted in Figure 11.12. At the end of the section, a scheme for reconstructing the original sampled signal $s_r[n]$ was given. The original continuous-time signal $s_c(t)$ in Figure 11.12 can, of course, be reconstructed from $s_r[n]$ by ideal bandlimited interpolation (ideal D/C conversion). Figure P11.31-1 shows a block diagram of the system for reconstructing a real continuous-time bandpass signal from a decimated complex signal. The complex bandpass filter $H_i(e^{j\omega})$ in the figure has a frequency response given by Eq. (11.80).

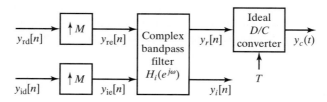

Figure P11.31-1

(a) Using the example depicted in Figure 11.13, show that the system of Figure P11.31-1 will reconstruct the original real bandpass signal (i.e., $y_c(t) = s_c(t)$) if the inputs to the reconstruction system are $y_{rd}[n] = s_{rd}[n]$ and $y_{id}[n] = s_{id}[n]$.

(b) Determine the impulse response $h_i[n] = h_{ri}[n] + jh_{ii}[n]$ of the complex bandpass filter in Figure P11.31-1.

(c) Draw a more detailed block diagram of the system of Figure P11.31-1 in which only real operations are shown. Eliminate any parts of the diagram that are not necessary to compute the final output.

(d) Now consider placing a complex linear time-invariant system between the system of Figure 11.12 and the system of Figure P11.31-1. This is depicted in Figure P11.31-2, where the frequency response of the system is denoted $H(e^{j\omega})$. Determine how $H(e^{j\omega})$ should be chosen if it is desired that

$$Y_c(j\Omega) = H_{\text{eff}}(j\Omega)S_c(j\Omega),$$

where

$$H_{\text{eff}}(j\Omega) = \begin{cases} 1, & \Omega_c < |\Omega| < \Omega_c + \Delta\Omega/2, \\ 0, & \text{otherwise}. \end{cases}$$

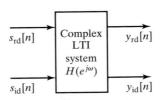

Figure P11.31-2

11.32. In Section 11.3, we mention that a causal complex cepstrum $\hat{x}[n]$ is equivalent to the minimum-phase condition of Section 5.4. Remember that $\hat{x}[n]$ is the inverse Fourier transform of $\hat{X}(e^{j\omega})$ as defined in Eq. (11.54). Note that because $X(e^{j\omega})$ and $\hat{X}(e^{j\omega})$ are defined, the region of convergence of both $X(z)$ and $\hat{X}(z)$ must include the unit circle.

    (a) Argue that the singularities (poles) of $\hat{X}(z)$ will occur wherever $X(z)$ has either poles or zeros. Use this fact to prove that if $\hat{x}[n]$ is causal, $x[n]$ is minimum phase.

    (b) Argue that if $x[n]$ is minimum phase, the constraints of the region of convergence require $\hat{x}[n]$ to be causal.

    We can examine this property for the case when $x[n]$ can be written as a superposition of complex exponentials. Specifically, consider a sequence $x[n]$ whose z-transform is

$$X(z) = A\frac{\displaystyle\prod_{k=1}^{M_i}(1 - a_k z^{-1})\prod_{k=1}^{M_o}(1 - b_k z)}{\displaystyle\prod_{k=1}^{N_i}(1 - c_k z^{-1})\prod_{k=1}^{N_o}(1 - d_k z)},$$

where $A > 0$ and a_k, b_k, c_k and d_k all have magnitude less than one.

    (c) Write an expression for $\hat{X}(z) = \log X(z)$.

    (d) Solve for $\hat{x}[n]$ by taking the inverse z-transform of your answer in Part (c).

    (e) Based on Part (d) and the expression for $X(z)$, argue that for sequences $x[n]$ of this form, a causal complex cepstrum is equivalent to having minimum phase.

A

RANDOM SIGNALS

In this appendix, we collect and summarize a number of results and establish the notation relating to the representation of stochastic signals. We make no attempt here to provide a detailed discussion of the difficult and subtle mathematical issues of the theory of random processes. Although our approach is not rigorous, we have summarized the important results and the mathematical assumptions implicit in their derivation.

A.1 DISCRETE-TIME RANDOM PROCESSES

The fundamental concept in the mathematical representation of stochastic signals is that of a *random process*. In our discussion of random processes as models for discrete-time signals, we assume that the reader is familiar with basic concepts of the theory of probability, such as random variables, probability distributions, and averages.

In utilizing the random-process model in practical signal-processing applications, we consider a particular sequence to be one of an ensemble of sample sequences. Given a discrete-time signal, the structure, i.e., the underlying probability law, of the corresponding random process is generally not known and must somehow be inferred. It may be possible to make reasonable assumptions about the structure of the process, or it may be possible to estimate the properties of a random-process representation from a finite segment of a typical sample sequence.

Formally, a random process is an indexed family of random variables $\{\mathbf{x}_n\}$ characterized by a set of probability distribution functions that, in general, may be a function of the index n. In using the concept of a random process as a model for discrete-time signals, the index n is associated with the time index. In other words, each sample value $x[n]$ of a random signal is assumed to have resulted from a mechanism that is governed by a probability law. An individual random variable $\mathbf{x}_n$ is described by the probability distribution function

$$P_{\mathbf{x}_n}(x_n, n) = \text{Probability}\,[\mathbf{x}_n \leq x_n], \tag{A.1}$$

where $\mathbf{x}_n$ denotes the random variable and x_n is a particular value of $\mathbf{x}_n$.[1] If $\mathbf{x}_n$ takes on a continuous range of values, it is equivalently specified by the *probability density function*

$$p_{\mathbf{x}_n}(x_n, n) = \frac{\partial P_{\mathbf{x}_n}(x_n, n)}{\partial x_n}, \tag{A.2}$$

or

$$P_{\mathbf{x}_n}(x_n, n) = \int_{-\infty}^{x_n} p_{\mathbf{x}_n}(x, n)dx. \tag{A.3}$$

The interdependence of two random variables $\mathbf{x}_n$ and $\mathbf{x}_m$ of a random process is described by the joint probability distribution function

$$P_{\mathbf{x}_n,\mathbf{x}_m}(x_n, n, x_m, m) = \text{Probability}\,[\mathbf{x}_n \leq x_n \text{ and } \mathbf{x}_m \leq x_m] \tag{A.4}$$

and by the joint probability density

$$p_{\mathbf{x}_n,\mathbf{x}_m}(x_n, n, x_m, m) = \frac{\partial^2 P_{\mathbf{x}_n,\mathbf{x}_m}(x_n, n, x_m, m)}{\partial x_n \partial x_m}. \tag{A.5}$$

Two random variables are *statistically independent* if knowledge of the value of one does not affect the probability density of the other. If all the random variables of a collection of random variables, $\{\mathbf{x}_n\}$, are statistically independent, then

$$P_{\mathbf{x}_n,\mathbf{x}_m}(x_n, n, x_m, m) = P_{\mathbf{x}_n}(x_n, n) \cdot P_{\mathbf{x}_m}(x_m, m) \qquad m \neq n. \tag{A.6}$$

A complete characterization of a random process requires the specification of all possible joint probability distributions. As we have indicated, these probability distributions may be a function of the time indices m and n. In the case where all the probability distributions are independent of a shift of time origin, the random process is said to be *stationary*. For example, the second-order distribution of a stationary process satisfies

$$P_{\mathbf{x}_{n+k},\mathbf{x}_{m+k}}(x_{n+k}, n + k, x_{m+k}, m + k) = P_{\mathbf{x}_n\mathbf{x}_m}(x_n, n, x_m, m) \quad \text{for all } k. \tag{A.7}$$

In many of the applications of discrete-time signal processing, random processes serve as models for signals in the sense that a particular signal can be considered a sample sequence of a random process. Although the details of such signals are unpredictable—making a deterministic approach to signal representation inappropriate—certain average properties of the ensemble can be determined, given the probability law of the process. These average properties often serve as a useful, although incomplete, characterization of such signals.

---

[1]In this appendix, boldface type is used to denote the random variables and regular type denotes dummy variables of probability functions.

A.2 AVERAGES

It is often useful to characterize a random variable by averages such as the mean and variance. Since a random process is an indexed set of random variables, we may likewise characterize the process by statistical averages of the random variables making up the random process. Such averages are called *ensemble averages*. We begin the discussion of averages with some definitions.

A.2.1 Definitions

The average, or mean, of a random process is defined as

$$m_{\mathbf{x}_n} = \mathcal{E}\{\mathbf{x}_n\} = \int_{-\infty}^{\infty} x p_{\mathbf{x}_n}(x, n) dx, \tag{A.8}$$

where $\mathcal{E}$ denotes an operator called *mathematical expectation*. In general, the mean (expected value) may depend on n. In addition, if $g(\cdot)$ is a single-valued function, then $g(\mathbf{x}_n)$ is a random variable, and the set of random variables $\{g(\mathbf{x}_n)\}$ defines a new random process. To compute averages of this new process, we can derive probability distributions of the new random variables, or it can be shown that

$$\mathcal{E}\{g(\mathbf{x}_n)\} = \int_{-\infty}^{\infty} g(x) p_{\mathbf{x}_n}(x, n) dx. \tag{A.9}$$

If the random variables are discrete—i.e., if they have quantized values—the integrals become summations over all possible values of the random variable. In that case $\mathcal{E}\{g(x)\}$ has the form

$$\mathcal{E}\{g(\mathbf{x}_n)\} = \sum_x g(x) \hat{p}_{\mathbf{x}_n}(x, n). \tag{A.10}$$

In cases where we are interested in the relationship between two (or more) random processes, we must be concerned with two sets of random variables $\{\mathbf{x}_n\}$ and $\{\mathbf{y}_m\}$. For example, the expected value of a function of two random variables is defined as

$$\mathcal{E}\{g(\mathbf{x}_n, \mathbf{y}_m)\} = \int_{-\infty}^{\infty} \int_{-\infty}^{\infty} g(x, y) p_{\mathbf{x}_n, \mathbf{y}_m}(x, n, y, m) dx \, dy, \tag{A.11}$$

where $p_{\mathbf{x}_n, \mathbf{y}_m}(x_m, n, y_m, m)$ is the joint probability density of the random variables $\mathbf{x}_n$ and $\mathbf{y}_m$.

The mathematical expectation operator is a linear operator; that is, it can be shown that

1. $\mathcal{E}\{\mathbf{x}_n + \mathbf{y}_m\} = \mathcal{E}\{\mathbf{x}_n\} + \mathcal{E}\{\mathbf{y}_m\}$; i.e., the average of a sum is the sum of the averages.
2. $\mathcal{E}\{a\mathbf{x}_n\} = a\mathcal{E}\{\mathbf{x}_n\}$; i.e., the average of a constant times $\mathbf{x}_n$ is equal to the constant times the average of $\mathbf{x}_n$.

In general, the average of a product of two random variables is not equal to the product of the averages. When this property holds, however, the two random variables are said to be *linearly independent* or *uncorrelated*. That is, $\mathbf{x}_n$ and $\mathbf{y}_m$ are linearly independent or uncorrelated if

$$\mathcal{E}\{\mathbf{x}_n \mathbf{y}_m\} = \mathcal{E}\{\mathbf{x}_m\} \cdot \mathcal{E}\{\mathbf{y}_m\}. \tag{A.12}$$

It is easy to see from Eqs. (A.11) and (A.12) that a sufficient condition for linear independence is

$$p_{\mathbf{x}_n, \mathbf{y}_m}(x_n, n, y_m, m) = p_{\mathbf{x}_n}(x_n, n) \cdot p_{\mathbf{y}_m}(y_m, m). \tag{A.13}$$

However, it can be shown that Eq. (A.13) is a stronger statement of independence than Eq. (A.12). As previously stated, random variables satisfying Eq. (A.13) are said to be *statistically independent*. If Eq. (A.13) holds for all values of n and m, the random processes $\{\mathbf{x}_n\}$ and $\{\mathbf{y}_m\}$ are said to be statistically independent. Statistically independent random processes are also linearly independent; but the converse is not true: Linear independence does not imply statistical independence.

It can be seen from Eqs. (A.9)–(A.11) that averages generally are functions of the time index. In the case of stationary processes, this is not true: For stationary processes, the mean is the same for all the random variables that constitute the process; i.e., the mean of a stationary process is a constant, which we denote simply m_x.

In addition to the mean of a random process, as defined in Eq. (A.8), a number of other averages are particularly important within the context of signal processing. These are defined next. For notational convenience, we assume that the probability distributions are continuous. Corresponding definitions for discrete random processes can be obtained by applying Eq. (A.10).

The *mean-square* value of $\mathbf{x}_n$ is the average of $|\mathbf{x}_n|^2$; i.e.,

$$\mathcal{E}\{|\mathbf{x}_n|^2\} = \text{mean square} = \int_{-\infty}^{\infty} |x|^2 p_{\mathbf{x}_n}(x, n) dx. \tag{A.14}$$

The mean-square value is sometimes referred to as the *average power*.

The *variance* of $\mathbf{x}_n$ is the mean-square value of $[\mathbf{x}_n - m_{x_n}]$; i.e.,

$$\text{var}[\mathbf{x}_n] = \mathcal{E}\{|(\mathbf{x}_n - m_{x_n})|^2\} = \sigma_{\mathbf{x}_n}^2. \tag{A.15}$$

Since the average of a sum is the sum of the averages, it can easily be shown that Eq. (A.15) can be written as

$$\text{var}[\mathbf{x}_n] = \mathbf{E}\{|\mathbf{x}_n|^2\} - |m_{x_n}|^2. \tag{A.16}$$

In general, the mean-square value and the variance are functions of time; however, they are constant for stationary processes.

The mean, mean square, and variance are simple averages that provide only a small amount of information about a process. A more useful average is the *autocorrelation sequence,* which is defined as

$$\begin{aligned} \phi_{xx}[n, m] &= \mathcal{E}\{\mathbf{x}_n \mathbf{x}_m^*\} \\ &= \int_{-\infty}^{\infty} \int_{-\infty}^{\infty} x_n x_m^* p_{\mathbf{x}_n, \mathbf{x}_m}(x_n, n, x_m, m) dx_n \, dx_m, \end{aligned} \tag{A.17}$$

where $*$ denotes complex conjugation. The autocovariance sequence of a random process is defined as

$$\gamma_{xx}[n, m] = \mathcal{E}\{(\mathbf{x}_n - m_{x_n})(\mathbf{x}_m - m_{x_m})^*\}, \tag{A.18}$$

which can be written as

$$\gamma_{xx}[n, m] = \phi_{xx}[n, m] - m_{x_n} m_{x_m}^*. \tag{A.19}$$

Note that, in general, both the autocorrelation and autocovariance are two-dimensional sequences, i.e., functions of two variables.

The autocorrelation sequence is a measure of the dependence between values of the random processes at different times. In this sense, it partially describes the time variation of a random signal. A measure of the dependence between two different random signals is obtained from the cross-correlation sequence. If $\{\mathbf{x}_n\}$ and $\{\mathbf{y}_m\}$ are two random processes, their cross-correlation is

$$\phi_{xx}[n, m] = \mathcal{E}\{\mathbf{x}_n\mathbf{y}_m^*\}$$
$$= \int_{-\infty}^{\infty} \int_{-\infty}^{\infty} xy^* p_{\mathbf{x}_n,\mathbf{y}_m}(x, n, y, m)dx\, dy, \tag{A.20}$$

where $p_{\mathbf{x}_n,\mathbf{y}_m}(x, n, y, m)$ is the joint probability density of $\mathbf{x}_n$ and $\mathbf{y}_m$. The cross-covariance function is defined as

$$\gamma_{xx}[n, m] = \mathcal{E}\{(\mathbf{x}_n - m_{x_n})(\mathbf{y}_m - m_{y_m})^*\}$$
$$= \phi_{xy}[n, m] - m_{x_n}m_{y_m}^*. \tag{A.21}$$

As we have pointed out, the statistical properties of a random process generally vary with time. However, a stationary random process is characterized by an equilibrium condition in which the statistical properties are invariant to a shift of time origin. This means that the first-order probability distribution is independent of time. Similarly, all the joint probability functions are also invariant to a shift of time origin; i.e., the second-order joint probability distributions depend only on the time difference $(m - n)$. First-order averages such as the mean and variance are independent of time; second-order averages, such as the autocorrelation $\phi_{xx}[n, m]$, are dependent on the time difference $(m - n)$. Thus, for a stationary process, we can write

$$m_x = \mathcal{E}\{\mathbf{x}_n\}, \tag{A.22}$$
$$\sigma_x^2 = \mathcal{E}\{|(\mathbf{x}_n - m_x)|^2\}, \tag{A.23}$$

both independent of n, and now if we denote the time difference by m, we have

$$\phi_{xx}[n + m, n] = \phi_{xx}[m] = \mathcal{E}\{\mathbf{x}_{n+m}\mathbf{x}_n^*\}. \tag{A.24}$$

That is, the autocorrelation sequence of a stationary random process is a one-dimensional sequence, a function of the time difference m.

In many instances, we encounter random processes that are not stationary in the *strict sense*—i.e., their probability distributions are not time invariant—but Eqs. (A.22)–(A.24) still hold. Such random processes are said to be *wide-sense stationary*.

A.2.2 Time Averages

In a signal-processing context, the notion of an ensemble of signals is a convenient mathematical concept that allows us to use the theory of probability to represent the signals. However, in a practical situation, we always have available at most a finite number of finite-length sequences rather than an infinite ensemble of sequences. For example, we might wish to infer the probability law or certain averages of the random-process representation from measurements on a single member of the ensemble. When the probability distributions are independent of time, it seems intuitively that the amplitude distribution (histogram) of a long segment of an individual sequence of samples should be approximately equal to the single probability density that describes each of the random variables of the random-process model. Similarly, the arithmetic average of a large number of samples of a single sequence should be very close to the mean of the process. To formalize these intuitive notions, we define the time average of a random process as

$$\langle \mathbf{x}_n \rangle = \lim_{L \to \infty} \frac{1}{2L+1} \sum_{n=-L}^{L} \mathbf{x}_n. \tag{A.25}$$

Similarly, the time autocorrelation sequence is defined as

$$\langle \mathbf{x}_{n+m}\mathbf{x}_n^* \rangle = \lim_{L \to \infty} \frac{1}{2L+1} \sum_{n=-L}^{L} \mathbf{x}_{n+m}\mathbf{x}_n^*. \tag{A.26}$$

It can be shown that the preceding limits exist if $\{\mathbf{x}_n\}$ is a stationary process with finite mean. As defined in Eqs. (A.25) and (A.26), these time averages are functions of an infinite set of random variables and thus are properly viewed as random variables themselves. However, under the condition known as *ergodicity*, the time averages in Eqs. (A.25) and (A.26) are equal to constants in the sense that the time averages of almost all possible sample sequences are equal to the same constant. Furthermore, they are equal to the corresponding ensemble average.[2] That is, for any single sample sequence $\{x[n]\}$ for $-\infty < n < \infty$,

$$\langle x[n] \rangle = \lim_{L \to \infty} \frac{1}{2L+1} \sum_{n=-L}^{L} x[n] = \mathcal{E}\{\mathbf{x}_n\} = m_x \tag{A.27}$$

and

$$\langle x[n+m]x^*[n] \rangle = \lim_{L \to \infty} \frac{1}{2L+1} \sum_{n=-L}^{L} x[n+m]x^*[n] = \mathcal{E}\{\mathbf{x}_{n+m}\mathbf{x}_n^*\} = \phi_{xx}[m]. \tag{A.28}$$

The time-average operator $\langle \cdot \rangle$ has the same properties as the ensemble-average operator $\mathcal{E}\{\cdot\}$. Thus, we generally do not distinguish between the random variable $\mathbf{x}_n$ and its value in a sample sequence, $x[n]$. For example, the expression $\mathcal{E}\{x[n]\}$ should be interpreted as $\mathcal{E}\{\mathbf{x}_n\} = \langle x[n] \rangle$. In general, a random process for which time averages equal ensemble averages is called an *ergodic process*.

In practice, it is common to assume that a given sequence is a sample sequence of an ergodic random process so that averages can be computed from a single sequence. Of course, we generally cannot compute with the limits in Eqs. (A.27) and (A.28), but instead the quantities

$$\hat{m}_x = \frac{1}{L} \sum_{n=0}^{L-1} x[n], \tag{A.29}$$

$$\hat{\sigma}_x^2 = \frac{1}{L} \sum_{n=0}^{L-1} |x[n] - \hat{m}_x|^2, \tag{A.30}$$

---

[2]A more precise statement is that the random variables $\langle \mathbf{x}_n \rangle$ and $\langle \mathbf{x}_{n+m}\mathbf{x}_n^* \rangle$ have means equal to m_x and $\phi_{xx}[m]$, respectively, and their variances are zero.

and

$$\langle x[n+m]x^*[n]\rangle_L = \frac{1}{L}\sum_{n=0}^{L-1} x[n+m]x^*[n] \qquad (A.31)$$

or similar quantities are often computed as *estimates* of the mean, variance, and autocorrelation. $\hat{m}_x$ and $\hat{\sigma}_x^2$ are referred to as the sample mean and sample variance, respectively. The estimation of averages of a random process from a finite segment of data is a problem of statistics, which we touched on briefly in Chapter 10.

A.3 PROPERTIES OF CORRELATION AND COVARIANCE SEQUENCES

Several useful properties of correlation and covariance functions follow in a simple way from the definitions. These properties are given in this section.

Consider two real stationary random processes $\{\mathbf{x}_n\}$ and $\{\mathbf{y}_n\}$ with autocorrelation, autocovariance, cross-correlation, and cross-covariance being given, respectively, by

$$\phi_{xx}[m] = \mathcal{E}\{\mathbf{x}_{n+m}\mathbf{x}_n^*\}, \qquad (A.32)$$

$$\gamma_{xx}[m] = \mathcal{E}\{(\mathbf{x}_{n+m} - m_x)(\mathbf{x}_n - m_x)^*\}, \qquad (A.33)$$

$$\phi_{xy}[m] = \mathcal{E}\{\mathbf{x}_{n+m}\mathbf{y}_n^*\}, \qquad (A.34)$$

$$\gamma_{xy}[m] = \mathcal{E}\{(\mathbf{x}_{n+m} - m_x)(\mathbf{y}_n - m_y)^*\}, \qquad (A.35)$$

where m_x and m_y are the means of the two processes. The following properties are easily derived by simple manipulations of the definitions:

Property 1

$$\gamma_{xx}[m] = \phi_{xx}[m] - |m_x|^2, \qquad (A.36a)$$

$$\gamma_{xy}[m] = \phi_{xy}[m] - m_x m_y^*. \qquad (A.36b)$$

These results follow directly from Eqs. (A.19) and (A.21), and they indicate that the correlation and covariance sequences are identical for zero-mean processes.

Property 2

$$\phi_{xx}[0] = E[|\mathbf{x}_n|^2] = \text{Mean-square value}, \qquad (A.37a)$$

$$\gamma_{xx}[0] = \sigma_x^2 = \text{Variance}. \qquad (A.37b)$$

Property 3

$$\phi_{xx}[-m] = \phi_{xx}^*[m], \qquad (A.38a)$$

$$\gamma_{xx}[-m] = \gamma_{xx}^*[m], \qquad (A.38b)$$

$$\phi_{xy}[-m] = \phi_{yx}^*[m], \qquad (A.38c)$$

$$\gamma_{xy}[-m] = \gamma_{yx}^*[m]. \qquad (A.38d)$$

Property 4

$$|\phi_{xy}[m]|^2 \leq \phi_{xx}[0]\phi_{yy}[0], \tag{A.39a}$$

$$|\gamma_{xy}[m]|^2 \leq \gamma_{xx}[0]\gamma_{yy}[0]. \tag{A.39b}$$

In particular,

$$|\phi_{xx}[m]| \leq \phi_{xx}[0], \tag{A.40a}$$

$$|\gamma_{xx}[m]| \leq \gamma_{xx}[0]. \tag{A.40b}$$

Property 5. If $\mathbf{y}_n = \mathbf{x}_{n-n_0}$, then

$$\phi_{yy}[m] = \phi_{xx}[m], \tag{A.41a}$$

$$\gamma_{yy}[m] = \gamma_{xx}[m]. \tag{A.41b}$$

Property 6. For many random processes, the random variables become uncorrelated as they become more separated in time. If this is true,

$$\lim_{m \to \infty} \gamma_{xx}[m] = 0, \tag{A.42a}$$

$$\lim_{m \to \infty} \phi_{xx}[m] = |m_x|^2, \tag{A.42b}$$

$$\lim_{m \to \infty} \gamma_{xy}[m] = 0, \tag{A.42c}$$

$$\lim_{m \to \infty} \phi_{xy}[m] = m_x m_y^*. \tag{A.42d}$$

The essence of these results is that the correlation and covariance are finite-energy sequences that tend to die out for large values of m. Thus, it is often possible to represent these sequences in terms of their Fourier transforms or z-transforms.

A.4 FOURIER TRANSFORM REPRESENTATION OF RANDOM SIGNALS

Although the Fourier transform of a random signal does not exist, the autocovariance and autocorrelation sequences of such a signal are aperiodic sequences for which the transform does exist. The spectral representation of the correlation functions plays an important role in describing the input–output relations for a linear time-invariant system when the input is a stochastic signal. Therefore, it is of interest to consider the properties of correlation and covariance sequences and their corresponding Fourier and z-transforms.

Let us define $\Phi_{xx}(e^{j\omega})$, $\Gamma_{xx}(e^{j\omega})$, $\Phi_{xy}(e^{j\omega})$, and $\Gamma_{xy}(e^{j\omega})$ as the Fourier transforms of $\phi_{xx}[m]$, $\gamma_{xx}[m]$, $\phi_{xy}[m]$, and $\gamma_{xy}[m]$, respectively. Since these functions are all discrete-time Fourier transforms of sequences, they must be periodic with period 2π. From Eqs. (A.36a) and (A.36b), it follows that, over one period $|\omega| \leq \pi$,

$$\Phi_{xx}(e^{j\omega}) = \Gamma_{xx}(e^{j\omega}) + 2\pi |m_x|^2 \delta(\omega), \qquad |\omega| < \pi, \tag{A.43a}$$

and

$$\Phi_{xy}(e^{j\omega}) = \Gamma_{xy}(e^{j\omega}) + 2\pi m_x m_y^* \delta(\omega), \qquad |\omega| < \pi. \tag{A.43b}$$

In the case of zero-mean processes ($m_x = 0$ and $m_y = 0$), the correlation and covariance functions are identical so that $\Phi_{xx}(e^{j\omega}) = \Gamma_{xx}(e^{j\omega})$ and $\Phi_{xy}(e^{j\omega}) = \Gamma_{xy}(e^{j\omega})$.

From the inverse Fourier transform equation, it follows that

$$\gamma_{xx}[m] = \frac{1}{2\pi} \int_{-\pi}^{\pi} \Gamma_{xx}(e^{j\omega}) e^{j\omega m} d\omega, \tag{A.44a}$$

$$\phi_{xx}[m] = \frac{1}{2\pi} \int_{-\pi}^{\pi} \Phi_{xx}(e^{j\omega}) e^{j\omega m} d\omega, \tag{A.44b}$$

and, consequently,

$$\mathcal{E}\{|x[n]|^2\} = \phi_{xx}[0] = \sigma_x^2 + |m_x|^2 = \frac{1}{2\pi} \int_{-\pi}^{\pi} \Phi_{xx}(e^{j\omega}) d\omega, \tag{A.45a}$$

$$\sigma_x^2 = \gamma_{xx}[0] = \frac{1}{2\pi} \int_{-\pi}^{\pi} \Gamma_{xx}(e^{j\omega}) d\omega. \tag{A.45b}$$

Sometimes it is notationally convenient to define the quantity

$$P_{xx}(\omega) = \Phi_{xx}(e^{j\omega}), \tag{A.46}$$

in which case Eqs.(A.45a) and (A.45b) are expressed as

$$\mathcal{E}\{|x[n]|^2\} = \frac{1}{2\pi} \int_{-\pi}^{\pi} P_{xx}(\omega) d\omega, \tag{A.47a}$$

$$\sigma_x^2 = \frac{1}{2\pi} \int_{-\pi}^{\pi} P_{xx}(\omega) d\omega - |m_x|^2. \tag{A.47b}$$

Thus, the area under $P_{xx}(\omega)$ for $-\pi \leq \omega \leq \pi$ is proportional to the average power in the signal. In fact, as we discussed in Section 2.10, the integral of $P_{xx}(\omega)$ over a band of frequencies is proportional to the power in the signal in that band. For this reason, the function $P_{xx}(\omega)$ is called the *power density spectrum,* or simply, the *power spectrum.* When $P_{xx}(\omega)$ is a constant independent of ω, the random process is referred to as a white-noise process, or simply, white noise. When $P_{xx}(\omega)$ is constant over a band and zero otherwise, we refer to it as bandlimited white noise.

From Eq. (A.38b), it follows that $P_{xx}(\omega)$ is always real valued, and for real random processes, $\phi_{xx}[m] = \phi_{xx}[-m]$, so in the real case, $P_{xx}(\omega)$ is both real and even; i.e.,

$$P_{xx}(\omega) = P_{xx}(-\omega). \tag{A.48}$$

An additional important property is that the power density spectrum is nonnegative. This point is discussed in Section 2.10.

The *cross power density spectrum* is defined as

$$P_{xy}(\omega) = \Phi_{xy}(e^{j\omega}). \tag{A.49}$$

This function is generally complex, and from Eq. (A.38c), it follows that

$$P_{xy}(\omega) = P_{yx}^*(\omega). \tag{A.50}$$

Finally, as shown in Section 2.10, if $x[n]$ is a random signal input to a linear time-invariant discrete-time system with frequency response $H(e^{j\omega})$, and if $y[n]$ is the corresponding output, then

$$\Phi_{yy}(e^{j\omega}) = |H(e^{j\omega})|^2 \Phi_{xx}(e^{j\omega}) \tag{A.51}$$

and

$$\Phi_{xy}(e^{j\omega}) = H(e^{j\omega})\Phi_{xy}(e^{j\omega}). \tag{A.52}$$

Example A.1   Noise Power Output of Ideal Lowpass Filter

Suppose that $x[n]$ is a zero-mean white-noise sequence with $\phi_{xx}[m] = \sigma_x^2 \delta[m]$ and power spectrum $\Phi_{xx}(e^{j\omega}) = \sigma_x^2$ for $|\omega| \leq \pi$, and furthermore, assume that $x[n]$ is the input to an ideal lowpass filter with cutoff frequency ω_c. Then from Eq. (A.51), it follows that the output $y[n]$ would be a bandlimited white noise process whose power spectrum would be

$$\Phi_{yy}(e^{j\omega}) = \begin{cases} \sigma_x^2, & |\omega| < \omega_c, \\ 0, & \omega_c < |\omega| \leq \pi. \end{cases} \tag{A.53}$$

Using the inverse Fourier transform, we obtain the autocorrelation sequence

$$\phi_{yy}[m] = \frac{\sin(\omega_c n)}{\pi n}. \tag{A.54}$$

Now, using Eq. (A.45a), we get for the average power of the output,

$$\mathcal{E}\{y^2[n]\} = \phi_{yy}[0] = \frac{1}{2\pi} \int_{-\omega_c}^{\omega_c} \sigma_x^2 d\omega = \sigma_x^2 \frac{\omega_c}{\pi}. \tag{A.55}$$

A.5 USE OF THE z-TRANSFORM IN AVERAGE POWER COMPUTATIONS

To carry out average power calculations using Eq. (A.45a), we must evaluate an integral of the power spectrum as was done in Example A.1. While the integral in that Example was easy to evaluate, such integrals in general are difficult to evaluate as real integrals. However, a result based on the z-transform makes the calculation of average output power straightforward in the important case of systems that have rational system functions.

In general, the z-transform can be used to represent the covariance function but not a correlation function. This is because when a signal has nonzero average value, its correlation function will contain an additive constant component that does not have a z-transform representation. When the average value is zero, however, the covariance and correlation functions are, of course, equal. If the z-transform of $\gamma_{xx}[m]$ exists, then since $\gamma_{xx}[-m] = \gamma_{xx}^*[m]$ it follows that in general

$$\Gamma_{xx}(z) = \Gamma_{xx}^*(1/z^*). \tag{A.56}$$

Furthermore, since $\gamma_{xx}[m]$ is two sided and conjugate-symmetric, it follows that the region of convergence of $\Gamma_{xx}(z)$ must be of the form

$$r_a < |z| < \frac{1}{r_a}$$

where necessarily $0 < r_a < 1$. In the important case when $\Gamma_{xx}(z)$ is a rational function of z, Eq. (A.56) implies that the poles and zeros of $\Gamma_{xx}(z)$ must occur in complex-conjugate reciprocal pairs.

The major advantage of the z-transform representation is that when $\Gamma_{xx}(z)$ is a rational function, the average power of the random signal can be computed easily using the relation

$$\mathcal{E}\{x^2[n]\} = \sigma_x^2 = \gamma_{xx}[0] = \left\{ \begin{array}{c} \text{Inverse } z\text{-transform} \\ \text{of } \Gamma_{xx}(z), \\ \text{evaluated for } m = 0 \end{array} \right\}. \tag{A.57}$$

It is straightforward to evaluate the right-hand side of this equation using a method based on the observation that when $\Gamma_{xx}(z)$ is a rational function of z, $\gamma_{xx}[m]$ can be computed for all m by employing a partial fraction expansion. Then to obtain the average power, we can simply evaluate $\gamma_{xx}[m]$ for $m = 0$.

The z-transform is also useful in determining the autocovariance and average power of the output of an LTI system when the input is a random signal. Generalizing Eq. (A.51) leads to

$$\Gamma_{yy}(z) = H(z)H^*(1/z^*)\Gamma_{xx}(z), \tag{A.58}$$

and from the properties of the z-transform and Eq. (A.58), it follows that the autocovariance of the output is the convolution

$$\gamma_{yy}[m] = h[m] * h^*[-m] * \gamma_{xx}[m]. \tag{A.59}$$

This result is particularly useful in quantization noise analysis where we need to compute the average output power when the input to a linear difference equation is a zero-mean white noise signal with average power σ_x^2. Since the autocovariance of such an input is $\gamma_{xx}[m] = \sigma_x^2\delta[m]$, it follows that the autocovariance of the output is $\gamma_{yy}[m] = \sigma_x^2(h[m] * h^*[-m])$, i.e., the covariance of the output is proportional to the deterministic autocorrelation of the impulse response of the LTI system. From this result it follows that

$$\mathcal{E}\{y^2[n]\} = \gamma_{yy}[0] = \sigma_x^2 \sum_{n=-\infty}^{\infty} |h[n]|^2. \tag{A.60}$$

As an alternative to computing the sum of squares of the impulse response sequence, which can be rather difficult for IIR systems, we can apply the method suggested in Eq. (A.57) to obtain $\mathcal{E}\{y^2[n]\}$ from a partial fraction expansion of $\Gamma_{yy}(z)$. Recall that for a white noise input with $\gamma_{xx}[m] = \sigma_x^2\delta[m]$, the z-transform is $\Gamma_{xx}(z) = \sigma_x^2$ so $\Gamma_{yy}(z) = \sigma_x^2 H(z)H^*(1/z^*)$. Therefore, Eq. (A.57) applied to the output of the system gives

$$\mathcal{E}\{y^2[n]\} = \gamma_{yy}[0] = \left\{ \begin{array}{c} \text{Inverse } z\text{-transform of} \\ \Gamma_{yy}(z) = H(z)H^*(1/z^*)\sigma_x^2, \\ \text{evaluated for } m = 0 \end{array} \right\}. \tag{A.61}$$

Now consider the special case of a stable and causal system having a rational system function of the form

$$H(z) = A \frac{\displaystyle\prod_{m=1}^{M}(1 - c_m z^{-1})}{\displaystyle\prod_{k=1}^{N}(1 - d_k z^{-1})} \qquad |z| > \max_k \{|d_k|\}, \qquad (A.62)$$

where $\max_k\{|d_k|\} < 1$ and $M < N$. Such a system function might describe the relationship between an internal round-off noise source and the output of a system implemented with fixed-point arithmetic. Substituting Eq. (A.62) for $H(z)$ in Eq. (A.58) gives

$$\Gamma_{yy}(z) = \sigma_x^2 H(z)H^*(1/z^*) = \sigma_x^2 |A|^2 \frac{\displaystyle\prod_{m=1}^{M}(1 - c_m z^{-1})(1 - c_m^* z)}{\displaystyle\prod_{k=1}^{N}(1 - d_k z^{-1})(1 - d_k^* z)}. \qquad (A.63)$$

Since we have assumed that $|d_k| < 1$ for all k, all of the original poles are inside the unit circle and therefore the other poles at $(d_k^*)^{-1}$ are at conjugate reciprocal locations outside the unit circle. The region of convergence for $\Gamma_{yy}(z)$ is therefore $\max_k |d_k| < |z| < \min_k |(d_k^*)^{-1}|$. For such rational functions, it can be shown that since $M < N$, the partial fraction expansion has the form

$$\Gamma_{yy}(z) = \sigma_x^2 \left(\sum_{k=1}^{N} \left(\frac{A_k}{1 - d_k z^{-1}} - \frac{A_k^*}{1 - (d_k^*)^{-1} z^{-1}} \right) \right), \qquad (A.64)$$

where the coefficients are found from

$$A_k = H(z)H^*(1/z^*)(1 - d_k z^{-1})\big|_{z=d_k}. \qquad (A.65)$$

Since the poles at $z = d_k$ are inside the inner boundary of the region of convergence, each of them corresponds to a right-sided sequence, while the poles at $z = (d_k^*)^{-1}$ each correspond to a left-sided sequence. Thus, the autocovariance function corresponding to Eq. (A.64) is

$$\gamma_{yy}[n] = \sigma_x^2 \sum_{k=1}^{N} (A_k (d_k)^n u[n] + A_k^* (d_k^*)^{-n} u[-n-1]),$$

from which it follows that we can obtain the average power from

$$\sigma_y^2 = \gamma_{yy}[0] = \sigma_x^2 \left(\sum_{k=1}^{N} A_k \right), \qquad (A.66)$$

where the quantities A_k are given by Eq: (A.65).

Thus, the computation of the total average power of the output of a system with rational system function and white noise input reduces to the straightforward problem of finding partial fraction expansion coefficients for the z-transform of the output autocorrelation function. The utility of this approach is illustrated by the following example.

Example A.2   Noise Power Output of a Second-Order IIR Filter

Consider a system with impulse response

$$h[n] = \frac{r^n \sin\theta(n+1)}{\sin\theta} u[n] \tag{A.67}$$

and system function

$$H(z) = \frac{1}{(1 - re^{j\theta}z^{-1})(1 - re^{-j\theta}z^{-1})}. \tag{A.68}$$

When the input is white noise with total average power σ_x^2, the *z*-transform of the autocovariance function of the output is

$$\Gamma_{yy}(z) = \sigma_x^2 \left(\frac{1}{(1 - re^{j\theta}z^{-1})(1 - re^{-j\theta}z^{-1})} \right) \left(\frac{1}{(1 - re^{-j\theta}z)(1 - re^{j\theta}z)} \right) \tag{A.69}$$

from which we obtain, using Eq. (A.65),

$$\mathcal{E}\{y^2[n]\} = \sigma_x^2 \left[\left(\frac{1}{(1 - re^{-j\theta}z^{-1})} \right) \left(\frac{1}{(1 - re^{-j\theta}z)(1 - re^{j\theta}z)} \right) \Bigg|_{z=re^{j\theta}} \right.$$
$$\left. + \left(\frac{1}{(1 - re^{j\theta}z^{-1})} \right) \left(\frac{1}{(1 - re^{-j\theta}z)(1 - re^{j\theta}z)} \right) \Bigg|_{z=re^{-j\theta}} \right]. \tag{A.70}$$

Making the indicated substitutions, placing both terms over a common denominator, and doing some algebra leads to

$$\mathcal{E}\{y^2[n]\} = \sigma_x^2 \left(\frac{1+r^2}{1-r^2} \right) \left(\frac{1}{1 - 2r^2\cos(2\theta) + r^4} \right). \tag{A.71}$$

Thus, using the partial fraction expansion of $\Gamma_{yy}(z)$ we have effectively evaluated the expression

$$\mathcal{E}\{y^2[n]\} = \sigma_x^2 \sum_{n=-\infty}^{\infty} |h[n]|^2 = \sigma_x^2 \sum_{n=0}^{\infty} \left| \frac{r^n \sin\theta(n+1)}{\sin\theta} \right|^2,$$

which would be difficult to sum in closed form, and the expression

$$\mathcal{E}\{y^2[n]\} = \frac{1}{2\pi} \int_{-\pi}^{\pi} \sigma_x^2 |H(e^{j\omega})|^2 d\omega = \frac{\sigma_x^2}{2\pi} \int_{-\pi}^{\pi} \frac{d\omega}{|(1 - re^{j\theta}e^{-j\omega})(1 - re^{-j\theta}e^{-j\omega})|^2},$$

which would be difficult to evaluate as a real integral.

The remarkably simple result of Example A.2 is an illustration of the power of the partial fraction method in evaluating average power formulas. In Chapter 6, we make use of this technique in the analysis of quantization effects in the implementation of digital filters.

B
CONTINUOUS-TIME FILTERS

The techniques discussed in Sections 7.1 and 7.2 for designing IIR digital filters rely on the availability of appropriate continuous-time filter designs. In this appendix, we briefly summarize the characteristics of several classes of lowpass filter approximations that we referred to in Chapter 7. More detailed discussions of these classes of filters appear in Guillemin (1957), Weinberg (1975), and Parks and Burrus (1987). Extensive design tables and formulas are found in Zverev (1967), and a variety of computer programs are available for computer-aided design of IIR digital filters based on analog filter approximations. (See, for example, Gray and Markel, 1976; DSP Committee 1979; Mersereau et al., 1984; and Mathworks, 1998.)

B.1 BUTTERWORTH LOWPASS FILTERS

Butterworth lowpass filters are defined by the property that the magnitude response is maximally flat in the passband. For an Nth-order lowpass filter, this means that the first $(2N - 1)$ derivatives of the magnitude-squared function are zero at $\Omega = 0$. Another property is that the magnitude response is monotonic in the passband and the stopband. The magnitude-squared function for a continuous-time Butterworth lowpass filter is of the form

$$|H_c(j\Omega)|^2 = \frac{1}{1 + (j\Omega/j\Omega_c)^{2N}}. \tag{B.1}$$

This function is plotted in Figure B.1.

As the parameter N in Eq. (B.1) increases, the filter characteristics become sharper: that is, they remain close to unity over more of the passband and become close to zero

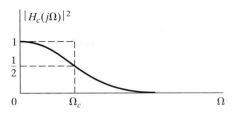

Figure B.1 Magnitude-squared function for continuous-time Butterworth filter.

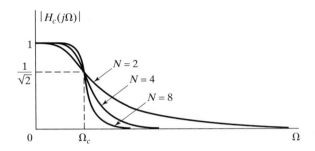

Figure B.2 Dependence of Butterworth magnitude characteristics on the order N.

more rapidly in the stopband, although the magnitude-squared function at the cutoff frequency Ω_c will always be equal to one-half because of the nature of Eq. (B.1). The dependence of the Butterworth filter characteristic on the parameter N is indicated in Figure B.2.

From the magnitude-squared function in Eq. (B.1), we observe by substituting $j\Omega = s$ that $H_c(s)H_c(-s)$ must be of the form

$$H_c(s)H_c(-s) = \frac{1}{1 + (s/j\Omega_c)^{2N}}.\tag{B.2}$$

The roots of the denominator polynomial (the poles of the magnitude-squared function) are therefore located at values of s satisfying $1 + (s/j\Omega_c)^{2N} = 0$; i.e.,

$$s_k = (-1)^{1/2N}(j\Omega_c) = \Omega_c e^{(j\pi/2N)(2k+N-1)}, \qquad k = 0, 1, \ldots, 2N - 1.\tag{B.3}$$

Thus, there are $2N$ poles equally spaced in angle on a circle of radius Ω_c in the s-plane. The poles are symmetrically located with respect to the imaginary axis. A pole never falls on the imaginary axis, and one occurs on the real axis for N odd, but not for N even. The angular spacing between the poles on the circle is π/N radians. For example, for $N = 3$, the poles are spaced by $\pi/3$ radians, or 60 degrees, as indicated in Figure B.3. To determine the system function of the analog filter to associate with the Butterworth magnitude-squared function, we must perform the factorization $H_c(s)H_c(-s)$. The poles of the magnitude-squared function always occur in pairs; i.e., if there is a pole at $s = s_k$, then a pole also occurs at $s = -s_k$. Consequently, to construct $H_c(s)$ from the magnitude-squared function, we would choose one pole from each such pair. To obtain a stable and causal filter, we should choose the poles on the left-half-plane part of the s-plane.

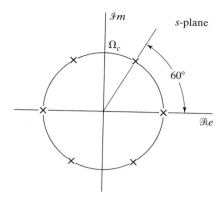

Figure B.3 *s*-plane pole locations for a third-order Butterworth filter.

B.2 CHEBYSHEV FILTERS

In a Butterworth filter, the magnitude response is monotonic in both the passband and the stopband. Consequently, if the filter specifications are in terms of maximum passband and stopband approximation error, the specifications are exceeded toward the low-frequency end of the passband and above the stopband cutoff frequency. A more efficient approach, which usually leads to a lower order filter, is to distribute the accuracy of the approximation uniformly over the passband or the stopband (or both). This is accomplished by choosing an approximation that has an equiripple behavior rather than a monotonic behavior. The class of Chebyshev filters has the property that the magnitude of the frequency response is either equiripple in the passband and monotonic in the stopband (referred to as a type I Chebyshev filter) or monotonic in the passband and equiripple in the stopband (a type II Chebyshev filter). The frequency response of a type I Chebyshev filter is shown in Figure B.4. The magnitude-squared function for this filter is of the form

$$|H_c(j\Omega)|^2 = \frac{1}{1 + \varepsilon^2 V_N^2(\Omega/\Omega_c)}, \tag{B.4}$$

where $V_N(x)$ is the Nth-order Chebyshev polynomial defined as

$$V_N(x) = \cos(N\cos^{-1} x). \tag{B.5}$$

For example, for $N = 0$, $V_0(x) = 1$; for $N = 1$, $V_1(x) = \cos(\cos^{-1} x) = x$; for $N = 2$, $V_2(x) = \cos(2\cos^{-1} x) = 2x^2 - 1$; and so on.

From Eq. (B.5), which defines the Chebyshev polynomials, it is straightforward to obtain a recurrence formula from which $V_{N+1}(x)$ can be obtained from $V_N(x)$ and $V_{N-1}(x)$. By applying trigonometric identities to Eq. (B.5), it follows that

$$V_{N+1}(x) = 2x V_N(x) - V_{N-1}(x). \tag{B.6}$$

From Eq. (B.5), we note that $V_N^2(x)$ varies between zero and unity for $0 < x < 1$. For $x > 1$, $\cos^{-1} x$ is imaginary, so $V_N(x)$ behaves as a hyperbolic cosine and consequently increases monotonically. Referring to Eq. (B.4), we see that $|H_c(j\Omega)|^2$ ripples between 1 and $1/(1 + \varepsilon^2)$ for $0 \leq \Omega/\Omega_c \leq 1$ and decreases monotonically for $\Omega/\Omega_c > 1$. Three parameters are required to specify the filter: ε, Ω_c, and N. In a typical design, ε is

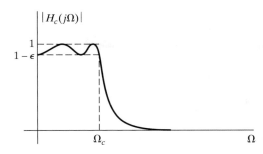

Figure B.4   Type I Chebyshev lowpass filter approximation.

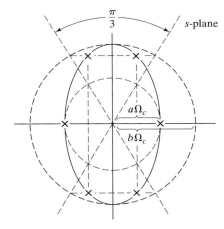

Figure B.5   Location of poles for a third-order type I lowpass Chebyshev filter.

specified by the allowable passband ripple and Ω_c is specified by the desired passband cutoff frequency. The order N is then chosen so that the stopband specifications are met.

The poles of the Chebyshev filter lie on an ellipse in the s-plane. As shown in Figure B.5, the ellipse is defined by two circles whose diameters are equal to the minor and major axes of the ellipse. The length of the minor axis is $2a\Omega_c$, where

$$a = \tfrac{1}{2}(\alpha^{1/N} - \alpha^{-1/N}) \tag{B.7}$$

with

$$\alpha = \varepsilon^{-1} + \sqrt{1 + \varepsilon^{-2}}. \tag{B.8}$$

The length of the major axis is $2b\Omega_c$, where

$$b = \tfrac{1}{2}(\alpha^{1/N} + \alpha^{-1/N}). \tag{B.9}$$

To locate the poles of the Chebyshev filter on the ellipse, we first identify the points on the major and minor circles equally spaced in angle with a spacing of π/N in such a way that the points are symmetrically located with respect to the imaginary axis and such that a point never falls on the imaginary axis and a point occurs on the real axis for N odd but not for N even. This division of the major and minor circles corresponds exactly to the manner in which the circle is divided in locating the poles of a Butterworth filter as in Eq. (B.3). The poles of a Chebyshev filter fall on the ellipse, with the ordinate specified by the points identified on the major circle and the abscissa specified by the points identified on the minor circle. In Figure B.5, the poles are shown for $N = 3$.

A type II Chebyshev lowpass filter can be related to a type I filter through a transformation. Specifically, if in Eq. (B.4) we replace the term $\varepsilon^2 V_N^2(\Omega/\Omega_c)$ by its reciprocal and also replace the argument of V_N^2 by its reciprocal, we obtain

$$|H_c(j\Omega)|^2 = \frac{1}{1 + [\varepsilon^2 V_N^2(\Omega_c/\Omega)]^{-1}}. \tag{B.10}$$

This is the analytic form for the type II Chebyshev lowpass filter. One approach to designing a type II Chebyshev filter is to first design a type I filter and then apply the preceding transformation.

B.3 ELLIPTIC FILTERS

If we distribute the error uniformly across the entire passband or across the entire stopband, as in the Chebyshev cases, we are able to meet the design specifications with a lower order filter than if we permit a monotonically increasing error in the passband, as in the Butterworth case. We note that in the type I Chebyshev and Butterworth approximations, the stopband error decreases monotonically with frequency, raising the possibility of further improvements if we distribute the stopband error uniformly across the stopband. This suggests the lowpass filter approximation in Figure B.6. Indeed, it can be shown (Papoulis, 1957) that this type of approximation (i.e., equiripple in the passband and the stopband) is the best that can be achieved for a given filter order N, in the sense that for given values of Ω_p, δ_1, and δ_2, the transition band $(\Omega_s - \Omega_p)$ is as small as possible.

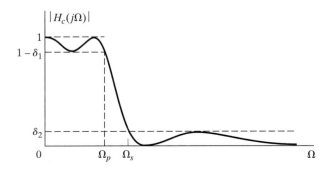

Figure B.6 Equiripple approximation in both passband and stopband.

This class of approximations, referred to as elliptic filters, has the form

$$|H_c(j\Omega)|^2 = \frac{1}{1 + \varepsilon^2 U_N^2(\Omega)}, \tag{B.11}$$

where $U_N(\Omega)$ is a Jacobian elliptic function. To obtain equiripple error in both the passband and the stopband, elliptic filters must have both poles and zeros. As can be seen from Figure B.6, such a filter will have zeros on the $j\Omega$-axis of the s-plane. A discussion of elliptic filter design, even on a superficial level, is beyond the scope of this appendix. The reader is referred to the texts by Guillemin (1957), Storer (1957), Gold and Rader (1969), and Parks and Burrus (1987) for more detailed discussions.

C

ANSWERS TO SELECTED BASIC PROBLEMS

This appendix contains the answers to the first 20 basic problems in each chapter.

Answers to Basic Problems in Chapter 2

2.1. **(a)** Always (2), (3), (5). If $g[n]$ is bounded, (1).
   (b) (3).
   (c) Always (1), (3), (4). If $n_0 = 0$, (2) and (5).
   (d) Always (1), (3), (4). If $n_0 \leq 0$, (2). If $n_0 = 0$, (5).
   (e) (1), (2), (4), (5).
   (f) Always (1), (2), (4), (5). If $b = 0$, (3).
   (g) (1), (3).
   (h) (1), (5).

2.2. **(a)** $N_4 = N_0 + N_2$, $N_5 = N_1 + N_3$.
   (b) At most $N + M - 1$ nonzero points.

2.3.

$$y[n] = \begin{cases} \dfrac{a^{-n}}{1-a}, & n < 0, \\ \dfrac{1}{1-a}, & n \geq 0. \end{cases}$$

2.4. $y[n] = 8[(1/2)^n - (1/4)^n]u[n]$.

2.5. **(a)** $y_h[n] = A_1(2)^n + A_2(3)^n$.
   (b) $h[n] = 2(3^n - 2^n)u[n]$.
   (c) $s[n] = [-8(2)^{(n-1)} + 9(3)^{(n-1)} + 1]u[n]$.

2.6. (a)

$$H(e^{j\omega}) = \frac{1 + 2e^{-j\omega} + e^{-j2\omega}}{1 - \frac{1}{2}e^{-j\omega}}.$$

(b) $y[n] + \frac{1}{2}y[n-1] + \frac{3}{4}y[n-2] = x[n] - \frac{1}{2}x[n-1] + x[n-3]$.

2.7. (a) Periodic, $N = 6$.
(b) Periodic, $N = 8$.
(c) Not periodic.
(d) Not periodic.

2.8. $y[n] = 3(-1/2)^n u[n] + 2(1/3)^n u[n]$.

2.9. (a)

$$h[n] = 2\left[\left(\frac{1}{2}\right)^n - \left(\frac{1}{3}\right)^n\right] u[n],$$

$$H(e^{j\omega}) = \frac{\frac{1}{3}e^{-j\omega}}{1 - \frac{5}{6}e^{-j\omega} + \frac{1}{6}e^{-j2\omega}},$$

$$s[n] = \left[-2\left(\frac{1}{2}\right)^n + \left(\frac{1}{3}\right)^n + 1\right] u[n].$$

(b) $y_h[n] = A_1(1/2)^n + A_2(1/3)^n$.
(c) $y[n] = 4(1/2)^n - 3(1/3)^n - 2(1/2)^n u[-n-1] + 2(1/3)^n u[-n-1]$.

2.10. (a)

$$y[n] = \begin{cases} a^{-1}/(1 - a^{-1}), & n \geq -1, \\ a^n/(1 - a^{-1}), & n \leq -2. \end{cases}$$

(b)

$$y[n] = \begin{cases} 1, & n \geq 3, \\ 2^{(n-3)}, & n \leq 2. \end{cases}$$

(c)

$$y[n] = \begin{cases} 1, & n \geq 0, \\ 2^n, & n \leq -1. \end{cases}$$

(d)

$$y[n] = \begin{cases} 0, & n \geq 9, \\ 1 - 2^{(n-9)}, & 8 \geq n \geq -1, \\ 2^{(n+1)} - 2^{(n-9)}, & -2 \geq n. \end{cases}$$

2.11. $y[n] = 2\sqrt{2}\sin(\pi(n-1)/4)$.

2.12. (a) $y[n] = n!u[n]$.
(b) The system is linear.
(c) The system is not time invariant.

2.13. **(a)**, **(b)**, and **(e)** are eigenfunctions of stable LTI systems.

2.14. **(a)** (iv).
   (b) (i).
   (c) (iii), $h[n] = (1/2)^n u[n]$.

2.15. **(a)** Not LTI. Inputs $\delta[n]$ and $\delta[n-1]$ violate TI.
   (b) Not causal. Consider $x[n] = \delta[n-1]$.
   (c) Stable.

2.16. **(a)** $y_h[n] = A_1(1/2)^n + A_2(-1/4)^n$.
   (b) Causal: $h_c[n] = 2(1/2)^n u[n] + (-1/4)^n u[n]$.
      Anticausal: $h_{ac}[n] = -2(1/2)^n u[-n-1] - (-1/4)^n u[-n-1]$.
   (c) $h_c[n]$ is absolutely summable, $h_{ac}[n]$ is not.
   (d) $y_p[n] = (1/3)(-1/4)^n u[n] + (2/3)(1/2)^n u[n] + 4(n+1)(1/2)^{(n+1)}u[n+1]$.

2.17. **(a)**

$$R(e^{j\omega}) = e^{-j\omega M/2}\frac{\sin\left(\omega\left(\frac{M+1}{2}\right)\right)}{\sin\left(\frac{\omega}{2}\right)}.$$

   (b) $W(e^{j\omega}) = (1/2)R(e^{j\omega}) + (1/4)R(e^{j(\omega-2\pi/M)}) + (1/4)R(e^{j(\omega+2\pi/M)})$.

2.18. Systems (a) and (b) are causal.

2.19. Systems (b), (c), (e), and (f) are stable.

2.20. **(a)** $h[n] = (1/a)^{(n-1)}u[n-1]$.
   (b) The system will be stable for $|a| > 1$.

Answers to Basic Problems in Chapter 3

3.1. **(a)** $\dfrac{1}{1 - \frac{1}{2}z^{-1}}, \quad |z| > \frac{1}{2}$.

   (b) $\dfrac{1}{1 - \frac{1}{2}z^{-1}}, \quad |z| < \frac{1}{2}$.

   (c) $\dfrac{-\frac{1}{2}z^{-1}}{1 - \frac{1}{2}z^{-1}}, \quad |z| < \frac{1}{2}$.

   (d) $1, \quad$ all z.

   (e) $z^{-1}, \quad z \neq 0$.

   (f) $z, \quad |z| < \infty$.

   (g) $\dfrac{1 - \left(\frac{1}{2}\right)^{10} z^{-10}}{1 - \frac{1}{2}z^{-1}}, \quad |z| \neq 0$.

3.2. $X(z) = \dfrac{z^{-1}(1 - z^{-N})}{(1 - z^{-1})^2}$.

3.3. (a) $X_a(z) = \dfrac{z^{-1}(\alpha - \alpha^{-1})}{(1 - \alpha z^{-1})(1 - \alpha^{-1} z^{-1})}$,   ROC: $|\alpha| < |z| < |\alpha^{-1}|$.

(b) $X_b(z) = \dfrac{1 - z^{-N}}{1 - z^{-1}}$,   ROC: $z \neq 0$.

(c) $X_c(z) = z^{-1} \dfrac{(1 - z^{-N})^2}{(1 - z^{-1})^2}$,   ROC: $z \neq 0$.

3.4. (a) $(1/3) < |z| < 2$, two sided.

(b) Two sequences. $(1/3) < |z| < 2$ and $2 < |z| < 3$.

(c) No. Causal sequence has $|z| > 3$, which does not include the unit circle.

3.5. $x[n] = 2\delta[n+1] + 5\delta[n] - 4\delta[n-1] - 3\delta[n-2]$.

3.6. (a) $x[n] = \left(-\frac{1}{2}\right)^n u[n]$, Fourier transform exists.

(b) $x[n] = -\left(-\frac{1}{2}\right)^n u[-n-1]$, Fourier transform does not exist.

(c) $x[n] = 4\left(-\frac{1}{2}\right)^n u[n] - 3\left(-\frac{1}{4}\right)^n u[n]$, Fourier transform exists.

(d) $x[n] = \left(-\frac{1}{2}\right)^n u[n]$, Fourier transform exists.

(e) $x[n] = -(a^{-(n+1)})u[n] + a^{-(n-1)}u[n-1]$, Fourier transform exists if $|a| > 1$.

3.7. (a) $H(z) = \dfrac{1 - z^{-1}}{1 + z^{-1}}$, $|z| > 1$.

(b) $\text{ROC}\{Y(z)\} = |z| > 1$.

(c) $y[n] = \left[-\frac{1}{3}\left(\frac{1}{2}\right)^n + \frac{1}{3}(-1)^n\right] u[n]$.

3.8. (a) $h[n] = \left(-\frac{3}{4}\right)^n u[n] - \left(-\frac{3}{4}\right)^{n-1} u[n-1]$.

(b) $y[n] = \frac{8}{13}\left(-\frac{3}{4}\right)^n u[n] - \frac{8}{13}\left(\frac{1}{3}\right)^n u[n]$.

(c) The system is stable.

3.9. (a) $|z| > (1/2)$.

(b) Yes. The ROC includes the unit circle.

(c) $X(z) = \dfrac{1 - \frac{1}{2}z^{-1}}{1 - 2z^{-1}}$.

(d) $h[n] = 2\left(\frac{1}{2}\right)^n u[n] - \left(-\frac{1}{4}\right)^n u[n]$.

3.10. (a) $|z| > \frac{3}{4}$.

(b) $0 < |z| < \infty$.

(c) $|z| < 2$.

(d) $|z| > 1$.

(e) $|z| < \infty$.

(f) $\frac{1}{2} < |z| < \sqrt{13}$.

3.11. (a) Causal.

(b) Not causal.

(c) Causal.

(d) Not causal.

3.12. **(a)**

Figure P3.12-1

(b)

Figure P3.12-2

(c)

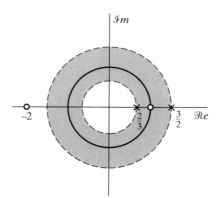

Figure P3.12-3

3.13. $g[11] = -\dfrac{1}{11!} + \dfrac{3}{9!} - \dfrac{2}{7!}$.

3.14. $A_1 = A_2 = 1/2$, $\alpha_1 = -1/2$, $\alpha_2 = 1/2$.

3.15. $h[n] = \left(\dfrac{1}{2}\right)^n (u[n] - u[n-10])$. The system is causal.

3.16. **(a)** $H(z) = \dfrac{1 - 2z^{-1}}{1 - \frac{2}{3}z^{-1}}$, $|z| > \frac{2}{3}$.

    (b) $h[n] = \left(\dfrac{2}{3}\right)^n u[n] - 2\left(\dfrac{2}{3}\right)^{(n-1)} u[n-1]$.

(c) $y[n] - \frac{2}{3}y[n-1] = x[n] - 2x[n-1]$.

(d) The system is stable and causal.

3.17. $h[0]$ can be 0, 1/3, or 1. To be painstakingly literal, $h[0]$ can also be 2/3, due to the impulse response $h[n] = (2/3)(2)^n u[n] - (1/3)(1/2)^n u[-n-1]$, which satisfies the difference equation but has no ROC. This noncausal system with no ROC can be implemented as the parallel combination of its causal and anticausal components.

3.18. **(a)** $h[n] = -2\delta[n] + \frac{1}{3}(-\frac{1}{2})^n u[n] + \frac{8}{3}u[n]$.

(b) $y[n] = \left(\dfrac{2}{\frac{3}{2} + \frac{j}{2}}\right) e^{j(\pi/2)n}$.

3.19. **(a)** $|z| > 1/2$.

(b) $1/3 < |z| < 2$.

(c) $|z| > 1/3$.

3.20. **(a)** $|z| > 2/3$.

(b) $|z| > 1/6$.

Answers to Basic Problems in Chapter 4

4.1. $x[n] = \sin(\pi n/2)$.

4.2. $\Omega_0 = 250\pi, 1750\pi$.

4.3. **(a)** $T = 1/12{,}000$.   **(b)** Not unique. $T = 5/12{,}000$.

4.4. **(a)** $T = 1/100$.   **(b)** Not unique. $T = 11/100$.

4.5. **(a)** $T \leq 1/10{,}000$.   **(b)** 625 Hz.   **(c)** 1250 Hz.

4.6. **(a)** $H_c(j\Omega) = 1/(a + j\Omega)$.

(b) $H_d(e^{j\omega}) = T/(1 - e^{-aT}e^{-j\omega})$.

4.7. **(a)**

$$X_c(j\Omega) = S_c(j\Omega)(1 + \alpha e^{-j\Omega\tau_d}),$$

$$X(e^{j\omega}) = \frac{1}{T}\sum_{k=-\infty}^{\infty} S_c\left(j\left(\frac{\omega}{T} - \frac{2\pi k}{T}\right)\right) + \frac{\alpha e^{-j\omega\tau_d/T}}{T}\sum_{k=-\infty}^{\infty} S_c\left(j\left(\frac{\omega}{T} - \frac{2\pi k}{T}\right)\right).$$

(b) $H(e^{j\omega}) = 1 + \alpha e^{-j\omega\tau_d/T}$.

(c) (i) $h[n] = \delta[n] + \alpha\delta[n-1]$.

(ii) $h[n] = \delta[n] + \alpha\frac{\sin(\pi(n-1/2))}{\pi(n-1/2)}$.

4.8. **(a)** $T \leq 1/20{,}000$.

(b) $h[n] = Tu[n]$.

(c) $X(e^{j\omega})|_{\omega=0}$.

(d) $T \leq 1/10{,}000$.

4.9. **(a)** $X\left(e^{j(\omega+\pi)}\right) = X\left(e^{j(\omega+\pi-\pi)}\right) = X\left(e^{j\omega}\right)$.

(b) $x[3] = 0$.

(c) $x[n] = \begin{cases} y[n/2], & n \text{ even}, \\ 0, & n \text{ odd}. \end{cases}$

4.10. (a) $x[n] = \cos(2\pi n/3)$.
    (b) $x[n] = -\sin(2\pi n/3)$.
    (c) $x[n] = \sin(2\pi n/5)/(\pi n/5000)$.

4.11. (a) $T = 1/40$, $T = 9/40$.
    (b) $t = 1/20$, unique.

4.12. (a)   (i) $y_c(t) = -6\pi \sin(6\pi t)$.
       (ii) $y_c(t) = -6\pi \sin(6\pi t)$.
    (b)   (i) Yes.
       (ii) No.

4.13. (a) $y[n] = \sin\left(\frac{\pi n}{2} - \frac{\pi}{4}\right)$.
    (b) Same $y[n]$.
    (c) $h_c(t)$ has no effect on T.

4.14. (a) No.
    (b) Yes.
    (c) No.
    (d) Yes.
    (e) Yes. (No information is lost; however, the signal cannot be recovered by the system in Figure 3.21.)

4.15. (a) Yes.
    (b) No.
    (c) Yes.

4.16. (a) $M/L = 5/2$, unique.
    (b) $M/L = 2/3$; also, $M/L = 7/3$.

4.17. (a) $\tilde{x}_d[n] = (4/3)\sin(\pi n/2)/(\pi n)$.
    (b) $\tilde{x}_d[n] = -\sin(3\pi n/4)$.

4.18. (a) $\omega_0 = 2\pi/3$.
    (b) $\omega_0 = 3\pi/5$.
    (c) $\omega_0 = \pi$.

4.19. $T \leq \pi/\Omega_0$.

4.20. (a) $F_s \geq 2000$ Hz.
    (b) $F_s \geq 4000$ Hz.

Answers to Basic Problems in Chapter 5

5.1. $x[n] = y[n]$, $\omega_c = \pi$.

5.2. (a) Poles: $z = 3, 1/3$, Zeros: $z = 0, \infty$.
    (b) $h[n] = -(3/8)(1/3)^n u[n] - (3/8)3^n u[-n - 1]$.

5.3. (a), (d) are the impulse responses.

5.4. (a) $H(z) = \dfrac{1 - 2z^{-1}}{1 - \frac{3}{4}z^{-1}}$, $|z| > 3/4$.
    (b) $h[n] = (3/4)^n u[n] - 2(3/4)^n u[n - 1]$.
    (c) $y[n] - (3/4)y[n - 1] = x[n] - 2x[n - 1]$.

(d) Stable and causal.

5.5. (a) $y[n]-(7/12)y[n-1]+(1/12)y[n-2] = 3x[n]-(19/6)x[n-1]+(2/3)x[n-2]$.

(b) $h[n] = 3\delta[n] - (2/3)(1/3)^{n-1}u[n-1] - (3/4)(1/4)^{n-1}u[n-1]$.

(c) Stable.

5.6. (a) $X(z) = \dfrac{1}{(1 - \frac{1}{2}z^{-1})(1 - 2z^{-1})}$, $\dfrac{1}{2} < |z| < 2$.

(b) $\frac{1}{2} < |z| < 2$.

(c) $h[n] = \delta[n] - \delta[n-2]$.

5.7. (a) $H(z) = \dfrac{1 - z^{-1}}{(1 - \frac{1}{2}z^{-1})(1 + \frac{3}{4}z^{-1})}$, $|z| > \dfrac{3}{4}$.

(b) $h[n] = -(2/5)(1/2)^n u[n] + (7/5)(-3/4)^n u[n]$.

(c) $y[n] + (1/4)y[n-1] - (3/8)y[n-2] = x[n] - x[n-1]$.

5.8. (a) $H(z) = \dfrac{z^{-1}}{1 - \frac{3}{2}z^{-1} - z^{-2}}$, $|z| > 2$.

(b) $h[n] = -(2/5)(-1/2)^n u[n] + (2/5)(2)^n u[n]$.

(c) $h[n] = -(2/5)(-1/2)^n u[n] - (2/5)(2)^n u[-n-1]$.

5.9.

$$h[n] = \left[-\frac{4}{3}(2)^{n-1} + \frac{1}{3}\left(\frac{1}{2}\right)^{n-1} \right] u[-n], |z| < \frac{1}{2},$$

$$h[n] = -\frac{4}{3}(2)^{n-1}u[-n] - \frac{1}{3}\left(\frac{1}{2}\right)^{n-1} u[n-1],   \frac{1}{2} < |z| < 2,$$

$$h[n] = \frac{4}{3}(2)^{n-1}u[n-1] - \frac{1}{3}\left(\frac{1}{2}\right)^{n-1} u[n-1], |z| > 2.$$

5.10. $H_i(z)$ cannot be causal and stable. The zero of a $H(z)$ at $z = \infty$ is a pole of $H_i(z)$. The existence of a pole at $z = \infty$ implies that the system is not causal.

5.11. (a) Cannot be determined.

(b) Cannot be determined.

(c) True.

(d) False.

5.12. (a) Stable.

(b)

$$H_1(z) = -9\frac{(1 + 0.2z^{-1})\left(1 - \frac{1}{3}z^{-1}\right)\left(1 + \frac{1}{3}z^{-1}\right)}{(1 - j0.9z^{-1})(1 + j0.9z^{-1})},$$

$$H_{ap}(z) = \frac{\left(z^{-1} - \frac{1}{3}\right)\left(z^{-1} + \frac{1}{3}\right)}{\left(1 - \frac{1}{3}z^{-1}\right)\left(1 + \frac{1}{3}z^{-1}\right)}.$$

5.13. $H_1(z)$, $H_3(z)$, and $H_4(z)$ are all-pass systems.

5.14. (a) 5.

(b) $\frac{1}{2}$.

5.15. (a) $\alpha = 1$, $\beta = 0$, $A(e^{j\omega}) = 1 + 4\cos(\omega n)$. The system is a generalized linear-phase system but not a linear-phase system, because $A(e^{j\omega})$ is not nonnegative for all ω.

(b) Not a generalized linear-phase or a linear-phase system.

(c) $\alpha = 1$, $\beta = 0$, $A(e^{j\omega}) = 3 + \cos(\omega n)$. Linear phase, since $|H(e^{j\omega})| = A(e^{j\omega}) \geq 0$ for all ω.

(d) $\alpha = 1/2$, $\beta = 0$, $A(e^{j\omega}) = 2\cos(\omega n/2)$. Generalized linear phase, because $A(e^{j\omega})$ is not nonnegative at all ω.

(e) $\alpha = 1$, $\beta = \pi/2$, $A(e^{j\omega}) = 2\sin(\omega n)$. Generalized linear phase, because $\beta \neq 0$.

5.16. $h[n]$ is not necessarily causal. Both $h[n] = \delta[n - \alpha]$ and $h[n] = \delta[n + 1] + \delta[n - (2\alpha + 1)]$ will have this phase.

5.17. $H_2(z)$ and $H_3(z)$ are linear-phase systems.

5.18. (a) $H_{\min}(z) = \dfrac{2\left(1 - \frac{1}{2}z^{-1}\right)}{1 + \frac{1}{3}z^{-1}}$.

(b) $H_{\min}(z) = 3\left(1 - \dfrac{1}{2}z^{-1}\right)$.

(c) $H_{\min}(z) = \dfrac{9}{4}\dfrac{\left(1 - \frac{1}{3}z^{-1}\right)\left(1 - \frac{1}{4}z^{-1}\right)}{\left(1 - \frac{3}{4}z^{-1}\right)^2}$.

5.19. $h_1[n] : 2$, $h_2[n] : 3/2$, $h_3[n] : 2$, $h_4[n] : 3$, $h_5[n] : 3$, $h_6[n] : 7/2$.

5.20. Systems $H_1(z)$ and $H_3(z)$ have a linear phase and can be implemented by a real-valued difference equation.

Answers to Basic Problems in Chapter 6

6.1. Network 1:

$$H(z) = \frac{1}{1 - 2r\cos\theta z^{-1} + r^2 z^{-2}}.$$

Network 2:

$$H(z) = \frac{r\sin\theta z^{-1}}{1 - 2r\cos\theta z^{-1} + r^2 z^{-2}}.$$

Both systems have the same denominators and thus the same poles.

6.2. $y[n] - 3y[n - 1] - y[n - 2] - y[n - 3] = x[n] - 2x[n - 1] + x[n - 2]$.

6.3. The system in Part (d) is the same as that in Part (a).

6.4. (a)

$$H(z) = \frac{2 + \frac{1}{4}z^{-1}}{1 + \frac{1}{4}z^{-1} - \frac{3}{8}z^{-2}}.$$

(b)

$$y[n] + \frac{1}{4}y[n - 1] - \frac{3}{8}y[n - 2] = 2x[n] + \frac{1}{4}x[n - 1].$$

6.5. (a)

$$y[n] - 4y[n-1] + 7y[n-3] + 2y[n-4] = x[n].$$

(b)

$$H(z) = \frac{1}{1 - 4z^{-1} + 7z^{-3} + 2z^{-4}}.$$

(c) Two multiplications and four additions.

(d) No. It requires at least four delays to implement a fourth-order system.

6.6.

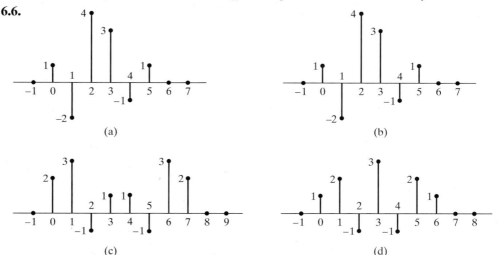

(a)   (b)

(c)   (d)

Figure P6.6-1

6.7.

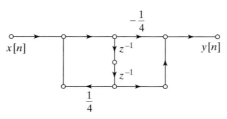

Figure P6.7-1

6.8. $y[n] - 2y[n-2] = 3x[n-1] + x[n-2].$

6.9. (a) $h[1] = 2.$

(b) $y[n] + y[n-1] - 8y[n-2] = x[n] + 3x[n-1] + x[n-2] - 8x[n-3].$

6.10. (a)

$$y[n] = x[n] + v[n-1].$$

$$v[n] = 2x[n] + \frac{1}{2}y[n] + w[n-1].$$

$$w[n] = x[n] + \frac{1}{2}y[n].$$

(b)

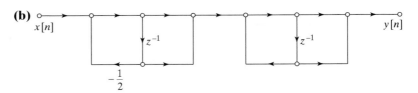

Figure P6.10

(c) The poles are at $z = -1/2$ and $z = 1$. Since the second pole is on the unit circle, the system is not stable.

6.11. (a)

Figure P6.11-1

(b)

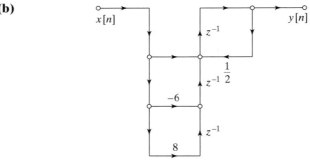

Figure P6.11-2

6.12. $y[n] - 8y[n-1] = -2x[n] + 6x[n-1] + 2x[n-2]$.

6.13.

Figure P6.13-1

6.14.

Figure P6.14-1

6.15.

Figure P6.15-1

6.16. (a)

Figure P6.16-1

(b) Both systems have the system function

$$H(z) = \frac{\left(1 - \frac{1}{2}z^{-1}\right)\left(1 - 2z^{-1} + 3z^{-2}\right)}{1 - \frac{1}{4}z^{-2}}.$$

6.17. (a)

Figure P6.17-1

(b)

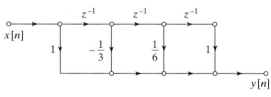

Figure P6.17-2

6.18. If $a = 2/3$, the overall system function is

$$H(z) = \frac{1 + 2z^{-1}}{1 + \frac{1}{4}z^{-1} - \frac{3}{8}z^{-2}}.$$

If $a = -2$, the overall system function is

$$H(z) = \frac{1 - \frac{2}{3}z^{-1}}{1 + \frac{1}{4}z^{-1} - \frac{3}{8}z^{-2}}.$$

6.19.

Figure P6.19-1

6.20.

Figure P6.20-1

Answers to Basic Problems in Chapter 7

7.1. (a)

$$H_1(z) = \frac{1 - e^{-aT}\cos(bT)z^{-1}}{1 - 2e^{-aT}\cos(bT)z^{-1} + e^{-2aT}z^{-2}}.$$

(b)

$H_2(z) = (1 - z^{-1})S_2(z)$, where

$$S_2(z) = \frac{a}{a^2 + b^2}\frac{1}{1 - z^{-1}} - \frac{1}{2(a + jb)}\frac{1}{1 - e^{-(a+jb)T}z^{-1}} - \frac{1}{2(a - jb)}\frac{1}{1 - e^{-(a-jb)T}}$$

(c) They are not equal.

7.2. (a)

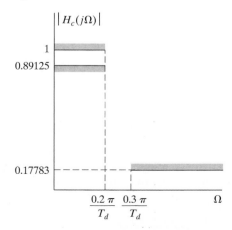

Figure P7.2-1

(b) $N = 6$, $\Omega_c T_d = 0.7032$.

(c) The poles in the s-plane are on a circle of radius $R = 0.7032/T_d$. They map to poles in the z-plane at $z = e^{s_k T_d}$. The factors of T_d cancel out, leaving the pole locations in the z-plane for $H(z)$ independent of T_d.

7.3. (a) $\hat{\delta}_2 = \delta_2/(1 + \delta_1)$.

(b)

$$\delta_2 = 0.18806$$

$$H(z) = \frac{0.3036 - 0.4723z^{-1}}{1 - 1.2971z^{-1} + 0.6949z^{-2}} + \frac{-2.2660 + 1.2114z^{-1}}{1 - 1.0691z^{-1} + 0.3699z^{-2}}$$

$$+ \frac{1.9624 - 0.6665z^{-1}}{1 - 0.9972z^{-1} + 0.2570z^{-2}}$$

(c) Use the same δ_2.

$$\frac{0.0007802(1 + z^{-1})^6}{(1 - 1.2686z^{-1} + 0.7051z^{-2})(1 - 1.0106z^{-1} + 0.3583z^{-2})(1 - 0.9044z^{-1} + 0.2155z^{-2})}.$$

7.4. (a)

$$H_c(s) = \frac{1}{s + 0.1} - \frac{0.5}{s + 0.2}.$$

The answer is not unique. Another possibility is

$$H_c(s) = \frac{1}{s + 0.1 + j2\pi} - \frac{0.5}{s + 0.2 + j2\pi}.$$

(b)

$$H_c(s) = \frac{2(1 + s)}{0.1813 + 1.8187s} - \frac{1 + s}{0.3297 + 1.6703s}.$$

This answer is unique.

7.5. (a) $M + 1 = 91$.

(b) $M/2 = 45$.

(c) $h_d[n] = \dfrac{\sin[0.625\pi(n - 45)]}{\pi(n - 45)} - \dfrac{\sin[0.3\pi(n - 45)]}{\pi(n - 45)}$.

7.6. (a) $\delta = 0.03$, $\beta = 2.181$.

(b) $\Delta\omega = 0.05\pi$, $M = 50$.

7.7.

$$0.99 \leq |H(e^{j\omega})| \leq 1.01, \qquad |\omega| \leq 0.2\pi,$$

$$|H(e^{j\omega})| \leq 0.01, \qquad 0.22\pi \leq |\omega| \leq \pi$$

7.8. (a) Six alternations. $L = 5$, so this does not satisfy the alternation theorem and is not optimal.

(b) Seven alternations, which satisfies the alternation theorem for $L = 5$.

7.9. $\omega_c = 0.4\pi$.

7.10. $\omega_c = 2.3842$ rad.

7.11. $\Omega_c = 2\pi(1250)$ rad/sec.

7.12. $\Omega_c = 2000$ rad/sec.

7.13. $T = 50\,\mu$s. This T is unique.

7.14. $T = 1.46$ ms. This T is unique.

7.15. The Hamming, Hanning, and Blackman windows may be used.

7.16. $\beta = 2.6524$, $M = 181$.

7.17.

$$\begin{aligned} |H_c(j\Omega)| &< 0.02, & |\Omega| &\le 2\pi(20) \text{ rad/sec}, \\ 0.95 < |H_c(j\Omega)| &< 1.05, & 2\pi(30) &\le |\Omega| \le 2\pi(70) \text{ rad/sec} \\ |H_c(j\Omega)| &< 0.001, & 2\pi(75) \text{ rad/sec} &\le |\Omega|. \end{aligned}$$

7.18.

$$\begin{aligned} |H_c(j\Omega)| &< 0.04, & |\Omega| &\le 726.5 \text{ rad/sec} \\ 0.995 < |H_c(j\Omega)| &< 1.005, & |\Omega| &\ge 1376.4 \text{rad/sec}. \end{aligned}$$

7.19. $T = 0.41667$ ms. This T is unique.

7.20. True.

Answers to Basic Problems in Chapter 8

8.1. **(a)** $x[n]$ is periodic with period $N = 6$.

(b) T will not avoid aliasing.

(c)

$$\tilde{X}[k] = 2\pi \begin{cases} a_0 + a_6 + a_{-6}, & k = 0, \\ a_1 + a_7 + a_{-5}, & k = 1, \\ a_2 + a_8 + a_{-4}, & k = 2, \\ a_3 + a_9 + a_{-3} + a_{-9}, & k = 3, \\ a_4 + a_{-2} + a_{-8}, & k = 4, \\ a_5 + a_{-1} + a_{-7}, & k = 5. \end{cases}$$

8.2. **(a)**

$$\tilde{X}_3 = \begin{cases} 3X[k/3], & \text{for } k = 3\ell, \\ 0, & \text{otherwise.} \end{cases}$$

(b)

$$\tilde{X}[k] = \begin{cases} 3, & k = 0, \\ -1, & k = 1. \end{cases}$$

$$\tilde{X}_3[k] = \begin{cases} 9, & k = 0, \\ 0, & k = 1, 2, 4, 5, \\ -3, & k = 3. \end{cases}$$

8.3. **(a)** $\tilde{x}_2[n]$.

(b) None of the sequences.

(c) $\tilde{x}_1[n]$ and $\tilde{x}_3[n]$.

8.4. **(a)**

$$X(e^{j\omega}) = \frac{1}{1 - ae^{-j\omega}}.$$

(b)

$$\tilde{X}[k] = \frac{1}{1 - ae^{-j(2\pi/N)k}}.$$

(c)

$$\tilde{X}[k] = X(e^{j\omega})|_{\omega=(2\pi k/N)}.$$

8.5. (a) $X[k] = 1.$
   (b) $X[k] = W_N^{kn_0}.$
   (c)

$$X[k] = \begin{cases} N/2, & k = 0, N/2, \\ 0, & \text{otherwise.} \end{cases}$$

(d)

$$X[k] = \begin{cases} N/2, & k = 0, \\ e^{-j(\pi k/N)(N/2-1)}(-1)^{(k-1)/2}\frac{1}{\sin(k\pi/N)}, & k \text{ odd}, \\ 0, & \text{otherwise.} \end{cases}$$

(e)

$$X[k] = \frac{1 - a}{1 - aW_N^k}.$$

8.6. (a)

$$X(e^{j\omega}) = \frac{1 - e^{j(\omega_0-\omega)N}}{1 - e^{j(\omega_0-\omega)}}.$$

(b)

$$X[k] = \frac{1 - e^{j\omega_0 N}}{1 - e^{j\omega_0}W_N^k}.$$

(c)

$$X[k] = \begin{cases} N, & k = k_0 \\ 0, & \text{otherwise.} \end{cases}$$

8.7.

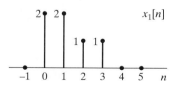

Figure P8.7-1

8.8.

$$y[n] = \begin{cases} \frac{1024}{1023}\left(\frac{1}{2}\right)^n, & 0 \le n \le 9, \\ 0, & \text{otherwise.} \end{cases}$$

8.9. (a) 1. Let $x_1[n] = \sum_m x[n + 5m].$
   2. Let $X_1[k]$ be the five-point FFT of $x_1[n].$ $M = 5.$
   3. $X_1[2]$ is $X(e^{j\omega})$ at $\omega = 4\pi/5.$
   (b) 1. Let $x_2[n]$ be $x[n]$ followed by seven zeros.
   2. Let $X_2[k]$ be the 27-point FFT of $x_2[n].$ $L = 27.$
   3. $X_2[5]$ is $X(e^{j\omega})$ at $\omega = 10\pi/27.$

8.10. $X_2[k] = (-1)^k X_1[k]$.

8.11.

Figure P8.11-1

8.12. (a)

$$X[k] = \begin{cases} 2, & k = 1, 3, \\ 0, & k = 2, 4. \end{cases}$$

(b)

$$H[k] = \begin{cases} 15, & k = 0, \\ -3 + j6, & k = 1, \\ -5, & k = 2, \\ -3 - j6, & k = 3. \end{cases}$$

(c) $y[n] = -3\delta[n] - 6\delta[n-1] + 3\delta[n-2] + 6\delta[n-3]$.
(d) $y[n] = -3\delta[n] - 6\delta[n-1] + 3\delta[n-2] + 6\delta[n-3]$.

8.13.

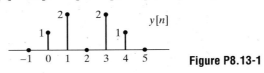

Figure P8.13-1

8.14. $x_3[2] = 9$.

8.15. $a = -1$. This is unique.

8.16. $b = 3$. This is unique.

8.17. $N = 9$.

8.18. $c = 2$.

8.19. $m = 2$. This is not unique. Any $m = 2 + 6\ell$ for integer ℓ works.

8.20. $N = 5$. This is unique.

Answers to Basic Problems in Chapter 9

9.1. If the input is $(1/N)X[((-n))_N]$, the output of the DFT program will be $x[n]$, the IDFT of $X[k]$.

9.2. (a) The gain is $-W_N^2$.
(b) There is one path. In general, there is only one path from any input sample to any output sample.
(c) By tracing paths, we see

$$X[2] = x[0] \cdot 1 + x[1]W_8^2 - x[2] - x[3]W_8^2 + \cdots$$
$$x[4] + x[5]W_8^2 - x[6] - x[7]W_8^2.$$

9.3. (a) Store $x[n]$ in $A[\cdot]$ in bit-reversed order, and $D[\cdot]$ will contain $X[k]$ in sequential (normal) order.

(b)

$$D[r] = \begin{cases} N, & r = 3, \\ 0, & \text{otherwise.} \end{cases}$$

(c)

$$C[r] = \begin{cases} 1, & r = 0, 1, 2, 3, \\ 0, & \text{otherwise.} \end{cases}$$

9.4. (a) $N/2$ butterflies with $2^{(m-1)}$ different coefficients.

(b) $y[n] = W_N^{2^{v-m}} y[n-1] + x[n]$.

(c) Period: 2^m, Frequency: $2\pi 2^{-m}$.

9.5.

$$X = AD - BD + CA - DA = AC - BD$$

$$Y = AD - BD + BC + BD = BC + AD.$$

9.6. Statement 1.

9.7. $\omega_k = 7\pi/16$.

9.8.

$$y[n] = X(e^{j\omega})|_{\omega=(2\pi/7)+(2\pi/21)(n-19)}.$$

9.9. (a) 2^{m-1}.

(b) 2^m.

9.10. $r[n] = e^{-j(2\pi/19)} W^{n^2/2}$ where $W = e^{-j(2\pi/10)}$.

9.11. $x[0], x[8], x[4], x[12], x[2], x[10], x[6], x[14], x[1], x[9], x[5], x[13], x[3], x[11], x[7], x[15]$.

9.12. False.

9.13. $m = 1$.

9.14.

$$r = \begin{cases} 0, & m = 0, \\ 0, 4, & m = 1, \\ 0, 2, 4, 6, & m = 2, \\ 0, 1, 2, 3, 4, 5, 6, 7, & m = 3. \end{cases}$$

9.15. $N = 64$.

9.16. $m = 3$ or 4.

9.17. Decimation-in-time.

9.18. 1021 is prime, so the program must implement the full DFT equations and cannot exploit any FFT algorithm. The computation time goes as N^2. Contrastingly, 1024 is a power of 2 and can exploit the $N \log N$ computation time of the FFT.

9.19.

$$a = -\sqrt{2}$$

$$b = -e^{-j(6\pi/8)}.$$

9.20.

$$y[n] = e^{j(2\pi/32)7} X^*(e^{j(7\pi/16)}).$$

Answers to Basic Problems in Chapter 10

10.1. **(a)** $f = 1500$ Hz.
     (b) $f = -2000$ Hz.

10.2. $N = 2048$ and 10000 Hz $< f < 10240$ Hz.

10.3. **(a)** 320 samples.
     (b) 400 DFT/second.
     (c) $N = 64$.
     (d) 250 Hz.

10.4. **(a)** $X[200] = 1 - j$.
     (b)

$$X(j2\pi(4000)) = 5 \times 10^{-5}(1 - j)$$
$$X(-j2\pi(4000)) = 5 \times 10^{-5}(1 + j).$$

10.5. **(a)** $T = 2\pi k_0/(N\Omega_0)$.
     (b) Not unique. $T = (2\pi/\Omega_0)(1 - k_0/N)$.

10.6.

$$X_c(j2\pi(4200)) = 5 \times 10^{-4}$$
$$X_c(-j2\pi(4200)) = 5 \times 10^{-4}$$
$$X_c(j2\pi(1000)) = 10^{-4}$$
$$X_c(-j2\pi(1000)) = 10^{-4}$$

10.7. $L = 1024$.

10.8. Rectangular, Hanning, Hamming, and Bartlett windows work.

10.9. $x_2[n]$ will have two distinct peaks.

10.10. $T > 1/1024$ sec.

10.11. $\Delta\Omega = 2\pi(2.44)$ rad/sec.

10.12. $N \geq 1600$.

10.13.

$$X_0[k] = \begin{cases} 18, & k = 3, 33, \\ 0, & \text{otherwise.} \end{cases}$$

$$X_1[k] = \begin{cases} 18, & k = 9, 27, \\ 0, & \text{otherwise.} \end{cases}$$

10.14. $x_2[n], x_3[n], x_6[n]$.

10.15. $\omega_0 = 0.25\pi$ rad/sample, $\lambda = \pi/80000$ rad/sample^2.

10.16. $\Delta f = 9.77$ Hz.

10.17. Methods 2 and 5 will improve the resolution.

10.18. The peaks will not have the same height. The peak from the rectangular window will be bigger.

10.19. $L = M + 1 = 124$.

10.20. **(a)** $A = 44.68$ dB.
     (b) Weak components will be visible if their amplitude exceeds 0.0058.

Answers to Basic Problems in Chapter 11

11.1. $\mathcal{I}m\{X(e^{j\omega})\} = 2a\sin\omega$.

11.2. $x[n] = (5/4)\delta[n] - \delta[n-1]$.

11.3.

$$x_1[n] = \delta[n] - \frac{1}{2}\delta[n-1]$$

$$x_2[n] = \delta[n] - \frac{1}{2}\delta[n+1].$$

11.4. $\mathcal{R}e\{X(e^{j\omega})\} = 1 - \cos(2\omega)$, $\mathcal{I}m\{X(e^{j\omega})\} = 0$.

11.5. (a) $x_i[n] = \sin\omega_0 n$.
    (b) $x_i[n] = -\cos\omega_0 n$.
    (c) $x_i[n] = (1 - \cos\omega_0 n)/(\pi n)$.

11.6. $x[n] = 5\delta[n] - 2\delta[n-1] + 3\delta[n-4]$.

11.7. (a) $x[n] = -\delta[n-1] - 2\delta[n-2]$.
    (b) Not unique. $x[n] = \delta[n] - \delta[n-1] - 2\delta[n-2]$ also satisfies the information given.

11.8. $X_{R2}(e^{j\omega})$ and $X_{R3}(e^{j\omega})$ are the answers.

11.9. $x[n] = -\delta[n] - 3\delta[n-1] - \delta[n-3]$ is the unique sequence satisfying the information given.

11.10. $h[n] = \pm(1/2)\left\{(-1/2)^n u[n] - (1/2)(-1/2)^{n-1}u[n-1]\right\}$.

11.11.

$$\mathcal{R}e\{X(e^{j\omega})\} = \begin{cases} 16\sin 3\omega, & 0 \le \omega \le \pi, \\ 0, & \text{otherwise.} \end{cases}$$

$$\mathcal{I}m\{X(e^{j\omega})\} = 0.$$

11.12. (a) $h[n] = \delta[n] - (1/3)\delta[n-1]$.
    (b) $h[n] = (1/3)\delta[n] - \delta[n-1]$.

11.13. $X_I(e^{j\omega}) = \cos\omega - \sin\omega - \cos 2\omega$.

11.14. $X_I(e^{j\omega}) = \sum_{k=0}^{\infty}(1/2)^k \sin k\omega$.

11.15. $x[n] = 4\delta[n] - \delta[n-1]$.

11.16. The facts are not consistent.

11.17. $x[n] = -\delta[n] + 3\delta[n-1]$ is the unique sequence satisfying the information given.

11.18. Two choices are $x[n] = 7\delta[n] + 2\delta[n-1]$ or $x[n] = 7\delta[n] + 2\delta[n-2]$.

11.19. $j X_I[k] = -j\delta[k-1] + j\delta[k-3]$.

11.20. $x_2[n]$ and $x_3[n]$ are consistent with the information given.

BIBLIOGRAPHY

AHMED, N., NATARAJAN, T., and RAO, K. R., "Discrete Cosine Transform," *IEEE Transactions on Computers,* Vol. C-23, pp. 90–93, Jan. 1974.

ALLEN, J. B., and RABINER, L. R., "A Unified Approach to Short-Time Fourier Analysis and Synthesis," *Proc. IEEE,* Vol. 65, pp. 1558–1564, Nov. 1977.

ANDREWS, H. C., and HUNT, B. R., *Digital Image Restoration,* Prentice Hall, Englewood Cliffs, NJ, 1977.

BARNES, C. W., and FAM, A. T., "Minimum Norm Recursive Digital Filters that are Free of Overflow Limit Cycles," *IEEE Trans. Circuits and Systems,* Vol. CAS-24, pp. 569–574, Oct. 1977.

BARTLETT, M. S., *An Introduction to Stochastic Processes with Special Reference to Methods and Applications,* Cambridge University Press, Cambridge, MA, 1953.

BENNETT, W. R., "Spectra of Quantized Signals," *Bell System Technical J.,* Vol. 27, pp. 446–472, 1948.

BLACKMAN, R. B., *Linear Data-Smoothing and Prediction in Theory and Practice,* Addison-Wesley Publishing Company, Reading, MA, 1965.

BLACKMAN, R. B., and TUKEY, J. W., *The Measurement of Power Spectra,* Dover Publications, New York, NY, 1958.

BLAHUT, R. E., *Fast Algorithms for Digital Signal Processing,* Addison-Wesley Publishing Company, Reading, MA, 1985.

BLUESTEIN, L. I., "A Linear Filtering Approach to the Computation of Discrete Fourier Transform," *IEEE Trans. Audio Electroacoustics,* Vol. AU-18, pp. 451–455, Dec. 1970.

BRACEWELL, R. N., "The Discrete Hartley Transform," *J. Optical Society of America,* Vol. 73, pp. 1832–1835, Dec. 1983.

BRACEWELL, R. N., "The Fast Hartley Transform," *Proc. IEEE,* Vol. 72, No. 8, pp. 1010–1018, Aug. 1984.

BRACEWELL, R. N., *The Fourier Transform and Its Applications,* 2nd ed. revised, McGraw-Hill Book Company, New York, NY, 1986.

BURRUS, C. S., "Efficient Fourier Transform and Convolution Algorithms," in *Advanced Topics in Signal Processing,* J. S. Lim and A. V. Oppenheim, Eds., Prentice Hall, Englewood Cliffs, NJ, 1988.

BURRUS, C. S., and Parks, T. W., *DFT/FFT and Convolution Algorithms Theory and Implementation,* Wiley, New York, NY, 1985.

CANDY, J. C., and TEMES, G. C., *Oversampling Delta-Sigma Data Converters,* IEEE Press, New York, NY, 1992.

CASTLEMAN, K. R., *Digital Image Processing,* Prentice Hall, Upper Saddle River, NJ, 1996.

CHAN, D. S. K., and RABINER, L. R., "An Algorithm for Minimizing Roundoff Noise in Cascade Realizations of Finite Impulse Response Digital Filters," *Bell System Technical J.,* Vol. 52, No. 3, pp. 347–385, Mar. 1973b.

CHAN, D. S. K., and RABINER, L. R., "Analysis of Quantization Errors in the Direct Form for Finite Impulse Response Digital Filters," *IEEE Trans. Audio Electroacoustics,* Vol. AU-21, pp. 354–366, Aug. 1973c.

CHELLAPPA, R., GIROD, B., MUNSON, D. C., TEKALP, A. M., and VETTERLI, M., "The Past, Present, and Future of Image and Multidimensional Signal Processing," *IEEE Signal Processing Magazine,* Vol. 15, No. 2, pp. 21–58, March 1998.

CHEN, W. H., SMITH C. H., and FRALICK, S. C., "A Fast Computational Algorithm for the Discrete Cosine Transform," *IEEE Trans. Commun.,* Vol. 25, pp. 1004–1009, Sept. 1977.

CHEN, X., and PARKS, T. W., "Design of FIR Filters in the Complex Domain," *IEEE Trans. Acoustics, Speech, and Signal Processing,* Vol. ASSP-35, pp. 144–153, 1987.

CHENEY, E. W., *Introduction to Approximation Theory,* 2nd ed., McGraw-Hill Book Company, New York, NY, 1982.

CHOW, Y., and CASSIGNOL, E., *Linear Signal Flow Graphs and Applications,* Wiley, New York, NY, 1962.

CHURCHILL, R. V., and BROWN, J. W., *Introduction to Complex Variables and Applications,* 5th ed., McGraw-Hill Book Company, New York, NY, 1990.

CLAASEN, T. A. C. M., MECKLENBRAUKER, W. F. G., and PEEK, J. B. H., "Second-Order Digital Filter with Only One Magnitude-Truncation Quantizer and Having Practically No Limit Cycles," *Electronics Letters,* Vol. 9, No. 2, pp. 531–532, Nov. 1973.

CLEMENTS, M. A., and PEASE, J., "On Causal Linear Phase IIR Digital Filters," *IEEE Trans. Acoustics, Speech, and Signal Processing,* Vol. ASSP-3, pp. 479–484, April 1989.

CONSTANTINIDES, A. G., "Spectral Transformations for Digital Filters," *Proc. IEE,* Vol. 117, No. 8, pp. 1585–1590, Aug. 1970.

COOLEY, J. W., LEWIS, P. A. W., and WELCH, P. D., "Historical Notes on the Fast Fourier Transform," *IEEE Trans. Audio Electroacoustics,* Vol. AU-15, pp. 76–79, June 1967.

COOLEY, J. W., and TUKEY, J. W., "An Algorithm for the Machine Computation of Complex Fourier Series," *Mathematics of Computation,* Vol. 19, pp. 297–301, Apr. 1965.

CROCHIERE, R. E., and OPPENHEIM, A. V., "Analysis of Linear Digital Networks," *Proc. IEEE,* Vol. 63, pp. 581–595, April 1975.

CROCHIERE, R. E., and RABINER, L. R., *Multirate Digital Signal Processing,* Prentice Hall, Englewood Cliffs, NJ, 1983.

DANIELS, R. W., *Approximation Methods for Electronic Filter Design,* McGraw-Hill Book Company, New York, NY, 1974.

DANIELSON, G. C., and LANCZOS, C., "Some Improvements in Practical Fourier Analysis and Their Application to X-Ray Scattering from Liquids," *J. Franklin Inst.,* Vol. 233, pp. 365–380 and 435–452, April and May 1942.

DAVENPORT, W. B., *Probability and Random Processes: An Introduction for Applied Scientists and Engineers,* McGraw-Hill Book Company, New York, NY, 1970.

DELLER, J. R., PROAKIS, J. G., and HANSEN, J. H. L., *Discrete-Time Processing of Speech Signals,* Macmillan Publishing Company, New York, NY, 1993.

DSP Committee, IEEE ASSP, Eds., *Programs for Digital Signal Processing,* IEEE Press, New York, NY, 1979.

DUDGEON, D. E., and MERSEREAU, R. M., *Two-Dimensional Digital Signal Processing,* Prentice Hall, Englewood Cliffs, NJ, 1984.

DUHAMEL, P., "Implementation of 'Split-Radix' FFT Algorithms for Complex, Real, and Real-Symmetric Data," *IEEE Trans. Acoustics, Speech, and Signal Processing,* Vol. ASSP-34, pp. 285–295, April 1986.

DUHAMEL, P., and HOLLMANN, H., "Split Radix FFT Algorithm," *Electronic Letters,* Vol. 20, pp. 14–16, Jan. 5, 1984.

EBERT, P. M., MAZO, J. E., and TAYLOR, M. C., "Overflow Oscillations in Digital Filters," *Bell System Technical J.,* Vol. 48, pp. 2999–3020, 1969.

ELLIOTT, D. F., and RAO, K. R., *Fast Transforms: Algorithms, Analysis, Applications,* Academic Press, New York, NY, 1982.

FETTWEIS, A., "Wave Digital Filters: Theory and Practice," *Proc. IEEE,* Vol. 74, No. 2, pp. 270–327, Feb. 1986.

FLANAGAN, J. L., *Speech Analysis, Synthesis and Perception,* 2nd ed., Springer-Verlag, New York, NY, 1972.

GARDNER, W. A., *Statistical Spectral Analysis: A Nonprobabilistic Theory,* Prentice Hall, Englewood Cliffs, NJ, 1988.

GENTLEMAN, W. M., and SANDE, G., "Fast Fourier Transforms for Fun and Profit," in *Proc. 1966 Fall Joint Computer Conf.,* AFIPS Conf. Proc., Vol. 29, pp. 563–578, Spartan Books, Washington, D.C., 1966.

GOERTZEL, G., "An Algorithm for the Evaluation of Finite Trigonometric Series," *American Math. Monthly,* Vol. 65, pp. 34–35, Jan. 1958.

GOLD, B., OPPENHEIM, A. V., and RADER, C. M., "Theory and Implementation of the Discrete Hilbert Transform," *Proc. Symp. Computer Processing in Communications,* Vol. 19, Polytechnic Press, New York, NY, 1970.

GOLD, B., and RADER, C. M., *Digital Processing of Signals,* McGraw-Hill Book Company, New York, NY, 1969.

GRAY, A. H., and MARKEL, J. D., "A Computer Program for Designing Digital Elliptic Filters," *IEEE Trans. Acoustics, Speech, and Signal Processing,* Vol. ASSP-24, pp. 529–538, Dec. 1976.

GROSSMAN, S. I., *Calculus Part 2,* 5th ed., Saunders College Publications, Fort Worth, TX, 1992.

GUILLEMIN, E. A., *Synthesis of Passive Networks,* Wiley, New York, NY, 1957.

HANNAN, E. J., *Time Series Analysis,* Methuen and Company, London, England, 1960.

HARRIS, F. J., "On the Use of Windows for Harmonic Analysis with the Discrete Fourier Transform," *Proc. IEEE,* Vol. 66, pp. 51–83, Jan. 1978.

HAYES, M. H., *Statistical Digital Signal Processing and Modeling,* Wiley, New York, NY, 1996.

HAYES, M. H., LIM, J. S., and OPPENHEIM, A. V., "Signal Reconstruction from Phase and Magnitude," *IEEE Trans. Acoustics, Speech, and Signal Processing,* Vol. ASSP-28, No. 6, pp. 672–680, Dec. 1980.

HAYKIN, S. S., *Adaptive Filter Theory,* 3rd ed., Prentice Hall, Upper Saddle River, NJ, 1996.

HEIDEMAN, M. T., JOHNSON, D. H., and BURRUS, C. S., "Gauss and the History of the Fast Fourier Transform," *IEEE ASSP Magazine,* Vol. 1, No. 4, pp. 14–21, Oct., 1984.

HELMS, H. D., "Fast Fourier Transform Method of Computing Difference Equations and Simulating Filters," *IEEE Trans. Audio Electroacoustics,* Vol. 15, No. 2, pp. 85–90, 1967.

HERRMANN, O., "On the Design of Nonrecursive Digital Filters with Linear Phase," *Elec. Lett.,* Vol. 6, No. 11, pp. 328–329, 1970.

HERRMANN, O., RABINER, L. R., and CHAN, D. S. K., "Practical Design Rules for Optimum Finite Impulse Response Lowpass Digital Filters," *Bell System Technical J.,* Vol. 52, No. 6, pp. 769–799, July–Aug. 1973.

HERRMANN, O., and SCHÜSSLER, W., "Design of Nonrecursive Digital Filters with Minimum Phase," *Elec. Lett.,* Vol. 6, No. 11, pp. 329–330, 1970a.

HERRMANN, O., and SCHÜSSLER, W., "On the Accuracy Problem in the Design of Nonrecursive Digital Filters," *Arch. Electronic Ubertragungstechnik,* Vol. 24, pp. 525–526, 1970b.

HEWES, C. R., BRODERSON, R. W., and BUSS, D. D., "Applications of CCD and Switched Capacitor Filter Technology," *Proc. IEEE,* Vol. 67, No. 10, pp. 1403–1415, Oct. 1979.

HNATEK, E. R., *A User's Handbook of D/A and A/D Converters,* Reprinted 1st ed., R. E. Krieger Publishing Co., Malabar, 1988.

HOFSTETTER, E., OPPENHEIM, A. V., and SIEGEL, J., "On Optimum Nonrecursive Digital Filters," *Proc. 9th Allerton Conf. Circuit System Theory,* Oct. 1971.

HWANG, S. Y., "On Optimization of Cascade Fixed Point Digital Filters," *IEEE Trans. Circuits and Systems,* Vol. CAS-21, No. 1, pp. 163–166, Jan. 1974.

JACKSON, L. B., "On the Interaction of Roundoff Noise and Dynamic Range in Digital Filters," *Bell System Technical J.,* Vol. 49, pp. 159–184, Feb. 1970a.

JACKSON, L. B., "Roundoff-Noise Analysis for Fixed-Point Digital Filters Realized in Cascade or Parallel Form," *IEEE Trans. Audio Electroacoustics,* Vol. AU-18, pp. 107–122, June 1970b.

JACKSON L. B., *Digital Filters and Signal Processing: with MATLAB Exercises,* 3rd ed., Kluwer Academic Publishers, Hingham, MA, 1996.

JAIN, A. K., *Fundamentals of Digital Image Processing,* Prentice Hall, Englewood Cliffs, NJ, 1989.

JAYANT, N. S., and Noll, P., *Digital Coding of Waveforms,* Prentice Hall, Englewood Cliffs, NJ, 1984.

JENKINS G. M., and WATTS, D. G., *Spectral Analysis and Its Applications,* Holden-Day, San Francisco, CA, 1968.

KAISER, J. F., "Digital Filters," Chap. 7 in *System Analysis by Digital Computer,* F. F. Kuo and J. F. Kaiser, Eds., Wiley, New York, NY, 1966.

KAISER, J. F., "Nonrecursive Digital Filter Design Using the I_0-Sinh Window Function," *Proc. 1974 IEEE International Symp. on Circuits and Systems,* San Francisco, CA, pp. 20–23, April 1974.

KAISER J. F., and HAMMING, R. W., "Sharpening the Response of a Symmetric Nonrecursive Filter by Multiple Use of the Same Filter," *IEEE Trans. Acoustics, Speech, and Signal Processing,* Vol. ASSP-25, No. 5, pp. 415–422, Oct. 1977.

KAISER, J. F., and SCHAFER, R. W., "On the Use of the I_0-Sinh Window for Spectrum Analysis," *IEEE Trans. Acoustics, Speech, and Signal Processing,* Vol. ASSP-28, No. 1, pp. 105–107, Feb. 1980.

KAN, E. P. F., and AGGARWAL, J. K., "Error Analysis of Digital Filters Employing Floating Point Arithmetic," *IEEE Trans. Circuit Theory,* Vol. CT-18, pp. 678–686, Nov. 1971.

KANEKO, T., and LIU, B., "Accumulation of Roundoff Error in Fast Fourier Transforms," *J. Assoc. Comput. Mach.,* Vol. 17, pp. 637–654, Oct. 1970.

KARAM, L. J., and McCLELLAN, J. H., "Complex Chebychev Approximation for FIR Filter Design," *IEEE Trans. Circuits and Systems,* Vol. 42, pp. 207–216, March 1995.

KAY, S. M., *Modern Spectral Estimation Theory and Application,* Prentice Hall, Englewood Cliffs, NJ, 1988.

KAY, S. M., and MARPLE, S. L., "Spectrum Analysis: A Modern Perspective," *Proc. IEEE,* Vol. 69, pp. 1380–1419, November 1981.

KNUTH, D. E., *The Art of Computer Programming; Seminumerical Algorithms,* 3rd ed., Addison-Wesley Publishing Co., Reading, MA, 1997.

KOOPMANNS, L. H., *Spectral Analysis of Time Series,* 2nd ed., Academic Press, New York, NY, 1995.

LAM, H. Y. F., *Analog and Digital Filters: Design and Realization,* Prentice Hall, Englewood Cliffs, NJ, 1979.

LIGHTHILL, M. J., *Introduction to Fourier Analysis and Generalized Functions,* Cambridge University Press, Cambridge, MA, 1958.

LIM, J. S., *Two-Dimensional Digital Signal Processing,* Prentice Hall, Englewood Cliffs, NJ, 1989.

LIU, B., and KANEKO, T., "Error Analysis of Digital Filters Realized in Floating-Point Arithmetic," *Proc. IEEE,* Vol. 57, pp. 1735–1747, Oct. 1969.

LIU, B., and PELED, A., "Heuristic Optimization of the Cascade Realization of Fixed Point Digital Filters," *IEEE Trans. Acoustics, Speech, and Signal Processing,* Vol. ASSP-23, pp. 464–473, 1975.

MACOVSKI, A., *Medical Image Processing,* Prentice Hall, Englewood Cliffs, NJ, 1983.

MAKHOUL, J., "Linear Prediction: A Tutorial Review," *Proc. IEEE,* Vol. 62, pp. 561–580, April 1975.

MAKHOUL, J., "A Fast Cosine Transform in One and Two Dimensions," *IEEE Trans. Acoustics, Speech, and Signal Processing,* Vol. 28, No. 1, Feb. 1980.

MARKEL, J. D., "FFT Pruning," *IEEE Trans. Audio and Electroacoustics,* Vol. AU-19, pp. 305–311, Dec. 1971.

MARKEL, J. D., and GRAY, A. H., Jr., *Linear Prediction of Speech,* Springer-Verlag, New York, NY, 1976.

MARPLE, S. L., *Digital Spectral Analysis with Applications,* Prentice Hall, Englewood Cliffs, NJ, 1987.

MARTUCCI, S. A., "Symmetrical Convolution and the Discrete Sine and Cosine Transforms," *IEEE Trans. Signal Processing,* Vol. 42, No. 5, pp. 1038–1051, May 1994.

MASON, S., and ZIMMERMANN, H. J., *Electronic Circuits, Signals and Systems,* Wiley, New York, NY, 1960.

Mathworks, *Signal Processing Toolbox Users Guide,* The Mathworks, Inc., Natick, MA, 1998.

McCLELLAN, J. H., and PARKS, T. W., "A Unified Approach to the Design of Optimum FIR Linear Phase Digital Filters," *IEEE Trans. Circuit Theory,* Vol. CT-20, pp. 697–701, Nov. 1973.

McCLELLAN, J. H., and RADER, C. M., *Number Theory in Digital Signal Processing,* Prentice Hall, Englewood Cliffs, NJ, 1979.

MERSEREAU, R. M., SCHAFER, R. W., BARNWELL, T. P., and SMITH, D. L., "A Digital Filter Design Package for PCs and TMS320s," *Proc. MIDCON,* Dallas, TX, 1984.

MILLS W. L., MULLIS, C. T., and ROBERTS, R. A., "Digital Filter Realizations Without Over-flow Oscillations," *IEEE Trans. Acoustics, Speech, and Signal Processing,* Vol. ASSP-26, pp. 334–338, Aug. 1978.

NAWAB, S. H., and QUATIERI, T. F., "Short-Time Fourier Transform," in *Advanced Topics in Signal Processing,* J. S. Lim and A. V. Oppenheim, Eds., Prentice Hall, Englewood Cliffs, NJ, 1988.

NYQUIST, H., "Certain Topics in Telegraph Transmission Theory," *AIEE Trans.,* pp. 617–644, 1928.

OETKEN, G., PARKS, T. W., and SCHÜSSLER, H. W., "New Results in the Design of Digital Interpola-tors," *IEEE Trans. Acoustics, Speech, and Signal Processing,* Vol. ASSP-23, pp. 301–309, June 1975.

OPPENHEIM, A. V., SCHAFER, R. W., and STOCKAM, T. G. Jr., "Nonlinear Filtering of Multiplied and Convolved Signals," *Proc. IEEE,* Vol. 56, No. 8, pp. 1264–1291, Aug. 1968.

OPPENHEIM, A. V., and JOHNSON, D. H., "Discrete Representation of Signals," *Proc. IEEE,* Vol. 60, No. 6, pp. 681–691, June 1972.

OPPENHEIM, A. V., and SCHAFER, R. W., *Digital Signal Processing,* Prentice Hall, Englewood Cliffs, NJ, 1975.

OPPENHEIM, A. V., and WILLSKY, A. S., with S. H. NAWAB, *Signals and Systems,* Second Edition, Prentice Hall, Upper Saddle River, NJ, 1997.

O'SHAUGHNESSY, D., *Speech Communication, Human and Machine,* Addison-Wesley Publishing Company, Reading, MA, 1987.

PAPOULIS, A., "On the Approximation Problem in Filter Design," *IRE Conv. Record,* Pt. 2, pp. 175–185, 1957.

PAPOULIS, A., *Probability, Random Variables and Stochastic Processes,* 3rd ed., McGraw-Hill Book Company, New York, NY, 1991.

PARKS, T. W., and BURRUS, C. S., *Digital Filter Design,* Wiley, New York, NY, 1987.

PARKS, T. W., and MCCLELLAN, J. H., "Chebyshev Approximation for Nonrecursive Digital Filters with Linear Phase," *IEEE Trans. Circuit Theory,* Vol. CT-19, pp. 189–194, Mar. 1972a.

PARKS, T. W., and MCCLELLAN, J. H., "A Program for the Design of Linear Phase Finite Impulse Response Filters," *IEEE Trans. Audio Electroacoustics,* Vol. AU-20, No. 3, pp. 195–199, Aug. 1972b.

PARSONS, T. *Voice and Speech Processing,* McGraw-Hill Book Company, New York, 1986.

PHILLIPS, C. L., and NAGLE, H. T., Jr., *Digital Control System Analysis and Design,* 3rd ed., Prentice Hall, Upper Saddle River, NJ, 1995.

PRATT W., *Digital Image Processing,* 2nd ed., Wiley, New York, NY, 1991.

RABINER, L. R., "The Design of Finite Impulse Response Digital Filters Using Linear Programming Techniques," *Bell System Technical J.,* pp. 1177–1198, July–Aug. 1972a.

RABINER, L. R., "Linear Program Design of Finite Impulse Response (FIR) Digital Filters," *IEEE Trans. Audio and Electroacoustics,* Vol. AU-20, No. 4, pp. 280–288, Oct. 1972b.

RABINER, L. R., and GOLD, B., *Theory and Application of Digital Signal Processing,* Prentice Hall, Englewood Cliffs, NJ, 1975.

RABINER, L. R., KAISER, J. F., HERRMANN, O., and DOLAN, M. T., "Some Comparisons Between FIR and IIR Digital Filters," *Bell System Technical J.,* Vol. 53, No. 2, pp. 305–331, Feb. 1974.

RABINER, L. R., and SCHAFER, R. W., *Digital Processing of Speech Signals,* Prentice Hall, Englewood Cliffs, NJ, 1978.

RABINER L. R., and SCHAFER, R. W., "On the Behavior of Minimax FIR Digital Hilbert Transformers," *Bell System Technical J.,* Vol. 53, No. 2, pp. 361–388, Feb. 1974.

RABINER, L. R., SCHAFER, R. W., and RADER, C. M., "The Chirp *z*-Transform Algorithm," *IEEE Trans. Audio Electroacoustics,* Vol. AU-17, pp. 86–92, June 1969.

RADER, C. M., "Discrete Fourier Transforms When the Number of Data Samples Is Prime," *Proc. IEEE,* Vol. 56, pp. 1107–1108, June 1968.

RADER, C. M., "An Improved Algorithm for High-Speed Autocorrelation with Applications to Spectral Estimation," *IEEE Trans. Audio Electroacoustics,* Vol. AU-18, pp. 439–441, Dec. 1970.

RADER, C. M., and Brenner, N. M., "A New Principle for Fast Fourier Transformation," *IEEE Trans. Acoustics, Speech, and Signal Processing,* Vol. ASSP-25, pp. 264–265, June 1976.

RADER, C. M., and GOLD, B., "Digital Filter Design Techniques in the Frequency Domain," *Proc. IEEE,* Vol. 55, pp. 149–171, Feb. 1967.

RAGAZZINI, J. R., and FRANKLIN, G. F., *Sampled Data Control Systems,* McGraw-Hill Book Company, New York, NY, 1958.

RAO, K. R., and YIP, P., *Discrete Cosine Transform: Algorithms, Advantages, Applications,* Academic Press, Boston, MA, 1990.

RAO, K. R., and HWANG, J. J., *Techniques and Standards for Image, Video, and Audio Coding,* Prentice Hall, Upper Saddle River, NJ, 1996.

RAO, S. K., and KAILATH, T., "Orthogonal Digital Filters for VLSI Implementation," *IEEE Trans. Circuits and Systems,* Vol. CAS-31, No. 11, pp. 933–945, Nov. 1984.

ROBINSON, E. A., and DURRANI, T. S., *Geophysical Signal Processing,* Prentice Hall, Englewood Cliffs, NJ, 1985.

ROBINSON, E. A., and TREITEL, S., *Geophysical Signal Analysis,* Prentice Hall, Englewood Cliffs, NJ, 1980.

RUNGE, C., "Über die Zerlegung Empirisch Gegebener Periodischer Functionen in Sinuswellen," *Z. Math. Physik,* Vol. 48, pp. 443–456, 1903; Vol. 53, pp. 117–123, 1905.

SANDBERG, I. W., "Floating-Point-Roundoff Accumulation in Digital Filter Realization," *Bell System Technical J.,* Vol. 46, pp. 1775–1791, Oct. 1967.

SCHAEFER, R. T., SCHAFER, R. W., and MERSEREAU, R. M., "Digital Signal Processing for Doppler Radar Signals," *Proc. 1979 IEEE Int. Conf. on Acoustics, Speech, and Signal Processing,* pp. 170–173, 1979.

SCHAFER, R. W., and RABINER, L. R., "A Digital Signal Processing Approach to Interpolation," *Proc. IEEE,* Vol. 61, pp. 692–702, June 1973.

SCHMID, H., *Electronic Analog/Digital Conversions,* Wiley, New York, NY, 1976.

SCHÜSSLER, H. W., and STEFFEN, P., "Some Advanced Topics in Filter Design," in *Advanced Topics in Signal Processing,* J. S. Lim and A. V. Oppenheim, Eds., Prentice Hall, Englewood Cliffs, NJ, 1988.

SHANNON, C. E., "Communication in the Presence of Noise," *Proc. IRE,* pp. 10–12, Jan. 1949.

SINGLETON, R. C., "An Algorithm for Computing the Mixed Radix Fast Fourier Transform," *IEEE Trans. Audio Electroacoustics,* Vol. AU-17, pp. 93–103, June 1969.

SKOLNIK, M. I., *Introduction to Radar Systems,* 2nd ed., McGraw-Hill Book Company, New York, NY, 1986.

SLEPIAN, D., LANDAU, H. T., and POLLACK, H. O., "Prolate Spheroidal Wave Functions, Fourier Analysis, and Uncertainty Principle (I and II)," *Bell System Technical J.,* Vol. 40, No. 1, pp. 43–80, Jan. 1961.

STEIGLITZ, K., "The Equivalence of Analog and Digital Signal Processing," *Information and Control,* Vol. 8, No. 5, pp. 455–467, Oct. 1965.

STOCKHAM, T. G., Jr., "High Speed Convolution and Correlation," *1966 Spring Joint Computer Conference,* AFIPS Proc., Vol. 28, pp. 229–233, 1966.

STORER, J. E., *Passive Network Synthesis,* McGraw-Hill Book Company, New York, NY, 1957.

TRIBOLET, J. M., "A New Phase Unwrapping Algorithm," *IEEE Trans. Acoustics, Speech, and Signal Processing,* Vol. ASSP-25, No. 2, pp. 170–177, April 1977.

TUKEY, J. W., *Exploratory Data Analysis,* Addison-Wesley Publishing Company, Reading, MA, 1977.

VAIDYANATHAN, P. P., *Multirate Systems and Filter Banks,* Prentice Hall, Englewood Cliffs, NJ, 1993.

VERNET, J. L., "Real Signals Fast Fourier Transform: Storage Capacity and Step Number Reduction by Means of an Odd Discrete Fourier Transform," *Proc. IEEE,* pp. 1531–1532, Oct. 1971.

VOLDER, J. E., "The CORDIC Trigonometric Computing Technique," *IRE Trans. Electronic Computers,* Vol. EC-8, pp. 330–334, Sept. 1959.

WEINBERG, L., *Network Analysis and Synthesis,* R. E. KREIGER, Huntington, NY, 1975.

WEINSTEIN, C. J., "Roundoff Noise in Floating Point Fast Fourier Transform Computation," *IEEE Trans. Audio Electroacoustics,* Vol. AU-17, pp. 209–215, Sept. 1969.

WEINSTEIN, C. J., and OPPENHEIM, A. V., "A Comparison of Roundoff Noise in Floating Point and Fixed Point Digital Filter Realizations," *Proc. IEEE,* Vol. 57, pp. 1181–1183, June 1969.

WELCH, P. D., "A Fixed-Point Fast Fourier Transform Error Analysis," *IEEE Trans. Audio Electroacoustics,* Vol. AU-17, pp. 153–157, June 1969.

WELCH, P. D., "The Use of the Fast Fourier Transform for the Estimation of Power Spectra," *IEEE Trans. Audio Electroacoustics,* Vol. AU-15, pp. 70–73, June 1970.

WIDROW, B., "Statistical Analysis of Amplitude-Quantized Sampled-Data Systems," *AIEE Trans. (Applications and Industry,)* Vol. 81, pp. 555–568, Jan. 1961.

WIDROW, B., "A Study of Rough Amplitude Quantization by Means of Nyquist Sampling Theory," *IRE Trans. Circuit Theory,* Vol. CT-3, pp. 266–276, Dec. 1956.

WIDROW, B., and STEARNS, S. D., *Adaptive Signal Processing,* Prentice Hall, Englewood Cliffs, NJ, 1985.

WINOGRAD, S., "On Computing the Discrete Fourier Transform," *Mathematics of Computation,* Vol. 32, No. 141, pp. 175–199, Jan. 1978.

ZVEREV, A. I., *Handbook of Filter Synthesis,* Wiley, New York, NY, 1967.

INDEX

Absolute summability, 51, 108
  for suddenly-applied
    exponential, 51–52
Accumulator, 19, 31, 34
  difference equation
    representation of, 34–35
  as time-invariant system, 20
Adaptive signal processing, 4
Additivity property, 18
Aliasing, 144–45
  prefiltering to avoid, 185–87
  in reconstruction of
    undersampled sinusoidal
    signal, 148–49
  sampling/reconstruction of
    sinusoidal signal, 147
Aliasing distortion, 144–45
All-pass decomposition, 280–82
All-pass systems, 274–79
  defined, 274
  first-and second-order, 275
  uses of, 278
Alternation theorem, 489–91
Amplitude spectrum, 49
Analog channel vocoder, 5
Analog signals, 8
  digital processing of, 185–201
    analog-to-digital (A/D)
      conversion, 187–93
    analysis of quantization
      errors, 193–97
    prefiltering to avoid aliasing,
      185–87
Analog-to-digital (A/D)
    conversion, 187–93
  oversampled:
    with direct quantization,
      201–5
    with noise shaping, 206–10
  sample-and-hold system,
    188–90
Analytic signal, 776
Antialiasing filter, 186
Aperiodic discrete-time
    sinusoids, 15
Approximation:
  Butterworth, 458–60
  Chebyshev, 460–63

elliptic, 463–65
  equiripple, 492–93
Attenuation, and group delay,
    effects of, 343–45
Autocorrelation, 65–66
Autocorrelation invariance, 526
Autocorrelation sequence:
  of $h[n]$, 67–68
  spectrum analysis of random
    signals using estimates of,
    743–48
Autocovariance sequence, 65–66
Autoregressive (AR) linear
    random process, 771
Autoregressive moving-average
    (ARMA) linear random
    process, 771
Auxiliary conditions, and
    linearity/time
    invariance/causality of a
    system, 39
Averaging of periodograms,
    737–38
  computation using the DFT,
    739

Backward difference, 21, 32,
    33–34
Bandlimited differentiator, ideal,
    159
Bandlimited signal,
    reconstruction of, from its
    samples, 150–53
Bandpass filters, FIR equiripple
    approximation of, 507–10
Bandpass sampling, 799–801
Bandpass signals, representation
    of, 796–99
Bandstop filters, frequency
    response of, 44
Bartlett windows, 468–71, 744,
    750–54
Basic sequences/sequence
    operations, 11–16
  combining, 13–16
  exponential sequences, 13–14
  unit sample sequence, 11
  unit step sequence, 11–13

Bilateral z-transform, 95
Bilinear transformation, 450–54,
    517
  Butterworth approximation,
    458–60
  of a Butterworth filter, 454–57
  Chebyshev approximation,
    460–63
  design examples, 454–65
  elliptic approximation, 463–65
  and warping, 453–54
Binary number system, 371–74
Bit-reversed order, 642–43
Blackman window, 468fn, 469
Block convolution, 585–87
  overlap-add method of, 723
  using time-dependent Fourier
    transform, 722–23
Block floating point, 668
Block diagram representation of
    linear constant-coefficient
    difference equations,
    341–48
Block processing, 693
Bounded-input bounded-output
    (BIBO), 21
Butterflies, 676
Butterworth approximation,
    458–60
Butterworth filter:
  bilinear transformation of,
    454–57
  impulse invariance with,
    446–49

Canonic form implementation,
    346, 347
Cascade connection, 29
Cascade form structures:
  coefficient quantization effects,
    379–82, 386–90
  FIR systems, 390, 412
  IIR systems, 356–59, 379–82,
    403–10
Cauchy integral theorem, 776
Cauchy principal values, 781–82
Cauchy-Riemann conditions, 776
Causal all-pass systems, 274–78

Causal generalized linear-phase
    systems, 297–311
  type I FIR linear-phase
    systems, 298
  type II FIR linear-phase
    systems, 298
  type III FIR linear-phase
    systems, 298
  type IV FIR linear-phase
    systems, 298–99
Causality, 21
  and auxiliary conditions, 39
  and region of convergence
    (ROC), 110–11
Causal sequences, 31, 129, 777–82
  defined, 778
  exponential sequence, 779–80
  finite-length sequence, 779
  sufficiency of Fourier
    transform for, 777–82
    exponential sequence,
      779–80
    finite-length sequence, 779
Characteristic of floating point
    number, 412
Charge-coupled devices (CCDs),
    2
Charge transport devices
    (CTDs), 2
Chebyshev approximation,
    460–63
Chirp transform algorithm
    (CTA), 656–61
  parameters, 661
Circular convolution:
  with delayed impulse
    sequence, 572–73
  and discrete Fourier transform,
    571–75
  N-point circular convolution,
    571
  of two rectangular pulses,
    573–75
Circular shift of a sequence,
    discrete Fourier
    transform, 564–67
Coefficient quantization, 377–39
  of an optimum FIR filter,
    386–90
  in FIR systems, 384–86
  in IIR systems, 377–82
  maintaining linear phase,
    390–91
  poles of quantized
    second-order sections,
    382–84

Commutative property of
    convolution, 29
Compensation for zero-order
    hold, 506–7
Complex-conjugate pole pair,
    direct form
    implementation of, 382–84
Complex cepstrum, 138, 788–89
Complex exponential sequences,
    14–15, 40
Complex sequences:
  defined, 789
  Hilbert transform relations for,
    789–801
Compressor, 20–21, 167–68
Computationally realizable
    systems, 242
Conjugate-antisymmetric
    sequence, 55–56
Conjugate-symmetric sequence,
    55–56
Conjugation property,
    z-transform, 123
Continuous phase, 343
Continuous-to-discrete-time
    (C/D) conversion, 141–49
Continuous-time complex
    exponentials and
    sinusoids, compared to
    discrete-time complex
    exponentials and
    sinusoids, 14
Continuous-time IIR filter
    design, 442–43
Continuous-time processing, of
    discrete-time signals,
    163–67
Continuous-time signals, 8
  aliasing, 144–45
    in reconstruction of
      undersampled sinusoidal
      signal, 148–49
    sampling/reconstruction of
      sinusoidal signal, 147
  bandlimited signal
    reconstruction, from its
    samples, 150–53
  continuous-time processing of
    discrete-time signals,
    163–67
    moving-average system with
      noninteger delay, 165–67
    noninteger delay, 164–65
  digital processing of analog
    signals, 185–201
    analog-to-digital (A/D)

conversion, 187–93
  analysis of quantization
    errors, 193–97
  digital-to-analog (D/A)
    conversion, 197–201
  prefiltering to avoid aliasing,
    185–87
discrete-time processing of,
    153–63
  changing sampling rate
    using, 167–78
  impulse invariance, 160–63
  linear time-invariant
    discrete-time systems,
    154–60
frequency-domain
    representation of, 142–49
multirate signal processing,
    179–84
  interchange of filtering and
    downsam-
    pling/upsampling,
    179–80
  polyphase decompositions,
    180–82
  polyphase implementation
    of decimation filters,
    182–83
  polyphase implementation
    of interpolation filters,
    183–84
Nyquist frequency/Nyquist
    rate, 146–47
periodic sampling, 140–42
sampling of, 140–239
sampling rate change by
    noninteger factor, 176–78
sampling rate increase by
    integer factor, 172–76
sampling rate reduction by
    integer factor, 167–72
unit impulse function, 142–43
Convergence:
  Fourier transform, 50–53
  z-transform, 96, 105–11
Convolution property,
    z-transform, 124–26
Convolution of sequences, 22–28
    124–25
  block, 585–88, 721–22
  circular, 571–75
  linear, 576–88
  periodic, 548–50
Convolution sum:
  analytic evaluation of, 26–28
  computation of, 25–26

defined, 23
Convolution-sum expression, 23–25
Convolution theorem, 60–61
　Fourier transform, 60–61
　z-transform, 124–25
CORDIC rotator algorithm, 677, 684
Correlation, computing using DFT, 746–48
Coupled form, for second-order system, 384
Coupled form oscillator, 425
Critically sampled filter bank, 770
Cross-correlation, 69–70

DB (decibels), 254
DCT-1, 590–93
　relationship between DFT and, 593–94
　See also Discrete cosine transform (DCT)
DCT-2, 590–93
　energy compaction property of, 595–98
　relationship between DFT and, 594–95
　See also Discrete cosine transform (DCT)
Dead bands, 416
Decimation, 172
Decimation filters, polyphase implementation of, 182–83
Decimation-in-frequency algorithms, 646–52
　alternate forms, 650–52
　in-place computation, 650
Decimation-in-time algorithms, 635–46
　alternate forms, 643–46
　in-place computations, 640–43
Decimator, 172
Decomposition:
　all-pass, 280–82
　of a linear-phase system, 308–11
Delay distortion, 242–43
Deterministic autocorrelation sequence, 67–68
DFS, *See* Discrete Fourier series (DFS)
DFT, *See* Discrete Fourier transform (DFT)
Difference equations, recursive computation of, 37–38
Differentiation in frequency, 60

Differentiation property, z-transform, 122–23
Differentiators, 516
　Kaiser window design of, 482–85
Digital filters, 439
Digital processing of analog signals, 185–201
　analog-to-digital (A/D) conversion, 187–93
　analysis of quantization errors, 193–97
　digital-to-analog (D/A) conversion, 197–201
　prefiltering to avoid aliasing, 185–87
Digital signal processing, 9
　early use of digital computers in, 5–6
Digital signals, 8
Digital-to-analog (D/A) conversion, 185, 197–201
　oversampling and noise shaping in, 210–13
Dirac delta function, 142–43
Direct form I implementation, of an LTI system, 346–48
Direct form II implementation, of an LTI system, 347–48
Discrete cosine transform (DCT), 589–99
　applications of, 598–99
　DCT-1, 590–93
　　relationship between DFT and, 593–94
　DCT-2, 590–93
　　energy compaction property of, 595–98
　　relationship between DFT and, 594–95
　definitions of, 589–90
　multiplication of, 599
Discrete Fourier series (DFS):
　duality in, 544–45
　of a periodic impulse train, 544
　of a periodic regular pulse train, 545–46
　properties of, 546–51
　　duality property, 547
　　linearity property, 546
　　periodic convolution, 548–51
　　shift of a sequence, 546–47
　　symmetry properties, 547
　representation of periodic sequences, 543–46

Discrete Fourier transform (DFT), 541–692
　autocorrelation sequence, spectrum analysis of random signals using estimates of, 743–48
　chirp transform algorithm (CTA), 656–61
　　parameters, 661
　computation of, 629–52
　　decimation-in-frequency algorithms, 646–52
　　decimation-in-time algorithms, 635–46
　　Goertzel algorithm, 633–35
　correlation, computing using DFT, 746–48
　defined, 541
　DFT analysis of sinusoidal signals, 697–714
　　oversampling/linear interpolation for frequency estimation, 713–14
　　spectral sampling, 703–14
　　with 32-point Kaiser window and zero-padding, 711–12
　　using a Kaiser window, 708–10
　　windowing, 698–703
　discrete cosine transform (DCT), 589–99
　　applications of, 598–99
　　DCT-1 and DCT-2, 590–93
　　definitions of, 589–90
　　energy compaction property of DCT-2, 595–98
　　multiplication of, 599
　　relationship between DFT and DCT-1, 593–94
　　relationship between DFT and DCT-2, 594–95
　discrete Fourier series (DFS), 543–46
　　duality in, 544–45
　　of a periodic impulse training, 544
　　of a periodic regular pulse train, 545–46
　　properties of, 546–51
　fast Fourier transform (FFT) algorithms, 629–92
　　algorithms for more general values of N, 655
　　coefficients, 654–55
　　indexing, 652–54

finite register length, effects of, 661–68

Fourier analysis of signals using, 693–97
relationship between DFT values, 697
as Fourier representation of finite-duration sequences, 559–64
implementation of, using convolution, 655–61
implementing linear time-invariant systems using, 582–88
linear convolution, 576–88
circular convolution as linear convolution with aliasing, 577–82
of two finite-length sequences, 277
periodic sequences, representation of, 542–46
of periodic signals, 551–55
periodogram, 730–42
averaging of, 737–39
defined, 732
explicit computation of, 732
modified, 732
periodogram analysis, 731, 739–42
power spectrum, 731–32, 734
properties of, 733–37
power spectrum estimation:
computing using DFT, 746–48
example based on estimation of autocorrelation sequence, 748–54
properties of, 564–76
circular convolution, 571–75
circular shift of a sequence, 564–67
duality property, 567–68
linearity property, 564
symmetry properties, 568–70
sampling the Fourier transform, 555–59
time-dependent Fourier analysis of nonstationary signals, 723–30
radar signals, 728–30
speech signals, 720–28
time-dependent Fourier transform, 714–22
block convolution using, 722–23

defined, 714
effect of the window in, 717–18
of a linear chirp signal, 715–17
sampling in time/frequency, 718–22
Winograd Fourier transform algorithm (WFTA), 655–56
Discrete Hilbert transforms, 775–810
analytic signal, 776, 789
bandpass sampling, 799–801
bandpass signals, representation of, 796–99
Cauchy principal values, 781–82
causal sequences:
defined, 778
exponential sequence, 779–80
finite-length sequence, 779
sufficiency of the Fourier transform for, 777–82
Hilbert transformers, 791
design of, 792–95
Kaiser window design of, 793–95
Hilbert transform relations, 775–801
for complex sequences, 789–801
magnitude, relationships between phase and, 788–89
Poisson's formulas, 776
Discrete-time differentiators, 482–85
Discrete-time filter, determining specifications for, 440–42
Discrete-time Fourier transform (DTFT), 48fn
Discrete-time impulse, 11
Discrete-time random signals, 65–70
Discrete-time signal processing, 2
continuous signals, 2
future of, 7
history of, 6–7
Discrete-time signals, 8–93
basic sequences/sequence operations, 11–16
combining basic sequences, 13–16

complex exponential sequence, 14
exponential sequences, 13–14
sinusoidal sequences, 13
unit sample sequence, 11
unit step sequence, 11–13
defined, 8
discrete-time systems, 16–22
graphic depiction of, 10
sampling frequency, 10
sampling period, 10
as sequences of numbers, 9–16
signal-processing systems, classification of, 8–9
Discrete-time sinusoids, periodic/aperiodic, 15
Discrete-time specifications, 443
Discrete-time systems, 16–22
causality, 21
coefficient quantization, 377–39
of an optimum FIR filter, 386–90
in FIR systems, 384–86
in IIR systems, 377–82
maintaining linear phase, 390–91
poles of quantized second-order sections, 382–84
defined mathematically, 16–17
discrete-time random signals, 65–70
autocorrelation/autocovariance sequence, 65–66
deterministic autocorrelation sequence, 67–68
power density spectrum, 68
random process, 65
white noise, 69
finite-precision numerical effects, 370–77
numerical representations, 371–74
quantization in implementing systems, 374–77
FIR systems, basic network structures for, 366–70
cascade form, 367–68
direct form, 367
linear-phase FIR system structures, 368–70

fixed-point realizations of IIR digital filters, zero-input limit cycles in, 413–18
Fourier transform theorems, 58–65
convolution theorem, 60–61
differentiation in frequency, 60
frequency shifting, 59
linearity of Fourier transform, 59
modulation or windowing theorem, 61–62
Parseval's theorem, 60
time reversal, 60
time shifting, 59
frequency-domain representation of, 40–48
eigenfunctions for linear time-invariant systems, 40–45, 61
frequency response of the ideal delay system, 41
frequency response of the moving-average system, 44–45
ideal frequency-selective filters, 43–44
sinusoidal response of linear time-invariant systems, 42–43
suddenly-applied complex exponential inputs, 46–48
ideal delay system, 17
IIR systems, basic structures for, 354–63
cascade form, 356–59
direct forms, 354–56
parallel form, 359–61
instability, 21–22
testing for, 22
limit cycles:
avoiding, 417–18
due to overflow, 416–17
due to round-off, 414–16
due to truncation, 414–16
in first-order system, 414–16
linear constant-coefficient difference equations, 34–40
block diagram representation of, 341–48
difference equation representation of the accumulator, 34–35
difference equation

representation of the moving-average system, 35–36
recursive computation of difference equations, 37–38
signal flow graph representation of, 348–53
linear systems, 18
accumulator system, 19
nonlinear system, 19
linear time-invariant systems, 22–28, 154–60
convolution sum, 23–28
eigenfunctions for, 40–45
properties of, 28–34
memoryless systems, 18
moving average, 17–18
and processing of continuous-time signals, 153
representation of sequences by Fourier transforms, 48–54
absolute summability for suddenly-applied exponential, 51–52
Fourier transform of complex exponential sequences, 54
Fourier transform of a constant, 53
inverse Fourier transform, 48
square-summability for the ideal lowpass filter, 52–53
round-off, limit cycles due to, 414–16
round-off noise, 391–413
analysis of cascade IIR structure, 403–10
analysis of direct-form FIR systems, 410–12
analysis of direct-form IIR structures, 391–99
first-order system, 396–97
floating-point realizations of discrete-time systems, 412–13
interaction between scaling and, 402–3
scaling in fixed-point implementations of IIR systems, 399–403
second-order system, 397
sinusoidal response of, 42–43
stability, 21–22
testing for, 22

structures for, 340–438
symmetric properties of Fourier transform, 55–58
conjugate-antisymmetric sequence, 55–56
conjugate-symmetric sequence, 55–56
even function, 55
even sequence, 55
illustration of, 57–58
odd function, 55
odd sequence, 55
time-invariant systems, 20–21
accumulator as, 20
compressor system, 20–21
transposed forms, 363–66
for a basic second-order section, 364–65
for first-order system with no zeros, 363–64
truncation, limit cycles due to, 414–16
Discrete-to-continuous-time (D/C) conversion, 152–53
Distortion compensation, 282–83
Doppler frequency, 729–30
Downsampling, 168–72
interchange of filtering and, 179–80
Duality property:
discrete Fourier series (DFS), 544–45, 547
discrete Fourier transform (DFT), 567–68

Eigenfunctions, 40
Eigenvalues, 40–41
See also Frequency response
Elliptic approximation, 463–65, 828–29
Elliptic continuous-time filters, 463–65, 828–29
Energy density spectrum, 60
Equiripple approximations, 492–93
Even function, 55
Even sequence, 55
Expander, 172
Exponential multiplication, 121–22
Exponential sequences, 13–14, 779–80
left-sided, 99–100
right-sided, 98
sum of, 100–101
Extraripple case, 493

Fast Fourier transform (FFT)
  algorithms, 2–3, 6, 576,
  629–92
algorithms for more general
  values of N, 655
coefficients, 654–55
decimation-in-frequency
  algorithms, 646–52
decimation-in-time FFT
  algorithms, 635–46
indexing, 652–54
Feedback, in IIR systems, 361–63
FFT, *See* Fast Fourier transforms
Filter design, 439–540
  bilinear transformation,
  450–65
discrete-time filter,
  determining specifications
  for, 440–42
discrete-time IIR filter design,
  from a continuous-time
  filters, 442–65
FIR discrete-time filter design,
  510–11
FIR equiripple approximation
  examples, 503–10
  bandpass filter, 507–10
  compensation for zero-order
  hold, 506–7
  lowpass filter, 503–6
IIR discrete-time filter design,
  510–11
impulse invariance, 443–49
  impulse invariance with a
  Butterworth filter, 446–49
optimum approximations of
  FIR filters, 486–502
  alternation theorem, 489–91
  characteristics of optimum
  FIR filters, 501–2
  equiripple approximations,
  492
  extraripple case, 493
  optimal type II lowpass
  filters, 497–98
  optimal type I lowpass
  filters, 491–97
  Parks-McClellan algorithm,
  498–501, 510
techniques, 439–540
windowing, 465–78
  Bartlett window, 468–71
  Blackman window, 468fn,
  469
  Hamming window, 468–71
  Hanning window, 468–71

incorporation of generalized
  linear phase, 469–73
Kaiser window method,
  474–78
linear-phase lowpass filter
  example, 472–73
properties of commonly
  used windows, 468–69
rectangular window, 468
Filtering, interchange of down-
  sampling/upsampling and,
  179–80
Filters:
  antialiasing, 186
  continuous-time IIR, 442–43
  finite impulse response (FIR),
  442
  frequency-selective, 439–540
  generalized linear-phase, 316
  ideal frequency-selective,
  43–44, 241–42
  linear-phase, 316
  optimal type II lowpass, 497–98
  optimal type I lowpass, 491–97
  sharp-cutoff antialiasing filters,
  186–87
  transversal, 367
Finite-duration sequences, 105,
  571
Fourier representation of,
  559–64
See also Discrete Fourier
  transform (DFT)
Finite impulse response (FIR)
  filters, 442
Finite-length sequences, 103–4,
  117, 779
  periodic sequence, 783–87
  and power expansion, 117
  sufficiency theorems for,
  782–87
Finite-precision numerical
  effects, 370–77
  numerical representations,
  371–74
  quantization in implementing
  systems, 374–77
Finite register length, effects of,
  661–68
FIR equiripple approximation
  examples, 503–10
  bandpass filter, 507–10
  compensation for zero-order
  hold, 506–7
  lowpass filter, 503–6
FIR filters:

optimum approximations of,
  486–502
  alternation theorem, 489–91
  characteristics of optimum
  FIR filters, 501–2
  equiripple approximations,
  492
  extraripple case, 493
  optimal type II lowpass
  filters, 497–98
  optimal type I lowpass
  filters, 491–97
First backward difference, 88, 524
First-order all-pass systems, 275
FIR linear-phase systems:
  examples of, 300–305
  locations of zeros for, 306–8
  relation of minimum-phase
  systems to, 308–11
  type III linear-phase system,
  302
  type II linear-phase system, 302
  type I linear-phase system,
  300–302
  type IV linear-phase system,
  302–5
FIR systems:
  basic network structures for,
  366–70
  cascade form, 367–68
  direct form, 367
  linear-phase FIR system
  structures, 368–70
  coefficient quantization in,
  384–86
  compensation of, 283–87
Fixed-point realizations of IIR
  digital filters, zero-input
  limit cycles in, 413–18
Floating-point realizations of
  discrete-time systems,
  412–13
Flow graph, determination of
  system function from,
  352–53
Flow graph
  reversal/transposition, 363
Forward difference, 21, 31, 33
Fourier transform, 63
  of complex exponential
  sequences, 54
  of a constant, 53
  linearity of, 59
  magnitude of, 49
  of a periodic impulse train, 552
  of periodic signals, 551–55

relationship between Fourier series coefficients and, 554–55
phase of, 49
sampling, 555–59
Frequency:
differentiation in, 60
Doppler, 729–30
Nyquist, 146–47
sampling, 10, 140–41
Frequency-domain representation:
of discrete-time signals/systems, 40–48
of sampling, 142–49
Frequency response, 40–41
determining impulse response from, 64
of the ideal delay system, 41
of the moving-average system, 44–45
Frequency-response compensation, 282–87
Frequency response for rational system functions, 253–70
multiple poles or zeros, 265–70
second-order FIR system, 268
second-order IIR system, 265–67
third-order IIR system, 268–70
single zero or pole, 258–265
Frequency-selective filters, 439–540
See also Filter design
Frequency shifting, 59
Fricative sounds, 724

Gain in *dB*, 254
Gain of the system, 241
Generalized linear phase, 295–97
linear systems with, 291–311
Generalized linear-phase filters, 316
Gibbs phenomenon, 468
Goertzel algorithm, 629, 633–35
Group delay:
and attenuation, effects of, 243–45
defined, 243

Hamming window, 468–71
Hanning window, 468–71
High-pass filters, 44

Kaiser window method design of, 479–82
Hilbert transformers, 791
design of, 792–95
Kaiser window design of, 793–95
See also Discrete Hilbert transforms
Hilbert transform relations, 775–76
for complex sequences, 789–801
See also Discrete Hilbert transforms
Homogeneous difference equation, 36
Homogeneous solution, 36

Ideal 90-degree phase shifter, 790–91
Ideal continuous-time bandlimited differentiator, discrete-time implementation of, 158–59
Ideal continuous-time lowpass filtering, using discrete-time lowpass filter, 155–57
Ideal continuous-to-discrete (C/D) converter, 141, 154, 185
Ideal delay system, 17, 31
frequency response of, 41
Ideal discrete-to-continuous (D/C) converter, 152–153, 185
Ideal frequency-selective filters, 43–44, 241–42
Ideal highpass filters:
frequency response of, 44, 242
impulse response of, 242
Ideal lowpass filters, 43
frequency response of, 43, 241
impulse response of, 242
with linear phase, 242–43
Ideal reconstruction filter, 150–51
Ideal signal reconstruction system, 150–51
IIR systems:
basic structures for, 354–63
cascade form, 356–59
direct forms, 354–56
parallel form, 359–61
coefficient quantization in, 377–82

feedback in, 361–63
Image processing, application of multidimensional digital processing to, 4
Impulse, 11
Impulse invariance, 160–63, 517
applied to continuous-time systems with rational system functions, 162–63
with a Butterworth filter, 446–49
discrete-time lowpass filter obtained by, 162
filter design by, 443–49
Impulse response:
determining from frequency response, 64
for a difference equation, determining, 64–65
Impulse responses, 31–32
Impulse train modulation, 142
Infinite-duration impulse response (IIR) systems, 32
Infinite impulse response (IIR) filters, 442
Infinite sum, expressing closed form, 97–98
Initial-rest conditions, 38
Initial-value theorem, 126
Innermost nonzero pole, 105
Inspection method for inverse *z*-transform, 111–12
Instability, 21–22
testing for, 22
Interpolation, 173
frequency-domain illustration of, 174
linear, 175–76
by filtering, 175
impulse response for, 175
Interpolator, 173
Inverse by partial fractions, 115–16
Inverse DFT algorithm, 676
Inverse Fourier transform, 48, 63
Inverse of non-rational *z*-transform, 122–23
Inverse systems:
defined, 33–34
system functions for, 248–50
Inverse transform, by a power series expansion, 117
Inverse *z*-transform, 111–18
inspection method, 111–12
inverse by partial fractions, 115–16

partial fraction expansion,
112–16
second-order z-transform,
113–14

Kaiser window, 474–82
DFT analysis of sinusoidal
signals using, 708–10
Kaiser window design, of Hilbert
transformers, 793–95
Kaiser window method, 474–78
examples of FIR filter design
by, 478–85
discrete-time differentiators,
482–85
highpass filter, 479–82
and lowpass filter design,
476–78
relationship to other windows,
478

Leakage, 701
Left-sided exponential sequence,
99–100
Left-sided sequence, 105
power series expansion for, 118
L'Hopital's rule, 151
Limit cycles:
avoiding, 417–18
due to overflow, 416–17
due to round-off, 414–16
due to truncation, 414–16
in first-order system, 414–16
Linear chirp signal,
time-dependent Fourier
transform of, 715–17
Linear constant-coefficient
difference equations,
34–40
block diagram representation
of, 341–48
impulse response for rational
system functions, 250–53
first-order IIR system,
251–52
simple FIR system, 252–53
inverse systems, 248–50
for first-order system, 249–50
with zero in the ROC, 250
region of convergence (ROC),
determining, 247–48
second-order system, 246–47
signal flow graph
representation of, 348–53
stability and causality, 247–48

system functions for systems
characterized by, 245–53
Linear convolution, 576–88
using discrete Fourier
transform (DFT), 576–88
Linearity:
and auxiliary conditions, 39
of Fourier transform, 59
Linearity property:
discrete Fourier series (DFS),
546
discrete Fourier transform
(DFT), 564
z-transform, 119
Linear noise models, 391–412
cascade form IIR systems,
403–410
direct form FIR systems,
410–12
direct form IIR systems,
391–403
parallel form IIR sytmes, 410
Linear phase, 291–311
causal generalized linear-phase
systems, 297–311
type I FIR linear-phase
systems, 298
type II FIR linear-phase
systems, 298
type III FIR linear-phase
systems, 298
type IV FIR linear-phase
systems, 298–99
decomposition of a
linear-phase system,
308–11
FIR linear-phase systems:
examples of, 300–305
locations of zeros for, 306–8
relation of minimum-phase
systems to, 308–11
type III linear-phase system,
302
type II linear-phase system,
302
type I linear-phase system,
300–302
type IV linear-phase system,
302–5
generalized linear phase,
295–97
ideal lowpass filter with,
293–95
systems with, 292–95
Linear-phase filters, 316

Linear-phase FIR system
structures, 368–70
Linear predictive coding (LPC), 4
Linear quantizers, 191
Linear system, 78–79
Linear systems, 18
accumulator system, 19
nonlinear system, 19
Linear time-invariant
discrete-time systems,
22–28, 154–60
all-pass systems, 274–79
causal all-pass systems,
274–78
defined, 274
first- and second-order
all-pass systems, 275
uses of, 278
convolution sum, 23–28
analytic evaluation of, 26–28
computation of, 25–26
defined, 23
direct form II implementation
of, 347–48
direct form I implementation
of, 346–47
eigenfunctions for, 40–45
frequency response of, 241–45
ideal frequency-selective
filter, 241–42
phase distortion and delay,
242–45
frequency response for rational
system functions, 253–70
multiple poles and zeros,
265–70
single zero or pole, 258–65
generalized linear phase, linear
systems with, 291–311
ideal continuous-time
bandlimited
differentiator,
discrete-time
implementation of, 158–59
ideal continuous-time lowpass
filtering, using
discrete-time lowpass
filter, 155–57
minimum-phase systems,
280–91
frequency-response
compensation, 282–87
minimum-phase and all-pass
decomposition, 280–82
properties of, 287–91
properties of, 28–34

cascade connection, 29
constraints of linearity/time invariance, 30–31
finite-duration impulse response (FIR) systems, 32–33
impulse responses, 31–32
infinite-duration impulse response (IIR) systems, 32
parallel connection, 30
relationship between magnitude and phase, 270–74
transform analysis of, 240–339
Log magnitude, 254
Long division, power series expansion by, 118

Magnitude:
Fourier transform, 49
relationships between phase and, 788–89
Magnitude distortions, 241
Magnitude response, 241
Magnitude spectrum, 49
Magnitude-squared function, 254, 258, 457
Mantissa, 412
Matlab, 3
Maximum entropy methods (MEM spectral analysis), 4
Maximum-phase sequence, 335
Memoryless systems, 18
MEM spectral analysis, 4
Microelectronics, 6–7
Microprocessors:
and discrete-time signals, 7
rapid evolution of, 2
Minimax criteriod, 489
Minimum energy-delay property, minimum-phase systems, 288–91
Minimum group-delay property, minimum-phase systems, 288
Minimum phase-lag property, minimum-phase systems, 287–88
Minimum-phase systems, 250, 280–91
frequency-response compensation, 282–87
minimum-phase and all-pass decomposition, 280–82
properties of, 287–91

minimum energy-delay property, 288–91
minimum group-delay property, 288
minimum phase-lag property, 287–88
relation of FIR linear-phase systems to, 308–11
Modified periodogram, 732
Modulation theorem, 61–62
Moving average, 17–18, 31
Moving-average (MA) linear random process, 771
Moving-average system:
difference equation representation of, 35–36
with noninteger delay, 165–67
Multidimensional signal processing, 3–4
Multiplication by an exponential sequence, 121–22
Multirate signal processing, 179–84
interchange of filtering and downsampling/upsampling, 179–80
polyphase decompositions, 180–82
polyphase implementation of decimation filters, 182–83
polyphase implementation of interpolation filters, 183–84
Multistage noise shaping (MASH), 210

Narrowband time-dependent Fourier analysis, 725
Nonanticipative system, 21
Noncausal window, 716
Noncomputable network, 363
Noninteger delay, 164–65
moving-average system with, 165–67
Nonlinear system, 19
Non-rational z-transform, inverse of, 122–23
Nonstationary signals:
time-dependent Fourier analysis of, 723–30
radar signals, 728–30
speech signals, 720–28
N-point circular convolution, 571
Nyquist frequency, 146–47
Nyquist rate, 146–47, 153

Nyquist sampling theorem, 146

Odd function, 55
Odd sequence, 55
Offset binary coding scheme, 191
One's complement, 371
One-sided z-transform, 95
Optimal type II lowpass filters, 497–98
Optimal type I lowpass filters, 491–97
Optimum approximations of FIR filters, 486–502
alternation theorem, 489–91
and polynomials, 290–91
characteristics of optimum FIR filters, 501–2
equiripple approximations, 492
extraripple case, 493
optimal type II lowpass filters, 497–98
optimal type I lowpass filters, 491–97
Parks-McClellan algorithm, 498–501
Outermost finite pole, 105
Overflow oscillations, in second-order system, 416–17
Overlap-add method, 586–87, 680
Overlap-save method, 587, 680
Oversampled A/D conversion:
with direct quantization, 201–5
with noise shaping, 206–10
Oversampled D/A conversion, and noise shaping, 210–13

Parallel connection, 30
Parallel form structures:
coefficient quantization effects, 382
IIR systems, 359–61
Roundoff noise effects, 410
Parks-McClellan algorithm, 487, 498–501, 506, 510
Parseval's relation:
discrete Fourier transform, 621
Fourier transform, 60
Parseval's theorem, 60, 400
Partial fractions, inverse by, 115–16
Periodic conjugate-antisymmetric components, 570
Periodic conjugate-symmetric components, 570

Periodic convolution:
  defined, 548
  discrete Fourier series (DFS),
    548–51
Periodic discrete-time sinusoids,
  15
Periodic impulse train, 143
Periodic sampling, 140–42
Periodic sequences,
    representation of, 542–46
Periodic signals, Fourier
    transform of, 551–55
Periodogram, 730–42
  averaging of, 737–39
  defined, 732
  explicit computation of, 732
  modified, 732
  periodogram analysis, 731,
    739–42
    example of, 739–42
  power spectrum, 731–32, 734
  properties of, 733–37
Periodogram analysis, 731,
    739–42
Phase:
  Fourier transform, 49
  relationships between
    magnitude and, 788–89
Phase distortion and delay,
    242–45
Phase response, 241
Phase shift, 241
Phase spectrum, 49
Plosive sounds, 724
Poisson's formulas, 776
Poles of X(z), 298
Polyphase decompositions,
    180–82
Polyphase implementation:
  of decimation filters, 182–83
  of interpolation filters, 183–84
Power density spectrum, 68
  of quantization noise, 207–9
Power series expansion, 116–18
  by long division, 118
  finite-length sequence, 117
  inverse transform by, 117
  for a left-sided sequence, 118
Power spectrum, 731–32, 734
  estimation:
    computing using DFT,
      746–48
    example based on estimation
      of autocorrelation
      sequence, 748–54

Prefiltering, to avoid aliasing,
    185–87
Product of two sequences, 11

Quantization errors, 372
  analysis of, 193–97
  for sinusoidal signal, 194–95
Quantization levels, 191–94
  coding of, 191
Quantization noise:
  example of, 195
  power density spectrum of,
    207–9

Radar signals:
  clutter, 729
  Doppler frequency, 729–30
  time-dependent Fourier
    analysis of, 728–30
Random process, 65
Random signals, 65–70, 811–23
  spectrum analysis of, 730–54
Rational system functions:
  frequency response for, 250–70
    multiple poles and zeros,
      265–70
    single zero or pole, 258–65
  impulse response for, 250–53
    first-order IIR system,
      251–52
    simple FIR system, 252–53
Reconstruction of sinusoidal
    signal, 147
  undersampled, aliasing in,
    148–49
Rectangular window, 471
Recursive computation of
    difference equations,
    37–38
Recursive representation, 35
Region of convergence (ROC),
    96, 105–11
  and causality, 110–11
  properties of, for z-transform,
    105–11
  and stability, 110–11
Region of convergence (ROC),
    96–97
Right-sided exponential
    sequence, 98
Right-sided sequence, 105, 108
Rotations, 684
Round-off, limit cycles due to,
    414–16

Sample-and-hold circuits, 188–89

Sampled-data Delta-Sigma
    modulator, 206
Sampling:
  bandpass, 799–801
  downsampling, 168–72, 179–80
  Fourier transform, 555–59
  frequency-domain
      representation of, 142–49
  periodic, 140–42
  spectral, 703–14
  in time/frequency, 718–22
  upsampling, 172–76, 179–80
Sampling the Fourier transform,
    555–59
Sampling frequency, 10, 140–41
Sampling period, 10, 140–42
Sampling rate change by
    noninteger factor, 176–78
Sampling rate compressor,
    167–68
Sampling rate expander, 172
Sampling rate increase by integer
    factor, 172–76
Sampling rate reduction by
    integer factor, 167–72
Second-order all-pass systems,
    275
Second-order z-transform,
    113–14
Sequence operations, 11–16
Sequences:
  autocorrelation, of h[n], 67–68
  autocovariance, 65–66
  causal, 31, 129, 777–82
  circular shift of, 564–67
  complex, 789–801
  complex exponential, 14
  conjugate-antisymmetric,
    55–56
  conjugate-symmetric, 55–56
  delayed impulse sequence,
    572–73
  deterministic autocorrelation,
    67–68
  even, 55
  exponential, 13–14, 779–80
  finite-duration, 105, 571
  finite-length, 103–4, 779
  left-sided, 99–100, 105, 118
  maximum-phase, 335
  odd, 55
  periodic, 542–46
  representation of, by Fourier
    transforms, 48–54
  right-sided, 98, 105, 108
  shifted exponential, 120–21

sinusoidal, 13
two-sided, 108–10
unit sample, 11
unit step, 11–13
Sharp-cutoff antialiasing filters,
    186–87
Shifted exponential sequence,
    120–21
Shift of sequences, 59
    circular, 564–67
    periodic, 546–47
Shifting:
    frequency, 59
    time, 59, 120–21
Shift of a sequence, discrete
    Fourier series (DFS),
    546–47
Short-time Fourier transform,
    See Time-dependent
    Fourier transform
Time-dependent Fourier
    transform, 714–22
Signal:
    defined, 8
    independent variable in
        mathematical
        representation of, 8
Signal expression processing, 3
Signal interpretation, 2–3
Signal modeling, 4
Signal processing, defined, 1–2
Signal processing
    microprocessors,
    processing capability of, 1
Signal-processing systems,
    classification of, 8–9
Sign bit, 371
Sign and magnitude, 371
Sink nodes, 349
Sinusoidal response, of linear
    time-invariant systems,
    42–43
Sinusoidal sequences, 13
Sinusoidal signals, DFT analysis
    of, 697–714
    oversampling/linear
        interpolation for
        frequency estimation,
        713–14
    spectral sampling, 703–14
    with 32-point Kaiser window
        and zero-padding, 711–12
    using a Kaiser window, 708–10
    windowing, 698–703
Source nodes, 349
Spectral analysis, 4, 6

Spectral sampling:
    effect of, 703–14
        illustration of, 703–6
    with frequencies matching
        DFT frequencies, 706–8
Spectrum:
    amplitude, 49
    energy density, 60
    magnitude, 49
    phase, 49
    power, 731–32, 734
    power density, 68, 207–9
Speech signals:
    time-dependent Fourier
        analysis of, 720–28
    narrowband spectrogram,
        728
    wideband spectrogram,
        725–28
Split-radix FFT (SRFFT), 683
Square-summability, for the ideal
    lowpass filter, 52–53
SRFFT, 683
Stability, 21–22, 30–31, 247–48
    and region of convergence,
        110–11
    testing for, 22
Stationary, use of term, 66
Stationary random signals,
    Fourier analysis of, 730–42
Steady-state response, 46
Stochastic signals, *See* Random
    signals
Structures for discrete-time
    systems, 340–418
    block diagram representation
        of linear
        constant-coefficient
        difference equations,
        341–48
    cascade form, *See* Cascade
        form structures
    direct form, *See* Direct form
        structures
    finite precision numerical
        effects, *See* Quantization
    FIR systems, 366–70
    IIR systems, 354–63
    parallel form, *See* Parallel form
        structures
    signal flow graph
        representation of linear
        constant-coefficient
        difference equations,
        348–53
    transposed forms, 363–66

Suddenly-applied complex
    exponential inputs, 46–48
Suddenly-applied exponential,
    absolute summability for,
    51–52
Sufficiency theorems for
    finite-length sequences,
    782–88
Summability:
    absolute, 51, 108
    for suddenly-applied
        exponential, 51–52
    square-, 52–53
Superposition, principal of, 22–24
Surface acoustic wave (SAW), 2
Switched-capacitor technologies,
    2
Symmetry properties:
    discrete Fourier series (DFS),
        547
    discrete Fourier transform,
        568–70
    discrete-time Fourier
        transform (DTFT), 55–58
Synthesis formula, 48
System function, 241
Systems, 374–77

Tapped delay line structure, 367
Telecommunications, and
    discrete-time signal
    processing, 7
Time-dependent Fourier analysis
    of nonstationary signals,
    723–30
    radar signals, 728–30
    speech signals, 720–28
Time-dependent Fourier
    transform, 693, 714–22
    block convolution using,
        722–23
    defined, 714
    effect of the window in, 717–18
    of a linear chirp signal, 715–17
    sampling in time/frequency,
        718–22
Time invariance, and auxiliary
    conditions, 39
Time-invariant systems, 20–21, 77
    accumulator as, 20
    compressor system, 20–21
Time-reversal property,
    z-transform, 123–24
Time reversal theorem, 60
Time-shifting property:
    discrete Fourier series, 546

discrete Fourier transform, 564–67
Fourier transform, 59
z-transform, 120–21
  shifted exponential sequence, 120–21
Time-shifting theorem, 59
Transient response, 46–47
Transposed forms, 363–66
  for a basic second-order section, 364–65
  for first-order system with no zeros, 363–64
Transversal filter structure, 367
Trapezoidal approximation, 525
Truncation, limit cycles due to, 414–16 "Twicing," 528
Two-dimensional signal processing techniques, 3–4
Two's complement, 371
Two-sided exponential sequence, 102
Two-sided sequences, 108–10
Type I FIR linear-phase systems, 298
Type II FIR linear-phase systems, 298
Type III FIR linear-phase systems, 298
Type IV FIR linear-phase systems, 298–99

Undersampled sinusoidal signal, reconstruction of, 148–49
Uniform convergence, of z-transform, 97
Unilateral z-transform, 95
Unit circle, 95–96
Unit impulse function, 142–43
  sifting property of, 143
Unit sample sequence, 11
Unit step sequence, 11–13
Unweighted approximation error, 503
Upsampling, 172–76

interchange of filtering and, 179–80

Video coding, 3
Vocal tract, 724
Vocoder simulations, 5
Voiced sounds, 724
Von Hann, Julius, 468fn

Warping, and bilinear transformation, 453–54
"Whitening" procedure, 333
White noise, 69
Wideband spectrogram, 725
Windowing, 465–78, 693
  Bartlett window, 468–71
  Blackman window, 468fn, 469
  effect of, 698–703
    on Fourier analysis of sinusoidal signals, 698–701
  Hamming window, 468–71
  Hanning window, 468–71
  incorporation of generalized linear phase, 469–73
  Kaiser window, 474–78
  linear-phase lowpass filter example, 472–73
  rectangular window, 468
Windowing theorem, 61–62
Winograd Fourier transform algorithm (WFTA), 655–56

Zero-input limit cycles, defined, 376, 413
Zero-order hold, compensation for, 506–7
Zero padding, 583
Zeros of $X(z)$, 98
z-plane, 96
z-transform, 94–139
  common pairs, 104
  defined, 95

evaluating a convolution using, 125
finite-duration sequence, 105
finite-length sequence, 103–4
infinite sum, expressing closed form, 97–98
inverse z-transform, 111–18
  inspection method, 111–12
  partial fraction expansion, 112–16
left-sided sequence, 99–100, 105
power series expansion, 116–18
  by long division, 118
  finite-length sequence, 117
  inverse transform by, 117
  for a left-sided sequence, 118
region of convergence properties for, 105–11
region of convergence (ROC), 96–97
right-sided sequence, 98, 105
sum of two exponential sequences, 100–101
transform properties, 119–26
  conjugation property, 123
  convolution property, 124–26
  differentiation property, 122–23
  exponential multiplication, 121–22
  initial-value theorem, 126
  inverse of non-rational z-transform, 122–23
  linearity property, 119
  time-reversal property, 123–24
  time-shifting property, 120–21
two-sided sequence, 102, 105
uniform convergence of, 97
unit circle, 95–96
z-transform operator Z, 95
z-transform pairs, 104